Environmental and Microbial Biotechnology

Series Editor

Ram Prasad, Department of Botany, Mahatma Gandhi Central University, Motihari, Bihar, India

Innovative and novel advances in microbial biotechnology are providing great understandings in to the machineries of nature, presenting fascinating prospects to apply principles of biology to different arenas of science. Sustainable elucidations are emerging to address the concerns on improving crop productivity through microbes, depleting natural resources, environmental pollution, microbial degradation of pollutants, nanomaterials, nanotoxicity & safety issues, safety of food & agricultural products etc. Simultaneously, there is an increasing demand for natural bio-products of therapeutic and industrial significance (in the areas of healthcare, environmental remediation, microbial biotechnology). Growing awareness and an increased attention on environmental issues such as climate change, energy use, and loss of non-renewable resources have carried out a superior quality for research that provides potential solutions to these problems. Emerging microbiome approaches potentially can significantly increase agriculture productivity & human healthcare and henceforth can contribute to meet several sustainable development goals.

The main objectives have provided an impetus for research on plants and microorganisms that produce novel bio-products with variable properties and understanding their mechanisms of action at cellular and molecular level. Hence, research activities of the environmental and microbial Biotechnology are comprehensively focused up on major sectors viz., bioresources, biorefining, bioremediation of organic and inorganic pollutants, environmental risk analysis of microorganisms, environmental assessment using microbiological indicators, enzymes for environment, food & industrial applications, nanomaterials & nanotoxicity, sustainable ecobiotechnology, biofertilizer, biocontrol agents for agriculture improvement and natural products for healthcare applications.

This book series is a state-of-the-art for a wide range of scientists, researchers, students, policy makers and academician involve in understanding and implementing the knowledge on environmental and microbial biotechnology to develop biologics for proper health care to continue life in smooth and sustainable strategy without any adverse effect.

More information about this series at http://www.springer.com/series/16324

Ajar Nath Yadav • Joginder Singh •
Chhatarpal Singh • Neelam Yadav
Editors

Current Trends in Microbial Biotechnology for Sustainable Agriculture

Springer

Editors
Ajar Nath Yadav
Department of Biotechnology
Dr. Khem Singh Gill Akal College of Agriculture
Eternal University, Baru Sahib
Sirmour, Himachal Pradesh, India

Chhatarpal Singh
Department of Environmental Microbiology
Babasaheb Ambedkar University
Lucknow, Uttar Pradesh, India

Joginder Singh
Department of Microbiology
Lovely Professional University
Phagwara, Punjab, India

Neelam Yadav
Veer Bahadur Singh Purvanchal University
Jaunpur, Uttar Pradesh, India

ISSN 2662-1681 ISSN 2662-169X (electronic)
Environmental and Microbial Biotechnology
ISBN 978-981-15-6948-7 ISBN 978-981-15-6949-4 (eBook)
https://doi.org/10.1007/978-981-15-6949-4

This Springer imprint is published by the registered company Springer Nature Singapore Pte Ltd.
The registered company address is: 152 Beach Road, #21-01/04 Gateway East, Singapore 189721, Singapore

Preface

Microbial biotechnology is an emerging field with greater applications in diverse sectors involving food security, human nutrition, plant protection, and overall basic research in the agricultural sciences. The environment has been sustaining the burden of mankind since decades, and indiscriminate use of the resources has led to the degradation of the environment, loss of soil fertility, and has created a need for sustainable strategies. The major focus in the coming decades would be on the green and clean environment by utilizing the soil- and plant-associated beneficial microbial communities. The plant–microbes interaction included the association of microbes with plant systems in the form of epiphytic, endophytic, and rhizospheric. The microbes associated with plant ecosystems play an important role in plant growth, development, and soil health. The soil and plant microbiomes promote plant growth directly or indirectly mechanisms by using diverse plant growth-promoting mechanisms, *viz*. releasing plant growth regulators; solubilization of phosphorus, potassium, and zinc; biological nitrogen fixation or by producing siderophores, ammonia, HCN, and other secondary metabolites. The PGP microbes belong to all three domains of archaea, bacteria, and eukarya. The most dominant and efficient plant growth-promoting microbes belong to different genera of *Arthrobacter, Azospirillum, Azotobacter, Bacillus, Burkholderia, Gluconoacetobacter, Methylobacterium, Paenibacillus, Pantoea, Penicillium, Piriformospora, Planomonospora, Pseudomonas, Rhizobium, Serratia,* and *Streptomyces*. These beneficial microbial communities represent a novel and promising solution for agro-environmental sustainability. Microbial communities possess a huge sink of capability by which they act as biofertilizers, bioprotectants, and biostimulants as well as for mitigation of different abiotic stress in plants. The utilization of beneficial soil and plant microbiomes will surely support sustainable agriculture.

The aim of the book ***"Current Trends in Microbial Biotechnology for Sustainable Agriculture"*** is to provide understanding of microbial diversity associated with plant systems and their role in plant growth and soil health. The book will be useful to scientists, research, and students related to microbiology, biotechnology, agriculture, molecular biology, environmental biology, and related subjects.

Sirmour, Himachal Pradesh, India — Ajar Nath Yadav
Phagwara, Punjab, India — Joginder Singh
Lucknow, Uttar Pradesh, India — Chhatarpal Singh
Jaunpur, Uttar Pradesh, India — Neelam Yadav

Acknowledgements

All the authors are sincerely acknowledged for contributing up-to-date information on the microbial biotechnology, plant–microbes interaction, biodiversity of soil and plant microbiomes, and their potential applications in plant growth promotion, enhanced crop production, control of phytopathogenic microbes, and role in soil health and fertility for agro-environmental sustainability. The editors are thankful to all the authors for their valuable contributions.

All editors would like to thank their families who were very patient and supportive during this journey. Our sincere thanks to the whole Springer team who were directly or indirectly involved in the compilation of this book. We are grateful to the many people who helped to bring this book to light. Our sincere thanks to Ms. Aakanksha Tyagi, Mr. Ashok Kumar and Ms. Beauty Christobel of Springer Nature, and Dr. Ram Prasad—Series editor of "Environmental and Microbial Biotechnology" for generous assistance and constant support.

The editor, Dr. Ajar Nath Yadav is grateful to his PhD research scholars Tanvir Kaur, Rubee Devi, Dr. Divjot Kour, Kusam Lata Rana, and colleagues for their support, love, and motivation in all his efforts during this project.

We are very sure that this book will be of great interest to the scientists, graduates, undergraduates, and postdocs who are investigating on "Microbial Biotechnology."

Contents

Editors and Contributors

About the Editors

Ajar Nath Yadav is an Assistant Professor (Senior Scale) and Assistant Controller of Examination at Eternal University, Baru Sahib, Himachal Pradesh, India. He has 5 years of teaching and 11 years of research experiences in the field of Microbial Biotechnology, Microbial Diversity, and Plant–Microbe Interactions. Dr. Yadav obtained doctorate in Microbial Biotechnology, jointly from IARI, New Delhi and BIT, Mesra, Ranchi, India; MSc (Biotechnology) from Bundelkhand University; and BSc (CBZ) from the University of Allahabad, India. Dr. Yadav has 197 publications, with h-index of 45, i10-index of 111, and 5151 citations (Google Scholar- on 29/09/2020). Dr. Yadav has published 115 research communications in different international and national conferences. In his credit one granted patent "Insecticidal formulation of novel strain of *Bacillus thuringiensis* AK 47" Dr. Yadav has got 12 Best Paper Presentation Awards, and 01 Young Scientist Award (NASI-Swarna Jayanti Puraskar). Dr. Yadav received "Outstanding Teacher Award" in 6th Annual Convocation 2018 by Eternal University, Baru Sahib, Himachal Pradesh. Dr. Yadav has a long-standing interest in teaching at the UG, PG, and PhD level and is involved in taking courses in microbiology and microbial biotechnology. Dr. Yadav is currently handling two projects. Dr. Yadav has guided 01 Ph.D. and 01 M.Sc. Scholar and presently he is guiding 05 scholars for PhD degree. He has been serving as an editor/editorial board member and reviewer for different national and international peer-reviewed journals. He has lifetime membership of Association of Microbiologists of India, and Indian Science Congress Council, India. Please visit https://sites.google.com/site/ajarbiotech/ for more details.

Joginder Singh is Professor at Department of Microbiology, Lovely Professional University, Punjab, India. Previously, he has worked as Young Scientist at Microbial Biotechnology and Biofertilizer Laboratory, Department of Botany, Jai Narain Vyas University funded by the DST, Govt. of India. His broad research interest is to explore different efficient strategies towards bioremediation and phytoremediation of pollutants and pesticides from water and soil. Presently, he is working on the designing and development of cleanup technology (biofilters) for in situ bioremediation of textile industrial effluents. He has more than 60 research and review articles in the peer-reviewed journals, edited six books, entitled and authored 30 chapters in edited books. Dr. Singh attended several International and National Seminars, Symposia, Conferences, and chaired technical sessions and presented papers in them. He serves as a reviewer for many prestigious journals. He is an active member of various scientific societies and organizations including Association of Microbiologists of India. He is an active member of various scientific societies and organizations including Association of Microbiologists of India, K. K. Nanda Foundation for Advancement of Plant Sciences, Indian Society for Radiation Biology, International Phytotechnology Society and European Federation of Biotechnology.

Chhatarpal Singh is a President of Agro Environmental Development Society (AEDS), Majhra Ghat, Rampur, Uttar Pradesh, India. Dr. Singh is currently working in the field of Agricultural and Environmental Microbiology; the research interest is agro-environmental development through various scientific and innovative approaches, *viz.* bio-formulation and crop residue-mediated biochar applications for the improvement of soil fertility and agriculture production. He has published various scientific research and review papers, book chapters, and magazine articles in the field of agricultural and environmental microbiology. He obtained his PhD degree in Environmental Microbiology with the specialization of Agricultural Microbiology from Babasaheb Bhimrao Ambedkar University Lucknow, Uttar Pradesh, India. He has been honored by various organizations for his outstanding contribution in the field of Agricultural Microbiology. Dr. Singh has organized various International Conference and Training Programs in the fields of agriculture, environment, and applied sciences.

Neelam Yadav is currently working on microbial diversity from diverse sources and their biotechnological applications in agriculture and allied sectors. She obtained her post-graduate degree from Veer Bahadur Singh Purvanchal University, Uttar Pradesh, India. She has research interest in the area of beneficial microbiomes and their biotechnological application in agriculture, environment, and allied sectors. To her credit, she has 65 publications in reputed international and national journals and with several publishers. She is the editor of nine books published by different international publishers. She is an editor/associate editor/reviewer of different international and national journals. She is a lifetime member of the Association of Microbiologist in India, Indian Science Congress Council, and National Academy of Sciences, India.

Contributors

Ahmed M. Abdel-Azeem Botany Department, Faculty of Science, University of Suez Canal, Ismailia, Egypt

Hebtallah H. Abo Mahas Zoology Department, Faculty of Science, University of Suez Canal, Ismailia, Egypt

Liliana Aguilar-Marcelino Centro Nacional de Investigación Disciplinaria en Salud Animal e Inocuidad, INIFAP, JiutepecMorelos, CP, México

Laith Khalil Tawfeeq Al-Ani School of Biology Science, Universiti Sains Malaysia, Minden, Pulau Pinang, Malaysia

Azka Asif School of Biological Sciences, University of the Punjab, Lahore, Pakistan

Muhammad Aslam Department of Agronomy, School of Agriculture, Lovely Professional University, Phagwara, Punjab, India

Diptanu Banik Department of Agronomy, School of Agriculture, Lovely Professional University, Phagwara, Punjab, India

Tanay Bhagwat Department of Chemical Engineering, Northeastern University, Boston, USA

Laxman Sonawane Bhushan Department of Entomology, School of Agriculture, Lovely Professional University, Phagwara, Punjab, India

Jairam Choudhary ICAR-Indian Institute of Farming Systems Research, Modipuram, Meerut, India

Ashim Debnath Department of Genetics and Plant Breeding, College of Agriculture, Narendra Deva University of Agriculture and Technology, Kumarganj, Ayodhya, Uttar Pradesh, India

Surajit De Mandal Key Laboratory of Bio-Pesticide Innovation and Application of Guangdong Province, College of Agriculture, South China Agricultural University, Guangzhou, China

Rubee Devi Department of Biotechnology, Dr. Khem Singh Gill Akal College of Agriculture, Eternal University, Baru Sahib, Sirmour, Himachal Pradesh, India

Shipa Rani Dey Department of Agronomy, School of Agriculture, Lovely Professional University, Jalandhar, Punjab, India

Vaishali Dhaka Department of Biotechnology, Lovely Professional University, Phagwara, Punjab, India

Daljeet Singh Dhanjal Department of Biotechnology, Lovely Professional University, Phagwara, Punjab, India

Neeraj Kumar Dubey Dept of Botany, Rashtriya Snatkottar Mahavidyalaya, Jaunpur, Uttar Pradesh, India

Padmanabh Dwivedi Department of Plant Physiology, I.Ag.Sci. BHU, Varanasi, Uttar Pradesh, India

Sara Atef Gezaf Botany Department, Faculty of Science, Arish University, North Sinai, Egypt

Sougata Ghosh Department of Microbiology, School of Science, RK University, Rajkot, Gujarat, India

Department of Chemical Engineering, Northeastern University, Boston, USA

Kapil Gupta Dept of Biotechnology, Central University of Rajasthan, Ajmer, Rajasthan, India

Rupal Gupta Amity Institute of Biotechnology, Amity University, Mumbai, Maharashtra, India

Shubhra Gupta Department of Biotechnology, Central University of Rajasthan, Ajmer, Rajasthan, India

Abd El-Latif Hesham Genetics Department, Faculty of Agriculture, Beni-Suef University, Beni-Suef, Egypt

Kashif Hussain Gyani Inder Singh Institute of Professional Studies, Dehradun, Uttarakhand, India

Touseef Hussain Department of Botany, Aligarh Muslim University, Aligarh, Uttar Pradesh, India

Durgesh Kumar Jaiswal Institute of Environment and Sustainable Development, Banaras Hindu University, Varanasi, Uttar Pradesh, India

Priyanka Jha Amity Institute of Biotechnology (AIBNK), Amity University, Kolkata Campus, Kolkata, West Bengal, India

Sudha Kannojiya Centre for Environment Science & Climate Resilient Agriculture, ICAR-Indian Agricultural Research Institute, New Delhi, India

Dhriti Kapoor Department of Botany, School of Bioengineering and Biosciences, Lovely Professional University, Phagwara, Punjab, India

Arun Karnwal Department of Microbiology, School of Bioengineering and Biosciences, Lovely Professional University, Phagwara, Punjab, India

Tanvir Kaur Department of Biotechnology, Dr. Khem Singh Gill Akal College of Agriculture, Eternal University, Baru Sahib, Sirmour, Himachal Pradesh, India

Zabish Khaliq School of Life Sciences, Forman Christian College (A Chartered University), Lahore, Pakistan

Jasmin Kumar Kheni Department of Biotechnology, Junagadh Agricultural University, Junagadh, Gujarat, India

Divjot Kour Department of Biotechnology, Dr. Khem Singh Gill Akal College of Agriculture, Eternal University, Baru Sahib, Sirmour, Himachal Pradesh, India

Ram Krishna ICAR-Directorate of Onion and Garlic Research, Pune, Maharashtra, India

Institute of Environment and Sustainable Development, Banaras Hindu University, Varanasi, Uttar Pradesh, India

Prasann Kumar Division of Research and Development, Lovely Professional University, Jalandhar, Punjab, India

Sandeep Kumar Centre for Environment Science & Climate Resilient Agriculture, ICAR-Indian Agricultural Research Institute, New Delhi, India

Sanjeev Kumar Department of Genetics and Plant Breeding, School of Agriculture, Lovely Professional University, Jalandhar, India

Vijay Kumar Department of Biotechnology, Lovely Faculty of Technology and Sciences, Lovely Professional University, Phagwara, Punjab, India

Vijay Kumar Department of Chemistry, Regional Ayurveda Research Institute for Drug Development, Gwalior, Madhya Pradesh, India

Monika Kundu Division of Agriculture Physics, ICAR-Indian Agricultural Research Institute, New Delhi, India

Lal Chand Malav ICAR-National Bureau of Soil Survey and Land Use Planning, Nagpur, India

Kauser Abdulla Malik School of Life Sciences, Forman Christian College (A Chartered University), Lahore, Pakistan

Samina Mehnaz School of Life Sciences, Forman Christian College (A Chartered University), Lahore, Pakistan

Hareem Mohsin Department of Allied Health Sciences, Superior University, Lahore, Pakistan

Salma Mukhtar School of Biological Sciences, University of the Punjab, Lahore, Pakistan

School of Life Sciences, Forman Christian College (A Chartered University), Lahore, Pakistan

Jayakumar Pathma Department of Entomology, School of Agriculture, Lovely Professional University, Phagwara, Punjab, India

Department of Trans-disciplinary Research, Division of Research and Development, Lovely Professional University, Phagwara, Punjab, India

Jyoti Bhimgonda Patil Department of Entomology, School of Agriculture, Lovely Professional University, Phagwara, Punjab, India

Ravishankar Patil Amity Institute of Biotechnology, Amity University, Mumbai, Maharashtra, India

Manoj Baliram Pohare Department of Infection, Immunity and Cardiovascular Diseases, University of Sheffield, Sheffield, UK

Shiv Prasad Centre for Environment Science & Climate Resilient Agriculture, ICAR-Indian Agricultural Research Institute, New Delhi, India

Yasir Rehman Department of Life Sciences, School of Science, University of Management & Technology (UMT), Lahore, Pakistan

Arpita Roy Department of Biotechnology, School of Engineering & Technology, Sharda University, Greater Noida, India

Hira Saleem Department of Microbiology & Molecular Genetics, University of the Punjab, Lahore, Pakistan

Samiksha Department of Biotechnology, School of Biosciences, Lovely Professional University, Jalandhar, India

Saurabh Saxena Department of Medical Laboratory Sciences, Lovely Faculty of Applied Medical Sciences (LFAMS), Lovely Professional University, Jalandhar, Punjab, India

Vinod Pravin Sharma Academy of Scientific and Innovative Research, CSIR-Indian Institute of Toxicology Research, Lucknow, Uttar Pradesh, India

Chhatarpal Singh School for Environmental Sciences, Babasaheb Bhimrao Ambedkar University, Lucknow, Uttar Pradesh, India

Jatinder Singh Department of Horticulture, School of Agriculture, Lovely Professional University, Phagwara, Punjab, India

Joginder Singh Department of Microbiology, Climate Mitigation and Sustainable Agriculture Research Lab (CMaSAR), Division of Research and Development, Lovely Professional University, Phagwara, Punjab, India

Department of Biotechnology, Lovely Professional University, Phagwara, Punjab, India

Major Singh ICAR-Directorate of Onion and Garlic Research, Pune, Maharashtra, India

Satendra Pal Singh Department of Botany, Y.D. (PG.) College, Lakhimpur, Uttar Pradesh, India

Simranjeet Singh Department of Biotechnology, Lovely Professional University, Jalandhar, Punjab, India

Punjab Biotechnology Incubators, Mohali, Punjab, India

Regional Advanced Water Testing Laboratory, Mohali, Punjab, India

Department of Microbiology, Climate Mitigation and Sustainable Agriculture Research Lab (CMaSAR), Division of Research and Development, Lovely Professional University, Phagwara, Punjab, India

Sonali Department of Biotechnology, Lovely Professional University, Jalandhar, Punjab, India

P. S. Soumia ICAR-Directorate of Onion and Garlic Research, Pune, Maharashtra, India

Rabia Tanvir University Diagnostic Lab (UDL), Department of Microbiology, University of Veterinary and Animal Sciences (UVAS), Lahore, Pakistan

Siddharth Thotapalli Department of Biotechnology, Lovely Professional University, Phagwara, Punjab, India

Varatharajalu Udayasuriyan Department of Plant Biotechnology, Centre for Plant Molecular Biology & Biotechnology, Tamil Nadu Agricultural Universityl, Coimbatore, India

Hina Upadhyay Department of Agronomy, School of Agriculture, Lovely Professional University, Phagwara, Punjab, India

Akash Varshney Department of Zoology, Y.D. (PG.) College, Lakhimpur, Uttar Pradesh, India

Jay Prakash Verma Institute of Environment and Sustainable Development, Banaras Hindu University, Varanasi, Uttar Pradesh, India

Sopan Ganpatrao Wagh Department of Bio-Science and Technology, Agribiotech College, VNMK University, Aurangabad, Maharashtra, India

Johnson Wahengbam Department of Entomology, School of Agriculture, Lovely Professional University, Phagwara, Punjab, India

Thomas J. Webster Department of Chemical Engineering, Northeastern University, Boston, USA

Ajar Nath Yadav Department of Biotechnology, Dr. Khem Singh Gill Akal College of Agriculture, Eternal University, Baru Sahib, Sirmour, Himachal Pradesh, India

Janardan Yadav Department of Soil Science, Institute of Agricultural Sciences, Banaras Hindu University, Varanasi, Uttar Pradesh, India

Neelam Yadav Veer Bahadur Singh Purvanchal University, Jaunpur, Uttar Pradesh, India

Soil Microbiomes for Healthy Nutrient Recycling

1

Shiv Prasad, Lal Chand Malav, Jairam Choudhary, Sudha Kannojiya, Monika Kundu, Sandeep Kumar, and Ajar Nath Yadav

Abstract

Nutrient cycling is a vital process in the ecosystem by which movement and exchange of nutrients in available forms from the environment into living organisms and then subsequently are recycled back into the atmosphere. Chemical elements such as C, O, H, S, N, and P are necessary to live. These elements must be recycled for organisms to live and to sustain plant growth and yield. In this context, microbes in the soil play a dynamic role. They help to release mineral nutrients through matter organic decomposition and mineral recycling. These mineralized nutrients are then absorbed by plant roots with water and used to make new organic material. They are also crucial to maintain soil structure and soil quality for sustainable plant growth. Currently, most of the world's soils are distinguished deficient in these nutrients, and there would be high demand for chemical fertilizers to meet the deficiency of nutrients. Synthetic chemical fertilizers are undoubtedly necessary for the healthy growth of plants. But, their

S. Prasad (✉) · S. Kannojiya · S. Kumar
Centre for Environment Science & Climate Resilient Agriculture, ICAR-Indian Agricultural Research Institute, New Delhi, India

L. C. Malav
ICAR-National Bureau of Soil Survey and Land Use Planning, Nagpur, India

J. Choudhary
ICAR-Indian Institute of Farming Systems Research, Modipuram, Meerut, India

M. Kundu
Division of Agriculture Physics, ICAR-Indian Agricultural Research Institute, New Delhi, India

A. N. Yadav
Department of Biotechnology, Dr. Khem Singh Gill Akal College of Agriculture, Eternal University, Baru Sahib, Sirmour, Himachal Pradesh, India

A. N. Yadav et al. (eds.), *Current Trends in Microbial Biotechnology for Sustainable Agriculture*, Environmental and Microbial Biotechnology,
https://doi.org/10.1007/978-981-15-6949-4_1

injudicious application is also harmful to the environment and living beings. However, the entire range of microbes associated with plants and their potential to replace synthetic farm inputs has only recently started. Accordingly, there is a need to explore the potent soil microbes for efficient nutrient recycling and identify alternative eco-friendly options for reducing chemical fertilizer's use and its adverse impacts. In this scenario, maintaining soil fertility and crop productivity using natural microbial diversity could be the best approach for enhancing the bioavailability of nutrients and improving soil health.

Keywords

Cycling · Microbes · Nutrient · Plants and soil quality · Soil · Sustainability

1.1 Introduction

Over the last decades, the global demand for food products has increased dramatically (Elferink and Schierhorn 2016). Global food demand is projected to enhance by 59–98% by 2050. In developing nations, food demand is also increasing, where the expansion of croplands resources is limited. In this scenario, for enhancing food production from existing land is hard to contribute to meet such an essential requirement (Bargaz et al. 2018). In order to address this problem, there is a need to enhance agricultural production sustainably through the use of efficient agro-bioresources, whereas soil microbial diversity can play an important role and also help to mitigate many problems associated with soil fertility, abiotic stress, insect pests, and diseases (Tilman et al. 2011; Utuk and Daniel 2015; Timmusk et al. 2017).

Soil serves as a plant growth medium and a major source of plant nutrients for quality food production. Nitrogen (N), phosphorus (P), potassium (K), and iron (Fe) are essential nutrients in crop production. Since most of the world's soils are known to lack in these nutrients, and there would be a high demand for chemical fertilizers to meet the deficiency of nutrients. Hence, there is an urgent need to explore the potential of soil microbes for proper nutrient recycling and to recognize alternative, sustainable, environment-friendly options for reducing the use and impacts of synthetic fertilizers (Malav et al. 2015). In this scenario, maintaining soil fertility and crop productivity through the use of natural microbial diversity could be a well-off approach for enhancing the bioavailability of nutrients and increasing soil health (Singh et al. 2015; Timmusk et al. 2017; Bargaz et al. 2018; Rana et al. 2020a, b).

Soils are regarded as home to a wide range of macro- and microorganisms of rhizospheric nature. Soil microbe diversity is the fundamental key component in regulating biogeochemical cycles (e.g., C, N, P, and many more). Biogeochemical cycling affects soil ecosystems, composition, and functions as well as the capacity of soils to provide readily available nutrients to plants and animals by converting dead organic matter into various nutrient forms and many auxiliary services to living beings (Aislabie et al. 2013). Biofertilizers and organic manure could be regarded as a better choice in the crop integrated nutrient management approach (Chaer et al. 2011; Kour et al. 2020b).

In this integration, soil microbes such as bacteria, archaea, and fungi play various crucial roles. Though very little is acknowledged about the small creature that is accountable for countless soil mechanisms in natural and managed agro-ecosystems (Yadav and Sidhu 2016; Sahu et al. 2017). Soil microbes have an immense impact on relations between soil and plant and microbe and play a vital role in sustaining soil fertility (Yadav et al. 2020c). Nutrient cycling is the most significant of these relationships. This chapter explains the potential of soil microbes for proper nutrient recycling, including diversity, abundance, and distribution, and their role in nutrient cycling of soil microbe organisms.

1.2 Soil Health and Sustainability

Soil health is defined as functional ability within agro-ecosystem boundaries that support biological productivity, promote plant and animal fitness, and sustain environmental quality (Doran and Parkin 1994). Healthy soil functions are to resist erosion, support water, and nutrient cycling, inactivate toxic pollutants, suppress pathogens, maintain soil organic matter, and enhance overall system productivity and sustainability (Singh et al. 2015; Dubey 2016; Sahu et al. 2017). The soil health directly or indirectly impact plant health, environmental health, and food safety and quality (Singh et al. 2020a; Takoutsing et al. 2016). The soil serves as a biological filter for removing unwanted solids and gaseous constituents from air and water (Singer and Ewing 2000; Sahu et al. 2017). Healthy soils produce nutritious crops that, in turn, nourish humans and animals. Certainly, soil quality is directly linked with food quality and quantity. Maintaining healthy soil implies managing land sustainably (FAO 2015). Managing soil health is not only necessary for agricultural sustainability but also for ecosystem function. However, erosion, deforestation, and intensive agriculture have led to the degradation of many soils. As we know, soils constitute the foundation for sustainable agricultural development. Therefore, keeping healthy is essential to maintain food production for future generations.

1.3 Soil Quality

Soil quality is the capability of the soil to perform functions that are crucial to agriculture and the environment. Soil Science Society of America established soil quality as the ability of particular kind of soil to function, within a natural or managed ecosystem, to support plants and animals productivity, maintain or improve quality of water and air, and promote human health and habitation (Carter et al. 1997). Soil quality is not limited to agricultural lands although most soil quality work has been done in agrarian systems. It is a blend of inherent and dynamic soil properties. Soil properties include soil organic matter, nutrient, soil structure, water infiltration rate, bulk density, and water holding capacity. Soil properties can change over months and years in response to land use.

Soil properties are dynamic and changed, depending on land management practices and the inherent properties of parental material (rocks). The soil quality is necessary for the integrity of ecosystems and sustainably supports human and animal health, plant growth (Pankhurst and Doube 1997). Declining soil quality is a vital concern worldwide (Singer and Ewing 2000). Healthy soils improve crop yields, drought and flood tolerance, and air and water quality and balance a range of other functions to satisfy the demands of both farmers and the community. Soil quality is a critical part and basic features of sustainable agro-ecosystem management, similar to water and air quality. The relationship between soil quality, environmental quality, and agricultural sustainability is shown in Fig. 1.1.

1.4 Soil Quality Indicators

Soil quality indicators are used to assess and identify soil properties that are responsive to management, affect, or associated with environmental consequences. There are three primary levels of soil indicators: chemical, physical, and biological. Soil quality integrates all of these indicators. Table 1.1 below shows the relationship between indicator type and soil function. Organic matter or soil carbon is itself an indicator of soil quality (Doran and Parkin 1996). It further affects other indicators like soil aggregate-stability (physical), nutrient availability (chemical), and nutrient cycling (biological). Chemical indicators give knowledge about equilibrium within soil solution (water and nutrients) and exchange sites (clay particles, soil organic

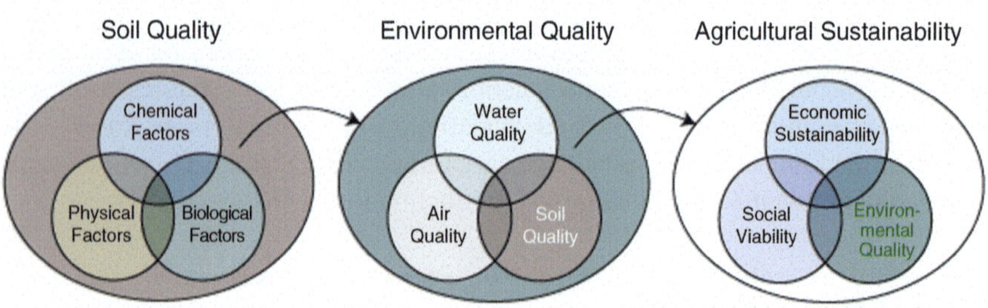

Fig. 1.1 Relationship between soil quality, environmental quality, and agricultural sustainability

Table 1.1 Indicator type and soil function relationship

Category of indicators	Related soil functions
Chemical	Cycling of nutrient, water relation, buffering capacity
Physical	Stability and physical support, water relation, habitats
Biological	Biodiversity, cycling of nutrient, filtering

matter), plant health, nutritional demands of plant and soil communities, levels of soil contaminants, and their availability for uptake by plants and animals.

Physical indicators give knowledge about soil hydrologic properties, such as water retention, that affects the availability of water to plants. Some indicators are related to nutrient availability by their impact on rooting volume and aeration status. Other measures tell us about the erosional situation. Biological indicators provide information regarding the organisms that form the soil food web that is responsible for organic matter decomposition and cycling of nutrients. Soil microbial respiration indicates the soil's ability to sustain plant growth (Doran and Parkin 1996).

1.5 Potential Role of Microbes for Soil Health

Soil microorganisms are responsible for making nutrient and organic matter cycling, in improving soil fertility, and leading to ecosystem productivity. Soil microbes form symbiotic relationships with plant roots (rhizobia, actinomycetes, mycorrhizal fungi, diazotrophic bacteria). They have the potential to improve nutrient mineralization and availability, produce plant growth hormones, and are antagonists of plant pests, fungi, or diseases (biocontrol agents). These organisms typically live in the soil, although, in some situations, they may help increase their communities by either inoculation or using several farm management techniques that improve their abundance and activity.

1.5.1 Soil as a Microbial Habitat

Soil represents a hospitable and dynamic habitat for microorganisms and is occupied by a wide range of microbial species. Microbes hold a fraction (<0.5%) of the total soil volume of topsoil. Usually, per gram of soil between one and ten million with a dominant number of bacteria and fungi is present (Fig. 1.2). The decomposition of organic residues and the cycling of nutrients is the significant role played by microbial species in soil (Pankhurst and Doube 1997). However, the soil also contains countless microorganisms capable of causing human disease (Rastegari et al. 2020b). Microbes are connected with the decay, and the nutrient cycling process is capable of sustaining and responding to quick changes in the environment. Hence, they rapidly adapt to environmental conditions, changes in microbial populations, and activities. Therefore, it can be considered as an excellent indicator of change in soil health (Singh et al. 2015; Sahu et al. 2017).

Soil microbes are classified as bacteria, actinomycetes, fungi, algae, and protozoa. The interactions of gases, water, organisms, and organic and inorganic constituents can be visualized in a per gram of soil (Fig. 1.3). Up to ten billion bacterial cells survive each gram of soil in and nearby plant roots, a sphere is known as the rhizosphere. Rhizobacteria are the most abundant group of soil microbes, both in absolute number and in diversity. They perform a vital role in nutrient cycling and decomposition of organic residues (Pankhurst and Doube 1997).

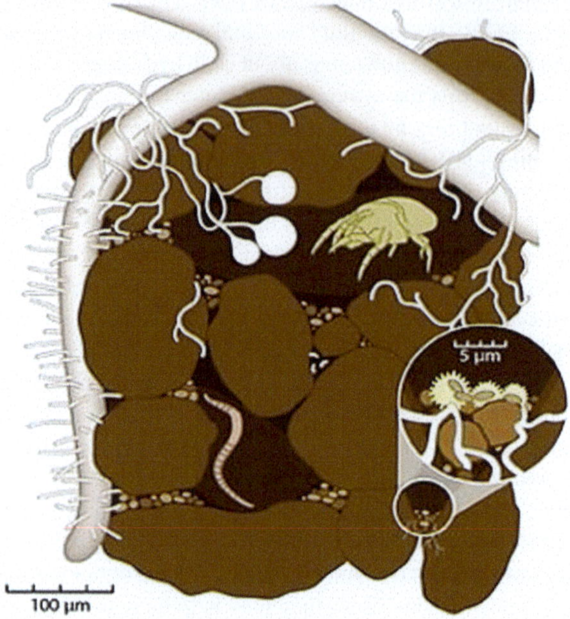

Fig. 1.2 A soil aggregate or ped is a naturally formed assemblage of sand, silt, clay, organic matter, root hairs, microorganisms, and their secretions, and resulting pores. (Source: Fortuna 2012)

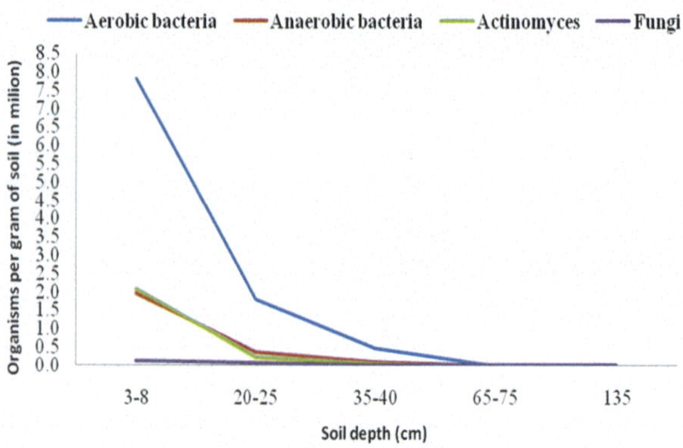

Fig. 1.3 Distribution of soil microorganisms with depth

1.5.2 Soil Microbes and Agro-Ecosystem Stability

Ecosystem stability is an essential part of sustainability, where microbes play a critical role. Stability of equilibrium of any system has two components: (i) Resistance—the ability of the ecosystem to continue to function without any change when stressed by disturbance (ii) Resilience—the strength of the ecosystem to recover after disturbance (Odum 1989; Seybold et al. 1999). Soil is the junction between the air, water, minerals, and organisms and is performing various functions in the natural and agro-ecosystem that we call ecosystem services. Soils play an essential role in the entire natural ecological cycles—C, N, oxygen, water, and nutrient, and also provide benefits through their contribution in several unique processes called ecosystem services. Suggested practices to increase agro-ecosystem stability and function are given in Table 1.2. Soil microbial biodiversity reflects the variability among living microorganisms extending from the countless of invisible microbes to more familiar macro-fauna like earthworms and termites. Soil microbes play an essential role in agro-ecosystem stability, including rich biodiversity, healthy biological cycles, and soil microbial activity; consequently, they are contributing to the build-up of stable soil agro-ecosystem (Doran and Parkin 1996; Pankhurst and Doube 1997; Madsen 2005; Fortuna 2012; Yadav et al. 2020a, b).

Table 1.2 Suggested practices to increase agroecosystem stability and function

Stability factor	Examples	Suggested factors
Disturbance (frequency and intensity)		
Chemical	Fertilizers and pesticides	Account for mineralization of organic amendments, be aware of non-target effects of pesticides
Biological	Introduction of exotic or weed species	
Physical	How often and what kind of tillage	Reduced, minimum or no-till practices
Diversity		
Species	Genetic resources (crop) or competition for water and nutrients (weeds)	Intercropping of varieties
Structure or habitat	Type of plant heights (e.g., to increase niche space among insect predators)	Intercropping of species
Temporal	Variety of plants through time	Rotations
Complexity		
Trophic groups	How many functions are represented	(For all types of complexity) Practices that improve habitat for soil
Redundancy	How many populations perform each function	Organisms, such as organic matter amendments, reduced disturbances
Food web structure	How do all of these groups interact	Increased diversity of resources and niches (habitat) (see above)
Nutrient or energy flux	How fast nutrients or energy move through the system Input: Output efficiencies	Low input and high organic matter Eliminate over-applications

Source: Doran and Parkin (1996)

1.5.3 Microorganisms and Soil Functions

Soil functions provide many benefits, such as cycling of nutrients, maintaining biodiversity and habitat, water relations, and maintaining water quality, acting as a biofilter and buffering providing physical stability and support crop production, and carbon sequestration. The summary of soil functions and its advantages for humans is given in Table 1.3. It is vital to maintain or improve soil quality over time and provide essential services in the face of disturbance, whether it is natural or human-induced. Typically, soil is not considered healthy if it is managed for short-term productivity at the cost of future degeneration (Doran and Parkin 1994).

The soil can store, govern the discharge and cycling of nutrients and elements. During these biogeochemical processes, similar to the water cycle, nutrients can be transformed into plant-available forms, contained in the soil, or even lost to air or water. Soil promotes the growth of a variety of plants, animals, and soil microorganisms, regularly by giving a different physical, chemical, and biological habitat. Soil acts as a filter to maintain water and air quality. Excess nutrients and toxic compounds can be degraded or otherwise made unavailable to plants and animals. Soil can maintain its porous structure to allow passage of air and water, resist erosive forces, and give a mechanism for plant roots. Soils also provide anchoring support for social structures (Doran and Parkin 1994).

Table 1.3 Summary of soil functions and advantages for humans

	Advantages for humans	
Soil function	On-site value	Off-site value
Cycling of nutrient	– Nutrients delivery to crops and plants – Storage of carbon to improves a kind of soil functions	– Improves air and water quality – N and C storage to reduce greenhouse gas emissions
Maintaining biodiversity and habitat	– Supports the growth of crops, rangeland plants, and trees – May increase resistance and resilience to stress – Reduces pesticide resistance	– Helps maintain genetic diversity Supports wild species and reduces extinction rates – Improves esthetics of landscape.
Water relations	– Provides erosion control – Allows on-site water recharge of streams and ponds – Makes water available for plants and animals	– Provides flood and sedimentation control – Groundwater recharge
Filtering and buffering	– Can maintain salt, metal, and micronutrient levels within range tolerable to plants and animals	– Improves water and air quality
Physical stability and support	– Acts as a medium for plant growth – Supports buildings and roads	– Stores archeological items – Stores garbage
Multiple functions	– Sustains productivity	– Maintains or improves air and water quality

Source: Doran and Parkin (1994)

Nutrient cycling and water regulation functions are natural soil processes occurring in each ecosystem. These functions provide various opportunities to humans for betterment of quality life, achieve sufficient food, quality water, flood control, and several more. Soil pollution can happen either because of anthropogenic activities or because of the natural process. However, it is mostly due to anthropogenic activities. A human can enhance the value of soil and take its maximum benefits because land management choices affect soil functions. Thus, it is necessary to realize what benefits we obtain from the earth. So we can have the greatness of achieving land management in a way that maintains essential soil functions. Several main benefits are long term or go beyond if the land is being managed properly. The community should respect the value of many off-site services and profits and the extent to which the landowner or community should pay to maintain these soil functions.

1.6 Role of Microorganisms in Nutrient Cycling

Soil microbes perform various functions in the pedosphere. They are essential in controlling biogeochemical processes (Table 1.4). The critical soil microbes regulated roles are: (i) soil organic matter formation and turnover which includes mineralization and carbon sequestration, (ii) nutrient cycling, (iii) disease dissemination and prevention, (iv) contaminant depletion, and (v) soil structure improvement (GHGs) (CO_2, CH_4, and N_2O, etc.) are the by-products of metabolic redox reactions of carbon and nitrogen compounds in soils (Madsen 2008). Nitrogen fertilizer application and cultural practices in soil management can stimulate microbial processes such as nitrification, denitrification, and mineralization that play a major role in the emission of GHG (Pathak et al. 2003; Rastegari et al. 2020a).

The quantity and composition of the microbial biomass depend on soil characteristics and the abundance of carbon (C) for energy and cell metabolism. Soil carbon inputs varied in chemical composition and nutrient content. Carbon recycling, degradation, and microbial function frequently contribute to an increased organic matter, which leads to soil aggregation. Various ecosystems have different types of potential to support biota and sequestration of soils C in organic matter. Soil organic carbon (SOC) is the backbone of organic matter, which is the source of energy for most of the soil biota. Microbiological decomposition of crop residues and organic matter provides access to carbon and the nutrients needed by most living species. Mineralization of organic-nitrogen into ammonium and the use of nitrogenous chemical fertilizers containing ammonium promote nitrification with the help of nitrifying bacteria and archaea that turn ammonium into nitrate. Therefore, nitrate undergoes a further microbially induced stage, denitrification (Maier et al. 2009; Fortuna 2012).

The food web present in the soil consists of various groups of microbes and helps for nutrients transfers and flow between the biotic and abiotic components (Sylvia et al. 2005). Mesofauna (collembolan and mites) perform a prominent role in nutrient cycling by slicing stocks into smaller pieces and directly helping to enhance the

Table 1.4 Examples of essential biogeochemical processes catalyzed by microorganisms in biosphere habitats

Process		Process	
Carbon cycle	**Nature of process**	***Nitrogen cycle***	**Nature of process**
Photosynthesis	Light-driven CO_2 fixation into biomass	N_2 fixation	N_2 gas becomes NH_3
C Respiration	Oxidation of organic C to CO_2	NH_4^+ oxidation	NH_3 becomes NO_2^-, NO_3^-
Cellulose decomposition	Depolymerization, respiration	Anaerobic NH_4^+oxidation	NO_2^- and NH_3 becomes N_2 gas
Methanogenesis	CH_4 production	Denitrification	NO_3^- is used as an electron acceptor and converted to N_2 gas
Aerobic CH_4 oxidation	CH_4 becomes CO_2		
Anaerobic CH_4 oxidation	CH_4 becomes CO_2	***Sulfur cycle*** S_2 oxidation	S^{2-} and S^0 become SO_4^{2-}
		SO_4^{2-} reduction	SO_4^{2-} is used as an electron acceptor and converted to N_2 gas
Biodegradation	**Nature of process**	***Other elements***	**Nature of process**
Synthetic organic compounds	Decomposition, CO_2 formation	H_2 oxidation	H_2 is oxidized to H^+; electrons reduce other substances
Petroleum hydrocarbons	Decomposition, CO_2 formation	Hg methylation and reduction	Organic Hg is formed and Hg^{2+} is converted to Hg
Fuel additives (MTBE)	Decomposition, CO_2 formation	(per)chlorate reduction	Oxidants in rocket fuel and other sources are converted to chloride
Nitroaromatics	Decomposition, CO_2 formation	U reduction	U oxyanion is used as an electron acceptor, therefore immobilized
Pharmaceuticals, personal care products	Decomposition	As reduction	As oxyanion is used as an electron acceptor; thus toxicity is diminished
Chlorinated solvents	Compounds are chlorinated through respiration in anaerobic habitats	Fe oxidation, acid mine drainage	FeS ores are oxidized, strong acidity is generated

As, arsenic; C, carbon; CH_4, methane; CO_2, carbon dioxide; Fe, iron; FeS, Iron sulphide; H, hydrogen; Hg, mercury; Hg^{2+}, mercuric ion; MTBE, methyl tertiary butyl ether; N_2, nitrogen; NH_3, ammonia; NH_4^+, ammonium; NO_2^-, nitrite; NO_3^-, nitrate; S_0, elemental sulfur; S^{2-}, sulfide; SO_4^{2-}, sulfate; and U, uranium

Source: Madsen (2005), Fortuna (2012)

surface area. It has greater exposure to microbes that are key to carbon cycling. All food webs include many trophic levels in a food chain. If organic carbon is derived from living animals, the term grazing is used. Soil microbes form an essential part of the detrital type of food chain because they obtain their organic carbon from dead substances. Elemental ratios of C:N:P:S relatively are constant in the biological systems and organisms. Those ratios and mass balance allow researchers to establish biochemical changes between species.

Most soil microbe members are chemo-heterotrophs, suggesting they receive carbon and energy by the oxidation of organic materials (Kumar et al. 2019b; Singh et al. 2020b). C-sequestration restricts the mineralization mechanism mediated by the CO_2 producing chemo-heterotrophs. Mineralization process by-products are metabolites, heat, and CO_2. CO_2 production can minimize concentrations of O_2 producing anoxic sites within micro-aggregates resulting in variation of micro-environments (Van Elsas et al. 2007; Sylvia et al. 2005). These microsites are habitats in which CO_2 is converted by archaea known as methanogens into CH_4 a GHG by anaerobic respiration. In neighboring microsites, methane can undergo oxidation into CO_2 with the help of a group of bacteria known as methanotrophs.

Microbes play a vital role in nutrient cycling and organic substances decomposition. This transforms the natural materials into biomass or mineralizes them to CO_2, water, and nutrients (Bloem et al. 1997; Pankhurst and Doube 1997; Malyan et al. 2019). Such effective microbes are also concerned with the production and oxidation of waste products, including organic industrial substances (Singh et al. 2016; Kumar et al. 2019a). The functions of these productive rhizospheric organisms own the potential to have a useful test of soil sustainability. This attribute cannot be obtained with higher organism diversity analysis and physical/chemical tests. Microorganisms respond quickly to environmental changes; hence, they adapt quickly to ambient conditions. This adaptation makes it possible for microbial studies to differentiate in the assessment of soil health. Thus, improvements in soil biota communities and activities can be an excellent predictor of soil health (Singh et al. 2015; Sindhu et al. 2016).

Soil is a diverse environment for several life forms and provides vegetation with mechanical assistance from which nutrients are derived. Soil microbes regularly interact with each other; at times, these relationships are beneficial to both parties (mutualism), symbiotic, and competitive. It increases soil health because the "healthy" soil biota may fight against the "poor" ones and also contribute significantly by degrading organic compounds to make nutrients available. Thus, the similar basic soil structure in the different geographical regions is found to support different biocommunities. Soils have different texture due to the percentage contribution of sand, silt, and clay, and that includes a diversity of microhabitats that sustain a wide variety of microbes. The atmosphere within soil shows less oxygen content from the above-ground due to the utilization of the available O_2 by soil biota and other metabolisms. Similarly, the concentration of CO_2 in the soil is higher than the level at the above-ground due to the generation of it as a by-product of microbial reactions (Sarkar et al. 2017; Kumar et al. 2017).

Microbes' reaction to environmental changes/stress is rapid relative to higher species, owing to their top surface to volume ratio. Those productive microbial communities may be regarded as soil architects (Rajendhran and Gunasekaran 2008). Several environmental functions, including plant growth, drinking water protection, or carbon sequestration, are strictly related to microbial service and its functional characteristics (Torsvik and Ovreas 2002; Lombard et al. 2011). A study on the development of abiotic and biotic interactions is very complex microbes function on a 3 μm scale and form biogeochemical soil interfaces (Totsche et al. 2010; Monier et al. 2011). Furthermore, most functional features, such as plant litter depletion or the formation of food web systems and nutrient cycling, are not the function of a single organism, but of closely associated microbial communities (Aneja et al. 2006; Sharma et al. 2012).

1.6.1 Organic Matter Decomposition

The decomposition of various forms of soil organic matter is one of the essential functions of soil biota. To make these organic compounds accessible to the autotrophic organisms, they must be processed into simple inorganic forms. Mineralization is the process of organic matter conversion into simpler inorganic forms, which is rendered primarily soil microbes, mostly fungus and bacteria (Gupta and Germida 1988; Xu et al. 2015). The organic substances that are brought into the soil are divided into three groups: the easily decomposable, moderately decomposable, and difficult to degrade, distinctly attached by various microbiota types. The consequence of microbial mineralization is, on the one hand, the release of energy, water, gases, etc., and, on the other, the creation of complex amorphous material humus through the process of humification.

1.6.2 Carbon Cycling

The balance between respiration and photosynthesis dominates terrestrial carbon cycling. Carbon is transferred into the soil from the atmosphere by autotrophic carbon-fixing species, primarily photosynthetic crop/plants and also photo- and chemoautotrophic microorganisms, which synthesize carbon dioxide (CO_2) in organic matter. Respiration is the primary process behind the transfer of carbon back to the atmosphere with the help of both autotrophic and heterotrophic organisms. The reverse pathway involves the decomposition of organic matter by heterotrophic carbon consuming bacteria, which use plant, animal, or microbial origin carbon as a metabolism base, retaining part of few carbons in their biomass and adding the remainder back into the environment as metabolites or as CO_2 (Gougoulias et al. 2014).

1.6.3 Nitrogen Cycle

All organisms require nitrogen because the protein and nucleic acids are essential elements. Animals derive nitrogen from organic sources, whereas plants derive inorganic nitrogen sources like NH_4^+ and NO_3- (Schimel and Bennett 2004). Nitrogen fixation is the reduction of atmospheric N_2 gas to NH_4^+. Nitrogen fixation is the only natural mechanism by which new nitrogen reaches the biosphere, and is thus necessary for the ecosystem's functioning. The enzyme nitrogenase catalyzes N-fixation. The ammonium generated by N-fixation is assimilated into amino acids and then converted into proteins. Under nitrogen-scarce conditions, N-fixing microorganisms have an advantage. Nitrogen fixation is carried out by free-living microorganisms such as *Azotobacter*, *Burkholderia*, *Clostridium,* and few methanogens, some of which may be kept associated with the rhizosphere of crops plants, and bacteria that shows symbiotic relationships with plants like *Rhizobium, Mesorhizobium,* and *Frankia* (Maier et al. 2009; Fortuna 2012; Santi et al. 2013). The nitrogen-fixing microbes have been reported from different habitats and host worldwide belonging to different genera of *Arthrobacter*, *Azoarcus*, *Azospirillum*, *Azotobacter*, *Bacillus*, *Enterobacter*, *Gluconoacetobacter*, *Herbaspirillum*, *Klebsiella*, *Pseudomonas*, and *Serratia* (Subrahmanyam et al. 2020; Suman et al. 2016; Yadav 2020).

Exudates from crop plants provide some of the energy needed to fasten nitrogen fixation. In agricultural soils, a significant source of N is rhizobia, which forms root nodules in symbiotic relationships with introduced legumes such as clover, lucerne, or lotus. Symbiotic interaction N-fixation levels are sometimes two to three orders of magnitude higher than free-living bacteria in the soil. Ammonia or ammonium ions are oxidized to nitrite and then to nitrate during nitrification. The two steps in nitrification—first are the formation of nitrite and then nitrate—are carried out by two distinct microbes. In soils, ammonia oxidation to nitrite is conducted by bacteria such as *Nitrosospira* and *Nitrosomonas*, while bacteria such as *Nitrobacter* and *Nitrospira* oxidize nitrite to nitrate. Nitrifying microbes utilize the energy derived from nitrification to assimilate CO_2. Nitrification is especially vital in soils as the degradation of nitrite and nitrate ions from ammonium to nitrite shifts their charge from positive into negative.

1.6.4 Siderophores Production

Iron is a vital nutrient, and part of many compounds that regulate and promote plant growth and development. In the soil, naturally, iron is present as a ferric ion (Fe^{3+}), which is too low to promote and facilitate soil microbial growth. It has been reported that some bacteria possess the ability to assimilate unavailable iron to overcome iron stress by producing ferric-specific ligands, referred to as siderophores, which are usually of low molecular weight (400–10,000) (Neiland and Nakamura 1997). Such microorganisms or bacteria are real iron scavengers as they have a high affinity for iron (Fe^{3+}) chelators that transfer iron to bacterial cells (Leong 1986). Soil microorganisms, especially rhizobacteria, are of great interest for siderophore production. Under iron stress conditions, these bacteria have a high chelating affinity for Fe^{3+} than Fe^{2+} ions, and Fe^{3+}ions are transferred to bacterial cells (Neiland 1995). Recent studies have indicated that biological control of different phytopathogenic organisms could be achieved using siderophore producing microorganism such as *Alcaligenes*, *Bacillus*, *Clavibacter*, *Curtobacterium*, *Flavobacterium*, *Kluyvera*, *Microbacterium,* and *Pseudomonas* (Verma et al. 2016, 2017; Yadav et al. 2017a).

1.6.5 Hormones Production

Microbial synthesis of the phytohormone has been known for a long time. Plant growth-promoting rhizobacteria (PGPR) is a group of microorganisms that colonize several plant species' rhizosphere and roots. They confer beneficial effects to plants by a variety of mechanisms, including indole-3-acetic acid synthesis of phytohormone auxin (IAA), which is essential for plant growth (Patten and Glick 1996; Kour et al. 2020a; Rana et al. 2020a, b). Eighty percent of microorganisms isolated from the rhizosphere of various crops have the potential to synthesize and release IAA as secondary metabolites (Patten and Glick 1996; Yadav et al. 2017b). The most common phytohormone produced by PGPR is indole-3-acetic acid, which participates in root growth and increases root surface area, thereby enabling plants to absorb more nutrients from the soil. Gibberellins associated with plant extension, mainly stem tissue, have been reported to be produced by *Bacillus pumilus* and *B. licheniformis* in the form of gibberellic acid. The phytohormone-producing rhizospheric microbes, when inoculated to crops, help plant growth promotion, enhance yield, and increase soil fertility for sustainable agriculture (Kumar et al. 2016; Singh and Yadav 2020; Yadav et al. 2018b).

1.6.6 Phosphate Solubilization

Phosphorus, after nitrogen, holds a second essential role in various critical processes in plant growth and development, including the division of cells, photosynthesis, and decomposition of sugar, energy, and nutrient conversion in a crop plant. Plants utilize phosphate ion the form of phosphate anions, but phosphate anions are incredibly reactive and get immobilized through precipitation with cations present in the soil such as Ca^{2+}, Mg^{2+}, Fe^{3+}, and Al^{3+}. Rhizobacteria help in the decomposition of organic compounds and make phosphorus available by the action of minerals and acids released by soil bacteria. Phosphorus mineralization is greatly affected by the microbial community, and phosphate-solubilizing bacteria such as species of *Bacillus* and *Paenibacillus* have been applied to soils to enhance the phosphorus status of plants specifically. *Pseudomonas, Bacillus,* and *Rhizobium* are the most potent phosphate solubilizers in the cropping system (Rodríguez and Fraga 1999; Kour et al. 2019).

Possible mechanisms for solubilization from organically bound phosphate involve either enzymes, namely C-P lyase, nonspecific phosphatases, and phytases. However, most of the bacterial genera solubilize phosphate through the production of organic acids such as gluconate, ketogluconate, acetate, lactate, oxalate, tartrate, succinate, citrate, and glycolate (Yadav et al. 2015). The rhizospheric phosphate utilizing bacteria could be a promising source for plant growth-promoting agent in agriculture (Rana et al. 2019; Yadav et al. 2018a). The rhizospheric phosphorus-solubilizing microbiomes may be used for mitigation of abiotic stress in plants such as high/low temperatures, alkaline/acidic, drought, and saline environments (Kour et al. 2018, 2019; Kumar et al. 2019c).

1.6.7 Manganese (Mn) Solubilizers

Redox condition and hydrogen ion concentration (pH) are two significant factors that influence the availability of Mn in the rhizosphere. Some rhizosphere bacteria such as *Bacillus, Pseudomonas,* and *Geobacter* can reduce oxidized Mn^{4+} to Mn^{2+} form, which is metabolically useful for crops (Wani et al. 2015). Consequently, Mn-reducer function in the rhizosphere is strongly favored. Products of organic matter can also help reduce Mn (Hue et al. 2014). *Gaeumannomyces graminis* is also an Mn oxidizer that impairs root lignification at infection sites. Effective rhizosphere Mn reducers like *Pseudomonas* sp. could have beneficial effects on plant nutrition and also help in biocontrol of pathogens. In comparison, Mn oxidization by rhizosphere bacteria supports plant growth in flooded soils where the abundance of Mn^{2+} can be high.

1.6.8 Iron Solubilizers

Iron dynamics in the rhizosphere is almost similar to that of manganese (Mn). Fe in the soil is a part of the structure of insoluble minerals Goethite (FeOOH) or hematite, in oxidized forms Fe^{3+}. Rhizosphere bacteria, such as *Bacillus, Pseudomonas, Geobacter, Alcaligenes, Clostridium,* and *Enterobacter,* can reduce oxidized Fe^{3+} to reduced Fe^{2+} form required by crop plants. Electrons and hydrogen ions are available in the rhizosphere, and consequently, Fe is diminished. However, it can be reprecipitated (Kaur et al. 2020; Wani et al. 2015).

1.6.9 Soil Enzymes

Soil enzymes act as a booster in the redox reaction through which plant residues decompose and make nutrients available. The material on that soil enzyme that has worked is considered the substrate. The enzymatic reaction releases a product, which may be a substrate-containing nutrient. There are so many sources of enzymes in the soil, such as living and dead microorganisms, soil animals, plant roots, and plant residues. Enzymes that are stable in the soil matrix retained or form complexes with humus, clay, and humus-clay compounds, which are no longer associated with sustainable cells. Stabilized enzymes contribute 40–60% of the total enzyme activity. It is believed that 40–60% of enzyme activity can come from stabilized enzymes. Thus, behavior is not strongly associated with microbial biomass or respiration. Enzyme activity is then the combined effect of long-term microbial development and viable sampling population activity.

Enzymes respond to changes in soil management long before more changes in soil quality indicators can be identified. Soil enzymes play a crucial part in the decomposition of organic matter and nutrient cycling (Table 1.5). There is no substantial evidence, apart from phosphatase activity, that directly relates enzyme activity to nutrient availability or crop production. The relation may be indirect

Table 1.5 Role of soil enzymes

Enzyme	Organic matter substances acted on	End product	Significance	Predictor of soil function
Beta glucosidase	Carbon compounds	Glucose (sugar)	Energy for microorganisms	Organic matter decomposition
FDA hydrolysis	Organic matter	Carbon and various nutrients	Energy and nutrients for microorganisms, measuremicrobial biomass	Organic matter decompositionnutrient cycling
Amidase	Carbon and nitrogen compounds	Ammonium (NH_4)	Plant-available NH_4	Nutrient cycling
Urease	Nitrogen (urea)	Ammonia (NH_3) and carbon dioxide (CO_2)	Plant-available NH_4	Nutrient cycling
Phosphatase	Phosphorus	Phosphate (PO_4)	Plant-available P	Nutrient cycling
Sulfatase	Sulfur	Sulfate (SO_4)	Plant-available S	Nutrient cycling

because nutrient mineralization is achieved with the contribution of enzyme activity to plant-available sources. Limited enzymatic activity (e.g., pesticide degrading enzymes) may contribute to dangerous chemical accumulation for the environment. Many of these toxic chemicals can also impede soil enzymatic activity.

Apart from these, as we know, plant growth and yield depend on the availability of nutrients and their efficient management. Therefore, it is essential to adopt the 4R Nutrient Stewardship concept of right nutrient application (i) Right source, (ii) Right rate, (iii) Right time, and (iv) Right place (Johnston and Bruulsema 2014). This concept integrates soil health with sustainable and precision farming practices. Right source means matching the source of the nutrient to the crop need and soil properties. A significant part of the source is balanced between the various nutrients, a considerable challenge globally in improving nutrient use efficiency. The right amount means balancing the nutrients added to the need for the seed as basic as that.

The applications of too much fertilizer contributes to excess soil nutrients and environmental degradation. Ultimately, striking a balance between the crop needs, environmental conditions, and the farmer's economic situation is required. Here, microbial biofertilizers can play a vital role in such cases. The right timing ensures that fertilizer nutrients are made available for the crop when needed. Efficiency in nutrient usage can be significantly improved when its supply is matched with crop demand. The right position means attempting to preserve nutrients so crops can use them. This is a question that presents the most significant challenge in smallholder farming systems, where most fertilizers are distributed, and in many cases, without incorporation (Johnston and Bruulsema 2014). Adaptation of 4R Nutrient

Stewardship concept of right nutrient application with the potential soil microbes helps to better nutrient recycling and long-term sustainability goal of our agriculture production system.

1.7 Conclusion and Future Perspectives

In soil processes, including nutrient cycling, soil organisms and their products play a crucial role. These mechanisms are essential for agriculture as well as water, air, and habitat quality protection. Agriculture is currently facing excessive pressure due to population development and related rises in urbanization, resource extraction, etc. However, cultural practices are influenced by the microbial activities. Therefore, it is necessary to consider the potential role of soil microbes for proper nutrient recycling and its impacts on soil health.

Furthermore, expanding our core knowledge regarding diversity and function of soil microbial component is a necessary task to alleviate the harmful effects of soil degradation. Research focusing more on the credentials of innovative microbial diversity in the soil remains essential practices. In the future, that would play a more critical role favorably for enhancing plant growth and yield as well as contribute towards a more environment-friendly alternative to support sustainable development.

Acknowledgments The authors are grateful to the ICAR-Indian Agricultural Research Institute (IARI), New Delhi, and the Indian Council of Agricultural Research for providing facilities and financial support to undertake these investigations. There are no conflicts of interest.

References

Aislabie J, Deslippe JR, Dymond J (2013) Soil microbes and their contribution to soil services. In: Ecosystem services in New Zealand—conditions and trends. Manaaki Whenua Press, Lincoln, New Zealand, pp 143–161

Aneja M, Sharma S, Schloter M, Munch JC (2006) Microbial degradation of beech litter—the influence of soil type, litter quality on the structure and function of microbial populations 15 involved in the turnover process. Microb Ecol 52:127–135

Bargaz A, Lyamlouli K, Chtouki M, Zeroual Y, Dhiba D (2018) Soil microbial resources for improving fertilizer efficiency in an integrated plant nutrient management system. Front Microbiol 9:1606–1606. https://doi.org/10.3389/fmicb.2018.01606

Bloem J, De Ruiter PC, Bouwman LA (1997) Soil food webs and nutrient cycling in agroecosystems. In: van Elsas JD (ed) Modern soil microbiology. Marcel Dekker, Inc., New York, pp 245–278

Carter MR, Gregorich EG, Anderson JW, Doran JW, Janzen HH, Pierce FJ (1997) Concepts of soil quality and significance. In: Gregorich EG, Carter MR (eds) Soil quality crop production ecosystem health. Elsevier, Amsterdam, pp 1–19

Chaer GM, Resende AS, Campello EFC, de Faria SM, Boddey RM (2011) Nitrogen-fixing legume tree species for the reclamation of severely degraded lands in Brazil. Tree Physiol 31(2):139–149

Doran JW, Parkin TB (1994) Defining and assessing soil quality. In: Doran JW, Coleman DC, Bezdicek DF, Stewart BA (eds) Defining soil quality for a sustainable environment. SSSA, Madison, pp 3–21

Doran JW, Parkin TB (1996) Quantitative indicators of soil quality: a minimum data set. In: Doran JW, Jones AJ (eds) Methods for assessing soil quality. SSSA, Inc., Madison, WI
Dubey RK (2016) Exploring rhizospheric interactions for agricultural sustainability, the need of integrative research on multi-trophic interactions. J Clean Prod 115:362–365
Elferink M, Schierhorn F (2016) Global demand for food is rising. Can we meet it? Harv Bus Rev 7:2016
FAO (2015) Food and Agriculture Organization (FAO), Healthy soils are the basis for healthy food production. http://www.fao.org/soils-2015/news/news-detail/en/c/277682/
Fortuna A (2012) The soil biota. Nat Educ Knowl 3(10):1
Gougoulias C, Clark JM, Shaw LJ (2014) The role of soil microbes in the global C-cycle: tracking the below-ground microbial processing of plant-derived C for manipulating carbon dynamics in agricultural systems. J Sci Food Agric 94(12):2362–2371
Gupta VSSR, Germida JJ (1988) Distribution of microbial biomass and its activity in different soil aggregate size calluses affected by cultivation. Soil Biol Biochem 2(6):777–786
Hue NV, Vega S, Silva J (2014) Manganese toxicity in a Hawaiian Oxisol affected by soil pH and organic amendments. Soil Sci Soc Am J 65(1):153–160
Johnston AM, Bruulsema TW (2014) 4R nutrient stewardship for improved nutrient use efficiency. Procedia Eng 83:365–370
Kaur T, Devi R, Rana KL, Kour D, Yadav AN (2019) Microbes with multifarious plant growth promoting attributes for sustainable agriculture. EU Voice 5:11–13
Kaur T, Rana KL, Kour D, Sheikh I, Yadav N, Yadav AN et al (2020) Microbe-mediated biofortification for micronutrients: present status and future challenges. In: Rastegari AA, Yadav AN, Yadav N (eds) Trends of microbial biotechnology for sustainable agriculture and biomedicine systems: perspectives for human health. Elsevier, Amsterdam, pp 1–17. https://doi.org/10.1016/B978-0-12-820528-0.00002-8
Kour D, Rana KL, Yadav AN (2018) Drought stress in plants and their mitigation by soil microbiomes. EU Voice 4:29–30
Kour D, Rana KL, Yadav AN, Sheikh I, Kumar V, Dhaliwal HS et al (2020a) Amelioration of drought stress in Foxtail millet (*Setaria italica* L.) by P-solubilizing drought-tolerant microbes with multifarious plant growth promoting attributes. Environ Sustain 3:23–34. https://doi.org/10.1007/s42398-020-00094-1
Kour D, Rana KL, Yadav AN, Yadav N, Kumar M, Kumar V et al (2020b) Microbial biofertilizers: bioresources and eco-friendly technologies for agricultural and environmental sustainability. Biocatal Agric Biotechnol 23:101487. https://doi.org/10.1016/j.bcab.2019.101487
Kour D, Rana KL, Yadav N, Yadav AN, Kumar A, Meena VS et al (2019) Rhizospheric microbiomes: biodiversity, mechanisms of plant growth promotion, and biotechnological applications for sustainable agriculture. In: Kumar A, Meena VS (eds) Plant growth promoting Rhizobacteria for agricultural sustainability: from theory to practices. Springer Singapore, Singapore, pp 19–65. https://doi.org/10.1007/978-981-13-7553-8_2
Kumar A, Chaturvedi AK, Yadav K, Arunkumar KP, Malyan SK, Raja P et al (2019a) Fungal phytoremediation of heavy metal-contaminated resources: current scenario and future prospects. In: Yadav AN, Singh S, Mishra S, Gupta A (eds) Recent advancement in white biotechnology through fungi, Perspective for sustainable environments, vol 3. Springer International Publishing, Cham, pp 437–461. https://doi.org/10.1007/978-3-030-25506-0_18
Kumar A, Maurya BR, Raghuwanshi R, Meena VS, Islam MT (2017) Co-inoculation with Enterobacter and Rhizobacteria on yield and nutrient uptake by wheat in the alluvial soil under Indo-Gangetic plain of India. J Plant Growth Regul 36(3):608–617
Kumar M, Kour D, Yadav AN, Saxena R, Rai PK, Jyoti A et al (2019b) Biodiversity of methylotrophic microbial communities and their potential role in mitigation of abiotic stresses in plants. Biologia 74:287–308. https://doi.org/10.2478/s11756-019-00190-6
Kumar V, Joshi S, Pant NC, Sangwan P, Yadav AN, Saxena A et al (2019c) Molecular approaches for combating multiple abiotic stresses in crops of arid and semi-arid region. In: Singh SP, Upadhyay SK, Pandey A, Kumar S (eds) Molecular approaches in plant biology and environmental challenges. Springer, Singapore, pp 149–170. https://doi.org/10.1007/978-981-15-0690-1_8

Kumar V, Yadav AN, Saxena A, Sangwan P, Dhaliwal HS (2016) Unravelling rhizospheric diversity and potential of phytase producing microbes. SM J Biol 2:1009

Leong J (1986) Siderophores: their biochemistry and possible role in the biocontrol of plant pathogens. Annu Rev Phytopathol 24:187–209

Lombard N, Prestat E, Elsas JD, Simonet P (2011) Soil-specific limitations for access and analysis of soil microbial communities by metagenomics. FEMS Microbiol Ecol 78(1):31–49

Madsen EL (2005) Identifying microorganisms responsible for ecologically critical biogeochemical processes. Nat Rev Microbiol 3(5):439

Madsen EL (2008) Microbial biogeochemistry: a grand synthesis in environmental microbiology: from genomes to biogeochemistry. Blackwell Publishing, Malden, MA, pp 281–299

Maier RM, Pepper IL, Gerba CP (2009) Environmental microbiology, vol 397. Academic Press, Burlington, MA

Malav LC, Khan SA, Gupta N (2015) Impacts of biogas slurry application on soil environment, yield and nutritional quality of baby corn. Soc Plant Res, 74

Malyan SK, Kumar A, Baram S, Kumar J, Singh S, Kumar SS et al (2019) Role of fungi in climate change abatement through carbon sequestration. In: Yadav AN, Singh S, Mishra S, Gupta A (eds) Recent advancement in white biotechnology through fungi, Perspective for sustainable environments, vol 3. Springer International Publishing, Cham, pp 283–295. https://doi.org/10.1007/978-3-030-25506-0_11

Monier J-M, Demanèche S, Delmont TO, Mathieu A, Vogel TM, Simonet P (2011) Metagenomic exploration of antibiotic resistance in soil. Curr Opin Microbiol 14:229–235. https://doi.org/10.1016/j.mib.2011.04.010

Neiland JB (1995) Siderophore: structure and function of microbial iron transport compounds. J Biol Chem 270:26723–26726

Neiland JB, Nakamura K (1997) Detection, determination, isolation, characterization, and regulation of microbial iron chelates. In: Winkelmann G (ed) Handbook of microbial iron chelates. CRC Press, Boca Raton, pp 1–14

Odum EP (1989) Ecology and our endangered life support systems. Sinauer Associates, Inc., Sunderland, MA

Pankhurst CE, Doube BM (1997) Biological indicators of soil health: a synthesis. In: Pankhurst CE, Doube BM, Gupta VVSR (eds) Biological indicators of soil health. CAB International, Wallingford, pp 419–435

Pathak H, Bhatia A, Prasad S, Singh S, Kumar S, Jain MC, Singh P (2003) Effect of DCD, FYM, and moisture regime on nitrous oxide emission from an Alluvial soil in rice-wheat cropping system. J Indian Soc Soil Sci 51(2):139–145

Patten CL, Glick BR (1996) Bacterial biosynthesis of indole-3-acetic acid. Can J Microbiol 42:207–220

Rajendhran J, Gunasekaran P (2008) Strategies for accessing soil metagenome for desired applications. Biotechnol Adv 26(6):576–590

Rana KL, Kour D, Kaur T, Devi R, Yadav AN, Yadav N et al (2020a) Endophytic microbes: biodiversity, plant growth-promoting mechanisms and potential applications for agricultural sustainability. Antonie Van Leeuwenhoek. https://doi.org/10.1007/s10482-020-01429-y

Rana KL, Kour D, Kaur T, Sheikh I, Yadav AN, Kumar V et al (2020b) Endophytic microbes from diverse wheat genotypes and their potential biotechnological applications in plant growth promotion and nutrient uptake. Proc Natl Acad Sci India B. https://doi.org/10.1007/s40011-020-01168-0

Rana KL, Kour D, Sheikh I, Dhiman A, Yadav N, Yadav AN et al (2019) Endophytic fungi: biodiversity, ecological significance and potential industrial applications. In: Yadav AN, Mishra S, Singh S, Gupta A (eds) Recent advancement in white biotechnology through fungi, Diversity and enzymes perspectives, vol 1. Springer, Cham, pp 1–62

Rastegari AA, Yadav AN, Yadav N (2020a) New and Future Developments in Microbial Biotechnology and Bioengineering: Trends of Microbial Biotechnology for Sustainable Agriculture and Biomedicine Systems: Diversity and Functional Perspectives. Elsevier, Amsterdam

Rastegari AA, Yadav AN, Yadav N (2020b) New and Future Developments in Microbial Biotechnology and Bioengineering: Trends of Microbial Biotechnology for Sustainable Agriculture and Biomedicine Systems: Perspectives for Human Health. Elsevier, Amsterdam

Rodríguez H, Fraga R (1999) Phosphate solubilizing bacteria and their role in plant growth promotion. Biotechnol Adv 17:319–339

Sahu N, Vasu D, Sahu A, Lal N, Singh SK (2017) Strength of microbes in nutrient cycling: a key to soil health. In: Meena VS, Mishra PK, Bisht JK, Pattanayak A (eds) Agriculturally important microbes for sustainable agriculture, Plant-soil-microbe nexus, vol I. Springer Singapore, Singapore, pp 69–86. https://doi.org/10.1007/978-981-10-5589-8_4

Santi C, Bogusz D, Franche C (2013) Biological nitrogen fixation in non-legume plants. Ann Bot 111(5):743–767

Sarkar A, Saha M, Meena VS (2017) Plant beneficial Rhizospheric microbes (PBRMs): prospects for increasing productivity and sustaining the resilience of soil fertility. In: Meena VS, Mishra PK, Bisht JK, Pattanayak A (eds) Agriculturally important microbes for sustainable agriculture, Plant-soil-microbe nexus, vol I. Springer Singapore, Singapore, pp 3–29. https://doi.org/10.1007/978-981-10-5589-8_1

Schimel JP, Bennett J (2004) Nitrogen mineralization: challenges of a changing paradigm. Ecol 85:591–602

Seybold CA, Herrick JE, Brejda JJ (1999) Soil resilience: a fundamental component of soil quality. Soil Sci 164(4):224–234

Sharma S, Mehta R, Gupta R, Schloter M (2012) Improved protocol for the extraction of 5 bacterial mRNA from soils. J Microbiol Methods 91:62–64

Sindhu SS, Phour M, Sehrawat A (2016) K-solubilizing microorganisms (KSM) and its effect on plant growth improvement. In: Meenav VS et al (eds) Potassium solubilizing microorganisms for sustainable agriculture. Springer, New Delhi, pp 171–185

Singer MJ, Ewing S (2000) Soil quality. In: Sumner ME (ed) Handbook of soil science. CRC Press, Boca Raton, pp G271–G298

Singh A, Kumari R, Yadav AN, Mishra S, Sachan A, Sachan SG (2020a) Tiny microbes, big yields: microorganisms for enhancing food crop production sustainable development. In: Rastegari AA, Yadav AN, Yadav N (eds) Trends of microbial biotechnology for sustainable agriculture and biomedicine systems: diversity and functional perspectives. Elsevier, Amsterdam, pp 1–16. https://doi.org/10.1016/B978-0-12-820526-6.00001-4

Singh B, Boukhris I, Pragya, Kumar V, Yadav AN, Farhat-Khemakhem A et al (2020b) Contribution of microbial phytases in improving plants growth and nutrition: a review. Pedosphere 30:295–313. https://doi.org/10.1016/S1002-0160(20)60010-8

Singh J, Yadav AN (2020) Natural bioactive products in sustainable agriculture. Springer, Singapore

Singh JS, Kumar A, Rai AN, Singh DP (2016) Cyanobacteria: a precious bioresource in agriculture, ecosystem, and environmental sustainability. Front Microbiol 7:529

Singh NP, Singh RK, Meena VS, Meena RK (2015) Can we use maize (*Zea mays*) rhizobacteria as plant growth promoter? Vegetos 28(1):8699

Subrahmanyam G, Kumar A, Sandilya SP, Chutia M, Yadav AN (2020) Diversity, plant growth promoting attributes, and agricultural applications of rhizospheric microbes. In: Yadav AN, Singh J, Rastegari AA, Yadav N (eds) Plant microbiomes for sustainable agriculture. Springer International Publishing, Cham, pp 1–52. https://doi.org/10.1007/978-3-030-38453-1_1

Suman A, Yadav AN, Verma P (2016) Endophytic microbes in crops: diversity and beneficial impact for sustainable agriculture. In: Singh D, Abhilash P, Prabha R (eds) Microbial inoculants in sustainable agricultural productivity, research perspectives. Springer-Verlag, New Delhi, pp 117–143. https://doi.org/10.1007/978-81-322-2647-5_7

Sylvia DM, Fuhrmann JJ, Hartel PG, Zuberer DA (2005) Principles and applications of soil microbiology. Prentice Hall, Upper Saddle River, NJ

Takoutsing B, Weber J, Aynekulu E, Martín JAR, Shepherd K, Sila A, Tchoundjeu Z, Diby L (2016) Assessment of soil health indicators for sustainable production of maize in smallholder farming systems in the highlands of Cameroon. Geoderma 276:64–73

Tilman D, Hill J, Befort BL (2011) Global food demand and the sustainable intensification of agriculture. Proc Natl Acad Sci U S A 108(50):20260–20264

Timmusk S, Behers L, Muthoni J, Muraya A, Aronsson AC (2017) Perspectives and challenges of microbial application for crop improvement. Front Plant Sci 8:49

Torsvik V, Ovreas L (2002) Microbial diversity and function in soil: from genes to ecosystems. Curr Opin Microbiol 5(3):240–245

Totsche KU, Rennert T, Gerzabek MH, Kogel-Knabner I, Smalla K, Spiteller M, Vogel HJ (2010) Biogeochemical interfaces in soil: the interdisciplinary challenge for soil science. J Plant Nutr Soil Sci 173:88–99

Utuk IO, Daniel EE (2015) Land degradation: a threat to food security: a global assessment. J Environ Earth Sci 5(8):13–21

Van Elsas JD, Jansson JK, Trevors JT (2007) Modern soil microbiology, 2nd edn. CRC Press, New York, NY

Verma P, Yadav AN, Khannam KS, Kumar S, Saxena AK, Suman A (2016) Molecular diversity and multifarious plant growth promoting attributes of Bacilli associated with wheat (*Triticum aestivum* L.) rhizosphere from six diverse agro-ecological zones of India. J Basic Microbiol 56:44–58

Verma P, Yadav AN, Kumar V, Singh DP, Saxena AK (2017) Beneficial plant-microbes interactions: biodiversity of microbes from diverse extreme environments and its impact for crop improvement. In: Singh DP, Singh HB, Prabha R (eds) Plant-microbe interactions in agro-ecological perspectives, Microbial interactions and agro-ecological impacts, vol 2. Springer Singapore, Singapore, pp 543–580. https://doi.org/10.1007/978-981-10-6593-4_22

Wani FS, Ahmad L, Ali T, Mushtaq A (2015) Role of microorganisms in nutrient mobilization and soil health—a review. J Pure Appl Microbiol 9:1401–1410

Xu Z, Yu G, Zhang X, Ge J, He N, Wang Q, Wang D (2015) The variations in soil microbial communities, enzyme activities. Appl Soil Ecol 86:19–29

Yadav AN (2020) Plant microbiomes for sustainable agriculture: current research and future challenges. In: Yadav AN, Singh J, Rastegari AA, Yadav N (eds) Plant microbiomes for sustainable agriculture. Springer International Publishing, Cham, pp 475–482. https://doi.org/10.1007/978-3-030-38453-1_16

Yadav AN, Kumar R, Kumar S, Kumar V, Sugitha T, Singh B et al (2017a) Beneficial microbiomes: biodiversity and potential biotechnological applications for sustainable agriculture and human health. J Appl Biol Biotechnol 5:45–57

Yadav AN, Kumar V, Prasad R, Saxena AK, Dhaliwal HS (2018a) Microbiome in crops: diversity, distribution and potential role in crops improvements. In: Prasad R, Gill SS, Tuteja N (eds) Crop improvement through microbial biotechnology. Elsevier, Los Angeles, pp 305–332

Yadav AN, Rastegari AA, Yadav N, Kour D (2020a) Advances in plant microbiome and sustainable agriculture: diversity and biotechnological applications. Springer, Singapore

Yadav AN, Rastegari AA, Yadav N, Kour D (2020b) Advances in plant microbiome and sustainable agriculture: functional annotation and future challenges. Springer, Singapore

Yadav AN, Sharma D, Gulati S, Singh S, Dey R, Pal KK et al (2015) Haloarchaea endowed with phosphorus solubilization attribute implicated in phosphorus cycle. Sci Rep 5:12293

Yadav AN, Singh J, Rastegari AA, Yadav N (2020c) Plant microbiomes for sustainable agriculture. Springer International Publishing, Cham

Yadav AN, Verma P, Kumar S, Kumar V, Kumar M, Singh BP et al. (2018b) Actinobacteria from rhizosphere: molecular diversity, distributions and potential biotechnological applications. In: Singh B, Gupta V, Passari A (eds) New and future developments in microbial biotechnology and bioengineering.Elsevier, Amsterdam, pp. 13–41. https://doi.org/10.1016/B978-0-444-63994-3.00002-3

Yadav AN, Verma P, Singh B, Chauhan VS, Suman A, Saxena AK (2017b) Plant growth promoting bacteria: biodiversity and multifunctional attributes for sustainable agriculture. Adv Biotechnol Microbiol 5:1–16

Yadav BK, Sidhu AS (2016) Dynamics of K and their bioavailability for plant nutrition. In: Meena VS et al (eds) Potassium solubilizing microorganisms for sustainable agriculture. Springer, New Delhi, pp 187–201

Soil Microbial Diversity: Calling Citizens for Sustainable Agricultural Development

2

Prasann Kumar and Shipa Rani Dey

Abstract

The significant use of land and climate leads to the projection of worldwide transformation and finally leads to an increased rate of extension for microbes. The equators of the earth and the countries residing on the same have threatened species, and their frequencies are too high on the same. It is reported that the pollution and fragmented lands coupled with previous are responsible for the loss of microbial diversity in the soil. This chapter describes the sustainable management of soil microbial diversity. A diverse group of microorganisms is found in plants that grow in metallic polluted soil effectively tolerating a high level of steel and providing various benefits to both soil and plant life. Rhizospheric bacteria are particularly well represented in the microorganisms involved in phytoremediation of heavy metal, as these can at the same time increase how to plant remediation takes place by changing soil bioavailability by modifying the pH of the soil, releasing the chelators and the reactions oxidation/reduction. In the same manner, in hyperaccumulators produced in metallic contaminated fields, steel-resistant fungus was frequently cited suggesting that this fungus progressed heavy metal resistance and could also be active in the phytoremediation. The microbe attached to the plant causes the metal to accumulate from the soil via the sorption mechanism. "The definition of biosorption is the microbial adsorption by metabolism dependent and active process of soluble/insoluble organic/inorganic metals". Some authors focused on the mechanism for bacterial absorption

P. Kumar (✉)
Divisions of Research and Development, Lovely Professional University, Jalandhar, Punjab, India
e-mail: prasann.21784@lpu.co.in

S. R. Dey
Department of Agronomy, School of Agriculture, Lovely Professional University, Jalandhar, Punjab, India

A. N. Yadav et al. (eds.), *Current Trends in Microbial Biotechnology for Sustainable Agriculture*, Environmental and Microbial Biotechnology,
https://doi.org/10.1007/978-981-15-6949-4_2

that reduced plant metal absorption. Research shows that metal bioavailability can be minimized through metal binding and/or metal bioavailability restricts the plant's root/shoot ratio.

Keywords

Agriculture · Biotic · Crop · Density · Future · Grass · Habitat · Microbes

2.1 Introduction

The existence of life and its diversity are the outstanding features of the earth. It leads to the extensive observation of the distribution of the natural habitat for plants, animals, and microbes. The biosphere of the earth consists of diverse biodiversity with special reference to the diversity of microorganisms, viz. bacteria, protozoa, fungi, unicellular algae (Yadav et al. 2020b). Its contribution to support the life form on the earth is extraordinary. The genetic information concerned with microbes is a virtually limitless pool for the information about its biodiversity. The soil has been reported as a source of the rich biodiversity of microbes. For example, the single pinch of the soil contains thousands of species of microbes. In terms of fungi diversity, a total 15,00,000 species have been reported. Limited knowledge is available for the fungi present in the soil concerning the pathogenic common fungi and mycorrhiza present in the soil (Yadav et al. 2019b). As we all know, agriculture is an artificial ecosystem, and it implies lots of implementation and practices, which leads to the destruction of diversity and promotes the smoothness of the areas in terms of its diversity. It has been reported that the grain crops of 12 species, 23 vegetable crops, and 35 fruit and nut crops have been cultivated mostly on the agricultural land species. In general, no more than 70 plant species spread over approximately 1440 million hectares of presently cultivated in the world. It is a general observation that a tropical rainforest contains over 100 species of trees in a given area. About 4–40 million species on the earth are unknown and unmonitored.

India is one of the world's top 12 mega-diversity countries, rich in biological diversity with about 81,000 species of animals and 47,000 species of plants. Myers et al. 2000, reported that the world harbors are full of plants, microorganisms, and animals, and its estimate is about ten million organisms, which includes algae, fungi, mosses, and higher plants. The total number of flowering plants is 15,000, algae 2000, fungi 15,000, bryophyte 2500, ferns 1000, gymnosperm 64 species of gymnosperm (Kumar and Dwivedi 2011). The key driver for biodiversity and ecosystem loss is the human transformation of land cover. The land cover and land use are two important acknowledged aspects in terms of microbial diversity. The physical surface of land comes under the land cover, and it includes vegetation and built structure. The shifting of one land cover to another form leads to habitat loss and finally leads to loss of habitat. The way to loss of habitat is different and the persistence of a species depends on factors, viz. the ability to migrate and adaptation evolution with interspecific interaction.

2.2 Soil Microbial Diversity

Soil microbial diversity is essential for the ecosphere's health, stability, and healthy and complete process. Everyday millions of people obtain their living from the soil. The effective biodiversity of agricultural system exists in the soil. Then the interaction of food web among the living things has a potential impact on crop quality, the interaction of soil-borne plant and animal pests and disease (affecting production levels) and the beneficial organisms. On this planet, 50% of the living protoplasm is microbial (Chakrabarti 2010). Microorganisms represent the richest repertoire of molecular and chemical diversity in nature, as they comprise the most diverse form of life and also they are the extraordinary reservoir of life in the biosphere (Biswas et al. 2018; Saxena et al. 2016; Verma et al. 2019). In nature, molecular and chemical diversity can be represented by the microorganism because it creates a huge source for the benefits of men.

They underlie the processes of the basic ecosystem which are the biogeochemical cycles and food chains, also maintain vital and often elegant relationships between themselves and higher organisms. The major sources for antimicrobial agents are microorganisms, and it produces a wide range of important medicinal compounds which are including with enzymes, enzyme inhibitors, antitumor agents, insecticides, vitamins, immunosuppressant, and immunomodulatory. To demonstrate the diversity of microorganisms which is extraordinary in terms of specialized metabolism, it is sensible to consider secondary metabolites of the genus *Streptomyces* with around 140 species or groups. Secondary metabolites have been recognized approximately 3500 antibiotics from the genus *Streptomyces* alone (Greene et al. 2000). In laboratory cultures, *Streptomyces griseus* can be induced to produce more than 50 antibiotics and over 180 secondary metabolites are produced by *Streptomyces hygroscopicus* alone. Given the endless combination of terrestrial, aquatic, and marine habitats and such enormous potential of secondary metabolite production in microorganisms and opportunities available for manipulation of the types and quantities produced in a laboratory, the biotechnology industry has a tremendous resource at hand for the discovery of new chemicals for biotechnological application.

2.2.1 The Indian Biodiversity Scenario

Biological diversity of the Indian subcontinent is found one of the richest in the world because of its vast geographical area, diverse topography along with climate with different biogeographically regions. For rich floral and faunal diversity in India, the popular place is Western Ghats, North-western, Eastern Himalayas, North-East Region, and Andaman and Nicobar Islands. The richness in diversity leads to the recognition of India on the 12 mega-diversity regions of the world. There are huge numbers of microbes as plant microbiomes (Kour et al. 2020b, c, d; Verma et al. 2016, 2015) and extremophilic microbes (Kumar et al. 2014; Pandey et al. 2013; Rajawat et al. 2020; Suman et al. 2015; Yadav et al. 2016, 2015d) have been isolated, identified, and characterized for potential applications in agriculture, medicine, and environment for sustainable development. Brussaad (2007) reported 72% of India's bio-wealth. The contribution of fungi, insects, and angiosperm in India's bio-wealth are 18, 40, and 13%, respectively. The world contribution of India's bio-wealth is around 8%. The most important mega-diversity centers are the Western Ghats, North-Eastern Hill regions, Bastar region inhabited by tribals, Andaman Nicobar Islands, mangrove forests of Sundarbans area, the Silent Valley of Kerala, Chilika Lake of Orissa, Sonar Lake of Maharashtra, and the Himalayan region.

Due to the richness of biodiversity, the Indo-Burma and Western Ghats/Sri Lanka considered as the hotspots of biodiversity in the subcontinent of India. The importance of biodiversity is commercial and the scientific lines are required for its management. The diverse population of microorganisms includes various places such as boiling waters (Kumar et al. 2014; Sahay et al. 2017), salt pans (Yadav et al. 2019a, 2015c), acid mine drainage, deep-sea vents (Yadav et al. 2017), and cold environment (Yadav et al. 2015a, b). The biodiversity and its protection along with conservation are considered as crucial for the living being (Tables 2.1, 2.2, and 2.3).

Table 2.1 Species richness of key soil eukaryotes

Taxonomic group	Species described	Species estimated
Protozoa	40,000	200,000
Fungi	70,000	1500,000
Nematodes	5000	20,000
Collembola	6500	15,000
Acari	20,000	80,000
Isoptera	2600	10,000
Earthworm	3700	8000

Source: Coleman (2001)

Table 2.2 Example of important heterotrophic surface soil bacteria

Organism	Characteristics	Function
Streptomyces	Gram +ve, aerobic, filamentous	Produce geosmins "earthy odor" and antibiotics
Bacillus	Gram +ve, aerobic, spore former	Carbon cycling, production of insecticides and antibiotics
Clostridium	Gram +ve, anaerobic, spore former	Carbon cycling, toxin production
Methylosinus	Aerobic	Methane oxidizer that can metabolize trichloroethane (TCE) using methane monooxygenase
Alcaligenes eutrophus	Gram −ve, aerobic	2,4-D degradation via plasmid pJP4
Rhizobium	Gram −ve, aerobic	Symbiotic nitrogen fixation with legumes
Frankia	Gram +ve, aerobic	Symbiotic nitrogen fixation with nonlegume
Agrobacterium	Gram +ve, aerobic	Important plant pathogens, cause crown gall disease

Source: Coleman (2001)

Table 2.3 Dominant cultural soil bacteria in surface soils

Organism	Characteristics	Function
Arthrobacter	Heterotrophic, aerobic, Gram variable. Up to 40% of culturable soil bacteria	Nutrient cycling and biodegradation
Streptomyces	Gram +ve, heterotrophic, aerobic actinomycete. 5–20% of culturable bacteria	Nutrient cycling and biodegradation, antibiotic production by *Streptomyces scabies*
Pseudomonas	Gram −ve, heterotroph, aerobic or facultatively anaerobic, possess wide array of enzyme systems, 10–20% of culturable bacteria	Nutrient cycling and biodegradation, including recalcitrant organics, a biocontrol agent
Bacillus	Gram +ve, aerobic heterotrophy, produces endospores, 2–10% of culturable soil bacteria	Nutrient cycling and biodegradation, biocontrol agent (*Bacillus thuringiensis*)

Source: Sharma (2011)

2.3 Soil Microbial Diversity and Its Impacts on Ecosystem Function

Studies have revealed the evidence of the significant relationship between processes and different components of plant diversity components, viz. the richness of species, richness of function, and composition of function. These were followed in natural and synthetically assembled groups of grassland species worldwide (Fig. 2.1; Tables 2.4 and 2.5) (Diaz and Cabido 2001). The range and more particularly the functional traits of plants (e.g., whether they harbor nitrogen-fixing symbionts, warm-season grasses, or rosette forbs) are generally strong drivers of ecosystem processes. These studies combined simplified microcosms and natural field sites, so extrapolation from them is limited. In terms of linkage, they are neither global nor simple, but its significant trends are found where the nitrogen-fixing symbionts. Although it is remarkable that a large portion of the research demonstrated that species abundance and functional composition had helpful effects over the ground biomass.

Ecosystem and biodiversity are the two faces of a single coin. On the one hand, creating a generation of biodiversity the ecosystem function exerts pressure while the other hand ecosystem of microbes is influenced by biodiversity (Loreau et al. 2001). The ecosystem is an unpredictable circumstance, which is an easy method that implies a functional system that covers the microorganisms and their ecosystem. Degradation of the environment occurs due to human exploitation. Well-being and prosperity of the environment have a direct impact on human well-being. A new biological specialty is emerging that supports the development of new diseases. The ecosystem functions and human health affected by environmental degradation, climate changes, and global warming. The biochemical and biogeochemical cycles of the microorganisms influence the physical, chemical conditions of the environment

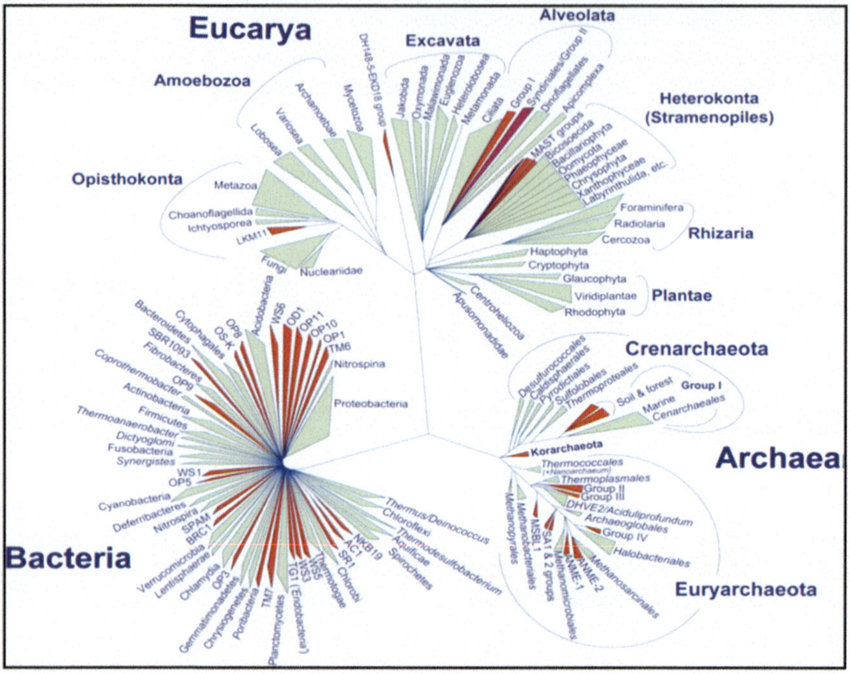

Fig. 2.1 Schematic phylogenetic tree of life based on current molecular knowledge (SSU rRNA and other molecular evidence). Green/light triangle represents phyla, divisions, or groups of high taxonomic rank for which one member has been cultivated and/or properly described (e.g., many protist species); the red/dark triangle represents high divergent lineage without cultivated or described species. (Adapted with permission from Lopez-Garcia and Moreira 2008)

Table 2.4 Diversity of microorganism (per gram) of typical garden soil at various depths

Depth (cm)	Bacteria	Actinomycetes	Fungi	Algae
3–8	9,750,000	2,080,000	119,000	25,000
20–25	2,179,000	245,000	50,000	5000
35–40	570,000	49,000	14,000	500
65–75	11,000	5000	6000	100
135–150	1400	–	3000	–

Source: Kumar and Dwivedi (2011)

contributing to the livelihood of the sustainable life (Naeem 2002). The microbial ecosystem manages with cellular interaction, the existence of fittest and terrestrial production. The ecosystem inside the human body keeps up a stable interior condition through symbiotic living (Table 2.4).

Table 2.5 Range of approximate biomass of each major component of the biota in a typical temperate grassland soil

Component of soil biota	Biomass (tons/ha)
Plant roots	Up to 90 but generally about 20
Bacteria	1–2
Actinomycetes	0–2
Fungi	2–5
Nematodes	0–0.2
Earthworms	0–2.5
Other soil animals	0–0.05
Viruses	Negligible

Source: Sharma (2011)

2.4 Soil Biodiversity and Its Role in Coping with Stress and Disturbances

Microorganisms play a crucial role in soil fertility and its maintenance. Human encroachments and disturbances such as the addition of pesticides affect the microbial components of an ecological niche (Magu 1998) and thus a simultaneous effect is observed on biotransformation reactions occurring in soil. For the assessment of adaptation of land use in a particular place and time, the microbes can be used as an indicator. Dilly and Blume 1998 reported that the combination of the concepts for suitability and improvement leads to the integrated of the ecophysiology of microbiota.

2.4.1 Abiotic Stress and Disturbance

Disturbances like rainfall, flood, fire, storms, nutrient availability, and soil erosion are natural occurrences in the environment, interrupting the development to a climax state and resulting in different patches of habitats at the landscape level. There are different abiotic stress including temperature (−2 to 20 °C—psychrophiles; 60 to 115 °C—thermophiles), salinity (2–5 M NaCl—halophiles) and pH (<4 acidophiles and >9—alkaliphiles), and drought. Microorganisms have been reported from diverse abiotic stress conditions and play significant roles in sustainable agriculture (Verma et al. 2017; Yadav et al. 2020a).

2.4.2 Biotic Stress and Disturbance

Plant production are detrimental by many soil organisms. It has been reported that, the living being viz., moles, rodents, snails, slugs, termites, ants, beetles and nematodes may significantly harm crops or become trouble in both rural and urban households. Numerous types of bacteria and actinomycetes can cause plant disease, but fungi are the most harmful cause for plant disease, which leads to maximum soil-borne crop diseases like wilts, blight, root rot, and club rot. Net primary productivity (NPP) is similar to the ecosystem nutrient use efficiency over soil nutrient supply. There is a proof which indicated that increasing trend with soil biodiversity, the nutrient use efficiency. Importance of the soil microbial biodiversity in increasing the water use efficiency studied by many scientists.

2.5 Dynamics of Microbial Communities in Metal-Polluted Areas

For the biological classification, the species are considered as the fundamental unit. For the measurement of biodiversity, spices can be taken as the measurement unit (Claridge et al. 1997). It is earlier described that microorganisms play vital roles in nutrient cycles and food webs. Microorganisms are present in high quantities in all kinds of habitats because of that reason microorganisms are selected as "test organisms." The increased value of the ratio between surface and volume represents the closest intimate and their interaction with the given climate. It would be the representation of higher sensitivity of microorganisms and their quick response. All the microbes are found reactive along with their respective given environment, viz. types of pollution, probiotics, xenobiotics, radioactivity, agricultural waste, and metals pollution.

The scientific results show that microbial population and its activity fall because of the presence of pollution which leads to the genetic and physiological characterization from standard communities. Studies reveal that heavy metal toxicity and microbial population can influence each other. The significant contribution of copper, zinc, iron, and nickel has been proved in the plant system for their optima activity (Kour et al. 2019; Malyan et al. 2019). But its higher concentration leads to the toxicity symptoms. The vital physiological process of microbes depends on these metals. Because of its trace in nature, its application is limited at the time of its application. The factors which influenced the metal toxicity are its total concentration, its availability to its organism. The mode of action depends on the types of organisms and metals, respectively. The mode of action includes the (1) binding with macromolecules like DNA, RNA, protein, etc., (2) malformation of enzymatic activity, (3) Reactive Oxygen Species formation, etc. For example, copper is an essential element, but its higher concentration leads to toxicity in plants. The formation of radicals leads to the damage of the cell. That is why the intercellular concentration of the copper and its level should be controlled. An ecosystem is calling the citizen for the mitigation of heavy metal toxicity in concern environment. A large number of studies has been conducted in terms of metal and ecosystem interaction (Kour et al. 2020a; Subrahmanyam et al. 2020). Even though, we all are facing the critical challenges for the discrimination between metals and its negative impact on the environment.

2.6 Bioinformatics in Soil Microbial Research

Bioinformatics is definite as an interdisciplinary field that creates software tools, databases, and methods to support genomic and post-genomic research. It includes the study of gene and protein expression, protein and DNA structure and function, protein production, genetic regulatory system, and clinical application. In many realms, biodiversity information is essential for decision-making, and it is critical to a governmental, scientific, and educational wide range of uses.

2.6.1 Biodiversity Database

The Global Invasive Species Database was developed by the IUCN/SSC Invasive Species Specialist Group (IISC) as part of the global initiative on invasive species led by the Global Invasive Species Programme (GISP). It delivers worldwide information to agencies, interested individuals, resource managers, and decision-makers on invasive alien species. The database covers invasive species and all groups of taxonomic from microorganisms to animals and plants which threaten biodiversity. Information on species is provided by skilled suppliers from all over the world, and it includes native and alien range, species biology, ecology, links and image, reference, and contacts. The biodiversity digitalization data includes various process, globally. Some of the important biodiversity databases are Phukan 2007.

2.6.2 Bacteria

The List of Bacterial Names with standing in Nomenclature (www.bacterio.cict.fr/).

2.6.3 Fungi

- National Fungus Collection, USDA (nt.ars-grin.gov/sbmlweb/collections/fungusCollection/Index.cfm)
- Oregon State University Mycological Collection (ocid.nacse.org/research/herbarium/myco/)
- University of Michigan Fungus Collection (www.herb.lsa.umich.edu/)

2.6.4 Viruses

- The Universal Virus Database (www.ncbi.nlm.gov/ICTVdb/)

2.6.5 Genetics

- National Microbial Germplasm and Invertebrate Genetic Resources Program (www.nscalliance.org/Bioinformatics/database.asp)

2.6.6 General All Biota

- Australian Biodiversity Information Facility (ABIF) (www.deh.gov.au/biodiversity/digir/)
- All Taxa Biodiversity Inventory (ATBI) (www.dlia.org/atbi/)
- Biodiversity and Biological Collections (biodiversity.uno.edu)
- European Natural History Specimen Information Network (ENHSIN) (www.nhm.ac.uk/science/rco/enhsin)
- Expert Center for Taxonomic Information (ETI) (www.eti.uva.nl/)
- Integrated Taxonomic Information System (ETIS) (www.eti.uva.nl/)
- National Biological Information Infrastructure (NBII) (www.itis.usda.gov/)
- National Biological Information Infrastructure (NBII) (www.nbii.gov/)
- World Biodiversity Information Network (REMIB) (www.conabio.gob.mx/remib_ingles/doctors/remib_ing.html)
- World Biodiversity Database (www.eti.uva.nl/Database/WBD.html)
- The Species Analyst (speciesanalyst.net)
- Species 2000 (www.sp2000.org)
- Nature Serve: An Online Encyclopedia of Life (www.natureserve.org)

In science, it is identified that we have defined the area where we can utilize information technology and computers together and it leads to significant social and scientific benefits. Its focuses are on biodiversity and ecosystem domains. Their synergistic opportunities fall into three major categories: Acquisition, Analysis and synthesis, and Dissemination.

2.7 Some Specific Opportunities

2.7.1 Modernizing the Biological Library

It seems to observe that the gathered data of biological information and data composed over the past 250 years is enormous. The organization, storing, and retrieving records are critical. New systems and devices must be produced for data extraction, content comprehension, and cross-lingual data recovery, making this a significant non-business application area for examining information incorporation, information purifying, information warehousing, and chronicling.

2.7.2 Digitizing the Biological Legacy

There is an earnest need to change over the documentation and new examples kept up in the exhibition halls and research centers overall which ranges more than billions of records, into metric-quality advanced arrangements. This gives a magnificent chance to propel explore on lossless picture pressure, 3D picture getting, apply autonomy, and the issue of coordinating physical ancient rarities into advanced libraries.

2.7.3 Multidimensional Observation and Recording

Endeavors are expected to empower the gathering of nitty-gritty data about the earth in various measurements and at different scales. This gives rich chances to investigate scaling sensor-combination strategies to huge fields, coordinated in situ nanosensing and creating and testing worldly spatial information get to techniques.

2.7.4 Mobile Computing

New instrumentation is expected to bring information and to gather, store, and transmit information from the field. Explicit open doors here incorporate the utilization of human—PC cooperation research to the multi-modular interface, sans hands frameworks, wearable PCs, remote nearness, mechanical autonomy, and human expansion.

Bioinformatics is a segment created by the merger of two hot areas: information technology and biotechnology. Without bioinformatics, new research in most fields of medicine and soil biodiversity would come to standstill. The technology is versatile and can be applied whenever gene, protein, and cell research are used for the diversity of soil microbial organisms. To foster biotechnology education and research in the country, the culture of DBT-supported courses have a huge impetus to the human resource generation.

2.8 Managing the Soil Biodiversity: Priorities

Soil animal and microbial diversity is a part of the biological resources of agro-ecosystems and must be considered in management decisions.

- The selection of plants and their spatial-transient association in the framework
- Adjustment of plant's protection from malady, or the nature of deposits (roots and shoots) created, through hereditary plant improvement
- Change in the sum or potentially nature of the natural deposits entering the dirt (outer or inward to the framework)

- Least soil unsettling influence and utilization of pesticides, water system, and manures
- Utilization of organic control rehearses.
- Vaccination of gainful soil life forms for illness control and soil fruitfulness improvement

2.9 Bioaugmentation Assisted Phytoextraction Mediated Through Microbes

The lack of natural capital becomes catastrophic during mankind's development and culture contributes to degradation by lethal deficiencies. The key casualties of the same were water and land resources. The biotic and abiotic pressure treatment with the human is the product of the anthropic operation. The SOIL, regarded as the essence of everlasting creation, seems to be the initial perpetrator, to one degree, of a multitude of waste products and chemical products (Aafi et al. (2012), arwidsson et al. (2010).

The addition of a substance to the soil can be called soil contaminants, leading to adverse effects on their function and ability. Contamination of heavy metals in India and abroad is of particular concern. The presence of heavy metals can lead to different diseases and disorders. All people and animals and live populations experience the same. The main cause is mercury, cadmium, plum, chromium, and arsenic. Cadmium is the seventh most harmful element causing cardiovascular and kidney diseases. Cr, Ni, and Pb may contribute to mutagenic, lung and brain injury. Such pathogens were blamed for the deaths. Based on the research, sorghum, in particular, the form of drilling for the scavenger of heavy metals from contaminated soil, is found to be responsible. So if we develop the same, it can contribute to polluted soil shaping strong scavenging metals (Azcón et al.(2010); Babu & Reddy (2011); Beolchini et al. (2009).

Different strategies are needed for the metals concerning organic contaminants. Metals cannot mineralize because of the fats. Thus, organic and inorganic pollutant remediation is generally different. Metal processing is one of the key methods to eliminate transmission and translocation in the atmosphere or food chain. Mulligan et al. (2001) document that such techniques can only be implemented in situ soil remediation following digging compared to soil flushing.

The soil recovery approach is only effective for point source exposure although non-point source contamination—i.e., low metal content with large ground emission—is less studied. The soil is contaminated due to the repeated use of fertilizers, trace metal pesticides, and atmospheric deposition. The concentration of metal in industrial sites is reported to be lower although enough to create a risk of damage to human beings and the environment through the food chain. Conversion through point source contamination due to concentration may also be feasible, depending upon several evidence, including metal species (Bubb et al. 1991; Santschi et al. 1997; Nguyen et al. 2005; Amaraneni 2006). The group in current technology usually calls phytoextraction green cures or phytoremediation. It is projected that a day

of pollution from a non-point source is now allocated (Fominar et al. (2004); Gadd (2004); Gonzalez-Chavez et al.(2004). This technique could be used to remove the very low concentration of pollutants from soil, according to Scientist. Where *Brassica juncea* is extracting Pb, it has been quantified by Blaylock (2000) that lead area is one of the main contaminants where only the metal concentration of no more than 1500 mg/kg can be obtained. Besides, as a result of low metal availability at one particular time, the main limit is the leniency of treatment (Baker et al. 2000). It is considered that a sensible remediation period is lower than 5 years (Khan et al. 2000) while the cleaning of the soil usually takes a lot longer than that (Baker et al. 2000; Dickinson and Pulford 2005). Generally, these should be less expensive than the physical ones if you have established ecosystems generated during the process (Glass 2000). There are many drawbacks in the phytoremediation process; a sluggish translocation speed from the roots to the flames. Smaller surface density, which is typically no greater than half a meter based upon the size of the field settlement (Hoberg et al.(2005); Hrynkiewicz et al.(2012); Joshi & Juwarkar (2009); Jurkevitch et al.(1988). The slowness of treatment was a major restriction reported by Baker et al. (2000).

A large number of chemical chelates, for instance, diethylenetriaminepentaacetic acid (DTPA), ethylenediamine tetra-acetic acid (EDTA), are already used to raise awareness of dangerous toxic metals found on the farm. The biodegradability of these compounds in soil, alternately low or may be considered to be the limiting factors for equal (Lombi et al. 2001) and additionally poisonous (Lasat 2002; McGrath et al. 2002; Romkens et al. 2002; Bouwman et al. 2005) for plants, microorganisms (biomass and diversity), and nematodes. When the soil value is taken into account, it certainly acts as a hazard (Chatterjee et al.(2009); Di Gregorio et al. (2011); Di Simine et al.(1998). Besides, it was shown that Pb mobilization by EDTA is quicker than using flora (Shen et al. 2002; Chen et al. 2004). With Pb leaching hazard in the soil, flora suggested no short-length extraction of high quantities of metals (Barona et al. 2001). Such compounds are ultimately costly, based on the amount used per hectare (Barona et al. 2001).

The alternative is to maximize a plant life and microorganism's synergistic effect (Glick 2003) through the application of coupling phytoextraction to soil biologic increase, also known as rhizoremediation (Kuiper et al. 2004). This method has been used extensively to remediate surface pollutant contaminants (Barac et al. 2004; Van Aken et al. 2004), but not now for metals. Two harmonizing ways are used to strengthen the overall use of metals through flowers; (1) an increase in soil metal mobility, followed by multiple plant metal concentrations. In that case, we simply deliver siderophores (Diels et al. 1999; Dubbin and Louise Ander 2003) and organic acids, through microorganisms that produce biosurfactants (Herman et al. 1995; Mulligan et al. 1999, 2001). These are used as a herbal chelating agent, together with the steel and organometallic co-ordinate compounds and/or types; and (2) increased vegetation productivity by connecting them with PGPR (Zhuang et al. 2007) and/or Arbuscular Mycorrhizal Fungi (AMF) plants (Khan 2006). This is the only way to improve plant biochemical effectiveness. For sorghum, the *Glomus* EM fungus has been applied to enhance cadmium absorption from contaminated soils and important repercussions have been found (Dimkpa et al.(2008, 2009a, b).

2.10 Metal Extraction and Its Mechanism from Soil by Microorganism-Assisted Plant

Many microorganisms decrease the metal toxicity in flowers through growing their aggregation through increasing the amount of metal collected from flora or with the use of certain items. Either plant life's biomass (dilution effects) or the knowledge of metals collected in plants are increased. Microorganisms simply limit the stress caused by additional metallic accumulation in plants.

$$+\text{Increasein Biomass of plant} = \frac{1}{\text{Concentration of metals per unit biomass}}$$

2.11 Significant Metal Accumulation by Plants

2.11.1 Bioavailability of Metals

One of the major factors that contribute to less metal extraction using plant life from the soil is the low concentration of metals in the soil solution. The scientist has generally said that the total amount of soil metallics are less than 1% (Whiting et al. 2001; Braud et al. 2006) and that the price for availability is influenced by a variety of physical and chemical soil characteristics, such as pH, CEC, and organic counts (Kayser et al. 2001). He said that, if we practice soil bioaugmentation, the superb results will be calculated. This is because more metal is found in the soil solution. It is a concern. Braud et al. (2006) have shown that to remediate heavy metallic lead, taking *Pseudomonas aerugino*sa and *Pseudomonas fluorescens*, it increases the awareness of lead in a soil solution by 11.3%. Therefore, it is, of course, best suitable for remediation because of the bio rise in crops. The binding of lead with Fe-Mn oxides and organic matter has been discovered (Duijff et al.(1991); Duss et al. (1986); Fasim et al.(2002). An extension of the extractable niche with the aid of an element up to 15 with an appreciation of Ni sensitivity in the soil has been shown by Abou-Shanab et al. (2006).

The presence of minerals frequently affects the essence of the soil's physicochemical and structural frameworks. The pH has an inverse relationship with the metals present in the soil while the acidity indicates an excellent relationship to the metals supply. Concentrations are up to 1.22 and 1.11, 1.33 and 1.33 times higher than those reported in non-bioaugmentable soil, respectively, in soils with the ectomycorrhizal fungus *Paxillus involutus*, depending on soil composition, in NH_4NO_3-extractable Cd, Cu, Pb, and Zn (Baum et al. 2006). Because the use of techniques is not equal to all microbial efficiency. Indeed, several extractants are used such as water (Chen et al. 2005; Di Gregorio et al. 2006; Wu et al. 2006a, b), $MgCl_2$ (Braud et al. 2006), NH_4NO_3 (Baum et al. 2006), DTPA (Chen et al. 2004; Di Gregorio et al. 2006), KNO_3 (Di Gregorio et al. 2006), and HCl (Wang et al. 2007). Bioavailability of metals is even divided into three unique swimming pools (Cao et al. 2007): right now soluble metals (with water), exchangeable metal forms (with KNO3), and complexes or adsorbed metallic structure (with EDTA).

Microbial siderophores, mainly localized in the rhizosphere (Bossier et al. 1988), enhance the Fe (III) mobility and additionally various other cations (Hofte et al. 1993; Diels et al. 2002). Principle factor analysis (PCA) shows shut fantastic relationships between the microorganisms that produce siderophores and the quantity of Cr and Pb in the exchangeable fraction (Braud et al. 2006). Bacteria such as *Azotobacter chroococcum* (N-fixing bacteria), *Bacillus megaterium* (P-solubilizer), *Bacillus mucilaginous* (K-solubilizer) (Wu et al. 2006a, b), and *Bacillus* sp. RJ 16 (Sheng and Xia 2006) can decrease the pH value, in all likelihood by using excreting low molecular weight natural acids, enhancing the bioavailability of Cd, Pb, and Zn (Chen et al. 2005).

In the bio-increased method, metal concentrations were shown to increase in the majority of genuinely extractable groups (Sheng and Xia 2006). However, after 6 months, in contrast with non-bioaugmented soil (Baum et al. 2006), concentrations of accessible metals are expanded inside the bio-amplified soil, which means that they remain alive and are metabolically active in the intention of inoculation microorganisms. Nevertheless, a host of other studies show that the reverse, i.e., the biological rise, results in a rejection of metal speciation. For example, Cd and Zn, Cr and Ni hypothesis were influenced without impact on Cd's and Zn's, Cr's or Ni's speciation, for example, by *Glomus Caledonian* (Chen et al. 2004) and *Glomus mosseae* (Citterio et al. 2005).

2.11.2 Metal Extraction by Plants

The metal sensibility of flowers in contrast to the engineered (non-augmented) ground, metal-dependent (Bi et al. 2003; Baum et al. 2006), and/or soil consciousness is enhanced and/or decreased (Chen et al. 2003). In response to heavy metal, PGPR reduces metallic toxicity by reducing the amount of ethylene produced from plants. On the other side, the more popular auxin synthesized by rhizobacteria was indole-3 acetic acid (IAA). In metallic absorption, the role of rhizobacteria is reported (Zaidi et al. 2006). Nevertheless, bio-increasing prices for metal harvested from plants will almost always rise. Simultaneous addition of rhizobacterial-synthesized EDTA and IAA in hydroponic conditions will increase Pb extraction by plant utilizing of EDTA (Lopez et al. 2005). The PGPR reduces the metallic toxicity in response to heavy metal by decreasing the amount of plant-generated ethylene. On the other hand, indole 3 acetic acid (IAA) was the most common auxin synthesized by rhizobacteria. It was shown to be involved in metal absorption (Souza et al. 1996) and by the fungi (Liao et al. 2003; Malcova et al. 2003; Leung et al. 2006).

PGPR such as *Agrobacterium, Alcaligenes, Arthrobacter, Azospirillum, Azotobacter, Bacillus, Burkholderia, Serratia, Pseudomonas*, and *Rhizobium* (O'Sullivan and O'Gara 1992; Hoflich et al. 1994; Carlot et al. 2002; Glick 2003) are mainly fascinating for metal extraction with the addition of vegetation.

Siderophores concerned with the alteration of metallic speciation in the soils are frequent productions of pseudomonads. Siderophores synthesized by *P. fluorescents* improve the intake of Fe with the absence of peanut chlorosis phenomena employing tomatoes (Duss etc., 1986), carnations and oats (Dujiff et al. 1997), vine and maize (Sharma and Johri 2003a, b). Some hydroxamate siderophores, for example, desferrioxamine B, can be complicated with Pb and potentially support Pb uptake through plants (Dubbin and Louise Ander 2003).

Nonetheless, Cd, Cu, and Zn complexation are considerably increased as opposed to desferrioxamine B (Neubauer et al. 2000) with Cd, Cu, and Zn nitrilotriacetate. Additionally, AMF can increase metal recovery from plants such as Cd in bean and maize (Guo et al. 1996). AMF can be used in soil characteristics and pH for alfalfa (El-Kherbawy et al. 1989), clover (Joner and Leyval 1997), and soybean (Heggo et al. 1990). The aggregation and translocation of Pb in plant lives have shown soil enhancement with *Glomus intraradices*, but the effect depends on plant species and steel locations. For example, the concentration of Pb in roots and *Zea mays* leaves are reduced but Pb in *Agrostis* capillaries is extended to the root (Malcova et al. 2003).

2.12 Plant-Associated Microbes Improve Heavy Metal Mobilization/Immobilization

A diverse group of microorganisms (Idris et al. 2004; Zarei et al. 2008, 2010) is found in plants that grow in metallic polluted soil effectively tolerating a high level of steel and providing various benefits to both soil and plant life. Rhizospheric bacteria are particularly well represented in the microorganisms involved in phytoremediation of heavy metal, as these can at the same time increase how to plant remediation takes place by changing soil bioavailability by modifying the pH of the soil, releasing the chelators and the reactions oxidation/reduction (Gadd 2000; Khan et al. 2009; Kidd et al. 2009; Ma et al. 2011). In the same manner, in hyperaccumulators produced in metallic contaminated fields, steel-resistant fungus was frequently cited suggesting that this fungus progressed heavy metal resistance and could also be active in the phytoremediation (Gohre and Paszkowski 2006; Miransari 2011). For example, cellulosimicrobe microorganism inoculation of Cr-resistant cellulose into unexperienced soil chilli grown in Cr (VI) lowered Cr uptake by 37% and root by 56% compared to uninoculated controls (Li et al. (2010); Majewska & Kurek (2005); Martino et al.(2003). This study shows that cell- and toxic Cr (IV) decreased to Cr in soil. Where (a) plant-associated microbes enhance plant nutrients and water intake. Microbial metabolites reduce the toxicity of steel; (b) metal biosorption; (c) metal reductions and reactions to complexing. Plant-associated microbes reduce the stress of heavy steel in plants utilizing (d) increasing the protection of antioxidants and/or producing ACC deaminases (Miguel (1999); Rajkumar et al. (2010); Saravanan et al. (2007); Sheng et al. (2008).

2.13 Metal Reduction and Oxidation

It has the ability for heavy metal to influence movement by certain microorganisms that are plant related and by corrosion or reduction reaction. Phytoextraction views show a great deal of curiosity in mostly metal corrosion by the microbiota of the rhizosphere. Many sulfur-oxidizing rhizosphere bacteria, for instance, may increase Cu mobilization as well as the intake in contaminated soils of plant tissue (Yang et al. (2012). The soil pH of rhizosphere via sulfur sulfate conversion is reduced by sulfur-oxidizing bacteria, which means that Cu is available to make the plant intake available. Similarly, Chen and Lin (2001) believed that through acidification reaction, the potential Fe/S oxidizing bacteria (Rajkumar et al. 2012) will increase metal bioavailability in soil (Shi et al. (2011); Tripathi et al. (2005), Venkatesh & Vedaraman (2012); Vivas et al.(2003).

2.14 Biosorption

The microbe attached to the plant causes the metal to accumulate from the soil via the sorption mechanism. "The definition of biosorption is the microbial adsorption by metabolism dependent and active process of soluble/insoluble organic/inorganic metals (Ma et al. 2011)." Some authors focused on the mechanism for bacterial absorption that reduced plant metal absorption. For example, in *Magnaporthe oryzae* and *Burkholderia* sp., Madhaiyan et al. (2007) report that deposition of Cd and Ni at shootings and roots of tomatoes are reduced. Research shows that metal bioavailability can be minimized through metal binding and/or metal bioavailability restricts the plant's root/shoot (Rajkumar et al. 2012). Filter barriers from roots to plant shoots can also play a role in mycorrhizal fungi, and these filtration barriers are also contrary to moving heavy metals (Rajkumar et al. 2012). Pine seedlings experimentation discovered that translocation of Pb, Zn, and Cd can reduce the EMF inoculation *Lactarius rufus*, *Scleroderma citrinum,* and *Amanita muscaria* from the plant roots to shoots as compared to the controls.

The mycelial on the outside and the inside surface has increased the amount of metal biosorption (Krupa and Kozdrój 2007). Mycorrhiza and roots combine to produce a wider surface area. This changes the root metal absorption rate. The fungal cells and their intracells link metals and control cell mobility (Meharg 2003). Though these investigations suggest the inoculation of plants for those metal-binding microbes, heavy metals, and phytostabilization metal-polluted soils, this is a good approach to plant protection. Some writers focused on the mechanism of microbial biosorption/bioaccumulation, which was not solely responsible for reducing metal accumulation and the translocation of plants. All of these results show that plant microorganisms associated with the use of metallurgical agents vary in their ability to modify the bioavailability of heavy metals and plants. The quality and type of metal deposition in the rhizosphere determines the microbial ability for colonization and survival. It occurs due to the physical, chemical, and biological properties of the earth. Examples include metal toxicity, indigenous microbial, changing pH levels, nutrient deficiencies, etc., which has the a greater potential to alter the microbe colonization in the soil. In the end, it changes the mobilization and immobilization of metals (Rajkumar et al. 2012). As all contaminated soils have a unique profile, the potential of plants can vary greatly to take up metals, their concentration, their microbial partner survival, and their potential for colonization, plant types, and conditions for growth (Rajkumar et al. 2012). Since plant microbes that may promote plant growth and/or mobilize/immobilize metals, there are several kinds of the interaction of plant microbes in soils that are contaminated by metals and concerns are evolved when the microbe interaction is manipulated.

Plant growth is promoted through the mobilization/immobilization of in vivo metal by microbial metabolites/processes but is unable to confer valuable traits on your host in soils contaminated with metal (Rajkumar et al. 2012). Alongside the isolation of a plant species linked to microbes and the reporting of its useful metabolites and processes, time is needed since it requires an analysis of over thousands of isolates (Rajkumar et al. 2012). For the option of single associated biomarkers with microbes, which can be used to efficiently help with phytoremediation by a microbe, impactful molecular research action is therefore required. Since the effect of the inoculated and necessary microbes is a significant factor in colonization and existence in the metal stress field environment, as well as the useful conduct for the plant growth and the total phytoremediation process in metal contaminated soils. Knowledge of different metal resistance, the existence, and adaptability of microbes can, therefore, be necessary to use their capabilities as a phytoremediation infection (Rajkumar et al. 2012). This information can also be used. Meanwhile, important progress in accepting the role of microbe plants in mobilizing/immobilizing mattresses and in the aim of such heavy metal phytoremediation methods has been made (Braud et al. 2009).

2.15 Conclusion and Prospects

The stability of the method, which still involves being reputable on the ground, easy to implement is like microbial leguminous seeds, used primarily in agriculture, and the scope for cleaning pollutant toxic soils varies between the pollutant and organic pollutant. Human actions that result in a loss of microbial diversity should be the central concern to scientists, the general public, regulatory agencies, and international organizations. The value of soil biodiversity is also to be recognized by society at large. We recommend that distinguishing the estimation of soil biodiversity as far as monetary advantages is an important stride in an examination program planned for supporting soil biodiversity, its utilization and as a feature of a wide procedure of preserving and utilizing agro-biodiversity. It is essential to underscore microbial decent variety as a wellspring of biotechnology just as proceeding to monitor, comprehend, and oversaw biodiversity while getting new data on assorted variety in the biosphere.

The majority of research on microbial biodiversity in soil has been concentrated on the soil of temperate regions. Knowledge of this respect for all tropical soils is poor. The soil in the tropics deserves particular attention. The future thrust areas of research may include the following aspects:

- Comparative study on the influence of chemical agriculture to that of organic agriculture.
- The rate intensification of agriculture in the tropic is greater than in other regions of the world. Some ecosystems are under particular threat of major changes or loss.
- Soil can be contaminated with a variety of inorganic pollutants, such as heavy metals through aerial deposition, fertilizers, and other human activities. Surprisingly, there is still a paucity of information on the effect(s), if any, of such contaminants on bacterial/microbial biodiversity under varying climatic conditions and land management practices. A decrease in bacterial/microbial diversity may result from a reduction in species richness due to pollutant toxicity.
- There is a paucity of information on the impact of crop rotation and monocultural cropping system on soil biodiversity.
- There is substantial information on the efficacy of biofertilizers in normal soils. Little is known about their efficacy in problem soils, i.e., acid and saline soils. The total inventory of N_2-fixing and phosphate-solubilizing microorganism in the saline tracts are desperately needed.

Acknowledgements P.K. and S.R.D. gratefully acknowledge the support provided by Lovely Professional University.

References

Aafi NE, Brhada F, Dary M, Maltouf AF, Pajuelo E (2012) Rhizostabilization of metals in soils using *Lupinus luteus* inoculated with the metal resistant *rhizobacterium serratia* sp. MSMC 541. Intl J Phytoreme 14:261–274

Abou-Shanab RAI, Angle JS, Chaney RL (2006) Bacterial inoculants affecting nickel uptake by *Alyssum murale* from low, moderate and high Ni soils. Soil Biol Biochem 38:2882–2889

Amaraneni SR (2006) Distribution of pesticides, PAHs and heavy metals in prawn ponds near Kolleru lake wetland, India. Environ Intl 32:294–302

Arwidsson Z, Johansson E, Von Kronhelm T, Allard B, Van Hees P (2010) Remediation of metal contaminated soil by organic metabolites from fungi production of organic acids. Water Air Soil Pollut 205:215–226

Azcón R, Perálvarez MDC, Roldán A, Barea JM (2010) Arbuscular mycorrhizal fungi, *Bacillus cereus*, and *Candida parapsilosis* from a multi-contaminated soil alleviate metal toxicity in plants. Microb Ecol 59:668–677

Babu AG, Reddy S (2011) Dual inoculation of arbuscular mycorrhizal and phosphate solubilizing fungi contribute in sustainable maintenance of plant health in fly ash ponds. Water Air Soil Pollut 219:3–10

Baker AJM, McGrath SP, Reeves RD, Smith JAC (2000) Metal hyperaccumulator plants: a review of the ecology and physiology of a biological resource for phytoremediation of metal-polluted soils. Lewis Publisher, Boca Raton, FL

Barac T, Taghavi S, Borremans B, Provoost A, Oeyen L, Colpaert JV, Vangronsveld J, van der Lelie D (2004) Engineered endophytic bacteria improve phytoremediation of water-soluble, volatile organic pollutants. Nat Biotechnol 22:583–588

Barona A, Aranguiz I, Elias A (2001) Metal associations in soils before and after EDTA extractive decontamination: implications for the effectiveness of further clean-up procedures. Environ Poll 113:79–85

Baum C, Hrynkiewicz K, Leinweber P, Meissner R (2006) Heavy-metal mobilization and uptake by mycorrhizal and nonmycorrhizal willows (*Salix dasyclados*). J Plant Nut Soil Sci 169:516–522

Beolchini F, Dell'Anno A, Propris LD, Ubaldini S, Cerrone F, Danovaro R (2009) Auto- and heterotrophic acidophilic bacteria enhance the bioremediation efficiency of sediments contaminated by heavy metals. Chemo 74:1321–1326

Bi YL, Li XL, Christie P (2003) Influence of early stages of arbuscular mycorrhiza on uptake of zinc and phosphorus by red clover from a low phosphorus soil amended with zinc and phosphorus. Chemo 50:831–837

Biswas S, Kundu D, Mazumdar S, Saha A, Majumdar B, Ghorai A et al (2018) Study on the activity and diversity of bacteria in a New Gangetic alluvial soil (Eutrocrept) under rice-wheat-jute cropping system. J Environ Biol 39:379–386

Blaylock MJ (2000) Field demonstrations of phytoremediation of lead-contaminated soils. In: Terry N, Banuelos G (eds) Phytoremediation of contaminated soil and water. Lewis Publishers, Boca Raton, FL, pp 1–12

Bossier P, Hofte M, Verstraete W (1988) Ecological significance of siderophores in soil. Adv Microbiol Ecol 10:385–414

Bouwman LA, Bloem J, Romkens PFAM, Japenga J (2005) EDGA amendment of slightly heavy metal loaded soil affects heavy metal solubility, crop growth and *microbivorous* nematodes but not bacteria and herbivorous nematodes. Soil Biol Biochem 37:271–278

Braud A, Jézéquel K, Bazot S, Lebeau T (2009) Enhanced phytoextraction of an agricultural Cr, Hg and Pb-contaminated soil by bioaugmentation with siderophore producing bacteria. Chemo 74:280–286

Braud A, Jezequel K, Vieille E, Tritter A, Lebeau T (2006) Changes in extractability of Cr and Pb in a polycontaminated soil after bioaugmentation with microbial producers of biosurfactants, organic acids and siderophores. Water, Air Soil Poll 6:261–279

Brussaad L, De Ruiter PC, Brown GG (2007) Soil biodiversity for agricultural sustainability. Agric Ecosystems Environ 121:233–244

Bubb JM, Rudd T, Lester JN (1991) Distribution of heavy metals in the River Yare and its associated broads II. Copper and cadmium. Sci Tot Environ 102:169–188

Cao A, Carucci A, Lai T, La Colla P, Tamburini E (2007) Effect of biodegradable chelating agents on heavy metals phytoextraction with *Mirabilis jalapa* and on its associated bacteria. Europ J Soil Biol 43:200–206

Carlot M, Giacomini A, Casella S (2002) Aspects of plant-microbe interactions in heavy metal polluted soil. Acta Biotechnol 22:13–20

Chakrabarti K (2010) Microbial diversity especially in agricultural soil. In Proceedings of national seminar on fertilizers and environment, Ghose K (ed), June 26, 2010, University of Calcutta, pp 36–37

Chatterjee S, Sau GB, Mukherjee SK (2009) Plant growth promotion by a hexavalent chromium reducing bacterial strain, *Cellulosimicrobium cellulans* KUCr3. World J Microbiol Biotechnol 25:1829–1836

Chen B, Shen H, Li X, Feng G, Christie P (2004) Effects of EDTA application and arbuscular mycorrhizal colonization on growth and zinc uptake by maize (*Zea mays* L.) in soil experimentally contaminated with zinc. Plant Soil 261:219–229

Chen BD, Li X, Tao HQ, Christie P, Wong MH (2003) The role of arbuscular mycorrhiza in zinc uptake by red clover growing in a calcareous soil spiked with various quantities of zinc. Chemo 50:839–846

Chen SY, Lin JG (2001) Effect of substrate concentration on bioleaching of metal-contaminated sediment. J Hazard Mater 82:77–89

Chen YX, Wang YP, Lin Q, Luo YM (2005) Effect of copper-tolerant rhizosphere bacteria on mobility of copper in soil and copper accumulation by *Elsholtzia splendens*. Environ Intl 31:861–866

Citterio S, Prato N, Fumagalli P, Aina R, Massa N, Santagostino A, Sgorbati S, Berta G (2005) The arbuscular mycorrhizal fungus *Glomus mosseae* induces growth and metal accumulation changes in *Cannabis sativa* L. Chemo 59:21–29

Claridge MF, Dawah HA, Wilson MR (1997) Practical approaches to species concepts for living organisms. In: Claridge MF, Dawah HA, Wilson MR (eds) Species: the units of biodiversity. Chapman & Hall, London, pp 3–15

Coleman DC (2001) Soil biota, soil systems and processes. In: Levin S (ed) Encyclopedia of biodiversity, vol 5. Academic Press, San Diego, pp 305–314

Di Gregorio S, Barbafieri M, Lampis S, Sanangelantoni AM, Tassi E, Vallini G (2006) Combined application of Triton X-100 and Sinorhizobium sp. Pb002 inoculum for the improvement of lead phytoextraction by *Brassica juncea* in EDTA amended soil. Chemo 63:293–299

Di Gregorio S, Lampis S, Vallini G (2011) Selenite precipitation by a rhizospheric strain of *Stenotrophomonas* sp isolated from the root system of *Astragalus bisulcatus*: a biotechnological and microbial community changes in rice rhizospheres affected by sulfur. Molecules 16:1409–1417

Di Simine CD, Sayer JA, Gadd GM (1998) Solubilization of zinc phosphate by a strain of *Pseudomonas fluorescens* isolated from a forest soil. Biol Fertil Soils 28:87–94

Diaz S, Cabido M (2001) Vive la difference: plant functional diversity matters to ecosystem processes. Trends Ecol Evol 16:646–655

Dickinson NM, Pulford ID (2005) Cadmium phytoextraction using short rotation coppice *Salix*: the evidence trail. Environ Intl 31:609–613

Diels L, De Smet M, Hooyberghs L, Corbisier P (1999) Heavy metals bioremediation of soil. Mol Biotechnol 12:154–158

Diels L, van der Lelie N, Bastiaens L (2002) New developments in treatment of heavy metal contaminated soils. Rev Environ Sci Biotechnol 1:75–82

Dilly O, Blume HP (1998) Indicators to assess sustainable land use with reference to soil microbiology. Adv Geoecol 31:29–36

Dimkpa CO, Merten D, Svatoš A, Buchel G, Kothe E (2009a) Siderophores mediate reduced and increased uptake of cadmium by *Streptomyces tendae* F4 and sunflower (*Helianthus annuus*), respectively. J Appl Microbiol 107:1687–1696

Dimkpa CO, Merten D, Svatoš A, Büchel G, Kothe E (2009b) Metal-induced oxidative stress impacting plant growth in contaminated soil is alleviated by microbial siderophores. Soil Biol Biochem 41:154–162

Dimkpa CO, Svatos A, Merten D, Buchel G, Kothe E (2008) Hydroxamate siderophores produced by *Streptomyces acidiscabies* E13 bind nickel and promote growth in cowpea (*Vigna unguiculata* L.) under nickel stress. Can J Microbiol 54:163–172

Dubbin WE, Louise Ander E (2003) Influence of microbial hydroxamate siderophores on Pb(II) desorption from a-FeOOH. Appl Geochem 18:1751–1756

Duijff BJ, Bakker PAHM, Schippers B (1991) Influence of pseudobactin- 358 on the iron nutrition of plants. Sixth International Fe Symposium, In

Dujiff BJ, Gianinazzi-Pearson V, Lemanceau P (1997) Involvement of the outer membrane lipopolysaccharides in the endophytic colonization of tomato roots by biocontrol Pseudomonas fluorescens strain WCS417r. New Phytol 135:325–334

Duss F, Mozafar A, Oertli JJ, Jaeggi W (1986) Effect of bacteria on the iron uptake by axenically-cultured roots of Fe-efficient and Fe-inefficient tomatoes (*Lycopersicon esculentum* Mill.). J Plant Nutr 9:587–598

El-Kherbawy M, Angle JS, Heggo A, Chaney RL (1989) Soil pH, rhizobia, and vesicular-arbuscular mycorrhizae inoculation effects on growth and heavy metal uptake of alfalfa (*Medicago sativa* L.). Biol Fert Soil 8:61–65

Fasim F, Ahmed N, Parsons R, Gadd GM (2002) Solubilization of zinc salts by a bacterium isolated from the air environment of a tannery. FEMS Microbiol Lett 213:1–6. Femsre.oxfordjournals.org

Fomina MA, Alexander IJ, Hillier S, Gadd GM (2004) Zinc phosphate and pyromorphite solubilization by soil plant-symbiotic fungi. Geomicrobiol J 21:351–366

Gadd GM (2000) Bioremedial potential of microbial mechanisms of metal mobilization and immobilization. Curr Opin Biotechnol 11:271–279. https://doi.org/10.1016/S0958-1669(00)00095-1
Gadd GM (2004) Bioremedial potential of microbial mechanisms of metal mobilization and immobilization. Curr Opin Biotechnol 11:271–279
Glass DJ (2000) Economic potential of phytoremediation. In: Raskin I, Ensley B (eds) Phytoremediation of toxic metals: using plants to clean up the environment. Wiley, New York, pp 15–31
Glick BR (2003) Phytoremediation: synergistic use of plants and bacteria to clean up the environment. Biotechnol Adv 21:383–393
Gohre V, Paszkowski U (2006) Contribution of the arbuscular mycorrhizal symbiosis to heavy metal phytoremediation. Planta 223:1115–1122
Gonzalez-Chavez MC, Carrillo-Gonzalez R, Wright SF, Nichols KA (2004) The role of glomalin, a protein produced by arbuscular mycorrhizal fungi, in sequestering potentially toxic elements. Environ Pollut 130:317–323
Greene EA, Kay JG, Jaber K, Stehmeier LG, Voordouw G (2000) Composition of soil microbial communities enriched on a mixture of aromatic hydrocarbons. Appl Environ Microbiol 66:5282
Guo Y, George E, Marschner H (1996) Contribution of an arbuscular mycorrhizal fungus to the uptake of cadmium and nickel in bean and maize plants. Plant Soil 184:195–205
Heggo A, Angle JS, Chaney RL (1990) Effects of vesicular-arbuscular mycorrhizal fungi on heavy metal uptake by soybean. Soil Biol Biochem 22:856–869
Herman D, Artiola J, Miller R (1995) Removal of cadmium, lead, and zinc from the soil by a rhamnolipid biosurfactant. Environ Sci Technol 29:2280–2285
Hoberg E, Marschner P, Lieberei R (2005) Organic acid exudation and pH changes by *Gordonia* sp. and *Pseudomonas fluorescens* grown with P adsorbed to goethite. Microbiol. Res 160:177–187
Hoflich G, Wiehe W, Kuhn G (1994) Plant growth stimulation by inoculation with symbiotic and association rhizosphere microorganisms. Experientia 50:23–28
Hofte M, Buysens S, Koedam N, Cornelis P (1993) Zinc affects siderophore-mediated high-affinity iron uptake systems in the rhizosphere *Pseudomonas aeruginosa* 7NSK2. Biometals 6:85–91
Hrynkiewicz K, Dabrowska G, Baum C, Niedojadlo K, Leinweber P (2012) Interactive and single effects of ectomycorrhiza formation and *Bacillus cereus* on metallothionein mt1 expression and phytoextraction of Cd and Zn by willows. Water Air Soil Pollut 223:957–968. http://en.wikipedia.org/wiki/Image.jpg
Idris R, Trifonova R, Puschenreiter M, Wenzel WW, Sessitsch A (2004) Bacterial communities associated with flowering plants of the Ni hyperaccumulator *Thlaspi goesingense*. Appl Environ Microbiol 70:2667–2677
Joner EJ, Leyval C (1997) Uptake of 109Cd by roots and hyphae of a *Glomus mosseae/Trifolium subterraneum* mycorrhiza from soil amended with high and low concentrations. New Phytol 135:353–360
Joshi PM, Juwarkar AA (2009) In vivo studies to elucidate the role of extracellular polymeric substances from Azotobacter in immobilization of heavy metals. Environ Sci Technol 43:5884–5889
Jurkevitch E, Hadar Y, Chen Y (1988) Involvement of bacterial siderophores in the remedy of lime-induced chlorosis in peanut. Soil Sci Soc Am J 52:1032–1037
Kayser G, Korckritz T, Markert B (2001) Bioleaching for the decontamination of heavy metals. Wasser Boden 53:54–58
Khan AG (2006) Mycorrhizremediation an enhanced form of phytoremediation. J Zhejiang Univ Sci B 7:503–514
Khan AG, Kuek C, Chaudhry TM, Khoo CS, Hayes WJ (2000) Role of plants, mycorrhizae and phytochelators in heavy metal contaminated land remediation. Chemo 41:197–207
Khan MS, Zaidi A, Wani PA, Oves M (2009) Role of plant growth promoting rhizobacteria in the remediation of metal contaminated soils. Environ Chem Lett 7:1–19
Kidd P, Barcelo J, Bernal MP, Navari-Izzo F, Poschenrieder C, Shilev S et al (2009) Trace element behaviour at the root–soil interface: implications in phytoremediation. Environ Exp Bot 67:243–259

Kour D, Kaur T, Devi R, Rana KL, Yadav N, Rastegari AA et al (2020a) Biotechnological applications of beneficial microbiomes for evergreen agriculture and human health. In: Rastegari AA, Yadav AN, Yadav N (eds) Trends of microbial biotechnology for sustainable agriculture and biomedicine systems: perspectives for human health. Elsevier, Amsterdam, pp 255–287. https://doi.org/10.1016/B978-0-12-820528-0.00019-3

Kour D, Rana KL, Kaur T, Sheikh I, Yadav AN, Kumar V et al (2020b) Microbe-mediated alleviation of drought stress and acquisition of phosphorus in great millet (*Sorghum bicolour* L.) by drought-adaptive and phosphorus-solubilizing microbes. Biocatal Agric Biotechnol 23:101501. https://doi.org/10.1016/j.bcab.2020.101501

Kour D, Rana KL, Kaur T, Singh B, Chauhan VS, Kumar A et al (2019) Extremophiles for hydrolytic enzymes productions: biodiversity and potential biotechnological applications. In: Molina G, Gupta VK, Singh B, Gathergood N (eds) Bioprocessing for biomolecules production. John Wiley & Sons, pp 321–372. https://doi.org/10.1002/9781119434436.ch16

Kour D, Rana KL, Sheikh I, Kumar V, Yadav AN, Dhaliwal HS et al (2020c) Alleviation of drought stress and plant growth promotion by *Pseudomonas libanensis* EU-LWNA-33, a drought-adaptive phosphorus-solubilizing bacterium. Proc Natl Acad Sci India Sec B Biol Sci. https://doi.org/10.1007/s40011-019-01151-4

Kour D, Rana KL, Yadav AN, Sheikh I, Kumar V, Dhaliwal HS et al (2020d) Amelioration of drought stress in Foxtail millet (*Setaria italica* L.) by P-solubilizing drought-tolerant microbes with multifarious plant growth promoting attributes. Environ Sustain 3:23–34. https://doi.org/10.1007/s42398-020-00094-1

Krupa P, Kozdrój J (2007) Ectomycorrhizal fungi and associated bacteria provide protection against heavy metals in inoculated pine (*Pinus sylvestris* L.) seedlings. Water Air Soil Pollut 182:83–90

Kuiper I, Lagendijk EL, Bloemberg GV, Lugtenberg BJJ (2004) Rhizoremediation: a beneficial plant microbe interaction. Mol Plant Microbe Interact 17:6–15

Kumar M, Yadav AN, Tiwari R, Prasanna R, Saxena AK (2014) Deciphering the diversity of culturable thermotolerant bacteria from Manikaran hot springs. Ann Microbiol 64:741–751

Kumar P, Dwivedi P (2011) Future habitat loss: greatest threat to the soil microbial biodiversity. J Funct Environ Bot 1(2):82–90

Lasat MM (2002) Phytoextraction of toxic metals: a review of biological mechanisms. J Environ Qual 31:109–120

Leung HM, Ye ZH, Wong MH (2006) Interactions of mycorrhizal fungi with *Pteris vittata* (As hyperaccumulator) in As-contaminated soils. Environ Poll 139:1–8

Li WC, Ye ZH, Wong MH (2010) Metal mobilization and production of short-chain organic acids by rhizosphere bacteria associated with a Cd/Zn hyperaccumulating plant. *Sedum alfredii*. Plant Soil 326:453–467

Liao JP, Lin XG, Cao ZH, Shi YQ, Wong MH (2003) Interactions between Arbuscular mycorrhizae and heavy metals under sand culture experiment. Chemo 50:847–853

Lombi E, Zhao FJ, Dunham SJ, Mcgrath SP (2001) Phytoremediation of heavy metal-contaminated soils: natural hyperaccumulation versus chemically enhanced phytoextraction. J Environ Qual 30:1916–1926

Lopez ML, Peralta-Videa JR, Benitez T, Gardea-Torresdey JL (2005) Enhancement of lead uptake by alfalfa (*Medicago sativa*) using EDTA and a plant growth promoter. Chemo 61:595–598

Lopez-Garcia P, Moreira D (2008) Tracking microbial biodiversity through molecular and genomic ecology. Res Microbiol 159:67–73

Loreau M, Naeem S, Inchausti P, Bengtsson J, Grime J, Hector A, Hooper DU, Huston MA, Raffaeli D, Schmid B, Tilman D, Wardle DA (2001) Biodiversity and ecosystem functioning: current knowledge and future challenges. Sci 294:804–808

Ma Y, Prasad MNV, Rajkumar M, Freitas H (2011) Plant growth promoting rhizobacteria and endophytes accelerate phytoremediation of metalliferous soils. Biotechnol Adv 29:248–258

Madhaiyan M, Poonguzhali S, Sa T (2007) Metal tolerating methylotrophic bacteria reduces nickel and cadmium toxicity and promotes plant growth of tomato (*Lycopersicon esculentum* L.). Chemosphere 69:220–228. https://doi.org/10.1016/j.chemosphere.2007.04.017

Magu SP (1998) Interaction of pesticides with soil microorganisms in relation to crop production. In: Kaushik BD (ed) Plant soil microbe interaction in relation to integrated nutrient management. Venus Printers & Publishers, New Delhi, pp 132–143

Majewska M, Kurek E (2005) Effect of microbial activity on Cd sorption/desorption processes in soil polluted with various Cd sources. Geophys Res Abs 7:04332

Malcova R, Vosatka M, Gryndler M (2003) Effects of inoculation with *Glomus intraradices* on lead uptake by *Zea mays* L. and *Agrostis capillaris* L. Appl Soil Ecol 23:55–67

Malyan SK, Kumar A, Baram S, Kumar J, Singh S, Kumar SS et al (2019) Role of fungi in climate change abatement through carbon sequestration. In: Yadav AN, Singh S, Mishra S, Gupta A (eds) Recent advancement in white biotechnology through fungi, Perspective for sustainable environments, vol 3. Springer International Publishing, Cham, pp 283–295. https://doi.org/10.1007/978-3-030-25506-0_11

Martino E, Perotto S, Parsons R, Gadd GM (2003) Solubilization of insoluble inorganic zinc compounds by ericoid mycorrhizal fungi derived from heavy metal polluted sites. Soil Biol Biochem 35:133–141

McGrath SP, Zhao J, Lombi E (2002) Phytoremediation of metals, metalloids, and radionuclides, vol 75. Adv Agron, In, pp 1–56

Meharg AA (2003) The mechanistic basis of interactions between mycorrhizal associations and toxic metal cations. Mycol Res 107:1253–1265

Miguel AA (1999) The ecological role of biodiversity in agroecosystems. Agric Ecosystems Environ 74:19–31

Miransari M (2011) Hyperaccumulators, arbuscular mycorrhizal fungi and stress of heavy metals. Biotechnol Adv 29:645–653

Mulligan CN, Yong RN, Gibbs BF (2001) Remediation technologies for metal-contaminated soils and groundwater: an evaluation. Eng Geol 60:193–207

Mulligan CN, Yong RN, Gibbs BF, James S, Bennett HPJ (1999) Metal removal from contaminated soils and sediments by biosurfactants surfactin. Environ Sci Technol 33:3812–3820

Myers N, Mittermelar RA, Mittermelar CG, da Fonseca GA, Kent J (2000) Biodiversity hotspots for conservation priorities. Nature 403:853–858

Naeem S (2002) Functioning of biodiversity. In: Munn T (ed) Encyclopedia of global environmental change. Wiley, New York, pp 20–36

Neubauer U, Nowack B, Furrer G, Schulin R (2000) Heavy metal sorption on clay minerals affected by the siderophore desferrioxamine B. Environ Sci Technol 34:2749–2755

Nguyen HL, Leermakers M, Elskens M, De Ridder F, Doan TH, Baeyens W (2005) Correlations, partitioning and bioaccumulation of heavy metals between different compartments of Lake Balaton. Sci. Total Environ 341:211–226

O'Sullivan DJ, O'Gara F (1992) Traits of fluorescent *Pseudomonas* spp. involved in suppression of plant root pathogens. Microbiol Rev 56:662–676

Pandey S, Singh S, Yadav AN, Nain L, Saxena AK (2013) Phylogenetic diversity and characterization of novel and efficient cellulase producing bacterial isolates from various extreme environments. Biosci Biotech Biochem 77:1474–1480

Phukan S (2007) Biodiversity information. In: Dwivedi P, Dwivedi SK, Kalita MC (eds) Biodiversity and environmental biotechnology. Scientific Publishers (India), Jodhpur, pp 251–269

Rajawat MVS, Singh R, Singh D, Yadav AN, Singh S, Kumar M et al (2020) Spatial distribution and identification of bacteria in stressed environments capable to weather potassium aluminosilicate mineral. Braz J Microbiol 51:751–764. https://doi.org/10.1007/s42770-019-00210-2

Rajkumar M, Ae N, Prasad MNV, Freitas H (2010) Potential of siderophore-producing bacteria for improving heavy metal phytoextraction. Trends Biotechnol 28:142–149

Rajkumar M, Sandhya S, Prasad MNV, Freitas H (2012) Perspectives of plant-associated microbes in heavy metal phytoremediation. Biotechnol Adv 30:1562–1574. https://doi.org/10.1016/j.biotechadv.2012.04.011

Romkens P, Bouwman L, Japenga J, Draaisma C (2002) Potentials and drawbacks of chelate-enhanced phytoremediation of soils. Environ Poll 116:109–121

Sahay H, Yadav AN, Singh AK, Singh S, Kaushik R, Saxena AK (2017) Hot springs of Indian Himalayas: potential sources of microbial diversity and thermostable hydrolytic enzymes. 3 Biotech 7:1–11

Santschi PH, Lenhart JJ, Honeyman BD (1997) Heterogeneous processes affecting trace contaminant distribution in estuaries: the role of natural organic matter. Marine Chem 58:99–125

Saravanan VS, Madhaiyan M, Thangaraju M (2007) Solubilization of zinc compounds by the diazotrophic, plant growth promoting bacterium *Gluconacetobacter diazotrophicus*. Chemo 66:1794–1798

Saxena AK, Yadav AN, Rajawat M, Kaushik R, Kumar R, Kumar M et al (2016) Microbial diversity of extreme regions: an unseen heritage and wealth. Indian J Plant Genet Resour 29:246–248

Sharma A, Johri BN (2003a) Combat of iron-deprivation through a plant growth promoting fluorescent Pseudomonas strain GRP3A in mung bean (*Vigna radiata* L. Wilzeck). Microbiol Res 158:77–81

Sharma A, Johri BN (2003b) Growth promoting influence of siderophore-producing *Pseudomonas* strains GRP3A and PRS9 in maize (*Zea mays* L.) under iron limiting conditions. Microbiol Res 158:243–248

Sharma PD (2011) Microbiology, 3rd edn. Rastogi Publication, Meerut

Shen ZG, Li XD, Wang CC, Chen HM, Chua H (2002) Lead phytoextraction from contaminated soil with high-biomass plant species. J Environ Qual 31:1893–1900

Sheng X, He L, Wang Q, Ye H, Jiang C (2008) Effects of inoculation of biosurfactant-producing *Bacillus* sp. J119 on plant growth and cadmium uptake in a cadmium-amended soil. J Hazard Mater 155:17–22

Sheng XF, Xia JJ (2006) Improvement of rape (*Brassica napus*) plant growth and cadmium uptake by cadmium-resistant bacteria. Chemo 64:1036–1042

Shi JY, Lin HR, Yuan XF, Chen XC, Shen CF, Chen YX (2011) Enhancement of copper availability and microbial community changes in rice rhizospheres affected by sulfur. Molecules 16:1409–1417

Souza MLP, Andreoli CV, Amaral MB, Domaszak SC (1996) Preliminary survey of heavy metal content in some soils of Paraná. Sanare. Revista Técnica da Sanepar 5:68–75

Subrahmanyam G, Kumar A, Sandilya SP, Chutia M, Yadav AN (2020) Diversity, plant growth promoting attributes, and agricultural applications of Rhizospheric microbes. In: Yadav AN, Singh J, Rastegari AA, Yadav N (eds) Plant microbiomes for sustainable agriculture. Springer International Publishing, Cham, pp 1–52. https://doi.org/10.1007/978-3-030-38453-1_1

Suman A, Verma P, Yadav AN, Saxena AK (2015) Bioprospecting for extracellular hydrolytic enzymes from culturable thermotolerant bacteria isolated from Manikaran thermal springs. Res J Biotechnol 10:33–42

Tripathi M, Munot HP, Shouche Y, Meyer JM, Goel R (2005) Isolation and functional characterization of siderophore-producing lead- and cadmium-resistant *Pseudomonas putida* KNP9. Curr Microbiol 50:233–237

Van Aken B, Yoon JM, Schnoor JL (2004) Biodegradation of nitrosubstituted explosives TNT, RDX, and HMX by a phytosymbiotic Methylobacterium sp. associated with poplar tissues (*Populus deltoides* -nigra DN34). Appl Environ Microbiol 70:508–517

Venkatesh NM, Vedaraman N (2012) Remediation of soil contaminated with copper using rhamnolipids produced from *Pseudomonas aeruginosa* MTCC 2297 using waste frying rice bran oil. Ann Microbiol 62:85–91

Verma P, Yadav AN, Khannam KS, Kumar S, Saxena AK, Suman A (2016) Molecular diversity and multifarious plant growth promoting attributes of Bacilli associated with wheat (*Triticum aestivum* L.) rhizosphere from six diverse agro-ecological zones of India. J Basic Microbiol 56:44–58

Verma P, Yadav AN, Khannam KS, Mishra S, Kumar S, Saxena AK et al (2019) Appraisal of diversity and functional attributes of thermotolerant wheat associated bacteria from the peninsular zone of India. Saud J Biol Sci 26:1882–1895. https://doi.org/10.1016/j.sjbs.2016.01.042

Verma P, Yadav AN, Khannam KS, Panjiar N, Kumar S, Saxena AK et al (2015) Assessment of genetic diversity and plant growth promoting attributes of psychrotolerant bacteria allied with wheat (*Triticum aestivum*) from the northern hills zone of India. Ann Microbiol 65:1885–1899

Verma P, Yadav AN, Kumar V, Singh DP, Saxena AK (2017) Beneficial plant-microbes interactions: biodiversity of microbes from diverse extreme environments and its impact for crop improvement. In: Singh DP, Singh HB, Prabha R (eds) Plant-microbe interactions in agro-ecological perspectives, Microbial interactions and agro-ecological impacts, vol 2. Springer Singapore, Singapore, pp 543–580. https://doi.org/10.1007/978-981-10-6593-4_22

Vivas A, Voros I, Biro B, Barea JM, Ruiz-Lozano JM, Azcon R (2003) Beneficial effects of indigenous Cd tolerant and Cd-sensitive *Glomus mosseae* associated with a Cd-adapted strain of *Brevibacillus* sp. in improving plant tolerance to Cd contamination. Appl Soil Ecol 24:177–186

Wang FY, Lin XG, Yin R (2007) Role of microbial inoculation and chitosan in phytoextraction of Cu, Zn, Pb and Cd by *Elsholtzia splendens* a field case. Environ Poll 147:248–255

Whiting SN, de Souza MP, Terry N (2001) Rhizosphere bacteria mobilize Zn for hyperaccumulation by *Thlaspi caerulescens*. Environ Sci Technol 35:3144–3150

Wu QT, Deng JC, Long XX, Morel JL, Schwartz C (2006a) Selection of appropriate organic additives for enhancing Zn and Cd phytoextraction by hyperaccumulators. J Environ Sci 18:1113–1118

Wu SC, Luo YM, Cheung KC, Wong MH (2006b) Influence of bacteria on Pb and Zn speciation, mobility and bioavailability in soil: a laboratory study. Environ Pollut 144:765–773

Yadav AN, Gulati S, Sharma D, Singh RN, Rajawat MVS, Kumar R et al (2019a) Seasonal variations in culturable archaea and their plant growth promoting attributes to predict their role in establishment of vegetation in Rann of Kutch. Biologia 74:1031–1043. https://doi.org/10.2478/s11756-019-00259-2

Yadav AN, Kumar R, Kumar S, Kumar V, Sugitha T, Singh B et al (2017) Beneficial microbiomes: biodiversity and potential biotechnological applications for sustainable agriculture and human health. J Appl Biol Biotechnol 5:45–57

Yadav AN, Mishra S, Singh S, Gupta A (2019b) Recent advancement in white biotechnology through fungi. Volume 1: diversity and enzymes perspectives. Springer International Publishing, Cham

Yadav AN, Rastegari AA, Yadav N, Kour D (2020a) Advances in plant microbiome and sustainable agriculture: diversity and biotechnological applications. Springer, Singapore

Yadav AN, Sachan SG, Verma P, Saxena AK (2015a) Prospecting cold deserts of north western Himalayas for microbial diversity and plant growth promoting attributes. J Biosci Bioeng 119:683–693

Yadav AN, Sachan SG, Verma P, Saxena AK (2016) Bioprospecting of plant growth promoting psychrotrophic Bacilli from cold desert of north western Indian Himalayas. Indian J Exp Biol 54:142–150

Yadav AN, Sachan SG, Verma P, Tyagi SP, Kaushik R, Saxena AK (2015b) Culturable diversity and functional annotation of psychrotrophic bacteria from cold desert of Leh Ladakh (India). World J Microbiol Biotechnol 31:95–108

Yadav AN, Sharma D, Gulati S, Singh S, Dey R, Pal KK et al (2015c) Haloarchaea endowed with phosphorus solubilization attribute implicated in phosphorus cycle. Sci Rep 5:12293

Yadav AN, Singh J, Rastegari AA, Yadav N (2020b) Plant microbiomes for sustainable agriculture. Springer International Publishing, Cham

Yadav AN, Verma P, Kumar M, Pal KK, Dey R, Gupta A et al (2015d) Diversity and phylogenetic profiling of niche-specific Bacilli from extreme environments of India. Ann Microbiol 65:611–629

Yang Q, Tu S, Wang G, Liao X, Yan X (2012) Effectiveness of applying arsenate reducing bacteria to enhance arsenic removal from polluted soils by *Pteris vittata* L. Int J Phytoremed 14:89–99

Zaidi S, Usmani S, Singh BR, Musarrat J (2006) Significance of Bacillus subtilis strain SJ-101 as a bioinoculant for concurrent plant growth promotion and nickel accumulation in *Brassica juncea*. Chemo 64:991–997

Zarei M, König S, Hempel S, Nekouei MK, Savaghebi G, Buscot F (2008) Community structure of Arbuscular mycorrhizal fungi associated to *Veronica rechingeri* at the Anguran zinc and lead mining region. Environ Pollut 156:1277–1283
Zarei M, Wubet T, Schäfer SH, Savaghebi GR, Salehi Jouzani G, Khayam Nekouei M, Buscot F (2010) Molecular diversity of arbuscular mycorrhizal fungi in relation to soil chemical properties and heavy metal contamination. Environ Pollut 158:2757–2765
Zhuang X, Chen J, Shim H, Bai Z (2007) New advances in plant growth promoting rhizobacteria for bioremediation. Environ Int 33:406–413

Metagenomics in Deciphering Microbial Communities Associated with Medicinal Plants

3

Hira Saleem, Hareem Mohsin, Azka Asif, Rabia Tanvir, and Yasir Rehman

Abstract

Medicinal plants hold a very important place in therapeutics. Plant growth is affected by a number of abiotic and biotic factors. Among these factors, microorganisms associated with these plants play an important role in the plant health and thus indirectly influence humans as well. Many of these microbes are known to be involved in the production of compounds that are not only useful for the host plant, but also have commercial importance. For in-depth analysis of these plant-associated microbiomes, metagenomic approaches provide the necessary platform of robust, high-throughput techniques. This chapter discusses the microbial communities associated with different medicinal plants, and how metagenomics can be helpful in studying their diversity and versatility.

H. Saleem
Department of Microbiology & Molecular Genetics, University of the Punjab, Lahore, Pakistan

H. Mohsin
Department of Allied Health Sciences, Superior University, Lahore, Pakistan

A. Asif
School of Biological Sciences, University of the Punjab, Lahore, Pakistan

R. Tanvir
University Diagnostic Lab (UDL), Department of Microbiology, University of Veterinary and Animal Sciences (UVAS), Lahore, Pakistan

Y. Rehman (✉)
Department of Life Sciences, School of Science, University of Management & Technology (UMT), Lahore, Pakistan
e-mail: Yasir.rehman@umt.edu.pk

A. N. Yadav et al. (eds.), *Current Trends in Microbial Biotechnology for Sustainable Agriculture*, Environmental and Microbial Biotechnology,
https://doi.org/10.1007/978-981-15-6949-4_3

Keywords

Diversity · Medicinal plants · Metagenomic approaches · Plant-associated microbiomes

3.1 Introduction

Plants are multicellular, autotrophic, and predominantly photosynthetic living organisms found on terrestrial as well as in aquatic ecosystem. The Kingdom Plantae has been divided on the basis of two factors: Flowering and Vasculature. On the basis of flowering, they can be classified into flowering plants such as Angiosperms and Gymnosperms and non-flowering plants including Bryophytes, Pteridophytes, and Thallophytes. On the basis of vasculature, they can be classified into vascular plants which contain a vascular system and non-vascular plants that do not have differentiation of xylem and phloem in them. They can also be classified on environmental basis, i.e., terrestrial and aquatic plants. Aquatic plants include diverse variety of algae while terrestrial include ferns mosses, gymnosperms, and angiosperms (flowering plants). Approximately 374,262 plant species are present worldwide (Christenhusz and Byng 2016).

3.2 Habitat-Based Diversity of Plants and Associated Microbes

On the basis of habitat, plant diversity has been classified into following categories with associated microbial flora.

3.2.1 Hydrophytes

These are wetland plants that particularly grow where plenty of water is available such as in wetlands and shallow rivers, lakes, ponds and marshes. They have adapted themselves according to such habitat. They are deprived of proper root systems and instead of stomata, they just have air spaces. Being residents of wetlands, they do not need to conserve water (Lefor 1999). They are classified as submerged, such as *Hydrilla verticillata,* which are rooted in mud of water bodies. The other is amphibious which is partly submerged; some part is present beneath the water while some above the water as in the case of *Limnophila heterophylla, Typha, Sagittaria,* etc. The last classification is free floating hydrophytes such as *Eichornia*, *Pistia*, *Wolffia*, and *Lemna* which float freely on the surface of water and are not present in rooted form. They are in contact with water as well with air. With the help of metagenomic approach such as clone libraries of the 16S rRNA genes, some epiphytic bacterial communities present on the *Hydrilla verticillate* have been identified; *Delta proteobacteria,*

Verrucomicrobiae, Armatimonadia and *Deinococci* (Gordon-Bradley et al. 2014). In a study done on microbial association with hydrophytes, it was seen that they are associated mostly with Cyanobacteria, Chlorophyta, and Bacillariophyta. Analysis also identified that the loosely attached algae included majority of cyanobacteria such as *Oscillatoria* spp., *Lyngbya* spp., *Microcystis* spp., and *Anabaena* spp., in addition to Chlorophyta members *Pediastrum* spp., *Scenedesmus* spp., *Quadrigula* spp., *Botryococcus,* and *Cladophora* spp. (Aboellil and Aboellil 2012).

3.2.2 Hygrophytes

These are plants which grow in shade and moist conditions. They have spongy root system in which their roots are adapted in the form of rhizomes. Their leaves have stomata, but the rate of transpiration is slow as they grow in a humid environment. Examples include aroids, ferns, bryophytes, *Begonias*, *Juncus,* and sundews (Stanford and Moran 1978). Microbial associations related to bryophytes are *Burkholderia, Serratia, Hafnia, Pantoea, Methanobacteria,* and *Methylobacteria* and were present abundantly inside and outside as well (Bragina et al. 2013). Several strains of *Pseudomonas putida, Xanthomonas* sp., *Serratia* sp., and *Bacillus* sp. are also known to be associated with moss (Opelt and Berg 2004).

3.2.3 Halophytes

These are unique plants which grow in hyper saline conditions. They have the ability to tolerate relatively high amounts of different salts especially sodium and magnesium salts such as NaCl, $MgSO_4$, and $MgCl_2$. To adapt in such harsh environment, they have a salt-expelling root system, named pneumatophores. *Sonneratia, Avicennia, Rhizophora, Ceriops,* and *Suaeda salsa* are a few examples of halophytes. Associated microbes are often seen to gain benefits from their hosts under stressful conditions. With the help of 16S rRNA-targeted metagenomics of *Suaeda salsa,* it was seen that microorganisms belonging to α-proteobacteria, Bacteroidetes, Verrucomicrobia and γ-proteobacteria were found associated with the plant. Moreover, analysis of soil rhizosphere of *Suaeda salsa* showed that *Curvularia protuberate, Fusarium culmorum, Microbulbifer* (*Alteromonadales*), *Pelagibius* (*Rhodospirillales*), *Halomonas* (*Oceanospirillales*) *Marinoscillum* (*Sphingobacteriales*), *Fulvivirga* (*Flexibacteraceae*), *Haloferula* (*Verrucomicrobiales*), *Pelagicoccus* (*Puniceicoccales*), and *Marinobacter* (*Alteromonadales)* genera were also abundantly found (Yuan et al. 2016).

3.2.4 Mesophytes

Mesophytes are plants which are present in moderate conditions of temperature as well as water. They have fully differentiated vascular system to transport the water from roots to the stem. They have herbaceous or woody stem and proper roots with root hairs to absorb water from the soil. They have stomata to transpire water into the environment. Mostly angiosperms fall into this category. Examples include *Jastropha curcas, Jastropha gossypifolia, Canna indica, and Zea mays.* When microbial associations of *Zea mays* was assessed through culture-dependent and culture-independent methods, various bacterial genera, i.e., *Enterobacter*, *Erwinia*, *Klebsiella*, *Pseudomonas*, and *Stenotrophomonas* were found. Culture-based analysis revealed that the predominant group was Firmicutes, mainly of *Bacillus* genus, while *Achromobacter*, *Lysinibacillus*, and *Paenibacillus* genera were rarely found in association with the roots (Pereira et al. 2011).

3.2.5 Xerophytes

These are the plants which are found in dry or xeric environment where water is scarce. These plants have adapted to store water in their stems, e.g., *Opuntia*, in leaves, e.g., *Bryophyllum* and *Aloe vera*, and in roots, e.g., *Asparagus*. That is why xerophytes are also known as succulents as they can retain water in different plant tissues during water scarcity. Metagenomics analysis of *Aloe vera* root revealed its microbiome predominantly consisting of *Proteobacteria* followed by *Firmicutes, Actinobacteria,* and *Bacteroidetes* (Akinsanya et al. 2015).

3.3 Microbes Associated with Medicinal Plants

Medicinal plants are important for their biologically active chemicals "phytochemicals." However, with the discovery of plant-associated microbiomes the focus has shifted to microorganisms and their interactions with their host plants (Huang et al. 2018). Phytotherapeutic compounds are not only important in medicine but also serve agricultural and industrial niches (Mohamad et al. 2019). Therefore, medicinal plants and their microbiomes are a reservoir of important compounds waiting to be explored.

Plant microenvironments are largely inhabited by bacteria which are crucial to plant health (Berg 2009; Hartmann et al. 2008). For instance, the nutrient-rich region of the rhizosphere contains about 10^{11} microorganisms per gram of root with ~30,000 species of prokaryotes (Berendsen 2012). Some of these are well studied and ubiquitous genera, *Pseudomonas* and *Bacillus* (Berg et al. 2011). However, some degree of specificity in microhabitats is observed owing to the varying factors of the microenvironments such as plant species, soil type, pedoclimate, pesticides, and various other biotic and abiotic factors (Berg et al. 2005; Berg and Smalla 2009; Fürnkranz et al. 2012; Köberl et al. 2013a; Singh and Mukerji 2006).

These microbiomes (containing both pathogens and commensals) are also known to be transmitted through generations by seeds and pollens (Fürnkranz et al. 2012; Hardoim et al. 2012; Hirsch and Mauchline 2012). For instance, take the case of ancestor land plants such as mosses; they are known to transfer a very diverse and primarily non-pathogenic microbiome from their sporophytes to gametophytes (Bragina et al. 2012).

In the case of medicinal plants, however, the microbiome possesses a higher degree of specificity owing to their host plant's distinctive exudates and divergent secondary metabolites (Qi et al. 2012; Singh and Mukerji 2006). Actinobacteria are a phylum of Gram-positive bacteria, known for their antibiotic production potential, for example, the genus *Streptomyces* has yielded many novel antibiotics over the years (Goodfellow and Fiedler 2010; Raaijmakers and Mazzola 2012).

Several studies done on various medicinal plants have reported that each of them hold a definitive microbiome majority of which are actinobacteria, producing compounds with antimicrobial as well as anticancer potential (Li et al. 2008; Zhao et al. 2012, 2011).

In retrospect, plant derivatives have been part of healthcare in traditional capacities, for instance, Australia's aboriginal medicine, traditional Chinese medicine (TCM), and the Mayan history are all rich with traditional methods of phytotherapeutics. These have also provided a platform for the discovery of new bioactive agents over the years (Huang et al. 2018). TCM is a hub for ethnopharmaceutical information for about 5000 species of plants and has been a basis for anticancer drugs as well (Miller et al. 2012b).

From 1981 to 2010, plant-derived medicines were about 26% of the entire pharmaceutical market, which boomed to 50% in 2010, and to 85% in 2017 (Ahn 2017; Newman and Cragg 2012). With the shift in research from plant extract compositions to their associated microorganisms instead, plant microbiomes particularly endomicrobiome are found to be directly involved in phytotherapeutic compounds (Chandra 2012; Egamberdieva et al. 2017).

However, with the increase in antibiotic resistance and the absence of new antibiotics to combat this situation, a decrease in the medicinally relevant microorganisms from the microbiome of plants has been observed (Miller et al. 2012a). Previously, phytotherapeutic compounds have yielded antimycotic, antiphlogistic, and hypertension treatment therapies (Abdel-Aziz et al. 2011; Li et al. 2003; Strobel et al. 2004). It is essential to study plant microbiomes and host–microbe interactions to understand and discover new bioactive compounds for chronic inflammations and infections (Nalini and Prakash 2017; Newman and Cragg 2012). Some of the many medicinal plants which will be discussed here are Dandelion (*Taraxacum*), Gingko (*Ginkgo biloba*), Turmeric (*Curcuma longa*), Evening primrose (*Oenothera*), Flax (*Linum usitatissimum*), Tea tree (*Melaleuca alternifolia*), Echinacea (*Echinacea*), Grapeseed extract (*Vitis vinifera*), Lavender (*Lavandula*), and Chamomile (*Matricaria chamomilla*).

3.3.1 *Taraxacum*

The genus *Taraxacum*, commonly known as dandelions, are plants rich in sugars and minerals and produces a lactone saturated white colored latex. The European pharmacopeia in 2005, along with the Committee on Herbal Medicinal Products of the EPA, European Medicines Agency, declared this genus as a medicinally important plant (Maggi 2019). Over the years, it has been used as a cholagogue, diuretic, choleretic, and as an appetizer. Its medicinal importance can be traced back to the tenth century A.D., when Arabs used it to treat liver and spleen disorders (Brown 2008). The extracts of dandelions have reportedly shown biological activities such as anti-inflammatory, antiangiogenic, hypoglycemic, antirheumatic, anticarcinogenic, and antinociceptive (Park et al. 2011; Shidoji and Ogawa 2004). To date, <1% of all species have been identified and studied which include *T. platycarpum*, *T. officinale*, *T. coreanum*, and *T. mongolicum* (Martinez et al. 2015). *Taraxacum* and its endobiome have been widely researched for new antimicrobials and antifungals (Machavariani et al. 2014).

A study done on *Taraxacum coreanum* showed extensive association with various fungal genus, many of which were new to Korea, isolated from roots, stems, and leaves of the plant. The endophytic fungi belonged to the *phylum Ascomycota* and *Basidiomycota*, out of which dominant species were *Alternaria*, *Cladosporium*, *Fusarium*, and *Phoma*. Novel isolates found belonged to the genera: *Apodus*, *Ceriporia*, *Dothideales*, *Leptodontidium*, *Nemania*, *Neoplaconema*, *Phaeosphaeria*, *Plectosphaerella*, and *Terfezia*. About 14% of the endophytic fungi showed promising antifungal activity against known plant pathogens such as *Botrytis cinereal*, *Phytophthora capsica, and Alternaria panax* (Paul et al. 2006). *Phoma* sp., an endophytic fungus isolated from *Taraxacum mongolicum*, is reported to produce an antimicrobial compound 2-hydroxy-6-methylbenzoic acid, particularly active for poultry and aquatic diseases (Wei-nan 2008; Zhang et al. 2013). In 2014, from the dandelion specie *Taraxacummongolicum*'s root, a novel actinomycete was isolated which was given the name *Micromonospora taraxaci* (Zhao et al. 2014a), noted to have potential bioactivity reported against bacterial pathogens (Boumehira et al. 2016; Carro et al. 2018).

3.3.2 *Ginkgo Bilboa*

Ginkgo Biloba is a native ancient Chinese tree, which has survived evolution as compared to its phylogenetic relatives, hence it is called a "living fossil" (Zhou 2009). It is used for illnesses like Alzheimer's (Rimbach et al. 2001), as a dietary supplement and is also widely researched for its flavonoids producing leaves (Ni et al. 2018a, b, 2017).

A spatial study done on *Ginkgo biloba* for bacterial communities on the plant showed a diversity of *Acidobacteria*, *Actinobacteria*, *Bacteroidetes*, *Proteobacteria*, and *Firmicutes* (Leff et al. 2015). Pandey and co-workers in 2009 also isolated a *Pseudomonas* sp. from roots of *Ginkgo biloba*, which took a significant part in plant growth by solubilizing tricalcium phosphate, increasing plan biomass in rice and barley (Pandey et al. 2009).

Endophytes from *G. Biloba* have been reported to show antimicrobial, cytotoxic, and antioxidant properties, for instance, the endophytic fungus *Chaetomium globosum*'s flavipins, are known to have antioxidant activity (Li et al. 2014; Ye et al. 2013;

Yuan et al. 2014). The cytotoxic, anticancer compounds found from the endophytic fungus studied by Li and co-workers are reported to be chaetoglobosins A, G, V, Vb, and C which were active against *Artemia salina* and *Mucor miehei* (Li et al. 2014; Qin et al. 2009). *Penicillium* sp. isolated from the plant produces adenosine, adenine, and 2-deoxyadenosine which are strong antioxidants (Yuan et al. 2014). Another fungus, *Xylaria* sp. is also reported to show antioxidant activity owing to its phenols and flavonoids (Liu et al. 2007). Compounds like 7-Amino-4-methylcoumarin extracted from *Xylaria* sp. YX-28, have shown antibacterial and antifungal activity as well (Liu et al. 2007). *Ginkgo biloba* also harbors *Aspergillus* sp. and is reported to produce Xanthoascin; a potent antimicrobial compound (Zhang et al. 2015).

3.3.3 *Curcuma longa*

Curcuma longa, which is colloquially known as turmeric, is a herb quite common as an ingredient in the southeast Asian cuisine. The desiccated rhizome of *C. longa* is known for its antipyretic and antiseptic properties owing to its curcuminoid and sesquiterpenoid compounds. Curcuminoid is used as an antimicrobial, antioxidant as well as anti-inflammatory agent (Aggarwal and Sung 2009; Jalgaonwala et al. 2010).

A study done on the bacterial endobiome of *C. longa* yielded *Pseudomonas putida*, *Clavibacter michiganensis*, and three *Bacillus* sp., namely *Bacillus pumilus*, *Bacillus thuringiensis*, and *Bacillus cereus*. Strains solubilized phosphate and produced indole 3 acetic acid (IAA) in addition to *P. putida's* siderophore production, all of which promote and regulate plant growth (Kumar et al. 2016). Strains were able to show antifungal as well as antibacterial activity against *Alternaria alternata* and *Fusarium solani*, and *Klebsiella pneumoniae* and *Escherichia coli*, respectively (Kumar et al. 2016). Another study reported *Paenibacillus* spp. colonization of the rhizome of *C. longa* which were active producers of IAA (Aswathy et al. 2013). *C. longa* also produces silver nanoparticles owing to endophytic species of *penicillium* which are used as an antibacterial particularly against *Staphylococcus aureus* and *Escherichia coli* (Singh et al. 2014). A fungal endophyte *Phoma herbarum* produces gentisyl alcohol which shows promising activity against plant leaf pathogen *Colletotrichum gloeosporioides* (Gupta et al. 2016). An Indonesian study done on *C. longa* screening for endophytic fungi with antioxidant activity revealed 44 associated strains, out of which six strains had DPPH scavenging activity significant enough to achieve inhibition more than 65% (Rachman et al. 2015). An endophytic specie of *Eurotium* has been associated with the production of asparaginase enzyme, which is an important component of medications against various types of leukemia (Jalgaonwala and Mahajan 2014).

3.3.4 *Oenothera biennis*

Oenothera biennis is a medicinal plant, the seed oil of which is mostly used for various health issues like asthma, eczema, arthritis, and other inflammations as well as premenstrual syndrome (Dante and Facchinetti 2011; Nikfarjam et al. 2016; Triantafyllidi et al. 2015). The seed oil contains gamma linoleic acid in addition to other phenolic compounds which constitute its potency for medical use (Mehmood et al. 2019; Munir et al. 2017). Some studies have associated the gamma linoleic acid production to endophytic fungus (Jiang et al. 2004).

A study done on roots of *O. biennis* showed a rich 88% of arbuscular mycorrhizae (Zubek and Błaszkowski 2009). Another study found a rich diversity in the root endobiome in varying percentages: *Arthrobacter*, *Variovorax*, *Rhizobium*, *Bradyrhizobium*, *Microbacterium*, *Agrobacterium*, *Bosea*, *Xanthomonas*, *Actinobacterium*, *Bacillus*, *Cellulomonas*, *Nocardioides*, *Paenibacillus*, *Caulobacter*, *Pseudoxanthomonas*, and *Sphingomonas* (Brannock 2004). O. *biennis* is known to grow at highly hydrocarbon contaminated places, and an endophyte study revealed colonization majorly by Alphaproteobacteria, Gammaproteobacteria (highest), or Actinobacteria most of which were IAA, siderophore, hydrogen cyanide and cellulase producers, and phosphate solubilizers with genes encoding hydrocarbon degradation potential (Pawlik et al. 2017). *Oenothera* sp. have also been reported to grow in gypsum-rich soil, with colonizations of Pleosporales, Sordariales, and Diaporthales in their root endospheres (Porras-Alfaro et al. 2014).

3.3.5 *Linum usitatissimum*

Linum usitatissimum, colloquially known as linseed, is rich in α-linolenic acid and the best dietary source of fiber mucilage (Cunnane et al. 1993). It is widely used to control diarrhea, gastrointestinal infections, and heart illnesses (Muir and Westcott 2003). *L. usitatissimum*'s endosphere is primarily composed of *proteobacteria*, *actinobacteria*, and *acidobacteria* in the same order of degree of colonization (Wijesinghe et al. 2015). Major members of the *proteobacteria* were found to be *Xanthomonadales*, *Pseudomonadales*, *Methylophilales*, and *Burkholderiales* (Wijesinghe et al. 2015).

3.3.6 *Melaleuca alternifolia*

Melaleuca alternifolia is the source of tea tree oil which is abundant in terpenes out of which terpinen-4-ol is the main antimicrobial component (Hart et al. 2000). Tea tree oil is also used for antifungal [*Botrytis* control (Abbey et al. 2019)], antiviral, and antiacne purposes (Brand et al. 2001; Miller et al. 2010). A study done for screening aluminum-resistant endophytic bacteria from tea tree showed 53 associated bacteria. Among these, *Burkholderia cepacia* showed phytohormone and siderophore production, regulating plant growth and seed germination (Zhao et al.

2014b). Given the potent antifungal and antibacterial capacity of tea tree oil, still not much research has been done on its endosphere.

3.3.7 *Echinacea*

Purple coneflowers, scientifically known as *Echinacea* is a member of commonly perceived medicinal plants used for healthcare. It is most known for its root's immunomodulatory properties and less-researched potential for the treatment of respiratory infections (Zhao et al. 2014b). Some of the common genera used in medicine are *Echinacea purpurea, Echinacea angustifolia,* and *Echinacea pallida.*

It also exhibits antifungal potentials owing to its diversity of endophytic fungi mostly belonging to the genera *Cladosporium*, *Colletotrichum*, *Fusarium*, *Glomerella*, *Ceratobasidium* and *Mycoleptodiscus*. Out of these the most abundant organisms were reported to be *Colletotrichum gloeosporioides*, *Fusarium oxysporum,* and *Cladosporium cladosporioides* (Rosa et al. 2012). Forty-one percent of these isolates had promising antifungal activity against plant pathogen *Colletotrichum* species (Rosa et al. 2012). A study of two species of medicinal importance from the purple coneflowers, *Echinacea purpurea* and *Echinacea angustifolia*, showed that these two plants also possessed a characteristic biome of endophytic bacteria. The most abundant were known to be *Pseudomonas, Actinobacteria and Bacillus* spp., *Staphylococcus, Curtobacterium, Mycobacterium, Arthrobacter, and Sphingomonas* (Chiellini et al. 2014). A study in 2015 focused on immunomodulatory functions of *Echinacea* extract and found that the ability to stimulate TNF-α by ethanolic extracts of *Echinacea purpurea* are due to LPS of endophytic bacteria present in the plant (Todd et al. 2015). Further studies showed that different plant tissues possess different bacterial communities, i.e., *Gammaproteobacteria*; *Pseudomonas* spp. in roots and rhizosphere and *Actinobacteria* in stem and leaves. *Pseudomonas* genera, which was ubiquitous in all plant parts, as well as other endophytes showed varying antibiotic resistance due to different interactions with the environment (Mengoni et al. 2014). *Echinacea* plant endophytes *Acinetobacter*, *Pseudomonas, Bacillus*, *Stenotrophomonas,* and *Wautersia* (*Ralstonia*) are also involved in IAA production (Lata et al. 2006). Fungal endophyte and entomopathogen *Beauveria bassiana* modulates *Echinacea purpurea*'s growth, pigment, and bioactive product formation (Gualandi 2010). Other endosphere studies on *Echinacea purpurea* show an abundance of *Colletotrichum dematium, Stagonosporopsis* sp. and *Alternaria alternata,* out of which a less prevalent fungus *Biscogniauxia mediterranea* produces fatty acid compounds (−)-5-methylmellein and (−)-(3*R*)-8-hydroxy-6-methoxy-3,5-dimethyl-3,4-dihydroisocoumarin with antifungal activities against *Phomopsis obscurans*, *P. viticola,* and *Fusarium oxysporum* (Carvalho et al. 2016).

3.3.8 *Vitis vinifera*

Extract of *Vitis vinifera* has been used for medicinal purposes since the Greek civilization. The extract is known to act as an antioxidant, as a remedy for obesity since it inactivates lipases, antimicrobial and antiulcer remedy (Mielnik et al. 2006; Moreno et al. 2003; Nuttall et al. 1998; Piccolo et al. 2016; Saito et al. 1998). Seed endophytes are important for seed germination and seedling development, initial plant growth, and some of them are even vertically transferred to the offspring (Truyens et al. 2015). *Vitis vinifera*'s seed is reported to be associated with firmicutes (Truyens et al. 2015) and the plant's growth is known to be modulated by a root-associated *Burkholderia phytofirmans* (Compant et al. 2008, 2005). A deeper analysis of different plant parts and their associated microbes revealed that seeds had populations of *Bacillus altitudinis, Bacillus altitudinis, Staphylococcus aureus* subsp. *Aureus*, and *Paenibacillus amylolyticus*, while pulp of the plant fruit also has *Bacillus weihenstephanensis* associated (Compant et al. 2005). Flowers of the plant were home to *Pseudomonas fulgida, Bacillus pumilus,* and *Bacillus thuringiensis* (Compant et al. 2005). The study also indicated that various stages of plant development harbor various microbial populations due to succession (Compant et al. 2005). The plant also hosts the fungus *Beauveria bassiana* in its endosphere which is a parasite to insects, protecting the plant from them (Rondot and Reineke 2018). Another variety of *Vitis vinifera* called "glera" is inhabited mostly by *Bacillus* along with *Microbacterium*, *Paenibacillus*, *Curtobacterium*, *Stenotrophomonas*, *Variovorax*, *Micrococcus*, and *Agrococcus* (Baldan et al. 2014). Another study seeking to find antagonistic fungi isolated 68 different taxa from *V. vinifera* with most of strains belonging to *Acremonium*, *Gibberella Alternaria*, *Fusarium*, *Botryotinia*, *Epicoccum*, *Penicillium*, *Nectria Cladosporium*, *Phoma*, *Aureobasidium* and *Trichoderma* species, out of which species of *Chaetomium*, *Phoma* (*P. glomerata*), and *Acremonium* were identified as antagonistic fungi (González and Tello 2011). Several species of *Nigrospora* and *Fusarium* are known to modulate grape characteristics such as flavonoids, reducing sugars and phenols (Yang et al. 2016).

3.3.9 *Lavandula*

Lavandula is a genus of flowering plants, mostly cultivated for their fragrance and oils. *Lavandula angustifolia* is the source of lavender oil used commercially and in medicines. The major uses of lavender oil include treatment of rheumatic pain, flatulent dyspepsia, as an antibacterial, and is used in aromatherapy as well (Evans 2009; Hammer et al. 1999).

Different tissues of *Lavandula angustifolia* were found mostly inhabited by *Pseudomonas* (51% of the total endosphere), followed by (in descending order) by *Stenotrophomonas*, *Rhizobium* and *Pantoea*, along with small quantities of Actinomycetes and *Bacillus* sp. Some of the strains showed strong antibacterial activities against *Burkholderia cepacia* (Emiliani et al. 2014). Arbuscular mycorrhizae are reported in *Lavandula angustifolia* indicating the presence of endophytic fungus in the root (Zubek and Błaszkowski 2009; Zubek et al. 2012). Upon research some fungal species were found to be: *Ambispora gerdemannii, Claroideoglomus claroideum, Glomus aureum, Funneliformis constrictum, Funneliformis mosseae, Paraglomus majewskii.* The cultivation conditions when shifted from a mineral fertilization regime to a manure fertilization regime showed an addition of a fungal specie, *Archaeosporatrappei*, to the set of fungal endobiome formerly discussed: (Zubek et al. 2012). A study done in Iran showed the association of *Planomicrobium chinense* with lavender root (Beiranvand et al. 2017).

3.3.10 *Matricaria chamomilla*

Chamomile is an eminent medicinal plant; its uses range from dry cough treatment, oils for therapy massages, digestive stimulant, aromatherapy, and colic treatment for infants (Singh et al. 2011; Weizman et al. 1993). Chamomile is known to possess bacterial endophytes with antifungal activities against *A. alternata, Chaetomium sp., P. variotii Byssochlamys sp., Aureobasidium sp., Fusarium sp.* (Goryluk-Salmonowicz et al. 2016). *Paenibacillus polymyxa* isolated from root of *M. chamomilla* reportedly showed antagonistic activity against phytopathogenic fungi, namely *Verticillium dahliae, Rhizoctonia solani,* and *Fusarium culmorum.* It is also detrimental to *Meloidogyne incognita,* a nematode, and acts as an antimicrobial to the human pathogen *Escherichia coli* (Köberl et al. 2013b, c, 2015). It was reported that volatile organic compounds, specifically pyrazine, from endophytic *Paenibacillus* play a role in the antibacterial and antifungal activities (Rybakova et al. 2016). Chamomile roots are also occupied by Sebacinales which help stimulate plant growth (Riess et al. 2014). *Matricaria recutita,* another member of the chamomile family, harbours rhizospheric microorganisms *Bacillus megaterium, Trichoderma harzianum,* and *Glomus intraradices* which communally modulate antioxidants in various ways for the plant, i.e., production of flavonoids and phenolics and scavenging free radicals (Gupta et al. 2017).

3.4 Metagenomics

The term metagenomics was introduced in 1998 where the importance of unculturable microorganisms and the potential of new metabolites and ecological pathways leading to knowledge beneficial to humans was discussed (Dubey et al. 2020; Handelsman et al. 1998; Shah et al. 2011; Thomas et al. 2012; Zarraonaindia et al. 2013). However, the concept was previously applied by Schmidt, when a λ phage library was constructed from seawater and analyzed for 16S rRNA genes (Schmidt et al. 1991). Metagenomics is the analysis of microbial sequence data sets obtained collectively from an environment (Handelsman et al. 1998). More precisely, "Metagenomics" provides a platform to study the DNA of all the microbial species present at a particular habitat independent of individual species' culturing and identification (Hugenholtz and Tyson 2008). Metagenomics analysis, or the culture-independent analysis, allows the study of DNA of microbial consortia isolated from different environments. This approach depends on the high-throughput sequencing techniques which are necessary for sequence analysis with both coverage and depth. This method of study not only allows the researcher to study the structure of microbial communities but also helps us to assign a particular function to a specific microbial community inhabiting different environments (Zhou et al. 2015). Various environments have been the focus of metagenomic studies including soil, feces, oral cavity, aquatic habitats, and hospital metagenomes for nosocomial infection studies (Coque et al. 2002). Soil metagenomes are particularly difficult to clone as compared to aquatic sources given the complex chemical composition of soil and the risks of unwanted molecules like polyphenolics clinging to DNA or interfering enzymes needed for cloning (Tsai and Olson 1992). However, with advances in the metagenomic techniques, today we have a deeper understanding of the soil microbial community (Rondon et al. 2000).

To put it crudely, metagenomics studies usually involve assembly, phylogenetic analysis, binning and analysis at community level. After environmental samples are collected, a metagenomic small-insert (<10 kb) library is constructed by extracting DNA, cloning it and transforming with a suitable vector into *E. coli* (Henne et al. 1999). Large insert libraries such as cosmid (pWE15 vector, 25–35 kb), bacterial artificial chromosomes (200 kb) and fosmids (40 kb) allow detection of operons as compared to small inserts (Béjà et al. 2002, 2000; Entcheva et al. 2001). Gram-negative hosts other than *E. coli* are also reported such as *Streptomyces lividans*, involved in the discovery of genes encoding novel antibiotics (Courtois et al. 2003). Direct sequencing is also possible courtesy of Next Generation Sequencing technologies (Roche 454, Illumina, ABISOLiD).

This metagenome can then be analyzed for a particular sequence by PCR or hybridization, screened for expression of a specific phenotype, or can be randomly sequenced to obtain microbial populations of a certain environment (Riesenfeld et al. 2004). Over the years, metagenomics has been used to study single genes and functional annotation such as in the discovery of novel cellulases, lipases, and proteases (Healy et al. 1995; Marco 2010), pathways such as antibiotic synthesis (Rondon et al. 2000), organisms, and phylogeny, for instance, in the discovery of archaeal enzymes which remain active only under extreme temperatures (Marco 2010; Stein et al. 1996), as well as communities (Tyson et al. 2004). Data analyzation and searching for functional genes is a cumbersome process (Aguiar-Pulido et al. 2016; Rodriguez-R and Konstantinidis 2014) and, although performed by sophisticated automated pipettors, it often takes 100,000 clones to be screened to achieve <10 active clones (Henne et al. 2000; Majerníks et al. 2001). Computational analysis usually involves either the functional metagenomics or the characterization of genes from millions of reads. Lack of effective transcription in the host due to codon usage bias, imperfect or weak translation, poor folding of protein due to lack of native cofactors and chaperons and poor secretion are some of the hindrances in metagenomic library analysis (Streit and Schmitz 2004). Sensitive screening method employing fluorogenic substrates, novel vectors, and host strains have been developed to overcome these difficulties (Streit and Schmitz 2004). Rapid analysis can now be done using microarray profiling of clones carrying conserved genes (Sebat et al. 2003). In addition, degenerate primers can also serve the same purpose.

For instance, by using robust techniques in metagenomics, genes encoding α-halocarboxylic acid degrading enzymes and novel hydrolases have been discovered (Bell et al. 2002; Marchesi and Weightman 2003). Numerous studies, for instance, biofilm studies like those carried out by European laboratories on metagenomics of highly diverse biofilms (Schmeisser et al. 2003) and of Tyson and co-workers studying low diversity acidophilic biofilms (Tyson et al. 2004) also take benefit of new and improved metagenomic techniques. Similarly, a sequence-based metagenomic study of the Sargasso Sea by Craig Venter was a large-scale project which fruited in many novel genes (Venter et al. 2004). Metagenomic studies have mostly focused on enzymes such as lipases and esterases (Schmeisser et al. 2003), oxidoreductases (Knietsch et al. 2003), nicotine amide (NAM)-dependent alcohol reductases (Hummel 1999), proteases (Santosa 2001), and nitrilases (DeSantis et al. 2002). Genes for vitamin biosynthesis (ascorbic acid) (Eschenfeldt et al. 2001), biotin biosynthesis (Streit and Entcheva 2003), and various novel therapeutic molecules such as antibiotics are also the focus of metagenomic research (Brady et al. 2002; Brady and Clardy 2003; MacNeil et al. 2001; Nikolouli and Mossialos 2012; Wang et al. 2000; Yan et al. 2014).

3.5 Approaches in Metagenomics

Current metagenomics approaches address taxonomic diversity by targeting and amplifying genes such as 16S rRNA, 18S rRNA, NifH, ribosomal internal transcribed region (ITS) prior to sequencing (Morgan and Huttenhower 2012). However, functional metagenomics permits us to study roles of microorganisms in a community, which is quite often not directly related to abundance (Vieites et al. 2008). For instance, if nitrogen fixers are 0.1% in a soil community, their role is far more crucial compared to other populations (Dinsdale et al. 2008).

Metagenomic approaches are divided broadly into two categories; whole genome shotgun sequencing and amplicon-based methods which include 16S rRNA sequencing for bacteria, 18S for eukaryotes, and ITS for fungi. Shotgun metagenomics identifies both culturable and unculturable organisms. Generally, biodiversity profile of the selected community is analyzed and then functionally annotated to lineages (Tringe et al. 2005). Therefore, shotgun sequencing can be of two types; sequence-based screening telling us about microbial diversity in an environmental sample or functional screens, identifying gene products without relating it to its parent organism (Madhavan et al. 2017). Metagenomic studies when launched must keep in mind the potential community to be found in the sample, for instance, complexity of soil sample vs. human skin would possibly be higher, for which more data for soil should be generated. As a result of deep sequence probing, novel and rare taxa could be identified (Sharpton 2014). This also makes shotgun sequencing expensive as compared to 16S (Quail et al. 2012).

16S rRNA gene sequencing relies on the variable regions V1-V9 of the bacterial ribosomal RNA gene for assigning bacteria to taxa (Chakravorty et al. 2007). Soil (Chong et al. 2012), human gut (Dethlefsen et al. 2008), and other environments' biodiversity can also be studied employing 16S rRNA sequencing. While assessing sequence similarities, divergence is permitted. Sequences with >97% similarity are clustered together into "Operational Taxonomical Units (OTUs)" or taxa (Morgan and Huttenhower 2012). However, strains or closely related species cannot be distinguished in this type of metagenomic analysis. For instance, strains O157: H7 and K-12 of *E. coli* cannot be differentiated with their 16S analysis (Weinstock 2012) nor *Shigella flexneri* from *E. coli* (Hilton et al. 2016). Therefore, taxa can be determined but specie level identification is blurry (Ranjan et al. 2016).

18S rRNA is fungal ribosomal component with conserved and variable regions mainly used for taxonomic analysis of fungi in microbial communities. ITS is found between 18S and 5.8S rRNA with higher variability and is used for studying fungal diversity in the environment (Bromberg et al. 2015). Pipelines for taxonomic and functional analysis include MG-RAST (Glass et al. 2010), Mothur (Schloss et al. 2009), and QIIME (Caporaso et al. 2010). QIIME uses UNITE database of ITS sequences of fungal rDNA (Kõljalg et al. 2005). Other genes are also checked by amplicon studies to identify secondary but pivotal functions like diversity of nitrogenase reductase (nifH) and nitrogen fixation activity (Igai et al. 2016). Metagenomic studies of fungal root communities revealed symbiotic arbuscular mycorrhizal fungi by analyzing SSU rRNA gene (Vasar et al. 2017).

3.6 Metagenomics and Diversity of Medicinal Plants

It has been reported by (Raynaud and Nunan 2014) that 1 g of soil holds 10^8–10^{10} microbial species. All these species are playing important roles especially in recycling of nutrients and bio-geochemical cycles and improving the productivity and biomass of the plant (Prakash et al. 2015). Metagenomics provides the platform for investigating how the microbial communities interact with each other and with their host plants that may result in healthier and high-yielding plants (Melcher et al. 2014). A huge diversity of microbes is associated with plants and is known to be involved in nitrogen fixation, enhancement of plant growth, and increasing resistance against different kinds of biotic and abiotic stresses (Lugtenberg and Kamilova 2009; Yang et al. 2009).

Medicinal plants are reported to have developed complex relations with the microbial communities of the rhizosphere, termed as the rhizomicrobiome where the bacterial species propagate by the phenomenon of quorum sensing. Such microbial communities are reported to have symbiotic (legume-rhizobium nitrogen fixation) and mycorrhizal associations. The medicinal plants together with microbial association are able to resist pathogenic strains and their compounds as well. Such communication highly influences the yield and health of the plant. Thus, omics approaches have allowed us to study the microbial consortia along with their functions and have added remarkable knowledge related to the economical and environment-friendly production of medicinal plants that will ultimately lead towards the reduced use of chemical-based plant promoting substances (fertilizers, herbicides, etc.). All this results in the overall improvement of medicinal plants, healthcare and quality of life (Hao and Xiao 2017).

3.6.1 *Cannabis* Microbiome

The plant microbiome can have harmful effects on human health, but when the plant health is considered, it is providing a number of benefits like stimulation of growth and conferring insect or microbial resistance (Turner et al. 2013). The microbiome of leaves and flowers of *Cannabis* residing on the exterior are called as the epiphytes and those within the tissues are called as endophytes. Endophytes usually gain entry within the plant via the rhizosphere and root junctions which are then translocated to other plant tissues via the xylem (Reinhold-Hurek and Hurek 2011). All these microbial communities whether they inhabit the part within the soil or above, are responsible for providing the optimum conditions for growth of *Cannabis* (Berendsen et al. 2012; Winston et al. 2014). A number of bacterial and fungal species have been identified in the endophytic microbiome of *Cannabis*. The fungal species include *Penicillium citrinum*, *Penicillium copticola,* and various species of *Aspergillus* (Kusari et al. 2013). The bacterial endophytic species include *Bacillus subtilis*, *Bacillus licheniformis,* and *Bacillus pumilus* which are known to be beneficial for the plant growth (McKernan et al. 2016; Shi et al. 2010).

3.6.2 *Ocimum sanctum* Microbiome

Ocimum sanctum commonly known as basil plant is known for its medicinal properties. This plant is also in association with endophytic bacterial communities of which the most abundant is *Bacillus pumilus*. This bacterial specie is reported to be beneficial as a bio-inoculant that can enhance the growth of plant. The microbial specie is also reported to be used as a probiotic (El-Badry 2016; Murugappan et al. 2013).

3.6.3 *Maytenus* spp. Microbiome

Maytenus is a shrub-like plant found in the tropical rainforests of Xishuangbanna in China. As reported by Qin et al. (2012), metagenomics approach was used to reveal the presence of *Actinomycetales* and newly reported *Acidimicrobiales* which were not reported before this particular study (Qin et al. 2012). The plants are reported to be used as a treatment for infectious and inflammatory diseases (Da Silva et al. 2011).

3.6.4 *Centella asiatica* Microbiome

Metagenomics analysis of *C. asiatica* was performed using PCR-DGGE analysis targeting the 16S rRNA gene. This study revealed the presence of novel endophytic *Actinobacteria* having medicinal potential due to their ability to produce bioactive metabolites. The plant species is also reported for its use in wound healing, diabetes, hypertension, etc. (Ernawati et al. 2016).

3.6.5 *Crocus sativus L* (Saffron) Microbiome

Culture-independent 16S rRNA analysis of plant rhizosphere and cormosphere showed the presence of 22 different genera isolated from rhizosphere whereas cormosphere was dominated by genus *Pantoea*. Statistical analysis on metagenomics data was also applied showing that the microbial load was diverse in different parts of the same plant (Ambardar et al. 2014).

3.6.6 *Ficus deltoidea* Microbiome

The diverse microbial community of *Actinobacteria* was studied using culture-dependent and culture-independent techniques. The culture-independent, i.e., metagenomics approach for 16S rRNA using DGGE showed the presence of a number of operational taxonomic units (OTUs) such as *Rhodococcus, Verrucosispora,* and *Streptomyces*. The associated microbial community is under research for production of bioactive compounds as the plant is medicinally important in treating cancer, diabetes, and cardiovascular diseases (JANATININGRUM et al. 2018).

3.6.7 *Tinospora crispa* Microbiome

Microbial diversity of plant *T. crispa* was usually done with culture-dependent techniques but the study conducted by Primanita et al. (2015) was performed using PCR-DGGE for metagenomics analysis. This study revealed the presence of endophytic *Actinomycetes* in abundance from various parts of the plant which include the stem, roots, and leaves. Metagenomics approach showed high percentage of *Actinomycetes* among which novel species were also identified showing less than 97% similarities from the already known species. The associated microbial consortia play an important role in the production of bioactive compounds which are used against several diseases (Primanita et al. 2015).

3.6.8 *Anoectochilus roxburghii* Microbiome

This wild plant is most commonly used in traditional Chinese medicine as a treatment of diabetes and tumors. Endophytic species associated with the plant are considered to be potential candidates for bioactive molecules. Metagenomics analysis revealed the presence of novel species like *Paenibacillus* spp. and *Brevibacillus* spp (Chen et al. 2014).

3.6.9 *Dendrobium officinale* Microbiome

D. officinale is considered to be a traditional and rare herbal plant found in China. Taxonomic classification of metagenomic data reveals the abundance of *Ascomycota, Glomerella, Cladosporium, Mycena, Colletotrichum,* and *Alternaria* as endophytic microbial species associated with the plant (Liu et al. 2017).

3.7 Conclusion

Metagenomics has led to the discovery of many novel microbes and microbial processes. Omics of medicinal plants has unveiled their microbiomes which are involved in conferring support to the plant and in the production of compounds of therapeutic importance. In the near future, with much more advanced and evolving technologies, scientists will be able to counter diseases which seem to be life-threatening today, and in doing so omics, especially metagenomics, will prove to be very useful.

3.8 Terminologies

- Annotation: Assigning functions to the genes in DNA sequence analysis.
- Arbuscular mycorrhizae: A symbiotic association between the fungus and a plant where the fungal hyphae penetrates the cortical root cells forming arbuscules.
- Arbuscules: Characteristic branched finger-like hyphae.
- Assembly: One major step in genomics/metagenomics analysis where DNA sequence reads are assembled together for contig formation.
- Bacterial artificial chromosomes (BAC): Vectors used for the insertion of gene of interest/large fragment of DNA in the host cell.
- Biofilms: A microbial consortium characterized by the production of exopolysaccharide layer for adherence of cells to surfaces.
- Bio-inoculant: Microorganisms introduced in the soil that make nitrogen and other nutrients available to the plants, thus reducing the use of nitrogen fertilizers.
- Corms/cormosphere: A swollen underground part of a plant that acts as a storage organ to fight against drastic climatic conditions. The microbial community associated with corm is referred to as cormosphere.
- Cosmid: A combination of plasmid having cos-sites integrated within. Cosmids are used in genetic engineering.
- Culture-dependent: Techniques which involve growing microorganisms in laboratories under artificial conditions.
- Culture-independent: Techniques that bypass the need of culturing microbes and allow the analysis of diverse microbial community within an environment or ecosystem.
- DGGE: Denaturing gradient gel electrophoresis. A culture-independent technique used in DNA fingerprinting.

- Endo-microbiome: A collective term used for all endophytic microbial strains residing in the plant tissues.
- Endophytes: Microbial communities residing within the plant tissues.
- Endo-spheres: All the associated microbial endophytes of a plant.
- Epiphytes: Any organism such as a plant or microorganism that inhabits the outer surface of plant.
- Fosmid: DNA constructs functioning in accordance with the replication and partitioning mechanisms of F plasmid to clone large DNA fragments.
- ITS: Internal transcribed spacer. A region of spacer DNA (non-coding DNA between genes) in small and large ribosomal subunits.
- Metagenomics: The study of environmental DNA or the study of DNA isolated directly from the environment.
- MG-RAST: Online metagenomic analysis tool for the phylogenetic and functional analysis of the metagenomic DNA.
- Mothur: Online tool for the analysis of metagenomic DNA obtained from environment.
- OTU: Operational taxonomic units which are used for the taxonomic classification of bacteria according to similarities.
- PCR: Molecular technique used for the amplification of small quantity of DNA.
- Pedoclimate: A micro-climate exhibiting the fusion of abiotic factors affecting the soil such as temperature, air, and water content.
- Pneumatophores: Also called as "air root" that grows vertically upwards out of soil or water for gaseous exchange.
- Probiotic: Bacteria inhabiting the digestive tract of humans, conferring a healthy system.
- QIIME: A next-generation online platform used for the analysis of microbiome from raw DNA reads.
- Quorum sensing: The release of chemicals called as auto-inducers which play a significant role in the cell density and biofilm formation.
- Rhizo-microbiome: An essential component of the plant ecosystem influencing plant health in physiological and pathological/stressful conditions.
- SSU rRNA gene: The gene encodes 16S rRNA used in the synthesis of small ribosomal subunit.
- UNITE: A database of ITS sequences of fungal rDNA.
- Vector (in terms of transformation): A molecule of DNA used as a vehicle carrying genetic material (foreign gene) to a host cell. A vector containing foreign DNA is known as recombinant DNA.

References

Abbey JA, Percival D, Abbey L, Asiedu SK, Prithiviraj B, Schilder A (2019) Biofungicides as alternative to synthetic fungicide control of grey mould (*Botrytis cinerea*)—prospects and challenges. Biocontrol Sci Tech 29:207–228

Abdel-Aziz H, Fawzy N, Ismail AI, El-Askary H (2011) Toxicological studies on a standardized extract of *Solanum indicum* ssp. *distichum*. Food Chem Toxicol 49:903–909

Aboellil A, Aboellil A (2012) Colonization abilities of microflora to attach aquatic plants. Global J Sci Front Res Biol Sci 12:21–27

Aggarwal BB, Sung B (2009) Pharmacological basis for the role of curcumin in chronic diseases: an age-old spice with modern targets. Trends Pharmacol Sci 30:85–94

Aguiar-Pulido V, Huang W, Suarez-Ulloa V, Cickovski T, Mathee K, Narasimhan G (2016) Metagenomics, metatranscriptomics, and metabolomics approaches for microbiome analysis. Evol Bioinform 12:5–16

Ahn K (2017) The worldwide trend of using botanical drugs and strategies for developing global drugs. BMB Rep 50:111–116. https://doi.org/10.5483/bmbrep.2017.50.3.221

Akinsanya MA, Goh JK, Lim SP, Ting ASY (2015) Metagenomics study of endophytic bacteria in Aloe vera using next-generation technology. Genom Data 6:159–163

Ambardar S, Sangwan N, Manjula A, Rajendhran J, Gunasekaran P, Lal R et al (2014) Identification of bacteria associated with underground parts of *Crocus sativus* by 16S rRNA gene targeted metagenomic approach. World J Microb Biotechnol 30:2701–2709

Aswathy AJ, Jasim B, Jyothis M, Radhakrishnan E (2013) Identification of two strains of *Paenibacillus* sp. as indole 3 acetic acid-producing rhizome-associated endophytic bacteria from *Curcuma longa*. 3 Biotech 3:219–224

Baldan E, Nigris S, Populin F, Zottini M, Squartini A, Baldan B (2014) Identification of culturable bacterial endophyte community isolated from tissues of *Vitis vinifera* "Glera". Plant Biosyst 148:508–516

Beiranvand M, Amin M, Hashemi-Shahraki A, Romani B, Yaghoubi S, Sadeghi P (2017) Antimicrobial activity of endophytic bacterial populations isolated from medical plants of Iran. Iran J Microbiol 9:11

Béjà O, Koonin EV, Aravind L, Taylor LT, Seitz H, Stein JL et al (2002) Comparative genomic analysis of archaeal genotypic variants in a single population and in two different oceanic provinces. Appl Environ Microbiol 68:335–345

Béjà O, Suzuki MT, Koonin EV, Aravind L, Hadd A, Nguyen LP et al (2000) Construction and analysis of bacterial artificial chromosome libraries from a marine microbial assemblage. Environ Microbiol 2:516–529

Bell PJ, Sunna A, Gibbs MD, Curach NC, Nevalainen H, Bergquist PL (2002) Prospecting for novel lipase genes using PCR. Microbiology 148:2283–2291

Berendsen RL, Pieterse CM (2012) The rhizosphere microbiome and plant health. Trends Plant Sci 17:478–48610

Berendsen RL, Pieterse CM, Bakker PA (2012) The rhizosphere microbiome and plant health. Trends Plant Sci 17:478–486

Berg G (2009) Plant–microbe interactions promoting plant growth and health: perspectives for controlled use of microorganisms in agriculture. Appl Microbiol Biotechnol 84:11–18

Berg G, Krechel A, Ditz M, Sikora RA, Ulrich A, Hallmann J (2005) Endophytic and ectophytic potato-associated bacterial communities differ in structure and antagonistic function against plant pathogenic fungi. FEMS Microbiol Ecol 51:215–229

Berg G, Smalla K (2009) Plant species and soil type cooperatively shape the structure and function of microbial communities in the rhizosphere. FEMS Microbiol Ecol 68:1–13

Berg G, Zachow C, Cardinale M, Müller H (2011) Ecology and human pathogenicity of plant-associated bacteria. In: Ehlers RU (ed) Regulation of biological control agents. Springer, Dordrecht, pp 175–189. https://doi.org/10.1007/978-90-481-3664-3

Boumehira AZ, El-Enshasy HA, Hacène H, Elsayed EA, Aziz R, Park EY (2016) Recent progress on the development of antibiotics from the genus *Micromonospora*. Biotechnol Bioproc Eng E21:199–223

Brady SF, Chao CJ, Clardy J (2002) New natural product families from an environmental DNA (eDNA) gene cluster. J Am Chem Soc 124:9968–9969

Brady SF, Clardy J (2003) Synthesis of long-chain fatty acid enol esters isolated from an environmental DNA clone. Org Lett 5:121–124

Bragina A, Berg C, Cardinale M, Shcherbakov A, Chebotar V, Berg G (2012) *Sphagnum* mosses harbour highly specific bacterial diversity during their whole lifecycle. ISME J 6:802–813

Bragina A, Berg C, Müller H, Moser D, Berg G (2013) Insights into functional bacterial diversity and its effects on Alpine bog ecosystem functioning. Sci Rep 3:1–8
Brand C, Ferrante A, Prager R, Riley T, Carson C, Finlay-Jones J et al (2001) The water-soluble components of the essential oil of *Melaleuca alternifolia* (tea tree oil) suppress the production of superoxide by human monocytes, but not neutrophils, activated *in vitro*. Inflamm Res 50:213–219
Brannock J (2004) Characterization of root endophytic bacteria. University of Cincinnati. http://rave.ohiolink.edu/etdc/view?acc_num=ucin1100806990
Bromberg JS, Fricke WF, Brinkman CC, Simon T, Mongodin EF (2015) Microbiota—implications for immunity and transplantation. Nat Rev Nephrol 11:342–353. https://doi.org/10.1038/nrneph.2015.70
Brown D (2008) RHS encyclopedia of herbs and their uses. Dorling Kindersley, London
Caporaso JG, Kuczynski J, Stombaugh J, Bittinger K, Bushman FD, Costello EK et al (2010) QIIME allows analysis of high-throughput community sequencing data. Nat Methods 7:335–336. https://doi.org/10.1038/nmeth.f.303
Carro L, Razmilic V, Nouioui I, Richardson L, Pan C, Golinska P et al (2018) Hunting for cultivable *Micromonospora* strains in soils of the Atacama Desert. Antonievan Leeuwenhoek 111:1375–1387
Carvalho CR, Wedge DE, Cantrell CL, Silva-Hughes AF, Pan Z, Moraes RM et al (2016) Molecular phylogeny, diversity, and bioprospecting of endophytic fungi associated with wild ethnomedicinal North American plant *Echinacea purpurea* (Asteraceae). Chem Biodivers 13:918–930
Chakravorty S, Helb D, Burday M, Connell N, Alland D (2007) A detailed analysis of 16S ribosomal RNA gene segments for the diagnosis of pathogenic bacteria. J Microbiol Methods 69:330–339
Chandra S (2012) Endophytic fungi: novel sources of anticancer lead molecules. Appl Microbiol Biotechnol 95:47–59
Chen Q, Liu B, Guan X, Tang J (2014) Analysis of *Anoectochilus roxburghii* root microbial diversity by metagenomic technology. J Agric Biotechnol 22:1441–1446
Chiellini C, Maida I, Emiliani G, Mengoni A, Mocali S, Fabiani A et al (2014) Endophytic and rhizospheric bacterial communities isolated from the medicinal plants *Echinacea purpurea* and *Echinacea angustifolia*. Int Microbiol 17:165–174
Chong CW, Pearce DA, Convey P, Yew WC, Tan IKP (2012) Patterns in the distribution of soil bacterial 16S rRNA gene sequences from different regions of Antarctica. Geoderma 181–182:45–55. https://doi.org/10.1016/j.geoderma.2012.02.017
Christenhusz MJ, Byng JW (2016) The number of known plants species in the world and its annual increase. Phytotaxa 261:201–217
Compant S, Kaplan H, Sessitsch A, Nowak J, Ait Barka E, Clément C (2008) Endophytic colonization of *Vitis vinifera* L. by *Burkholderia phytofirmans* strain PsJN: from the rhizosphere to inflorescence tissues. FEMS Microbiol Ecol 63:84–93
Compant S, Reiter B, Sessitsch A, Nowak J, Clément C, Barka EA (2005) Endophytic colonization of *Vitis vinifera* L. by plant growth-promoting bacterium Burkholderia sp. strain PsJN. Appl Environ Microbiol 71:1685–1693
Coque TM, Oliver A, Pérez-Díaz JC, Baquero F, Cantón R (2002) Genes encoding TEM-4, SHV-2, and CTX-M-10 extended-spectrum β-lactamases are carried by multiple *Klebsiella pneumoniae* clones in a single hospital (Madrid, 1989 to 2000). Antimicrob Agents Chemother 46:500–510
Courtois S, Cappellano CM, Ball M, Francou F-X, Normand P, Helynck G et al (2003) Recombinant environmental libraries provide access to microbial diversity for drug discovery from natural products. Appl Environ Microbiol 69:49–55
Cunnane SC, Ganguli S, Menard C, Liede AC, Hamadeh MJ, Chen Z-Y et al (1993) High α-linolenic acid flaxseed (*Linum usitatissimum*): some nutritional properties in humans. Br J Nutr 69:443–453
Da Silva G, Serrano R, Silva O (2011) *Maytenus heterophylla* and *Maytenus senegalensis*, two traditional herbal medicines. J Nat Sci Biol Med 2:59

Dante G, Facchinetti F (2011) Herbal treatments for alleviating premenstrual symptoms: a systematic review. J Psychosom Obstet Gynecol 32:42–51

DeSantis G, Zhu Z, Greenberg WA, Wong K, Chaplin J, Hanson SR et al (2002) An enzyme library approach to biocatalysis: development of nitrilases for enantioselective production of carboxylic acid derivatives. J Am Chem Soc 124:9024–9025

Dethlefsen L, Huse S, Sogin ML, Relman DA (2008) The pervasive effects of an antibiotic on the human gut microbiota, as revealed by deep 16S rRNA sequencing. PLoS Biol 6:e280. https://doi.org/10.1371/journal.pbio.0060280

Dinsdale EA, Edwards RA, Hall D, Angly F, Breitbart M, Brulc JM et al (2008) Functional metagenomic profiling of nine biomes. Nature 452:629–632. https://doi.org/10.1038/nature06810

Dubey RK, Tripathi V, Prabha R, Chaurasia R, Singh DP, Rao CS et al (2020) Unravelling the soil microbiome: perspectives for environmental sustainability. Springer, Cham

Egamberdieva D, Wirth S, Behrendt U, Ahmad P, Berg G (2017) Antimicrobial activity of medicinal plants correlates with the proportion of antagonistic endophytes. Front Microbiol 8:199

El-Badry M (2016) Bacterial community metagenomic and variation of some medicinal plant rhizosphere collected form Sinai. SCIREA J Agric 1:16

Emiliani G, Mengoni A, Maida I, Perrin E, Chiellini C, Fondi M et al (2014) Linking bacterial endophytic communities to essential oils: clues from *Lavandula angustifolia* Mill. Evid Based Complement Alternat Med 2014:650905. https://doi.org/10.1155/2014/650905

Entcheva P, Liebl W, Johann A, Hartsch T, Streit WR (2001) Direct cloning from enrichment cultures, a reliable strategy for isolation of complete operons and genes from microbial consortia. Appl Environ Microbiol 67:89–99

Ernawati M, Solihin DD, Lestari Y (2016) Community structures of endophytic actinobacteria from medicinal plant *Centella asiatica* L. urban-based on metagenomic approach. Int J Pharm Pharm Sci 8:292–297

Eschenfeldt WH, Stols L, Rosenbaum H, Khambatta ZS, Quaite-Randall E, Wu S et al (2001) DNA from uncultured organisms as a source of 2, 5-diketo-D-gluconic acid reductases. Appl Environ Microbiol 67:4206–4214

Evans WC (2009) Trease and Evans Pharmacognosy, International Edition E-Book. Elsevier Health Sciences, Saunders Ltd., London

Fürnkranz M, Lukesch B, Müller H, Huss H, Grube M, Berg G (2012) Microbial diversity inside pumpkins: microhabitat-specific communities display a high antagonistic potential against phytopathogens. Microb Ecol 63:418–428

Glass EM, Wilkening J, Wilke A, Antonopoulos D, Meyer F (2010, 2010) Using the metagenomics RAST server (MG-RAST) for analyzing shotgun metagenomes. Cold Spring Harbor Protocols:pdb.prot5368

González V, Tello ML (2011) The endophytic mycota associated with *Vitis vinifera* in central Spain. Fungal Divers 47:29–42

Goodfellow M, Fiedler H-P (2010) A guide to successful bioprospecting: informed by actinobacterial systematics. Antonie Van Leeuwenhoek 98:119–142

Gordon-Bradley N, Lymperopoulou DS, Williams HN (2014) Differences in bacterial community structure on *Hydrilla verticillata* and *Vallisneria americana* in a freshwater spring. Microbes Environ 29(1):67–73

Goryluk-Salmonowicz A, Piorek M, Rekosz-Burlaga H, Studnicki M, Błaszczyk M (2016) Endophytic detection in selected European herbal plants. Pol J Microbiol 65:369–375

Gualandi Jr RJ (2010) Fungal endophytes enhance growth and production of natural products in *Echinacea purpurea* (Moench.) Master's Thesis, University of Tennessee. https://trace.tennessee.edu/utk_gradthes/713

Gupta R, Saikia SK, Pandey R (2017) Bioconsortia augments antioxidant and yield *in Matricaria recutita* L. against *Meloidogyne incognita* (Kofoid and White) Chitwood infestation. Proc Natl Acad Sci India Sect B Biol Sci 87:335–342

Gupta S, Kaul S, Singh B, Vishwakarma RA, Dhar MK (2016) Production of gentisyl alcohol from Phoma herbarum endophytic in *Curcuma longa* L. and its antagonistic activity towards leaf spot pathogen *Colletotrichum gloeosporioides*. Appl Biochem Biotechnol 180:1093–1109

Hammer KA, Carson CF, Riley TV (1999) Antimicrobial activity of essential oils and other plant extracts. J Appl Microbiol 86:985–990

Handelsman J, Rondon MR, Brady SF, Clardy J, Goodman RM (1998) Molecular biological access to the chemistry of unknown soil microbes: a new frontier for natural products. Chem Biol 5:R245–R249

Hao D-c, Xiao P-g (2017) Rhizosphere microbiota and microbiome of medicinal plants: from molecular biology to omics approaches. Chin Herb Med 9:199–217

Hardoim PR, Hardoim CC, Van Overbeek LS, Van Elsas JD (2012) Dynamics of seed-borne rice endophytes on early plant growth stages. PLoS One 7(2):e30438

Hart P, Brand C, Carson C, Riley T, Prager R, Finlay-Jones J (2000) Terpinen-4-ol, the main component of the essential oil of *Melaleuca alternifolia* (tea tree oil), suppresses inflammatory mediator production by activated human monocytes. Inflamm Res 49:619–626

Hartmann A, Rothballer M, Schmid M (2008) Lorenz Hiltner, a pioneer in rhizosphere microbial ecology and soil bacteriology research. Plant Soil 312:7–14

Healy F, Ray R, Aldrich H, Wilkie A, Ingram L, Shanmugam K (1995) Direct isolation of functional genes encoding cellulases from the microbial consortia in a thermophilic, anaerobic digester maintained on lignocellulose. Appl Microbiol Biotechnol 43:667–674

Henne A, Daniel R, Schmitz RA, Gottschalk G (1999) Construction of environmental DNA libraries in *Escherichia coli* and screening for the presence of genes conferring utilization of 4-Hydroxybutyrate. Appl Environ Microbiol 65:3901–3907

Henne A, Schmitz RA, Bömeke M, Gottschalk G, Daniel R (2000) Screening of environmental DNA libraries for the presence of genes conferring lipolytic activity on *Escherichia coli*. Appl Environ Microbiol 66:3113–3116

Hilton SK, Castro-Nallar E, Pérez-Losada M, Toma I, McCaffrey TA, Hoffman EP et al (2016) Metataxonomic and metagenomic approaches vs. culture-based techniques for clinical pathology. Front Microbiol 7:484

Hirsch PR, Mauchline TH (2012) Who's who in the plant root microbiome? Nat Biotechnol 30:961

Huang W, Long C, Lam E (2018) Roles of plant-associated microbiota in traditional herbal medicine. Trends Plant Sci 23:559–562

Hugenholtz P, Tyson GW (2008) Metagenomics. Nature 455:481–483

Hummel W (1999) Large-scale applications of NAD (P)-dependent oxidoreductases: recent developments. Trends Biotechnol 17:487–492

Igai K, Itakura M, Nishijima S, Tsurumaru H, Suda W, Tsutaya T et al (2016) Nitrogen fixation and *nifH* diversity in human gut microbiota. Sci Rep 6:1–11

Jalgaonwala R, Mohite B, Mahajan R (2010) Evaluation of endophytes for their antimicrobial activity from indigenous medicinal plants belonging to North Maharashtra region India. Int J Pharm Biomed Res 1:136–141

Jalgaonwala RE, Mahajan RT (2014) Production of anticancer enzyme asparaginase from endophytic *Eurotium* sp. isolated from rhizomes of *Curcuma longa*. Eur J Exp Biol 4:36–43

Janatiningrum I, Solihin D, Meryandini A, Lestari Y (2018) Comparative study on the diversity of endophytic actinobacteria communities from *Ficus deltoidea* using metagenomic and culture-dependent approaches. Biodiversitas 19:1514–1520

Jiang M, Zhang Y, Hu X, Huang Q (2004) Endophytic fungus producing gamma-linolenic acid from evening primrose (*Oenothera biennis* L.). Chinese J Oil Crop Sci 26:78–81

Knietsch A, Waschkowitz T, Bowien S, Henne A, Daniel R (2003) Metagenomes of complex microbial consortia derived from different soils as sources for novel genes conferring formation of carbonyls from short-chain polyols on *Escherichia coli*. J Mol Microbiol Biotechnol 5:46–56

Köberl M, Ramadan EM, Adam M, Cardinale M, Hallmann J, Heuer H et al (2013b) *Bacillus* and *Streptomyces* were selected as broad-spectrum antagonists against soilborne pathogens from arid areas in Egypt. FEMS Microbiol Lett 342:168–178

Köberl M, Schmidt R, Ramadan EM, Bauer R, Berg G (2013a) The microbiome of medicinal plants and its potential for biocontrol and promotion of plant growth and quality. Front Microbiol 4:400

Köberl M, Schmidt R, Ramadan EM, Bauer R, Berg G (2013c) The microbiome of medicinal plants: diversity and importance for plant growth, quality and health. Front Microbiol 4:400

Köberl M, White RA, Erschen S, El-Arabi TF, Jansson JK, Berg G (2015) Draft genome sequence of *Paenibacillus polymyxa* strain Mc5Re-14, an antagonistic root endophyte of *Matricaria chamomilla*. Genome Announc 3:e00861–e00815

Kõljalg U, Larsson K-H, Abarenkov K, Nilsson RH, Alexander IJ, Eberhardt U et al (2005) UNITE: a database providing web-based methods for the molecular identification of ectomycorrhizal fungi. New Phytol 166:1063–1068

Kumar A, Singh R, Yadav A, Giri D, Singh P, Pandey KD (2016) Isolation and characterization of bacterial endophytes of *Curcuma longa* L. 3 Biotech 6:60

Kusari P, Kusari S, Spiteller M, Kayser O (2013) Endophytic fungi harbored in *Cannabis sativa* L.: diversity and potential as biocontrol agents against host plant-specific phytopathogens. Fungal Divers 60:137–151

Lata H, Li X, Silva B, Moraes R, Halda-Alija L (2006) Identification of IAA-producing endophytic bacteria from micropropagated *Echinacea* plants using 16S rRNA sequencing. Plant Cell Tissue Organ Cult 85:353–359

Leff JW, Del Tredici P, Friedman WE, Fierer N (2015) Spatial structuring of bacterial communities within individual *Ginkgo biloba* trees. Environ Microbiol 17:2352–2361

Lefor MW (1999) Hydrophyte. In: Knödel, Klaus, Lange, Gerhard, Voigt, Hans-Jürgen Environmental Geology. Springer, Dordrecht, pp 338–338. https://doi.org/10.1007/1-4020-4494-1_184

Li H, Xiao J, Gao Y-Q, Tang JJ, Zhang A-L, Gao J-M (2014) Chaetoglobosins from *Chaetomium globosum*, an endophytic fungus in *Ginkgo biloba*, and their phytotoxic and cytotoxic activities. J Agric Food Chem 62:3734–3741

Li J, Zhao GZ, Chen HH, Wang HB, Qin S, Zhu WY et al (2008) Antitumour and antimicrobial activities of endophytic streptomycetes from pharmaceutical plants in rainforest. Lett Appl Microbiol 47:574–580

Li RW, Myers SP, Leach DN, Lin GD, Leach G (2003) A cross-cultural study: anti-inflammatory activity of Australian and Chinese plants. J Ethnopharmacol 85:25–32

Liu W, Zhou Z, Liu Y, Hu X, Guo Y, Li J (2017) Application of high-throughput internal transcribed spacer rRNA metagenomics analysis in deciphering endophytic fungi diversity of *Dendrobium Officinale*. J Biobaased Mater Bioenergy 11:106–118. https://doi.org/10.1166/jbmb.2017.1647

Liu X, Dong M, Chen X, Jiang M, Lv X, Yan G (2007) Antioxidant activity and phenolics of an endophytic *Xylaria* sp. from *Ginkgo biloba*. Food Chem 105:548–554

Lugtenberg B, Kamilova F (2009) Plant-growth-promoting rhizobacteria. Ann Rev Microbiol 63:541–556

Machavariani NG, Ivankova TD, Sineva ON, Terekhova LP (2014) Isolation of endophytic actinomycetes from medicinal plants of the Moscow region, Russia. World Appl Sci J 30:1599–1604

MacNeil I, Tiong C, Minor C, August P, Grossman T, Loiacono K et al (2001) Expression and isolation of antimicrobial small molecules from soil DNA libraries. J Mol Microbiol Biotechnol 3:301–308

Madhavan A, Sindhu R, Parameswaran B, Sukumaran RK, Pandey A (2017) Metagenome analysis: a powerful tool for enzyme bioprospecting. Appl Biochem Biotechnol 183:636–651

Maggi F (2019) Dandelion. In: Nabavi SM, Silva AS (eds) Nonvitamin and nonmineral nutritional supplements. Elsevier, San Diego, pp 203–204. https://doi.org/10.1016/C2016-0-03546-5

Majerník A, Gottschalk G, Daniel R (2001) Screening of environmental DNA libraries for the presence of genes conferring Na+ (Li+)/H+ antiporter activity on *Escherichia coli*: characterization of the recovered genes and the corresponding gene products. J Bacteriol 183:6645–6653

Marchesi JR, Weightman AJ (2003) Comparing the dehalogenase gene pool in cultivated α-halocarboxylic acid-degrading bacteria with the environmental metagene pool. Appl Environ Microbiol 69:4375–4382

Marco D (2010) Metagenomics. Theory, methods and applications. Caister Academic Press, Norfolk

Martinez M, Poirrier P, Chamy R, Prüfer D, Schulze-Gronover C, Jorquera L et al (2015) *Taraxacum officinale* and related species—an ethnopharmacological review and its potential as a commercial medicinal plant. J Ethnopharmacol 169:244–262

McKernan K, Spangler J, Helbert Y, Lynch RC, Devitt-Lee A, Zhang L et al (2016) Metagenomic analysis of medicinal Cannabis samples; pathogenic bacteria, toxigenic fungi, and beneficial microbes grow in culture-based yeast and mold tests. F1000Res 5:2471–2471. https://doi.org/10.12688/f1000research.9662.1

Mehmood Z, Khan MS, Qais FA, Ahmad I (2019) Herb and modern drug interactions: efficacy, quality, and safety aspects. In: New look to phytomedicine. Elsevier, London, pp 503–520

Melcher U, Verma R, Schneider WL (2014) Metagenomic search strategies for interactions among plants and multiple microbes. Front Plant Sci 5:268

Mengoni A, Maida I, Chiellini C, Emiliani G, Mocali S, Fabiani A et al (2014) Antibiotic resistance differentiates *Echinacea purpurea* endophytic bacterial communities with respect to plant organs. Res Microbiol 165:686–694

Mielnik M, Olsen E, Vogt G, Adeline D, Skrede G (2006) Grape seed extract as antioxidant in cooked, cold stored Turkey meat. LWT-Food Sci Technol 39:191–198

Miller KI, Ingrey SD, Alvin A, Sze MYD, Roufogalis BD, Neilan BA (2010) Endophytes and the microbial genetics of traditional medicines. Microbiol Aust 31:60–63

Miller KI, Qing C, Sze DMY, Neilan BA (2012b) Investigation of the biosynthetic potential of endophytes in traditional Chinese anticancer herbs. PLoS One 7(5):e35953

Miller KI, Qing C, Sze DM-Y, Roufogalis BD, Neilan BA (2012a) Culturable endophytes of medicinal plants and the genetic basis for their bioactivity. Microb Ecol 64:431–449

Mohamad OAA, Ma J-B, Liu Y-H, Li L, Hatab S, Li W-J (2019) Medicinal plant-associated microbes as a source of protection and production of crops. In: Egamberdieva D, Tiezzi A (eds) Medically important plant biomes: source of secondary metabolites. Springer, Singapore, pp 239–263. https://doi.org/10.1007/978-981-13-9566-6

Moreno DA, Ilic N, Poulev A, Brasaemle DL, Fried SK, Raskin I (2003) Inhibitory effects of grape seed extract on lipases. Nutrition 19:876–879

Morgan XC, Huttenhower C (2012) Chapter 12: Human microbiome analysis. PLoS Comput Biol 8:e1002808

Muir AD, Westcott ND (2003) Flax: the genus *Linum*. CRC Press, Boca Raton, FL

Munir R, Semmar N, Farman M, Ahmad NS (2017) An updated review on pharmacological activities and phytochemical constituents of evening primrose (genus *Oenothera*). Asian Pac J Trop Biomed 7:1046–1054

Murugappan R, Begum SB, Roobia RR (2013) Symbiotic influence of endophytic *Bacillus pumilus* on growth promotion and probiotic potential of the medicinal plant *Ocimum sanctum*. Symbiosis 60:91–99

Nalini M, Prakash H (2017) Diversity and bioprospecting of actinomycete endophytes from the medicinal plants. Lett Appl Microbiol 64:261–270

Newman DJ, Cragg GM (2012) Natural products as sources of new drugs over the 30 years from 1981 to 2010. J Nat Prod 75:311–335

Ni J, Dong L, Jiang Z, Yang X, Chen Z, Wu Y et al (2018a) Comprehensive transcriptome analysis and flavonoid profiling of ginkgo leaves reveals flavonoid content alterations in day–night cycles. PLoS One 13(3):e0193897

Ni J, Dong L, Jiang Z, Yang X, Sun Z, Li J et al (2018b) Salicylic acid-induced flavonoid accumulation in *Ginkgo biloba* leaves is dependent on red and far-red light. Ind Crop Prod 118:102–110

Ni J, Hao J, Jiang Z, Zhan X, Dong L, Yang X et al (2017) NaCl induces flavonoid biosynthesis through a putative novel pathway in post-harvest Ginkgo leaves. Front Plant Sci 8:920

Nikfarjam M, Bahmani M, Heidari-Soureshjani S (2016) Phytotherapy for depression: a review of the most important medicinal plants of flora of Iran effective on depression. J Chem Pharm Sci 9:1242–1247

Nikolouli K, Mossialos D (2012) Bioactive compounds synthesized by non-ribosomal peptide synthetases and type-I polyketide synthases discovered through genome-mining and metagenomics. Biotechnol Lett 34:1393–1403

Nuttall S, Kendall M, Bombardelli E, Morazzoni P (1998) An evaluation of the antioxidant activity of a standardized grape seed extract, Leucoselect®. J Clin Pharm Ther 23:385–389

Opelt K, Berg G (2004) Diversity and antagonistic potential of bacteria associated with bryophytes from nutrient-poor habitats of the Baltic Sea Coast. Appl Environ Microbiol 70:6569–6579

Pandey G, Dorrian SJ, Russell RJ, Oakeshott JG (2009) Biotransformation of the neonicotinoid insecticides imidacloprid and thiamethoxam by *Pseudomonas* sp. 1G. Bioch Biophys Res Commun 380:710–714

Park CM, Park JY, Noh KH, Shin JH, Song YS (2011) *Taraxacum officinale* Weber extracts inhibit LPS-induced oxidative stress and nitric oxide production via the NF-κB modulation in RAW 264.7 cells. J Ethnopharmacol 133:834–842

Paul NC, Kim WK, Woo SK, Park MS, Yu SH (2006) Diversity of endophytic fungi associated with *Taraxacum coreanum* and their antifungal activity. Mycobiology 34:185–190

Pawlik M, Cania B, Thijs S, Vangronsveld J, Piotrowska-Seget Z (2017) Hydrocarbon degradation potential and plant growth-promoting activity of culturable endophytic bacteria of *Lotus corniculatus* and *Oenothera biennis* from a long-term polluted site. Environ Sci Pollut Res Int 24:19640–19652

Pereira P, Ibáñez F, Rosenblueth M, Etcheverry M, Martínez-Romero E (2011) Analysis of the bacterial diversity associated with the roots of maize (*Zea mays* L.) through culture-dependent and culture-independent methods. ISRN Ecol 2011:938546

Piccolo S, Alfonzo A, Burruano S, Moschetti G (2016) Detection of bacterial endophytes in *Vitis vinifera* L. and antibiotic activity against grapevine fungal pathogens. In: Biocontrol of major grapevine diseases: leading research. CABI, Boston, pp 182–190

Porras-Alfaro A, Raghavan S, Garcia M, Sinsabaugh RL, Natvig DO, Lowrey TK (2014) Endophytic fungal symbionts associated with gypsophilous plants. Botany 92:295–301

Prakash O, Sharma R, Rahi P, Karthikeyan N (2015) Role of microorganisms in plant nutrition and health. In: Nutrient use efficiency: from basics to advances. Springer, New Delhi, pp 125–161

Primanita M, Wahyudi A, Lestari Y (2015) 16S rRNA-based metagenomic analysis of endophytic *Actinomycetes* diversity from *Tinospora crispa* L. Miers. Microbiol Indones 9:4

Qi X, Wang E, Xing M, Zhao W, Chen X (2012) Rhizosphere and non-rhizosphere bacterial community composition of the wild medicinal plant *Rumex patientia*. World J Microbiol Biotechnol 28:2257–2265

Qin J-C, Zhang Y-M, Gao J-M, Bai M-S, Yang S-X, Laatsch H et al (2009) Bioactive metabolites produced by *Chaetomium globosum*, an endophytic fungus isolated from *Ginkgo biloba*. Bioorg Med Chem Lett 19:1572–1574

Qin S, Chen HH, Zhao GZ, Li J, Zhu WY, Xu LH et al (2012) Abundant and diverse endophytic actinobacteria associated with medicinal plant *Maytenus austro yunnanensis* in Xishuangbanna tropical rainforest revealed by culture-dependent and culture-independent methods. Environ Microbiol Rep 4:522–531

Quail MA, Smith M, Coupland P, Otto TD, Harris SR, Connor TR et al (2012) A tale of three next generation sequencing platforms: comparison of Ion Torrent, Pacific biosciences and Illumina MiSeq sequencers. BMC Genomics 13:341

Raaijmakers JM, Mazzola M (2012) Diversity and natural functions of antibiotics produced by beneficial and plant pathogenic bacteria. Annu Rev Phytopathol 50:403–424

Rachman F, Septiana E, Lekatompessy SJ, Widowati T, Sukiman HI, Simanjuntak P (2015) Screening for endophytic fungi from turmeric plant (*Curcuma longa* L.) of Sukabumi and Cibinong with potency as antioxidant compounds producer. Pak J Biol Sci 18:42

Ranjan R, Rani A, Metwally A, McGee HS, Perkins DL (2016) Analysis of the microbiome: advantages of whole genome shotgun versus 16S amplicon sequencing. Biochem Biophys Res Commun 469:967–977

Raynaud X, Nunan N (2014) Spatial ecology of bacteria at the microscale in soil. PLoS One 9:e87217

Reinhold-Hurek B, Hurek T (2011) Living inside plants: bacterial endophytes. Curr Opin Plant Biol 14:435–443

Riesenfeld CS, Schloss PD, Handelsman J (2004) Metagenomics: genomic analysis of microbial communities. Annu Rev Genet 38:525–552

Riess K, Oberwinkler F, Bauer R, Garnica S (2014) Communities of endophytic sebacinales associated with roots of herbaceous plants in agricultural and grassland ecosystems are dominated by *Serendipita herbamans* sp. nov. PLoS One 9:e94676–e94676

Rimbach G, Gohil K, Matsugo S, Moini H, Saliou C, Virgili F et al (2001) Induction of glutathione synthesis in human keratinocytes by *Ginkgo biloba* extract (EGb761). Biofactors 15:39–52

Rodriguez-R LM, Konstantinidis KT (2014) Bypassing cultivation to identify bacterial species. Microbe 9:111–118

Rondon MR, August PR, Bettermann AD, Brady SF, Grossman TH, Liles MR et al (2000) Cloning the soil metagenome: a strategy for accessing the genetic and functional diversity of uncultured microorganisms. Appl Environ Microbiol 66:2541–2547

Rondot Y, Reineke A (2018) Endophytic *Beauveria bassiana* in grapevine *Vitis vinifera* (L.) reduces infestation with piercing-sucking insects. Biol Control 116:82–89

Rosa LH, Tabanca N, Techen N, Wedge DE, Pan Z, Bernier UR et al (2012) Diversity and biological activities of endophytic fungi associated with micropropagated medicinal plant *Echinacea purpurea* (L.) Moench. Am J Plant Sci 3:1105–1114

Rybakova D, Cernava T, Köberl M, Liebminger S, Etemadi M, Berg G (2016) Endophytes-assisted biocontrol: novel insights in ecology and the mode of action of *Paenibacillus*. Plant Soil 405:125–140

Saito M, Hosoyama H, Ariga T, Kataoka S, Yamaji N (1998) Antiulcer activity of grape seed extract and procyanidins. J Agric Food Chem 46:1460–1464

Santosa DA (2001) Rapid extraction and purification of environmental DNA for molecular cloning applications and molecular diversity studies. Mol Biotechnol 17:59–64

Schloss PD, Westcott SL, Ryabin T, Hall JR, Hartmann M, Hollister EB et al (2009) Introducing mothur: open-source, platform-independent, community-supported software for describing and comparing microbial communities. Appl Environ Microbiol 75:7537–7541

Schmeisser C, Stöckigt C, Raasch C, Wingender J, Timmis K, Wenderoth D et al (2003) Metagenome survey of biofilms in drinking-water networks. Appl Environ Microbiol 69:7298–7309

Schmidt TM, DeLong EF, Pace NR (1991) Analysis of a marine picoplankton community by 16S rRNA gene cloning and sequencing. J Bacteriol 173:4371–4378

Sebat JL, Colwell FS, Crawford RL (2003) Metagenomic profiling: microarray analysis of an environmental genomic library. Appl Environ Microbiol 69:4927–4934

Shah N, Tang H, Doak TG, Ye Y (2011) Comparing bacterial communities inferred from 16S rRNA gene sequencing and shotgun metagenomics. In: Altman RS (ed) Biocomputing 2011. World Scientific, Kohala Coast, HI, pp 165–176. https://doi.org/10.1142/8021

Sharpton TJ (2014) An introduction to the analysis of shotgun metagenomic data. Front Plant Sci 5:209

Shi Y, Lou K, Li C (2010) Growth and photosynthetic efficiency promotion of sugar beet (*Beta vulgaris* L.) by endophytic bacteria. Photosynth Res 105:5–13

Shidoji Y, Ogawa H (2004) Natural occurrence of cancer-preventive geranylgeranoic acid in medicinal herbs. J Lipid Res 45:1092–1103

Singh D, Rathod V, Ninganagouda S, Hiremath J, Singh AK, Mathew J (2014) Optimization and characterization of silver nanoparticle by endophytic fungi *Penicillium* sp. isolated from *Curcuma longa* (turmeric) and application studies against MDR*E. coli* and *S. aureus*. Bioinorg Chem Appl 2014:408021

Singh G, Mukerji KG (2006) Root exudates as determinant of rhizospheric microbial biodiversity. In: Mukerji KG, Manoharachary C, Singh J (eds) Microbial activity in the rhizosphere. Springer, Berlin, Heidelberg, pp 39–53. https://doi.org/10.1007/3-540-29420-1

Singh O, Khanam Z, Misra N, Srivastava MK (2011) Chamomile (*Matricaria chamomilla* L.): an overview. Pharmacogn Rev 5:82–95

Stanford QH, Moran W (1978) Geography: a study of its physical elements. Oxford University Press, Toronto

Stein JL, Marsh TL, Wu KY, Shizuya H, DeLong EF (1996) Characterization of uncultivated prokaryotes: isolation and analysis of a 40-kilobase-pair genome fragment from a planktonic marine archaeon. J Bacteriol 178:591–599

Streit W, Entcheva P (2003) Biotin in microbes, the genes involved in its biosynthesis, its biochemical role and perspectives for biotechnological production. Appl Microbiol Biotechnol 61:21–31

Streit WR, Schmitz RA (2004) Metagenomics—the key to the uncultured microbes. Curr Opin Microbiol 7:492–498

Strobel G, Daisy B, Castillo U, Harper J (2004) Natural products from endophytic microorganisms. J Nat Prod 67:257–268

Thomas T, Gilbert J, Meyer F (2012) Metagenomics—a guide from sampling to data analysis. Microb Inform Exp 2(1):3

Todd DA, Gulledge TV, Britton ER, Oberhofer M, Leyte-Lugo M, Moody AN et al (2015) Ethanolic *Echinacea purpurea* extracts contain a mixture of cytokine-suppressive and cytokine-inducing compounds, including some that originate from endophytic bacteria. PLoS One 10(5):e0124276

Triantafyllidi A, Xanthos T, Papalois A, Triantafillidis JK (2015) Herbal and plant therapy in patients with inflammatory bowel disease. Ann Gastroenterol 28(2):210–220

Tringe SG, von Mering C, Kobayashi A, Salamov AA, Chen K, Chang HW et al (2005) Comparative metagenomics of microbial communities. Science 308:554

Truyens S, Weyens N, Cuypers A, Vangronsveld J (2015) Bacterial seed endophytes: genera, vertical transmission and interaction with plants. Environ Microbiol Rep 7:40–50

Tsai Y-L, Olson BH (1992) Rapid method for separation of bacterial DNA from humic substances in sediments for polymerase chain reaction. Appl Environ Microbiol 58:2292–2295

Turner TR, James EK, Poole PS (2013) The plant microbiome. Genome Biol 14:209

Tyson GW, Chapman J, Hugenholtz P, Allen EE, Ram RJ, Richardson PM et al (2004) Community structure and metabolism through reconstruction of microbial genomes from the environment. Nature 428:37–43

Vasar M, Andreson R, Davison J, Jairus T, Moora M, Remm M et al (2017) Increased sequencing depth does not increase captured diversity of arbuscular mycorrhizal fungi. Mycorrhiza 27:761–773

Venter JC, Remington K, Heidelberg JF, Halpern AL, Rusch D, Eisen JA et al (2004) Environmental genome shotgun sequencing of the Sargasso Sea. Science 304:66–74

Vieites JM, Guazzaroni M-E, Beloqui A, Golyshin PN, Ferrer M (2008) Metagenomics approaches in systems microbiology. FEMS Microbiol Rev 33:236–255

Wang G-Y-S, Graziani E, Waters B, Pan W, Li X, McDermott J et al (2000) Novel natural products from soil DNA libraries in a streptomycete host. Org Lett 2:2401–2404

Wei-nan L (2008) Primary studies on isolation and detection of three strains endophytic fungi from Dandelion (*Taraxacum mongolicum Hand-Mazz*) and its antimicrobial activity of poultry pathogenic. J Anhui Agric Sci 22:9540–9542

Weinstock GM (2012) Genomic approaches to studying the human microbiota. Nature 489:250–256

Weizman Z, Alkrinawi S, Goldfarb D, Bitran C (1993) Efficacy of herbal tea preparation in infantile colic. J Pediatr 122:650–652

Wijesinghe M, Germida J, Walley F (2015) Root-associated bacterial communities in flax (*Linum usitatissimum*) and their response to arbuscular mycorrhizal (AM) inoculation. In: Soils and crops workshop. https://harvest.usask.ca/bitstream/handle/10388/8902/M.%20Wijesingh%20et%20al.,%202015.pdf?sequence=1

Winston ME, Hampton-Marcell J, Zarraonaindia I, Owens SM, Moreau CS, Gilbert JA et al (2014) Understanding cultivar-specificity and soil determinants of the *Cannabis* microbiome. PLoS One 9(6):e99641

Yan X, Tang X-X, Chen L, Yi Z-W, Fang M-J, Wu Z et al (2014) Two new cytotoxic indole alkaloids from a deep-sea sediment derived metagenomic clone. Mar Drugs 12:2156–2163

Yang J, Kloepper JW, Ryu C-M (2009) Rhizosphere bacteria help plants tolerate abiotic stress. Trends Plant Sci 14:1–4

Yang M-Z, Ma M-D, Yuan M-Q, Huang Z-Y, Yang W-X, Zhang H-B et al (2016) Fungal endophytes as a metabolic fine-tuning regulator for wine grape. PLoS One 11(9):e0163186

Ye Y, Xiao Y, Ma L, Li H, Xie Z, Wang M et al (2013) Flavipin in *Chaetomium globosum* CDW7, an endophytic fungus from *Ginkgo biloba*, contributes to antioxidant activity. Appl Microbiol Biotechnol 97:7131–7139

Yuan Y, Tian J-M, Xiao J, Shao Q, Gao J-M (2014) Bioactive metabolites isolated from *Penicillium* sp. YY-20, the endophytic fungus from *Ginkgo biloba*. Nat Prod Res 28:278–281

Yuan Z, Druzhinina IS, Labbé J, Redman R, Qin Y, Rodriguez R et al (2016) Specialized microbiome of a halophyte and its role in helping non-host plants to withstand salinity. Sci Rep 6:1–13

Zarraonaindia I, Smith DP, Gilbert JA (2013) Beyond the genome: community-level analysis of the microbial world. Biol Philos 28:261–282

Zhang H, Xiong Y, Zhao H, Yi Y, Zhang C, Yu C et al (2013) An antimicrobial compound from the endophytic fungus *Phoma* sp. isolated from the medicinal plant *Taraxacum mongolicum*. J Taiwan Inst Chem E 44:177–181

Zhang W, Wei W, Shi J, Chen C, Zhao G, Jiao R et al (2015) Natural phenolic metabolites from endophytic *Aspergillus* sp. IFB-YXS with antimicrobial activity. Bioorg Med Chem Lett 25:2698–2701

Zhao J, Guo L, He H, Liu C, Zhang Y, Li C et al (2014a) *Micromonospora taraxaci* sp. nov., a novel endophytic actinomycete isolated from dandelion root (*Taraxacum mongolicum Hand.-Mazz.*). Antonievan Leeuwenhoek 106:667–674

Zhao K, Penttinen P, Chen Q, Guan T, Lindström K, Ao X et al (2012) The rhizospheres of traditional medicinal plants in Panxi, China, host a diverse selection of actinobacteria with antimicrobial properties. Appl Microbiol Biotechnol 94:1321–1335

Zhao K, Penttinen P, Guan T, Xiao J, Chen Q, Xu J et al (2011) The diversity and anti-microbial activity of endophytic actinomycetes isolated from medicinal plants in Panxi plateau, China. Curr Microbiol 62:182–190

Zhao X, Song P, Feng L, Hong W, WuC YR et al (2014b) Isolation and identification of a growth-promoting and aluminum-resistant endophytic bacterium from tea tree. Acta Agric Univ Jiangxiensis 36:407–412

Zhou J, He Z, Yang Y, Deng Y, Tringe SG, Alvarez-Cohen L (2015) High-throughput metagenomic technologies for complex microbial community analysis: open and closed formats. MBio 6:e02288–e02214

Zhou Z-Y (2009) An overview of fossil Ginkgoales. Palaeoworld 18:1–22

Zubek S, Błaszkowski J (2009) Medicinal plants as hosts of arbuscular mycorrhizal fungi and dark septate endophytes. Phytochem Rev 8:571

Zubek S, Stefanowicz AM, Błaszkowski J, Niklińska M, Seidler-Łożykowska K (2012) Arbuscular mycorrhizal fungi and soil microbial communities under contrasting fertilization of three medicinal plants. Appl Soil Ecol 59:106–115

Role of Metagenomics in Deciphering the Microbial Communities Associated with Rhizosphere of Economically Important Plants

4

Priyanka Jha and Vijay Kumar

Abstract

The physiological development of the plant system is significantly affected by microbial communities. Selected members of the microbial community are advantageous while few are unfavorable to the plant growth. Few pathogens rather colonize the rhizosphere to break the defense mechanism of the plant to trigger disease whereas others propagate through the various aerial parts of the plant to spread disease pertaining to human. However, the significance of microbial communities related to rhizosphere has been widely recognized recently. To improve the plant growth and development, it will be beneficial to know the microbial structure present in the rhizospheric microbiome. This can enhance the present situation of sustainable growth of agro-ecosystem related to soil microbiome by enhancing the final yields. In this chapter, we summarize the community structure of the microbes in the rhizosphere of various economic important plants through metagenomic approach.

Keywords

Agro-ecosystem · Metagenomics · Microbiome · Plants · Rhizosphere

---

P. Jha (✉)
Amity Institute of Biotechnology (AIBNK), Amity University, Kolkata Campus, Kolkata, West Bengal, India
e-mail: pjha@kol.amity.edu

V. Kumar
Department of Biotechnology, Lovely Faculty of Technology and Sciences, Lovely Professional University, Phagwara, Punjab, India

A. N. Yadav et al. (eds.), *Current Trends in Microbial Biotechnology for Sustainable Agriculture*, Environmental and Microbial Biotechnology,
https://doi.org/10.1007/978-981-15-6949-4_4

4.1 Introduction

The role of microbial diversity in plant growth and development is always considered of special interest to biologists. Plant rhizosphere is colonized with huge population of microorganisms that attain higher densities at a short span of time. This results in higher number of microbial genes in the microbiome than the plant genes. Several reports on plant–microbe interaction has been reported which suggests the influence of the interaction on seed germination, seedling growth, and plant productivity. Predominantly found microbes in rhizosphere are bacteria, nematodes, worms, fungi, and algae (Bonkowski et al. 2009). The microbes active in the rhizospheric zone survive predominantly on the nutrients released by the plant roots. Plant roots mostly secrete mucilage and exudates, also known as rhizodeposits which regulate the microbial diversity of the zone. A study by Cook et al. (1995), suggested that plant cells may regulate the microbial diversity of the rhizosphere by alternating the rhizodeposits as per their benefit. Microbial diversity related to nitrogen fixers, mycorrhiza present in the rhizospheric region has been reported to have advantageous influence on plant growth and development. However, pathogenic bacteria, fungi, and worms account as non-beneficial to the plant development. Also, proliferation of human pathogens inside and on the surface of the leaf epidermis has been reported by several studies (Kaestli et al. 2012). Hence, understanding the structural and functional diversity of microbial world in the rhizospheric zone is an important step to improve the plant growth, health, and productivity. Metagenomics approach has allowed examining and identifying microbial diversity and richness from culturable to the unculturable forms leading to enhanced analysis of microbial genome evolution and heterogeneity.

The latest sequencing technologies have made way for unraveling the microbial diversity and the genomic content in the rhizosphere (Table 4.1). The rhizospheric microbial population has been reported with capabilities of nitrogen fixation and helping in plant growth and development (Lugtenberg and Kamilova 2009; Yang et al. 2009). Suitable literature reports have found that microbes invade the root and tissue of the healthy plant. Such microbial growth affects the plant health by induction of certain phenotypes in favor of the plant development and disease resistance (Ryu et al. 2003). However, studies are required for better understanding and higher specificity to learn about the relationship between plant and microbial interactions. To combat such scenario, metagenomic approach was employed in several studies relating soils from different geographical region for comparing microbial populations (Taghavi et al. 2009; Fierer et al. 2012; Unno and Shinano 2013; Luo et al. 2014). Report by Fierer et al. (2012) suggested low abundance of biotic genes in desert soil microbiome thus indicating the predominance of genes related to abiotic conditions in such microbiome. In past few decades, with the advent of metagenomics approach, biological sciences relating to study of microbial interactions have reached new heights. This technique has permitted extension to the soil microbial interactions in rhizospheric zone with more understanding towards novel genes, hidden genetic features, and newer metabolic pathways. The main purpose of this review is to provide brief insights on the current advances of metagenomics

Table 4.1 Case studies relating metagenomics approach to microbial interactions

Sl. No.	Case studies	References
1	The study introduces a new term known as metaphenome which combines genetic potential of the microbiome and available resources	Jansson and Hofmockel (2018)
2	Samples from reference, rehabilitated and non-rehabilitated mine site of Urucum massif, Brazil, is compared. Results suggested similar microbiome composition for reference and rehabilitated sites samples	Gastauer et al. (2019)
3	The microbial populations present in the rhizospheric zone of the barley and alfalfa planted soil contaminated are predominated by Proteobacteria, Bacteroidetes, and Actinobacteria phylum	Kumar et al. (2018)
4	16S gene profiling and metagenomic shotgun sequencing was done to find out the microbial interactions in the wild and domesticated roots of barley. Bacterial families such as *Commamonadaceae, Flavobacteriaceae,* and *Rhizobiaceae* predominated the rhizospheric zone of the barley plant	Bulgarelli et al. (2015)
5	Amplicon sequencing approach was utilized to find the microbial population present in the roots of *Vitis vinifera*. Predominance of Alphaproteobacteria, Rhizobiales, Acidobacteria-GP4 was found	Gupta et al. (2019)

approaches in deciphering the different microbial communities associated to rhizosphere of different economic important plant.

4.2 Achievements with Metagenomics in Economic Important Plant and Microbial Interactions

Metagenomics is a promising tool which provides fundamental knowledge on microbe–microbe and plant–microbe interactions and has remarkable potential to enhance sustainable plant productivity (Bramhachari et al. 2017). Several reports have been documented on a metagenomic approach to explore the microbial diversity in plants (Unno and Shinano 2013; Mendes et al. 2014; Yadav et al. 2015). The first metagenomic library was prepared from the samples of picoplankton by Schmidt et al. (1991) which was followed by Healy et al. (1995) for generating metagenomic libraries for a variety of cellulase from cellulose digesters. In the year 2000, a term metagenomic library was named to bacterial artificial chromosomes with DNA inserts from soil sample (Rondon et al. 2000). Huge volume of metagenomic data is generated utilizing this approach which influences the findings of novel genes and enzymes, metabolic pathways, and microbial interaction. In this section, we will highlight recent achievements in scientific studies relating metagenomic approach for microbial interactions in various economic important plants.

4.2.1 Medicinal Plants

Medicinal plants are mostly used as herbal remedies and play an essential role in traditional healthcare in most of the developing countries. The word allelopathy is commonly used to define the chemical involvements of two or more plant species possibly due to the release of biochemicals in the rhizosphere (Duke 2010). With reference to allelopathy, allelopathic toxicity is described as a phenomenon where a particular plant variety negatively affects itself due to repeated plantation in the same soil. Huge number of medicinal plants (approx. 70%) suffers due to allelopathic toxicity. Wu et al. (2018) reported the change in microbial population in the rhizosphere due to allelopathic toxicity of *Rehmannia glutinosa*, a traditional Chinese herbal plant. The toxicity phenomenon resulted in reduction of bacterial population of Burkholderiaceae and Pseudomonadaceae and an increase in Sphingomonadaceae and Streptomycetaceae. Also, comparative metagenomics suggested a reduction in abundance of *Azotobacter*, *Pseudomonas*, *Burkholderia,* and *Lysobacter* in 2-year monocultured soil.

Similarly, McKernan and his group studied the microbiome composition of the rhizosphere of medicinally important *Cannabis* sp. (McKernan et al. 2016). Metagenomic analysis suggested the abundance of pathogenic bacterial and fungal species such as *E. coli, Salmonella enterica, Penicillium citrinum, Pseudomonas aeruginosa,* and *Clostridium botulinum*. As found in the results of the study, *P. citrinum* is a growth-promoting endophyte for *Cannabis* sp. and also had higher abundance in the study. However, this pathogenic strain produces nephrotoxin citrinin which in case present in the plant extracts might pose for a serious health threat.

Tian and Zhang (2017) studied the microbial diversity of the rhizosphere related to the halophyte *Messerschmidia sibirica*. The 16S rRNA gene sequencing was accomplished using Illumina HiSeq platform to identify the bacterial diversity relating the halophyte. The halophyte *M. sibirica* has traditional medicinal values along with significant commercial and ecological importance. In addition, they also help in soil improvement, phytoremediation. The metagenomic throughput study indicated the predominance of Proteobacteria and Actinobacteria. The genera observed to be abundant were *Pseudomonas*, *Bacillus*, *Sphingomonas*, and *Rhizobium*.

4.2.2 Plants Producing Cereals

The next-generation throughput sequencing technique has been applied widely to examine the gut microbiome and interactions between host and microbial metabolism (Broderick 2015, Martin et al. 2014; Table 4.2). Ina report by Mendes et al. (2014), the microbial population from the soil reservoir used for soybean farming in Amazon forest soils is examined. With the help of shotgun sequencing, predominance of Deltaproteobacteria, Sphingobacteria, and Chloroflexi along with Gammaproteobacteria and Solibacteres were found. In another study by Unno and Shinano (2013), the metagenomic study indicated a variation in the abundance of bacterial community that improves the phytic acid utilization and the plant health.

Table 4.2 Examples for microbial community diversity in rhizospheric zone for host plants

Sl. No.	Host plant	Microbial community	References
1	Rice	Proteobacteria, Acidobacteria, Firmicutes, Bacteroidetes	Arjun and Harikrishnan (2011)
2	*Arabidopsis*	Proteobacteria, Actinobacteria, Bacteroidetes	Bodenhausen et al. (2013)
3	Wheat	*Pseudomonas aeruginosa, Bacillus cereus, Alcaligenes faecalis*	Egamberdiyeva et al. (2008)
4	*Arachis hypogaea*	Proteobacteria, Actinobacteria, Firmicutes, Bacteroidetes, Acidobacteria	Haldar and Sengupta (2015)
5	Wheat	Proteobacteria, Archaea, Firmicutes, Actinobacteria, Fungi	Hernandez-Leon et al. (2012)
6	Rice	Firmicutes, Actinobacteria, Gammaproteobacteria, Methanobacteriales, Methanomicrobiales, Methanosarcinales	Knief et al. (2012)

Table 4.3 List of rhizodeposits released by host plant in rhizospheric zone

Sl. No.	Rhizodeposit	Host plant	References
1	Strigolactone (plant hormone)	Rice	Cardoso et al. (2014)
2	Malic acid, citric acid	Tomato	de Weert et al. (2002)
3	Sugar (fructose, maltose), amino acid, organic acid	Maize	Carvalhais et al. (2011)
4	Salicylic acid, gamma-amibobutyric acid	Arabidopsis	Badri et al. (2013)
5	Mugineic acid	Barley	Takagi et al. (1984)
6	Mucilage	Maize	Iijima et al. (2000)

The microbial community included Bacteroidetes, Betaproteobacteria, Chlorobi, and Methanobacteria. In a recent study by Kumar et al. (2018), metagenomic analysis showed the rhizospheric microbial structure in alfalfa and barley planted oil contaminated soil samples. The sequencing study revealed the abundance of Proteobacteria, Bacteroidetes, and Actinobacteria of approx. 46%, 21.4%, and 10.4%, respectively. Plants can significantly alter the microbial structure by producing rhizodepositions (Table 4.3). The 16S rRNA amplicon sequencing was employed to reveal the rhizospheric microbial structure for *Brassica oleracea* (O'Brien et al. 2018). The rhizosphere of the organic fertilized 12-wee- old cabbages showed increased abundance of *Thiobacillus* and reduced abundance of cyanobacteria *Phormidium* in synthetic fertilized soils.

4.2.3 Leguminous Plants

In the late nineteenth century, scientists grouped mycorrhiza and bacterial communities present in root nodules of the leguminous plant as root symbionts (Morton 1981). The symbiotic nitrogen fixation is one of the most significant mutualistic functions among microbes present in close vicinity of leguminous rhizosphere. The mutualistic approach by the microbe allows the conversion of atmospheric nitrogen into nutrient which can be further taken by the plant (van der Heijden et al. 2006). In a recent study by Dinnage et al. (2019), metagenomic sequencing was done to isolate the rhizospheric microbial population from the bulk soil near *Acacia acuminata*. Results suggested the predominance of Bradyrhizobiaceae clade along with Rhizobiaceae.

Since Beijerinck's experimentation with *Rhizobium* and leguminous plant symbiosis which further resulted in root nodule formation, biochemistry-relating nitrogen fixation made substantial progress (Quispel 1974). In leguminous plants, formation of root nodules is somehow significantly influenced by flavonoid pathway which attracts rhizobial microbes and activates the *nod* gene expression. Several reports have been found with respect to the flavonoid pathway influence on *nod* gene expression induction (Fig. 4.1). The mycorrhizal interaction with nodules permit increased nitrogen fixation which in turn allows higher nutrient uptake. However, signaling pathways for various plant–microbe interactions varies. While a lot many studies have inferred by analyzing the mutant plant varieties that the candidate genes are related to metabolite transfer to rhizospheric zone (Carvalhais et al. 2015; Foo et al. 2013; Zhang et al. 2009).

The flavonoid metabolism pathway is one of the most studied biosynthetic metabolisms (Fig. 4.1). Flavonoid metabolism is primarily initiated by the phenylpropanoid metabolites which are formed from malonyl CoA and p-coumaroyl CoA (Stafford 1990). There are certain types of flavonoids which are produced from CoA ester compounds such as cinnamic acid. The flavonoid diversity ranges due to various basal flavonoid structures such as flavonols, flavones, flavonones, and many more (Fig. 4.1).

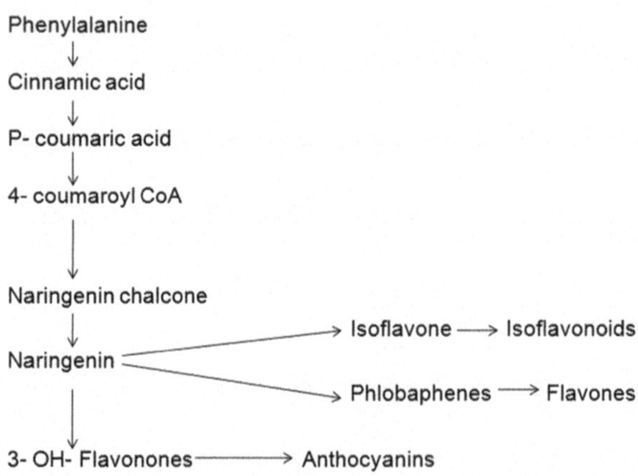

Fig. 4.1 The flavonoid metabolism pathway in plant systems

4.2.4 Essential Oil-Bearing Plants

A plant resource comprises several types of natural products including essential oils. Essential oils (non-toxic compounds) are natural bioactive products and have potential applications and economic value. Recent study by Shaikh et al. (2018) revealed the rhizospheric study of *Mentha arvensis* via metagenomic approach. *M. arvensis* is widely cultivated aromatic plant mostly for essential oil utilized to treat skin ache and pain. The metagenomic study resulted in predominance of *Aspergillus niger* followed by *Rhizopus stolonifer* and *Rhizopus nigricans.* Abundance hit for *Aspergillus flavus*, *A. terricola*, *Trichoderma viride*, and *Zygorhynchus molleri* was also found. The study revealed a hike by 0.88% when treated with *Trichoderma viride* whereas when treated with *A. niger* an increase of 0.78% in essential oil production was observed. The experimental set inoculated with *T. viride* showed highest production of menthol (approx. 98%) when compared with the set inoculated with rhizospheric fungi. As per a recent study by Xu et al. (2018), rhizospheric microbiome has a significant role in enhancing the fruit quality and health of citrus plant. The amplicon sequencing revealed the predominance of Proteobacteria, Actinobacteria, Bacteroidetes, and Acidobacteria as taxa. Also, the highly abundant microbes found in the rhizosphere are *Pseudomonas*, *Agrobacterium*, *Burkholderia*, *Rhizobium*, and *Bradyrhizobium* which later were observed to be core microbes in the near vicinity of the plant root often helping for stress tolerance and disease resistance. In the year 2010, Kaewkla and Franco (2010) isolated *Pseudonocardia eucalypti* from the roots of native Australian eucalyptus tree. The *P. eucalypti* is an endophytic actinobacteria-bearing Gram-positive characteristics. The 16S rRNA phylogenetic studies also showed sequences having 96.1% and 96.3% similarity to *Pseudonocardia acaciae* and *Pseudonocardia spinosispora,* respectively. Metagenomic study of *Aloe vera* microbiome suggested the presence of Proteobacteria, Firmicutes, Bacteroidetes, and Actinobacteria in a study by Akinsanya et al. (2015).

In 2017, Pereira et al. (2017) revealed the variation in microbiome composition in the rhizosphere of *Eucalyptus grandis* and *Acacia mangium*. Quantitative PCR was performed on soil samples with a depth of 0–800 cm. Results reported predominance of Proteobacteria in subsurface layers of soil with a depth of 0–300 cm and Acidobacteria in surface layer. A change in microbial composition was possibly due to the rhizospheric effect of *A. mangium* in mixed stands with *E. grandis*. Also, *A. mangium* increases the acidification of rhizosphere by absorption of cations and simultaneous release of H^+ ions leading to microbiome composition variation. The monospecific cultivation of *E. grandis* suggested an abundance of Firmicutes and Proteobacteria.

4.3 Insight on Plant Growth-Promoting Rhizobacteria

Plant growth-promoting rhizobacteria (PGPRs), an important group of microorganism, belongs to rhizosphere bacteria and involved in to promote plant growth and health via different mechanism (Ali et al. 2015; Yadav et al. 2020b). The PGPRs play a vital role in plant-based agricultural system, mostly as biofertilizers for the replacement of pesticides and chemical fertilizers which often contaminate the environment (Kour et al. 2020; Singh et al. 2020b). Recent advances with regard to utilizing metagenomics approaches expanded overall understanding to characterize soil microbial communities and plant–microbe interactions in the rhizosphere (Souza et al. 2015; Goel et al. 2017).

A range of microbial population share mutualistic relationship with leguminous plants by producing biological minerals, and hence improving soil fertility and plant growth (Rosenberg and Rosenberg 2016; Verma et al. 2017). Simultaneous evolution of plants with rhizospheric microbial community is necessary to withstand biotic and abiotic stresses and improving the sustainability (Khan et al. 2016; Kumar et al. 2019; Singh et al. 2016). The PGPRs show positive association with plant hosts. The PGPRs help in improving barren lands to fertile zones for cultivation, soil quality, and enriches the plant health (Bhardwaj et al. 2014; Yadav et al. 2018). The soil fertility along with plant health can be improved by various ways such as by producing certain compounds in vicinity of plant roots which enhance plant health, nitrogen fixation, phosphate, and potassium solubilization, enhance hormone production and by reducing the harmful effects of pathogenic microbes by inducing systemic resistance, enzyme, and volatile organic compound production and antibiosis (Tripathi et al. 2012; Rastegari et al. 2020; Yadav et al. 2020a) (Table 4.4). The PGPRs act as biofertilizers which can increase the nutrient uptake from rhizosphere. The direct mechanism of PGPRs invades the host root and improves plant health by nitrogen fixation, production of siderophores, indole-3-acetic acid (IAA); by degradation of environmental pollutants and mitigation of different abiotic stress such as temperature, pH, drought, radiation, and salinity (Ma et al. 2011; Tank and Saraf 2010; Kour et al. 2019; Singh et al. 2020a; Singh and Yadav 2020). Furthermore, the applications of metagenomics not only provide insights to microbial/taxonomic diversity but also access to metabolic diversity of genes. In addition, metagenomics can be utilized for the development of next-generation fungicides and pesticides to improve the organic agriculture efficiency. However, a detailed metagenomic approach of the PGPRs is highly recommended to explore the new rhizospheric flora to improve the sustainable agricultural/plant productivity.

Table 4.4 Role of plant growth-promoting rhizobacteria in enhancing plant growth and development

Sl. No.	Name of the microbe	Important role	References
1	*Azotobacter aceae*	Nitrogen fixation	Bhattacharyya and Jha (2012)
2	*Bacillus circulans*	Phosphate solubilization	Oteino et al. (2015)
3	*Azospirillim brasilence*	Polycyclic aromatic hydrocarbon degradation	Orlandini et al. (2014)
4	*Azospirillim brasilence*	Indole acetic acid synthesis	Orlandini et al. (2014)
5	*Burkholderia sp.*	Induction of ethylene production	Islam et al. (2016)
6	*Bacillus subtilis*	Nickel accumulation	Prathap and Ranjitha (2015)
7	*Pseudomonas putida*	Ethylene, salicylic acid production	Tiwari et al. (2016)
8	*Bacillus amyloliquefaciens*	Maintenance of elicitors, secondary metabolite production	Srivastava et al. (2016)
9	*Pseudomonas fluorescens*	Degradation of trichloroethylene and resistance against halo blight	Ramadan et al. (2016)
10	*Bacillus mucilaginosus*	Enhanced potassium intake	Liu et al. (2012)

4.4 Biotechnological Impact of Next-Generation Sequencing Technologies

In the year 1986, a group of scientists (Pace et al. 1986) initially coined the concept of DNA cloning directly from environmental samples to understand the structural diversity of microbial communities. This strategy was a modified version of shotgun cloning of DNA extracted from natural samples based on 16S rRNA genes. However, in the year 1998, Handelsman along with the scientific group proposed the term "metagenome" through a study based on extracting soil microbial communities and its underlying importance as initial source of novel natural compounds (Handelsman et al. 1998). The study proved that metagenomics approach had significant contribution in new chemical compound mining from uncultured microorganisms.

Metagenomics approach can be further classified into two major sections known as structural and functional metagenomics, which focus on various outlooks of determining the microbial community linked to a particular microbiome. The structural metagenomics allows the examination of the structural diversity of uncultured microbial population for reconstruction of metabolic pathways (Handelsman 2005). In this way, the microbial community study will allow to examine underlying connections of various microbes in a specific ecosystem with respect to different biotic and abiotic stresses. However, the functional metagenomic approach focuses to identify genes related to a specific function. This technique involves preparation of gene expression libraries associating different metagenomic clones based on activity-based screening.

Table 4.5 Novel gene discovery through metagenomics

Sl. No.	Target gene identified	Screening technique	Functions	References
1	Naphthalene dioxygenase	Functional metagenomics	Applicable in heavy metal/ oil contaminated soil samples	Ono et al. (2007)
2	Salt resistance genes	Functional metagenomics	Aids in resisting varying salt concentrations	Mirete et al. (2015)
3	Nickel resistance genes	Functional metagenomics	Helps in resisting nickel present in near vicinity	Mirete et al. (2007)
4	Antimicrobial molecules	Functional metagenomics	Confers antibiotic attributes	Ia et al. (2001)
5	Dioxygenase-degrading cluster	Sequencing-based metagenomics	Degradation of phenylalkanoic acid (PAA)	Zaprasis et al. (2009)
6	Cellulase	Functional metagenomics	Thermotolerant enzyme	Garg et al. (2016)

The Sanger sequencing technology was one of the initial sequencing platforms employed for metagenomics (Sanger et al. 1977). However, with evolution of time, the rise of next-generation sequencing platforms allowed higher sequencing capacities at lower cost (Klindworth et al. 2013). Also, current versions of NGS platform have higher capacities of up to 5000 Mb of DNA sequences per day in comparison to 6 Mb data generated by Sanger sequencing (Kircher and Kelso 2010).

The advent of metagenomics has a significant contribution in the field of biotechnology with deciphering the microbial roles in commercially available enzymes, production of antibiotics, and in biochemical transformations for biotechnological advancements (Fernández-Arrojo et al. 2010). Structural and functional-based metagenomics has been widely employed (Table 4.5) for identifying the new genes providing resistance against harsh conditions, antibiotics, salinity, and heavy metals. Also, functional metagenomics can provide the deeper understanding towards biochemical pathways employed by the microbes in varying biotic and abiotic stresses (Table 4.5). Thus, the similar property can be utilized further to enhance the survival capacity of the microbes being employed in industry.

4.5 Conclusion and Future Prospects

Over the last few years with the advent of metagenomics, significant advancements have been made. To deduce the interlinking biochemical processes that control the microbial organization in the subsurface, an elaborated examination of the microbial structure, diversity, and its dynamics is a prerequisite. The findings presented reveal that metagenomic approaches have emerged as a modern tool that controls diverse aspects of microbial communities and has potential application in the plant–microbes interaction. The microbiome structure beneath the surface significantly helps in plant growth and development. These microbial communities have undisputable contribution towards soil fertility, enhanced nutrient accession to plant, resisting from biotic and abiotic stresses and phytoremediation of organic compounds. In some cases, microbial community has been observed to vary with soil depth; however, in deeper horizon, no distinct population has been detected. In majority of the studies, Proteobacteria has been observed to be in high abundance which indicates active nitrogen fixation, nutrient uptake capacity, and polycyclic aromatic hydrocarbon degradation. Apart from Proteobacteria, Bacteroidetes, Firmicutes, and Acidobacteria have also been obtained in higher number. Also, integrated omics technology such as next-generation sequencing and metagenomics has made it possible to unlock the rich microbial potential from the subsurface horizons. However, more elaborated research needs to be conducted along with other omics analyses such as meta-transcriptomics and meta-proteomics to discover the in situ functions associated with the microbial structures. In addition, the use of novel approaches will certainly elucidate the novel and diverse mechanisms of PGPRs activity as well as new PGPRs identity which will provide a new look on applications of PGPRs biology. Further research will provide novel insights for better understanding of metagenomic approach related to microbes. Ideally, this will inevitably improve the modern metagenomic approaches for microbial communities to associate with economically important plant.

Acknowledgement P.J. acknowledge Amity Institute of Biotechnology (AIBNK), Amity University, Kolkata, West Bengal, India, for providing the infrastructure facilities. V.K. would like to acknowledge the Department of Science and Technology (DST), SERB, Govt. of India under Startup Research Scheme (File No. SRG/2019/001279).

References

Akinsanya MA, Goh JK, Lim SP, Ting AS (2015) Metagenomics study of endophytic bacteria in *Aloe vera* using next-generation technology. Genom Data 6:159–163

Ali GS, Norman D, El-Sayed AS (2015) Soluble and volatile metabolites of plant growth-promoting rhizobacteria (PGPRs): role and practical applications in inhibiting pathogens and activating induced systematic resistance. Adv Bot Res 75:241–284

Arjun JK, Harikrishnan K (2011) Metagenomic analysis of bacterial diversity in the rice rhizosphere soil microbiome. Biotechnol Bioinformatics Bioeng 1:361–367

Badri DV, Chaparro JM, Zhang R, Shen Q, Vivanco JM (2013) Application of natural blends of phytochemicals derived from the root exudates of *Arabidopsis* to the soil reveal that phenolic-related compounds predominantly modulate the soil microbiome. J Biol Chem 288:4502–4512

Bhardwaj D, Ansari MW, Sahoo RK, Tuteja N (2014) Biofertilizers function as key player in sustainable agriculture by improving soil fertility, plant tolerance and crop productivity. Microbial Cell Fact 13:1–10

Bhattacharyya PN, Jha DK (2012) Plant growth-promoting rhizobacteria (PGPR): emergence in agriculture. World J Microbial Biotechnol 28:1327–1350

Bodenhausen N, Horton MW, Bergelson J (2013) Bacterial communities associated with the leaves and the roots of *Arabidopsis thaliana*. PLoS One 8:e56329

Bonkowski M, Villenave C, Griffiths B (2009) Rhizosphere fauna: the functional and structural diversity of intimate interactions of soil fauna with plant roots. Plant Soil 321:213–233

Bramhachari PV, Nagaraju GP, Kariali E (2017) Metagenomic approaches in understanding the mechanism and function of PGPRs: perspectives for sustainable agriculture. In: Meena VS,

Mishra PK, Bisht JK, Pattanayak A (eds) Agriculturally important microbes for sustainable agriculture. Springer, Singapore

Broderick NA (2015) A common origin for immunity and digestion. Front Immunol 6:72

Bulgarelli D, Garrido-Oter R, Münch PC, Weiman A, Dröge J, Pan Y, McHardy AC, Schulze-Lefert P (2015) Structure and function of the bacterial root microbiota in wild and domesticated barley. Cell Host Microbe 17:392–403

Cardoso C, Zhang Y, Jamil M, Hepworth J, Charnikhova T, Dimkpa SO, Meharg C, Wright MH, Liu J, Meng X, Wang Y, Li J, McCouch SR, Leyser O, Price AH, Bouwmeester HJ, Ruyter-Spira C (2014) Natural variation of rice strigolactone biosynthesis is associated with the deletion of two MAX1 orthologs. Proc Natl Acad Sci U S A 111:2379–2384

Carvalhais LC, Dennis PG, Badri DV, Kidd BN, Vivanco JM, Schenk PM (2015) Linking jasmonic acid signaling, root exudates, and rhizosphere microbiomes. Mol Plant-Microbe Interact 28:1049–1058

Carvalhais LC, Dennis PG, Fedoseyenko D, Hajirezaei MR, Borriss R, von Wirén N (2011) Root exudation of sugars, amino acids, and organic acids by maize as affected by nitrogen, phosphorus, potassium, and iron deficiency. J Plant Nutr Soil Sci 174:3–11

Cook RJ, Thomashow LS, Weller DM, Fujimoto D, Mazzola M, Bangera G, Kim DS (1995) Molecular mechanisms of defense by rhizobacteria against root disease. Proc Natl Acad Sci U S A 92:4197

de Weert S, Vermeiren H, Mulders IH, Kuiper I, Hendrickx N, Bloemberg GV, Vanderleyden J, Mot RD, Lugtenberg BJJ (2002) Flagella-driven chemotaxis towards exudate components is an important trait for tomato root colonization by *Pseudomonas fluorescens*. Mol Plant-Microbe Interact 15:1173–1180

Dinnage R, Simonsen AK, Cardillo M, Thrall PH, Barrett LG, Prober SM (2019) Larger legume plants host a greater diversity of symbiotic nitrogen-fixing bacteria. J Ecol 107:977–991

Duke SO (2010) Allelopathy: current status of research and future of the discipline: a commentary. Allelopathy J 25:17–30

Egamberdiyeva D, Kamilova F, Validov S, Gafurova L, Kucharova Z, Lugtenberg B (2008) High incidence of plant growth-stimulating bacteria associated with the rhizosphere of wheat grown in salinated soil in Uzbekistan. Environ Microbiol 10:1–9

Fernández-Arrojo L, Guazzaroni ME, López-Cortés N, Beloqui A, Ferrer M (2010) Metagenomic era for biocatalyst identification. Curr Opin Biotechnol 21:725–733

Fierer N, Leff JW, Adams BJ, Nielsen UN, Bates ST, Lauber CL, Owens S, Gilbert JA, Wall DH, Caporaso JG (2012) Cross-biome metagenomic analyses of soil microbial communities and their functional attributes. Proc Natl Acad Sci U S A 109:21390–21395

Foo E, Yoneyama K, Hugill CJ, Quittenden LJ, Reid JB (2013) Strigolactones and the regulation of pea symbioses in response to nitrate and phosphate deficiency. Mol Plant 6:76–87

Garg R, Srivastava R, Brahma V, Verma L, Karthikeyan S, Sahni G (2016) Biochemical and structural characterization of a novel halotolerant cellulase from soil metagenome. Sci Rep 6:1–15

Gastauer M, Vera MPO, de Souza KP, Pires ES, Alves R, Caldeira CF, Ramos SJ, Oliveira G (2019) A metagenomic survey of soil microbial communities along a rehabilitation chronosequence after iron ore mining. Sci Data 6:190008

Goel R, Suyal DC, Narayan DB, Soni R (2017) Soil metagenomics: a tool for sustainable agriculture. In: Kalia V, Shouche Y, Purohit H, Rahi P (eds) Mining of microbial wealth and metagenomics. Springer, Singapore. https://doi.org/10.1007/978-981-10-5708-3_13

Gupta VVSR, Bramley RGV, Greenfield P, Yu J, Herderich MJ (2019) Vineyard soil microbiome composition related to rotundone concentration in Australian cool climate 'peppery' shiraz grapes. Front Microbiol 10:1607

Haldar S, Sengupta S (2015) Impact of plant development on the rhizobacterial population of *Arachis hypogaea*: a multifactorial analysis. J Basic Microbiol 55:922–928

Handelsman J (2005) Metagenomics: application of genomics to uncultured microorganisms. Microbiol Mol Biol Rev 69:195–195

Handelsman J, Rondon MR, Brady SF, Clardy J, Goodman RM (1998) Molecular biological access to the chemistry of unknown soil microbes: a new frontier for natural products. Chem Biol 5:R245–R249

Healy FG, Ray RM, Aldrich HC, Wilkie AC, Ingram LO, Shanmugam KT (1995) Direct isolation of functional genes encoding cellulases from the microbial consortia in a thermophilic, anaerobic digester maintained on lignocellulose. Appl Microbiol Biotechnol 43:667–674

Hernandez-Leon R, Martinez-Trujillo M, Valencia-Cantero E (2012) Construction and characterization of a metagenomic DNA library from the rhizosphere of wheat (*Triticum aestivum*). Phyton Int J Exp Bot 81:12

Ia MN, Tiong CL, Minor C, August PR, Grossman TH, Loiacono KA, Lynch BA, Phillips T, Narula S, Sundaramoorthi R, Tyler A, Aldredge T, Long H, Gilman M, Holt D, Osburne MS (2001) Expression and isolation of antimicrobial small molecules from soil DNA libraries. J Mol Microbiol Biotechnol 3:301–308

Iijima M, Griffiths B, Bengough AG (2000) Sloughing of cap cells and carbon exudation from maize seedling roots in compacted sand. New Phytol 145:477–482

Islam S, Akanda AM, Prova A, Islam Md T, Md H (2016) Isolation and identification of plant growth promoting rhizobacteria from cucumber rhizosphere and their effect on plant growth promotion and disease suppression. Front Microbiol 6:1–12

Jansson JK, Hofmockel KS (2018) The soil microbiome—from metagenomics to metaphenomics. Curr Opin Microbiol 43:162–168

Kaestli M, Schmid M, Mayo M, Rothballer M, Harrington G, Richardson L, Hill A, Hill J, Tuanyok A, Keim P, Hartmann A, Currie BJ (2012) Out of the ground: aerial and exotic habitats of the melioidosis bacterium *Burkholderia pseudomallei* in grasses in Australia. Environ Microbiol 14:2058–2070

Kaewkla O, Franco CMM (2010) *Pseudonocardia eucalypti* sp. nov., an endophytic actinobacterium with a unique knobby spore surface, isolated from roots of a native Australian eucalyptus tree. Int J Syst Evol Microbiol 61:742–746

Khan Z, Rho H, Firrincieli A, Hung H, Luna V, Masciarelli O, Kim SH, Doty SL (2016) Growth enhancement and drought tolerance of hybrid poplar upon inoculation with endophyte consortia. Curr Plant Biol 6:1–10

Kircher M, Kelso J (2010) High-throughput DNA sequencing—concepts and limitations. BioEssay 32:524–536

Klindworth A, Pruesse E, Schweer T, Peplies J, Quast C, Horn M, Glockner FO (2013) Evaluation of general 16S ribosomal RNA gene PCR primers for classical and next-generation sequencing-based diversity studies. Nucleic Acids Res 41:e1–e11

Knief C, Delmotte N, Chaffron S, Stark M, Innerebner G, Wassmann R, von Mering C, Vorholt JA (2012) Metaproteogenomic analysis of microbial communities in the phyllosphere and rhizosphere of rice. ISME J 6:1378–1390

Kour D, Rana KL, Yadav AN, Yadav N, Kumar M, Kumar V et al (2020) Microbial biofertilizers: bioresources and eco-friendly technologies for agricultural and environmental sustainability. Biocatal Agric Biotechnol 23:101487. https://doi.org/10.1016/j.bcab.2019.101487

Kour D, Rana KL, Yadav N, Yadav AN, Singh J, Rastegari AA et al (2019) Agriculturally and industrially important fungi: current developments and potential biotechnological applications. In: Yadav AN, Singh S, Mishra S, Gupta A (eds) Recent advancement in white biotechnology through fungi, Volume 2: perspective for value-added products and environments. Springer International Publishing, Cham, pp 1–64. https://doi.org/10.1007/978-3-030-14846-1_1

Kumar V, AlMomin S, Al-Aqeel H, Al-Salameen F, Nair S, Shajan A (2018) Metagenomic analysis of rhizosphere microflora of oil- contaminated soil planted with barley and alfalfa. PLoS One 13:e0202127

Kumar V, Joshi S, Pant NC, Sangwan P, Yadav AN, Saxena A et al (2019) Molecular approaches for combating multiple abiotic stresses in crops of arid and semi-arid region. In: Singh SP, Upadhyay SK, Pandey A, Kumar S (eds) Molecular approaches in plant biology and environmental challenges. Springer, Singapore, pp 149–170. https://doi.org/10.1007/978-981-15-0690-1_8

Liu D, Lian B, Dong H (2012) Isolation of *Paenibacillus* sp. and assessment of its potential for enhancing mineral weathering. J Geom 29:413–421

Lugtenberg B, Kamilova F (2009) Plant-growth-promoting rhizobacteria. Ann Rev Microbiol 63:541–556

Luo C, Rodriguez-R LM, Johnston ER, Wu L, Cheng L, Xue K, Tu Q, Deng Y, He Z, Shi JZ, Yuan MM, Sherry RA, Li D, Luo Y, Schuur EA, Chain P, Tiedje JM, Zhou J, Konstantinidis KT (2014) Soil microbial community responses to a decade of warming as revealed by comparative metagenomics. Appl Environ Microbiol 80:1777–1786

Ma Y, Rajkumar M, Luo YM, Freitas H (2011) Inoculation of endophytic bacteria on host and nonhost plants-effects on plant growth and Ni uptake. J Hazard Mater 195:230–237

Martin R, Miquel S, Langella P, Bermudez-Humaran LG (2014) The role of metagenomics in understanding the human microbiome in health and disease. Virulence 5:413–423

McKernan K, Spangler J, Helbert Y, Lynch RC, Devitt-Lee A, Zhang L, Orphe W, Warner J, Foss T, Hudalla CJ, Silva M, Smith DR (2016) Metagenomic analysis of medicinal Cannabis samples; pathogenic bacteria, toxigenic fungi, and beneficial microbes grow in culture-based yeast and mold tests. F1000Res 5:2471

Mendes LW, Kuramae EE, Navarrete AA, van Veen JA, Tsai SM (2014) Taxonomical and functional microbial community selection in soybean rhizosphere. ISME J 8:1577–1587

Mirete S, De Figueras CG, González-Pastor JE (2007) Novel nickel resistance genes from the rhizosphere metagenome of plants adapted to acid mine drainage. Appl Environ Microbiol 73:6001–6011

Mirete S, Mora-Ruiz MR, Lamprecht-Grandío M, de Figueras CG, Rosselló-Móra R, González-Pastor JE (2015) Salt resistance genes revealed by functional metagenomics from brines and moderate-salinity rhizosphere within a hypersaline environment. Front Microbiol 6:1121

Morton AG (1981) History of botanical science. An account of the development of botany from ancient times to the present day. Academic Press, London

O'Brien FJM, Dumont MG, Webb JS, Poppy GM (2018) Rhizosphere bacterial communities differ according to fertilizer regimes and cabbage (*Brassica oleracea* var. *capitata* L.) harvest time, but not aphid herbivory. Front Microbiol 9:1620

Ono A, Miyazaki R, Sota M, Ohtsubo Y, Nagata Y, Tsuda M (2007) Isolation and characterization of naphthalene-catabolic genes and plasmids from oil-contaminated soil by using two cultivation-independent approcahes. Appl Microbiol Biotechnol 74:501–510

Orlandini V, Emiliani G, Fondi M, Maida E, Perrin E, Fani R (2014) Network analysis of plasmidomes: the *Azospirillum brasilense* Sp245 case. Int J Evol Biol 2014:951035

Oteino N, Lally RD, Kiwanuka S, Lloyd A, Ryan D, Germaine KJ, Dowling DN (2015) Plant growth promotion induced by phosphate solubilizing endophytic *Pseudomonas* isolates. Front Microbiol 6:745

Pace NR, Stahl DA, Lane DJ, Olsen GJ (1986) The analysis of natural microbial populations by ribosomal RNA sequences. In: Cou MK (ed) Advances in microbial ecology. Springer, Boston, MA, pp 1–55

Pereira APA, Andrade PAM, Bini D, Durrer A, Robin A, Bouillet JP, Andreote FD, Cardoso EJBN (2017) Shifts in the bacterial community composition along deep soil profiles in monospecific and mixed stands of *Eucalyptus grandis* and *Acacia mangium*. PLoS One 12:e0180371

Prathap M, Ranjitha KBD (2015) A critical review on plant growth promoting rhizobacteria. J Plant Pathol Microbiol 6:1–4

Quispel ANN (ed) (1974) The biology of nitrogen fixation. North-Holland Press, Amsterdam

Ramadan EM, AbdelHafez AA, Hassan EA, Saber FM (2016) Plant growth promoting rhizobacteria and their potential for biocontrol of phytopathogens. Afr J Microbiol Res 10:486–504

Rastegari AA, Yadav AN, Yadav N (2020) New and future developments in microbial biotechnology and bioengineering: Trends of microbial biotechnology for sustainable agriculture and biomedicine systems: diversity and functional perspectives. Elsevier, Amsterdam

Rondon MR, August PR, Bettermann AD, Brady SF, Grossman TH, Liles MR, Loiacono KA, Lynch BA, MacNeil IA, Minor C, Tiong CL, Gilman M, Osburne MS, Clardy J, Handelsman

J, Goodman RM (2000) Cloning the soil metagenome: a strategy for accessing the genetic and functional diversity of uncultured microorganisms. Appl Environ Microbiol 66:2541–2547

Rosenberg E, Rosenberg IZ (2016) Microbes drive evolution of animals and plants: the hologenome concept. Microbial Biol 7:1–8

Ryu CM, Farag MA, Hu CH, Reddy MS, Wei HX, Paré PW, Kloepper JW (2003) Bacterial volatiles promote growth in Arabidopsis. Proc Natl Acad Sci U S A 100:4927–4932

Sanger F, Air GM, Barrell BG, Brown NL, Coulson AR, Fiddes JC, Hutchison CA III, Slocombe PM, Smith M (1977) Nucleotide sequence of bacteriophage phi X174 DNA. Nature 265:687–695

Schmidt TM, DeLong EF, Pace NR (1991) Analysis of a marine picoplankton community by 16S rRNA gene cloning and sequencing. J Bacteriol 173:4371–4378

Shaikh MN, Kasabe UI, Mokat DN (2018) Influence of rhizosphere fungi on essential oil production and menthol content in *Mentha arvensis* L. J Essent Oil Bear Plant 21:1076–1081

Singh A, Kumari R, Yadav AN, Mishra S, Sachan A, Sachan SG (2020a) Tiny microbes, big yields: microorganisms for enhancing food crop production sustainable development. In: Rastegari AA, Yadav AN, Awasthi AK, Yadav N (eds) Trends of microbial biotechnology for sustainable agriculture and biomedicine systems: diversity and functional perspectives. Elsevier, Amsterdam, pp 1–16. https://doi.org/10.1016/B978-0-12-820526-6.00001-4

Singh B, Boukhris I, Pragya KV, Yadav AN, Farhat-Khemakhem A et al (2020b) Contribution of microbial phytases in improving plants growth and nutrition: a review. Pedosphere 30:295–313. https://doi.org/10.1016/S1002-0160(20)60010-8

Singh J, Yadav AN (2020) Natural bioactive products in sustainable agriculture. Springer, Singapore

Singh RN, Gaba S, Yadav AN, Gaur P, Gulati S, Kaushik R et al (2016) First, high quality draft genome sequence of a plant growth promoting and cold active enzymes producing psychrotrophic *Arthrobacter agilis* strain L77. Stand Genomic Sci 11:54. https://doi.org/10.1186/s40793-016-0176-4

Souza RC, Hungria M, Cantão ME, Vasconcelos ATR, Nogueira MA, Vicente VA (2015) Metagenomic analysis reveals microbial functional redundancies and specificities in a soil under different tillage and crop-management regimes. Appl Soil Ecol 86:106–112

Srivastava S, Bist V, Srivastava S, Singh PC, Trivedi PK, Asif MH, Chauhan PS, Nautiyal CS (2016) Unraveling aspects of *Bacillus amyloliquefaciens* mediated enhanced production of rice under biotic stress of *Rhizoctonia solani*. Front Plant Sci 7:587

Stafford HA (1990) Flavonoid metabolism. CRC Press, Boca Raton, FL

Taghavi S, Garafola C, Monchy S, Newman L, Hoffman A, Weyens N, Barac T, Vangronsveld J, van der Lelie D (2009) Genome survey and characterization of endophytic bacteria exhibiting a beneficial effect on growth and development of poplar trees. Appl Environ Microbiol 75:748–757

Takagi S, Nomoto K, Takemoto T (1984) Physiological aspect of mugineic acid, a possible phytosiderophore of graminaceous plants. J Plant Nutr 7:469–477

Tank N, Saraf M (2010) Salinity-resistant plant growth promoting rhizobacteria ameliorates sodium chloride stress on tomato plants. J Plant Interact 5:51–58

Tian XY, Zhang CS (2017) Illumina-based analysis of endophytic and rhizosphere bacterial diversity of the coastal halophyte *Messerschmidia sibirica*. Front Microbiol 8:2288

Tiwari S, Lata C, Chauhan PS, Nautiyal CS (2016) *Pseudomonas putida* attunes morphophysiological, biochemical and molecular responses in *Cicer arietinum* L. during drought stress and recovery. Plant Physiol Biochem 99:108–117

Tripathi DK, Singh VP, Kumar D, Chauhan DK (2012) Impact of exogenous silicon addition on chromium uptake, growth, mineral elements, oxidative stress, antioxidant capacity, and leaf and root structures in rice seedlings exposed to hexavalent chromium. Acta Physiol Plant 34:279–289

Unno Y, Shinano T (2013) Metagenomic analysis of the rhizosphere soil microbiome with respect to phytic acid utilization. Microbes Environ 28:120–127 van der Heijden MGA, Bakker R, Verwaal J, Scheublin TR, Rutten M, Van Logtestijn R, Staehelin C (2006) Symbiotic bacteria as a determinant of plant community structure and plant productivity in dune grassland. FEMS Microbiol Ecol 56:178–187

Verma P, Yadav AN, Kumar V, Singh DP, Saxena AK (2017) Beneficial plant-microbes interactions: biodiversity of microbes from diverse extreme environments and its impact for crop improvement. In: Singh DP, Singh HB, Prabha R (eds) Plant-microbe interactions in agro-ecological perspectives, Microbial interactions and agro-ecological impacts, vol 2. Springer Singapore, Singapore, pp 543–580. https://doi.org/10.1007/978-981-10-6593-4_22

Wu L, Wang J, Wu H, Chen J, Xiao Z, Qin X, Zhang Z, Lin W (2018) Comparative metagenomic analysis of rhizosphere microbial community composition and functional potentials under *Rehmannia glutinosa* consecutive monoculture. Int J Mol Sci 19:2394

Xu J, Zhang Y, Zhang P, Trivedi P, Riera N, Wang Y, Liu X, Fan G, Tang J, Coletta-Filho HD, Cubero J, Deng X, Ancona V, Lu Z, Zhong B, Roper MC, Capote N, Catara V, Pietersen G, Vernière C, Al-Sadi AM, Li L, Yang F, Xu X, Wang J, Yang H, Jin T, Wang N (2018) The structure and function of the global citrus rhizosphere microbiome. Nat Commun 9:4894

Yadav AN, Kumar V, Prasad R, Saxena AK, Dhaliwal HS (2018) Microbiome in crops: diversity, distribution and potential role in crops improvements. In: Prasad R, Gill SS, Tuteja N (eds) Crop improvement through microbial biotechnology. Elsevier, New York, pp 305–332

Yadav AN, Mishra S, Kour D, Yadav N, Kumar A (2020a) Agriculturally important fungi for sustainable agriculture, volume 1: perspective for diversity and crop productivity. Springer International Publishing, Cham

Yadav AN, Sachan SG, Verma P, Saxena AK (2015) Prospecting cold deserts of north western Himalayas for microbial diversity and plant growth promoting attributes. J Biosci Bioeng 119:683–693

Yadav AN, Singh J, Rastegari AA, Yadav N (2020b) Plant microbiomes for sustainable agriculture. Springer International Publishing, Cham

Yang J, Kloepper JW, Ryu CM (2009) Rhizosphere bacteria help plants tolerate abiotic stress. Trend Plant Sci 14:1–4

Zaprasis A, Liu YJ, Liu SJ, Drake HL, Horn MA (2009) Abundance of novel and diverse tfdA-like genes, encoding putative phenoxyalkanoic acid herbicide-degrading dioxygenases, in soil. Appl Environ Microbiol 76:119–128

Zhang J, Subramanian S, Stacey G, Yu O (2009) Flavones and flavonols play distinct critical roles during nodulation of *Medicago truncatula* by *Sinorhizobium meliloti*. Plant J 57:171–183

Plant–Microbe Association for Mutual Benefits for Plant Growth and Soil Health

5

Surajit De Mandal, Sonali, Simranjeet Singh,
Kashif Hussain, and Touseef Hussain

Abstract

The beneficial associations of plants and microbes exemplify a complex and multi-organ system composed of participatory organisms and the environmental forces acting on them. Current knowledge of plant-microbe symbiosis involves a series of associations with varying degrees of intimacy and mutual dependence. Generally, rhizosphere microbes can help the plant by maintaining nutrient recycling, hormones production, preventing microbial infections and improving tolerance towards potentially hazardous compounds. Symbiotic relationships are known to be extremely beneficial for the enhancement of overall plant growth, especially in those soils that are deprived of certain minerals like P or N. However, in case of well-fertilized arable soils, symbiotic microbial growth is found to reduce significantly due to the improved bioavailability of nutrients in the soil. In addition to the vast benefits of symbiotic microbial growth in the rhizosphere, it also offers an overall increase in crop productivity, therefore making it an essential area of research.

Keywords

Microbes · PGPR · Plant symbiotic · Rhizosphere · Soil health

S. De Mandal
Key Laboratory of Bio-Pesticide Innovation and Application of Guangdong Province, College of Agriculture, South China Agricultural University, Guangzhou, China

Sonali · S. Singh
Department of Biotechnology, Lovely Professional University, Jalandhar, Punjab, India

K. Hussain
Gyani Inder Singh Institute of Professional Studies, Dehradun, Uttarakhand, India

T. Hussain (✉)
Department of Botany, Aligarh Muslim University, Aligarh, Uttar Pradesh, India

A. N. Yadav et al. (eds.), *Current Trends in Microbial Biotechnology for Sustainable Agriculture*, Environmental and Microbial Biotechnology,
https://doi.org/10.1007/978-981-15-6949-4_5

5.1 Introduction

Plants are well-known to co-operate with the large microbial community in nature. This plant-associated microbiota involves various groups of organisms like bacteria, archaea, fungi, *etc.* and acts as symbiont or pathogen (Berendsen et al. 2012; Vorholt 2012; Hussain and Khan 2020). These complex interactions between plants and microbes have a significant impact on plant growth and productivities (Hussain et al. 2020a; Kumar et al. 2020a, b; Singh et al. 2020a, b, c). Several studies highlighted the beneficial activities of these microbial communities in plant health such as increasing nutrient availability (De Mandal et al. 2018; Yadav et al. 2018), adaptation to environmental variations and abiotic stresses (Garbaye 1994), disease suppression (Haney et al. 2015), stimulating plant hormone (Laskar et al. 2018), priming of the plant immune system, establishment of mycorrhizal associations (Rolli et al. 2015), and induced systemic resistance (Van der Ent et al. 2009; Van Der Heijden et al. 2016; Zamioudis et al. 2015). In turn, the host plant also secretes secondary metabolites that favour the growth of specialized microorganisms (Hassani et al. 2018; Thrall et al. 2007). Plant root can synthesize and secrete several compounds in the form of root exudates, which plays a significant role in mediating the interaction between the plant and the microorganisms (Rastegari et al. 2020a; Singh and Yadav 2020; Yadav et al. 2020f). Two different types of compounds are identified in the root executes, *i.e.*, low molecular weight compounds such as amino acids, sugars, phenolics, organic acids, secondary metabolites, and high molecular weight compounds like proteins, mucilage, *etc.* (Bais et al. 2002; Weir et al. 2004).

Rhizospheric microorganisms influenced by these metabolites and, in turn, interacts with the plant roots in a positive (symbiotic), negative (*e.g.*, parasitic/pathogenic) or neutral ways and affect plant physiology (Kour et al. 2019; Yadav et al. 2020e). It has been stated that plant synthesizes the root exudates, which allows the aggregation of the particular microbial community in the rhizosphere, and the microbial densities were 100 times more in the rhizospheres as compared to the bulk soil. These further illustrate the significance of root exudates in shaping the soil microbiota (Ciccazzo et al. 2014; Faure et al. 2009; Lareen et al. 2016).

The growing demand for crop production encourages the use of sustainable agricultural practices around the world. Several research have been undertaken to meet environmental and economic sustainability (Rastegari et al. 2020b; Yadav et al. 2020b, d). The exploitation of rhizosphere microorganisms is considered as an important way for sustainable and healthy crop production. However, the interaction of the plants and microbes is affected by several ecological factors (Yadav et al. 2020c). Optimization of the root-associated microorganisms by improving their abilities to supply nutrients, protection against pathogens, and tolerance in hostile environments improve the overall agricultural productivity (Barea 2015; Zolla et al. 2013).

Plants were known to evolve with their adaptation to survive in the abiotic and biotic stresses. However, they often rely on other partners to combat the pathogens (Turner et al. 2013). In nature, plants have been associated with microbial communities for millions of years. The mycorrhizal fungi have evolved with the plants for more than 400 MY. During this process, they acquired several mechanisms that modulate plant–microbe interactions to survive in hostile environments (Oldroyd 2013; Pirozynski and Malloch 1975). There have been several studies on plant microbiota as well as hosts; however, the underlying mechanisms of plant–microbes interaction have not yet been fully revealed (Kumar et al. 2019b; Subrahmanyam et al. 2020). This chapter is an overview of the different microbial associations between microbial groups and host plants. We discuss and review the progress of recent research on microbial services for the benefit of plant species. This knowledge will be useful for the detailed understanding of plant–microbiome interactions that can be exploited for the improvement of agricultural practices.

5.2 Plants–Microbes Association

5.2.1 Endophytic Microbiome

Plant-associated microbiota may present in the rhizospheric soil or rhizoplane and within the tissues (endophytes) of the plant. The endophytes can escape the immune protection of the plant defense system and colonize inside without causing any disease. They produce several bioactive molecules which play an imperative role in plant development and protection against various pathogens and hostile environments (Datta et al. 2020; Suman et al. 2016). However, most of the endophytes are unculturable, and thus their interaction mechanism with the plant is mostly studied using the molecular-based approach, while some microorganisms colonize the interior of plants, such as mycorrhizal fungi, rhizobia, and pathogens. They are not considered among the core groups of endophytes because they transfer nutrients from external sources like atmosphere or appearance of symptoms of disease in the host plant (Barea 2015; Brader et al. 2014; Mercado-Blanco 2015) (Table 5.1).

5.2.1.1 Bacterial Endophytes

The bacterial endophytes enter the plant via rhizosphere through root epidermis and cortex. They are further divided into three categories such as passenger endophytes (limited to root cortex), opportunistic endophytes (limited to particular tissues in roots of plants like root cortex and also show root proliferating properties), and competent endophytes (that have the capability to spread to vascular tissues or other tissue of the plant) (Hardoim et al. 2008; Rana et al. 2019a). Once vascular tissue is invaded, the endophytes can spread and colonize to the vegetative parts like fruits, flowers, and seeds. Selection for the colonization of endophytes in the seeds could provide beneficial bacteria to the next generation (Compant et al. 2010; Jambon et al. 2018; Truyens et al. 2015).

Rhizosphere colonizing microbes are known to promote the growth of plant growth, in addition to that confer them the ability to adapt to extreme environmental conditions. It has been reported that rice seed is colonized by diverse endophytic bacteria that serve as a source of beneficial bacterial communities in the growing

Table 5.1 Plant–microbe association and PGP attributes of different microbes

Microbes	PGP attributes	Host/association
Archaea		
Halobacterium sp., *Halococcus hamelinensis, Haloferax alexandrinus, Haloterrigena thermotolerans, Methanobacterium bryantii, Methanosarcina, Methanospirillum* sp., *Natrialba* sp., *Natronoarchaeum annanilyticum, Nitrosomonas communis*	P-solubilization, IAA, siderophore, nitrogen fixation	Abutilon, cressa, maize, rice, sporobolus, Suaeda nudiflora
Actinobacteria		
Arthrobacter humicola, A. methylotrophus, Arthrobacter sp., *Cellulosimicrobium* sp., *Kocuria, Micrococcus luteus, Streptomyces*	P-solubilization, IAA, biocontrol	Cowpea, millet, mustard, wheat
Bacteroidetes		
Flavobacterium psychrophilum, Flavobacterium sp., *Sphingobacterium* sp.	P-solubilization, K-solubilization	Barley, millet, wheat
Proteobacteria		
Achromobacter piechaudii, Acinetobacter sp., *Advenella* sp., *Agrobacterium larrymoorei, Alcaligenes* sp., *Azotobacter tropicalis, Bradyrhizobium* sp., *Enterobacter* sp., *Methylobacterium phyllosphaerae, M. radiotolerans, Nitrinicola lacisaponensis, Pantoea agglomerans* sp., *Providencia rustigianii, Pseudomonas cedrina, P. fluorescens, P. gessardii, P. putida, P. rhodesiae, P. thivervalensis, Serratia marcescens, Tetrathiobacter* sp., *Variovorax*	Multifunction PGP attributes including solubilization of P, K, Zn; production of ammonia, HCN siderophore, and biocontrol	Amaranth, barley, buckwheat, cotton, cowpea, gram, maize, millet, mustard, oat, rice, sunflower, tomato, wheat
Firmicutes		
Bacillus aerophilus, B. alcalophilus, B. altitudinis, B. amyloliquefaciens, B. cereus, B. circulans, B. endophyticus, B. flexus, B. fusiformis, B. licheniformis, B. megaterium, B. methylotrophicus, B. mojavensis, B. pumilus, B. solisalsi, B. sphaericus, B. tequilensis, B. thuringiensis, Exiguobacterium acetylicum, Lysinibacillus, Paenibacillus alvei, P. dendritiformis, P. polymyxa, P. xylanexedens, Planococcus salinarum, Staphylococcus	Multifunction PGP attributes	Amaranth, apple, barley, buckwheat, maize, mustard, oat, pepper, rice, sorghum, sunflower, tomato, wheat
Fungi		
Gliocladium, Leptosphaeria, Metarhizium, Penicillium, Piriformospora indica, Sporotrichum thermophile, Trichoderma, T. longibrachiatum, Williopsis saturnus	IAA, siderophore P-solubilization, biocontrol	Amaranth, barley, buckwheat cotton, maize, oat, rice, sorghum, soybean, wheat

Sources: Verma et al. (2017b)

plant and help in plant growth and development (Walitang et al. 2019). Under both stress and normal conditions, these endophytes help in seed germination and seedling development (Bent and Chanway 1998; Gond et al. 2015a, b). These seed endophytes also showed antifungal activity against various plant pathogens. Endophytic bacteria isolated from the seeds of commercial wheat cultivar demonstrated high biocontrol activities against *Fusarium graminearum* (Herrera et al. 2016). Similarly, seed endophytic bacteria such as *Bacillus* and *Pseudomonas* were found to have antagonistic effects on *F. oxysporum* f. sp. *lycopersici* (Fol.) (Gagne-Bourgue et al. 2013; Sundaramoorthy and Balabaskar 2013). Similarly, colonization of the endophytic bacteria in the root causes enhanced expression and activity of vacuolar proton pumps H^+-ATPase (V-PPase) that confer drought tolerance in pepper plants (Vigani et al. 2019).

Endophytic bacteria derived from the halophytes helps to alleviate stress induced by salinity in plants by regulating the plant hormones, assisting in the uptake of nutritional compounds and modulating the synthesis of ROS via various mechanisms such as enhancing the solubilization of phosphate compounds, increase the process of nitrogen fixation, improving the catalytic activity of enzyme 1-aminocyc lopropane-1-carboxylic acid deaminase and elevating the production of compounds like siderophores, abscisic acid (ABA), volatiles, and indole-3-acetic acid (IAA) (Kour et al. 2020a; Rana et al. 2020a). Under the saline conditions, plant growth-promoting endophytic bacteria can be involved in the growth stimulation, nutrient acquisition, symbiotic performance, and stress tolerance in chickpea. It was found that the plants inoculated with *Bacillus subtilis* (BERA 71) showed enhanced stability of the membranes when subjected to saline conditions, which can be attributed to the suppression of lipid peroxidation, reduction in the production of ROS and the accumulation of proline (Abd_Allah et al. 2018).

5.2.1.2 Fungal Endophytes

The fungal endophytes may be present inside the roots, leaves, stems, and forms an association with plants which can be neutral, mutualistic, or antagonistic (Chadha et al. 2014). They are divided into two groups, clavicipitaceous and the non-clavicipitaceous endophytes. These non-clavicipitaceous endophytes can be further subdivided into three classes: Class 2 endophytes (grow in rhizomes, roots, and shoots), class 3 endophytes (reside only in shoots of plants), and class 4 endophytes (present only in the roots of plants) (Rodriguez et al. 2009; Yadav 2019). Similar to bacterial endophytes, fungal endophytes also contribute to the plant fitness in biotic (plant pathogens, insects, and nematodes) and abiotic stress (drought, extreme pH, nutrient limitation, salination, temperature) (Rana et al. 2020a, b, c). They can produce several bioactive secondary metabolites, including volatile organic compounds that act as a defense substance against pests and pathogens. These metabolites also act in specific interaction and communication within the host (Lugtenberg et al. 2016; Rana et al. 2019b).

The sugarcane endophyte *Epicoccum nigrum* was reported to possess biocontrol activities against several pathogens such as *Sclerotinia sclerotiorumin, Pythium,* and *Monilinia* spp. (de Lima Favaro et al. 2012). Fungal endophytes also involve the reduction of the growth of pathogen through fungal–fungal interactions. For example, the secondary metabolites released from the endophyte antagonistic *Fusarium verticillioides* break down the plant compounds that suppress the growth of the plant pathogen *Ustilago maydis* (Estrada et al. 2012). However, it has been proved that specific physiological and environmental conditions are needed to express the secondary metabolites by endophytic fungi. For example, the genome of the host is essential for the optimum expression of the secondary metabolites by Epichloe (Brakhage 2013; Lugtenberg et al. 2016; Netzker et al. 2015; Schardl and Panaccione 2005). Endophytic fungi also involved in the higher yield and quality medicinal plants. The endophytic fungus AL12 (*Gilmaniella* sp.) promotes plant growth by improving the primary metabolism of plants by enhancing the rate of glycolysis, photosynthesis, and the TCA cycle, which provides sufficient energy and carbon for the synthesis of sesquiterpenoid in the traditional Chinese herb *Atractylodes lancea*. This further explains the importance of the interaction between plant- endophytic fungal (Yuan et al. 2016).

5.2.2 Plant Growth-Promoting Rhizobacteria

The narrow zone of soil near the root system is termed as the rhizosphere (Walker et al. 2003). The bacterial group inheriting in the rhizosphere is named as "rhizobacteria" (Kloeppe et al. 1999; Koul et al. 2019; Ahmad et al. 2019; Kumar et al. 2020a, b). These colonized bacterial communities residing nearby root aid in plant growth are also known as Plant Growth-Promoting Rhizobacteria (PGPR) (Beneduzi et al. 2012; Kaur et al. 2018). Other than this, these bacterial communities also serve as the ecological method for managing plant diseases (Compant et al. 2005). Moreover, PGPR acting as BCA has an additional advantage over traditional chemical practices as they are non-toxic naturally occurring microbes (Rai et al. 2020; Singh et al. 2020a). Numerous reports have suggested the use of PGPR as a control method for regulating root diseases (Lucy et al. 2004; Whipps 2001). Diverse bacterial species have been isolated, which could act as a potential biocontrol agent for cereals. For instance, *Azotobacter*, *Bacillus*, and *Pseudomonas* isolated from the root of the plant have been accorded to show antagonistic activity against phytopathogens and can act as an effective disease controlling system (Berg and Smalla 2009).

Many reports on various crops like chickpea, tomato, and wheat have highlighted the biocontrol ability of both Bacillus spp. and Pseudomonas sp. against soil-dwelling pathogenic microbes (Perez-Montano et al. 2014). Moreover, *Bacillus* species like *Bacillus cereus, Bacillus thuringiensis,* and *Bacillus licheniformis* has also been comprehended as biocontrol agents (Thakur et al. 2020). For instance, one study reported *Bacillus spp.* from the rhizosphere of chickpea, which was found to be effective in regulating the growth of *Fusarium oxysporum* responsible for causing Fusarium wilt disease. Another study reported the *Bacillus* strains isolated from the rhizosphere soil of wild grass and sorghum in South Africa and Ethiopia, respectively. The isolated *Bacillus* strains showed antagonistic activity against *F. oxysporum* responsible for root rot disease and *Pythium ultimum* responsible for crown rot disease (Idris et al. 2007).

Furthermore, PGPR involves different mechanisms to reduce the phytopathogens and induce system resistance in the plant via antibiotic, by secreting toxic bio-surfactant and volatile compounds and cell-wall degrading enzymes (Van Loon et al. 1998; Whipps 2001; Compant et al. 2005; Perez-Montano et al. 2014). The primitive proposed mechanism discussed only the siderophores, which degrade the iron and eradicate the plant pathogens (Raaijmakers et al. 2002).

5.2.3 Breeding Microbe-Optimized Plants

There are certain microbial communities which can interact with different types of plant. For example, inoculation of *Pseudomonas simiae* WCS417r in the soil used for growing Arabidopsis thaliana showed a four-fold increase in its yield (Wintermans et al. 2016). It also illustrated the potential of microbial interaction, which influences the gene expression of the plant (Smith et al. 1999). Therefore, breeding horticultural plants aids in optimizing and maintaining the microbial community beneficial for both, which is the chief intention of this approach (Kumar et al. 2015; Kumar et al. 2018; Singh et al. 2016, 2018). So far, such attempts have not been initiated in this direction, but to initiate to overcome the cumbersome process of breeding, we must understand how beneficial microbes are attracted and established by the plants (Yadav et al. 2020a). Further, genetically modified plants and their breeding allows us to generate the microbe-optimized which exudates to attract the particular rhizospheric microbes to colonize at either root or leaf at the right time (Trivedi et al. 2017).

5.2.4 Engineering Microbiome, Plant-Optimized Microbiomes

The method involves the genetic amendment of either consortium of beneficial microbes or individual microorganisms, which helps in developing an optimized plant/soil environment. These genetically engineered microbes can be used as the inoculum for several horticultural crops growing in a diverse type of soil to improve the crop yield. This approach has not been implemented in agricultural fields yet, but the available literature supports the adaptation of engineered soil microbes to crops with time, which would result in improved plant–microbe interactions (Berendsen et al. 2012). The evidence suggests that naturally occurring plant microbiomes are significant in the development and progression of disease in plants (Bulgarelli et al. 2013). Therefore, it is necessary to study the detailed mechanism of attraction of microbes to the rhizospheric region and their colonization in the roots.

5.2.5 Pairing Microbe-Optimized Plant Seed with the Optimal Microbiome

Still, researchers are making continuous efforts to microbes that allow or improve the yield of a particular crop. One of the approaches to optimize plant–microbe interaction is to coat the seeds with suitable microbes keeping the type of soil into consideration. While considering the transient nature of the microbiomes, this approach is considered far better than the other methods of application like root soaks or sprays. The consortium of microbes used as inoculum enhances nutrient absorption in plant and also act the biocontrol agent against phytopathogens and pest. To ensure the effectiveness and viability of the beneficial soil microorganisms, certain amendments in the soil become necessary.

Rhizobium, beneficial microbes for legume, is now commercially available to improve the yield. In addition, to assist in the formation of root nodules by nitrogen-fixing bacteria in leguminous plants for improved growth, they also aid in the suppression of disease-causing microbes and limit nutrient availability and assimilation. Studies have demonstrated the potential application of microorganism-derived growth-promoting compounds in the production of effective vermicompost formulations (both aqueous extracts and granular), which can stabilize and increase the shelf life of bio formulations (Kalra et al. 2010). Improved nodulation of soybean was noticed when soybean was co-inoculated with *Bradyrhizobium* and *B. megaterium* (Liu and Sinclair 1990). Lately, it was established that concoctions of PGPR improve the biocontrol ability against multiple phytopathogens and promote the growth of the plant (Liu et al. 2018).

5.3 Current Scenario and the Need for Adopting of Biocontrol Agents in India

Securing food has become the top priority around the world (Porter et al. 2014). It is due to an exponential increase in the population in developing countries, so to fulfill the demand of the growing population, there is a need for advancement in the approaches to substantially increase the crop yield. Although chemical fertilizers and pesticides have been employed to improve the yield, excessive use has induced a detrimental effect on both environment and health.

Despite the awareness about environmental issues, most of the countries has not started the use of biocontrol agents in agricultural practices. Hence, it has become prominent to discover and assess the biocontrol ability of PGPR against various phytopathogens. Pilot studies involving mass production of *Trichoderma* have already been initiated to inhibit the growth of phytopathogens (Korolev et al. 2008; Cumagun 2014). *Trichoderma* has been reported to modify the signalling pathway to contend *Botrytis cinerea*, Cucumber mosaic virus (CMV), and *Fusarium* (Elad et al. 1998; Wang et al. 2005; Vitti et al. 2015).

Another study reported about the biocontrol potential of *Pseudomonas fluorescens* against *Ralstonia solanacearum* responsible for causing in wilting in tomato (Vanitha et al. 2009). Therefore, BCA holds a special place and can be used to decrease the dependency on pesticides and other chemical agents. This approach requires the government to bring awareness and supports research and development for studying the biocontrol agents in collaboration with industries and research institutes.

5.4 Plant–Microbe Interactions at the Post-genomic Era

The advance of modern technologies, including the advancement in sequencing technologies along with advanced bioinformatics tools, has impressively accelerated the studies of plant–microbe interactions. This method allows us to produce a massive amount of sequencing data in less time and with a low cost and thus improve our understanding of gene, genome, pathways, regulatory network of plants, microbes, and their associations. The bioinformatics tools and comparative analysis of genomic DNA have unveiled the information regarding the biological pathway, gene function, genome make-up, regulatory networks, and phylogenetic variation among the microbes, which has substantially improved our knowledge about the metabolism of microbes. Several NGS methods such as shotgun, amplicon, whole genome, transcriptomic, and metatranscriptomic sequencing has been introduced to analyze the plant microbial interactions.

The most used NGS method is the amplicon sequencing-based approach, where a particular marker gene is sequenced from all the microbial species present in a microbial community and frequently used to study the plant–microbe interaction. Several studies used this technique to analyze the rhizosphere and phyllosphere microbial communities, which play a significant role in plants by interfering the fitness, growth, protection, and other traits. These phyllosphere microbes can influence plant biogeography as well as ecosystem function by regulating plant systems under different ecological conditions (Friesen et al. 2011; Meyer and Leveau 2012). High-throughput sequencing was used to study the functional biogeography of plants and plant–microbe interactions. Kembel et al. (2014) investigate the poorly understood association between bacterial biodiversity on leaves vs. host tree attributes. They showed that bacterial leaf communities were highly correlated with the host evolutionary relatedness as well as functional traits (Kembel et al. 2014). This technique is used to reveal how the interaction between plant and microbes ecosystem processes during early succession (Knelman et al. 2012).

Shotgun sequencing used to study the microbial communities as well as their functional aspects by sequencing all the genes present inside the metagenome—several studies focusing the sequencing of the entire microbial communities instead of a single species using NGS approaches, whereas very few metagenomic-related studies have been conducted to assess the microbial community surrounding the plant via shotgun sequencing. Analysis of WGS is now growing interest day by day and used to answer various biological answers. The halotolerant endophyte *Bacillus flexus* KLBMP 4941 of the halophyte *Limonium sinense* can improve host seedling growth under salt stress conditions. Analysis of the complete genome of this bacteria identified the presence of genes associated with plant growth promotion (PGP) including nitrogen fixation, siderophore, spermidine, and acetoin synthesis as well as high salinity tolerance (Na^+/H^+ antiporter, glycine betaine transporter, and betaine-aldehyde dehydrogenase) were identified (Wang et al. 2017). *Bacillus paralicheniformis* KMS 80 (MTCC No. 12704) plays a vivacious role in the biological nitrogen fixation and growth promotion in *Oryza sativa* L. WGS analysis revealed

21 genes for nitrogen metabolism pathway and two main transcriptional factor glnR and tnrA that regulates the nitrogen fixation (Annapurna et al. 2018).

Transcriptomic and metatranscriptomic approaches are used to study the protein-coding genes of any organism and have been widely used for the study of plant–microbe interaction. This approach was used to study the role of rhizosphere microbial communities in different developmental stages of the plants. This study identifies the potential role of microbial genes in the regulation of various metabolic pathways (Chaparro et al. 2014; Kumari et al. 2017). The mass spectrometry-based quantitative proteomic analysis was performed to study the interaction between the endophytic plant growth-promoting *Gluconacetobacter diazotrophicus* and sugarcane. This experiment shows that plants associated with *G. diazotrophicus* have higher nitrogen fixation ability as well as the overexpression of signal cascade proteins (Lery et al. 2011). The metabolomic approach was used to study the complex nodulation process in Soybean by *Bradyrhizobium japonicum*. It was found that 166 metabolites significantly regulated during bacterial inoculation, and trehalose was the most strongly induced metabolite (Brechenmacher et al. 2010). It has been found that lipophilic secondary metabolites produced by *Macrophomina phaseolina* might play a significant role in the plant–fungus interactions responsible for severe diseases of *E. globules* (Salvatore et al. 2020).

5.5 Importance of Microbes in Agriculture Farming

Plant growth-promoting bacteria are most widely used for sustainable agriculture all over the world. PGPRs are widely used inoculants in agriculture soils. They can enhance the nutrient uptake and inhibit the growth of various phytopathogens by producing secondary metabolites and other substances (Ahemad and Kibret 2014). Based on the effects on plants, PGPR roles can be divided into direct and indirect impacts, illustrated in Fig. 5.1.

5.5.1 The Direct Impact of PGP Microbes on Plant Nutritions

Nutrients are one of the major limiting factors for plants as some are required in small quantities or some in large. These elements also play a vital role in plant metabolism. PGP bacteria provide nutrients to the plant by solubilization of minerals as well as help in various hormones production.

5.5.1.1 Nitrogen Fixation

Nitrogen is another vital element for the growth and development of the plants. Plant species are unable to convert atmospheric nitrogen into ammonia. This process is mediated by different microorganisms using a complex nitrogen system known as nitrogenase (Babalola 2010; Backer et al. 2018; Sharma et al. 2019). PGPR bacteria perform the nitrogen fixation by making the symbiotic and non-symbiotic relationship with plants bacterium (Kapoor et al. 2019; Singh et al. 2019).

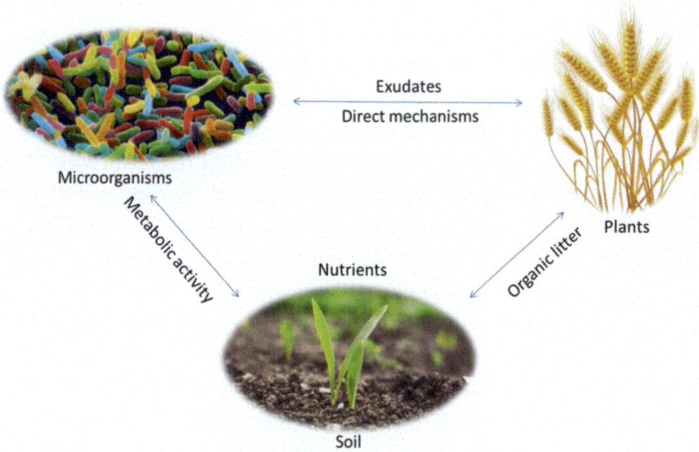

Fig. 5.1 Mutual relationship between plants, microbes, and soil

Twenty nif genes have been reported which are classified into eight operons: *nif*USVWZ, *nif*ENX, *nif*J, *nif*HDKTY, *nif*BQ, *nif*LA, *nif*F, and *nif*M. The *nif*D and *nif*K genes encode the FeMo-protein, and nifH encodes the Fe-protein (Gupta et al. 2017). Example of symbiotic bacteria: *Rhizobium* with the leguminous plant, *Frankia* with the non-leguminous plant.

5.5.1.2 Phosphorus Solubilization

After nitrogen, phosphorus is one of the essential minerals for the plant. Phosphorus plays a vital role in various metabolic and biochemical pathways such as photosynthesis, Adenosine Triphosphate (ATP). Phosphorus fertilizers are used by the farmer to provide these nutrients but this source is a costly, less productive and environmentally unsafe method. To compensate for this problem, PGPRs are used to offer an eco-friendly approach, by improved uptake of phosphorus from soil either by the release of phosphorus by substrate degradation, biochemical phosphate mineralization (Gupta et al. 2017; Singh et al. 2020b). Some examples are *B. megaterium* from chickpea, *B. licheniformis* from both wheat and spinach, *Enterobacter agglomerans* from tomato, *P. chlororaphis* as well as *P. putida* from soybean, (Abd-Alla 1994; Ahemad and Khan 2011; Rajkumar et al. 2008). There are many reports of P-solubilization by diverse groups of microorganisms including archaea, fungi, and bacteria, and these potential P-solubilizing microbes could be used as bio-inoculant to fulfill the requirement of chemical fertilizers for sustainable agriculture (Singh et al. 2020b; Verma et al. 2016, 2015; Yadav et al. 2015).

5.5.1.3 Potassium Solubilization

Similar to nitrogen and phosphorus, potassium is also important for the high yield of crops. Potassium (K) is the most abundant inorganic cation in plants comprising up to 10% of dry weight plants and is not assimilated into the organic matter rather remains in its ionic form only throughout its "life" in the plant. Potassium plays an essential role in root development, stomata opening and closing, and plant growth (Abd-Alla 1994; Khan et al. 2002; Backer et al. 2018). Examples: phosphate solubilization was mediated by the bacterial member *Bacillus mucilaginosus* and *Azotobacter chroococcum* associated with the wheat plants and *Rhizobium, Bacillus edaphicus* in cotton. A wide range of potassium-solubilizing microbes have been reported, viz. *Acidithiobacillus, Agrobacterium, Arthrobacter, Aspergillus, Bacillus, Burkholderia, Enterobacter Pantoea, Flectobacillus, Klebsiella, Microbacterium, Myroides, Paenibacillus, Pseudomonas,* and *Stenotrophomonas* (Rajawat et al. 2020; Verma et al. 2017a; Yadav et al. 2017).

Verma et al. (2014) reported several plant growth-promoting bacterial members such as *Paenibacillus dendritiformis, Bacillus megaterium, Paenibacillus amylolyticus, Duganella violaceusniger, Pseudomonas thivervalensis, Psychrobacter fozii, Stenotrophomonas maltophilia, Pseudomonas monteilii, Pseudomonas lini* that can solubilize K and phosphorus and zinc; producte IAA, siderophores, GA, HCN, ammonia, ACC and perform nitrogen fixation as well as biocontrol activities. Verma et al. (2015) reported *Bacillus megaterium, Bacillus horikoshii, Bacillus amyloliquefaciens, Exiguobacterium antarcticum, Achromobacter piechaudii, Stenotrophomonas maltophilia, Klebsiella* sp. as K-solubilizers which showed other plant growth-promoting attributes including phosphorus and zinc solubilization, production of IAA, siderophores, HCN, ammonia, ACC, GA, nitrogen fixation as well as biocontrol activity. Verma et al. (2016) reported potassium-solubilizing *Bacillus aerophilus, Bacillus atrophaeus, Bacillus cereus, Bacillus circulans, Bacillus endophyticus, Bacillus horikoshii, Bacillus licheniformis, Bacillus megaterium, Bacillus mojavensis, Bacillus pumilus, Bacillus sphaericus, Exiguobacterium antarcticum, Paenibacillus amylolyticus, Paenibacillus dendritiformis, Paenibacillus polymyxa, Planococcus citreus,* and *Planococcus salinarum* which also showed the production of GA, IAA, ACC, siderophores, ammonia, HCN, chitinase, protease, lipase, β-glucanase, solubilization of phosphorus and zinc as well as nitrogen fixation capability and biocontrol activity.

Potassium-solubilizing bacteria are an essential constituent of soil microbial community as they play a significant part in the K cycle (Kour et al. 2020d; Kumar et al. 2019a). The mechanism behind solubilization of K from the soil is a complex phenomenon as various factors affect this process, such as involvement of microbes, the nutritional quality of soil, amount and type of mineral available in the soil, and various other environmental factors. One of the most efficient ways of utilizing K from the soil is the use of K-solubilizing microbes that can utilize a reservoir of K from the soil and make available to the plants as various literature have reported about the use of KSM has proved to be useful (Kour et al. 2020c).

5.5.1.4 Siderophores Production

Siderophores are low molecular weight iron-chelating compounds produced by PGPR, which transport elements into the cells (Kumar et al. 2019). As Fe^{3+} form of iron is abundantly found in soil but its insoluble nature restricts its use by the plants

(Burd et al. 2000). Examples: *Phyllobacterium* strain (siderophore producing). Siderophores are responsible for iron solubilization and transportation into the bacterial cells. Bacteria produce either hydroxamate or catecholate type of siderophores. Under conditions of iron limitations, siderophore producers are able to bind and transport iron-siderophore complex by the expression of specific proteins. The siderophores production is favorable for plants as it can inhibit the growth of plant pathogen. Siderophores influence the growth of the plants both directly and indirectly. Siderophores and their substituted derivatives have varied applications in agricultural, environmental, and medical sciences (Sharaff et al. 2020).

5.5.2 The Indirect Impact of PGP Microbes on Plant Nutritions

PGPR has shown to produce numerous volatile compounds that show antagonist effects towards the disease-causing microorganism (Whipps 2001). Several biocontrol agents have been included: decyl alcohol, 3,5,5-trimethylhexanol, kanosamine, 2, 4 diacetylphloroglucinol (2, 4-DAPG), xanthobaccin, phenazine-1-carboxylic acid, pyrrolnitrin, zwittermycin A, viscosinamide, etc (Rezzonico et al. 2007; Whipps 2001). Antibiotics produced by these PGPRs help in the inhibition of growth of phytopathogens and lead to the improvement of the total yields. *Bacillus* sp. such as *Bacillus* strain D13 produces decyl alcohol which inhibits the growth of *Xanthomonas* species (Whipps 2001). *Pseudomonas* is one of the most common soil bacteria that produces 2,4-diacetyl phloroglucinol (2,4-DAPG) (Rezzonico et al. 2007). 2,4-DAPG is the most effective antibiotic and has a broad species spectrum and shows antifungal and antibacterial activities (Saraf et al. 2014). Several *Pseudomonas* species were associated with the production of rhamnolipids, for example, *P. fluorescens* DR54 and DSS7 producing CLPs showing antimicrobial and surfactant properties (Din et al. 2019). PCA and pyrrolnitrin have proved to be effective against various classes of bacteria and fungi including ascomycete, deuteromycete, and basidiomycete. The use of these biocontrol agents also results in the destruction of plant-promoting pathogens such as arbuscular mycorrhizal (AM) fungi and hence requires further research before their application (Bhale et al. 2018).

5.5.2.1 Enzymes Production

The plant produces enzymes in response to biotic and abiotic stresses which includes ethylene, responsible for the stunted root growth and ageing effects on plants (Karnwal et al. 2019; Singh et al. 2019; Sidhu, et al. 2019). Some of the PGPRs including *Aspergillus*, rhizobacteria, *Pseudomonas,* and *Bacillus* sp. have shown to produce 1-aminocyclopropane-1-carboxylate (ACC) deaminase results in stimulation of plant growth and reduction in the ethylene production in plants (Zain et al. 2019). PGPRs producing chitinases and lytic enzymes can efficiently reduce phytopathogens leading to the removal of biotic stress (Chowdhury and Bagchi 2017; Kour et al. 2020b; Mondal et al. 2020).

5.5.2.2 Hydrogen Cyanide Production

Apart from the above-mentioned antibiotics and enzymes, some of the plant-promoting bacteria also helps in plant growth by inducing HCN production in plants (Chowdhury and Bagchi 2017). HCN is mainly associated with inhibition of the growth of plant pathogens and produced from the glycine via the use of HCN synthase enzyme which is present on the surface of PGPRs lipid bilayer (Siddiqui et al. 2006). PGPRs which can produce HCN include *Bacillus, Bacillus subtilis HussainT-AMU, Pseudomonas, Rhizobium, etc.* and are associated with growth of major disease-causing nematodes, for example, *Meloidogyne javanica, M. incognita* and *Thielaviopsis basicola* and thereby disease such as "root-knot galling" and black rot (Van 2006; Hussain et al. 2020c).

5.5.2.3 Induced System Resistance

Induced systemic resistance (ISR) or immunization of plants against a pathogen is a recent term and involves the development of resistance mechanism in plants against a disease either by stimulating with chemicals or root colonization with PGPRs (Hussain et al. 2020c). The following term known as ISR and PGPR can stimulate root colonization of the bacteria such as *Pseudomonas* sp., *Bacillus* sp. etc. (Chalam et al. 1997; Hussain et al. 2020b; Singh et al. 2020a). The induction of ISR can be achieved with various parts and products of PGPRs like flagella, salicylic acid, LPS, and siderophores.

5.5.2.4 Emerging Biocontrol Strategies

Implementation of Plant Exudates to Attract Beneficial Biocontrol Microbes

The exudates discharged by plant plays an imperative role in determining the composition of soil and functioning of the microbial community. Exudates attract certain group of the microbial community by performing a specific function (Rahman et al. 2017). For example, legumes release flavonoids to attract the definite nitrogen-fixing rhizobacteria (Cooper 2007) and in return, these microbial species help in activating defense system of the plant to fight against foliar diseases (Ryu et al. 2004). Moreover, now soil microbial community are extensively explored and used in agricultural practices to improve nutrient uptake in plant and generate resistance against plant diseases (Cao et al. 2011; Kavoo-Mwangi et al. 2013; Singh et al. 2018). The interlink between microbial diversity and different exudate has been well establishing in hormone-treated plants (Carvalhais et al. 2013, 2015). Additionally, strigolactone has been reported to attract Mycorrhiza and other microbes which were having the ability to improve water availability, defense system, and phosphate solubilization ability (Rahman et al. 2017). Other organic composites like fumarate, malate, and succinate have been found effective in attracting *Pseudomonas fluorescens,* which acts as an effective BCA against various phytopathogens (Oku et al. 2014). Considering the evidence about the using of plant exudates to attract useful microbes is the viable solution to combat pathogens responsible for various plant diseases. In addition, the microbial community residing in rhizosphere could be influenced by treating plants with signalling chemicals to attract useful microbes (Carvalhais et al. 2015; Wintermans et al. 2016).

Use of Substrates to Maintain Beneficial Biocontrol Microbes

One of the essential components for effective growth, metabolic activity as well as the functioning of microbes is "substrate." The substrate plays a vital role in culturing the beneficial biocontrol microbes. Most of these microbes can be cultured in in vitro conditions via traditional culturing procedures (Bai et al. 2015). This approach allows us to isolate beneficial microbes from their natural environment and maintain the microbial population in the rhizosphere to control as well as regulate the growth of plant pathogens, by providing the suitable substrate. Additionally, nutrition also allows the microbes to adapt themselves to survive in a varied environment.

Phyllosphere Biocontrol

The fungi responsible for causing foliar diseases have been reported to affect different types of crops (Madden and Nutter 1995). Six of the fungus has been comprehended throughout the world to the causative agent of foliar diseases (Dean et al. 2012). Therefore, extensive knowledge about these foliar causing agents will be a decisive step for protecting crops. The use of microbes as BCA has emerged as an eco-friendly substitute for synthetic chemical (Maksimov et al. 2011). Additionally, spraying of BCA formulation has been found effective in curbing the foliar diseases (Heydari and Pessarakli 2010). Moreover, the liquid formulation was also tested on the avocado plant affected stem-end rot pathogen (Demoz and Korsten 2006). Other than this, various bacteria having antagonistic potential were isolated to cease the growth of *Erwinia chrysanthemi* responsible for causing stem rot disease in tomato (Aysan et al. 2003). A study reported about serenade compound, which was obtained from *B. subtilis* strain and was exhibiting the antagonistic activity against fungi affecting the blueberries (Scherm et al. 2004). Additionally, plants also synthesize antimicrobial agents on their leaf surface to defend themselves and produce exudates to attract the growth-promoting microbes (Vorholt 2012). There is numerous evidence available in previously published literature highlight the leaf-colonizing microbes which aid in developing defense mechanism to cease the progression of foliar disease in plants (Morris and Monier 2003). Strategies like niche occupation and pre-emptive colonization have also been proposed as an effective way to protect crops from pathogens (Lindow 1987). It is believed that pioneer strategies like phyllosphere microbiome profiling (Vorholt 2012) and plant and microbial interaction could pave new opportunities to improve plants defense system and meet the demand for food security.

***Fungi* as Biocontrol Agents**

Presently, fungi have emerged as effective BCA and predominantly used to improve the yield of crops (Malyan et al. 2019). In 2019, Adnan and his colleagues reported about the *Trichoderma* species which acted as the effective BCA against phytopathogen. The antagonistic activity of *Trichoderma* spp. was because of the synthesis of a bioactive molecule having antagonistic potential against both *Pythium ultimum* and *Rhizoctonia solani* (Harman and Nelson 1994). *Pochonia chlamydosporia*, another fungal isolated exhibiting antagonistic potential against root-knot nematodes of different crops conditions (Manzanilla-Lopez et al. 2013). And, endophytic colonization of *P. chlamydosporia* in the plant has been found to be effective antagonistic potential against pathogen and also found to improve the plant growth (Maciá-Vicente et al. 2009).

Various studies have proved the potential of mycorrhizal associations in conferring resistance to plant against numerous pathogenic diseases and have been employed to provide the plant roots with lasting protection against pathogens (Akhtar and Siddiqui 2008). Mycorrhizal associations are the predominant type of fungal associations found in roots of most plants. The mycorrhizal association can be described as the establishment of the beneficial fungi in the root cortical tissue during growth and development of plant which creates an unfavorable microenvironment that inhibits the proliferation of pathogens. The utility of these mycorrhizal fungal associations as a biocontrol agent to prevent the onset of pathogenic diseases in plants is noble and eco-friendly. Several studies have proved that tree seedlings with mycorrhizal associations exhibit more resistance to feeder roots against pathogenic fungi/bacteria/nematodes than non-mycorrhizal roots (Schouteden et al. 2015).

Ectomycorrhizae grow on the surface of roots and do not penetrate inside the root cells producing a net-like structure known as the Hartig net. They prevent the pathogenic attack through various mechanisms like the synthesis of antifungal agents, antibiosis, and development of fungal mantle that obstructs the entry into plant roots. (Duchesne 1994).

Vesicular arbuscular mycorrhizal fungi (VAM) is another important part of the microbial soil community which provides significant benefit for plants (Sukhada et al. 2011). VAM fungi not only benefits the plant by improving overall development and growth but also confers resistance to host plant against pathogenic microbes. (Ziedan et al. 2011). The application of VAM fungi belonging to the genus *Glomus* is more pronounced than others and includes various species like *G. mosseae, G. fasciculatum, G. monosporum, G. constrictum,* and *G. macrosporum.* They enhance the natural defense system of host plants and restrict the entry of soil-borne pathogens into the roots, thus preventing the infection. Moreover, studies have reported a decrease in the incidence of root-knot infection caused by nematodes in plants (Linderman 1994). The infection caused by *Pseudomonas syringae* in tomatoes results in huge productivity loss, which can be overcome to a significant level by establishing the mycorrhizal fungi in the roots of the host plant (Song et al. 2015). The mycorrhizal association serves a physical barrier and involves certain chemical reactions inducing some direct as well as indirect effects (Fitter and Garbaye 1994). The indirect effect involves the enhancement of nutrient uptake potential in plants, elevating the lignification in roots, mitigating the environmental stresses and altering the microenvironment in the mycorrhizospheric zone thereby promoting the growth of other beneficial microbes (Tripathi et al. 2008; Linderman 1994).

Arbuscular mycorrhizal fungi (AMF) are well established in the treatment of several plant diseases, especially those infecting the root system of plants (Xavier and Boyetchko 2004). They are known to provide systemic resistance to host plants which aid in suppressing most pathogenic diseases (Pozo and Azcón-Aguilar 2007). Studies have reported that the mycorrhizal induced resistance (MIR) is the outcome of exhaustion of intermediate compounds formed in the salicylic acid (SA)-dependent defense pathway of active depletion of components in the SA-dependent defense pathway, which causes systemic priming of jasmonic acid-dependent defenses (Pozo and Azcón-Aguilar 2007). However, the actual mechanism and role of jasmonates in mycorrhizal induced resistance are still unclear (Hause et al. 2007) and the long-distance signals supervising MIR remain to be resolved.

5.6 Conclusion and Future Prospects

The advancement in the field of agriculture considerably relies on the progressive development in the biotechnology, especially considering the conventional breeding practices and genetic modification concerning to improve interactions among plants and microbial communities. In conventional plant biotechnology, the plant breeding approach mainly emphasizes on the inheritance of beneficial traits but did not consider the benefit of plant–microbe interaction. In contrast, the genetic approach focuses on enhancing the colonization potential of beneficial of rhizospheric microbes within a soil microbial community and improving the plant–microbe interactions by alteration of certain factors in the plants as well as microbes. Considering the hazardous effects of fertilizers on the environment, it becomes a necessity to prioritize the research focus towards the plant–microbe interactions involved in the uptake of nutrients to enhance crop productivity in the specified arable agricultural land. However, the specific mechanisms that stimulate the colonization of rhizospheric microorganisms and their regulation according to the nutritional status of the plant are surpassingly complicated and difficult to predict. Conserving the microscopic diversity of the soil is considered overly beneficial in all aspects. Various research studies could be conducted to observe the variations in the diversity in response to specific treatments, which would confer resistance to plants towards various treatments. Moreover, design management strategies may allow the maintenance of oil diversity and productivity.

References

Abd_Allah EF, Alqarawi AA, Hashem A, Radhakrishnan R, Al-Huqail AA, Al-Otibi FON et al (2018) Endophytic bacterium *Bacillus subtilis* (BERA 71) improves salt tolerance in chickpea plants by regulating the plant defense mechanisms. J Plant Interact 13:37–44

Abd-Alla M (1994) Solubilization of rock phosphates by *Rhizobium* and *Bradyrhizobium*. Folia Microbiol 39:53–56

Adnan M, Islam W, Shabbir A, Khan KA, Ghramh HA, Huang Z et al (2019) Plant defense against fungal pathogens by antagonistic fungi with *Trichoderma* in focus. Microb Pathog 129:7–18

Ahmad G, Nishat Y, Haris M, Danish M, Hussain T (2019) Efficiency of soil, plant and microbes for healthy plant immunity and sustainable agricultural system. In: Varma A, Tripathi S, Prasad R (eds) Plant microbe interface. Springer International Publishing, Cham, pp 325–346. https://doi.org/10.1007/978-3-030-19831-2_15

Ahemad M, Khan M (2011) Toxicological assessment of selective pesticides towards plant growth promoting activities of phosphate solubilizing *Pseudomonas aeruginosa*. Acta Microbiol Immunol Hung 58(3):169–187

Ahemad M, Kibret M (2014) Mechanisms and applications of plant growth promoting rhizobacteria: current perspective. J King Saud Univ Sci 26(1):1–20

Akhtar MS, Siddiqui ZA (2008) Arbuscular mycorrhizal fungi as potential bioprotectants against plant pathogens. In: Siddiqui ZA, Akhtar MS, Futai K (eds) Mycorrhizae: sustainable agriculture and forestry. Springer, Dordrecht, pp 61–97. https://doi.org/10.1007/978-1-4020-8770-7_3

Annapurna K, Govindasamy V, Sharma M, Ghosh A, Chikara SK (2018) Whole genome shotgun sequence of *Bacillus paralicheniformis* strain KMS 80, a rhizobacterial endophyte isolated from rice (*Oryza sativa* L.). 3 Biotech 8(5):223

Aysan Y, Karatas A, Cinar O (2003) Biological control of bacterial stem rot caused by Erwinia chrysanthemi on tomato. Crop Prot 22:807–811

Babalola OO (2010) Beneficial bacteria of agricultural importance. Biotechnol Lett 32(11):1559–1570

Backer R, Rokem JS, Ilangumaran G, Lamont J, Praslickova D, Ricci E et al (2018) Plant growth-promoting rhizobacteria: context, mechanisms of action, and roadmap to commercialization of biostimulants for sustainable agriculture. Front Plant Sci 9:1473

Bai Y, Müller DB, Srinivas G, Garrido-Oter R, Potthoff E, Rott M et al (2015) Functional overlap of the *Arabidopsis* leaf and root microbiota. Nature 528:364

Bais HP, Walker TS, Schweizer HP, Vivanco JM (2002) Root specific elicitation and antimicrobial activity of rosmarinic acid in hairy root cultures of *Ocimum basilicum*. Plant Physiol Biochem 40(11):983–995

Barea J (2015) Future challenges and perspectives for applying microbial biotechnology in sustainable agriculture based on a better understanding of plant-microbiome interactions. J Soil Sci Plant Nutr 15(2):261–282

Beneduzi A, Ambrosini A, Passaglia LM (2012) Plant growth-promoting rhizobacteria (PGPR): their potential as antagonists and biocontrol agents. Genet Mol Biol 35(4):1044–1051

Bent E, Chanway CP (1998) The growth-promoting effects of a bacterial endophyte on lodgepole pine are partially inhibited by the presence of other rhizobacteria. Can J Microbiol 44(10):980–988

Berendsen RL, Pieterse CM, Bakker PA (2012) The rhizosphere microbiome and plant health. Trends Plant Sci 17(8):478–486

Berg G, Smalla K (2009) Plant species and soil type cooperatively shape the structure and function of microbial communities in the rhizosphere. FEMS Microbiol Ecol 68(1):1–13

Bhale UN, Bansode SA, Singh S (2018) Multifactorial role of arbuscular mycorrhizae in agroecosystem. In: Gehlot P, Singh J (eds) Fungi and their role in sustainable development: current perspectives. Springer Singapore, Singapore, pp 205–220. https://doi.org/10.1007/978-981-13-0393-7_12

Brader G, Compant S, Mitter B, Trognitz F, Sessitsch A (2014) Metabolic potential of endophytic bacteria. Curr Opin Biotechnol 27:30–37

Brakhage AA (2013) Regulation of fungal secondary metabolism. Nat Rev Microbiol 11(1):21–32

Brechenmacher L, Lei Z, Libault M, Findley S, Sugawara M, Sadowsky MJ et al (2010) Soybean metabolites regulated in root hairs in response to the symbiotic bacterium *Bradyrhizobium japonicum*. Plant Physiol 153(4):1808–1822

Bulgarelli D, Schlaeppi K, Spaepen S, Van Themaat EVL, Schulze-Lefert P (2013) Structure and functions of the bacterial microbiota of plants. Annu Rev Plant Biol 64:807–838

Burd GI, Dixon DG, Glick BR (2000) Plant growth-promoting bacteria that decrease heavy metal toxicity in plants. Can J Microbiol 46(3):237–245

Cao Y, Zhang Z, Ling N, Yuan Y, Zheng X, Shen B, Shen Q (2011) *Bacillus subtilis* SQR 9 can control Fusarium wilt in cucumber by colonizing plant roots. Biol Fertil Soils 47:495–506
Carvalhais LC, Dennis PG, Badri DV, Kidd BN, Vivanco JM, Schenk PM (2015) Linking jasmonic acid signaling, root exudates, and rhizosphere microbiomes. Mol Plant-Microbe Interact 28:1049–1058
Carvalhais LC, Dennis PG, Badri DV, Tyson GW, Vivanco JM, Schenk PM (2013) Activation of the jasmonic acid plant defence pathway alters the composition of rhizosphere bacterial communities. PLoS One 8:e56457
Chadha N, Mishra M, Prasad R, Varma A (2014) Root endophytic fungi: research update. J Biol Life Sci USA 5(2):135–158
Chalam A, Sasikala C, Ramana CV, Uma N, Rao PR (1997) Effect of pesticides on the diazotrophic growth and nitrogenase activity of purple nonsulfur bacteria. Bulletin Environ Cont Toxicol 58(3):463–468
Chaparro JM, Badri DV, Vivanco JM (2014) Rhizosphere microbiome assemblage is affected by plant development. ISME J 8(4):790–803
Chowdhury N, Bagchi A (2017) Structural insight into the gene expression profiling of the HCN operon in *Pseudomonas aeruginosa*. Appl Biochem Biotechnol 182(3):1144–1157
Ciccazzo S, Esposito A, Rolli E, Zerbe S, Daffonchio D, Brusetti L (2014) Different pioneer plant species select specific rhizosphere bacterial communities in a high mountain environment. SpringerPlus 3(1):391
Compant S, Clément C, Sessitsch A (2010) Plant growth-promoting bacteria in the rhizo-and endosphere of plants: their role, colonization, mechanisms involved and prospects for utilization. Soil Biol Biochem 42(5):669–678
Compant S, Duffy B, Nowak J, Clément C, Barka EA (2005) Use of plant growth-promoting bacteria for biocontrol of plant diseases: principles, mechanisms of action, and future prospects. Appl Environ Microbiol 71(9):4951–4959
Cooper J (2007) Early interactions between legumes and rhizobia: disclosing complexity in a molecular dialogue. J Appl Microbiol 103:1355–1365
Cumagun CJR (2014) Advances in formulation of *Trichoderma* for biocontrol. In: Biotechnology and biology of *Trichoderma*. Elsevier, Waltham, MA, pp 527–531
Datta S, Singh S, Kumar V, Dhanjal DS, Sidhu GK, Amin DS et al (2020) Endophytic bacteria in xenobiotic degradation. In: Kumar A, Singh VK (eds) Microbial endophytes. Woodhead Publishing, Cambridge, MA, pp 125–156. https://doi.org/10.1016/B978-0-12-818734-0.00006-1
de Lima Favaro LC, de Souza Sebastianes FL, Araújo WL (2012) *Epicoccum nigrum* P16, a sugarcane endophyte, produces antifungal compounds and induces root growth. PLoS One 7(6):e36826
De Mandal S, Singh SS, Kumar NS (2018) Analyzing plant growth promoting *Bacillus* sp. and related genera in Mizoram, Indo-Burma biodiversity Hotspot. Biocat Agric Biotechnol 15:370–376
Dean R, Van Kan JA, Pretorius ZA, Hammond-Kosack KE, Di Pietro A, Spanu PD, Rudd JJ, Dickman M, Kahmann R, Ellis J (2012) The Top 10 fungal pathogens in molecular plant pathology. Mol Plant Pathol 13:414–430
Demoz BT, Korsten L (2006) Bacillus subtilis attachment, colonization, and survival on avocado flowers and its mode of action on stem-end rot pathogens. Biol Control 37:68–74
Din BU, Sarfraz S, Xia Y, Kamran MA, Javed MT, Sultan T et al (2019) Mechanistic elucidation of germination potential and growth of wheat inoculated with exopolysaccharide and ACC-deaminase producing *Bacillus* strains under induced salinity stress. Ecotoxicol Environ Saf 183:109466
Duchesne LC (1994) Role of ectomycorrhizal fungi in biocontrol. In: Pfleger FL, Linderman RG (eds) Mycorrhizae and plant health. APS Press, St. Paul, MN, pp 27–45
Elad Y, Kirshner B, Yehuda N, Sztejnberg A (1998) Management of powdery mildew and gray mold of cucumber by *Trichoderma harzianum* T39 and Ampelomyces quisqualis AQ10. BioControl 43(2):241–251

Estrada AER, Jonkers W, Kistler HC, May G (2012) Interactions between *Fusarium verticillioides, Ustilago maydis*, and *Zea mays*: an endophyte, a pathogen, and their shared plant host. Fungal Genet Biol 49(7):578–587
Faure D, Vereecke D, Leveau JH (2009) Molecular communication in the rhizosphere. Plant Soil 321(1–2):279–303
Fitter A, Garbaye J (1994) Interactions between mycorrhizal fungi and other soil organisms. Plant Soil 159(1):123–132
Friesen ML, Porter SS, Stark SC, Von Wettberg EJ, Sachs JL, Martinez-Romero E (2011) Microbially mediated plant functional traits. Annu Rev Ecol Evol Syst 42:23–46
Gagne-Bourgue F, Aliferis K, Seguin P, Rani M, Samson R, Jabaji S (2013) Isolation and characterization of indigenous endophytic bacteria associated with leaves of switchgrass (*Panicum virgatum* L.) cultivars. J Appl Microbiol 114(3):836–853
Garbaye J (1994) Tansley review no. 76 helper bacteria: a new dimension to the mycorrhizal symbiosis. New Phytol 128(2):197–210
Gond S, Torres M, Bergen M, Helsel Z, White J Jr (2015a) Induction of salt tolerance and up-regulation of aquaporin genes in tropical corn by rhizobacterium *Pantoea agglomerans*. Lett Appl Microbiol 60(4):392–399
Gond SK, Bergen MS, Torres MS, White JF Jr (2015b) Endophytic *Bacillus* spp. produce antifungal lipopeptides and induce host defence gene expression in maize. Microbiol Res 172:79–87
Gupta G, Snehi SK, Singh V (2017) Role of PGPR in biofilm formations and its importance in plant health. John Wiley & Sons Ltd, Hoboken, NJ, pp 27–42. https://doi.org/10.1002/9781119246329.ch2
Haney CH, Samuel BS, Bush J, Ausubel FM (2015) Associations with rhizosphere bacteria can confer an adaptive advantage to plants. Nat Plants 1(6):15051
Hardoim PR, van Overbeek LS, van Elsas JD (2008) Properties of bacterial endophytes and their proposed role in plant growth. Trends Microbiol 16(10):463–471
Harman G, Nelson E (1994) Mechanisms of protection of seed and seedlings by biological seed treatments: implication for practical disease control. In: Maude R (ed) Seed treatment, progress and prospects. British Crop Protection Council, Hampshire, pp 383–392
Hassani MA, Durán P, Hacquard S (2018) Microbial interactions within the plant holobiont. Microbiome 6(1):58
Hause B, Mrosk C, Isayenkov S, Strack D (2007) Jasmonates in arbuscular mycorrhizal interactions. Phytochemistry 68(1):101–110
Herrera SD, Grossi C, Zawoznik M, Groppa MD (2016) Wheat seeds harbour bacterial endophytes with potential as plant growth promoters and biocontrol agents of Fusarium graminearum. Microbiol Res 186:37–43
Heydari A, Pessarakli M (2010) A review on biological control of fungal plant pathogens using microbial antagonists. J Biol Sci 10(4):273–290
Hussain T, Khan AA (2020) *Bacillus subtilis Hussain T-AMU* and its antifungal activity against Potato black scurf caused by *Rhizoctonia solani*. Biocatal Agric Biotechnol 23:101433
Hussain T, Akthar N, Aminedi R, Danish M, Nishat Y, Patel S (2020a) Role of the potent microbial based bioagents and their emerging strategies for the ecofriendly management of agricultural phytopathogens. In: Singh J, Yadav AN (eds) Natural bioactive products in sustainable agriculture. Springer Singapore, Singapore, pp 45–66. https://doi.org/10.1007/978-981-15-3024-1_4
Hussain T, Haris M, Shakeel A, Khan AA, Khan MA (2020c) Bio-nematicidal activities by culture filtrate of *Bacillus subtilis Hussain T-AMU*: new promising biosurfactant bioagent for the management of Root Galling caused by *Meloidogyne incognita*. Vegetos 33:229–238. https://doi.org/10.1007/s42535-020-00099-5
Hussain T, Singh S, Danish M, Pervez R, Hussain K, Husain R (2020b) Natural metabolites: an eco-friendly approach to manage plant diseases and for better agriculture farming. In: Singh J, Yadav AN (eds) Natural bioactive products in sustainable agriculture. Springer Singapore, Singapore, pp 1–13. https://doi.org/10.1007/978-981-15-3024-1_1
Idris HA, Labuschagne N, Korsten L (2007) Screening rhizobacteria for biological control of Fusarium root and crown rot of sorghum in Ethiopia. Biol Control 40(1):97–106

Jambon I, Thijs S, Weyens N, Vangronsveld J (2018) Harnessing plant-bacteria-fungi interactions to improve plant growth and degradation of organic pollutants. J Plant Interact 13(1):119–130
Kalra A, Chandra M, Awasthi A, Singh AK, Khanuja SPS (2010) Natural compounds enhancing growth and survival of rhizobial inoculants in vermicompost-based formulations. Biol Fertil Soils 46(5):521–524
Kapoor D, Singh S, Kumar V, Romero R, Prasad R, Singh J (2019) Antioxidant enzymes regulation in plants in reference to reactive oxygen species (ROS) and reactive nitrogen species (RNS). Plant Gene 19:100182
Karnwal A, Singh S, Kumar V, Sidhu GK, Dhanjal DS, Datta S et al (2019) Fungal enzymes for the textile industry. In: Yadav AN, Mishra S, Singh S, Gupta A (eds) Recent advancement in White biotechnology through fungi, Diversity and enzymes perspectives, vol 1. Springer International Publishing, Cham, pp 459–482. https://doi.org/10.1007/978-3-030-10480-1_14
Kaur P, Singh S, Kumar V, Singh N, Singh J (2018) Effect of rhizobacteria on arsenic uptake by macrophyte Eichhornia crassipes (Mart.) Solms. Int J Phytoremediation 20(2):114–120
Kavoo-Mwangi A, Kahangi E, Ateka E, Onguso J, Mukhongo R, Mwangi E, Jefwa J (2013) Growth effects of microorganisms based commercial products inoculated to tissue cultured banana cultivated in three different soils in Kenya. Appl Soil Ecol 64:152–162
Kembel SW, O'Connor TK, Arnold HK, Hubbell SP, Wright SJ, Green JL (2014) Relationships between phyllosphere bacterial communities and plant functional traits in a neotropical forest. Proc Natl Acad Sci U S A 111(38):13715–13720
Khan M, Zaidi A, Aamil M (2002) Biocontrol of fungal pathogens by the use of plant growth promoting rhizobacteria and nitrogen fixing microorganisms. Ind J Bot Soc 81:255–263
Kloeppe J, Rodriguez-Kabana R, Zehnder A, Murphy J, Sikora E, Fernandez C (1999) Plant root-bacterial interactions in biological control of soilborne diseases and potential extension to systemic and foliar diseases. Australas Plant Pathol 28(1):21–26
Knelman JE, Legg TM, O'Neill SP, Washenberger CL, González A, Cleveland CC et al (2012) Bacterial community structure and function change in association with colonizer plants during early primary succession in a glacier forefield. Soil Biol Biochem 46:172–180
Korolev N, David DR, Elad Y (2008) The role of phytohormones in basal resistance and *Trichoderma*-induced systemic resistance to *Botrytis cinerea* in *Arabidopsis thaliana*. BioControl 53(4):667–683
Koul B, Singh S, Dhanjal DS, Singh J (2019) Plant growth-promoting rhizobacteria (PGPRs): a fruitful resource. In: Singh DP, Prabha R (eds) Microbial interventions in agriculture and environment, Soil and crop health management, vol 3. Springer Singapore, Singapore, pp 83–127. https://doi.org/10.1007/978-981-32-9084-6_5
Kour D, Kaur T, Devi R, Rana KL, Yadav N, Rastegari AA et al (2020a) Biotechnological applications of beneficial microbiomes for evergreen agriculture and human health. In: Rastegari AA, Yadav AN, Yadav N (eds) Trends of microbial biotechnology for sustainable agriculture and biomedicine systems: perspectives for human health. Elsevier, Amsterdam, pp 255–287. https://doi.org/10.1016/B978-0-12-820528-0.00019-3
Kour D, Kaur T, Yadav N, Rastegari AA, Singh B, Kumar V et al (2020b) Phytases from microbes in phosphorus acquisition for plant growth promotion and soil health. In: Rastegari AA, Yadav AN, Yadav N (eds) Trends of microbial biotechnology for sustainable agriculture and biomedicine systems: diversity and functional perspectives. Elsevier, Amsterdam, pp 157–176. https://doi.org/10.1016/B978-0-12-820526-6.00011-7
Kour D, Rana KL, Kaur T, Devi R, Yadav N, Halder SK et al (2020c) Potassium solubilizing and mobilizing microbes: biodiversity, mechanisms of solubilization and biotechnological implication for alleviations of abiotic stress. In: Rastegari AA, Yadav AN, Yadav N (eds) Trends of microbial biotechnology for sustainable agriculture and biomedicine systems: diversity and functional perspective. Elsevier, Amsterdam, pp 177–202. https://doi.org/10.1016/B978-0-12-820526-6.00012-9
Kour D, Rana KL, Yadav AN, Yadav N, Kumar M, Kumar V et al (2020d) Microbial biofertilizers: bioresources and eco-friendly technologies for agricultural and environmental sustainability. Biocatal Agric Biotechnol 23:101487. https://doi.org/10.1016/j.bcab.2019.101487

Kour D, Rana KL, Yadav N, Yadav AN, Kumar A, Meena VS et al (2019) Rhizospheric microbiomes: biodiversity, mechanisms of plant growth promotion, and biotechnological applications for sustainable agriculture. In: Kumar A, Meena VS (eds) Plant growth promoting rhizobacteria for agricultural sustainability: from theory to practices. Springer Singapore, Singapore, pp 19–65. https://doi.org/10.1007/978-981-13-7553-8_2

Kumar A, Devi S, Agrawal H, Singh S, Singh J (2020a) Rhizoremediation: a unique plant microbiome association of biodegradation. In: Varma A, Tripathi S, Prasad R (eds) Plant microbe symbiosis. Springer International Publishing, Cham, pp 203–220. https://doi.org/10.1007/978-3-030-36248-5_11

Kumar M, Kour D, Yadav AN, Saxena R, Rai PK, Jyoti A et al (2019a) Biodiversity of methylotrophic microbial communities and their potential role in mitigation of abiotic stresses in plants. Biologia 74:287–308. https://doi.org/10.2478/s11756-019-00190-6

Kumar V, Joshi S, Pant NC, Sangwan P, Yadav AN, Saxena A et al (2019b) Molecular approaches for combating multiple abiotic stresses in crops of arid and semi-arid region. In: Singh SP, Upadhyay SK, Pandey A, Kumar S (eds) Molecular approaches in plant biology and environmental challenges. Springer, Singapore, pp 149–170. https://doi.org/10.1007/978-981-15-0690-1_8

Kumar V, Shahi SK, Singh S (2018) Bioremediation: an eco-sustainable approach for restoration of contaminated sites. In: Singh J, Sharma D, Kumar G, Sharma NR (eds) Microbial bioprospecting for sustainable development. Springer Singapore, Singapore, pp 115–136. https://doi.org/10.1007/978-981-13-0053-0_6

Kumar V, Singh S, Upadhyay N (2019) Effects of organophosphate pesticides on siderophore producing soils microorganisms. Biocatal Agric Biotechnol 21:101359

Kumar V, Singh S, Singh J, Upadhyay N (2015) Potential of plant growth promoting traits by bacteria isolated from heavy metal contaminated soils. Bull Environ Contam Toxicol 94(6):807–814

Kumar V, Singh S, Singh S, Datta S, Dhanjal DS, Singh J (2020b) Methods and techniques for the chemical profiling and quality control of natural products and natural product-derived drugs. In: Singh J, Meshram V, Gupta M (eds) Bioactive natural products in drug discovery. Springer Singapore, Singapore, pp 585–598. https://doi.org/10.1007/978-981-15-1394-7_20

Kumari A, Sumer S, Jalan B, Nongbri PL, Laskar MA (2017) Impact of next-generation sequencing technology in plant–microbe interaction study. In: Microbial applications, vol 1. Springer, Cham, pp 269–294

Lareen A, Burton F, Schäfer P (2016) Plant root-microbe communication in shaping root microbiomes. Plant Mol Biol 90(6):575–587

Laskar F, Das Purkayastha S, Sen A, Bhattacharya MK, Misra BB (2018) Diversity of methanogenic archaea in freshwater sediments of lacustrine ecosystems. J Basic Microbiol 58(2):101–119

Lery LM, Hemerly AS, Nogueira EM, von Krüger WM, Bisch PM (2011) Quantitative proteomic analysis of the interaction between the endophytic plant-growth-promoting bacterium *Gluconacetobacter diazotrophicus* and sugarcane. Mol Plant-Microbe Interact 24(5):562–576

Linderman RG (1994) Role of VAM fungi in biocontrol. In: Pfleger FL, Linderman RG (eds) Mycorrhizae and plant health. APS Press, St. Paul, MN

Lindow SE (1987) Competitive exclusion of epiphytic bacteria by ice-*Pseudomonas syringae* mutants. Appl Environ Microbiol 53:2520–2527

Liu ZL, Sinclair JB (1990) Enhanced soybean plant growth and nodulation by *Bradyrhizobium* in the presence of strains of *Bacillus megaterium*. Phytopathology 80:1024

Liu K, McInroy JA, Hu CH, Kloepper JW (2018) Mixtures of plant-growth-promoting rhizobacteria enhance biological control of multiple plant diseases and plant-growth promotion in the presence of pathogens. Plant Dis 102(1):67–72

Lucy M, Reed E, Glick BR (2004) Applications of free living plant growth-promoting rhizobacteria. Antonie Van Leeuwenhoek 86(1):1–25

Lugtenberg BJ, Caradus JR, Johnson LJ (2016) Fungal endophytes for sustainable crop production. FEMS Microbiol Ecol 92(12):fiw194

Macia-Vicente J, Rosso L, Ciancio A, Jansson HB, Lopez-Llorca L (2009) Colonisation of barley roots by endophytic *Fusarium equiseti* and *Pochonia chlamydosporia*: effects on plant growth and disease. Ann Appl Biol 155(3):391–401

Madden LV, Nutter FW Jr (1995) Modeling crop losses at the field scale. Can J Plant Pathol 17(2):124–137

Maksimov I, Abizgil'Dina R, Pusenkova L (2011) Plant growth promoting rhizobacteria as alternative to chemical crop protectors from pathogens. Appl Biochem Microbiol 47:333–345

Malyan SK, Kumar A, Baram S, Kumar J, Singh S, Kumar SS et al (2019) Role of fungi in climate change abatement through carbon sequestration. In: Yadav AN, Singh S, Mishra S, Gupta A (eds) Recent advancement in White biotechnology through fungi, Perspective for sustainable environments, vol 3. Springer International Publishing, Cham, pp 283–295. https://doi.org/10.1007/978-3-030-25506-0_11

Manzanilla-Lopez RH, Esteves I, Finetti-Sialer MM, Hirsch PR, Ward E, Devonshire J et al (2013) *Pochonia chlamydosporia*: advances and challenges to improve its performance as a biological control agent of sedentary endo-parasitic nematodes. J Nematol 45(1):1

Mercado-Blanco J (2015) Life of microbes inside the plant. In: Principles of plant-microbe interactions. Springer, Cham, pp 25–32

Meyer KM, Leveau JH (2012) Microbiology of the phyllosphere: a playground for testing ecological concepts. Oecologia 168(3):621–629

Mondal S, Halder SK, Yadav AN, Mondal KC (2020) Microbial consortium with multifunctional plant growth promoting attributes: future perspective in agriculture. In: Yadav AN, Rastegari AA, Yadav N, Kour D (eds) Advances in plant microbiome and sustainable agriculture, volume 2: functional annotation and future challenges. Springer, Singapore, pp 219–254. https://doi.org/10.1007/978-981-15-3204-7_10

Morris CE, Monier JM (2003) The ecological significance of biofilm formation by plant-associated bacteria. Annu Rev Phytopathol 41:429–453

Netzker T, Fischer J, Weber J, Mattern DJ, König CC, Valiante V et al (2015) Microbial communication leading to the activation of silent fungal secondary metabolite gene clusters. Front Microbiol 6:299

Oku S, Komatsu A, Nakashimada Y, Tajima T, Kato J (2014) Identification of Pseudomonas fluorescens chemotaxis sensory proteins for malate, succinate, and fumarate, and their involvement in root colonization. Microbes Environ, ME14128

Oldroyd GE (2013) Speak, friend, and enter: signalling systems that promote beneficial symbiotic associations in plants. Nat Rev Microbiol 11(4):252–263

Perez-Montano F, Alías-Villegas C, Bellogín R, Del Cerro P, Espuny M, Jiménez-Guerrero I et al (2014) Plant growth promotion in cereal and leguminous agricultural important plants: from microorganism capacities to crop production. Microbiol Res 169(5–6):325–336

Pirozynski K, Malloch D (1975) The origin of land plants: a matter of mycotrophism. Biosystems 6(3):153–164

Porter JR, Xie L, Challinor AJ, Cochrane K, Howden SM, Iqbal MM, Lobell DB, Travasso MI (2014) Food security and food production systems. Cambridge University Press, Cambridge, pp 485–533

Pozo MJ, Azcón-Aguilar C (2007) Unraveling mycorrhiza-induced resistance. Curr Opin Plant Biol 10(4):393–398

Raaijmakers JM, Vlami M, De Souza JT (2002) Antibiotic production by bacterial biocontrol agents. Antonie Van Leeuwenhoek 81(1–4):537

Rahman SF, Singh E, Pieterse CMJ, Schenk PM (2017) Emerging microbial biocontrol strategies for plant pathogens. Plant Sci 267:102–111

Rai PK, Singh M, Anand K, Saurabhj S, Kaur T, Kour D et al (2020) Role and potential applications of plant growth promotion rhizobacteria for sustainable agriculture. In: Rastegari AA, Yadav AN, Yadav N (eds) Trends of microbial biotechnology for sustainable agriculture and biomedicine systems: diversity and functional perspectives. Elsevier, Amsterdam, pp 49–60. https://doi.org/10.1016/B978-0-12-820526-6.00004-X

Rajawat MVS, Singh R, Singh D, Yadav AN, Singh S, Kumar M et al (2020) Spatial distribution and identification of bacteria in stressed environments capable to weather potassium aluminosilicate mineral. Braz J Microbiol 51:751–764. https://doi.org/10.1007/s42770-019-00210-2

Rajkumar M, Ma Y, Freitas H (2008) Characterization of metal-resistant plant-growth promoting Bacillus weihenstephanensis isolated from serpentine soil in Portugal. J Basic Microbiol 48(6):500–508

Rana KL, Kour D, Kaur T, Devi R, Negi C, Yadav AN et al (2020a) Endophytic fungi from medicinal plants: biodiversity and biotechnological applications. In: Kumar A, Radhakrishnan EK (eds) Microbial endophytes. Woodhead Publishing, Cambridge, pp 273–305. https://doi.org/10.1016/B978-0-12-819654-0.00011-9

Rana KL, Kour D, Kaur T, Devi R, Yadav AN, Yadav N et al (2020b) Endophytic microbes: biodiversity, plant growth-promoting mechanisms and potential applications for agricultural sustainability. Antonie Van Leeuwenhoek. https://doi.org/10.1007/s10482-020-01429-y

Rana KL, Kour D, Kaur T, Sheikh I, Yadav AN, Kumar V et al (2020c) Endophytic microbes from diverse wheat genotypes and their potential biotechnological applications in plant growth promotion and nutrient uptake. Proc Natl Acad Sci India B. https://doi.org/10.1007/s40011-020-01168-0

Rana KL, Kour D, Sheikh I, Yadav N, Yadav AN, Kumar V et al (2019a) Biodiversity of endophytic fungi from diverse niches and their biotechnological applications. In: Singh BP (ed) Advances in Endophytic fungal research: present status and future challenges. Springer International Publishing, Cham, pp 105–144. https://doi.org/10.1007/978-3-030-03589-1_6

Rana KL, Kour D, Yadav AN (2019b) Endophytic microbiomes: biodiversity, ecological significance and biotechnological applications. Res J Biotechnol 14:142–162

Rastegari AA, Yadav AN, Yadav N (2020a) New and future developments in microbial biotechnology and bioengineering: Trends of microbial biotechnology for sustainable agriculture and biomedicine systems: diversity and functional perspectives. Elsevier, Amsterdam

Rastegari AA, Yadav AN, Yadav N (2020b) New and future developments in microbial biotechnology and bioengineering: Trends of microbial biotechnology for sustainable agriculture and biomedicine systems: perspectives for human health. Elsevier, Amsterdam

Rezzonico F, Zala M, Keel C, Duffy B, Moënne-Loccoz Y, Défago G (2007) Is the ability of biocontrol fluorescent pseudomonads to produce the antifungal metabolite 2, 4-diacetylphloroglucinol really synonymous with higher plant protection? New Phytol 173(4):861–872

Rodriguez R, White J Jr, Arnold AE, Redman RS (2009) Fungal endophytes: diversity and functional roles. New Phytol 182(2):314–330

Rolli E, Marasco R, Vigani G, Ettoumi B, Mapelli F, Deangelis ML et al (2015) Improved plant resistance to drought is promoted by the root-associated microbiome as a water stress-dependent trait. Environ Microbiol 17(2):316–331

Ryu CM, Farag MA, Hu CH, Reddy MS, Kloepper JW, Paré PW (2004) Bacterial volatiles induce systemic resistance in Arabidopsis. Plant Physiol 134:1017–1026

Salvatore MM, Félix C, Lima F, Ferreira V, Naviglio D, Salvatore F et al (2020) Secondary metabolites produced by *Macrophomina phaseolina* isolated from Eucalyptus globulus. Agriculture 10(3):72

Saraf M, Pandya U, Thakkar A (2014) Role of allelochemicals in plant growth promoting rhizobacteria for biocontrol of phytopathogens. Microbiol Res 169(1):18–29

Schardl CL, Panaccione DG (2005) Biosynthesis of ergot and loline alkaloids. In: Roberts CA, West CP, Spiers DE (eds) Neotyphodium in cool-season grasses. Blackwell, Ames, IA, pp 73–92

Scherm H, Ngugi H, Savelle A, Edwards J (2004) Biological control of infection of blueberry flowers caused by *Monilinia vaccinii-corymbosi*. Biol Control 29:199–206

Schouteden N, De Waele D, Panis B, Vos CM (2015) Arbuscular mycorrhizal fungi for the biocontrol of plant-parasitic nematodes: a review of the mechanisms involved. Front Microbiol 6:1280

Sharaff MS, Subrahmanyam G, Kumar A, Yadav AN (2020) Mechanistic understanding of root-microbiome interaction for sustainable agriculture in polluted soils. In: Rastegari AA, Yadav AN, Yadav N (eds) Trends of microbial biotechnology for sustainable agriculture and biomedicine systems: diversity and functional perspectives. Elsevier, Amsterdam, pp 61–84. https://doi.org/10.1016/B978-0-12-820526-6.00005-1

Sharma K, Sharma S, Prasad SR (2019) PGPR: renewable tool for sustainable agriculture. Int J Curr Microbiol App Sci 8(1):525–530

Siddiqui IA, Shaukat SS, Sheikh IH, Khan A (2006) Role of cyanide production by *Pseudomonas fluorescens* CHA0 in the suppression of root-knot nematode, *Meloidogyne javanica* in tomato. World J Microbiol Biotechnol 22(6):641–650

Singh A, Kumar R, Yadav AN, Mishra S, Sachan S, Sachan SG (2020c) Tiny microbes, big yields: microorganisms for enhancing food crop production sustainable development. In: Rastegari AA, Yadav AN, Yadav N (eds) Trends of microbial biotechnology for sustainable agriculture and biomedicine systems: diversity and functional perspectives. Elsevier, Amsterdam, pp 1–16. https://doi.org/10.1016/B978-0-12-820526-6.00001-4

Singh S, Kumar V, Chauhan A, Datta S, Wani AB, Singh N, Singh J (2018) Toxicity, degradation and analysis of the herbicide atrazine. Environ Chem Lett 16(1):211–237

Singh S, Kumar V, Kapoor D, Kumar S, Singh S, Dhanjal DS, Datta S, Samuel J, Dey P, Wang S, Prasad R, Singh J (2019) Revealing on hydrogen sulfide and nitric oxide signals coordination for plant growth under stress conditions. Physiol Plant 168(2):13002

Singh J, Yadav AN (2020) Natural bioactive products in sustainable agriculture. Springer, Singapore

Singh S, Kumar V, Singh S, Dhanjal DS, Datta S, Singh J (2020a) Global scenario of plant–microbiome for sustainable agriculture: current advancements and future challenges. In: Plant microbiomes for sustainable agriculture. Springer, Cham, pp 425–443

Singh S, Kumar V, Singla S, Sharma M, Singh DP, Prasad R, Thakur VK, Singh J (2020b) Kinetic study of the biodegradation of acephate by indigenous soil bacterial isolates in the presence of humic acid and metal ions. Biomol Ther 10(3):433

Singh S, Singh N, Kumar V, Datta S, Wani AB, Singh D, Singh K, Singh J (2016) Toxicity, monitoring and biodegradation of the fungicide carbendazim. Environ Chem Lett 14:317–329

Smith KP, Handelsman J, Goodman RM (1999) Genetic basis in plants for interactions with disease-suppressive bacteria. Proc Natl Acad Sci 96:4786–4790

Song Y, Chen D, Lu K, Sun Z, Zeng R (2015) Enhanced tomato disease resistance primed by arbuscular mycorrhizal fungus. Front Plant Sci 6:786

Subrahmanyam G, Kumar A, Sandilya SP, Chutia M, Yadav AN (2020) Diversity, plant growth promoting attributes, and agricultural applications of rhizospheric microbes. In: Yadav AN, Singh J, Rastegari AA, Yadav N (eds) Plant microbiomes for sustainable agriculture. Springer International Publishing, Cham, pp 1–52. https://doi.org/10.1007/978-3-030-38453-1_1

Sukhada M, Manjula R, Rawal R (2011) Evaluation of arbuscular mycorrhiza and other biocontrol agents against *Phytophthora parasitica* var. nicotianae infecting papaya (*Carica papaya* cv. Surya) and enumeration of pathogen population using immunotechniques. Biol Control 58(1):22–29

Suman A, Yadav AN, Verma P (2016) Endophytic microbes in crops: diversity and beneficial impact for sustainable agriculture. In: Singh D, Abhilash P, Prabha R (eds) Microbial inoculants in sustainable agricultural productivity, research perspectives. Springer-Verlag, New Delhi, pp 117–143. https://doi.org/10.1007/978-81-322-2647-5_7

Sundaramoorthy S, Balabaskar P (2013) Evaluation of combined efficacy of *Pseudomonas fluorescens* and *Bacillus subtilis* in managing tomato wilt caused by *Fusarium oxysporum f.* sp. lycopersici (Fol). Plant Pathol J 12(4):154–161

Thakur N, Kaur S, Tomar P, Thakur S, Yadav AN (2020) Microbial biopesticides: current status and advancement for sustainable agriculture and environment. In: Rastegari AA, Yadav AN, Yadav N (eds) Trends of microbial biotechnology for sustainable agriculture and biomedicine systems: diversity and functional perspectives. Elsevier, Amsterdam, pp 243–282. https://doi.org/10.1016/B978-0-12-820526-6.00016-6

Thrall PH, Hochberg ME, Burdon JJ, Bever JD (2007) Coevolution of symbiotic mutualists and parasites in a community context. Trends Ecol Evol 22(3):120–126

Tripathi S, Kamal S, Sheramati I, Oelmuller R, Varma A (2008) Mycorrhizal fungi and other root endophytes as biocontrol agents against root pathogens. In: Mycorrhiza. Springer, Berlin, pp 281–306

Trivedi P, Schenk PM, Wallenstein MD, Singh BK (2017) Tiny microbes, big yields: enhancing food crop production with biological solutions. Microbiol Biotech 10:999–1003

Truyens S, Weyens N, Cuypers A, Vangronsveld J (2015) Bacterial seed endophytes: genera, vertical transmission and interaction with plants. Environ Microbiol Rep 7(1):40–50
Turner TR, James EK, Poole PS (2013) The plant microbiome. Genome Biol 14(6):209
Van der Ent S, Van Hulten M, Pozo MJ, Czechowski T, Udvardi MK, Pieterse CM et al (2009) Priming of plant innate immunity by rhizobacteria and β-aminobutyric acid: differences and similarities in regulation. New Phytol 183(2):419–431
Van Der Heijden MG, De Bruin S, Luckerhoff L, Van Logtestijn RS, Schlaeppi K (2016) A widespread plant-fungal-bacterial symbiosis promotes plant biodiversity, plant nutrition and seedling recruitment. ISME J 10(2):389–399
Van LC (2006) Plant responses to plant growth-promoting bacteria. Eur J Plant Pathol 119:243–254
Van Loon L, Bakker P, Pieterse C (1998) Systemic resistance induced by rhizosphere bacteria. Annu Rev Phytopathol 36(1):453–483
Vanitha SC, Niranjana SR, Mortensen CN, Umesha S (2009) Bacterial wilt of tomato in Karnataka and its management by Pseudomonas fluorescens. BioControl 54(5):685–695
Verma P, Yadav AN, Kazy SK, Saxena AK, Suman A (2014) Evaluating the diversity and phylogeny of plant growth promoting bacteria associated with wheat (Triticum aestivum) growing in central zone of India. Int J Curr Microbiol Appl Sci 3:432–447
Verma P, Yadav AN, Khannam KS, Kumar S, Saxena AK, Suman A (2016) Molecular diversity and multifarious plant growth promoting attributes of Bacilli associated with wheat (*Triticum aestivum* L.) rhizosphere from six diverse agro-ecological zones of India. J Basic Microbiol 56:44–58
Verma P, Yadav AN, Khannam KS, Panjiar N, Kumar S, Saxena AK et al (2015) Assessment of genetic diversity and plant growth promoting attributes of psychrotolerant bacteria allied with wheat (*Triticum aestivum*) from the northern hills zone of India. Ann Microbiol 65:1885–1899
Verma P, Yadav AN, Khannam KS, Saxena AK, Suman A (2017a) Potassium-solubilizing microbes: diversity, distribution, and role in plant growth promotion. In: Panpatte DG, Jhala YK, Vyas RV, Shelat HN (eds) Microorganisms for green revolution, Microbes for sustainable crop production, vol 1. Springer Singapore, Singapore, pp 125–149. https://doi.org/10.1007/978-981-10-6241-4_7
Verma P, Yadav AN, Kumar V, Singh DP, Saxena AK (2017b) Beneficial plant-microbes interactions: biodiversity of microbes from diverse extreme environments and its impact for crop improvement. In: Singh DP, Singh HB, Prabha R (eds) Plant-microbe interactions in agro-ecological perspectives, Microbial interactions and agro-ecological impacts, vol 2. Springer Singapore, Singapore, pp 543–580. https://doi.org/10.1007/978-981-10-6593-4_22
Vigani G, Rolli E, Marasco R, Dell'Orto M, Michoud G, Soussi A et al (2019) Root bacterial endophytes confer drought resistance and enhance expression and activity of a vacuolar H+-pumping pyrophosphatase in pepper plants. Environ Microbiol 21(9):3212–3228
Vitti A, La Monaca E, Sofo A, Scopa A, Cuypers A, Nuzzaci M (2015) Beneficial effects of *Trichoderma harzianum* T-22 in tomato seedlings infected by Cucumber mosaic virus (CMV). BioControl 60(1):135–147
Vorholt JA (2012) Microbial life in the phyllosphere. Nat Rev Microbiol 10(12):828–840
Walitang D, Samaddar S, Choudhury AR, Chatterjee P, Ahmed S, Sa T (2019) Diversity and plant growth-promoting potential of bacterial endophytes in rice in plant growth promoting rhizobacteria (PGPR): prospects for sustainable agriculture. Springer, Singapore, pp 3–17
Walker TS, Bais HP, Grotewold E, Vivanco JM (2003) Root exudation and rhizosphere biology. Plant Physiol 132(1):44–51
Wang H, Chang KF, Hwang SF, Turnbull GD, Howard RJ, Blade SF, Callan NW (2005) Fusarium root rot of coneflower seedlings and integrated control using *Trichoderma* and fungicides. BioControl 50(2):317–329
Wang TT, Ding P, Chen P, Xing K, Bai JL, Wan W et al (2017) Complete genome sequence of endophyte Bacillus flexus KLBMP 4941 reveals its plant growth promotion mechanism and genetic basis for salt tolerance. J Biotechnol 260:38–41
Weir TL, Park SW, Vivanco JM (2004) Biochemical and physiological mechanisms mediated by allelochemicals. Curr Opin Plant Biol 7(4):472–479

Whipps JM (2001) Microbial interactions and biocontrol in the rhizosphere. J Exp Botany 52(suppl_1):487–511

Wintermans PC, Bakker PA, Pieterse CM (2016) Natural genetic variation in *Arabidopsis* for responsiveness to plant growth-promoting rhizobacteria. Plant Mol Biol 90:623–634

Xavier LJ, Boyetchko SM (2004) Arbuscular mycorrhizal fungi in plant disease control. In: Fungal biotechnology in agricultural, food, and environmental applications. CRC Press, Boca Raton, FL, pp 183–194

Yadav AN (2019) Endophytic fungi for plant growth promotion and adaptation under abiotic stress conditions. Acta Sci Agric 3:91–93

Yadav AN, Kour D, Kaur T, Devi R, Gukeria G, Rana KL et al (2020a) Microbial biotechnology for sustainable agriculture: current research and future challenges. In: Rastegari AA, Yadav AN, Yadav N (eds) Trends of microbial biotechnology for sustainable agriculture and biomedicine systems: diversity and functional perspectives. Elsevier, Amsterdam, pp 331–343. https://doi.org/10.1016/B978-0-12-820526-6.00020-8

Yadav AN, Kumar R, Kumar S, Kumar V, Sugitha T, Singh B et al (2017) Beneficial microbiomes: biodiversity and potential biotechnological applications for sustainable agriculture and human health. J Appl Biol Biotechnol 5:45–57

Yadav AN, Mishra S, Kour D, Yadav N, Kumar A (2020b) Agriculturally important fungi for sustainable agriculture, volume 1: perspective for diversity and crop productivity. Springer International Publishing, Cham

Yadav AN, Mishra S, Kour D, Yadav N, Kumar A (2020c) Agriculturally important fungi for sustainable agriculture, volume 2: functional annotation for crop protection. Springer International Publishing, Cham

Yadav AN, Rastegari AA, Yadav N (2020d) Microbiomes of extreme environments: biodiversity and biotechnological applications. CRC Press, Taylor & Francis, Boca Raton

Yadav AN, Rastegari AA, Yadav N, Kour D (2020e) Advances in plant microbiome and sustainable agriculture: diversity and biotechnological applications. Springer, Singapore

Yadav AN, Sharma D, Gulati S, Singh S, Dey R, Pal KK et al (2015) Haloarchaea endowed with phosphorus solubilization attribute implicated in phosphorus cycle. Sci Rep 5:12293

Yadav AN, Singh J, Rastegari AA, Yadav N (2020f) Plant microbiomes for sustainable agriculture. Springer International Publishing, Cham

Yadav AN, Verma P, Kumar S, Kumar V, Kumar M, Singh BP et al (2018) Actinobacteria from Rhizosphere: molecular diversity, distributions and potential biotechnological applications. In: Singh B, Gupta V, Passari A (eds) New and future developments in microbial biotechnology and bioengineering. Elsevier, Los Angeles, CA, pp 13–41. https://doi.org/10.1016/B978-0-444-63994-3.00002-3

Yuan J, Zhou JY, Li X, Dai CC (2016) The primary mechanism of endophytic fungus *Gilmaniella* sp. AL12 promotion of plant growth and sesquiterpenoid accumulation in *Atractylodes lancea.* Plant Cell Tiss Organ Culture 125(3):571–584

Zain M, Yasmin S, Hafeez FY (2019) Isolation and characterization of plant growth promoting antagonistic bacteria from cotton and sugarcane plants for suppression of Phytopathogenic *Fusarium* species. Iran J Biotechnol 17(2):61–70

Zamioudis C, Korteland J, Van Pelt JA, van Hamersveld M, Dombrowski N, Bai Y et al (2015) Rhizobacterial volatiles and photosynthesis-related signals coordinate MYB 72 expression in *Arabidopsis* roots during onset of induced systemic resistance and iron-deficiency responses. Plant J 84(2):309–322

Ziedan ES, Elewa I, Mostafa M, Sahab A (2011) Application of mycorrhizae for controlling root diseases of sesame. J Plant Protect Res 51(4):355–361

Zolla G, Bakker MG, Badri DV, Chaparro JM, Sheflin AM, Manter DK et al (2013) Understanding root–microbiome interactions. Mol Microb Ecol Rhizosphere 1:743–754

Deciphering and Harnessing Plant Microbiomes: Detangling the Patterns and Process—A Clean, Green Road to Sustainable Agriculture

6

Jayakumar Pathma, Ashim Debnath,
Jyoti Bhimgonda Patil, and Laxman Sonawane Bhushan

Abstract

Plant microbiome refers to the diverse microbial counterparts that are associated with plants and plays a crucial role in host biology, ecology, and evolution. Though plant microbiomes have history of co-evolution with the host plants, certain members other than the core-microbiomes are shaped by various factors including plant genotype, plant age, associated host plant tissue or organ, other interacting microbial associates, arthropods, various environmental factors such as soil physio-chemistry, and human inference such as crop domestication, intensive and extensive cultivation, and use of agrochemicals especially in case of agro-ecosystems. Classical knowledge based on microbial culturing techniques and biochemical analysis prejudiced that when a plant interacts with a microbial partner the relationship could be detrimental as with pathogen interaction or promote plant growth in case of symbiotic associations. Advances in molecular techniques such as culture-independent approaches, next-generation sequencing, and high-throughput screening methods helped us to understand the robust nature of plant-associated microbiomes and their crucial role in plant fitness, environmental protection, and human health. This chapter gives a glimpse

J. Pathma (✉)
Department of Entomology, School of Agriculture, Lovely Professional University, Phagwara, Punjab, India

Department of trans-disciplinary research, Division of Research and Development, Lovely Professional University, Phagwara, Punjab, India

A. Debnath
Department of Genetics and Plant Breeding, College of Agriculture, Narendra Deva University of Agriculture and Technology, Kumarganj, Ayodhya, Uttar Pradesh, India

J. B. Patil · L. S. Bhushan
Department of Entomology, School of Agriculture, Lovely Professional University, Phagwara, Punjab, India

A. N. Yadav et al. (eds.), *Current Trends in Microbial Biotechnology for Sustainable Agriculture*, Environmental and Microbial Biotechnology,
https://doi.org/10.1007/978-981-15-6949-4_6

of patterns of plant microbiome associations and their importance in plant health and emphasise the importance of both basic and applied research which will enlighten us with deeper insights on the plant microbiomes. This will help us identify economical, eco-friendly, and effective strategies of manipulating the plant-associated microbiomes which can open up new avenues in maintaining plant health and ecological fitness and sustain crop production in a clean green way preserving the nature's serenity and human health.

Keywords

Agrochemical · Eco-friendly · Microbiome · Plants · Sustainable agriculture

6.1 Introduction

Population growth is predicted to be 9.8 billion in 2050 and reach 11.2 billion by 2100. Shrinkage of land, depletion of non-renewable resources, over utilisation of renewable resources, and climate change due to increased human activities pose a threat to global food security and life subsistence. Though certain technologies such as precision agriculture, use of genetically modified (GM) crops that tolerate stress and give higher yield have been formulated, there are objections for use of GM crops from many parts of the world. Problems such as environmental pollution, pesticide resistance, and pest resurgence that arise due to use of agrochemicals and growing awareness on health benefits of organic farming, etc. had changed preference of consumers to organically grown produce which resulted in the shift of interest of the agrochemical industries to search for better performing microbial inoculants and their products. The compound annual growth rate (CAGR) of global biopesticide and biostimulant market is estimated to be 17% and 10.9%, respectively. Microbial products were accounted to occupy more than 60% of the bioformulations (Arora et al. 2020; Kour et al. 2020f). The predicted estimate of biopesticides market is 7–11 billion $ by 2025 while that of biostimulants is $ 3.12 billion in 2022 (Sessitsch et al. 2018). This had insisted the global investors to invest in plant microbiome research so as to find better commercial alternatives to chemical molecules. Plant microbiomes apart from playing a key role in crop production also have diverse applications in the field of medicine and health care and bioremediation of polluted environment (Rastegari et al. 2020a; Yadav et al. 2020g). Above all, the plant microbiomes are repositories of biodiversity and play an important role in overall fitness of plants, healthy self-sustaining environment, and ecological functions, with wide applications in food and agriculture industry as well as human health care thereby contributing to the Nation's economic growth (Rastegari et al. 2020b).

6.2 Plant Microbiomes

Plant microbiomes are dynamic and are controlled by various biotic and abiotic factors and their interactions. An individual's phenotypic expression is said to be the combined expression of complex interactions between the genomes of the host and their microbial associates (hologenome) and the individual is referred to as 'holobiont' (Bordenstein and Theis 2015). Lynn Margulis in 1991 introduced the term 'holobiont' to refer to the host and its inherited single microbial symbiont and later the term was extended to refer to the microbial community associated with the host (Margulis and Fester 1991; Yadav et al. 2020a). This relationship between plants and microbes is predicted to have prevailed from the time of evolution of terrestrial plants 450 million years ago. The selection pressure from various biotic and abiotic components would have shaped the structure of the holobiont with 'microbe–microbe' interactions playing a significant role in architecturing the microbial community structure associated with the host plant (Singh et al. 2020b; Yadav et al. 2020g). Holobiont research has brought a paradigm shift in the way we see a living organism and the living world and helps us understand how to optimise ecological interactions to reap the benefits of ecosystem services.

The first International Conference on 'Holobionts' was held in Paris in April 2017 (Faure et al. 2018). Plant microbiome comprises of diverse gene pool with respect to its origin and function and may include those from prokaryotes and eukaryotes associated with the host environment (agro-ecosytem, forest ecosystem, etc.) and is majorly influenced by the host plant genotype though many other factors also play a notable role in shaping the associated microbes. Host plant tissue (vegetative parts like root, stem, leaf, and reproductive parts like flowers, fruits, and seeds) harbouring the microbiota plays a significant role in structuring a habitat-specific microbiome and had evidenced variation of associated microbial community within the same plant. The microbiome in turn imparts a significant impact on the respective host plant tissue and the plant as a whole by influencing the physiology and biochemistry of the associated plant tissue.

In addition to host plant, the physiochemical characteristics of soil including soil type, nutritional status, and the interactional effect of root exudates with the soil chemistry in the rhizosphere region co-ordinates the microbiome assemblage especially the 'rhizosphere microbes'. Similarly, the microbial community composition above ground is determined by environmental conditions such as radiation, precipitation, microclimate, physiological conditions, nutrient availability, and phytochemistry of the above ground plant parts (the phyllosphere) which might vary at a large scale between plant species and at a lesser scale within plant parts (Andrews and Harris 2000; Mercier and Lindow 2000; Bednarek and Osbourn 2009). The plant immune system is another important factor sculpturing the associated microbiomes (Yadav 2020; Yadav et al. 2020e). In general, microbes activate plants immune response by jasmonic acid (JA) and ethylene (ET) signalling pathways and plants recognise both pathogenic and beneficial microbe as a non-self, but still how they recruit and assemble beneficial counterparts needs further research (van Wees et al. 2008).

Certain members represent core-microbiomes with long history of association with the host plants since their evolution. Tailoring of host microbiomes may be influenced by ecological and evolutionary factors and are shaped by four major process, viz. dispersal (organisms are moved across space), speciation (new species creation), selection (reveals deterministic fitness among species), and drift (change in species abundance stochastically) (Vellend 2010). 'Hologenome evolution theory' defines 'holobiont' as a unit of evolutionary selection and states that the genomes of the host and that of associated microbiota behaves together as a consortium and tackles environmental changes. The diverse microbial partners co-evolve as well as aid in the survival of holobiont and provide necessary time for the evolvement of the host genome (Zilber-Rosenberg and Rosenberg 2008; Rosenberg and Zilber-Rosenberg 2018).

Plants offer an exclusive habitat to diverse microbes including bacteria, archaea, fungi, oomycete, and viruses (Yadav et al. 2017a, b). The outcome of plant–microbe interaction under the influence of abiotic factors prevailing in the environment might be mutualistic (beneficial to both), commensal (silent- no pain or gain), or pathogenic (detrimental to the host plant) (Rodriguez et al. 2009). Based on their ecological niche, plant microbiomes could be classified as above ground dwellers (those colonising phyllosphere: leaves, stem, flower, fruits, and seeds) below ground dwellers (those which colonise the rhizosphere) (Yadav et al. 2020b). Based on their type of association with the plant tissues, they can be classified as ephiphytes which dwell on the plant surface (phylloplane-leaf surface, cauloplane-stem surface, anthoplane-flower surface, carpoplane-fruit surface, and rhizoplane-root surface) and endophytes that dwell within the plant tissues (Fig. 6.1). Few investigations show that certain rhizosphere microbes enter the plant tissues and establish themselves as endophytes both below and above ground plant parts. However, each plant part harbours distinct microbial community profiles with respect to diversity and

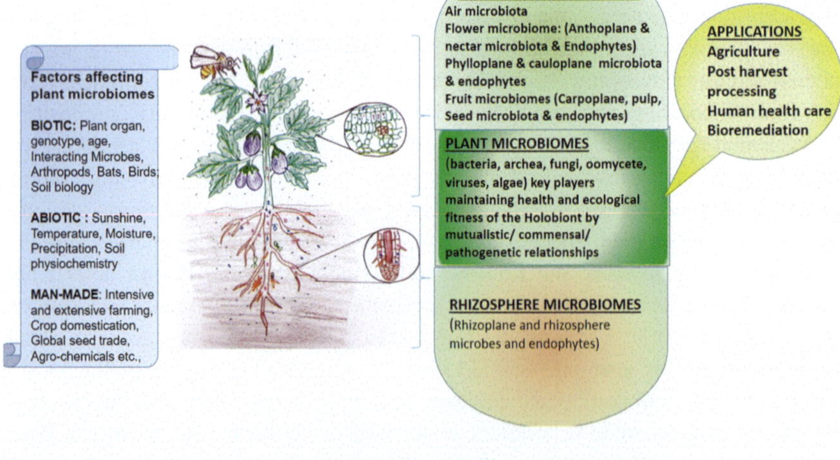

Fig. 6.1 Ecological habitats of plant microbiome and factors shaping them

abundance. Endophytes of roots may differ from those from shoots of the same plant species. This is because few members especially endophytes are transferred vertically from parent to the progeny via seeds or plant parts used for propagation while others get horizontal entry into the plant during its life cycle (Rosenblueth and Martinez-Romero 2006; Ryan et al. 2008; Yadav et al. 2020f). Members of plant microbiomes play an important role in plant growth promotion and biocontrol of diseases and are of great commercial value as biofertilisers and biopesticides (Table 6.1). Additionally, plant-associated microbiomes are important components of multi-tropic interactions and influence a myriad of interactions in the ecology (Mondal et al. 2020; Rai et al. 2020). One such example is induction of volatiles (allelochemicals) that initiate plant–arthropod interactions (Beck and Vannette 2017). The interaction might positively affect the host plant as in case of pollination (Rering et al. 2018) and attraction of natural enemies of insect pests, while in some cases be detrimental to the host plant by attracting herbivorous insects that feed on plants or deter natural enemies of the pest (Pineda et al. 2017; Sugio et al. 2015) or attract herbivores with vector potential to spread phytopathogenic disease (Jimenez-Martınez et al. 2004; Mann et al. 2012). Thus, understanding the complex interactions can help us harness the existing microbiomes so as to favour sustainable crop production.

6.2.1 Rhizosphere Microbiome

Rhizosphere microbes refer to the microbes inhabiting the root surface (rhizoplane) as epiphytes, root tissues internally as endophytes as well as those inhabiting the thin layer of soil adhering to the roots (rhizosphere) which is a continuum of the rhizoplane (Mwajita et al. 2013; Subrahmanyam et al. 2020; Verma et al. 2017). Microbial community structure of rhizosphere is rich both in diversity and abundance as compared to the bulk soil (Reinhold-Hurek et al. 2015). Rhizosphere is a nutritionally rich region attributed by plants root exudates, mucilage factors, and signalling molecules curating microbial associations (Kent and Triplett 2002; Lebeis et al. 2015). Though reduced oxygen availability, high osmotic pressure and extreme variations in moisture, water content, availability of minerals, and certain heavy metals are certain glitches to microbes interacting with rhizosphere, rhizospheric soil is the richest reservoir of immense microbial diversity that plays an important role in plant health as well as support human health by acting as source of novel biomolecules of therapeutic properties supporting human health. Approximately 10^{11} microbial cells with 30,000 different species were reported to inhabit 1 g of root (Berendsen et al. 2012). They include bacteria, fungi, protozoans, and algae with bacterial population dominating other microbes (Saharan and Nehra 2011; Yadav et al. 2018a; Rana et al. 2020d).

Root microbiome plays a key role in plant health and ecological fitness. Numerous factors including plant genotype, age, soil physio-chemistry as well as human interference in terms of crop domestication and fertilisation for yield enhancement shapes root microbiomes (Pathma et al. 2019a). Profiling of the root microbiome of

Table 6.1 Elite microbes from plant microbiomes of commercial value

Microorganisms	Plant/plants species	Beneficial effects	References
Serratia plymuthica HRO-C48	Strawberry	Growth promotion	Berg (2009)
Methylobacterium extorquens	Strawberry	Flavour biosynthesis and enhancement	Zabetakis (1997)
Alcaligenes xylosoxidans, Bacillus pumilus, Pseudomonas marginalis, P. brassicacearum, P. putida, P. oryzihabitans, Rhodococcus spp.	Indian mustard	Improved root growth under heavy metal toxicity	Belimov et al. (2005)
Pseudomonas aeruginosa, Serratia liquefaciens	Broad beans	Bioremediation by associated bean plants grown in oil-polluted sand	Radwan et al. (2005)
Pseudomonas fluorescens 89B-61, *Bacillus amyloliquefaciens* IN937a, *B. subtilis* GB03, *B. pasteurii* C9, *Paenibacillus polymyxa* E681, *S. marcescens* 90–16	*Arabidopsis*	Increased fresh weight (foliar growth)	Ryu et al. (2005)
Bradyrhizobium sp.	Green gram	Improved nodulation along with increased root and shoot growth	Shaharoona et al. (2006)
Serratia marcescens NBRI1213	Betelvine	Increase in shoot and root length and dry weight, biocontrol of *Phytophthora*	Lavania et al. (2006)
Pseudomonas spp.	Wild plants	Increased shoot and root length, total dry weight, and total microbial activity	Ahn et al. (2007)
Azospirillum amazonense	Rice	Increase in number of panicles, grain dry matter and nitrogen at grain maturation stage	Rodrigues et al. (2008)
Acinetobacter spp., *Alcaligenes faecalis, Bacillus cereus, Pseudomonas aeruginosa, Enterobacter hormaechei, Pantoae* spp.	Wheat	Improved plant growth and nutrition under salt stress	Egamberdieva (2008)
Pseudomonas fluorescens G (ACC-5), *P. fluorescens* (ACC-14), *P. putida* A (Q-7)	Pea	Increased root and shoot length, fresh and dry weight, number of leaves per plant and water use efficiency under drought	Zahir et al. (2008)

(continued)

Table 6.1 (continued)

Microorganisms	Plant/plants species	Beneficial effects	References
Azospirillum spp.	Paddy, millets, oilseeds, fruits, sugarcane, banana	Biopromoter	Berg (2009)
Bacillus megaterium M-3, *Pseudomonas putida* BA 8, *Burkholderia gladii* BA7, *Agrobacterium rubi* A16	Radish	Increased seed germination percentage under saline conditions	Kaymak et al. (2009)
Bacillus cepacia OSU-7	Stored potatoes	Biocontrol of *Fusarium*	Recep et al. (2009)
Bacillus subtilis FZB24	Potatoes, vegetables	Plant growth promotion and disease control	Berg (2009)
Bacillus OSU-142, *Bacillus* M-3, *Pseudomonas* BA-8	Strawberry	Increased total sugar, reduced sugar, and total soluble solids.	Pirlak and Kose (2009)
Bradyrhizobium japonicum	Soybean	Nitrogen fixation and plant growth promotion	Berg (2009)
Ampelomyces quisqualis M-10	Apples, grapes, cucurbits, and tomatoes	Biocontrol of powdery mildew	Berg (2009)
Bacillus amyloliquefaciens IARI-HHS2-30	Wheat	Plant growth promoting and mitigation of cold stress	Verma et al. (2015b)
Halolamina pelagica CDK2	*Suaeda nudiflora*	P-solubilisation	Gaba et al. (2017)
Acinetobacter guillouiae EUB2RT.R1	Wheat	Plant growth promotion and nutrient uptake	Rana et al. (2020c)
Pseudomonas libanensis EU-LWNA-33	Wheat, maize, rice, sorghum, and finger millet	Alleviation of drought stress and plant growth promotion	Kour et al. (2020d)
Streptomyces laurentii EU-LWT$_3$-69, *Penicillium* sp. EU-DSF-10	*Amaranthus*, buck wheat, millets, and maize	Microbe-mediated alleviation of drought stress and acquisition of phosphorus in great millet (*Sorghum bicolor* L.)	Kour et al. (2020c)
Acinetobacter calcoaceticus EU- LRNA-72 *Penicillium* sp. EU-FTF-6	Wheat, maize, foxtail millet, and finger millet	Amelioration of drought stress in Foxtail millet (Setaria italica L.)	Kour et al. (2020e)

Arabidopsis by culture-independent technique documented a consistency in the composition of associated microbiomes among different lineages of *Arabidopsis* as well as among those grown in diverse soils across different continents evidencing evolutionary conservation of the associated microbiomes (Beattie 2015). Rhizosphere microbiomes of tomato varied between cultivars and transfer of a

Flavobacterium TRM1 from the rhizosphere of tomato variety Hawaii 7996 resistant to wilt pathogen *Ralstonia solanacearum*, to a susceptible variety Moneymaker conferred resistance in Moneymarker evidencing the use of the beneficial bacteria as probiotics for plants to maintain plant health (Kwak et al. 2018). Until last few decades, rhizosphere microbiome lured the attention of microbiologists due to their potent role in plant growth promotion and biocontrol of phytopathogens. Rhizosphere microbes include plant growth-promoting rhizobacteria (PGPR) belonging to various genera such as *Pseudomonas, Bacillus, Microbacterium, Azotobacter, Azospirillum, Rhizobium,* and *Streptomyces* as well as numerous beneficial fungi (*Trichoderma*) and mycorrhiza (*Globus, Gigaspora,* etc.) (Kent and Triplett 2002; Mwajita et al. 2013; Pathma et al. 2019b; Kour et al. 2019).

PGPR promote plant growth directly by aiding nutrient availability for host plant (mechanisms include fixing atmospheric nitrogen, solubilising and mobilising phosphorous, potassium, silica, and other essential mineral nutrients) and phytohormone production (IAA, cytokinins, ACC deaminase) (Kour et al. 2020a, b; Singh et al. 2020c).

PGPR indirectly support plant growth by protecting host plants from phytopathogens (mechanisms include competition, parasitism, production of antibiotics, siderophores, hydrogen cyanide, fungal cell wall degrading enzymes, and induction of systemic resistance in plants) (Kent and Triplett 2002; Pathma et al. 2010b, 2011, 2019a, b, Mwajita et al. 2013). Bioformulations of *Azospirillum, Azotobacter,* and *Rhizobium* have been used as biofertiliser for decades (Mohanram and Kumar 2019; Kour et al. 2020f). Similarly, numerous species of soil dwelling *Trichoderma* especially *T. harzianum* and *T. viride* which are highly interactive with the plants roots have been reported to play an active role in protecting plants against a wide range of soil and seed-borne fungal pathogens, viz. *Fusarium, Phytophthora, Sclerotium, Rhizoctonia, Ustilago, Sclerotinia,* and *Pythium* causing diseases on various species of agriculturally important plants (Rana et al. 2020a, b, c; Yadav et al. 2018b).

Numerous biocontrol mechanisms including mycoparasitism, competition, production of fungal cell wall degrading enzymes, antibiotics, and siderophores have been reported (Singh et al. 2014). *Trichoderma* was also reported to enhance nutrient uptake and produce hormones involved in phytostimulation (Fiorentino et al. 2018; Kaur et al. 2020). Harman et al. (2019) documented that endophytic *Trichoderma* has the potential to increase the photosynthetic ability of the plant. Thus, *Trichoderma* also received greater attention and commercial value as biopesticide. Similarly, *Pseudomonas* spp. and *Bacillus* spp. in the rhizosphere region due to their potent plant growth promotion and biocontrol properties have also been commercialised for use in eco-friendly crop production (Van Peer and Schippers 1988; Vessey 2003; Araujo et al. 2005; Pathma et al. 2010a, 2019b; Sharma et al. 2019).

Apart from protecting the plant from phytopathogens, certain rhizobacterial species have evidenced to deter pest infestation by inducing production of secondary metabolites or defensive compounds in plants that deters insect herbivores. For instance, aphid infestation on barley was impacted by plant biochemistry induced by associated rhizobacteria (Tetard-Jones et al. 2012; Pieterse and Dicke 2007).

Badri et al. (2013) showed that the rhizosphere soil microbiomes apart from impacting the plant growth patterns also influenced the plant phytochemistry especially the leaf metabolome which deterred the feeding preference of the insect herbivore *Trichoplusia ni* on *Arabidopsis*. Investigations showed that *Pseudomonas fluorescens* WCS417r-induced resistance in *Arabidopsis* and reduced the performance of generalist feeder *Spodoptera exigua* but unaffected specialist feeder, *Pieris rapae* (Van Oosten et al. 2008). Cosme et al. (2016) showed that a rice root endophytic fungi *Piriformospora indica* induced GA signalling in plants making it tolerant to rice root feeding weevil *Lissorhoptrus oryzophilus*.

6.2.2 Phyllosphere Microbiomes

Phyllosphere in a broader sense includes all the above ground plant parts, viz. leaves, stems, buds, flowers, fruits, and seeds offering a diverse habitat for microbes. Phyllosphere is claimed to be a largest habitat for microbes next to soil as leaf surface area of terrestrial plants alone exceeds 6.4×10^8 km^2 worldwide with a rough estimate to support bacterial population of around 10^{26} cells (Lindow and Brandl 2003; Vorholt 2012). The phyllosphere is an ephemeral environment in contrast to rhizosphere, and hence microbe-inhabiting phyllosphere should be able to adapt the phyllosphere by establishing itself in newly expanding niches. Based on the area of colonisation, microbes can be classified as epiphytes colonising surface of aerial plant parts (phylloplane) and endophytes living within tissues of the aerial plant parts (endosphere) (Singh et al. 2020a). The term phyllosphere was initially coined by a plant pathologist Last in 1955 who worked on the effect of seasons on the relative abundance of *Sporobolomyces* sp. (pink yeast) and *Erysiphe graminis* (powdery mildew pathogen) on cereals during which he enumerated and documented the characteristic microflora inhabiting the phyllosphere. As compared to rhizosphere, phyllosphere especially leaves are less rich in nutrients as well as subjected to environmental pressure including radiation and extremities of temperature and moisture which significantly impacts the associated phyllosphere microbiomes (Kumar et al. 2019a, b). The phyllosphere microbiomes included bacteria, archaea, fungi, oomycetes, viruses, and nematodes (Koskella 2013; Vorholt 2012; Lindow and Brandl 2003) with bacteria exceeding other groups in numbers and diversity (Andrews and Harris 2000).

6.2.2.1 Leaf and Stem Microbiomes

Phyllosphere microenvironment comprises of leaf surface and internal leaf tissues as they are connected by natural openings, viz. stomata and hydathodes which pave entry sites for associated microbes (Morris 2002). Doan and Leveau (2015) described that phyllosphere includes the 'phylloplane' (leaf surface landscape) and 'phyllotelma' (leaf surface waterscape). Phyllosphere microbial communities (PMCs) attracted the interest of microbiologists and pathologists in 1950s when they realised the economic importance of the microbes inhabiting this micro-habitat. Broadly, phyllosphere microbes were classified as residential microbial community (native to healthy plant leaves) and casual community (introduced accidently) and foliar pathogens were considered as a part of this complex community (Leben 1965). Resident communities were hypothesised to produce antagonistic compounds that directly hinder the growth of phytopathogens or indirectly hamper the phytopathogens by competing for nutrient, alteration of plant physiology, etc. (Last and Deighton 1965; Leben 1965). The leaf inhabiting microbiomes apart from establishing in newly formed leaves should also adapt themselves to diurnal cycles of light and plant metabolism as well as withstand UV-radiations to which they are exposed constantly.

Apart from fighting leaf morphological defences like the waxy cuticle, etc., they also need to withstand the plant defences triggered by the entry of these ephiphytes into plant apoplast for their successful establishment as endophytes (Rana et al. 2019; Yadav et al. 2020d). Mechaber et al. (1996) showed that environmental variation (morphology and microclimate) across the leaves can affect the distribution pattern of phyllosphere microbes. Bacteria belonging to families Pseudomonadaceae, Enterobacteriaceae, and Microbacteriaceae are found to be common inhabitants of the microbial communities of leaf surfaces irrespective of the plant species across time and space (Ercolani 1991; Thompson et al. 1993; Yang et al. 2001; Lindow and Brandl 2003; Krimm et al. 2005; Ostman et al. 2010). Phyllosphere supports bacterial abundance with an average number from 10^6–10^7 cells/cm^2 leaf area (Lindow and Brandl 2003; Leveau 2006). Certain phyllosphere microbes are detrimental foliar phytopathogens while a few have beneficial effects on plant growth which depicts their overall effect on plant growth and ecological fitness (Vacher et al. 2016). *Beijerinckia* and *Azotobacter* inhabiting the phyllosphere were reported to fix atmospheric nitrogen which was mobilised by rain resulting in plant growth in tropical and temperate plant species (Ruinen 1965; Jones 1970). Presence of diazotropic nitrogen-fixing proteobacteria (*Klebsiella*) and Cyanobacteria (*Scytonema*, *Nostoc*, and *Stigonema*) were reported from the phyllosphere which were beneficial to host plants (Malyan et al. 2020). Rico et al. (2014) reported that apart from provision of nitrogen to plants the presence of these N-fixing bacteria increased the drought tolerance and environmental adaptability of host plants.

Advent of meta-omics and advancement in techniques of fingerprinting as well as big data analysis helped us realise the complex nature of phyllosphere microbes with respect to taxonomic and functional diversity (Jumpponen and Jones 2009; Delmotte et al. 2009; Vacher et al. 2016). Leaf morphology, viz. cuticular layer, wax layer, specialised cells, trichomes, stomatal openings, hydathodes, and leaf physiology including respiration, photosynthesis, and evapotranspiration modulates the O_2 and CO_2 fluxes, water vapour, production of ROS (in response to PAMPs), etc. and plays an important role in microbial colonisation (Torres 2010). By-products of plant cell growth, namely methanol which is released through stomata has been documented as a carbon source for growth of some phyllosphere microbes (Vacher et al. 2016). Iguchi et al. (2015) reported the presence of beneficial *Methylobacterium* spp. in abundance on the phyllosphere of rice plants which induced plant growth by N_2 fixation and production of phytohormones such as ACC deaminase in addition to induction of systemic resistance in plants against phytopathogens (Madhaiyan et al.

2004; Maliti et al. 2005; Abanda-Nkpwatt et al. 2006; De Costa et al. 2008; Chinnadurai et al. 2009).

Apart from plant growth-promoting activity, certain phyllosphere microbes are found to degrade organic pollutants and can be commercially exploited for bioremediation (Van Aken et al. 2004; Ilori et al. 2006; Wei et al. 2017; Sharaff et al. 2020). Phyllosphere bacteria induced plant defence signalling which not only evaded foliar phytopathogens but also influenced insect herbivory (Humphrey et al. 2014). *Bacillus amyloliquefaciens*, an endophyte reported from vanilla was found to possess potent biocontrol and plant growth-promoting properties and played an important role in the health of host plant (White et al. 2014). Studies showed that epiphytic nitrogen-fixing bacteria *Herbaspirillum seropedicae* could penetrate the leaf tissues of pineapple through the stomata and establish themselves as endophytes aiding plant nutrition (Baldotto et al. 2011). Studies on spray application of a diazotroph, *Azospirillum brasilense* on wheat and maize plants evidenced their colonisation and endophytic growth into leaf tissues via stomata though they failed to establish as epiphytes (Fukami et al. 2016).

Among phyllosphere microbes, many studies focussed on an epiphytic phytopathogenic bacterium *Pseudomonas syringae* documented across many plant species (Mansfield et al. 2012). It can grow endophytically into plant tissues and is also prevalent in waterbodies, snow, rain, and clouds. Morris et al. (2014) stated that this cosmopolitan distribution is the nature of phyllosphere ephiphytes including *P. syringae* which are swept and absorbed into atmosphere and catalyse ice and cloud formation and are involved in rainfall induction. Hirano and Upper (2000) documented the presence of conserved genes '*ice*' or '*ina*' that leads to ice formation and this was claimed to be a nutrient access mechanism by inducing frost damage in phyllosphere (Morris et al. 2013).

P. syringae was used as a model to study microbial interactions in phyllosphere (Melotto et al. 2008; Innerebner et al. 2011) and to decipher microbial adaptations to epiphytic regime (Burch et al. 2014). Leaf morphology (cuticular wax, trichomes), physiology, and biochemistry (nutrient composition and metabolites) vary among plant species and genotypes, and these are certain factors which aid the selection of associated phyllosphere microbiomes (Inacio et al. 2010; Kembel and Mueller 2014; Kembel et al. 2014; Hunter et al. 2015; Mason et al. 2015). Also leaf phosphorous and aluminium content influenced associated bacterial and fungal communities, respectively (Kembel et al. 2014; Kembel and Mueller 2014). Also leaf microclimate including temperature (Bernard et al. 2013) and water content (Morris 2002; Yadav et al. 2005) influenced the phyllosphere microbes by modifying the pH and water availability.

Among prokaryotes, bacteria, viz. *Methylobacterium, Pseudomonas, Sphingomonas, Azotobacter, Beijerinckia,* and *Klebsiella*; cyanobacteria, viz. *Nostoc, Scytonema,* and *Stigonema* (Delmotte et al. 2009; Kembel et al. 2014; Vacher et al. 2016) were found in abundance. Among eukaryotes, fungi, viz. *Cladosporium, Aureobasidium,* and *Taphrina* belonging to Ascomycota (Coince et al. 2014; Kembel and Mueller 2014) and yeasts genera *Sporobolomyces* and *Cryptococcus* belonging to Basidiomycota (Cordier et al. 2012; Ottesen et al. 2013) were common phyllosphere microbes. The community composition was dynamic and season dependent (Penuelas et al. 2012). Density of yeast was as high as 500 cells/cm^2 with minor variation among the plant species (Inacio et al. 2010). *Aureobasidium pullulans* was found to have antagonistic potential against phytopathogens (Cordier et al. 2012). Rodriguez et al. (2009) reported that certain endophytic fungi lives inside leaf tissues without causing noticeable symptoms and has an epiphytic stage at times of horizontal transmission between mature plants prior to entry into leaf tissues. These endophytic fungi benefit plant growth by inducing resistance in plants against biotic (pathogens, herbivores) and abiotic (extreme temperature, drought and salinity) stress (Rodriguez et al. 2009; Porras-Alfaro and Bayman 2011; Kennedy et al. 2020). Certain phyllosphere dwelling endophytic fungi acts as latent pathogens as in case of *Diplodia mutila* which is triggered by strong light to generate ROS which is detrimental to the host, the tropical palm tree *Iriartea deltoidea* at the seedling stage (Alvarez-Loayza et al. 2011). Certain phyllosphere fungi also acts as phytopathogens and includes *Erysiphe, Podosphaera, Phyllactinia, Blumeria, Colletotrichum, Helminthosporium, Puccinia, Septoria, Pseudocercosporella, Pyrenopeziza, Leptosphaeria, Botrytis*, etc. (McCartney and Fitt 1998; Glawe 2008).

Leaves of raw eaten green vegetables and herbs, viz. *Lepidium sativum, Cichorium endivia*, and *Thymus vulgaris* were reported to harbour beneficial bacteria, viz. *Bifidobacterium, Lactobacillus,* and *Streptococcus* in a good concentration (10^{-5} CFU per gram of plant tissue) which are probiotic in nature and cause health benefits to humans comparable to that caused by administering probiotic supplements. Additionally, bacteria belonging to genera, viz. *Enterococcus, Pediococcus, Leuconostoc, Bacillus, Propionibacterium, Akkermansia, Staphylococcus,* and *Clostridium* were also reported from phyllosphere of different plant species utilised as herbs and eaten raw (Patz et al. 2019).

Culture-dependent and metagenomic approach to study the impact of nitrogen fertilisation on phyllosphere microbiomes of spinach and rocket showed the presence of genera *Bacillus, Pseudomonas, Phyllobacterium, Exiguobacterium, Pantoea,* and unclassified Enterobacteriaceae in both the host plants; however, it evidenced variation in species diversity and their population percentage (Kumar et al. 2019a). Additionally, the study evidenced that nitrogen fertilisation impacted the microbial population in a characteristic pattern in each hosts. Core fungal microbiome included members of Ascomycota and Basidiomycota in both the crops; however, the dominance of specific members varied. This study also revealed that the microbial diversity was influenced by plant species, leaf nutrition, and time (Darlison et al. 2019). Studies on bacterial community composition of leaf and petals of plants, viz. *Lotus corniculatus* and *Saponaria officinalis* showed the presence of similar taxa in the leaves of the plants tested and similar results were obtained with the leaf microbiomes.

However, there was a considerable variation among the plant parts of the same species indicating the habitat specificity of microbes. Flowers had less bacterial diversity as compared to leaves (Junker et al. 2011). Studies involving comparison on sugarcane microbiomes in different plant parts showed a variation in species diversity and their relative abundance. de Souza et al. (2016) showed that Rhizobiales dominated rhizosphere while Saprospirales and Rhodospirillales intensively colonised roots and young shoots, respectively, as endophytes. Enterobacteriales colonised sugarcane stalks and leaves both as endophytes and epiphytes at a higher extent as compared to roots, young shoots and soil and sugarcane stalks had abundance of endophytic Pseudomonadales. Among the fungal communities, Polyporales were enriched in roots, bulk soil, and young shoots while stalks and leaves were dominated by Capnodiales. Saccharomycetales intensively colonised stalks both as endophytes and epiphytes. Armanhi et al. (2018) reported that members of Moraxellaceae, Pseudomonadaceae, and Enterobacteriaceae occurred as endophytes of sugarcane stalks and leaves, while Sphingomonadaceae and Rhizobiaceae occurred in leaves. Additionally, Rhizobiaceae also occurred as epiphytes in stalk and that many of these members were reported to possess plant growth-promoting activity.

6.2.2.2 Floral Microbiomes

Flowers are important reproductive structures, and their microbial inhabitants play an important role in the reproductive success in terms of yield, food safety as well as conservation of the plant species (Aleklett et al. 2014). Floral (anthosphere) microbiomes were found to have certain distinct members which were not present in vegetative parts (Ottesen et al. 2013). Floral structures, viz. sepals, petals, ovaries, stigma, style, nectaries, stamens, and pollens differ considerably in anatomy, physiology, and biochemistry and provide a unique habitat for the microbial cells (Junker et al. 2011). For instance, the petals lack lignin which increases its colonisation by fungal endophytes as compared to the sepals which are rich in lignin (Ngugi and Scherm 2006). Also the sepals contain trichomes and few oil glands while petals are rich in conical cells and bright pigments which affect the surface temperature and wettability and in turn select microbes preferring the microclimate. Similarly, the physio-chemistry of the stigma plays an important role recruiting the microbes.

Stigma has been identified as a primary site of entry for the pathogen *Erwinia amylovora* causing fire blight. Stigma–style pathway is considered as one of the routes exploited by fungus to colonise the ovary (Ngugi and Scherm 2006; Aleklett et al. 2014). Analyses of epiphytic bacterial community of apple blossoms showed the presence of microbial agonists, viz. *Pantoea agglomerans*, *Pseudomonas* spp. and *Cryptococcus* spp. which deterred the growth of *Erwinia* (Pusey et al. 2009). Few investigations have documented certain novel species of yeasts, *Wickerhamomyces* and *Candida* in flowers (Groenewald et al. 2011; Jindamorakot et al. 2008; Rosa et al. 2007). Various studies documented that fungal communities belonging to Ascomycota dominated the floral microbiome, and this was followed by bacterial members. Very few studies represented archaea and viruses as a part of floral microbiomes. *Metschnikowia, Cryptococcus, Pseudomonas,* and *Acinetobacter* were found to be common inhabitants of anthosphere over a large geographical range. Ottesen et al. (2013) documented that an unidentified fungi was the most prevalent member in tomato flowers by 18S rRNA sequences analyses. Analyses of 16S rRNA sequence of apple floral microbiota showed the dominance of

Deinococcus-Thermus and TM7-affiliated bacteria (Shade et al. 2013). Comparative analyses of microbiomes of leaves and petals of *Lotus corniculatus* and *Saponaria officinalis* evidenced that microbial community composition has higher degree of habitat specificity as compared to host plant genotype specificity and that flowers relatively harboured less bacterial diversity. The floral scents emitted by the flower volatiles played a critical role in tailoring the associated microbes. This reflects the natural defence mechanisms of avoiding microbes that are pathogenic to plants and affect their reproductive ability (Singh and Yadav 2020). Thus, floral scents apart from playing a role in attraction of pollinators also play an equally important role in deterring factors detrimental to plant health and fitness (Junker et al. 2011). Several bee species have been documented to play an important role in transferring naturally occurring antagonistic bacteria that evade pathogens across plants and thereby shape up establishment of bacterial communities in flowers (Johnson et al. 1993; Maccagnani et al. 2009).

Nectar Microbiome

Nectar biochemistry is strongly impacted by nectar microbiomes. Nectar microbiomes are influenced by host plants genetics, intra-specific variation of the floral traits, microbial members of the community, their order of succession, herbivory, and other environmental conditions (Tucker and Fukami 2014; Samuni-Blank et al. 2014). The role of nectar microbiomes in plant-pollinator mutualism and plant fitness is inevitable. Sandhu and Waraich (1985) reported that the nectar microflora are transferred among flowers and between other plant parts by means of nectar consumers such as insects, birds, and bats. As compared to other plant parts, floral nectar is reported to nurture relatively less microbes comprising of bacteria, yeasts, and fungi that can tolerate high osmotic pressure, toxic secondary metabolites, lytic enzymes, H_2O_2, and ROS. The bacterial and yeast population inhabiting the nectar is estimated to be approximately $>10^7$ cells/mm^3 and $>10^5$ cells/mm^3, respectively (de Vega et al. 2009; Herrera et al. 2009; Fridman et al. 2012). Along with the host plant, nectar microbes play an important role in composing the nectar biochemistry, viz. composition of various sugars, volatile compounds, etc. as well as the volume of nectar produced which alters the foraging behaviour of the pollinators, thereby impacting seed setting and plant health (Vannette and Fukami 2018; Alvarez-Perez et al. 2019).

Lenaerts et al. (2017) showed that bacteria altered the nectar chemistry which affected the life history of a generalist parasitoid of aphid. It has been hypothesised that nectar biochemistry as influenced by the inhabiting microbes also plays an important role in recruiting specialist pollinators and repelling nectar robbers thereby altering the pollination behaviour (Gonzalez-Teuber and Heil 2009; Richardson et al. 2016). Good et al. (2014) showed that honey bee, *Apis mellifera* avoided floral nectar inhabited by bacteria *Lactobacillus kunkeei*, *Asaia astilbes* and *Erwinia tasmaniensis* but the preference was unaffected by the presence of yeast *Metschnikowia reukaufii*. The preference was not based on the presence or absence of microbe but was depended on the secreted microbial metabolite which alters the floral scent and acts as chemical cues. Rering et al. (2018) evidenced that *M.*

reukaufii produced distinct compounds that were more attractive to bees. He also reported that fungi present in floral nectar microbiomes were known to produce volatiles 2-ethyl-1-hexanol and 2-nonanone which attracted bees thereby improving pollination. Profiling the nectar microbiomes of different plant species of the Mediterranean region showed the frequent occurrence of certain bacteria–yeast associations such as *Leuconostoc* sp. with *M. reukaufii*, *Acinetobacter* spp. with *Metschnikowia gruessii* and *M. reukaufii* (Alvarez-Perez and Herrera 2013).

Structuring of microbiomes of nectar followed specific patterns. Certain hypothesis assumes floral nectar to be sterile and that microbial communities are introduced initially by the visitors recruited by the nectar chemistry. Studies on nectar microbiome of dioecious shrub *Eurya emarginata* showed a variation in the microbial composition of male and female flowers. It also evidenced that reduced visit of the pollinators in male flower reduced the occurrence of yeasts which in turn increased the bacterial abundance (Tsuji and Fukami 2018). Thus, the microbial community composition is affected by the order of microbial colonisation termed as 'priority effects'. Studies showed that prior introduction of bacterium *Neokomagataea* sp. in monkey-flower (*Diplacus aurantiacus*) nectar caused its domination across multiple generations and even eliminated a common yeast inhabitant *M. reukaufii* (Toju et al. 2018). Alvarez-Perez et al. (2019) documented that opportunistic bacteria and yeasts inhabit the floral nectar and the yeast–bacterium interactions influence the mutualistic interactions between host plants and their floral visitors. Interaction mechanisms include cross talk signalling, physical complex formation, nutrition exchange, antibiosis, and horizontal gene transfer. Certain nectar dwelling species of yeasts such as *Metschnikowia* and bacteria of genera *Pseudomonas* and *Pantoea* are known to produce antimicrobial compounds that inhibit the growth of plant pathogens (Parret and De Mot 2002; Duffy et al. 2006; Pusey et al. 2009; Walterson and Stavrinides 2015). *Acinetobacter baylyi* reported from floral nectar is reported to aid in horizontal gene transfer by killing and extracting genes from the co-existing species (Fridman et al. 2012; Cooper et al. 2017). Fridman et al. (2012) reported that members of Gammaproteobacterium, viz. *Acinetobacter* spp., *Erwinia* spp., *Pantoea* spp., and *Pseudomonas* spp. dominated nectar microbiomes of three different plants, viz. *Amygdalus communis*, *Citrus paradise,* and *Nicotiana glauca*. The microbiomes also included members of other species, namely *Bacillus* spp., *Paenibacillus* spp., *Staphylococcus* spp., *Chryseobacterium* sp., *Arthrobacter* spp., *Curtobacterium* spp., *Kocuria* spp., *Asaia* sp., and *Bartonella* sp. at comparatively lower frequencies. Thus, microbial members of the nectar produce a variety of semiochemicals including volatiles, metabolites, and quorum-sensing molecules that inhibits, repels, or attracts other microbial species, arthropods, and bird species inhabiting the ecosystem. Certain compounds of microbial origin such as Farnesol, an important quorum-sensing molecule found in various fungal species is also a constituent of insect pheromone which when present in the floral nectar triggers a behavioural responses in the insect attracting it towards the flowers (Sobhy et al. 2018; Rering et al. 2018). Better understanding of the factors and mechanisms orchestrating the nectar microbiomes and their effects on plant-pollinator mutualisms as well as plant–pest interactions can help us forecast the future as well as tailor microbiomes to improve pollination and plant health.

6.2.2.3 Fruit Microbiomes

Microbes inhabit the fruits both externally on the surface (carpoplane) as epiphytes and internally as endophytes as they do with other plant parts. Ottesen et al. (2013) made a comparative study on the microbiomes harboured in different tissues of tomato by sequencing 18S and 16SrRNA moieties and shotgun metagenomics which evidenced that the microbial communities present in each plant organ was distinct; however, the flowers and fruits shared few common members of *Brachybacterium, Chryseomonas, Microvirga, Microbacterium, Microbacteriaceae, Microvirga, Microbacteriaceae, Paracocccus, Rhizobium,* and *Sphingomonas* which were not found in any other plant tissue. Research shows that these microbiomes play an important role in the fruit physiology and health, and their role becomes essential in fruits after harvest. Setati et al. (2012) documented the presence of antagonistic yeast-like fungi *Aureobasidium pullulans* previously reported from phyllosphere on fruit surfaces which could be used as potential biocontrol agent. Studies on the fungal microbiome of apple fruits managed organically and conventionally showed that members of Ascomycota were dominant followed by Basidiomycota, Chytridiomycota, and unidentified fungi. Organic apples had members of Ascomycota, viz. *Phaeoramularia, Phaeosphaeria, Stagonospora,* and unidentified Mycosphaerellaceae in abundance while members of Basidiomycota, viz. *Cystofilobasidium, Leucosporidiella,* and *Guehomyces*, were dominant in conventionally grown apples (Abdelfattah et al. 2016). Wassermann et al. (2019) in an attempt to study microbiomes associated with apple fruit by employing 16S rRNA gene analyses, q-PCR, and (FISH-CLSM) identified that each part of the fruit, viz. peel, fruit pulp, seeds, calyx, and stem harboured distinct microbiomes among which fruit pulp and seed showed more bacterial colonisation while the peel has less.

Bacterial members belonging to Proteobacteria dominated followed by Bacteroidetes, Actinobacteria, and Firmicutes with differences in bacteria genera, viz. *Ralstonia*, *Sphingomonas*, *Pseudomonas*, *Massilia*, *Methylobacterium*, *Burkholderia*, *Pantoea,* and *Hymenobacter* among the fruit tissues. The investigation also showed that apples harboured bacterial members supporting human health and that the bacterial community composition varied between conventionally and organically grown apples. Organic apples had abundance of *Methylobacterium*, *Spirosoma*, *Hymenobacter,* and *Zymomonas* while conventional apples had *Burkholderia*, *Acinetobacter*, *Erwinia,* and *Pantoea* in abundance. Thus, knowledge on fruit microbiomes will help us to understand the beneficial microbes, and the role played by them in avoiding post-harvest damages including physiological disorders and microbial contaminations by food-borne pathogens. This will enable us to design microbial consortia for biological control of post-harvest diseases as well as enhance the quality, palatability, and preserve the nutritional value of the fruits and vegetables after harvest (Droby and Wisniewski 2018).

Seed Microbiomes

Seeds play the most important role to initiate the life cycle, reproduce the species, and also facilitate dispersal, adaptation, and persistence of the plant species in new environments (Fenner and Thompson 2005). Seed-to-seedling transition is one of the most important bottlenecks in a plant's life cycle as various biotic (soil and seed-borne pathogens, granivores) and abiotic stress (drought, nutrient deficiencies, heavy metal toxicities, soil temperature, etc.) affects seed germination and seedling growth (Leck et al. 2008; Bever et al. 2015). Microbes with a co-evolutionary history with plants interact with seeds at all stages leading to the development of microbiome which will be essential for the overall development and performance of plants (Hardoim et al. 2015). Symbiotic microbial communities (primarily bacteria and fungi) are present in seeds which are essential for nutrient acquisition in seedlings, modulation of plant development, and defence from pathogen. Microbes occur in various parts of seeds including embryonic tissues, endosperm, and seed coat (Suman et al. 2016; Yadav et al. 2020c). Ephiphytes found on the seed surface (on seed coat) are transmitted vertically and horizontally while endophytic microbes colonising the internal seed tissues (embryo and endosperm) are transmitted vertically to the progeny. At times, the ephiphytes also enter into endophytic life style and vice versa (Rodriguez et al. 2009; Barret et al. 2016). Though many studies document endophytes as mutualistic or silent partners of host plant, there are exceptions where endophytes (bacteria, fungi, virus) act as pathogens to the host plants emphasising that mutualism and pathogenicity are not strictly inherent microbial properties but are environment-specific expressions (Hume et al. 2016). Certain studies reported that endophytic microbes dwelling in seeds are distinct from those present in the soil substrate proposing that they might be recruited from the parent plant while few others report that rhizosphere microbes can colonise the plants systematically and enter the flowers and contribute partially to the seed microbiomes (Compant et al. 2008, 2010, 2011).

Johnston-Monje and Raizada (2013) showed that few bacterial endophytes harboured in the seeds are capable of colonising the seedlings systemically as the seeds germinate and grow thereby expanding their territory into different plant organs as well as exit into rhizosphere via roots. In general, eukaryotes belonging to Ascomycete and Basidiomycete and prokaryotic bacteria of the phyla Proteobacteria, Actinobacteria, Firmicutes, and Bacteroidetes were commonly found associated with seed tissues (Barret et al. 2015; Johnston-Monje et al. 2016). However, the seed inhabiting microbial species and microbial community composition varied with plant species (Links et al. 2014), genotype (Barret et al. 2015), seed developmental stage (Liu et al. 2013), eco-geographical location (Klaedtke et al. 2016), and counteracting phytopathogens (Rezki et al. 2016). Thus, seed microbiomes are outcomes of numerous interlinked factors and processes (Frank et al. 2017).

Johnston-Monje and Raizada (2011) and Liu et al. (2012) documented that seeds of different maize genotypes including hybrids as well its wild ancestor teosinte harboured similar bacterial genera evidencing the long-term conservation of seed endophytic microbial community. These experiments proved that the seed bacterial community of maize was unaffected by ecology, evolutionary boundary, and ethnography but was determined by the host genetics. Seeds of *Brassica* plants were dominated by Ascomycetes (class: Dothideomycetes, Eurotiomycetes, Leotiomycetes, Sordariomycetes) and Basidiomycetes (class: Tremellomycetes) (Barret et al. 2015). Bacteria belonging to genera *Pseudomonas, Bacillus, Paenibacillus, Acinetobacter, Pantoea, Micrococcus,* and *Staphylococcus* were found to be the common inhabitants of seeds as endophytes of which many of them were beneficial to host plant (Truyens et al. 2015). Many studies showed that epiphytic and endophytic bacterial communities of seeds were similar in diversity. However, this was not true in case of seed inhabiting fungal communities where the ephiphytes were dominated by phytopathogenic species of *Fusarium, Alternaria, Leptosphaeria, Phoma,* and *Pyrenophora*. This was in line with research investigations involving *Centaurea stoebe* and its ancestral relative *Centaurea jacea* where majority of the seed endophytes were found to be pathogenic in nature while its roots harboured non-pathogenic endophytes. This is because of the inability of pathogenic seed endophytes of *Centaurea* to systemically colonise the plants which evaded the risk of the seeds acting as vectors of these pathogens (Geisen et al. 2017).

Microbes inhabiting seeds have been evidenced to provide services essential for seed germination and survival which includes breaking of seed dormancy by cytokinin production; induction of seed germination and growth by phytohormone production and provision of macro- and micronutrients (by N_2 fixation, P, K, S solubilisation); protecting seeds from abiotic stress such as heavy metal toxicity and from biotic stress caused by phytopathogens and pests (Thakur et al. 2020; Tiwari et al. 2020). Evidence of bacterial occurrence as endophytes on surface sterilised seeds of different plant species, namely paddy, maize, barley, annual ryegrass, common bean, pumpkin, grapevine, alfalfa, coffee, tobacco, quinoa, giant cardon cactus, several eucalyptus species, Norway spruce, etc. have been reported and reviewed (Frank et al. 2017). Seed inhabiting fungal endophytes *Epichloe* sp. and *Neotyphodium* sp. associated with poaceae plants was evidenced to play a significant role in protection of plants from phytopathogenic infections (Perez et al. 2016).

Thus, knowledge on seed microbiome and their role can help us understand and nurture them in a way to benefit crop production. However, globalisation and global seed trade which had centralised seed production and distribution had posed a potential risk to seed microbiomes as it might lead to loss in heterogeneous microbiomes as structured by the plant genotype and environmental interactions. This will slowly favour establishment of highly homogenous plant microbiomes posing danger to some key microbial players which might be endangered (Berg and Raaijmakers 2018). Thus, ways to preserve the heterogeneous nature of seed-associated microbiomes should be an objective of prime focus in case of centralised seed production programmes to improve the ecological fitness of the seed material produced.

6.3 Tools in Microbiome Analysis

Metagenomics revealed the identity and deciphered the role of the majority (>90%) of unculturable microbes harboured in the host plants which was not possible with conventional microbial culturing techniques. Meta-analysis and high-throughput screening unveiled the fact that differences occur in pattern of microbiome assemblage and their activity in host plants with respect to plant species, genotypes, plant age, plant tissue, soil biochemistry, and geography. Metagenomics and proteomics untangled large microbial metabolic clusters, their tissue-specific interactions with host plants, etc. and help us appreciate nature's creation and identify novel lead compounds and molecules with improved bio-efficacy and high degree of target specificity that could be harnessed and utilised for sustainable crop production. Advances in next-generation sequencing and bio-informatics tools had opened up pathways to study the molecular taxonomic and functional diversity of the plant microbiota and has elucidated how plants coordinate with core microbial partners belonging to certain phyla, viz. Proteobacteria, Firmicutes, Bacteroidetes, and Actinobacteria which have been found across a variety of host species including arabidopsis, brassica, soya bean, grapevine, and few tree species studied. Advanced molecular biology techniques widened the traditional perspective that host–microbe interactions not only refers to specific pathogenic interactions leading to disease or symbiotic interactions leading to nitrogen fixation, etc. but has much more beyond. Whole genome sequencing provides deeper insights on the impact of microbiome functions on the host plant that is engrained at different strata, viz. species, subspecies and strain (Beattie 2015). Tools used to decipher plant–microbiome interaction is summarised (Table 6.2).

Sequencing of marker genes and internal transcribed spacer (ITS) regions are efficient tools for quantitative surveys and reveals the patterns of microbiome assemblage and community compositions in different plant parts over diversified environmental conditions (Hacquard et al. 2017). Mitter et al. (2017) profiled seed endophyte community of crops such as wheat and soya bean using 16S rRNA gene sequences and analysed the bacterial colonisation by q-PCR and fluorescence in situ hybridisation using double labelling of oligonucleotide probes (DOPE-FISH/CSLM) microscopy. Use of appropriate techniques such as metataxonomic primer-based amplicon sequencing (16S rRNA for bacteria and archaea or intergenic spacer (ITS) region for eukaryotes), short gun sequencing of entire genome (metagenomic, metagenomic, and metaproteomic studies) coupled with use of liquid chromatography mass spectrometry (LC-MSMS) and nuclear magnetic resonance (NMR) for detection of metabolites (metabolomics) for exometabolic profiling of plant exudates (Zhalnina et al. 2018), breeding techniques such as multi-generation plant trait selection experiments (Panke-Buisse et al. 2015), crop mutant line experiments (Senga et al. 2017) coupled with microscopy (Rybakova et al. 2017) will provide us better insights of the taxonomic and functional diversity of the members of the microbiomes (Bulgarelli et al. 2015; Sergaki et al. 2018).

Table 6.2 Molecular tools in microbiome analysis

Dominant members of host plant-associated microbiomes	Host plant(s)	Ecological niche	Tools/ approaches in diversity analysis	Insights	References
Methylobacterium, Sphingomonas	Clover, soybean, Arabidopsis	Phyllosphere	Metaproteomics	Physiology and mechanism of interactions (bacteria–bacteria/ bacteria–plant) and identification of enzymatic functions playing a key role in nitrogen and carbon cycles	Delmotte et al. (2009)
Azorhizobium, Sinorhizobium, Rhodopseudomonas, Methylobacterium, Streptomyces, Sphingomonas	Paddy	Phyllosphere	Metaproteogenomics	Taxonomical diversity of microbes, physiological adaptations of rhizosphere and phyllosphere bacteria to supplement functions of the harboured niche	Knief et al. (2012)
Rhizobium, Bradyrhizobium, Azospirillum, Acidovorax, Dechloromonas, Methylococcus, Anaeromyxobacter, Herbaspirillum, Rhodopseudomonas, Magnetospirillum, Syntrophobacter, Methylobacterium, Magnetosprillum, Methylosinus, Geobacter, Desulfovibrio	Paddy	Rhizosphere	Metaproteogenomics	Taxonomical diversity of microbes, physiological adaptations of rhizosphere and phyllosphere bacteria to supplement functions of the harboured niche	Knief et al. (2012)
Proteobacteria, Gammaproteobacteria Alphaproteobacteria, Firmicutes, Planctomycetes, Verrumicrobia and Fusobacteria	Paddy	Root endosphere	Metagenomics	Microbial metabolic capacity inside rice roots	Sessitsch et al. (2012)

Azotobacter tropicalis, Bacillus aerophilus, Lysinibacillus fusiformis, Methylobacterium phyllosphaerae, Planococcus salinarum, Pseudomonas chlororaphis, Pseudomonas fluorescens, Pseudomonas rhodesiae and Serratia marcescens	Wheat from Southern hills zone of India	Epiphytic, endophytic, and rhizospheric	16S rRNA gene, amplified rDNA restriction analysis	Wheat-associated microbial diversity, functional characterisation for plant growth promotion, abiotic stress of acidic	Verma et al. (2013)
Arthrobacter, Bacillus, Corynebacterium, Duganella, Exiguobacterium, Lysinibacillus, Methylobacterium, Paenibacillus, Pantoea, Pseudomonas, Psychrobacter, Serratia, Stenotrophomonas	Wheat from Central zone of India	Epiphytic, endophytic and rhizospheric	16S rRNA gene, amplified rDNA restriction analysis	Wheat-associated microbial diversity, functional characterisation for plant growth promotion, abiotic stress of high temperature	Verma et al. (2014)
Alphaproteobacteria, Betaproteobacteria Gammaproteobacteria, Deinococcus-Thermus, Actinobacteria, Bacteroidetes and Firmicutes	Arabidopsis and their wild relatives	Root and leaf microbiomes	Culture-independent techniques	Evolutionary conservation of the associated microbiomes and functional overlap of root and leaf microbiota	Bai et al. (2015)
Flavobacteriaceae, Rhizobiaceae and Comamonadaceae,	Domesticated and wild varieties of barley	Root microbiomes	Combined use of 16S rRNA and shotgun metagenome analysis	Shows that microbiota assemblage in root-soil interface is driven by combined action of host–microbe and microbe–microbe interactions	Bulgarelli et al. (2015)

(continued)

Table 6.2 (continued)

Dominant members of host plant-associated microbiomes	Host plant(s)	Ecological niche	Tools/ approaches in diversity analysis	Insights	References
Achromobacter, Arthrobacter, Bacillus, Bordetella, Brevundimonas, Enterobacter, Exiguobacterium, Flavobacterium, Kluyvera, Kocuria, Leclercia, Methylobacterium, Pantoea, Providencia, Pseudomonas, Staphylococcus, Stenotrophomonas	Wheat from Northern hills zone of India	Epiphytic, endophytic, and rhizospheric	16S rRNA gene, amplified rDNA restriction analysis	Wheat-associated microbial diversity, functional characterisation for plant growth promotion, abiotic stress of cold stress	Verma et al. (2015a)
Haloarcula argentinensis, Halobacterium, Halococcus hamelinensis, Halococcus, Haloferax, Halolamina, Halostagnicola kamekurae, Haloterrigena thermotolerans, Natrialba, Natrinema, Natronoarchaeum mannanilyticum	*Halophytic Plants, Abutilon, Dichanthium, Sporobolus, Cenchrus, Suaeda nudiflora*	Rhizospheric microbiomes	16S rRNA gene, amplified rDNA restriction analysis	Phosphorus solubilising haloarchaea	Yadav et al. (2015)
Bacillus aerophilus, Bacillus amyloliquefaciens, Bacillus horikoshii, Bacillus megaterium, Bacillus sphaericus, Bacillus thuringiensis, Exiguobacterium acetylicum, Lysinibacillus fusiformis, Paenibacillus dendritiformis, Paenibacillus lautus, Planococcus citreus, Planomicrobium okeanokoites, Staphylococcus arlettae	Wheat from different agro-ecological zone of India	Rhizospheric	16S rRNA gene, amplified rDNA restriction analysis	Wheat-associated microbial diversity form different agro-ecological zone of India, functional characterisation for plant growth promotion, abiotic stress of temperature, pH, salinity, and drought	Verma et al. (2016)
Bradyrhizobiaceae, Bacillaceae, Burkholderiaceae, Caulobacteraceae, Comamonadaceae, Enterobacteriaceae, Rhizobiaceae, Xanthomonadaceae, Sphingomonadaceae, Moraxellaceae, Pseudomonadaceae, Oxalobacteraceae, Paenibacillaceae	Sugarcane	Roots and stalks	Community-based culture collection (CBC) approach	Identification and isolation of novel plant growth-promoting microbes	Armanhi et al. (2018)

Bacillus, *Glomus*	Grapevine	Rhizosphere microbiome	16S rRNA gene and ITS high-throughput amplicon sequencing (HTAS)	Host genetic control of plant microbiome	Berlanas et al. (2019)
Pseudomonas, Flavobacterium, Herbaspirillum, Curtobacterium, Microbacterium, Delftia, Enterobacter, Pantoea, Sphingomonas, Stenotrophomonas, Xanthomonas,	Paddy (cross-breed and inbreed varieties grown in different locations)	Seed endophytes	Terminal restriction fragment length polymorphism (TRFLP) analysis	Conservation and transmission of core microbial endophytes of paddy seeds is unaltered by cross-breeding, inbreeding and cultivation in different eco-geographic locations.	Walitang et al. (2019)
Haloarcula argentinensis, Halobacterium, Halococcus hamelinensis, Halococcus, Haloferax, Halolamina, Halostagnicola kamekurae, Haloterrigena thermotolerans, Natrialba, Natrinema, Natronoarchaeum mannanilyticum	*Halophytic Plants, Abutilon, Dichanthium, Sporobolus, Cenchrus, Suaeda nudiflora*	Rhizospheric microbiomes	16S rRNA gene, amplified rDNA restriction analysis	Seasonal variations of rhizospheric microbiomes from halophytic plants and their PGP attributes	Yadav et al. (2019)
Achromobacter spanius, Bacillus amyloliquefaciens, Bacillus siamensis, Delftia acidovorans, Methylobacterium mesophilicum, Paenibacillus amylolyticus, Pseudomonas aeruginosa, Pseudomonas japonica, Rhodobacter capsulatus, Rhodobacter sphaeroides, Salmonella bongori, Staphylococcus succinus	Wheat from Peninsular zone of India	Epiphytic, endophytic, and rhizospheric	16S rRNA gene, amplified rDNA restriction analysis	Wheat-associated microbial diversity, functional characterisation for plant growth promotion, abiotic stress of drought and heat	Verma et al. (2019)

6.4 Engineering Plant Microbiomes for Eco-Friendly, Sustainable Crop Production

Engineering a 'healthy microbiome' requires in-depth knowledge on the host–microbiome interactions, their co-evolutionary signatures aiding their assemblage and functioning. Success of microbiome research involves large culture collections, cataloguing the molecular and functional diversity of the microbes associated with plants, identifying the responsible genes, creation of data banks, affordable genome sequencing techniques, bio-informatics tools, molecular docking studies, gnotobiotic reconstitution system, economic and high quality meta-analysis and high-throughput systems that will provide information with high fidelity.

Ardanov et al. (2012) showed that the microbial community composition of vegetative plant parts could be modified by infecting the plant with endophytes. Infection of potato plants with endophytic, *Methylobacterium* induced resistance against a phytopathogen P*ectobacterium atrosepticum* causing black leg in potato. Similar experiments carried out on different plants with different pathogen combinations evidenced that certain endophytic bacteria could directly or indirectly protect plants against phytopathogens and hence could be introduced to various plant species to induce plant defences. However, intensive research is required to assess the compatibility of the introduced endophyte with the host plant so as to establish and exhibit its maximum potential in plant protection. In an attempt to colonise gnotobiotic *Arabidopsis* plants devoid of microbes with synthetic microbial communities colonising the root and phyllosphere of the test plants and evaluation of the established communities by gene sequencing evidenced the consistent assemblage of microbiomes resembling native microbiomes as that of wild *Arabidopsis* rather than being influenced by applied strains thereby indicating the robust mechanisms involved in microbiome assemblage (Bai et al. 2015). Mitter et al. (2017) attempted a new method of modifying the seed microbiomes of wheat by introducing an endophytic bacterium *Paraburkholderia phytofirmans* PsJN through flowers of the parent plant. The results evidenced that the bacterium introduced through flowers not only established themselves in the seeds produced by the plant but also altered the microbial community structure by altering the proportion of the members belonging to different taxa in the treated samples.

Plant microbiomes can be used as models to unveil the mechanisms underlying microbiome assembly and functioning. With advanced molecular techniques that flood us with data and methodologies for big data analyses now the challenge is on the conceptual networking of the results with underlying evolutionary and ecological mechanisms (Vacher et al. 2016). This will provide us better understanding of their ecological role. Hence, extensive research and region-specific field trials are required to engineer health microbiomes with a foresight to avoid perturbing synthetically engineered microbial communities and maintain harmony with nature. Progress in DNA sequencing techniques and high-throughput screening technologies such as metagenomics, metatranscriptomics, metaproteomics, and metametabolomics has revolutionised the field of microbiome research and had provided us deeper insights on host–microbiome associations and interactions which will enable us tailor beneficial microbiomes in plants by using simple techniques and support crop production sustainably without the risk of introducing GMO, ecological dilapidation, and environmental pollution.

6.5 Conclusion and Future Perspective

Population explosion, expansion of demographic activities and climate change had put agriculture under pressure. Addressing the demand by establishing sustainable production systems is a great challenge to agriculturists and human community. Plant microbiomes play an important role in hosts nutrient assimilation, growth promotion, and biocontrol of phytopathogens. They influence the chemical ecology of the host plant and thereby its interaction with other living organisms at different tropic levels including pollinators, insect pests, and their natural enemies. Plant microbiomes acting as an interface between plant and environment also act as an interface between humans and natural microbiomes. They act as a nodal route of exposure of humans to antibiotic resistance through direct contact, food chain, and globalisation (Chen et al. 2019). Thus, plant microbiomes apart from influencing the plant health and fitness also tremendously impact human health. Extensive research on microbiomes associated with different plants and plant tissues specifically and community profiling using both culturable and non-culturable methods will help us identify the key microbial players that shape the host plant microbiomes as well as impact the host plant fitness and in turn human health. In-depth studies will reveal the possibilities of engineering the plant microbiomes and laying a clean green road which utilises nature's assets so as to improve crop protection and production in an economical and eco-friendly manner. Deciphering plant microbiomes and devising simple methods to efficiently harness them so as to provide a profound effect on plant growth and plant protection will open up new avenues of breeding plants by introducing required beneficial traits without controversial genetic manipulation of the plants or use of pollution causing agrochemicals thereby laying a clean green road to sustainable agriculture.

Acknowledgement The authors wish to acknowledge the support and guidance provided by Dr. R. Nagarajaprakash, Group leader, Chemical Sciences Research Group, Division of Research and Development, Lovely Professional University, Phagwara, Punjab, India, for improving the chapter.

References

Abanda-Nkpwatt D, Tschiersch MM, Boettner JM et al (2006) Molecular interaction between *Methylobacterium extorquens* and seedlings: growth promotion, methanol consumption, and localization of the methanol emission site. J Exp Bot 57:4025–4032

Abdelfattah A, Wisniewski M, Droby S et al (2016) Spatial and compositional variation in the fungal communities of organic and conventionally grown apple fruit at the consumer point-of-purchase. Hortic Res 3:16047

Ahn TS, Ka JO, Lee GH et al (2007) Microcosm study for revegetation of barren land with wild plants by some plant growth-promoting rhizobacteria. J Microbiol Biotechnol 17:52–57

Aleklett K, Hart M, Shade A et al (2014) The microbial ecology of flowers: an emerging frontier in phyllosphere research. Botany 92:253–266

Alvarez-Loayza P, White JFJ, Torres MS et al (2011) Light converts endosymbiotic fungus to pathogen, influencing seedling survival and niche-space filling of a common tropical tree, *Iriartea deltoidea*. PLoS One 6:e16386

Alvarez-Perez S, Herrera CM (2013) Composition, richness and non-random assembly of culturable bacterial–microfungal communities in floral nectar of Mediterranean plants. FEMS Microbiol Ecol 83:685–699

Alvarez-Perez S, Lievens B, Fukami T (2019) Yeast–bacterium interactions: the next frontier in nectar research. Trends Plant Sci 24(5):393–401. https://doi.org/10.1016/j.tplants.2019.01.012

Andrews JH, Harris RF (2000) The ecology and biogeography of microorganisms on plant surfaces. Annu Rev Phytopathol 38:145–180

Araujo FF, Henning AA, Hungria M (2005) Phytohormones and antibiotics produced by *Bacillus subtilis* and their effects on seed pathogenic fungi and on soybean root development. World J Microbiol Biotechnol 21:1639–1645

Ardanov P, Sessitsch A, Haggman H (2012) *Methylobacterium*-induced endophyte community changes correspond with protection of plants against pathogen attack. PLoS One 7:e46802

Armanhi JSL, de Souza RSC, Damasceno NB et al (2018) A community-based culture collection for targeting novel plant growth-promoting bacteria from the sugarcane microbiome. Front Plant Sci 8:2191. https://doi.org/10.3389/fpls.2017.02191

Arora NK, Fatima T, Mishra I, Verma S (2020) Microbe-based inoculants: role in next green revolution: environmental concerns and sustainable development. Springer, Singapore, pp 191–246

Badri DV, Zolla G, Bakker MG et al (2013) Potential impact of soil microbiomes on the leaf metabolome and on herbivore feeding behaviour. New Phytol 198:264–273

Bai Y, Muller DB, Srinivas G et al (2015) Functional overlap of the *Arabidopsis* leaf and root microbiota. Nature 528:364–369

Baldotto LEB, Olivares FL, Bressan-Smith R (2011) Structural interaction between GFP-labeled diazotrophic endophytic bacterium *Herbaspirillum seropedicae* RAM10 and pineapple plantlets "Vitoria". Braz J Microbiol 42:114–125

Barret M, Briand M, Bonneau S et al (2015) Emergence shapes the structure of the seed microbiota. Appl Environ Microbiol 81:1257–1266

Barret M, Guimbaud J-F, Darrasse A, Jacques M-A (2016) Plant microbiota affects seed transmission of phytopathogenic micro-organisms. Mol Plant Pathol 17:791–795

Beattie GA (2015) Microbiomes: curating communities from plants. Nature 528:340–341

Beck JJ, Vannette RL (2017) Harnessing insect–microbe chemical communications to control insect pests of agricultural systems. J Agric Food Chem 65:23–28

Bednarek P, Osbourn A (2009) Plant–microbe interactions: chemical diversity in plant defense. Science 324:746–748

Belimov AA, Hontzeas N, Safronova VI (2005) Cadmium-tolerant plant growth-promoting bacteria associated with the roots of Indian mustard (*Brassica juncea* L. Czern.). Soil Biol Biochem 37:241–250

Berendsen RL, Pieterse CM, Bakker PA (2012) The rhizosphere microbiome and plant health. Trends Plant Sci 17:478–486

Berg G (2009) Plant–microbe interactions promoting plant growth and health: perspectives for controlled use of microorganisms in agriculture. Appl Microbiol Biotechnol 84:11–18

Berg G, Raaijmakers JM (2018) Saving seed microbiomes. ISME J 12:1167–1170

Berlanas C, Berbegal M, Elena G (2019) The fungal and bacterial rhizosphere microbiome associated with grapevine rootstock genotypes in mature and young vineyards. Front Microbiol 10:1142. https://doi.org/10.3389/fmicb.2019.01142

Bernard F, Sache I, Suffert F et al (2013) The development of a foliar fungal pathogen does react to leaf temperature. New Phytol 198:232–240

Bever JD, Mangan SA, Alexander HM (2015) Maintenance of plant species diversity by pathogens. Annu Rev Ecol Evol Systemat 46:305–325

Bordenstein SR, Theis KR (2015) Host biology in light of the microbiome: ten principles of holobionts and hologenomes. PLoS Biol 13:e1002226

Bulgarelli D, Garrido-Oter R, Munch PC et al (2015) Structure and function of the bacterial root microbiota in wild and domesticated barley. Cell Host Microbe 17(3):392–403. https://doi.org/10.1016/j.chom.2015.01.011

Burch AY, Zeisler V, Yokota K et al (2014) The hygroscopic biosurfactant syringafactin produced by Pseudomonas syringae enhances fitness on leaf surfaces during fluctuating humidity. Environ Microbiol 16:2086–2098

Chen QL, Cui HL, Su JQ et al (2019) Antibiotic resistomes in plant microbiomes. Trends Plant Sci 24:530–541. https://doi.org/10.1016/j.tplants.2019.02.010

Chinnadurai C, Balachandar D, Sundaram SP (2009) Characterization of 1-aminocyclopropane-1-carboxylate deaminase producing methylobacteria from phyllosphere of rice and their role in ethylene regulation. World J Microbiol Biotechnol 25:1403–1411

Coince A, Cordier T, Lengelle J et al (2014) Leaf and root-associated fungal assemblages do not follow similar elevational diversity patterns. PLoS One 9:e100668

Compant S, Clement C, Sessitsch A (2010) Plant growth promoting bacteria in the rhizo- and endosphere of plants: their role, colonization, mechanisms involved and prospects for utilization. Soil Biol Biochem 42:669–678

Compant S, Kaplan H, Sessitsch A et al (2008) Endophytic colonization of *Vitisvinifera* L. by *Burkholderia phytofirmans* strain PsJN: from the rhizosphere to inflorescence tissues. FEMS Microbiol Ecol 63:84–93

Compant S, Mitter B, Colli-Mull JG et al (2011) Endophytes of grapevine flowers, berries, and seeds: identification of cultivable bacteria, comparison with other plant parts, and visualization of niches of colonization. Microb Ecol 62:188–197

Cooper RM, Tsimring L, Hasty J (2017) Inter-species population dynamics enhance microbial horizontal gene transfer and spread of antibiotic resistance. elife 6:e25950

Cordier T, Robin C, Capdevielle X et al (2012) The composition of phyllosphere fungal assemblages of European beech (*Fagussyl vatica*) varies significantly along an elevation gradient. New Phytol 196:510–519

Cosme M, Lu J, Erb M et al (2016) A fungal endophyte helps plants to tolerate root herbivory through changes in gibberellin and jasmonate signaling. New Phytol 211:1065–1076

Darlison J, Mogren L, Rosberg AK et al (2019) Leaf mineral content govern microbial community structure in the phyllosphere of spinach (*Spinacia oleracea*) and rocket (*Diplotaxis tenuifolia*) Alsanius. Sci Total Environ 675:501–512

De Costa DM, Samarasinghe SST, Dias HRD et al (2008) Control of rice sheath blight by phyllosphere epiphytic microbial antagonists. Phytoparasitica 36:52–65

de Souza RSC, Okura VK, Armanhi JSL et al (2016) Unlocking the bacterial and fungal communities assemblages of sugarcane microbiome. Sci Rep 6:28774

de Vega C, Herrera CM, Johnson SD (2009) Yeasts in floral nectar of some South African plants: quantification and associations with pollinator type and sugar concentration. S Afr J Bot 75:798–806

Delmotte N, Knief C, Chaffron S et al (2009) Community proteogenomics reveals insights into the physiology of phyllosphere bacteria. Proc Natl Acad Sci U S A 106:16428–16433

Doan HK, Leveau JHJ (2015) Artificial surfaces in phyllosphere microbiology. Phytopathology 105:1036–1042

Droby S, Wisniewski M (2018) The fruit microbiome: a new frontier for postharvest biocontrol and postharvest biology. Postharvest Biol Technol 140:107–112

Duffy B, Voglesanger J, Schoch B et al (2006) Biocontrol of *Erwinia amylovora* using a commercial yeast strain mixture. Acta Hortic 704:363–366

Egamberdieva D (2008) Plant growth promoting properties of rhizobacteria isolated from wheat and pea grown in loamy sand soil. Turk J Biol 32:9–15

Ercolani GL (1991) Distribution of epiphytic bacteria on olive leaves and the influence of leaf age and sampling time. Microb Ecol 21:35–48

Faure D, Simon JC, Heulin T (2018) Holobiont: a conceptual framework to explore the eco-evolutionary and functional implications of host-microbiota interactions in all ecosystems. New Phytol 132:1–4

Fenner M, Thompson K (2005) The ecology of seeds. Cambridge University press, Cambridge

Fiorentino N, Ventorino V, Woo SL et al (2018) *Trichoderma*- based biostimulants modulate rhizosphere microbial populations and improve N uptake efficiency, yield, and nutritional quality of leafy vegetables. Front Plant Sci 9:743

Frank AC, Guzman JPS, Shay JE (2017) Transmission of bacterial endophytes. Microorganisms 5:70

Fridman S, Gerchman Y, Halpern M et al (2012) Bacterial communities in floral nectar. Environ Microbiol Rep 4:97–104

Fukami J, Nogueira MA, Araujo RS et al (2016) Accessing inoculation methods of maize and wheat with *Azospirillum brasilense*. AMB Express 6:3

Gaba S, Singh RN, Abrol S, Yadav AN, Saxena AK, Kaushik R (2017) Draft Genome Sequence of *Halolamina pelagica* CDK2 isolated from natural Salterns from Rann of Kutch, Gujarat, India. Genome Announc 5:1–2

Geisen S, Kostenko O, Cnossen MC et al (2017) Seed and root endophytic fungi in a range expanding and a related plant species. Front Microbiol 8:1645

Glawe DA (2008) The powdery mildews: a review of the world's most familiar (yet poorly known) plant pathogens. Annu Rev Phytopathol 46:27–51

Gonzalez-Teuber M, Heil M (2009) Nectar chemistry is tailored for both attraction of mutualists and protection from exploiters. Plant Signal Behav 4:809–813

Good AP, Gauthier M-PL, Vannette RL et al (2014) Honey bees avoid nectar colonized by three bacterial species, but not by a yeast species, isolated from the bee gut. PLoS One 9:e86494

Groenewald M, Robert V, Smith MT (2011) Five novel Wickerhamomyces and Metschnikowia related yeast species, *Wickerhamomyces chaumierensis* sp. nov., *Candida pseudoflosculorum* sp. nov., *Candida danieliae* sp. nov., *Candida robnettiae* sp. nov. and *Candida eppingiae* sp. nov., isolated from plants. Int J Syst Evol Microbiol 61:2015–2022

Hacquard S, Spaepen S, Garrido-Oter R et al (2017) Interplay between innate immunity and the plant microbiota. Annu Rev Phytopathol 55:565–589

Hardoim PR, Van Overbeek S, Berge G et al (2015) The hidden world within plants: ecological and evolutionary considerations for defining functioning of microbial endophytes. Microbiol Mol Biol Rev 79:293–320

Harman GE, Doni F, Khadka RB et al (2019) Endophytic strains of Trichoderma increase plants' photosynthetic capability. J Appl Microbiol. https://doi.org/10.1111/jam.14368

Herrera CM, de Vega C, Canto A et al (2009) Yeasts in floral nectar: a quantitative survey. Ann Bot 103:1415–1423

Hirano SS, Upper CD (2000) Bacteria in the leaf ecosystem with emphasis on Pseudomonas syringae—a pathogen, ice nucleus, and epiphyte. Microbiol Mol Biol Rev 64:624–653

Hume DE, Ryan GD, Gibert A et al (2016) Epichloe fungal endophytes for grassland ecosystems. In: Lichtfouse E (ed) Sustainable agriculture reviews, vol 19. Springer International Publishing, Cham, pp 233–305

Humphrey PT, Nguyen TT, Villalobos MM et al (2014) Diversity and abundance of phyllosphere bacteria are linked to insect herbivory. Mol Ecol 23:1497–1515

Hunter PJ, Pink DAC, Bending GD (2015) Cultivar-level genotype differences influence diversity and com-position of lettuce (*Lactuca* sp.) phyllosphere fungal communities. Fungal Ecol 17:183–186

Iguchi H, Yurimoto H, Sakai Y (2015) Interactions of methylotrophs with plants and other heterotrophic bacteria. Microorganisms 3:137–151

Ilori MO, Amund OO, Ezeani CK et al (2006) Occurrence and growth potentials of hydrocarbon degrading bacteria on the phylloplane of some tropical plants. Afr J Biotechnol 5:542–545

Inacio J, Ludwig W, Spencer-Martins I et al (2010) Assessment of phylloplane yeasts on selected Mediterranean plants by FISH with group- and species-specific oligonucleotide probes. FEMS Microbiol Ecol 71:61–72

Innerebner G, Knief C, Vorholt JA (2011) Protection of Arabidopsis thaliana against leaf-pathogenic *Pseudomonas syringae* by Sphingomonas strains in a controlled model system. Appl Environ Microbiol 77:3202–3210

Jimenez-Martınez ES, Bosque-Perez NA, Berger PH et al (2004) Volatile cues influence the response of *Rhopalosiphum padi* (Homoptera: Aphididae) to Barley yellow dwarf virus-infected transgenic and untransformed wheat. Environ Entomol 33:1207–1216

Jindamorakot S, Limtong S, Yongmanitchai W et al (2008) *Candida ratchasimensis* sp. nov. and *Candida khaoyaiensis* sp. nov., two anamorphic yeast species isolated from flowers in Thailand. FEMS Yeast Res 8:955–960

Johnson KB, Stockwell VO, Burgett DM, Sugar D, Loper JE (1993) Dispersal of *Erwinia amylovora* and *Pseudomonas fluorescens* by honey bees from hives to apple and pear blossoms. Phytopathology 83:479–484

Johnston-Monje D, Lundberg DS, Lazarovits G et al (2016) Bacterial populations in juvenile maize rhizospheres originate from both seed and soil. Plant Soil 405:337–355

Johnston-Monje D, Raizada MN (2011) Conservation and diversity of seed associated endophytes in Zea across boundaries of evolution, ethnography and ecology. PLoS One 6:e20396

Johnston-Monje D, Raizada MN (2013) Surveying diverse Zea seed for populations of bacterial endophytes. In: Molecular microbial ecology of the rhizosphere. John Wiley & Sons Inc, Hoboken, NJ, pp 445–455

Jones K (1970) Nitrogen fixation in the phyllosphere of the douglas fir, *Pseudotsuga douglasii*. Ann Bot 34:239–244

Jumpponen A, Jones KL (2009) Massively parallel 454 sequencing indicates hyper diverse fungal communities in temperate *Quercusma crocarpa* phyllosphere. New Phytol 184:438–448

Junker RR, Loewel C, Gross R et al (2011) Composition of epiphytic bacterial communities differs on petals and leaves. Plant Biol 13:918–924

Kaur T, Rana KL, Kour D, Sheikh I, Yadav N, Kumar V et al (2020) Microbe-mediated biofortification for micronutrients: present status and future challenges. In: Rastegari AA, Yadav AN, Yadav N (eds) Trends of microbial biotechnology for sustainable agriculture and biomedicine systems: perspectives for human health. Elsevier, Amsterdam, pp 1–17. https://doi.org/10.1016/B978-0-12-820528-0.00002-8

Kaymak HC, Guvenc I, Yarali F et al (2009) The effects of bio-priming with PGPR on germination of radish (*Raphanus sativus* L.) seeds under saline conditions. Turk J Agric 33:173–179

Kembel SW, Mueller RC (2014) Plant traits and taxonomy drive host associations in tropical phyllosphere fungal communities. Botany 92:303–311

Kembel SW, O'Connor TK, Arnold HK (2014) Relationships between phyllosphere bacterial communities and plant functional traits in a neotropical forest. Proc Natl Acad Sci U S A 111:13715–13720

Kennedy RK, Nagarajaprakash R, Pathma J (2020) Therapeutic potency of bioactive compounds from fungal endophytes. In: Gelhot P, Panwar JS (eds) New and future developments in microbial bioetechnology and bioengineering—recent advances in application of Fungi and fungal metabolites: applications in healthcare. Elsevier, Amsterdam

Kent AD, Triplett EW (2002) Microbial communities and their interactions in soil and rhizosphere ecosystems. Annu Rev Microbiol 56:211–236

Klaedtke S, Jacques MA, Raggi L et al (2016) Terroir is a key driver of seed-associated microbial assemblages. Environ Microbiol 18:1792–1804

Knief C, Delmotte N, Chaffron S et al (2012) Metaproteogenomic analysis of microbial communities in the phyllosphere and rhizosphere of rice. ISME J 6:1378–1390

Koskella B (2013) Phage-mediated selection on microbiota of a long-lived host. Curr Biol 23:1256–1260

Kour D, Kaur T, Yadav N, Rastegari AA, Singh B, Kumar V et al (2020a) Phytases from microbes in phosphorus acquisition for plant growth promotion and soil health. In: Rastegari AA, Yadav

AN, Yadav N (eds) Trends of microbial biotechnology for sustainable agriculture and biomedicine systems: diversity and functional perspectives. Elsevier, Amsterdam, pp 157–176. https://doi.org/10.1016/B978-0-12-820526-6.00011-7

Kour D, Rana KL, Kaur T, Devi R, Yadav N, Halder SK et al (2020b) Potassium solubilizing and mobilizing microbes: biodiversity, mechanisms of solubilization and biotechnological implication for alleviations of abiotic stress. In: Rastegari AA, Yadav AN, Yadav N (eds) Trends of microbial biotechnology for sustainable agriculture and biomedicine systems: diversity and functional perspective. Elsevier, Amsterdam, pp 177–202. https://doi.org/10.1016/B978-0-12-820526-6.00012-9

Kour D, Rana KL, Kaur T, Sheikh I, Yadav AN, Kumar V et al (2020c) Microbe-mediated alleviation of drought stress and acquisition of phosphorus in great millet (*Sorghum bicolor* L.) by drought-adaptive and phosphorus-solubilizing microbes. Biocatal Agric Biotechnol 23:101501. https://doi.org/10.1016/j.bcab.2020.101501

Kour D, Rana KL, Sheikh I, Kumar V, Yadav AN, Dhaliwal HS et al (2020d) Alleviation of drought stress and plant growth promotion by *Pseudomonas libanensis* EU-LWNA-33, a drought-adaptive phosphorus-solubilizing bacterium. Proc Natl Acad Sci India Sec B Biol Sci. https://doi.org/10.1007/s40011-019-01151-4

Kour D, Rana KL, Yadav AN, Sheikh I, Kumar V, Dhaliwal HS et al (2020e) Amelioration of drought stress in Foxtail millet (*Setaria italica* L.) by P-solubilizing drought-tolerant microbes with multifarious plant growth promoting attributes. Environ Sustain 3:23–34. https://doi.org/10.1007/s42398-020-00094-1

Kour D, Rana KL, Yadav AN, Yadav N, Kumar M, Kumar V et al (2020f) Microbial biofertilizers: bioresources and eco-friendly technologies for agricultural and environmental sustainability. Biocatal Agric Biotechnol 23:101487. https://doi.org/10.1016/j.bcab.2019.101487

Kour D, Rana KL, Yadav N, Yadav AN, Singh J, Rastegari AA et al (2019) Agriculturally and industrially important fungi: current developments and potential biotechnological applications. In: Yadav AN, Singh S, Mishra S, Gupta A (eds) Recent advancement in White biotechnology through Fungi, Perspective for value-added products and environments, vol 2. Springer International Publishing, Cham, pp 1–64. https://doi.org/10.1007/978-3-030-14846-1_1

Krimm U, Abanda-Nkpwatt D, Schwab W et al (2005) Epiphytic microorganisms on strawberry plants (*Fragaria ananassa* cv. Elsanta): identification of bacterial isolates and analysis of their interaction with leaf surfaces. FEMS Microbiol Ecol 53:483–492

Kumar M, Kour D, Yadav AN, Saxena R, Rai PK, Jyoti A et al (2019a) Biodiversity of methylotrophic microbial communities and their potential role in mitigation of abiotic stresses in plants. Biologia 74:287–308. https://doi.org/10.2478/s11756-019-00190-6

Kumar M, Saxena R, Rai PK, Tomar RS, Yadav N, Rana KL et al (2019b) Genetic diversity of methylotrophic yeast and their impact on environments. In: Yadav AN, Singh S, Mishra S, Gupta A (eds) Recent advancement in White biotechnology through Fungi, Perspective for sustainable environments, vol 3. Springer International Publishing, Cham, pp 53–71. https://doi.org/10.1007/978-3-030-25506-0_3

Kwak M-J, Kong HG, Choi K (2018) Rhizosphere microbiome structure alters to enable wilt resistance in tomato. Nat Biotechnol 36(11):1100–1116

Last FT, Deighton FC (1965) The non-parasitic microflora on the surfaces of living leaves. Trans Br Mycol Soc 48:83–99

Lavania M, Chauhan PS, Chauhan SVS et al (2006) Induction of plant defense enzymes and phenolics by treatment with plant growth-promoting rhizobacteria Serratiamarcescens NBRI1213. Curr Microbiol 52:363–368

Lebeis SL, Paredes SH, Lundberg DS et al (2015) Salicylic acid modulates colonization of the root microbiome by specific bacterial taxa. Science 349:860–864

Leben C (1965) Epiphytic microorganisms in relation to plant disease. Annu Rev Phytopathol 3:209–230

Leck MA, Parker VT, Simpson R (2008) Seedling ecology and evolution. Cambridge University Press, Cambridge

Lenaerts M, Goelen T, Paulussen C et al (2017) Nectar bacteria affect life history of a generalist aphid parasitoid by altering nectar chemistry. Funct Ecol 31:2061–2069

Leveau JHJ (2006) Microbial communities in the phyllosphere. In: Riederer M, Mueller C (eds) Biology of the plant cuticle. Blackwell Publishing, Oxford, pp 334–367

Lindow SE, Brandl MT (2003) Microbiology of the phyllosphere. Appl Environ Microbiol 69:1875–1883

Links MG, Demeke T, Grafenhan T et al (2014) Simultaneous profiling of seed associated bacteria and fungi reveals antagonistic interactions between microorganisms within a shared epiphytic microbiome on *Triticum* and *Brassica* seeds. New Phytol 202:542–553

Liu Y, Zuo S, Xu L et al (2012) Study on diversity of endophytic bacterial communities in seeds of hybrid maize and their parental lines. Arch Microbiol 194:1001–1012

Liu Y, Zuo S, Zou YY et al (2013) Investigation on diversity and population succession dynamics of endophytic bacteria from seeds of maize (*Zea mays* L., Nongda 108) at different growth stages. Ann Microbiol 63:71–79

Maccagnani B, Giacomello F, Fanti M et al (2009) *Apis mellifera* and *Osmia cornuta* as carriers for the secondary spread of *Bacillus subtilis* on apple flowers. BioControl 54:123–133

Madhaiyan M, Poonguzhali S, Senthilkumar M et al (2004) Growth promotion and induction of systemic resistance in rice cultivar Co-47 (*Oryza sativa* L.) by *Methylobacterium* spp. Bot Bull Acad Sin 45:315–324

Maliti CM, Basile DV, Corpe WA (2005) Effects of *Methylobacterium* spp. strains on rice *Oryza sativa* L. callus induction, plantlet regeneration, and seedlings growth *in vitro*. J Torrey Bot Soc 132:355–367

Malyan SK, Singh S, Bachheti A, Chahar M, Sah MK, Narender et al. (2020) Cyanobacteria: a perspective paradigm for agriculture and environment. In: Rastegari AA, Yadav AN, Yadav N (eds) Trends of microbial biotechnology for sustainable agriculture and biomedicine systems: diversity and functional perspectives. Elsevier, Amsterdam pp 215–224. https://doi.org/10.1016/B978-0-12-820526-6.00014-2

Mann RS, Ali JG, Hermann SL et al (2012) Induced release of a plant-defense volatile 'deceptively' attracts insect vectors to plants infected with a bacterial pathogen. PLoS Pathog 8:e1002610

Mansfield J, Genin S, Magori S et al (2012) Top 10 plant pathogenic bacteria in molecular plant pathology. Mol Plant Pathol 13:614–629

Margulis L, Fester R (eds) (1991) Symbiosis as a source of evolutionary innovation: speciation and morphogenesis. MIT Press, Cambridge

Mason CJ, Pfammatter JA, Holeski LM et al (2015) Foliar bacterial communities of trembling aspen in a common garden. Can J Microbiol 61:143–149

McCartney HA, Fitt BDL (1998) Dispersal of foliar fungal plant pathogens: mechanisms, gradients and spatial patterns. In: Jones DG (ed) The epidemiology of plant diseases. Springer, Dordrecht, pp 138–160

Mechaber WL, Marshall DB, Mechabers RA et al (1996) Mapping leaf surface landscapes. Proc Natl Acad Sci U S A 93:4600–4603

Melotto M, Underwood W, He SY (2008) Role of stomata in plant innate immunity and foliar bacterial diseases. Annu Rev Phytopathol 46:101–122

Mercier J, Lindow SE (2000) Role of leaf surface sugars in colonization of plants by bacterial epiphytes. Appl Environ Microbiol 66:369–374

Mitter B, Pfaffenbichler N, Flavell R et al (2017) New approach to modify plant microbiomes and traits by introducing beneficial bacteria at flowering into progeny seeds. Front Microbiol 8:1–11

Mohanram S, Kumar P (2019) Rhizosphere microbiome: revisiting the synergy of plant-microbe interactions. Ann Microbiol 69:307–320

Mondal S, Halder SK, Yadav AN, Mondal KC (2020) Microbial consortium with multifunctional plant growth promoting attributes: future perspective in agriculture. In: Yadav AN, Rastegari AA, Yadav N, Kour D (eds) Advances in plant microbiome and sustainable agriculture, Functional annotation and future challenges, vol 2. Springer, Singapore, pp 219–254. https://doi.org/10.1007/978-981-15-3204-7_10

Morris CE (2002) Phyllosphere. eLS. https://doi.org/10.1038/npg.els.0000400

Morris CE, Conen F, Huffman JA et al (2014) Bio precipitation: a feed back cycle linking Earth history, ecosystem dynamics and land use through biological ice nucleators in the atmosphere. Glob Change Biol 20:341–351

Morris CE, Monteil CL, Berge O (2013) The life history of *Pseudomonas syringae*: linking agriculture to earth system processes. Annu Rev Phytopathol 51:85–104

Mwajita M, Murage H, Tani A et al (2013) Evaluation of rhizosphere, rhizoplane and phyllosphere bacteria and fungi isolated from rice in Kenya for plant growth promoters. SpringerPlus 2:606

Ngugi HK, Scherm H (2006) Biology of flower-infecting fungi. Annu Rev Phytopathol 44:261–282

Oosten V, Bodenhausen VR, Reymond N et al (2008) Differential effectiveness of microbially induced resistance against herbivorous insects in *Arabidopsis*. Mol Plant-Microbe Interact 21:919–930

Ostman O, Drakare S, Kritzberg ES et al (2010) Regional invariance among microbial communities. Ecol Lett 13:118–127

Ottesen AR, Gonzalez Pena A, White JR et al (2013) Baseline survey of the anatomical microbial ecology of an important food plant: *Solanum lycopersicum* (tomato). BMC Microbiol 13:114

Panke-Buisse K, Poole AC, Goodrich JK et al (2015) Selection on soil microbiomes reveals reproducible impacts on plant function. ISME J 9:980–989

Parret AH, De Mot R (2002) Bacteria killing their own kind: novel bacteriocins of *Pseudomonas* and other γ-proteobacteria. Trends Microbiol 10:107–112

Pathma J, Ayyadurai N, Sakthivel N (2010a) Assessment of genetic and functional relationship of antagonistic fluorescent pseudomonads of rice rhizosphere by repetitive sequence, protein coding sequence and functional gene analyses. J Microbiol 48:715–727

Pathma J, Kamaraj Kennedy R, Sakthivel N (2010b) Mechanisms of fluorescent pseudomonads that mediate biological control of phytopathogens and plant growth promotion of crop plants. In: Maheshwari DK (ed) Bacteria in agro-biology: plant growth. Springer-Verlag, Berlin Heidelberg, pp 77–105

Pathma J, Rahul GR, Kamaraj Kennedy R, Subashri R, Sakthivel N (2011) Secondary metabolite production by bacterial antagonists. J Biol Control 25:165–181

Pathma J, Raman G, Kennedy RK et al (2019a) Recent advances in plant-microbe interaction. In: Sharma SG, Sharma M, Sharma NR (eds) Microcosm: microbial diversity, interventions and scope. Springer, Singapore

Pathma J, Raman G, Sakthivel N (2019b) Microbiome of rhizospheric soil and vermicompost and their applications in soil fertility, pest and pathogen management for sustainable agriculture. In: Panpatte DG, Jhala YK (eds) Soil fertility management for sustainable development. Springer, Singapore, pp 189–210

Patz S, Witzel K, Ann-Christin S et al (2019) Culture dependent and independent analysis of potential probiotic bacterial genera and species present in the phyllosphere of raw eaten produce. Int J Mol Sci 20:3661

Penuelas J, Rico L, Ogaya R et al (2012) Summer season and long-term drought increase the richness of bacteria and fungi in the foliar phyllosphere of Quercusilexin a mixed Mediterranean forest. Plant Biol 14:565–575

Perez LI, Gundel PE, Omacini M (2016) Can the defensive mutualism between grasses and fungal endophytes protect non-symbiotic neighbours from soil pathogens? Plant Soil 405:289–298

Pieterse CMJ, Dicke M (2007) Plant interactions with microbes and insects: molecular mechanisms to ecology. Trends Plant Sci 12:564–569

Pineda A, Kaplan I, Bezemer TM (2017) Steering soil microbiomes to suppress aboveground insect pests. Trends Plant Sci 22:770–778

Pirlak L, Kose M (2009) Effects of plant growth promoting rhizobacteria on yield and some fruit properties of strawberry. J Plant Nutr 32:1173–1184

Porras-Alfaro A, Bayman P (2011) Hidden fungi, emergent properties: endophytes and microbiomes. Annu Rev Phytopathol 49:291–315

Pusey PL, Stockwell VO, Mazzola M (2009) Epiphytic bacteria and yeasts on apple blossoms and their potential as antagonists of *Erwinia amylovora*. Phytopathology 99:571–581

Radwan SS, Dashti N, El-Nemr IM (2005) Enhancing the growth of *Vicia faba* plants by microbial inoculation to their phytoremediation potential for oily desert areas. Int J Phytoremediation 7:19–32

Rai PK, Singh M, Anand K, Saurabhj S, Kaur T, Kour D et al (2020) Role and potential applications of plant growth promotion rhizobacteria for sustainable agriculture. In: Rastegari AA, Yadav AN, Yadav N (eds) Trends of microbial biotechnology for sustainable agriculture and biomedicine systems: diversity and functional perspectives. Elsevier, Amsterdam, pp 49–60. https://doi.org/10.1016/B978-0-12-820526-6.00004-X

Rana KL, Kour D, Kaur T, Devi R, Yadav AN, Yadav N et al (2020a) Endophytic microbes: biodiversity, plant growth-promoting mechanisms and potential applications for agricultural sustainability. Antonie Van Leeuwenhoek. https://doi.org/10.1007/s10482-020-01429-y

Rana KL, Kour D, Kaur T, Devi R, Yadav N, Rastegari AA et al (2020b) Biodiversity, phylogenetic profiling and mechanisms of colonization of seed microbiomes. In: Rastegari AA, Yadav AN, Yadav N (eds) Trends of microbial biotechnology for sustainable agriculture and biomedicine systems: diversity and functional perspectives. Elsevier, Amsterdam, pp 99–125. https://doi.org/10.1016/B978-0-12-820526-6.00007-5

Rana KL, Kour D, Kaur T, Devi R, Yadav N, Subrahmanyam G et al (2020c) Biotechnological applications of seed microbiomes for sustainable agriculture and environments. In: Rastegari AA, Yadav AN, Yadav N (eds) Trends of microbial biotechnology for sustainable agriculture and biomedicine systems: diversity and functional perspectives. Elsevier, Amsterdam, pp 127–143. https://doi.org/10.1016/B978-0-12-820526-6.00008-7

Rana KL, Kour D, Kaur T, Sheikh I, Yadav AN, Kumar V et al (2020d) Endophytic microbes from diverse wheat genotypes and their potential biotechnological applications in plant growth promotion and nutrient uptake. Proc Natl Acad Sci India Sect B Biol Sci. https://doi.org/10.1007/s40011-020-01168-0

Rana KL, Kour D, Yadav AN (2019) Endophytic microbiomes: biodiversity, ecological significance and biotechnological applications. Res J Biotechnol 14:142–162

Rastegari AA, Yadav AN, Yadav N (2020a) New and future developments in microbial biotechnology and bioengineering: Trends of microbial biotechnology for sustainable agriculture and biomedicine systems: diversity and functional perspectives. Elsevier, Amsterdam

Rastegari AA, Yadav AN, Yadav N (2020b) New and future developments in microbial biotechnology and bioengineering: Trends of microbial biotechnology for sustainable agriculture and biomedicine systems: perspectives for human health. Elsevier, Amsterdam

Recep K, Fikrettin S, Erkol D et al (2009) Biological control of the potato dry rot caused by *Fusarium* species using PGPR strains. Biol Control 50:194–198

Reinhold-Hurek B, Bunger W, Burbano CS et al (2015) Roots shaping their microbiome: global hotspots for microbial activity. Annu Rev Phytopathol 53:403

Rering CC, Beck JJ, Hall GW et al (2018) Nectar-inhabiting microorganisms influence nectar volatile composition and attractiveness to a generalist pollinator. New Phytol 220:750–759

Rezki S, Campion C, Preveaux A et al (2016) Differences in stability of seed-associated microbial assemblages in response to invasion by phytopathogenic microorganisms. Peer J 4:e1923

Richardson LL, Bowers MD, Irwin RE et al (2016) Nectar chemistry mediates the behavior of parasitized bees: consequences for plant fitness. Ecology 97:325–337

Rico L, Ogaya R, Terradas J et al (2014) Community structures of N_2-fixing bacteria associated with the phyllosphere of a Holm oak forest and their response to drought. Plant Biol 16:586–593

Rodrigues EP, Rodrigues LS, de Oliveira ALM et al (2008) *Azospirillum amazonense* inoculation: effects on growth, yield and N2 fixation of rice (*Oryza sativa* L.). Plant Soil 302:249–261

Rodriguez RJ, White JF Jr, Arnold AE et al (2009) Fungal endophytes: diversity and functional roles. New Phytol 182:314–330

Rosa CA, Pagnocca FC, Lachance M-A et al (2007) *Candida flosculorum* sp. nov. and *Candida floris* sp. nov., two yeast species associated with tropical flowers. Int J Syst Evol Microbiol 57:2970–2974

Rosenberg E, Zilber-Rosenberg I (2018) The hologenome concept of evolution after 10 years. Microbiome 6:78

Rosenblueth M, Martinez-Romero E (2006) Bacterial endophytes and their interactions with hosts. Mol Plant-Microbe Interact 19:827–837

Ruinen J (1965) The phyllosphere. III. Nitrogen fixation in the phyllosphere. Plant Soil 22:375–394

Ryan RP, Germaine K, Franks A et al (2008) Bacterial endophytes: recent developments and applications. FEMS Microbiol Lett 278:1–9

Rybakova D, Mancinelli R, Wikstrom M et al (2017) The structure of the Brassica napus seed microbiome is cultivar-dependent and affects the interactions of symbionts and pathogens. Microbiome 5:104

Ryu CM, Hu CH, Locy RD et al (2005) Study of mechanisms for plant growth promotion elicited by rhizobacteria in *Arabidopsis thaliana*. Plant Soil 268:285–292

Saharan BS, Nehra V (2011) Plant growth promoting rhizobacteria: a critical review. Life Sci Med Res 21:1–30

Samuni-Blank M, Izhaki I, Laviad S et al (2014) The role of abiotic environmental conditions and herbivory in shaping bacterial community composition in floral nectar. PLoS One 9:e99107

Sandhu DK, Waraich MK (1985) Yeasts associated with pollinating bees and flower nectar. Microb Ecol 11:51–58

Senga RA, Terrazas SA, Balbirnie K et al (2017) Root hair mutations displace the barley rhizosphere microbiota. Front Plant Sci 8:1094

Sergaki C, Lagunas B, Lidbury I et al (2018) Challenges and approaches in microbiome research: from fundamental to applied. Front Plant Sci 9:1205

Sessitsch A, Brader G, Pfaffenbichler N et al (2018) The contribution of plant microbiota to economy growth. Microb Biotechnol 11:1–5

Sessitsch A, Hardoim P, Doring J et al (2012) Functional characteristics of an endophyte community colonizing rice roots as revealed by metagenomic analysis. Mol Plant-Microbe Interact 25:28–36

Setati ME, Jacobson D, Andong U-C et al (2012) The vineyard yeast microbiome, a mixed model microbial map. PLoS One 7:e52609

Shade MM, Handelsman J (2013) Unexpected diversity during community succession in the apple flower microbiome. MBio 4. https://doi.org/10.1128/mBio.00602-12

Shaharoona B, Bibi R, Arshad M et al (2006) 1-Aminocylopropane-1-carboxylate (ACC) deaminase rhizobacteria extenuates ACC-induced classical triple response in etiolated pea seedlings. Pak J Bot 38:1491–1499

Sharaff MS, Subrahmanyam G, Kumar A, Yadav AN (2020) Mechanistic understanding of root-microbiome interaction for sustainable agriculture in polluted soils. In: Rastegari AA, Yadav AN, Yadav N (eds) Trends of microbial biotechnology for sustainable agriculture and biomedicine systems: diversity and functional perspectives. Elsevier, Amsterdam, pp 61–84. https://doi.org/10.1016/B978-0-12-820526-6.00005-1

Sharma S, Kour D, Rana KL, Dhiman A, Thakur S, Thakur P et al (2019) *Trichoderma*: biodiversity, ecological significances, and industrial applications. In: Yadav AN, Mishra S, Singh S, Gupta A (eds) Recent advancement in White biotechnology through fungi, Diversity and enzymes perspectives, vol 1. Springer International Publishing, Cham, pp 85–120. https://doi.org/10.1007/978-3-030-10480-1_3

Singh A, Kumar R, Yadav AN, Mishra S, Sachan S, Sachan SG (2020a) Tiny microbes, big yields: microorganisms for enhancing food crop production sustainable development. In: Rastegari AA, Yadav AN, Yadav N (eds) Trends of microbial biotechnology for sustainable agriculture and biomedicine systems: diversity and functional perspectives. Elsevier, Amsterdam, pp 1–15. https://doi.org/10.1016/B978-0-12-820526-6.00001-4

Singh A, Kumari R, Yadav AN, Mishra S, Sachan A, Sachan SG (2020b) Tiny microbes, big yields: microorganisms for enhancing food crop production sustainable development. In: Rastegari AA, Yadav AN, Awasthi AK, Yadav N (eds) Trends of microbial biotechnology for sustainable agriculture and biomedicine systems: diversity and functional perspectives. Elsevier, Amsterdam, pp 1–16. https://doi.org/10.1016/B978-0-12-820526-6.00001-4

Singh B, Boukhris I, Pragya, Kumar V, Yadav AN, Farhat-Khemakhem A et al. (2020c) Contribution of microbial phytases to the improvement of plant growth and nutrition: a review. Pedosphere 30:295–313. https://doi.org/10.1016/S1002-0160(20)60010-8

Singh J, Yadav AN (2020) Natural bioactive products in sustainable agriculture. Springer, Singapore
Singh V, Upadhyay RS, Sarma BK et al (2014) Seed bio-priming with *Trichoderma asperellum* effectively modulate plant growth promotion in pea. IJAEB 9:361–365
Sobhy IS, Baets D, Goelen D et al (2018) Sweet scents: nectar specialist yeasts enhance nectar attraction of a generalist aphid parasitoid without affecting survival. Front Plant Sic 9:1009
Subrahmanyam G, Kumar A, Sandilya SP, Chutia M, Yadav AN (2020) Diversity, plant growth promoting attributes, and agricultural applications of rhizospheric microbes. In: Yadav AN, Singh J, Rastegari AA, Yadav N (eds) Plant microbiomes for sustainable agriculture. Springer International Publishing, Cham, pp 1–52. https://doi.org/10.1007/978-3-030-38453-1_1
Sugio A, Dubreuil G, Giron D et al (2015) Plant–insect interactions under bacterial influence: ecological implications and underlying mechanisms. J Exp Bot 66:467–478
Suman A, Yadav AN, Verma P (2016) Endophytic microbes in crops: diversity and beneficial impact for sustainable agriculture. In: Singh D, Abhilash P, Prabha R (eds) Microbial inoculants in sustainable agricultural productivity, research perspectives. Springer-Verlag, New Delhi, pp 117–143. https://doi.org/10.1007/978-81-322-2647-5_7
Tetard-Jones C, Kertesz MA, Preziosi RF (2012) Identification of plant quantitative trait loci modulating a rhizobacteria-aphid indirect effect. PLoS One 7:e41524
Thakur N, Kaur S, Tomar P, Thakur S, Yadav AN (2020) Microbial biopesticides: current status and advancement for sustainable agriculture and environment. In: Rastegari AA, Yadav AN, Yadav N (eds) Trends of microbial biotechnology for sustainable agriculture and biomedicine systems: diversity and functional perspectives. Elsevier, Amsterdam, pp 243–282. https://doi.org/10.1016/B978-0-12-820526-6.00016-6
Thompson IP, Bailey MJ, Fenlon JS et al (1993) Quantitative and qualitative seasonal changes in the microbial community from the phyllosphere of sugar-beet (*Beta vulgaris*). Plant Soil 150:177–191
Tiwari P, Bajpai M, Singh LK, Mishra S, Yadav AN (2020) Phytohormones producing fungal communities: metabolic engineering for abiotic stress tolerance in crops. In: Yadav AN, Mishra S, Kour D, Yadav N, Kumar A (eds) Agriculturally important fungi for sustainable agriculture, volume 1: perspective for diversity and crop productivity. Springer, Cham. https://doi.org/10.1007/978-3-030-45971-0_8
Toju H, Vannette RL, Dhami MK et al (2018) Priority effects can persist across floral generations in nectar microbial metacommunities. Oikos 127:345–352
Torres MA (2010) ROS in biotic interactions. Physiol Plant 138:414–429
Truyens S, Weyens N, Cuypers A et al (2015) Bacterial seed endophytes: genera, vertical transmission and interaction with plants. Environ Microbiol Rep 7:40–50
Tsuji K, Fukami T (2018) Community-wide consequences of sexual dimorphism: evidence from nectar microbes in dioecious plants. Ecology 99:2476–2484
Tucker CM, Fukami T (2014) Environmental variability counteracts priority effects to facilitate species coexistence: evidence from nectar microbes. Proc Biol Sci 281:20132637
Vacher C, Hampe A, Porte A et al (2016) The phyllosphere: microbial jungle at the plant-climate interface. Ann Rev Ecol Evol Systemat 47:1–24
Van Aken B, Yoon JM, Schnoor JL (2004) Biodegradation of nitro-substituted explosives 2,4,6-trinitrotoluene, hexahydro-1,3,5-triazine, an octahydro-1,3,5,7- tetranitro-1,3,5-tetrazocine by a photosymbiotic *Methylobacterium* sp associated with poplar tissues (*Populus deltoids* x nigra DN34). Appl Environ Microb 70:508–517
Van Peer R, Schippers B (1988) Plant growth responses to bacterization with selected pseudomonas Spp. strains and rhizosphere microbial development in hydroponic cultures. Can J Microbiol 35:456–463
Van Wees SC, van der Ent S, Pieterse CM (2008) Plant immune responses triggered by beneficial microbes. Curr Opin Plant Biol 11:443–448
Vannette RL, Fukami T (2018) Contrasting effects of yeasts and bacteria on floral nectar traits. Ann Bot 121:1343–1349
Vellend M (2010) Conceptual synthesis in community ecology. Q Rev Biol 85:183–206

Verma P, Yadav AN, Kazy SK, Saxena AK, Suman A (2013) Elucidating the diversity and plant growth promoting attributes of wheat (*Triticum aestivum*) associated acidotolerant bacteria from southern hills zone of India. Natl J Life Sci 10:219–227

Verma P, Yadav AN, Kazy SK, Saxena AK, Suman A (2014) Evaluating the diversity and phylogeny of plant growth promoting bacteria associated with wheat (*Triticum aestivum*) growing in central zone of India. Int J Curr Microbiol Appl Sci 3:432–447

Verma P, Yadav AN, Khannam KS, Kumar S, Saxena AK, Suman A (2016) Molecular diversity and multifarious plant growth promoting attributes of Bacilli associated with wheat (*Triticum aestivum* L.) rhizosphere from six diverse agro-ecological zones of India. J Basic Microbiol 56:44–58

Verma P, Yadav AN, Khannam KS, Mishra S, Kumar S, Saxena AK et al (2019) Appraisal of diversity and functional attributes of thermotolerant wheat associated bacteria from the peninsular zone of India. Saudi J Biol Sci 26:1882–1895. https://doi.org/10.1016/j.sjbs.2016.01.042

Verma P, Yadav AN, Khannam KS, Panjiar N, Kumar S, Saxena AK et al (2015a) Assessment of genetic diversity and plant growth promoting attributes of psychrotolerant bacteria allied with wheat (*Triticum aestivum*) from the northern hills zone of India. Ann Microbiol 65:1885–1899

Verma P, Yadav AN, Kumar V, Singh DP, Saxena AK (2017) Beneficial plant-microbes interactions: biodiversity of microbes from diverse extreme environments and its impact for crop improvement. In: Singh DP, Singh HB, Prabha R (eds) Plant-microbe interactions in agro-ecological perspectives, Microbial interactions and agro-ecological impacts, vol 2. Springer Singapore, Singapore, pp 543–580. https://doi.org/10.1007/978-981-10-6593-4_22

Verma P, Yadav AN, Shukla L, Saxena AK, Suman A (2015b) Alleviation of cold stress in wheat seedlings by *Bacillus amyloliquefaciens* IARI-HHS2-30, an endophytic psychrotolerant K-solubilizing bacterium from NW Indian Himalayas. Natl J Life Sci 12:105–110

Vessey JK (2003) Plant growth promoting rhizobacteria as biofertilizers. Plant Soil 255:571–586

Vorholt JA (2012) Microbial life in the phyllosphere. Nat Rev Microbiol 10:828–840

Walitang DI, Kim CG, Jeon S et al (2019) Conservation and transmission of seed bacterial endophytes across generations following crossbreeding and repeated inbreeding of rice at different geographic locations. MicrobiologyOpen 8:e00662

Walterson AM, Stavrinides J (2015) Pantoea: insights into a highly versatile and diverse genus within the Enterobacteriaceae. FEMS Microbiol Rev 39:968–984

Wassermann B, Muller H, Berg G (2019) An apple a day: which bacteria do we eat with organic and conventional apples? Front Microbiol 10:1629

Wei X, Lyu S, Yu Y et al (2017) Phylloremediation of air pollutants: exploiting the potential of plant leaves and leaf-associated microbes. Front Plant Sci 8:1318

White JF, Torres MS, Sullivan RF et al (2014) Occurrence of *Bacillus amyloliquefaciens* as a systemic endophyte of vanilla orchids: *B. amyloliquefaciens* as a systemic endophyte of Vanilla orchids. Microsc Res Tech 7:874–885

Yadav AN (2020) Plant microbiomes for sustainable agriculture: current research and future challenges. In: Yadav AN, Singh J, Rastegari AA, Yadav N (eds) Plant microbiomes for sustainable agriculture. Springer International Publishing, Cham, pp 475–482. https://doi.org/10.1007/978-3-030-38453-1_16

Yadav AN, Gulati S, Sharma D, Singh RN, Rajawat MVS, Kumar R et al (2019) Seasonal variations in culturable archaea and their plant growth promoting attributes to predict their role in establishment of vegetation in Rann of Kutch. Biologia 74:1031–1043. https://doi.org/10.2478/s11756-019-00259-2

Yadav AN, Kaur T, Kour D, Rana KL, Yadav N, Rastegari AA et al (2020a) Saline microbiome: biodiversity, ecological significance and potential role in amelioration of salt stress in plants. In: Rastegari AA, Yadav AN, Yadav N (eds) Trends of microbial biotechnology for sustainable agriculture and biomedicine systems: diversity and functional perspectives. Elsevier, Amsterdam, pp 283–309. https://doi.org/10.1016/B978-0-12-820526-6.00018-X

Yadav AN, Kour D, Kaur T, Devi R, Gukeria G, Rana KL et al (2020b) Microbial biotechnology for sustainable agriculture: current research and future challenges. In: Rastegari AA, Yadav AN, Yadav N (eds) Trends of microbial biotechnology for sustainable agriculture and biomedicine systems: diversity and functional perspectives. Elsevier, Amsterdam, pp 331–343. https://doi.org/10.1016/B978-0-12-820526-6.00020-8

Yadav AN, Kour D, Kaur T, Devi R, Yadav N (2020c) Agriculturally important fungi for crop productivity: current research and future challenges. In: Yadav AN, Mishra S, Kour D, Yadav N, Kumar A (eds) Agriculturally important fungi for sustainable agriculture, volume 1: perspective for diversity and crop productivity. Springer, Cham. https://doi.org/10.1007/978-3-030-45971-0_12

Yadav AN, Kumar R, Kumar S, Kumar V, Sugitha T, Singh B et al (2017a) Beneficial microbiomes: biodiversity and potential biotechnological applications for sustainable agriculture and human health. J Appl Biol Biotechnol 5:45–57

Yadav AN, Kumar V, Prasad R, Saxena AK, Dhaliwal HS (2018a) Microbiome in crops: diversity, distribution and potential role in crops improvements. In: Prasad R, Gill SS, Tuteja N (eds) Crop improvement through microbial biotechnology. Elsevier, Cambridge, MA, pp 305–332

Yadav AN, Mishra S, Kour D, Yadav N, Kumar A (2020d) Agriculturally important fungi for sustainable agriculture, volume 1: perspective for diversity and crop productivity. Springer International Publishing, Cham

Yadav AN, Rastegari AA, Yadav N, Kour D (2020e) Advances in plant microbiome and sustainable agriculture: diversity and biotechnological applications. Springer, Singapore

Yadav AN, Rastegari AA, Yadav N, Kour D (2020f) Advances in plant microbiome and sustainable agriculture: functional annotation and future challenges. Springer, Singapore

Yadav AN, Sharma D, Gulati S, Singh S, Dey R, Pal KK et al (2015) Haloarchaea endowed with phosphorus solubilization attribute implicated in phosphorus cycle. Sci Rep 5:12293

Yadav AN, Singh J, Rastegari AA, Yadav N (2020g) Plant microbiomes for sustainable agriculture. Springer International Publishing, Cham

Yadav AN, Verma P, Kumar V, Sangwan P, Mishra S, Panjiar N et al (2018b) Biodiversity of the genus *Penicillium* in different habitats. In: Gupta VK, Rodriguez-Couto S (eds) New and future developments in microbial biotechnology and bioengineering, *Penicillium* system properties and applications. Elsevier, Amsterdam, pp 3–18. https://doi.org/10.1016/B978-0-444-63501-3.00001-6

Yadav AN, Verma P, Singh B, Chauhan VS, Suman A, Saxena AK (2017b) Plant growth promoting bacteria: biodiversity and multifunctional attributes for sustainable agriculture. Adv Biotechnol Microbiol 5:1–16

Yadav RKP, Karamanoli K, Vokou D (2005) Bacterial colonization of the phyllosphere of Mediterranean perennial species as influenced by leaf structural and chemical features. Microb Ecol 50:185–196

Yang CH, Crowley DE, Borneman J et al (2001) Microbial phyllosphere populations are more complex than previously realized. Proc Natl Acad Sci U S A 98:3889–3894

Zabetakis I (1997) Enhancement of flavour biosynthesis from strawberry (*Fragariax ananassa*) callus cultures by *Methylobacterium* species. Plant Cell Tissue Organ Cult 50:179–183

Zahir ZA, Munir A, Asghar HN et al (2008) Effectiveness of rhizobacteria containing ACC-deaminase for growth promotion of pea (*Pisum sativum*) under drought conditions. J Microbiol Biotechnol 18:958–963

Zhalnina K, Louie KB, Hao Z et al (2018) Dynamic root exudate chemistry and microbial substrate preferences drive patterns in rhizosphere microbial community assembly. Nat Microbiol 3:470–480

Zilber-Rosenberg I, Rosenberg E (2008) Role of microorganisms in the evolution of animals and plants: the hologenome theory of evolution. FEMS Microbiol Rev 32:723–735

Rhizosphere Biology: A Key to Agricultural Sustainability

7

Simranjeet Singh, Vijay Kumar, Daljeet Singh Dhanjal, Sonali, Vaishali Dhaka, Siddharth Thotapalli, Joginder Singh, Laith Khalil Tawfeeq Al-Ani, and Liliana Aguilar-Marcelino

Abstract

The interactions of plant–microbe enable various types of transformations in the rhizosphere, which might be harmful, neutral, or beneficial. These interactions are proved helpful to plants for enhancing the biological, chemical, and physical properties of soil by facilitating the nutrients balance of the soil. Mutualistic plant–microbe interaction in the rhizosphere can enhance the nutrient uptake from roots, improve the biomass productivity and potentially, the ability to tolerate environmental stress. The microbial communities present in the rhizosphere influences the development of phytopathogens, the fitness of the ecological plants,

S. Singh
Department of Biotechnology, Lovely Professional University, Phagwara, Punjab, India

Punjab Biotechnology Incubators, Mohali, Punjab, India

Regional Advanced Water Testing Laboratory, Mohali, Punjab, India

V. Kumar
Department of Chemistry, Regional Ayurveda Research Institute for Drug Development, Gwalior, Madhya Pradesh, India

D. S. Dhanjal · V. Dhaka · S. Thotapalli · J. Singh
Department of Biotechnology, Lovely Professional University, Phagwara, Punjab, India

Sonali
Department of Microbiology, Lovely Professional University, Phagwara, Punjab, India

L. K. T. Al-Ani
School of Biology Science, Universiti Sains Malaysia, Pulau Pinang, Malaysia

L. Aguilar-Marcelino (✉)
Centro Nacional de Investigación Disciplinaria en Salud Animal e Inocuidad, INIFAP, Morelos, Mexico
e-mail: aguilar.liliana@inifap.gob.mx

A. N. Yadav et al. (eds.), *Current Trends in Microbial Biotechnology for Sustainable Agriculture*, Environmental and Microbial Biotechnology,
https://doi.org/10.1007/978-981-15-6949-4_7

and resistance of heavy metals and acquisition of nutrients. For improving the yields, varieties, and sustainability of the crops, the plant–microbe interaction is now getting considered as a valuable asset. Bioprospecting, the rhizospheric microorganisms with the ability to confer tolerance towards stress to host plant and using their symbiotic interaction with plants to improve the overall plant growth and crop productivity, could significantly aid in decreasing the adverse effects of stress on plants. The emerging field of engineering of ecosystems and rhizosphere marks a promising opportunity to fill critical research gaps and to develop sustainable solutions. Exploration of plant–microbe interactions is the key to understand the mechanism of rhizosphere priming, management of the carbon cycle in soil, and improve the crop productivity under current and future climatic conditions.

Keywords

Agricultural sustainability · Environment · Nutrient uptake · Roots exudates · Soil microbiomes

7.1 Introduction

In the year 1904, Hiltner coined the term "rhizosphere". It is referred to the soil zone present around the legume roots, which supports the bacterial activity. The rhizosphere is divided into three different types of regions (Lynch and de Leij 2012). These include the ecto-rhizosphere, rhizoplane, and endo-rhizosphere zone. The root tissue, which includes the layers of cortical and endodermis, is known as endo-rhizosphere. The rhizoplane includes the root surface area with the polysaccharide layer of mucilaginous and along with epidermis layer, whereas ecto-rhizosphere is defined as the region soil, which is adjacent to the root (Linderman 1991). Since various organic compounds get accumulated and released by roots exudation in the rhizosphere, this region is enriched with the nutrients (Ligaba et al. 2004). These nutrients are utilized by the microorganisms occurring in these regions as the sources of energy and carbon to increase their microbial activity and growth (Lugtenberg and Kamilova 2009). The microbial communities present in the rhizosphere influences the development of phytopathogens (Nehl et al. 1997), the fitness of the ecological plants (Barriuso et al. 2008), resistance of heavy metals (Kuffner et al. 2008), and acquisition of nutrients (Lynch 1990; Kour et al. 2020c).

The different types of organisms are found in the rhizosphere, namely archaea, nematodes, bacteria, protozoa, algae, fungi, arthropods, and oomycetes (Raaijmakers et al. 2009; Kour et al. 2019b; Yadav et al. 2018). The released nutrients from the plants are utilized by the different groups of the rhizospheric microbiome. It has been observed that in the regulation of plant roots activity and microbial diversity, the rhizodeposits (i.e. exudates) provides the major driving force to them. The pathogenic fungi, nematodes, oomycetes, bacteria, and fungi are the deleterious rhizosphere organisms (Van Baarlen et al. 2007; Tyler and Triplett 2008; Thakur et al. 2020). The defence of the frontline for the roots of plants against the pathogens of soil-borne attack is provided by the rhizosphere (Cook et al. 1995). This book chapter covers different aspects of plant–microbe interactions; new, improved engineering methods for bio-formulations. Efforts have also made to summarize the use of recombinant DNA technology to modify rhizosphere populations and their possible role of rhizospheric microbes in agricultural sustainability.

7.2 Plant–Microbe Interaction

The bacteria which are associated with the plant and capable of colonizing the roots are known as "rhizobacteria". They are classified into three groups, namely: (1) neutral, (2) beneficial, and (3) deleterious depending on their effects on plant growth. The bacteria stimulating the growth of plant referred to as beneficial rhizobacteria or also known as plant growth-promoting rhizobacteria (PGPR) (Kour et al. 2020b; Singh et al. 2020a). PGPR enhances crops growth indicating their potential in the agriculture field as biofertilizers (Timmusk et al. 1999; Kour et al. 2020f). The rhizospheric microorganisms are capable of forming the NH_4^+ by decomposing the proteins into amino acids via the ammonification process. The nitrification (NO_3^- formation) occurs after the ammonification at a rapid rate in most soils; hence, both NH_4^+ and NO_3^- are available for the plants but majorly NO_3^- is the main nitrogen source for the plants (Sylvia et al. 1999; Marschner 2011).

According to the root exudates quantity and quality, microbes associated with the rhizosphere are often transient (Biswas et al. 2018; Rana et al. 2020a). The rhizosphere-associated microbe's variation depends on the parameters influencing the chemical and biological aspects of the root (Yang and Crowley 2000; Morgan et al. 2005). The interactions of plant–microbe enable various types of transformations in the rhizosphere; for example, nutrient cycling mainly the sequestration of carbon and nitrogen (Philippot et al. 2013). The interaction between the plant and microbe might be harmful, neutral, or beneficial. The plant–microbe interaction is considered as a valuable asset due to their capabilities to improve the yields, varieties, and sustainability of the crop (Gopal and Gupta 2016). The primary factors which are involved in the inhibition or attraction of microbe's proliferation in the rhizosphere are the root exudates (Moore et al. 2014). Positive and beneficial interactions among rhizospheric microorganisms are favourable for good practices of agriculture. These interactions are not only important for the plant growth and development but also enhances the biological, chemical, and physical properties of soil by facilitating the nutrient balance of soil via biogeochemical cycles (Velmourougane et al. 2017). There are many ecological benefits due to this interaction, such as the availability of nutrients to the plants and promoting the plant growth (Boddey and Dobereiner 1995; Yadav et al. 2020c). The rhizospheric microbiome is able to protect the plant against the abiotic and biotic stress (Verma et al. 2017; Yadav et al. 2019).

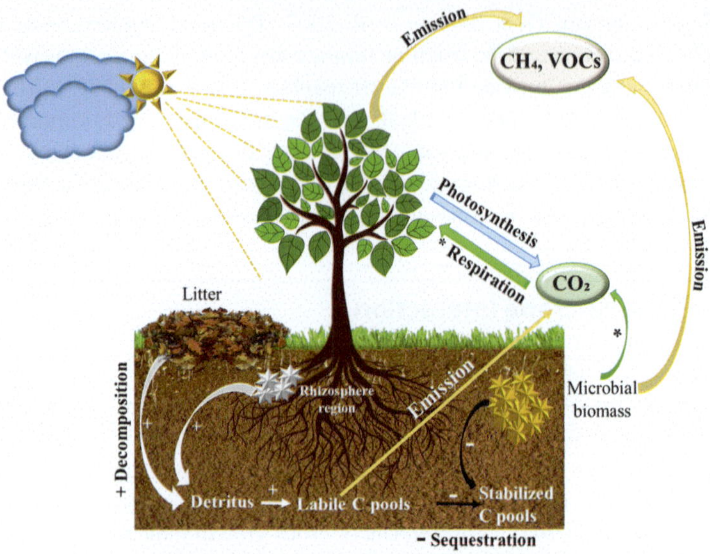

Fig. 7.1 Interaction of rhizosphere region with the other components of the plant ecosystem

The belowground diversity of the plant may perform as insurance under the different conditions of the environment for maintaining the productivity of the plants (Wagg et al. 2011). The rhizospheric microbes are considered as the soil quality bioindicators for the plants (Schnitzer et al. 2011; Yadav et al. 2020b). These rhizospheric microorganisms protect plants from the attack of the phytopathogens (Lugtenberg and Kamilova 2009). These include abiotic stress and disease control, root growth stimulation, biofertilization, and rhizoremediation (Kumar et al. 2019a, c). They can also facilitate the trace elements uptake, i.e. iron. In soil, iron is an abundant element under the conditions of alkaline and neutral (Andrews et al. 2003; Buckling et al. 2007). The interaction of the rhizosphere region with the other components of the plant ecosystem is illustrated in Fig. 7.1.

7.3 Engineering of Rhizosphere

Plant preservation is essential because of various reasons as it provides feed, food, fuel, aid in regulating carbon as well as the water cycle, climate, nutrition entrapment, and serve as habitat for wildlife. Considering, the massive diversity in the genotype of collected as well as generated plant species, the assessment of their genetic diversity of these plants has become highly important (Shishido et al. 2019). It could maintain the plant ecosystem and its values by stabilizing and generating stress tolerance in both cultivated and native ecosystem, and by retaining both cultivation and functioning of the ecosystem. These opinions direct that the selection of both species and genotypes should be taken into consideration while designing the breeding programme (Turnbull et al. 2016).

Hence, plant ecosystems can be engineered to improve carbon storage involving the allocated carbon in both above and belowground biomass for separating into the structural form or transport them to the soil for the conversion of recalcitrant minerals like calcite (Nogia et al. 2016). In 2010, Jansson and his colleagues comprehended and reviewed the potential of engineered plants in enhancing the carbon storage capacity and also introduced the term "phytosequestration"(Jackson and Baker 2010), whereas another group of scientists discussed the potential of terrestrial ecosystems in improving carbon storage. In the long run, storage of carbon in soil will become necessary. Therefore, a better understanding of the metabolic processes of microbial communities in rhizosphere and their interaction with the host plant and mechanism involved in carbon deposition is required (Dignac et al. 2017).

7.4 Plant Metabolism Through Rhizosphere Engineering

The conventional approach of plant breeding and advanced plant genetic engineering has been a success to accumulate desirable genes associated with stress response and tolerance in the plant genome. Most commonly employed strategy by plants to modify the rhizosphere is by altering exudation potential of roots; in view of this, researchers have attempted to develop transgenic plants that can alter the rhizospheric region by regulating the efflux of organic anions and H^+ in roots (Backer et al. 2018). Since the identification of several genes involved in root exudation, it has become possible to regulate the expression of those genes in plants for the incorporation of new features in the redesigned rhizosphere (Mark et al. 2005). For example, insertion of Arabidopsis vacuolar H^+ pyrophosphatase gene AVP1 in tomato and rice plants resulted in enhanced malate and citrate efflux, approximately 50%, on treatment with $AlPO_4$. This can be attributed to the increase of the tolerance in Al^{+3}-induced stress conditions and enhance the utilization of the insoluble form of phosphorus (Pasapula et al. 2011; Singh et al. 2020b). However, rhizosphere engineering is a complex process depending on several factors such as (1) inactivation of the engineered trait of the plant in the soil; (2) inability of the low rate of root exudation to affect the rhizosphere; (3) limited information about the composition of root exudates; and (4) variation in concentration and release time of root exudates during the development of plant and external stimuli.

Another approach involves exploring genetically diverse crops with desirable characters for partitioning and allocation of carbon (Canarini et al. 2019). It is debatable that increased distribution of photosynthate in rhizosphere will occur at the expense of carbon partitioning into harvestable compounds. However, reports suggest that inadequate sink demand can inhibit the process of photosynthesis through feedback response and make it sink limited. Thus, there is an immense potential for belowground allocation of carbon for long-term storage without imperilling crop productivity (Kaiser et al. 2015).

7.5 Genetic Modification of Rhizospheric Microbes

Genetic modification of microorganisms presents a unique opportunity to promote plant growth, confer resistance towards various diseases, and induce stress tolerance. Till now, numerous bacterial species have been identified to possess many advantageous effects but selecting and engineering a sustainable organism remains a challenge (Ortíz-Castro et al. 2009). For example, considering the inhabitation of two microbes in a niche, there can be six broad ecological interactions between them, namely commensalism, competition, predation, amensalism, cooperation, and null interaction. With the increase in microbial species in a niche, the perplexity of the ecological interactions among them increases linearly (Mougi 2016). The major challenge is to maximize positive interactions like cooperation and eliminate negative interactions like competition and parasitism. In view of this, it is an arduous task to minimize the competition between two strain co-cultures. The rate of plant growth, rate of seeding, sensitivity to pathogenic organisms, stabilization in adverse conditions, and sustainability of the microbiota are greatly influenced by the environmental factors such as pH, temperature, availability of nutrients, and exudates of the host plant (Bashey 2015). Besides these challenges, knowledge about interactions of natural soil microorganisms, including PGPR, can be exploited to develop a synthetic microbial community with desirable traits.

Numerous rhizosphere colonizing microorganisms have been identified as belonging to a wide range of genera whose genome sequences are publicly available, which are amenable to genetic modifications (Devi et al. 2020; Jacoby et al. 2017). These genera comprise of *Pseudomonas, Streptomyces, Rhizobium,* and *Bacillus.* Complete genetic sequences are available for *Streptomyces* spp., especially the ones used as PGPR. Still, they have certain limitations such as they have large genomes and possess mobile components which pose difficulty in engineering. *Bacillus* species are considered as an ideal organism to develop the synthetic microbial community as it is comparatively easy to modify genetically, has detailed information on genome sequences, contains many strains that promote plant growth, and are currently utilized as biocontrol agents (Vurukonda et al. 2018; Subrahmanyam et al. 2020). A consortium comprising of three different microbes, genetically modified *Bacillus* spp. and two other nitrogen-fixing microbes (natural or engineered) like *Bradyrhizobium, Pseudomonas,* and *Rhizobium* can provide many of the advantages of the complex natural microbiota of rhizosphere (de Souza et al. 2015; Yadav 2020).

To promote cooperation over competition, each strain can be engineered to make it deficient in certain essential genes such as elimination of gene synthesizing an essential enzyme or co-factor that is required by all strains (Hibbing et al. 2010). For instance, this could be understood as the system where *Bacillus* requires a co-factor produced by *Pseudomonas*, on the contrary, the *Pseudomonas* depends on the genes of *Rhizobium*, and *Bacillus* has the ability to remediate the waste generated by *Rhizobium* and recycle it for mutual use. This functional interaction among the strains on subsequent addition of the other strains as a consortium of three strains will have >729 predicted interaction, whereas a consortium of four strains will have about 531,441 predicted interaction (Gupta and Diwan 2017).

Hence, there is a need to limit the strain number to three in synthetic microbial community system so that their interaction among each other and with host plant could be controlled. In order to design the microbial consortium for an engineered rhizosphere, some critical realms need to be followed for their competence (McCarty and Ledesma-Amaro 2019; Mondal et al. 2020). Numerous traits need to be assessed prior to their selection for developing engineered microbial consortium: (a) Proficiency of microbes on colonizing the host plant roots in the rhizosphere, (b) Do the microbes colonize effectively on the host plant? (c) Are the microbes capable of surviving as well as competing with the other microbes in the consortium? (d) Is the adherence of microbes with the surface of root effective? (e) Does the microbe aid in promoting the plant growth or enhancing the growth of member of the consortium? (f) Do the microbes multiply themselves to reach the desired density? (g) Do the strains involved in consortium enable them to survive under abiotic stress? (Compant et al. 2019). The most important factor is the growth density irrespective of the reason that microbes will have a positive effect on the plant or not.

For instance, *Pseudomonas* spp. requires the growth density about 10^5–10^6 CFU/g of root to save the plant pathogens like *G. tritici* as well as *Pythium* spp. (Kwak and Weller 2013). If these standards are taken into consideration, then these microbial consortia could be used in the engineered rhizosphere, and these microbial consortia will help the plant in tolerating the effects induced by fertilizers, herbicides, and pesticides without losing their beneficial effects (Woo and Pepe 2018).

7.6 Molecular Mechanisms in the Rhizosphere

Previous studies mentioned the potential of PGPR in improving the growth of plants under stress conditions. Even advancement in molecular techniques has unveiled information regarding the genetic basis of PGPR that is showing the advantageous effect on plants (Shivakumar and Bhaktavatchalu 2017). Some of the studies that provide information regarding the molecular basis of PGPR have been comprehended in Table 7.1. Therefore, screening of the mechanism regulating the activities of PGPR will open the new avenue for genetic modifications of the microbe and host plant to improve their plant growing ability, especially under stress conditions.

In a study reported by Wang and collaborators, a microarray-based study was conducted to expand their knowledge about biochemical and physiological changes that take place in the plant. For this, they inoculated *Pseudomonas fluorescens* strain FPT9601-T5 (PGPR) in *Arabidopsis* plant. The result obtained on the analysis revealed that 200 genes out of 22,810 genes of *Arabidopsis* plant were showing different expression, i.e. two-fold increase in expression in PGPR-treated plant (Wang et al. 2005). Later, the majority of genes were found to be involved in different cellular processes like metabolic processes, stress response, and signal transduction. Moreover, upregulation of auxin-regulated genes, as well as nodulin-like genes and downregulation of ethylene-responsive genes, was observed (Markakis et al. 2012).

Table 7.1 The molecular studies involving PGPR under stress conditions

Species of plant	Microbial species	Molecular method used	References
Arabidopsis thaliana	*Bacillus megaterium* BP17	Microarray	Vibhuti et al. (2017)
Arabidopsis thaliana	*Bacillus subtilis* GB03	RT-PCR	Zhang et al. (2010)
Arabidopsis thaliana	*Pseudomonas fluorescens* FPT9601-T5	Microarray	Wang et al. (2005)
Arabidopsis thaliana	*Pseudomonas fluorescens* strain SS101	Microarray and LC-QTOF-MS	van de Mortel et al. (2012)
Arabidopsis thaliana	*Pseudomonas putida* MTCC5279	Microarray	Srivastava et al. (2012)
Cucumis sativus	*Acinetobacter calcoaceticus* SE370 and *Burkholderia cepacia* SE4	GC and enzyme-based assay	Khan et al. (2014)
Abelmoschus esculentus	*Enterobacter* sp. UPMR18	RT-PCR	Habib et al. (2016)
Piper nigrum	*Bacillus licheniformis* K11	2D-PAGE and PCR	Lim and Kim (2013)
Oryza sativa	*Azospirillum brasilense* Sp245	qRT-PCR	Vargas et al. (2012)
Saccharum officinarum	*Gluconacetobacter diazotrophicus* PAL5	Illumina sequencing	Vargas et al. (2014)
Triticum aestivum	*Dietzianatronolimnaea* STR1	qRT-PCR	Bharti et al. (2016)
Triticum aestivum	*Acinetobacter guillouiae* EU-B2RT.R1	16S rRNA-PCR	Rana et al. (2020b)
Triticum aestivum	*Pseudomonas libanensis* EULWNA-33	16S rRNA-PCR	Kour et al. (2020d)
Setaria italica	*Acinetobacter calcoaceticus* EU-LRNA-72	16S rRNA-PCR	Kour et al. (2020e)
Sorghum bicolor	*Streptomyces laurentii* EU-LWT$_3$–69	16S rRNA-PCR	Kour et al. (2020c)

Whereas another group of researchers with the help of RNA-Seq technology, i.e. Illumina, revealed that the inoculation of *Gluconacetobacter diazotrophicus* strain PAL5 in sugarcane triggered the ABA-dependent signalling genes and made its resistance to drought (Vargas et al. 2014). In 2015, Kim and his group showed that VOCs synthesized by *Bacillus subtilis* strain JS influenced the gene expression profiles of the tobacco. The upregulation in genes related to photosynthesis pathways was observed, signifying the VOC-mediated improvement in the growth of the plant (Tahir et al. 2017).

Other than the previous studies discussing gene expression profiles, proteomic analysis has also been conducted to gather more information about proteins as well as pathways triggered during host–PGPR interaction. As recognition of candidate protein among different PGPR could serve as a valuable resource for promoting the growth of the targeted plant in the near future (Singh et al. 2017). In 2008, Buensanteai and collaborators conducted an experiment on *Bacillus amyloliquefaciens* strain KPS46 inoculated in soybean plant to investigate the role of synthesized extracellular protein in improving plant growth and inducing systemic resistance (Radhakrishnan et al. 2017). For the separation of extracellular proteins synthesized by strain KPS46 (wild-type), KPS46 (mutant-type), N19G1, the methods like mass spectrometry (MS), two-dimensional polyacrylamide gel electrophoresis (2D–PAGE), and exploring of protein database were employed. The results obtained showed the presence of 20 extracellular proteins which could have a role in inducing resistance and plant development (Atshan et al. 2015). Another study revealed the presence of six different stress proteins on the molecular assessment of the pepper plant inoculated with *Bacillus licheniformis* strain K11 under drought stress. Even though there are technical constraints of using proteomic techniques for assessing the PGPR–host interaction but advancement in molecular techniques involving top-down proteomics and MALDI-TOF promises to extend our knowledge about the molecular basis for PGPR–host plant interaction in the near future (Lim and Kim 2013).

Furthermore, metabolic profiling of bacteria and plant is an alternative approach to understand the mechanism of symbiotic interactions. For instance, GC–MS analysis of drought-stressed wheat seedlings revealed the presence of seven stress-related VOCs in the rhizosphere and secondary metabolites were found to be β-pinene, benzaldehyde, and geranyl acetone. These three VOCs are likely to be considered as a promising candidate for rapid assessment of crop under drought stress. Hence, the deep insight about the genes, secondary metabolites, and proteins involved in plant–PGPR interaction and are responsible for abiotic stress resistance can be used for developing engineered plants. These engineered plants will harbour genes that control stress or microbes that alleviate the stress (Vaishnav et al. 2017).

7.7 Role of Rhizospheric Microbes for Agricultural Sustainability

7.7.1 Mutual Plant–Microbe Interactions

To overcome the adverse effects caused by environmental stresses, various strategies have been demonstrated. Transcriptome engineering is one such method to develop crops tolerant to abiotic stress (Cohen and Leach 2019). To date, the commonly used strategy to combat environmental stress in plants is to overexpress the single genes that encode for enzymes involved in the transportation of ions and scavenging of ROS. The application of this approach is limited due to the resultant pleiotropic effects on growth of the plant and comprehended multiple pathways in response to environmental stress (Xie et al. 2019). Utilization of agrochemicals is another method to enhance crop productivity in boosting crop productivity, but it is cost-intensive and has adverse effects on the environment on long-term use (Aktar et al. 2009). Employment of beneficial microbes in the rhizosphere of plants is another strategy to reduce the harmful effects of climatic fluctuations on the growth of plants and crop productivity. Mutualistic plant–microbe interaction in the

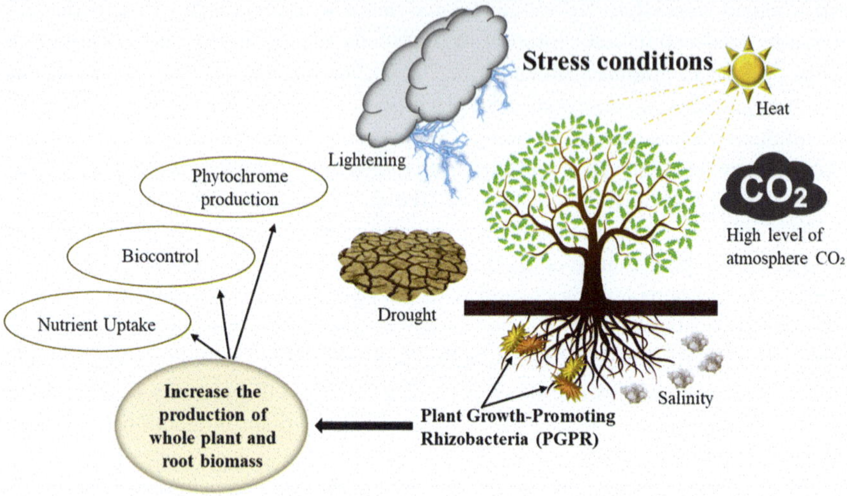

Fig. 7.2 Role of PGPR under the stress conditions

rhizosphere can enhance the nutrient uptake from roots, improve the biomass productivity and potentially, the ability to tolerate environmental stress (Igiehon and Babalola 2018). Bioprospecting, the rhizospheric microorganisms with the ability to confer tolerance towards stress to host plant and using their symbiotic interaction with plants to improve the overall plant growth and crop productivity, could significantly aid in decreasing the adverse effects of stress on plants. This approach has several advantages such as the ability of PGPR to confer multiple environmental stress tolerance to host plant, their application to diverse plant hosts and enhanced crop productivity as illustrated in Fig. 7.2 (Odelade and Babalola 2019).

7.7.2 Mitigation of Drought Stress

Among the environmental factors, drought is considered as the most critical factor that hampers plant growth and threatens crop productivity. Drought stress can be attributed to climatic changes, agronomic and edaphic factors (Rastegari et al. 2020a). Researchers predict that in the future, drought stress will worsen if the global supply of freshwater and climatic hitches remain a hurdle (Nadeem et al. 2019). In view of fluctuations in precipitation and global temperature, drought will hinder the production of biomass, feed, and most importantly, food. Thus, to ensure food security, the development of drought-tolerant crops becomes a necessity for a sustainable future. Most bioenergy crops used for biofuel production are tolerant towards drought conditions like poplar, miscanthus, etc. Therefore, there is an urgent need to enhance the tolerance of bioenergy crops towards drought and significantly improve their water use efficiency (WUE) for sustainable production of biomass in semi-arid and arid regions (Von Cossel et al. 2019).

Genetic engineering techniques have been extensively used to induce drought tolerance in plants, despite the efforts, there has been slow progress owing to the involvement of numerous genes and sophistication associated with the traits (Khan et al. 2019a, b; Rastegari et al. 2020b). It has been observed that the rhizosphere and microbiota associated with it play a vital role in constraining the capability of plants to manage the drought stress (Kour et al. 2019a; Verma et al. 2014, 2019; Yadav and Yadav 2018). The rhizosphere of plants is colonized by diverse microorganisms including plant growth-promoting rhizobacteria (PGPR) which provides them with the ability to cope with drought by aiding in the production of exopolysaccharides (EPS), phytohormones, and volatile organic compounds (VOCs) (Naseem et al. 2018; Tiwari et al. 2020). They also help in accumulating various antioxidants and osmolytes. Moreover, they can also alter the morphology of root in response to stress and regulate the stress-responsive genes (Sharma et al. 2019). For instance, it has been observed that the drought tolerance of wheat plant was enhanced by the inoculation of indole acetic acid (IAA) producing *Azospirillum* species which improved the growth of roots and induced lateral roots formation (Vurukonda et al. 2016). Similarly, the growth of *Lavandula dentata* in drought was stimulated by IAA producing plant growth-promoting bacteria, *Bacillus thuringiensis* that increased nutrient availability and improved the metabolic activities of the plant (Armada et al. 2016). In another study, grapevine and *Arabidopsis* plants were able to adapt to drought conditions when they were inoculated with GFP-labelled *Pseudomonas species* and *Acinetobacter species* which induced a water-stress mechanism to cope with drought (Rolli et al. 2015).

Upon inoculation of leaves of *Platycladus orientalis* with *Bacillus subtilis,* an increase in ABA concentration in shoots and stomatal conductance was observed, that provided drought resistance to the plant. Due to increased ABA levels, the water content in leaves enhanced, water potential improved, and cytokinin levels increased drastically (Liu et al. 2013). In another study, an isolate from the rhizosphere of *Brassica napus*, *Phyllobacterium brassicacearum* strain STM196 inoculated in *Arabidopsis* plants aided in acclimation of drought stress by enhancing ABA concentrations, reducing transpiration in leaves and increasing tolerance towards osmotic stress (Ahkami et al. 2017). Also, an inoculation of soybean plants with gibberellin-producing rhizobacterium, *Pseudomonas putida* strain H-2–3, an increase in fresh weight and length of shoots under drought conditions was reported (Kang et al. 2014b). In response to drought stress, they produced more chlorophyll, abscisic acid, and salicylic acid in comparison to control plants (Radhakrishnan et al. 2014).

7.7.3 Mitigation of Salinity Stress

Salinity is another major environmental factor that adversely affects the productivity of plants globally. Presence of salt in excess in the soil creates ionic imbalance and ion toxicity in plants which further triggers water deficiency in plants due to hyperosmotic stress and induces an imbalance in the metabolic activities (Shrivastava and Kumar 2015; Rajawat et al. 2020; Yadav et al. 2015; Kang et al. 2014a). Plants cope with stress due to salinity in various ways such as by producing polyamines and osmolytes, triggering defence mechanisms, preventing deposition of reactive oxygen species and regulating the transport of ions (Khan et al. 2019a, b; Gaba et al. 2017; Yadav et al. 2020a).

A study demonstrated that uptake of Na^+ ions by the plant was reduced significantly and the production of biomass enhanced when the wheat seedlings were subjected to the application PGPR like *Paenibacillus, Enterobacter, Bacillus,* etc. that synthesized exopolysaccharides (EPS) under highly saline conditions (Egamberdieva et al. 2019). In another study, PGPR inoculation in tomato plants reduced the adverse effects of ethylene, released under stress conditions, on the growth of roots by the activity of enzyme ACC deaminase which resulted in improved plant growth in water-deficit and saline conditions (Ilangumaran and Smith 2017). A recent study described the use of *Dietzia natronolimnaea* strain STR1, i.e. carotenoid producing and halotolerant, in combating the effects of salinity in wheat plants. Wheat plants inoculated with halotolerant PGPR showed higher levels of proline and production of numerous antioxidants that conferred salinity tolerance to the plants. Moreover, application of PGPR activated certain pathways in a plant-like ABA signalling, Fe transport, SOS pathways, etc. (Bharti et al. 2016).

In comparison to the uninoculated peanut seedlings, the inoculated peanut seedlings showed enhanced ion homeostasis, less accumulation of ROS, and improved growth under saline conditions. Another study showed the synergistic action of *Bacillus drentensis* and *Enterobacter cloacae* to aid in withstanding salinity in mung beans with foliar application of silicon (Ahkami et al. 2017). Moreover, when peanut seedlings inoculated with *Haererohalobacter*, *Brachybacterium saurashtrense,* and *Brevibacterium casei* were subjected to highly saline conditions by incorporation of 100 MNaCl, grown plants showed overall improved growth (Shukla et al. 2012).

7.7.4 Mitigation of Heavy Metals Stress

Heavy metals like Ni, As, Cr, Cd, Cu, Pb, Zn, etc. at low concentrations are essential to microbes and plants for the growth and metabolic activities but can present a major challenge if the concentration exceeds the tolerance limits (Singh et al. 2011). The presence of toxic heavy metals in soil greatly influence the characteristics of the plant and phytoremediation potentials; however, bacteria present in soil can significantly enhance the phytoremediation potential of the plant through synergistic action and hence the term, microbe-assisted phytoremediation (Ojuederie and Babalola 2017; Sharaff et al. 2020).

Reports suggest that PGPR also aid in protecting host plant from ill effects of toxicity caused by heavy metals. PGPR are known to possess this ability to cover a wide range of genera such as *Bradyrhizobium, Mesorhizobium, Sinorhizobium, Rhizobium, Pseudomonas, Azotobacter,* and *Bacillus* (Wani et al. 2008; Rai et al. 2020). For instance, a study showed that application of *Bacillus licheniformis* could significantly improve the germination of rice plant seed and enhance the biochemical characteristics of rice when subjected to stress induced by Ni. Therefore, highlighting the potential of the strain in protecting the rice plant from heavy metal toxicity (Jamil et al. 2014). Like most microorganisms, PGPR has also evolved in certain unique ways to tolerate heavy metals such as mobilization, immobilization, and transformation of heavy metals into either inactive form or less toxic utilizable form (Tiwari and Lata 2018). PGPR are known to follow five mechanisms broadly to increase heavy metal resistance: (1) Extrusion of heavy metals by transportation through efflux pumps; (2) Exclusion of heavy metals by direct removal from target sites; (3) Inactivation of heavy metals through the formation of complexes like the formation of thiol-containing complex structures; (4) Biotransformation of heavy metals from a toxic oxidation state to a less toxic oxidation state such as the conversion of highly toxic Cr^{+4} into less toxicCr^{+6}; and (5) Addition or removal of methyl from heavy metals, i.e. methylation and demethylation (Ma et al. 2016).

Similarly, plants also possess various mechanisms to cope with heavy metal resistance; however, the process by which microbes and plants interact at the molecular level to combat heavy metal toxicity remains unclear. Furthermore, increasing the knowledge about plant–microbe interactions, genes involved, and mechanisms of regulation, it would be possible to engineer plants for enhanced growth heavy metals contaminated sites (Mishra et al. 2017).

7.7.5 Mitigation of Heat Stress

Temperature is one of the abiotic stresses which negatively impact the growth, homeostasis, and metabolic activities of plants and microorganisms. Bioprospecting PGPR with the ability to promote plant growth at alleviated temperatures would possibly enhance global crop productivity, especially concerning the increased rate of global warming (Kour et al. 2020a). The experimental evidence supporting the effect of PGPR isolates in enhancing crop production at high temperatures is less. Till now, thermostable PGPR isolates stable even at 60 °C (Rodriguez et al. 2008) have been reported in the literature, but they lack the ability to provide thermostability to host plant. Nonetheless, some studies have shown the application of PGPR isolates to cope with the negative impacts due to low temperature-induced stress (Barka et al. 2006; Dimkpa et al. 2009). Low temperature-induced stress has resulted in enhanced synthesis of certain compounds like proline, sugar, anthocyanin, etc. (Dimkpaet al. 2009). In a study, grapevine plants inoculated with *Burkholderia phytofirmans* lead to increased production of carbohydrates, proline, and phenols along with the improved accumulation of starch (Barkaet al. 2006; Kumar et al. 2019b). However, PGPR-inoculated grapevine plant showed reduced biomass production and imbalance of electrolytes when subjected to low temperature (4 °C).

7.7.6 Combating Elevation CO_2 Levels

The process of photosynthesis plays a significant role in the uptake of atmospheric CO_2 and its conversion to organic carbon in plants biomass. The rise in CO_2 levels in atmosphere enhances the photosynthetic process in C3 plants, helping the proliferation of rhizospheric bacteria with enhanced localization of photosynthate in soil. Climatic fluctuations greatly influence the composition of plants as well as the diversity that threatens the soil microbes and edaphic characteristics of soil, including quality and quantity of organic matter in the soil. It also has a negative impact on various nutrient cycles like the carbon cycle, methane cycle, nitrogen cycle, and terrestrial ecosystem climates (Dorrepaal et al. 2009; Malyan et al. 2019). The PGPR utilization has enhanced the grassland management technology (Antoun et al. 1998; Van Der Heijden et al. 2006), restoration of the ecosystem (Requena et al. 2001), and reforestation (Chanway 1997). The PGPR have a remarkable ability to improve the accumulation of carbon in terrestrial systems by enhancing crop productivity and reducing the carbon loss through respiration in microbial systems at alleviated atmospheric CO_2 levels (Nie et al. 2015). However, the possibility of escalation of atmospheric CO_2 concentrations in future will broaden the horizon of PGPR application. The impact of microorganisms on the host plant through plant–microbe interactions is well known, but the mechanisms involved at the molecular level still remain unclear. Thus, it becomes important to study the plant growth dynamics and mechanism of rhizobacteria colonization to exploit the potential of PGPR further.

7.8 Conclusion and Future Prospects

Increasing crop productivity has become a global necessity. There is a need to improve environmental management practices, revert the effects of changing climate, and forecast the interaction and impact of plant ecosystems on atmospheric processes. To meet the ecological requirements, there is a need to understand plant ecosystem dynamics in stressful environments. The emerging field of engineering of ecosystems and rhizosphere marks a promising opportunity to fill critical research gaps and to develop solutions. The interactions within ectophytic and endophytic microbial communities along with mycorrhizal–rhizospheric relationship to promote plant growth and enhance nutrient uptake still remain unknown. Plant–microbe interactions is the key to understand the mechanism of rhizosphere priming, management of the carbon cycle in soil, and improve the crop productivity under current and future climatic conditions. Recent advancement in genetic engineering offers an exciting opportunity to fulfil the research gaps. Future studies will explore the synthetic approaches, which improves the production of bioenergy crops under abiotic and biotic conditions.

References

Ahkami AH, Allen White R, Handakumbura PP, Jansson C (2017) Rhizosphere engineering: enhancing sustainable plant ecosystem productivity. Rhizosphere 3:233–243

Aktar W, Sengupta D, Chowdhury A (2009) Impact of pesticides use in agriculture: their benefits and hazards. Interdiscip Toxicol 2:1–12. https://doi.org/10.2478/v10102-009-0001-7

Andrews SC, Robinson AK, Rodríguez-Quiñones F (2003) Bacterial iron homeostasis. FEMS Microbiol Rev 27:215–237

Antoun H, Beauchamp CJ, Goussard N et al (1998) Potential of *Rhizobium* and *Bradyrhizobium* species as plant growth promoting rhizobacteria on non-legumes: effect on radishes (*Raphanus sativus* L.). Plant Soil 204:57–67. https://doi.org/10.1023/A:1004326910584

Armada E, Probanza A, Roldán A, Azcón R (2016) Native plant growth promoting bacteria *Bacillus thuringiensis* and mixed or individual mycorrhizal species improved drought tolerance and oxidative metabolism in *Lavandula dentata* plants. J Plant Physiol 192:1–12. https://doi.org/10.1016/j.jplph.2015.11.007

Atshan SS, Shamsudin MN, Sekawi Z et al (2015) Comparative proteomic analysis of extracellular proteins expressed by various clonal types of *Staphylococcus aureus* and during planktonic growth and biofilm development. Front Microbiol 6:524. https://doi.org/10.3389/fmicb.2015.00524

Backer R, Rokem JS, Ilangumaran G et al (2018) Plant growth-promoting rhizobacteria: context, mechanisms of action, and roadmap to commercialization of biostimulants for sustainable agriculture. Front Plant Sci 9:1473

Barka EA, Nowak J, Clément C (2006) Enhancement of chilling resistance of inoculated grapevine plantlets with a plant growth-promoting rhizobacterium, *Burkholderia phytofirmans* strain PsJN. Appl Environ Microbiol 72:7246–7252. https://doi.org/10.1128/AEM.01047-06

Barriuso J, Ramos Solano B, Santamaría C et al (2008) Effect of inoculation with putative plant growth-promoting rhizobacteria isolated from *Pinus* spp. on *Pinuspinea* growth, mycorrhization and rhizosphere microbial communities. J Appl Microbiol 105:1298–1309. https://doi.org/10.1111/j.1365-2672.2008.03862.x

Bashey F (2015) Within-host competitive interactions as a mechanism for the maintenance of parasite diversity. Philos Trans R Soc B Biol Sci 370:20140301

Bharti N, Pandey SS, Barnawal D et al (2016) Plant growth promoting rhizobacteria *Dietzia natronolimnaea* modulates the expression of stress responsive genes providing protection of wheat from salinity stress. Sci Rep 6:34768. https://doi.org/10.1038/srep34768

Biswas S, Kundu D, Mazumdar S, Saha A, Majumdar B, Ghorai A et al (2018) Study on the activity and diversity of bacteria in a New Gangetic alluvial soil (Eutrocrept) under rice-wheat-jute cropping system. J Environ Biol 39:379–386

Boddey RM, Dobereiner J (1995) Nitrogen fixation associated with grasses and cereals: recent progress and perspectives for the future. In: Nitrogen economy in tropical soils. Springer, Dordrecht, pp 241–250

Buckling A, Harrison F, Vos M et al (2007) Siderophore-mediated cooperation and virulence in *Pseudomonas aeruginosa*. FEMS Microbiol Ecol 62:135–141

Canarini A, Kaiser C, Merchant A et al (2019) Root exudation of primary metabolites: mechanisms and their roles in plant responses to environmental stimuli. Front Plant Sci 10:157

Chanway CP (1997) Inoculation of tree roots with plant growth promoting soil bacteria: an emerging technology for reforestation. For Sci 43:99–112. https://doi.org/10.1093/forestscience/43.1.99

Cohen SP, Leach JE (2019) Abiotic and biotic stresses induce a core transcriptome response in rice. Sci Rep 9:1–11. https://doi.org/10.1038/s41598-019-42731-8

Compant S, Samad A, Faist H, Sessitsch A (2019) A review on the plant microbiome: ecology, functions, and emerging trends in microbial application. J Adv Res 19:29–37

Cook RJ, Thomashow LS, Weller DM et al (1995) Molecular mechanisms of defense by rhizobacteria against root disease. Proc Natl Acad Sci U S A 92:4197–4201. https://doi.org/10.1073/pnas.92.10.4197

de Souza R, Ambrosini A, Passaglia LMP (2015) Plant growth-promoting bacteria as inoculants in agricultural soils. Genet Mol Biol 38:401–419

Devi R, Kaur T, Kour D, Rana KL, Yadav A, Yadav AN (2020) Beneficial fungal communities from different habitats and their roles in plant growth promotion and soil health. Microb Biosystems 5:21–47. https://doi.org/10.21608/mb.2020.32802.1016

Dignac MF, Derrien D, Barré P et al (2017) Increasing soil carbon storage: mechanisms, effects of agricultural practices and proxies. A review. Agron Sustain Dev 37:1–27

Dimkpa C, Weinand T, Asch F (2009) Plant-rhizobacteria interactions alleviate abiotic stress conditions. Plant Cell Environ 32:1682–1694. https://doi.org/10.1111/j.1365-3040.2009.02028.x

Dorrepaal E, Toet S, Van Logtestijn RSP et al (2009) Carbon respiration from subsurface peat accelerated by climate warming in the subarctic. Nature 460:616–619. https://doi.org/10.1038/nature08216

Egamberdieva D, Wirth S, Bellingrath-Kimura SD et al (2019) Salt-tolerant plant growth promoting rhizobacteria for enhancing crop productivity of saline soils. Front Microbiol 10:2791. https://doi.org/10.3389/fmicb.2019.02791

Gaba S, Singh RN, Abrol S, Yadav AN, Saxena AK, Kaushik R (2017) Draft genome sequence of *Halolamina pelagica* CDK2 isolated from natural Salterns from Rann of Kutch, Gujarat, India. Genome Announc 5:1–2. https://doi.org/10.1128/genomeA.01593-16

Gopal M, Gupta A (2016) Microbiome selection could spur next-generation plant breeding strategies. Front Microbiol 7:1971. https://doi.org/10.3389/fmicb.2016.01971

Gupta P, Diwan B (2017) Bacterial exopolysaccharide mediated heavy metal removal: a review on biosynthesis, mechanism and remediation strategies. Biotechnol Rep 13:58–71

Habib SH, Kausar H, Saud HM (2016) Plant growth-promoting rhizobacteria enhance salinity stress tolerance in okra through ROS-scavenging enzymes. Biomed Res Int 2016:6284547. https://doi.org/10.1155/2016/6284547

Hibbing ME, Fuqua C, Parsek MR, Peterson SB (2010) Bacterial competition: surviving and thriving in the microbial jungle. Nat Rev Microbiol 8:15–25

Igiehon NO, Babalola OO (2018) Rhizosphere microbiome modulators: contributions of nitrogen fixing bacteria towards sustainable agriculture. Int J Environ Res Public Health 15:574

Ilangumaran G, Smith DL (2017) Plant growth promoting rhizobacteria in amelioration of salinity stress: a systems biology perspective. Front Plant Sci 8:1768

Jackson RB, Baker JS (2010) Opportunities and constraints for forest climate mitigation. Bioscience 60:698–707. https://doi.org/10.1525/bio.2010.60.9.7

Jacoby R, Peukert M, Succurro A et al (2017) The role of soil microorganisms in plant mineral nutrition—current knowledge and future directions. Front Plant Sci 8. https://doi.org/10.3389/fpls.2017.01617

Jamil M, Zeb S, Anees M et al (2014) Role of *Bacillus licheniformis* in phytoremediation of nickel contaminated soil cultivated with rice. Int J Phytoremediation 16:554–571. https://doi.org/10.1080/15226514.2013.798621

Kaiser C, Kilburn MR, Clode PL et al (2015) Exploring the transfer of recent plant photosynthates to soil microbes: mycorrhizal pathway vs direct root exudation. New Phytol 205:1537–1551. https://doi.org/10.1111/nph.13138

Kang SM, Khan AL, Waqas M et al (2014a) Plant growth-promoting rhizobacteria reduce adverse effects of salinity and osmotic stress by regulating phytohormones and antioxidants in *Cucumis sativus*. J Plant Interact 9:673–682. https://doi.org/10.1080/17429145.2014.894587

Kang SM, Radhakrishnan R, Khan AL et al (2014b) Gibberellin secreting rhizobacterium, *Pseudomonas putida* H-2-3 modulates the hormonal and stress physiology of soybean to improve the plant growth under saline and drought conditions. Plant Physiol Biochem 84:115–124. https://doi.org/10.1016/j.plaphy.2014.09.001

Khan IU, Gannon V, Jokinen CC, Kent R, Koning W, Lapen DR, et al (2014) A national investigation of the prevalence and diversity of thermophilic *Campylobacter* species in agricultural watersheds in Canada. Water Res 61: 243–252. https://doi.org/10.1016/j.watres.2014.05.027

Khan A, Khan AL, Muneer S et al (2019a) Silicon and salinity: crosstalk in crop-mediated stress tolerance mechanisms. Front Plant Sci 10:1429

Khan S, Anwar S, Yu S et al (2019b) Development of drought-tolerant transgenic wheat: achievements and limitations. Int J Mol Sci 20:3350. https://doi.org/10.3390/ijms20133350

Kour D, Kaur T, Devi R, Rana KL, Yadav N, Rastegari AA et al (2020a) Biotechnological applications of beneficial microbiomes for evergreen agriculture and human health. In: Rastegari AA, Yadav AN, Yadav N (eds) Trends of microbial biotechnology for sustainable agriculture and biomedicine systems: perspectives for human health. Elsevier, Amsterdam, pp 255–279. https://doi.org/10.1016/B978-0-12-820528-0.00019-3

Kour D, Kaur T, Yadav N, Rastegari AA, Singh B, Kumar V et al (2020b) Phytases from microbes in phosphorus acquisition for plant growth promotion and soil health. In: Rastegari AA, Yadav AN, Yadav N (eds) Trends of microbial biotechnology for sustainable agriculture and biomedicine systems: diversity and functional perspectives. Elsevier, Amsterdam, pp 157–176. https://doi.org/10.1016/B978-0-12-820526-6.00011-7

Kour D, Rana KL, Kaur T, Sheikh I, Yadav AN, Kumar V et al (2020c) Microbe-mediated alleviation of drought stress and acquisition of phosphorus in great millet (*Sorghum bicolour* L.) by drought-adaptive and phosphorus-solubilizing microbes. Biocatal Agric Biotechnol 23:101501. https://doi.org/10.1016/j.bcab.2020.101501

Kour D, Rana KL, Sheikh I, Kumar V, Yadav AN, Dhaliwal HS et al (2020d) Alleviation of drought stress and plant growth promotion by *Pseudomonas libanensis* EU-LWNA-33, a drought-adaptive phosphorus-solubilizing bacterium. Proc Natl Acad Sci India B. https://doi.org/10.1007/s40011-019-01151-4

Kour D, Rana KL, Yadav AN, Sheikh I, Kumar V, Dhaliwal HS et al (2020e) Amelioration of drought stress in foxtail millet (*Setaria italica* L.) by P-solubilizing drought-tolerant microbes with multifarious plant growth promoting attributes. Environ Sustain 3:23–34. https://doi.org/10.1007/s42398-020-00094-1

Kour D, Rana KL, Yadav AN, Yadav N, Kumar M, Kumar V et al (2020f) Microbial biofertilizers: bioresources and eco-friendly technologies for agricultural and environmental sustainability. Biocatal Agric Biotechnol 23:101487. https://doi.org/10.1016/j.bcab.2019.101487

Kour D, Rana KL, Yadav AN, Yadav N, Kumar V, Kumar A et al (2019a) Drought-tolerant phosphorus-solubilizing microbes: biodiversity and biotechnological applications for alleviation of drought stress in plants. In: Sayyed RZ, Arora NK, Reddy MS (eds) Plant growth promoting rhizobacteria for sustainable stress management, Rhizobacteria in abiotic stress management, vol 1. Springer, Singapore, pp 255–308. https://doi.org/10.1007/978-981-13-6536-2_13

Kour D, Rana KL, Yadav N, Yadav AN, Kumar A, Meena VS et al (2019b) Rhizospheric microbiomes: biodiversity, mechanisms of plant growth promotion, and biotechnological applications for sustainable agriculture. In: Kumar A, Meena VS (eds) Plant growth promoting rhizobacteria for agricultural sustainability: from theory to practices. Springer, Singapore, pp 19–65. https://doi.org/10.1007/978-981-13-7553-8_2

Kuffner M, Puschenreiter M, Wieshammer G et al (2008) Rhizosphere bacteria affect growth and metal uptake of heavy metal accumulating willows. Plant Soil 304:35–44. https://doi.org/10.1007/s11104-007-9517-9

Kumar A, Chaturvedi AK, Yadav K, Arunkumar KP, Malyan SK, Raja P et al (2019a) Fungal phytoremediation of heavy metal-contaminated resources: current scenario and future prospects. In: Yadav AN, Singh S, Mishra S, Gupta A (eds) Recent advancement in white biotechnology through fungi, Perspective for sustainable environments, vol 3. Springer International Publishing, Cham, pp 437–461. https://doi.org/10.1007/978-3-030-25506-0_18

Kumar M, Kour D, Yadav AN, Saxena R, Rai PK, Jyoti A et al (2019b) Biodiversity of methylotrophic microbial communities and their potential role in mitigation of abiotic stresses in plants. Biologia 74:287–308. https://doi.org/10.2478/s11756-019-00190-6

Kumar V, Joshi S, Pant NC, Sangwan P, Yadav AN, Saxena A et al (2019c) Molecular approaches for combating multiple abiotic stresses in crops of arid and semi-arid region. In: Singh SP, Upadhyay SK, Pandey A, Kumar S (eds) Molecular approaches in plant biology and environmental challenges. Springer, Singapore, pp 149–170. https://doi.org/10.1007/978-981-15-0690-1_8

Kwak YS, Weller DM (2013) Take-all of wheat and natural disease suppression: a review. Plant Pathol J 29:125–135

Ligaba A, Shen H, Shibata K et al (2004) The role of phosphorus in aluminium-induced citrate and malate exudation from rape (*Brassica napus*). Physiol Plant 120:575–584. https://doi.org/10.1111/j.0031-9317.2004.0290.x

Lim JH, Kim SD (2013) Induction of drought stress resistance by multi-functional PGPR *Bacillus licheniformis* K11 in pepper. Plant Pathol J 29:201–208. https://doi.org/10.5423/PPJ.SI.02.2013.0021

Linderman RG (1991) Mycorrhizal interactions in the rhizosphere. In: The rhizosphere and plant growth. Springer, Dordrecht, pp 343–348

Liu F, Xing S, Ma H et al (2013) Cytokinin-producing, plant growth-promoting rhizobacteria that confer resistance to drought stress in *Platycladus orientalis* container seedlings. Appl Microbiol Biotechnol 97:9155–9164. https://doi.org/10.1007/s00253-013-5193-2

Lugtenberg B, Kamilova F (2009) Plant-growth-promoting rhizobacteria. Annu Rev Microbiol 63:541–556. https://doi.org/10.1146/annurev.micro.62.081307.162918

Lynch JM (1990) The rhizosphere. John Wiley, Chichester; New York

Lynch JM, de Leij F (2012) Rhizosphere. In: eLS. John Wiley & Sons, Ltd, Chichester

Ma Y, Oliveira RS, Freitas H, Zhang C (2016) Biochemical and molecular mechanisms of plant-microbe-metal interactions: relevance for phytoremediation. Front Plant Sci 7:918. https://doi.org/10.3389/fpls.2016.00918

Malyan SK, Kumar A, Baram S, Kumar J, Singh S, Kumar SS et al (2019) Role of fungi in climate change abatement through carbon sequestration. In: Yadav AN, Singh S, Mishra S, Gupta A (eds) Recent advancement in White biotechnology through fungi, Perspective for sustainable environments, vol 3. Springer International Publishing, Cham, pp 283–295. https://doi.org/10.1007/978-3-030-25506-0_11

Mark GL, Dow JM, Kiely PD et al (2005) Transcriptome profiling of bacterial responses to root exudates identifies genes involved in microbe-plant interactions. Proc Natl Acad Sci U S A 102:17454–17459. https://doi.org/10.1073/pnas.0506407102

Markakis MN, De Cnodder T, Lewandowski M et al (2012) Identification of genes involved in the ACC-mediated control of root cell elongation in *Arabidopsis thaliana*. BMC Plant Biol 12:208. https://doi.org/10.1186/1471-2229-12-208

Marschner P (2011) Marschner's mineral nutrition of higher plants, 3rd edn. Elsevier Inc., Amsterdam

McCarty NS, Ledesma-Amaro R (2019) Synthetic biology tools to engineer microbial communities for biotechnology. Trends Biotechnol 37:181–197

Mishra J, Singh R, Arora NK (2017) Alleviation of heavy metal stress in plants and remediation of soil by rhizosphere microorganisms. Front Microbiol 8:1706. https://doi.org/10.3389/fmicb.2017.01706

Mondal S, Halder SK, Yadav AN, Mondal KC (2020) Microbial consortium with multifunctional plant growth promoting attributes: future perspective in agriculture. In: Yadav AN, Rastegari AA, Yadav N, Kour D (eds) Advances in plant microbiome and sustainable agriculture, Functional annotation and future challenges, vol 2. Springer, Singapore, pp 219–254. https://doi.org/10.1007/978-981-15-3204-7_10

Moore BD, Andrew RL, Külheim C, Foley WJ (2014) Explaining intraspecific diversity in plant secondary metabolites in an ecological context. New Phytol 201:733–750

Morgan JAW, Bending GD, White PJ (2005) Biological costs and benefits to plant-microbe interactions in the rhizosphere. J Exp Bot 56:1729–1739. https://doi.org/10.1093/jxb/eri205

Mougi A (2016) The roles of amensalistic and commensalistic interactions in large ecological network stability. Sci Rep 6:29929. https://doi.org/10.1038/srep29929

Nadeem M, Li J, Yahya M et al (2019) Research progress and perspective on drought stress in legumes: a review. Int J Mol Sci 20:799

Naseem H, Ahsan M, Shahid MA, Khan N (2018) Exopolysaccharides producing rhizobacteria and their role in plant growth and drought tolerance. J Basic Microbiol 58:1009–1022

Nehl DB, Allen SJ, Brown JF (1997) Deleterious rhizosphere bacteria: an integrating perspective. Appl Soil Ecol 5:1–20. https://doi.org/10.1016/S0929-1393(96)00124-2

Nie M, Bell C, Wallenstein MD, Pendall E (2015) Increased plant productivity and decreased microbial respiratory C loss by plant growth-promoting rhizobacteria under elevated CO_2. Sci Rep 5:9212. https://doi.org/10.1038/srep09212

Nogia P, Sidhu GK, Mehrotra R, Mehrotra S (2016) Capturing atmospheric carbon: biological and nonbiological methods. Int J Low-Carbon Technol 11:266–274. https://doi.org/10.1093/ijlct/ctt077

Odelade KA, Babalola OO (2019) Bacteria, fungi and archaea domains in rhizospheric soil and their effects in enhancing agricultural productivity. Int J Environ Res Public Health 16:3873

Ojuederie OB, Babalola OO (2017) Microbial and plant-assisted bioremediation of heavy metal polluted environments: a review. Int J Environ Res Public Health 14:1504

Ortíz-Castro R, Contreras-Cornejo HA, Macías-Rodríguez L, López-Bucio J (2009) The role of microbial signals in plant growth and development. Plant Signal Behav 4:701–712. https://doi.org/10.4161/psb.4.8.9047

Pasapula V, Shen G, Kuppu S et al (2011) Expression of an *Arabidopsis* vacuolar H+-pyrophosphatase gene (AVP1) in cotton improves drought- and salt tolerance and increases fibre yield in the field conditions. Plant Biotechnol J 9:88–99. https://doi.org/10.1111/j.1467-7652.2010.00535.x

Philippot L, Raaijmakers JM, Lemanceau P, Van Der Putten WH (2013) Going back to the roots: the microbial ecology of the rhizosphere. Nat Rev Microbiol 11:789–799

Raaijmakers JM, Paulitz TC, Steinberg C et al (2009) The rhizosphere: a playground and battlefield for soil-borne pathogens and beneficial microorganisms. Plant Soil 321:341–361. https://doi.org/10.1007/s11104-008-9568-6

Radhakrishnan R, Hashem A, Abd Allah EF (2017) *Bacillus*: a biological tool for crop improvement through bio-molecular changes in adverse environments. Front Physiol 8:667. https://doi.org/10.3389/fphys.2017.00667

Radhakrishnan R, Kang SM, Baek IY, Lee IJ (2014) Characterization of plant growth-promoting traits of *Penicillium* species against the effects of high soil salinity and root disease. J Plant Interact 9:754–762. https://doi.org/10.1080/17429145.2014.930524

Rai PK, Singh M, Anand K, Saurabhj S, Kaur T, Kour D et al (2020) Role and potential applications of plant growth promotion rhizobacteria for sustainable agriculture. In: Rastegari AA, Yadav AN, Yadav N (eds) Trends of microbial biotechnology for sustainable agriculture and biomedicine systems: diversity and functional perspectives. Elsevier, Amsterdam, pp 49–60. https://doi.org/10.1016/B978-0-12-820526-6.00004-X

Rajawat MVS, Singh R, Singh D, Yadav AN, Singh S, Kumar M et al (2020) Spatial distribution and identification of bacteria in stressed environments capable to weather potassium aluminosilicate mineral. Braz J Microbiol 51:751–764. https://doi.org/10.1007/s42770-019-00210-2

Rana KL, Kour D, Kaur T, Devi R, Yadav AN, Yadav N et al (2020a) Endophytic microbes: biodiversity, plant growth-promoting mechanisms and potential applications for agricultural sustainability. Antonie Van Leeuwenhoek. https://doi.org/10.1007/s10482-020-01429-y

Rana KL, Kour D, Kaur T, Sheikh I, Yadav AN, Kumar V et al (2020b) Endophytic microbes from diverse wheat genotypes and their potential biotechnological applications in plant growth promotion and nutrient uptake. Proc Natl Acad Sci India B. https://doi.org/10.1007/s40011-020-01168-0

Rastegari AA, Yadav AN, Yadav N (2020a) New and future developments in microbial biotechnology and bioengineering: Trends of microbial biotechnology for sustainable agriculture and biomedicine systems: diversity and functional perspectives. Elsevier, Amsterdam

Rastegari AA, Yadav AN, Yadav N (2020b) New and future developments in microbial biotechnology and bioengineering: Trends of microbial biotechnology for sustainable agriculture and biomedicine systems: perspectives for human health. Elsevier, Amsterdam

Requena N, Perez-Solis E, Azcón-Aguilar C et al (2001) Management of indigenous plant-microbe symbioses aids restoration of desertified ecosystems. Appl Environ Microbiol 67:495–498. https://doi.org/10.1128/AEM.67.2.495-498.2001

Rodriguez RJ, Henson J, Van Volkenburgh E et al (2008) Stress tolerance in plants via habitat-adapted symbiosis. ISME J 2:404–416. https://doi.org/10.1038/ismej.2007.106

Rolli E, Marasco R, Vigani G et al (2015) Improved plant resistance to drought is promoted by the root-associated microbiome as a water stress-dependent trait. Environ Microbiol 17:316–331. https://doi.org/10.1111/1462-2920.12439

Schnitzer SA, Klironomos JN, Hille Ris Lambers J et al (2011) Soil microbes drive the classic plant diversity-productivity pattern. Ecology 92:296–303. https://doi.org/10.1890/10-0773.1

Sharaff MS, Subrahmanyam G, Kumar A, Yadav AN (2020) Mechanistic understanding of root-microbiome interaction for sustainable agriculture in polluted soils. In: Rastegari AA, Yadav AN, Yadav N (eds) Trends of microbial biotechnology for sustainable agriculture and biomedicine systems: diversity and functional perspectives. Elsevier, Amsterdam, pp 61–84. https://doi.org/10.1016/B978-0-12-820526-6.00005-1

Sharma A, Shahzad B, Kumar V et al (2019) Phytohormones regulate accumulation of osmolytes under abiotic stress. Biomol Ther 9:285

Shishido R, Akimoto M, Htut T et al (2019) Assessment of genetic diversity and genetic structure of wild rice populations in Myanmar. Breed Sci 69:471–477. https://doi.org/10.1270/jsbbs.18165

Shivakumar S, Bhaktavatchalu S (2017) Role of plant growth-promoting rhizobacteria (PGPR) in the improvement of vegetable crop production under stress conditions. In: Microbial strategies for vegetable production. Springer International Publishing, Cham, pp 81–97

Shrivastava P, Kumar R (2015) Soil salinity: a serious environmental issue and plant growth promoting bacteria as one of the tools for its alleviation. Saudi J Biol Sci 22:123–131

Shukla PS, Agarwal PK, Jha B (2012) Improved salinity tolerance of *Arachis hypogaea* (L.) by the interaction of halotolerant plant-growth-promoting rhizobacteria. J Plant Growth Regul 31:195–206. https://doi.org/10.1007/s00344-011-9231-y

Singh A, Kumar R, Yadav AN, Mishra S, Sachan S, Sachan SG (2020a) Tiny microbes, big yields: microorganisms for enhancing food crop production sustainable development. In: Rastegari AA, Yadav AN, Yadav N (eds) Trends of microbial biotechnology for sustainable agriculture and biomedicine systems: diversity and functional perspectives. Elsevier, Amsterdam, pp 1–15. https://doi.org/10.1016/B978-0-12-820526-6.00001-4

Singh B, Boukhris I, Pragya, Kumar V, Yadav AN, Farhat-Khemakhem A et al (2020b) Contribution of microbial phytases to the improvement of plant growth and nutrition: a review. Pedosphere 30:295–313. https://doi.org/10.1016/S1002-0160(20)60010-8

Singh R, Gautam N, Mishra A, Gupta R (2011) Heavy metals and living systems: an overview. Indian J Pharmacol 43:246–253

Singh RP, Runthala A, Khan S, Jha PN (2017) Quantitative proteomics analysis reveals the tolerance of wheat to salt stress in response to *Enterobacter cloacae* SBP-8. PLoS One 12:e0183513. https://doi.org/10.1371/journal.pone.0183513

Srivastava S, Chaudhry V, Mishra A et al (2012) Gene expression profiling through microarray analysis in *Arabidopsis thaliana* colonized by *Pseudomonas putida* MTCC 5279, a plant growth promoting rhizobacterium. Plant Signal Behav 7:235–245. https://doi.org/10.4161/psb.18957

Subrahmanyam G, Kumar A, Sandilya SP, Chutia M, Yadav AN (2020) Diversity, plant growth promoting attributes, and agricultural applications of rhizospheric microbes. In: Yadav AN, Singh J, Rastegari AA, Yadav N (eds) Plant microbiomes for sustainable agriculture. Springer, Cham, pp 1–52. https://doi.org/10.1007/978-3-030-38453-1_1

Sylvia DM, Fuhrmann JJ, Hartel PG, Zuberer DA (1999) Habitat and organisms. In: Principles and applications of soil microbiology. Prentice Hall, Upper Saddle River, NJ, pp 72–92

Tahir HAS, Gu Q, Wu H et al (2017) Plant growth promotion by volatile organic compounds produced by *Bacillus subtilis* SYST2. Front Microbiol 8:171. https://doi.org/10.3389/fmicb.2017.00171

Thakur N, Kaur S, Tomar P, Thakur S, Yadav AN (2020) Microbial biopesticides: current status and advancement for sustainable agriculture and environment. In: Rastegari AA, Yadav AN, Yadav N (eds) Trends of microbial biotechnology for sustainable agriculture and biomedicine systems: diversity and functional perspectives. Elsevier, Amsterdam, pp 243–282. https://doi.org/10.1016/B978-0-12-820526-6.00016-6

Timmusk S, Nicander B, Granhall U, Tillberg E (1999) Cytokinin production by *Paenibacillus polymyxa*. Soil Biol Biochem 31:1847–1852. https://doi.org/10.1016/S0038-0717(99)00113-3

Tiwari P, Bajpai M, Singh LK, Mishra S, Yadav AN (2020) Phytohormones producing fungal communities: metabolic engineering for abiotic stress tolerance in crops. In: Yadav AN, Mishra S, Kour D, Yadav N, Kumar A (eds) Agriculturally important fungi for sustainable agriculture, volume 1: perspective for diversity and crop productivity. Springer, Cham. https://doi.org/10.1007/978-3-030-45971-0_8

Tiwari S, Lata C (2018) Heavy metal stress, signaling, and tolerance due to plant-associated microbes: an overview. Front Plant Sci 9:452

Turnbull LA, Isbell F, Purves DW et al (2016) Understanding the value of plant diversity for ecosystem functioning through niche theory. Proc R Soc B Biol Sci 283:20160536

Tyler HL, Triplett EW (2008) Plants as a habitat for beneficial and/or human pathogenic bacteria. Annu Rev Phytopathol 46:53–73. https://doi.org/10.1146/annurev.phyto.011708.103102

Vaishnav A, Varma A, Tuteja N, Choudhary DK (2017) Characterization of bacterial volatiles and their impact on plant health under abiotic stress. In: Volatiles and food security: role of volatiles in agro-ecosystems. Springer, Singapore, pp 15–24

Van Baarlen P, Van Belkum A, Summerbell RC et al (2007) Molecular mechanisms of pathogenicity: how do pathogenic microorganisms develop cross-kingdom host jumps? FEMS Microbiol Rev 31:239–277

van de Mortel JE, de Vos RCH, Dekkers E et al (2012) Metabolic and transcriptomic changes induced in arabidopsis by the rhizobacterium *Pseudomonas fluorescens* SS101. Plant Physiol 160:2173–2188. https://doi.org/10.1104/pp.112.207324

Van Der Heijden MGA, Bakker R, Verwaal J et al (2006) Symbiotic bacteria as a determinant of plant community structure and plant productivity in dune grassland. FEMS Microbiol Ecol 56:178–187. https://doi.org/10.1111/j.1574-6941.2006.00086.x

Vargas L, Brígida ABS, Mota Filho JP et al (2014) Drought tolerance conferred to sugarcane by association with *Gluconacetobacter diazotrophicus*: a transcriptomic view of hormone pathways. PLoS One 9:e114744. https://doi.org/10.1371/journal.pone.0114744

Vargas L, de Carvalho TLG, Ferreira PCG et al (2012) Early responses of rice (*Oryza sativa* L.) seedlings to inoculation with beneficial diazotrophic bacteria are dependent on plant and bacterial genotypes. Plant Soil 356:127–137. https://doi.org/10.1007/s11104-012-1274-8

Velmourougane K, Prasanna R, Singh S et al (2017) Modulating rhizosphere colonization, plant growth, soil nutrient availability and plant defense enzyme activity through *Trichoderma viride-Azotobacter chroococcum* biofilm inoculation in chickpea. Plant Soil 421:157–174. https://doi.org/10.1007/s11104-017-3445-0

Verma P, Yadav AN, Kazy SK, Saxena AK, Suman A (2014) Evaluating the diversity and phylogeny of plant growth promoting bacteria associated with wheat (*Triticum aestivum*) growing in central zone of India. Int J Curr Microbiol Appl Sci 3:432–447

Verma P, Yadav AN, Khannam KS, Mishra S, Kumar S, Saxena AK et al (2019) Appraisal of diversity and functional attributes of thermotolerant wheat associated bacteria from the peninsular zone of India. Saudi J Biol Sci 26:1882–1895. https://doi.org/10.1016/j.sjbs.2016.01.042

Verma P, Yadav AN, Kumar V, Singh DP, Saxena AK (2017) Beneficial plant-microbes interactions: biodiversity of microbes from diverse extreme environments and its impact for crop improvement. In: Singh DP, Singh HB, Prabha R (eds) Plant-microbe interactions in agro-ecological perspectives, Microbial interactions and agro-ecological impacts, vol 2. Springer, Singapore, pp 543–580. https://doi.org/10.1007/978-981-10-6593-4_22

Vibhuti M, Kumar A, Sheoran N et al (2017) Molecular basis of endophytic *Bacillus megaterium*-induced growth promotion in *Arabidopsis thaliana*: revelation by microarray-based gene expression analysis. J Plant Growth Regul 36:118–130. https://doi.org/10.1007/s00344-016-9624-z

Von Cossel M, Wagner M, Lask J et al (2019) Prospects of bioenergy cropping systems for a more social-ecologically sound bioeconomy. Agronomy 9:605

Vurukonda SSKP, Giovanardi D, Stefani E (2018) Plant growth promoting and biocontrol activity of *Streptomyces* spp. as endophytes. Int J Mol Sci 19:952

Vurukonda SSKP, Vardharajula S, Shrivastava M, SkZ A (2016) Enhancement of drought stress tolerance in crops by plant growth promoting rhizobacteria. Microbiol Res 184:13–24

Wagg C, Jansa J, Schmid B, van der Heijden MGA (2011) Below ground biodiversity effects of plant symbionts support aboveground productivity. Ecol Lett 14:1001–1009. https://doi.org/10.1111/j.1461-0248.2011.01666.x

Wang Y, Ohara Y, Nakayashiki H et al (2005) Microarray analysis of the gene expression profile induced by the endophytic plant growth-promoting rhizobacteria, *Pseudomonas fluorescens* FPT9601-T5 in Arabidopsis. Mol Plant-Microbe Interact 18:385–396. https://doi.org/10.1094/MPMI-18-0385

Wani PA, Khan MS, Zaidi A (2008) Effect of metal-tolerant plant growth-promoting rhizobium on the performance of pea grown in metal-amended soil. Arch Environ Contam Toxicol 55:33–42. https://doi.org/10.1007/s00244-007-9097-y

Woo SL, Pepe O (2018) Microbial consortia: promising probiotics as plant biostimulants for sustainable agriculture. Front Plant Sci 9:1801. https://doi.org/10.3389/fpls.2018.01801

Xie X, He Z, Chen N et al (2019) The roles of environmental factors in regulation of oxidative stress in plant. Biomed Res Int 2019:9732325. https://doi.org/10.1155/2019/9732325

Yadav AN (2020) Plant microbiomes for sustainable agriculture: current research and future challenges. In: Yadav AN, Singh J, Rastegari AA, Yadav N (eds) Plant microbiomes for sustainable agriculture. Springer International Publishing, Cham, pp 475–482. https://doi.org/10.1007/978-3-030-38453-1_16

Yadav AN, Gulati S, Sharma D, Singh RN, Rajawat MVS, Kumar R et al (2019) Seasonal variations in culturable archaea and their plant growth promoting attributes to predict their role in establishment of vegetation in Rann of Kutch. Biologia 74:1031–1043. https://doi.org/10.2478/s11756-019-00259-2

Yadav AN, Kaur T, Kour D, Rana KL, Yadav N, Rastegari AA et al (2020a) Saline microbiome: biodiversity, ecological significance and potential role in amelioration of salt stress in plants. In: Rastegari AA, Yadav AN, Yadav N (eds) Trends of microbial biotechnology for sustainable agriculture and biomedicine systems: diversity and functional perspectives. Elsevier, Amsterdam, pp 283–309. https://doi.org/10.1016/B978-0-12-820526-6.00018-X

Yadav AN, Rastegari AA, Yadav N, Kour D (2020b) Advances in plant microbiome and sustainable agriculture: diversity and biotechnological applications. Springer, Singapore

Yadav AN, Sharma D, Gulati S, Singh S, Dey R, Pal KK et al (2015) Haloarchaea endowed with phosphorus solubilization attribute implicated in phosphorus cycle. Sci Rep 5:12293

Yadav AN, Singh J, Rastegari AA, Yadav N (2020c) Plant microbiomes for sustainable agriculture. Springer, Cham

Yadav AN, Verma P, Kumar S, Kumar V, Kumar M, Singh BP et al (2018) Actinobacteria from rhizosphere: molecular diversity, distributions and potential biotechnological applications. In: Singh B, Gupta V, Passari A (eds) New and future developments in microbial biotechnology and bioengineering. Elsevier, Amsterdam, pp 13–41. https://doi.org/10.1016/B978-0-444-63994-3.00002-3

Yadav AN, Yadav N (2018) Stress-adaptive microbes for plant growth promotion and alleviation of drought stress in plants. Acta Sci Agric 2:85–88

Yang CH, Crowley DE (2000) Rhizosphere microbial community structure in relation to root location and plant iron nutritional status. Appl Environ Microbiol 66:345–351. https://doi.org/10.1128/AEM.66.1.345-351.2000

Zhang H, Murzello C, Sun Y et al (2010) Choline and osmotic-stress tolerance induced in arabidopsis by the soil microbe *Bacillus subtilis* (GB03). Mol Plant-Microbe Interact 23:1097–1104. https://doi.org/10.1094/MPMI-23-8-1097

Rhizosphere Microbiomes and Their Potential Role in Increasing Soil Fertility and Crop Productivity

8

Sara Atef Gezaf, Hebtallah H. Abo Mahas, and Ahmed M. Abdel-Azeem

Abstract

Plant microbiome in rhizosphere plays the most critical role in plant growth promoting (PGP), development, and fertilization of soil. Plants and rhizospheric soil are natural resources that harbor microorganisms, and this plays important roles in the maintenance of nutrient balance and ecosystem function. The diverse group of microbes is significant components of soil plant systems, where they are bound in an intense network of interactions within the (rhizosphere-phyllospheric-endophytic). The microbes with PGP attributes have emerged as an important and promising tool for sustainable agriculture. PGP microbes promote plant growth directly or indirectly either by releasing plant growth phytohormones; solubilization of phosphorus, potassium, and zinc; and biological process such as nitrogen fixation or by producing siderophore, ammonia, and other secondary metabolites which have antagonistic activity against pathogenic microbes. The PGP microbes belong to different phylum of archaea (Euryarchaeota), bacteria (Acidobacteria, Actinobacteria, Bacteroidetes, Deinococcus-Thermus, Firmicutes, and Proteobacteria), and fungi (Ascomycota and Basidiomycota).

Keywords

Abiotic stress resistance · Biodiversity · PGPR · Phytohormone · Plant microbes · Rhizosphere · Soil fertility

S. A. Gezaf
Botany Department, Faculty of Science, Arish University, North Sinai, Egypt

H. H. Abo Mahas
Zoology Department, Faculty of Science, University of Suez Canal, Ismailia, Egypt

A. M. Abdel-Azeem (✉)
Botany Department, Faculty of Science, University of Suez Canal, Ismailia, Egypt

A. N. Yadav et al. (eds.), *Current Trends in Microbial Biotechnology for Sustainable Agriculture*, Environmental and Microbial Biotechnology,
https://doi.org/10.1007/978-981-15-6949-4_8

8.1 Introduction

Microbes are very important for the maintenance of life on Earth, until now we still understand little about the majority of microbes in environments such as soils, oceans, atmosphere, and even those living on and in our bodies (Turner et al. 2013). The plant microbiomes (phyllospheric, endophytic, and rhizospheric) and microbiomes of extreme habitat (acidophilic, alkaliphilic, psychrophilic, halophilic, thermophilic, and xerophilic) are natural bioresearches, which play vital roles in the maintenance of global nutrient stability and ecosystem (Yadav 2017; Kour et al. 2020). Plant–microbe interaction resulted in adaptation, plant growth promotion, uptake of micronutrient, and production of different groups of secondary metabolites and bioactive compounds with potential applications in agriculture, medicine, and industry (Yadav 2019; Kumar et al. 2019a; Yadav et al. 2020).

In general, there are three types of plant–microbe interactions. They are epiphytic, endophytic, and rhizospheric. The rhizosphere is the region of soil influenced by roots through the release of substrates that affect microbial activity such as rhizodeposition of exudates, mucilage, and sloughed cells. Root exudates contain a variety of compounds, predominately organic acids and sugars, but also amino acids, fatty acids, vitamins, growth factors, hormones, and antimicrobial compounds (Turner et al. 2013; Yadav et al. 2017a). The phyllosphere or aerial surface of a plant is a common niche for synergism between microbes and areal parts of plant is a much more dynamic environment than the rhizosphere (Verma et al. 2016a, b).

The PGP microbes could be applied as biofertilizers instead of the chemical fertilizers and for the improvement of different abiotic stresses in crops including salinity, temperature, and drought (Yadav 2017; Kour et al. 2019a, b; Verma et al. 2017). Rhizobacteria showed beneficial traits for the development and growth promotion of plants by means of direct and indirect ways and referred to as plant growth-promoting rhizobacteria (PGPR) with their holistic association with plants (Dheeman et al. 2017; Subrahmanyam et al. 2020). Among the fungal groups arbuscular mycorrhizal (AM) fungi are known to promote activities which can improve agricultural developments. In exchange for the AM fungi providing all of these nutrients, the plant in turn provides the mycorrhizae with carbon and other nutrients (Yadav 2019). The endophytic microbes are referred to as microorganisms which infect plant parts without causing any symptoms to their host, which colonize in the interior of the plant parts such as root, stem, or seeds (Zabalgogeazcoa 2008). Endophytic fungi are agriculturally important as they can enhance plant growth; improve plant nutrition through different direct and indirect PGP attributes including solubilization of phosphorus, potassium, and zinc; production of phytohormones (indole acetic acids, gibberellic acids, and cytokinin) conditions (Rana et al. 2020a, b, 2019). Also due to protection of plant against biotic and abiotic stresses, they are considered as ecofriendly bioresources (Yadav 2019).

8.2 The Plant Microbiomes

Plants are exposed to huge numbers of microorganisms that are present in the top soil and are found on leaves and stems (Sivakumar and Thamizhiniyan 2012). Plant–microbe interactions play a vital role to ensure sustainability in agriculture and ecosystem restoration (Badri et al. 2009). In the past, the interaction of microbes with plants was simply thought of as being an effect, but today it is considered as a process with a high level of complexity in which at least different types of microbes share information without sharing the same spaces from a cellular perspective (Yadav et al. 2017b). Plant–microbe interaction is a mode of communication between plants and microbes which is initiated by the secretion of different signaling molecules (Rastegari et al. 2020; Singh et al. 2020). Plants have evolved unique and sophisticated defense mechanism that involves innate immune system consisting of two classes of immune receptors. These receptors can recognize the presence of oneself molecules both inside and outside of host cells to distinguish a microbial mutualist from pathogens (Jones and Dangl 2006).

Microbial communities affect the plant physiology directly or indirectly, in a positive or negative manner, by various interactions like mutualism, commensalism, amensalism, and pathogenic consequences. In plants, commensalism or mutualism is one of the most common interactions found. The interactions may be categorized as positive, negative, or neutral which largely depend on the nature of microorganisms associating the host (Abhilash et al. 2012; Rai et al. 2020). Positive interactions stimulate plant growth by conferring abiotic and/or biotic stress tolerance and help the plants for the revitalization of nutrient-deficient and contaminated soils. Negative interactions involve host–pathogen interactions resulting in many plant diseases and adverse effects and host life (Akram et al. 2017). Moreover, some microbes reside in the soil surrounding the plant roots just to obtain their nutrition from root exudates. They do not influence the plant growth or physiology in a positive or negative way, thus forming neutral interactions (Akram et al. 2017).

The biodiversity of plant microbiomes ranged between archaea (Euryarchaeota), bacteria (Actinobacteria, Acidobacteria, Bacteroidetes, Deinococcus-Thermus, Proteobacteria, and Firmicutes), and fungi (Ascomycota and Basidiomycota) has been characterized genetically for its beneficial attributes for human welfare (Kumar et al. 2019b; Sharaff et al. 2020). All types of microorganisms (fungi, bacteria, and actinobacteria) have been discovered as endophytes. The most frequently encountered endophytes are fungi (Khare et al. 2018). There are very few reports of halophilic archaea as PGP including rhizospheric as well as endophytic (Yadav et al. 2015; Gaba et al. 2017).

8.3 The Rhizosphere of Plant Microbiomes

The rhizosphere is the narrow zone of soil that is directly influenced by roots through the release of substrates that affect microbial activity and root exudates, can contain up to one thousand microbial cells per gram of root (Egamberdieva et al. 2008), and have prokaryotic species more than 30,000 (Mendes et al. 2011; Qiu et al. 2014). It is considered as an important and active zone for microbial colonization and activity depending on the distance away from plant roots and forms a system especially suitable for obtaining culturable beneficial microbes (Hartmann et al. 2008; Yadav et al. 2017b). The collective rhizosphere microbes' genome is much larger than that of the plant and is also referred to as the plant's second genome (Berendsen et al. 2012; Qiu et al. 2014). The rhizospheric microbes are influenced by several factors such as soil type and moisture, temperature, PH, age, and conditions of plants (Verma et al. 2015).

There are a variety of microbes that can be found growing in rhizosphere microhabitats. It is universally considering that members of any microbial group can develop important functions in the ecosystem (Giri et al. 2005; Barea et al. 2005). Most studies on rhizosphere microbiology, especially those describing co-operative microbial interactions, are focused on bacteria and fungi (Bowen and Rovira 1999; Barea et al. 2005). Bacteria and fungi have very different living habits and a variety of saprophytic and symbiotic relationships, both detrimental (pathogenic) and beneficial (mutualistic) (Barea et al. 2004). Barea et al. (2005) concluded that detrimental microbes included both the major plant pathogens and the minor parasitic and non-parasitic deleterious rhizosphere, bacteria and fungi. Beneficial saprophytes, from a diversity of microbial groups, are able to promote plant growth and health. These include:

(i) Decomposing microbes of organic debris.
(ii) Plant growth-promoting rhizobacteria (PGPR).
(iii) Fungal and bacterial with antagonistic activity of root pathogens.

Most of the plant–microbe interaction research within the past has focused on the traditional symbiosis between plants and arbuscular mycorrhizae (Parniske 2008), nitrogen fixation by rhizobia within the nodules of legume roots (Oldroyd et al. 2011). However, the role of endophytes that reside in plants is yet to be explored to its fullest potential. Endophytic microorganisms and their role in crop health are now attracting great interest from researchers (Jain and Pundir 2017).

Among the sustainable efforts, the role of root-associated microbes especially arbuscular mycorrhizal (AM) fungi in imparting stress tolerance has been exploited by many researchers in the recent years (Garg and Singla 2012). Mycorrhizal symbioses are ubiquitous system of green technology. In these symbioses, the fungal mycelia scavenge through soil for resources (often phosphorus or nitrogen) and provide these resources to plants in exchange of organic carbon. The associations are mutualistic frequently but sometimes exist as parasitism depending upon fungal nature (Prasad 2017). They are ubiquitous soil-borne fungi, whose origin and divergence dates back to over 450 million years (Gutjahr and Parniske 2013).

Besides mycorrhizal endophytes, non-mycorrhizal endophytes (hereafter referred to as endophytes) have been recovered from most plants. Fungal endophytes are microfungi that internally infect living plant tissues without causing disease or any harm to plant and live in mutualistic association with plants for at least a part of their life cycle (Lugtenberg et al. 2016). Lynch and Whipps (1991) suggest the three different zones of the rhizospheric soil: (a) Endorhizosphere means the endodermis and cortical layers inside the roots. (b) Rhizoplane means the root surface with mucilaginous polysaccharide layer. (c) Ectorhizosphere means the soil particles past the root surface that are impacted by root exudates.

The zone of endorhizosphere is internal root colonization because it is a physical location inside the plant (Fig. 8.1) Compared to bulk soil, the endorhizosphere is abundant in various nutrients due to an aggregation of root exudates (Dakora and Phillips 2002), including sugars, amino acids, vitamins, organic acids, and enzymes (Gray and Smith 2005). Root exudates release water, oxygen, and ions, but most importantly include carbon-containing compounds (Uren 2000). Some root exudates act as repellents against pathogens while others function as attractants that aggregate beneficial microbes (Ahemad and Kibret 2014) based on the physiological status, species of plants, and microorganisms (Kang et al. 2010).

In the rhizosphere, various interactions occur between rhizobacteria and plant root. For example, interactions of signal molecules between plant roots and rhizobacteria are important and occur in the rhizosphere (Werner 2000); and these interactions influence in plant growth and crop production (Shaikh et al. 2018). The role of the rhizosphere is critical for PGP, nutrition, and crop quality (Berg and Smalla 2009; Hassan et al. 2019a, b) In addition, the rhizosphere is where plant roots communicate with beneficial rhizobacteria for energy and nutrition. Plant growth-promoting rhizobacteria (PGPR) may affect plant growth, development, and disease

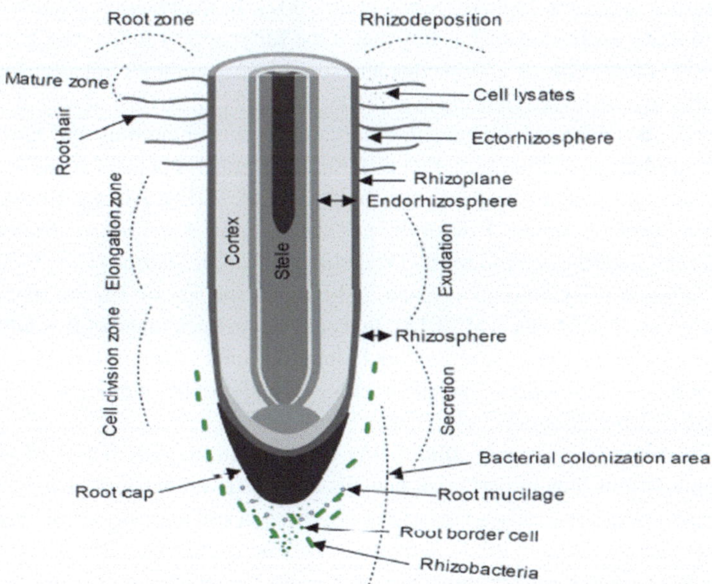

Fig. 8.1 Types of different root zones in the rhizosphere where rhizodeposition, root exudates, and root border cells provide nutrients for PGPR growth and root colonization (Bertin et al. 2003; Prashar et al. 2014; Liu et al. 2017; Hassan et al. 2019a, b)

suppression by one or more direct or indirect mechanisms. Bacterial genera such as *Bacillus* and *Pseudomonas* have been extensively studied and utilized as biocontrol agents, biofertilizers, and also have been shown to trigger induced systemic resistance (ISR) (Kloepper et al. 2004; Takishita et al. 2018; Hassan et al. 2019a, b).

8.4 Plant Growth Promoting and Rhizospheric Microbiomes

Plant related with microbes have been shown to be beneficial by promoting plant growth (PGP) either directly, for example, by fixation of atmospheric nitrogen, solubilization of minerals such as phosphorus, potassium, and zinc; production of sideropores and plant growth hormones such as cytokinins, gibberellins, and auxins, or indirectly, by production of antagonistic substances by inducing resistance against plant pathogens (Tilak et al. 2005; Verma et al. 2016a; Kour et al. 2017).

8.4.1 Improving Soil Fertility

Rhizosphere is a field where microbes are under the influence of plant roots. Roots are the sites for uptake of mineral elements and exudation of organic compounds that act as carbon and energy sources for the indigenous microflora (Hinsinger et al. 2009). Plant–microbe interactions in the rhizosphere are critical for regulating biogeochemical recycling of mineral elements and maintaining the microbial community structure in the rhizosphere (Singh et al. 2007). In many cases, plant–microbe interactions are evolved in such a way that some fungi appear to live non-pathologically inside plant roots as endophytes and some form symbiotic relationship with roots called mycorrhizae (Gehring et al. 2006; Suman et al. 2016; Yadav et al. 2018).

Mineral phosphate-solubilizing microorganisms are ubiquitous and have variable cell numbers in different soil that differ in their mineral phosphate-solubilizing ability from one medium to another (Chauhan et al. 2014). Among rhizospheric fungi the most common P-solubilizing strains are *Aspergillus*, *Penicillium*, *Trichoderma*, and *Rhizoctonia solani* (Wakelin et al. 2004; Sharma et al. 2013); species of *Glomus* of AM fungi (Prasad 2017); and species of *Pseudomonas* and *Bacillus* in bacteria (Mehta et al. 2015). In recent years, huge range of P-solubilizing endophytic fungi have been identified including the genera *Penicillium*, *Aspergillus*, *Piriformospora*, *Curvularia*, and other class of endophytic symbionts AM fungi (Mehta et al. 2019; Mondal et al. 2020).

In soil, although phosphate-solubilizing fungi constitute only 0.1–0.5% of total fungal populations they impart great benefits towards plant nutrition. Unlike bacteria, fungal hyphae can easily go over long distances in soil and release more organic acids than bacteria (Kucey 1983). Among many benefits provided by AMF, the most significant one is to improve phosphorus nutrition of the host plant with low phosphate levels that is achieved by the large surface area of their hyphae and their high affinity P mobilization mechanisms (Van der Heijden et al. 2006). The AMF are probably the most abundant fungi commonly present in agricultural soils, and the arbuscules are the main sites for the exchange of P, N, and other minerals mobilized by the thin fungal hyphae in soils (Rashid et al. 2016). *Trichoderma harzianum* can solubilize P by chelating and reducing molecules (Altomare et al. 1999). Wahid and Mehana (2000) reported that there is an increase of more than 30% in response to P-solubilizing fungal inoculation in soil containing rock phosphate and superphosphate. Srivastav et al. (2004) reported significantly higher solubilization of rock phosphate under in vitro conditions by fungal strains of *Aspergillus niger*, *Curvularia lunata, Rhizoctonia solani,* and *Fusarium oxysporum.* A dark septate root endophytic fungus *Curvularia geniculata* isolated from *Parthenium hysterophorus* roots was known to enhance plant growth through P-solubilization and phytohormones production (Priyadharsini and Muthukumar 2017).

In study reported by Efthymiou et al. (2018), the wheat crop inoculated with *Penicillium aculeatum* significantly increased the shoot biomass and P content of wheat. These results led to the development of novel bioinoculants containing phosphate-solubilizing *Penicillium* fungi to increase the fertility value of P-rich biochar. The species of *Rhizobium* and *Bradyrhizobium* establish symbiotic associations with roots in leguminous plants such as soybean, pea, peanut, and alfalfa, convert N_2 into ammonia, and make it available to the plants as a source of N (Badawi et al. 2011). In agriculture, 80% of the biologically fixed N comes from *Rhizobium, Bradyrhizobium, Sinorhizobium, Azorhizobium, Mesorhizobium,* and *Allorhizobium* of the family Rhizobiaceae in association with the leguminous plants. A lot of studies have shown *P. indicia* as phosphorus mobilizer (Singh et al. 2000).

A variety of nitrogen-fixing microbes like *Arthrobacter, Azoarcus, Azospirillum, Azotobacter, Bacillus, Enterobacter, Gluconoacetobacter, Herbaspirillum,* and *Klebsiella, Pseudomonas* are isolated from the rhizosphere of varied crops, which contribute fixed nitrogen to the associated plants (Suman et al. 2016; Niste et al. 2013; Olivares et al. 2013). A study by Barazani et al. (2005) confirmed the growth increase in *Nicotiana tobaccum* due to *P. indica* and showed that the growth promotion was related to better aptness, as enhanced seed production was observed in treated plants. Nath et al. (2012) studied *Penicillium* sp. isolated from tea leaves as phosphate solubilizer. *Penicillium* sp. Significantly increased plant biomass, related growth parameters, assimilation of essential nutrients such as potassium, calcium, magnesium, and reduced the sodium toxicity in cucumber plants under salinity and drought stress, when compared with control plants (Lata et al. 2018).

Aspergillus terreus also produces siderophores which chelate the iron and activate the plant defense mechanism (Chhipa and Deshmukh 2019). In the rhizosphere of the many plant species, diverse rhizobacterial species with the potential to enhance plant growth, crop production, and biological control activity were recorded by many researchers. PGPR genera present in the rhizosphere include *Azotobacter, Azospirillum, Agrobacterium, Arthrobacter, Bacillus, Burkholderia, Caulobacter, Chromobacterium, Erwinia, Flavobacterium, Pseudomonas, Micrococcus,* and *Serratia* (Gray and Smith 2005; Duy et al. 2015; Hassan et al. 2019a, b).

8.4.2 Phytohormones Producing Microbes

Endophytic bacteria provide a large array of beneficial effects to their host plant. It promotes plant growth by producing plant growth-enhancing substances such as indole acetic acid (IAA) (Naveed et al. 2015; Tiwari et al. 2020), cytokinins (CK) (Garcia de Salamone et al. 2001), gibberellic acid (GA) (Uma Maheswari et al. 2013), and improving nutrient absorption, including nitrogen fixation (Mirza et al. 2001). PGP activities of endophytes are attributed to the production of iron-chelating agents, siderophores as in rice by *Enterobacter* spp. and *Burkholderia* spp. (Souza et al. 2013), indole acetic acid (IAA), and other growth hormones as in cashew by *Staphylococcus saprophyticus* and *Escherichia coli* (Lins et al. 2014). Endophytic *Azospirillum* spp. is reported to accumulate the abscisic acid (ABA) in mitigating water stress tolerance in maize. Plant growth-promoting hormones IAA and gibberellins further enhance the effect (Cohen et al. 2009). Few of the soil-borne pathogens like *Fusarium oxysporum, Pythium* spp., *Phytophthora* spp., *Aphanomyces* spp., *Sclerotium rolfsii, Gaeumannomyces graminis, Rhizoctonia solani, Verticillium* spp., and *Thielaviopsis basicola* are found to be negatively affected by PGPR (Sahu et al. 2017).

Secondary metabolite like colletotric acid, isolated from the endophytic fungus *Colletotrichum gloeosporioides*, dwelling in *Artemisia annua* (Zou et al. 2000). A study by Sirrenberg et al. (2007) noted the production of indole acetic acid in submerged culture of *Piriformospora indica* when colonized with Arabidopsis thaliana. *P. indica* can synthesize a hormone and release it into the root tissue, influence a phytohormone level by interfering with its synthesis, degradation, or modification, or interfere with a phytohormone signaling pathways, or any combination of these possibilities (Oelmüller et al. 2009). The different species of Aspergillus genera were also identified as gibberellin producers, such as *Aspergillus flavus, A. niger*, which also induce the production of defense hormone salicylic and jasmonic acid (Hasan 2002; Khan et al. 2011). *Penicillium* sp. from cucumber roots has been found to synthesize GA and IAA. Inoculating these strains in cucumber plants under drought stress has shown a significant increase in plant biomass, growth parameters, and assimilation of essential nutrients and reduced sodium toxicity (Waqas et al. 2012). Direct effects of alkaloids by endophytes in host plants are a standard phenomenon as in *Fescue* (by the endophytes *Neotyphodium* spp. and *Epichloë* spp.), wherein the host plant leaves are shielded from herbivores by the assembly of alkaloid, loline, produced by mutualistic fungal endophytes (Roberts and Lindow 2014).

8.4.3 Abiotic Stress Resistance Microbes

Rhizosphere microorganisms with their intrinsic metabolic and genetic capabilities contribute to reduce abiotic stresses in plants (Gopalakrishnan et al. 2015). Rhizosphere microorganisms also increase tolerance to low non-freezing temperatures resulting in higher and faster accumulation of stress-related proteins and metabolites (Theocharis et al. 2012; Mohanram and Kumar 2019). Several microorganisms of the genera *Pseudomonas, Bacillus, Achromobacter, Enterobacter, Azotobacter, Methylobacterium,* and *Trichoderma* have been widely studied in PGP by reduction of multiple kinds of abiotic stresses (Atieno et al. 2012; Sorty et al. 2016; Meena et al. 2017). Treatment of Indian mustard (*Brassica juncea*) with the fungus, *Trichoderma harzianum*, improved the uptake of essential nutrients and enhanced accumulation of antioxidants and decreased Na^+ uptake under saline conditions (Ahmad et al. 2015). Better root colonizing capability of *Pseudomonas* sp. along with its ability to produce exopolysaccharides led to enhanced tolerance towards salinity (Sen and Chandrasekhar 2014).

Novel stress-tolerant bacteria such as *Brachybacterium saurashtrense, Zhihengliuella* sp., and *Brevibacterium casei* have also been reported from plant rhizospheres (Jha et al. 2012; Mohanram and Kumar 2019). The plant root colonized with *P. indica* showed tolerance in different abiotic stresses like extreme temperature, salinity, drought, and heavy metals (Chhipa and Deshmukh 2019). *P. indica* readily colonizes the *A. thaliana* and increases the yield and salt tolerance in *Hordeum vulgare* (Waller et al. 2005); and barley plant (Chadha et al. 2014). *P. indica* showed drought and salt tolerance in cacao, barley, and Chinese cabbage plants (Abo Nouh 2019). Tolerance to abiotic stress was induced in *A. thaliana*; overall growth and biomass production was achieved in herbaceous mono- and dicots, medicinal plants, and other important crops (Chadha et al. 2014). *P. brevicompactum* isolated from wild barley species was helpful in drought tolerance improvement of barley plant in drought condition (Abo Nouh 2019). *Curvularia* sp. also confers heat and drought stress to *Lycopersicum esculentum* (Rodriguez and Redman 2008). *Curvularia* sp. confers thermos tolerance to grasses and also provides thermos tolerance ability to other plants like tomato, watermelon, and wheat (Abo Nouh 2019). *Colletotrichum magna* and *C. protuberata* are well reported for water stress tolerance in wheat (*Triticum* sp.), tomato (*Solanum lycopersicum*), and watermelon (*Citrullus lanatus*) plants (Raghuwanshi 2018). *Penicillium* sp. and *Phomaglomerata* significantly increased plant biomass, related growth parameters, assimilation of essential nutrients such as potassium, calcium, and magnesium and reduced the sodium toxicity in cucumber plants under salinity and drought stress, when compared with control plants (Abo Nouh 2019). Epichloë species may enhance the eco-physiology of host plants and enable plants to counter abiotic stresses such as drought and metal contamination (Rodriguez et al. 2009). *Penicillium brevicompactum* isolated from wild barley species was helpful in drought tolerance improvement of barley plant in drought condition (Chhipa and Deshmukh 2019).

8.4.4 Plant Pathogen Resistance

In the early 1970s, several researchers identified microbial populations in the rhizosphere a form the first barrier to pathogen infection (Barea et al. 2005). Currently, it is well known that some soils are naturally suppressive to some soil-borne plant pathogens including *Fusarium, Rhizoctonia, Pythium,* and *Phytophthora* (Thakur et al. 2020). Although this suppression relates to both physicochemical and microbiological features of the soil, in most systems the biological elements are the first factors in disease suppression and therefore the topic of biological control of plant pathogens (Weller et al. 2002). Among the prokaryotes, a broad range of bacteria such as *Agrobacterium, Bacillus* spp. (e.g., *B. cereus, B. pumilis,* and *B. subtilis*), and *Streptomyces* have been shown their ability against soil-borne pathogens.

Most bacteria studied as biocontrol are *Pseudomonas* spp., such as *P. aeruginosa* and *P. fluorescens* which may be among the most effective root colonizing bacteria (Barea et al. 2005). Some strains of Actinomycete genera such as *Streptomyces, Streptosporangium, Thermobifida,* and *Micromonospora* display biological control activity against some root fungal pathogens (Franco-Correa et al. 2010). Between the eukaryotes, there are a variety of fungal species that display antagonistic properties and have been applied in biocontrol, but the ubiquitous *Trichoderma species* clearly dominate. In addition, fungi non-pathogenic species such as *Pythium* and *Fusarium* are receiving increasing interest as antagonists (Barea et al. 2005).

Pandey and Upadhyay (2000) reported that rhizosphere of healthy pigeon pea plant was heavily colonized by resident *Trichoderma* and *Gliocladium* which were highly antagonistic to the pathogen. *T. viride* formed loops and coiling and ruptured the cell wall of the pathogen. Some pathogenic diseases of plant controlled by AMF: Verticillium wilt caused by *Verticillium dahliae* (Karagiannidis et al. 2002), root-knot nematode caused by *Meloidogyne incognita* (Momotaz et al. 2015) in Tomato. White rot onion caused by *Sclerotium cepivorum* in onion (Torres-Barragán et al. 1996). Root rot caused by *Aphanomyces euteiches* in pea (Larsen and Bødkar 2001). Verticillium wilt caused by *Verticillium* sp. in cotton (Kobra et al. 2011). *Aspergillus terreus* enhanced the sunflower growth (*Helianthus annuus L.)* and disease resistivity against the stem rot caused by Sclerotium rolfsii (Waqas et al. 2015). *P. indica*, which can regulate development, is also able to act as a biofertilizer and also is a good candidate to improve commercial plant production and might be especially useful in agroforestry and flori-horticulture applications (Varma et al. 1999). In barley, the root endophyte *P. indica* confers disease resistance by a different mechanism (Waller et al. 2005).

P. indica showed as a biocontrol agent against plant pathogen in maize, tomato, wheat, and barley (Kumar et al. 2009). *P. indica* showed the reduced severity of Verticillium wilt by 30% in tomato, caused by *Verticillium dahliae*, and increased leaf biomass by 20% (Fakhro et al. 2010). Colonization of *P. indica* controlled various plant diseases such as powdery mildew, eyespot, Rhizoctonia root rot, Fusarium wilt, black root rot, yellow leaf mosaic, Verticillium wilt, cyst nematode, and leaf blight in barley, wheat, maize, tomato, and *Arabidopsis* plants (Chhipa and Deshmukh 2019).

Trichoderma is widely used as biocontrol agent against phytopathogenic fungi and as a biofertilizer (Saba et al. 2012). *Trichoderma* sp. uses several mechanisms such as antibiosis, mycoparasitism and competition for nutrients and space and is also able to promote growth and development of plant and induce the defense response of plants (Talapatra et al. 2017). *Trichoderma* sp. has used as BCA against plant pathogenic fungi like Botrytis *cinerea, Fusarium* spp., *Pythium* spp., and *Rhizoctonia* spp. (Park et al. 2018). *Trichoderma* endophyte from an essential medicinal plant of Assam, *Rauwolfia serpentina*, showed antagonistic activity against *Fusarium oxysporum* and *Phytophthora* spp. (Doley and Jha 2010).

Mycoparasitic *Trichoderma* species are used commercially as biological control agents against plant-pathogenic fungi such as *Rhizoctonia solani, Botrytis cinerea, Sclerotium rolfsii, Sclerotinia sclerotiorum, Pythium* spp., *and Fusarium* spp. *T. harzianum* protected bean seedlings against pre-emergence damping off infection, reduced the disease severity, and increased the plant growth in the presence of *R. solani* pathogen (Paula et al. 2001). *P. indica* increased the resistance in barley against root rot causing agent *Fusarium culmorum* and *Blumeria graminis* (Waller et al. 2005). *P. citrinum* endophyte enhances the sunflower growth (*Helianthus annuus L.*) and disease resistivity against the stem rot caused by *Sclerotium rolfsii* and leaf spot and blight caused by *Alternaria alternata* (Waqas et al. 2015). *P. brevicompactum* has been reported to suppress various seed-borne pathogens including *Rhynchosporium, Pyrenophora, Fusarium,* and *Cochliobolus* and soil-borne pathogen *Gaeumannomyces graminis var. tritici* (Murphy et al. 2015). *Curvularia* sp. endophyte from an essential medicinal plant of Assam, *Rauwolfia serpentina*, showed antagonistic activity against *Fusarium oxysporum* and *Phytophthora* spp. (Li et al. 2000; Doley and Jha 2010).

8.5 Conclusion

Plant microbiome in rhizosphere plays the most critical role in plant growth promoting (PGP), development, and fertilization of soil. The diverse group of microbes is significant components of soil plant systems. The microbes with PGP attributes have emerged as an important and promising tool for sustainable agriculture. PGP microbes promote plant growth and directly or indirectly development; either by releasing plant growth phytohormones; solubilization of phosphorus, potassium, and zinc; and biological process such as nitrogen fixation or by producing siderophore, ammonia, and other secondary metabolites which have antagonistic activity against pathogenic microbes. In coming time, biofertilizers will not only act as potential alternative for feeding the emerging population but also will improve productivity and support the growth of the plants during stress conditions. Therefore, it is crucial to realize the importance of biofertilizers and their implementation in modern agriculture. Sustainable agriculture should change rather from growing plants, cultivation of plant–microbial communities must be done, which will ultimately lead to high productivity with negligible energy and chemical investments simultaneously with minimum pressures on the environment.

References

Abhilash P, Powell JR, Singh HB, Singh BK (2012) Plant–microbe interactions: novel applications for exploitation in multipurpose remediation technologies. Trends Biotechnol 30:416–420

Abo Nouh FA (2019) Endophytic fungi for sustainable agriculture. Microb Biosyst 4(1):31–44. https://doi.org/10.21608/MB.2019.38886

Ahemad M, Kibret M (2014) Mechanisms and applications of plant growth promoting rhizobacteria: current perspective. J King Saud Univ Sci 26:1–20

Ahmad P, Hashem A, Abd-Allah EF, Alqarawi AA, John R, Egamberdieva D, Gucel S (2015) Role of *Trichoderma harzianum* in mitigating NaCl stress in Indian mustard (*Brassica juncea* L) through antioxidative defense system. Front Plant Sci 6:868

Akram MS, Shahid M, Tahir M, Mehmood F, Ijaz M (2017) Plant-microbe interactions: current perspectives of mechanisms behind symbiotic and pathogenic associations. In: Singh DP, Singh HB, Prabha R (eds) Plant-microbe interactions in agro-ecological perspectives. Springer International Publishing AG, Singapore, pp 97–196. https://doi.org/10.1007/978-981-10-5813-4

Altomare C, Norvell WA, Björkman T, Harman GE (1999) Solubilization of phosphates and micronutrients by the plant growth-promoting and biocontrol fungus *Trichoderma harzianum* Rifai 1295-22. Appl Environ Microbiol 65:2926–2933

Atieno M, Hermann L, Okalebo R, Lesueur D (2012) Efficiency of different formulations of *Bradyrhizobium japonicum* and effect of coinoculation of *Bacillus subtilis* with two different strains of Bradyrhizobium japonicum. World J Microbiol Biotechnol 28:2541–2550

Badawi F, Sh F, Biomy AMM, Desoky AH (2011) Peanut plant growth and yield as influenced by co-inoculation with *Bradyrhizobium* and some rhizo-microorganisms under sandy loam soil conditions. Ann Agrar Sci 56:17–25

Badri DV, Weir TL, van der Lelie D, Vivanco JM (2009) Rhizosphere chemical dialogues: plant–microbe interactions. Curr Opin Biotechnol 20:642–650

Barazani O, Benderoth M, Groten K, Kuhlemeier C, Baldwin IT (2005) *Piriformospora indica* and *Sebacina vermifera* increase growth performance at the expense of herbivore resistance in *Nicotiana attenuata*. Oecologia 146:234–243

Barea JM, Azcón R, Azcón-Aguilar C (2004) Mycorrhizal fungiand plant growth promoting rhizobacteria. In: Varma A, Abbott L, Werner D, Hampp R (eds) Plant surface microbiology. Springer-Verlag, Heidelberg, pp 351–371

Barea JM, Pozo MJ, Azcon R, Azcon Aguilar C (2005) Microbial co-operation in the rhizosphere. J Exp Bot 56(417):1761–1778

Berendsen RL, Pieterse CMJ, Bakker P (2012) The rhizosphere microbiome and plant health. Trends Plant Sci 17:478–486

Berg G, Smalla K (2009) Plant species and soil type cooperatively shape the structure and function of microbial communities in the rhizosphere. FEMS Microbiol Ecol 68:1–13

Bertin C, Yang X, Weston LA (2003) The role of root exudates and allelochemicals in the rhizosphere. Plant Soil 256:67–83

Bowen GD, Rovira AD (1999) The rhizosphere and its management to improve plant growth. Adv Agron 66:1–102

Chadha N, Mishra M, Prasad R, Varma A (2014) Root endophytic fungi: research update. J Biol Life Sci USA 5:135–158

Chauhan A, Guleria S, Walia A, Mahajan R, Verma S, Shirkot CK (2014) Isolation and characterization of *Bacillus* sp. with their effect on growth of tomato seedlings. Indian J Agric Biochem 27(2):193–201

Chhipa H, Deshmukh SK (2019) Fungal endophytes: rising tools in sustainable agriculture production. In: Jha S (ed) Endophytes and secondary metabolites. Springer International Publishing AG, Cham, pp 1–24

Cohen AC, Travaglia CN, Bottini R, Piccoli PN (2009) Participation of abscisic acid and gibberellins produced by endophytic *Azospirillum* in the alleviation of drought effects in maize. Botany 87:455–462

Dakora FD, Phillips DA (2002) Root exudates as mediators of mineral acquisition in low-nutrient environments. In: Food security in nutrient-stressed environments: exploiting plants' genetic capabilities. Springer, Dordrecht, pp 201–213

Dheeman S, Maheshwari DK, Baliyan N (2017) Bacterial endophytes for ecological intensification of agriculture. In: Maheshwari DK (ed) Endophytes: biology and biotechnology, sustainable development and biodiversity, vol 15. Springer International Publishing AG, Cham, pp 193–231. https://doi.org/10.1007/978-3-319-66541-2_1

Doley P, Jha DK (2010) Endophytic fungal assemblages from ethnomedicinal plant *Rauwolfia serpentina* (L) Benth. J Plant Pathol Microbiol 40(1):44–48

Duy MV, Hoi NT, Ve NB, Thuc LV, Trang NQ (2015) Influence of *Cellulomonas flavigena, Azospirillum* sp. and *Pseudomonas* sp. on rice growth and yield grown in submerged soil amended with rice straw. Recent trends PGPR research for sustainable crop productivity 4th Asian PGPR conference Proceeding, HaNoi, VietNam

Efthymiou A, Grønlund M, Müller-Stover DS, Jakobsen I (2018) Augmentation of the phosphorus fertilizer value of biochar by inoculation of wheat with selected *Penicillium* strains. Soil Biol Biochem 116:139–147

Egamberdieva D, Kamilova F, Validov S, Gafurova L, Kucharova Z, Lugtenberg B (2008) High incidence of plant growth-stimulating bacteria associated with the rhizosphere of wheat grown on salinated soil in Uzbekistan. Environ Microbiol 10:1–9

Fakhro A, Andrade-Linares DR, von Bargen S, Bandte M, Büttner C, Grosch R, Schwarz D, Franken P (2010) Impact of *Piriformospora indica* on tomato growth and on interaction with fungal and viral pathogens. Mycorrhiza 20(3):191–200

Franco-Correa M, Quintana A, Duque C, Suarez C, Rodríguez MX, Barea JM (2010) Evaluation of actinomycete strains for key traits related with plant growth promotion and mycorrhiza helping activities. Appl Soil Ecol 45:209–217

Gaba S, Singh RN, Abrol S, Yadav AN, Saxena AK et al (2017) Draft genome sequence of *Halolamina pelagica* CDK2 isolated from natural salterns from Rann of Kutch, Gujarat, India. Genome Announc 5(6):1–2

Garcia de Salamone IE, Hynes RK, Nelson LM (2001) Cytokinin production by plant growth promoting rhizobacteria and selected mutants. Can J Microbiol 47(5):404–411

Garg N, Singla P (2012) The role of *Glomus mosseae* on key physiological and biochemical parameters of pea plants grown in arsenic contaminated soil. Sci Hortic 143:92–101

Gehring CA, Mueller RC, Whitham TG (2006) Environmental and genetic effects on the formation of ectomycorrhizal and arbuscular mycorrhizal associations in cottonwoods. Oecologia 149:158–164

Giri B, Giang PH, Kumari R, Prasad R, Varma A (2005) Microbial diversity in soils. In: Buscot F, Varma S (eds) Microorganisms in soils: roles in genesis and functions. Springer-Verlag, Heidelberg, pp 195–212

Gopalakrishnan S, Sathya A, Vijayabharathi R, Varshney RK, Gowda CL, Krishnamurthy L (2015) Plant growth promoting rhizobia: challenges and opportunities. 3 Biotech 5:355–377

Gray E, Smith D (2005) Intracellular and extracellular PGPR: commonalities and distinctions in the plant-bacterium signaling processes. Soil Biol Biochem 37:395–412

Gutjahr C, Parniske M (2013) Cell and developmental biology of the arbuscular mycorrhiza symbiosis. Annu Rev Cell Dev Biol 29:593–617

Hartmann A, Rothballer M, Schmid M (2008) Lorenz Hiltner, a pioneer in rhizosphere microbial ecology and soil bacteriology research. Plant Soil 312:7–14

Hasan HAH (2002) Gibberellin and auxin production by plant root fungi and their biosynthesis under salinity-calcium interaction. Rostlinná Výroba 48:101–106

Hassan MK, McInroy JA, Jones J, Shantharaj D, Liles MR, Kloepper JW (2019a) Pectin-rich amendment enhances soybean growth promotion and nodulation mediated by *Bacillus Velezensis* strains. Plan Theory 8:120

Hassan MK, Mcinroy JA, Kloepper JW (2019b) The interactions of rhizo deposits with plant growth-promoting rhizobacteria in the rhizosphere: a review. Agric J 9:142

Hinsinger P, Bengough AG, Vetterlein D, Young IM (2009) Rhizosphere: biophysics, biogeochemistry and ecological relevance. Plant Soil 321:117–152

Jain P, Pundir RK (2017) Potential role of endophytes in sustainable agriculture-recent developments and future prospects. In: Maheshwari DK (ed) Endophytes: biology and biotechnology, sustainable development and biodiversity, vol 15. Springer International Publishing AG, Cham, pp 145–160. https://doi.org/10.1007/978-3-319-66541-2_1

Jha B, Gontia I, Hartmann A (2012) The roots of the halophyte *Salicornia brachiata* are a source of new halo tolerant diazotrophic bacteria with plant growth-promoting potential. Plant Soil 356:265–277

Jones JDG, Dangl JL (2006) The plant immune system. Nature 444:323–329

Kang BG, Kim WT, Yun HS, Chang SC (2010) Use of plant growth-promoting rhizobacteria to control stress responses of plant roots. Plant Biotechnol Rep 4:179–183

Karagiannidis N, Bletsos F, Stavropoulos N (2002) Effect of Verticillium wilt (*Verticillium dahliae* Kleb.) and mycorrhiza (*Glomus mosseae*) on root colonization, growth and nutrient uptake in tomato and eggplant seedlings. Sci Hortic 94:145–156

Khan AL, Hamayun M, Ahmad N, Waqas M, Kang SM, Kim YH, Lee IJ (2011) *Exophiala* sp. LHL08 reprograms *Cucumis sativus* to higher growth under abiotic stresses. Physiol Plant 143(4):329–343

Khare E, Mishra J, Arora NK (2018) Multifaceted interactions between endophytes and plant: developments and prospects. Front Microbiol 9(2732):1–12

Kloepper JW, Ryu CM, Zhang S (2004) Induced systemic resistance and promotion of plant growth by *Bacillus* spp. Phytopathology 94:1259–1266

Kobra N, Jalil K, Youbert G (2011) Arbuscular mycorrhizal fungi and biological control of *Verticillium* wilted cotton plants. Arch Phytopathol Plant Protect 44(10):933–942

Kour D, Rana KL, Verma P, Yadav AN, Kumar V (2017) Biofertilizers: eco-friendly technologies and bioresources for sustainable agriculture. In Proceeding of International Conference on Innovative Research in Engineering Science and Technology, IREST/PP/014

Kour D, Rana KL, Yadav AN, Yadav N, Kumar M, Kumar V et al (2020) Microbial biofertilizers: bioresources and eco-friendly technologies for agricultural and environmental sustainability. Biocatal Agric Biotechnol 23:101487. https://doi.org/10.1016/j.bcab.2019.101487

Kour D, Rana KL, Yadav AN, Yadav N, Kumar V, Kumar A et al (2019a) Drought-tolerant phosphorus-solubilizing microbes: biodiversity and biotechnological applications for alleviation of drought stress in plants. In: Sayyed RZ, Arora NK, Reddy MS (eds) Plant growth promoting rhizobacteria for sustainable stress management, Rhizobacteria in abiotic stress management, vol 1. Springer, Singapore, pp 255–308. https://doi.org/10.1007/978-981-13-6536-2_13

Kour D, Rana KL, Yadav N, Yadav AN, Kumar A, Meena VS et al (2019b) Rhizospheric microbiomes: biodiversity, mechanisms of plant growth promotion, and biotechnological applications for sustainable agriculture. In: Kumar A, Meena VS (eds) Plant growth promoting Rhizobacteria for agricultural sustainability: from theory to practices. Springer, Singapore, pp 19–65. https://doi.org/10.1007/978-981-13-7553-8_2

Kucey RMN (1983) Phosphate solubilizing bacteria and fungi in various cultivated and virgin Alberta soils. Can J Soil Sci 63:671–678

Kumar M, Kour D, Yadav AN, Saxena R, Rai PK, Jyoti A et al (2019a) Biodiversity of methylotrophic microbial communities and their potential role in mitigation of abiotic stresses in plants. Biologia 74:287–308. https://doi.org/10.2478/s11756-019-00190-6

Kumar M, Yadav V, Tuteja N (2009) Antioxidant enzyme activities in maize plants colonized with *Piriformospora indica*. Microbiology 155:780–790

Kumar V, Joshi S, Pant NC, Sangwan P, Yadav AN, Saxena A et al (2019b) Molecular approaches for combating multiple abiotic stresses in crops of arid and semi-arid region. In: Singh SP, Upadhyay SK, Pandey A, Kumar S (eds) Molecular approaches in plant biology and environmental challenges. Springer, Singapore, pp 149–170. https://doi.org/10.1007/978-981-15-0690-1_8

Larsen J, Bødker L (2001) Interactions between pea root-inhabiting fungi examined using signature fatty acids. New Phytol 149:487–493

Lata R, Chowdhury S, Gond S, White JF (2018) Induction of abiotic stress tolerance in plants by endophytic microbes. Appl Microbiol 66(4):268–276

Li JY, Strobel GA, Harper JK, Lobkovsky E, Clardy J (2000) Cryptocin, a potent tetramic acid antimycotic from the endophytic fungus *Cryptosporiopsis quercina*. Org Lett 2:767–770

Lins MRCR, Fontes JM, Vasconcelos NM, Santos DMS, Ferreira OE, Ribeiro MRC, Azevedo JL, Araújo JM, GMS L (2014) Plant growth promoting potential of endophytic bacteria isolated from cashew leaves. Afr J Biotechnol 13:3360–3365

Liu H, Carvalhais LC, Crawford M, Singh E, Dennis PG, Pieterse CMJ, Schenk PM (2017) Inner plant values: diversity, colonization and benefits from endophytic bacteria. Front Microbiol 8:2552

Lugtenberg BJJ, Caradus JR, Johnson LJ (2016) Fungal endophytes for sustainable crop production. FEMS Microbiol Ecol 92(12):1–17

Lynch J, Whipps J (1991) Substrate flow in the rhizosphere. In: The rhizosphere and plant growth. Springer, Dordrecht, pp 15–24

Meena KK, Sorty AM, Bitla UM, Choudhary K, Gupta P, Pareek A, et al. (2017) Abiotic stress responses and microbe mediated mitigation in plants: the omics strategies. Front Plant Sci 8:172

Mehta P, Walia A, Shirkot CK (2015) Functional diversity of phosphate solubilizing plant growth promoting Rhizobacteria isolated from apple trees in the trans Himalayan Region of Himachal Pradesh, India. Biol Agri Hort 31(4):265–288

Mehta P, Sharma P, Putatunda C, Walia A (2019) Endophytic fungi: role in phosphate solubilization. In: Singh BP (ed) Advances in endophytic fungal research. Cham, Springer Nature Switzerland AG, pp 183–209. https://doi.org/10.1007/978-3-030-03589-1

Qiu M, Li S, Zhou X, Cui X, Vivanco JM, Zhang N, Shen Q, Zhang R (2014) De-coupling of root microbiome associations followed by antagonist inoculation improves rhizosphere soil suppressiveness. Biol Fertil Soils 50:217–224. https://doi.org/10.1007/s00374-013-0835-1

Mendes R, Kruijt M, de Bruijn I, Dekkers E, van der Voort M, Schneider JH, et al. (2011) Deciphering the rhizosphere microbiome for disease-suppressive bacteria. Science 332:1097–1100

Mirza MS, Ahmad W, Latif F, Haurat J, Bally R, Normand P, Malik KA (2001) Isolation partial characterization and the effect of plant growth-promoting bacteria (PGPB) on micro propagated sugarcane in vitro. Plant Soil 237(1):47–54

Mohanram S, Kumar P (2019) Rhizosphere microbiome: revisiting the synergy of plant-microbe interactions. Ann Microbiol 69:307–320

Momotaz R, Alam MM, Islam MN, Alam KM, Rahman MZ (2015) Management of the root-knot nematode of tomato by inoculation with Arbuscular Mycorrhizal fungi. Int J Sustain Crop Prod 10(2):48–54

Mondal S, Halder SK, Yadav AN, Mondal KC (2020) Microbial consortium with multifunctional plant growth promoting attributes: future perspective in agriculture. In: Yadav AN, Rastegari AA, Yadav N, Kour D (eds) Advances in plant microbiome and sustainable agriculture, Functional annotation and future challenges, vol 2. Springer, Singapore, pp 219–254. https://doi.org/10.1007/978-981-15-3204-7_10

Murphy BR, Doohan FM, Hodkinson TR (2015) Fungal root endophytes of a wild barley species increase yield in a nutrient-stressed barley cultivar. Symbiosis 65(1):1–7

Nath R, Sharma GD, Barooah M (2012) Efficiency of tricalcium phosphate solubilization by two different endophytic Penicillium sp. isolated from tea (*Camelia sinensis* L) Er. J Exp Biol 2(4):1354–1358

Naveed M, Qureshi MA, Zahir ZA, Hussain MB, Sessitsch A, Mitter B (2015) L-tryptophan-dependent biosynthesis of indole-3-acetic acid (IAA) improves plant growth and colonization of maize by *Burkholderia phytofirmans* PsJN. Ann Microbiol 65:1391–1389

Niste M, Vidican R, Pop R, Rotar I (2013) Stress factors affecting symbiosis activity and nitrogen fixation by *Rhizobium* cultured in vitro. ProEnvironment 6(13):42–45

Oelmüller R, Sherameti I, Tripathi S, Varma A, Jena F, Botanik A, Str D (2009) *Piriformospora indica*, a cultivable root endophyte with multiple biotechnological applications. Symbiosis 49:1–17

Oldroyd GE, Murray JD, Poole PS, Downie JA (2011) The rules of engagement in the legume rhizobial symbiosis. Annu Rev Genet 45:119–144

Olivares J, Bedmar EJ, Sanjuán J (2013) Biological nitrogen fixation in the context of global change. Mol Plant Microb Int 26(5):486–494

Pandey KK, Upadhyay JP (2000) Microbial population from rhizosphere and non-rhizosphere soil of pigeon pea. J Mycol Plant Pathol 30(1):7–10

Park Y, Mishra RC, Yoon S, Kim H, Park C, Seo S, Bae H (2018) Endophytic *Trichoderma citrinoviride* isolated from mountain-cultivated ginseng (*Panax ginseng*) has great potential as a biocontrol agent against ginseng pathogens. J Ginseng Res 43:408–420

Parniske M (2008) Arbuscular mycorrhiza: the mother of plant root endosymbioses. Nat Rev Microbiol 6(10):763–775

de Paula TJ, Rotter C, Han B (2001) Effect of soil moisture and panting date on Rhizoctonia root rot of beans and its control Journal of American Science by *Trichoderma harizanum.* Bulletin OILB/SROP 24(3):99–10

Prasad K (2017) Biology, diversity and promising role of mycorrhizal endophytes for Green Technology. In: Maheshwari DK (ed) Endophytes: biology and biotechnology, sustainable development and biodiversity, vol 15. Springer International Publishing AG, Cham, pp 267–302. https://doi.org/10.1007/978-3-319-66541-2_1

Prashar P, Kapoor N, Sachdeva S (2014) Rhizosphere: its structure, bacterial diversity and significance. Rev Environ Sci Biotechnol 13:63–77

Priyadharsini P, Muthukumar T (2017) The root endophytic fungus *Curvularia geniculate* from *Parthenium hysterophorus* roots improves plant growth through phosphate solubilization and phytohormone production. Fungal Ecol 27:69–77

Raghuwanshi R (2018) Fungal community in mitigating impacts of drought in plants. In: Gehlot P, Singh J (eds) Fungi and their role in sustainable development: current perspectives. Springer Nature Singapore Pvt Ltd, Singapore, pp 267–382. https://doi.org/10.1007/978-981-13-0393-7_15

Rai PK, Singh M, Anand K, Saurabhj S, Kaur T, Kour D et al (2020) Role and potential applications of plant growth promotion rhizobacteria for sustainable agriculture. In: Rastegari AA, Yadav AN, Yadav N (eds) Trends of microbial biotechnology for sustainable agriculture and biomedicine systems: diversity and functional perspectives. Elsevier, Amsterdam, pp 49–60. https://doi.org/10.1016/B978-0-12-820526-6.00004-X

Rana KL, Kour D, Kaur T, Devi R, Yadav AN, Yadav N et al (2020a) Endophytic microbes: biodiversity, plant growth-promoting mechanisms and potential applications for agricultural sustainability. Antonie Van Leeuwenhoek. https://doi.org/10.1007/s10482-020-01429-y

Rana KL, Kour D, Kaur T, Sheikh I, Yadav AN, Kumar V et al (2020b) Endophytic microbes from diverse wheat genotypes and their potential biotechnological applications in plant growth promotion and nutrient uptake. Proc Natl Acad Sci India B. https://doi.org/10.1007/s40011-020-01168-0

Rana KL, Kour D, Yadav AN (2019) Endophytic microbiomes: biodiversity, ecological significance and biotechnological applications. Res J Biotechnol 14:142–162

Rashid MA, Mujawar LH, Shahzad T, Almeelbi T, Ismail IMI, Oves M (2016) Bacteria and fungi can contribute to nutrients bioavailability and aggregate formation in degraded soils. Microbiol Res 183:26–41

Rastegari AA, Yadav AN, Yadav N (2020) New and future developments in microbial biotechnology and bioengineering: Trends of microbial biotechnology for sustainable agriculture and biomedicine systems: diversity and functional perspectives. Elsevier, Amsterdam

Roberts E, Lindow S (2014) Loline alkaloid production by fungal endophytes of *Fescue* species select for particular epiphytic bacterial microflora. ISME J 8:359–368

Rodriguez R, Redman R (2008) More than 400 million years of evolution and some plants still can't make it on their own : plant stress tolerance via fungal symbiosis. J Exp Botany 59(5):1109–1114

Rodriguez RJ, White JF, Arnold AE, Redman RS et al (2009) Fungal endophytes: diversity and functional roles. New Phytol 182:314–330

Saba H, Vibhash D, Manisha M, Prashant KS, Farhan H, Tauseef A (2012) *Trichoderma* a promising plant growth stimulator and biocontrol agent. Mycosphere 3(4):524–531. https://doi.org/10.5943/mycosphere/3/4/14

Sahu PK, Gupta A, Lavanya G, Bakade R, Singh DP (2017) Bacterial endophytes: potential candidates for plant growth promotion. In: Singh DP, Singh HB, Prabha R (eds) Plant-microbe interactions in agro-ecological perspectives. Springer International Publishing AG, Singapore, pp 611–632. https://doi.org/10.1007/978-981-10-5813-4

Sen S, Chandrasekhar CN (2014) Effect of PGPR on growth promotion of rice (Oryza sativa L) under salt stress. Asian J Plant Sci Res 4:62–67

Shaikh S, Wani S, Sayyed R (2018) Impact of interactions between rhizosphere and rhizobacteria: a review. J Bacteriol Mycol 5:1058

Sharaff MS, Subrahmanyam G, Kumar A, Yadav AN (2020) Mechanistic understanding of root-microbiome interaction for sustainable agriculture in polluted soils. In: Rastegari AA, Yadav AN, Yadav N (eds) Trends of microbial biotechnology for sustainable agriculture and biomedicine systems: diversity and functional perspectives. Elsevier, Amsterdam, pp 61–84. https://doi.org/10.1016/B978-0-12-820526-6.00005-1

Sharma SB, Sayyed RZ, Trivedi MH, Gobi TA (2013) Phosphate solubilizing microbes: sustainable approach for managing phosphorus deficiency in agricultural soils. SpringerPlus 2:587

Singh A, Kumar R, Yadav AN, Mishra S, Sachan S, Sachan SG (2020) Tiny microbes, big yields: microorganisms for enhancing food crop production sustainable development. In: Rastegari AA, Yadav AN, Yadav N (eds) Trends of microbial biotechnology for sustainable agriculture and biomedicine systems: diversity and functional perspectives. Elsevier, Amsterdam, pp 1–15. https://doi.org/10.1016/B978-0-12-820526-6.00001-4

Singh BK, Munro S, Potts JM, Millard P (2007) Influence of grass species and soil type on rhizosphere microbial community structure in grassland soils. Appl Soil Ecol 36:147–155. https://doi.org/10.1016/j.apsoil.2007.01.004

Singh A, Sharma J, Rexer KH, Varma A (2000) Plant productivity determinants beyond minerals, water and light: *Piriformospora indica*—a revolutionary plant growth promoting fungus. Curr Sci 79:1548–1554

Sirrenberg A, Göbel C, Grond S, Czempinski N, Ratzinger A, Karlovsky P, et al. (2007) *Piriformospora indica* affects plant growth by auxin production. Physiol Plant 131(4):581–589

Sivakumar PV, Thamizhiniyan P (2012) Enhancement in growth and yield of tomato by using AM fungi and *Azospirillum*. Int J Environ Biol 2(3):137–141

Sorty AM, Meena KK, Choudhary K, Bitla UM, Minhas PS, Krishnani KK (2016) Effect of plant growth promoting bacteria associated with halophytic weed (*Psoralea corylifolia* L) on germination and seedling growth of wheat under saline conditions. Appl Biochem Biotechnol 180:872–882

Souza R, Beneduzi A, Ambrosini A, Costa PB, Meyer J, Vargas LK, Schoenfeld R, Passaglia LMP (2013) The effect of plant growth-promoting rhizobacteria on the growth of rice (*Oryza sativa* L) cropped in southern Brazilian fields. Plant Soil 366(1):585–603

Srivastav S, Yadav KS, Kundu BS (2004) Prospects of using phosphate solubilizing *Pseudomonas* as biofungicide. Indian J Microbiol 44:91–94

Subrahmanyam G, Kumar A, Sandilya SP, Chutia M, Yadav AN (2020) Diversity, plant growth promoting attributes, and agricultural applications of rhizospheric microbes. In: Yadav AN, Singh J, Rastegari AA, Yadav N (eds) Plant microbiomes for sustainable agriculture. Springer, Cham, pp 1–52. https://doi.org/10.1007/978-3-030-38453-1_1

Suman A, Yadav AN, Verma P (2016) Endophytic microbes in crops: diversity and beneficial impact for sustainable agriculture. In: Singh DP, Abhilash PC, Prabha R (eds) Microbial inoculants in sustainable agricultural productivity, research perspectives. Springer-Verlag, New Delhi, pp 117–143

Takishita Y, Charron JB, Smith DL (2018) Biocontrol rhizobacterium *Pseudomonas* sp. 23S induces systemic resistance in tomato (*Solanum lycopersicum* L) against bacterial canker *Clavibacter michiganensis* subsp michiganensis. Front Microbiol 9:2119

Talapatra K, Das AR, Saha AK, Das P (2017) In vitro antagonistic activity of a root endophytic fungus towards plant pathogenic fungi. J Appl Biol Biotechnol 5(2):68–71

Thakur N, Kaur S, Tomar P, Thakur S, Yadav AN (2020) Microbial biopesticides: current status and advancement for sustainable agriculture and environment. In: Rastegari AA, Yadav AN, Yadav N (eds) Trends of microbial biotechnology for sustainable agriculture and biomedicine systems: diversity and functional perspectives. Elsevier, Amsterdam, pp 243–282. https://doi.org/10.1016/B978-0-12-820526-6.00016-6

Theocharis A, Bordiec S, Fernandez O, Paquis S, Dhondt-Cordelier S, Baillieul F, Clément C, Barka EA (2012) *Burkholderia phytofirmans* PsJN primes Vitis vinifera L and confers a better tolerance to low nonfreezing temperatures. Mol Plant-Microbe Interact 25(2):241–249

Tilak K, Ranganayaki N, Pal KK, De R, Saxena AK (2005) Diversity of plant growth and soil health supporting bacteria. Curr Sci 89(1):136–150

Tiwari P, Bajpai M, Singh LK, Mishra S, Yadav AN (2020) Phytohormones producing fungal communities: metabolic engineering for abiotic stress tolerance in crops. In: Yadav AN, Mishra S, Kour D, Yadav N, Kumar A (eds) Agriculturally important fungi for sustainable agriculture, Perspective for diversity and crop productivity, vol 1. Springer, Cham, pp 56–86. https://doi.org/10.1007/978-3-030-45971-0_8

Torres-Barragán A, Zavale-Tamejia E, Gonzalez-Chavez C, Ferrera-Cerrato R (1996) The use of arbuscular mycorrhizae to control onion white rot (*Sclerotium cepivorum*) under field conditions. Mycorrhiza 6:253–257

Turner TR, James EK, Poole PS (2013) The plant microbiome. Genome Biol 14:209

Uma Maheswari T, Anbukkarasi K, Hemalatha T, Chendrayan K (2013) Studies on phytohormone producing ability of indigenous endophytic bacteria isolated from tropical legume crops. Int J Curr Microbiol App Sci 2(6):127–136

Uren NC (2000) Types amounts and possible functions of compounds released into the rhizosphere by soil-grown plants in the rhizosphere. CRC Press, Boca Raton, FL, pp 35–56

Van der Heijden MGA, Streitwolf-Engel RR, Siegrist S, Neudecker A, Ineichen K, Boller T, Wiemken A, Sanders IR (2006) The mycorrhizal contribution to plant productivity, plant nutrition and soil structure in experimental grassland. New Phytol 172:739–752

Varma A, Sudha S, Franken P (1999) *Piriformospora indica*-a cul-tivable plant growth promoting root endophyte with similari-ties to arbuscular mycorrhizal fungi. Appl Environ Microbiol 65:2741–2744

Verma P, Yadav AN, Khannam KS et al (2016a) Appraisal of diversity and functional attributes of thermotolerant wheat associated bacteria from the peninsular zone of India. Saudi J Biol Sci 26:1882–1895. https://doi.org/10.1016/j.sjbs.2016.01.042

Verma P, Yadav AN, Khannam KS, Kumar S, Saxena AK et al (2016b) Molecular diversity and multifarious plant growth promoting attributes of Bacilli associated with wheat (*Triticum aestivum* L.) rhizosphere from six diverse agro-ecological zones of India. J Basic Microbiol 56(1):44–58

Verma P, Yadav AN, Khannam KS et al (2016c) Appraisal of diversity and functional attributes of thermotolerant wheat associated bacteria from the peninsular zone of India. Saud J Biol Sci 26:1882–1895. https://doi.org/10.1016/j.sjbs.2016.01.042

Verma P, Yadav AN, Khannam KS, Panjiar N, Kumar S et al (2015) Assessment of genetic diversity and plant growth promoting attributes of psychotolerant bacteria allied with wheat (*Triticum aestivum*) from the northern hills zone of India. Ann Microbiol 65(4):1885–1899

Verma P, Yadav AN, Kumar V, Singh DP, Saxena AK (2017) Beneficial plant-microbes interactions: biodiversity of microbes from diverse extreme environments and its impact for crop improvement. In: Singh DP, Singh HB, Prabha R (eds) Plant-microbe interactions in agro-ecological perspectives, Microbial interactions and agro-ecological impacts, vol 2. Springer, Singapore, pp 543–580. https://doi.org/10.1007/978-981-10-6593-4_22

Wahid OA, Mehana TA (2000) Impact of phosphate solubilizing fungi on the yield and phosphorus uptake by wheat and faba bean plants. Microbiol Res 155:221–227

Wakelin SA, Warren RA, Harvey PR, Ryder MH (2004) Phosphate solubilization by *Penicillium* sp. closely associated with wheat roots. Biol Fertil Soils 40:36–43

Waller F, Achatz B, Baltruschat H (2005) The endophytic fungus *Piriformospora indica* reprograms barley to salt-stress tolerance, disease resistance, and higher yield. Proc Natl Acad Sci U S A 102:13386–13391

Waqas M, Khan AL, Hamayun M, Shahzad R, Kang SM, Kim JG, Lee IJ (2015) Endophytic fungi promote plant growth and mitigate the adverse effects of stem rot: an example of *Penicillium citrinum* and *Aspergillus terreus*. J Plant Interact 10(1):280–287

Waqas M, Khan AL, Kamran M, Hamayun M, Kang SM, Kim YH, Lee IJ (2012) Endophytic fungi produce gibberellins and indoleacetic acid and promotes hostplant growth during stress. Molecules 17:10754–10773

Weller DM, Raaijmakers JM, Gardener BBM, Thomashow LS (2002) Microbial populations responsible for specific soil suppressiveness to plant pathogens. Annu Rev Phytopathol 40:309–348

Werner D (2000) Organic signals between plants and microorganisms in the rhizosphere: biochemistry and organic substances at the soil–plant interface. Marcel Dekker, New York, NY, pp 197–222

Yadav AN (2019) Endophytic fungi for plant growth promotion and adaptation under abiotic stress conditions. Acta Sci Agric 3(1):91–93

Yadav AN (2017) Agriculturally important microbiomes: biodiversity and multifarious PGP attributes for amelioration of diverse abiotic stresses in crops for sustainable agriculture. Biomed J Sci Tech Res 1(4):861–864. https://doi.org/10.26717/BJSTR.2017.01.000321

Yadav AN, Sharma D, Gulati S, Singh S, Kaushik R (2015) Haloarchaea endowed with phosphorus solubilization attribute implicated in phosphorus cycle. Sci Rep 5:12293

Yadav AN, Singh J, Rastegari AA, Yadav N (2020) Plant microbiomes for sustainable agriculture. Springer, Cham

Yadav AN, Verma P, Kaushik R, Dhaliwal HS, Saxena AK (2017a) Archaea endowed with plant growth promoting attributes. EC Microbiol 8(6):294–298

Yadav AN, Verma P, Kour D, Rana K, Kumar V et al (2017b) Plant microbiomes and its beneficial multifunctional plant growth promoting attributes. Int J Environ Sci Nat Resour 3(1):1–8

Yadav AN, Verma P, Kumar S, Kumar V, Kumar M, Kumari Sugitha TC et al (2018) Actinobacteria from rhizosphere: molecular diversity, distributions, and potential biotechnological applications. In: Singh BP, Gupta VK, Passari AK (eds) New and future developments in microbial biotechnology and bioengineering. Elsevier, Amsterdam, pp 13–41. https://doi.org/10.1016/B978-0-444-63994-3.00002-3

Zabalgogeazcoa I (2008) Fungal endophytes and their interaction with plant pathogens. Span J Agric Res 6:138–146

Zou WX, Meng JC, Lu H, Chen GX, Shi GX, Zhang TY, Tan RX (2000) Metabolites of *Colletotrichum gloeosporioides*, an endophytic fungus in *Artemisia mongolica*. J Nat Prod 63:1529–1530

Plant Growth-Promoting Rhizobacteria (PGPR): Current and Future Prospects for Crop Improvement

9

Kapil Gupta, Neeraj Kumar Dubey, Satendra Pal Singh, Jasmin Kumar Kheni, Shubhra Gupta, and Akash Varshney

Abstract

Population growth and high food demand is the biggest problem of the world. It is necessary to find and apply new techniques in agricultural to enhance the productivity. The chemicals used in agriculture increase yield, kill pathogens, pests, and weeds but harm the ecosystem severely. With increasing concerns about the agrochemicals side effects, a better alternative can be used of microorganisms to the plants and rhizosphere microbial populations. Use of plant growth-promoting rhizobacteria (PGPR) has been found to be a potential alternative and promising technique compared to old routinely used technique which is increasing the burden of pollution to the soil in agriculture. Naturally occurring soil microflora present in the rhizosphere adheres to the surface of the plant roots and imparts beneficial effect on plant growth and production. PGPR are known as biofertilizers and are used for soil quality improvement; they are key players for improvement in agriculture yields. Phytopathogens affect plant health, which is a major threat to sustainable agriculture worldwide. PGPRs apply different mechanisms to protect plants from disease also help plant to grow healthy under environmental stresses. In this chapter, PGPR mediated different mechanisms are discussed

K. Gupta (✉) · S. Gupta
Department of Biotechnology, Central University of Rajasthan, Ajmer, Rajasthan, India

N. K. Dubey
Department of Botany, Rashtriya Snatkottar Mahavidyalaya, Jaunpur, Uttar Pradesh, India

S. P. Singh
Department of Botany, Y.D. (PG.) College, Lakhimpur, Uttar Pradesh, India

J. K. Kheni
Department of Biotechnology, Junagadh Agricultural University, Junagadh, Gujarat, India

A. Varshney
Department of Zoology, Y.D. (PG.) College, Lakhimpur, Uttar Pradesh, India

A. N. Yadav et al. (eds.), *Current Trends in Microbial Biotechnology for Sustainable Agriculture*, Environmental and Microbial Biotechnology,
https://doi.org/10.1007/978-981-15-6949-4_9

that help plants in healthy growth. This approach to improve sustainable agriculture with the use of PGPRs can be commercialized by using PGPR with global applicability.

Keywords

Biofertilizers · Microorganism · Microflora · PGPR · Rhizobacteria · Stress management

9.1 Introduction

Plant growth-promoting rhizobacteria (PGPR) are important resources due to their ability to enhance productivity, profitability, and sustainability at the same time and also food security and rural livelihood is achieved. Use of PGPRs or their by-products is gaining more attention and their use in agriculture could help farmers by providing the technology which is low-cost and environmentally safe. Agriculture is facing arable land reduction and its expansion is impossible, which leads to impart pressure for over-production of crops; therefore, an improved farming technology is required to improve the fertility of the soil and crop production. For example, some techniques being used are sustainable management practices, agricultural intensification, and some other techniques like use of cultivars having disease resistance, salt tolerance, drought tolerance, heavy metal tolerance, and better nutritional values.

Rhizosphere, the soil zone surrounding the plant roots are reservoir of the microbes. Bacteria present in the rhizosphere can be categorized as symbiotic or non-symbiotic, depending on whether they are beneficial for the plant or not (Kundan et al. 2015). Rhizosphere is under control of plant roots through the release of chemical substrates which affects the microbial activity (Barea et al. 2005; Yadav et al. 2017). Rhizobacteria has to compete with other rhizospheric microbes for the nutrients and other compounds secreted by the host plant roots.

Soil microbes are also beneficial to plants in stresses and have been reported in sustained crop production (Khan et al. 2016; Compant et al. 2016). PGPR are the best examples of plant–microbes association, which leads to enhance plant growth and crop production (Yadav et al. 2020d). PGPR establishment with plants affect soil characteristics and convert unfertile land into fertile and improved land. PGPR work as biofertilizers by increasing the availability and uptake of nutrients from poor nutrient containing soil (Kour et al. 2020g). Neutralization of plant stress is also achieved through PGPR for biotic (insects, disease) and abiotic stress (water, salt, light, temperature, etc.) that plants face in the environment (Fasciglione et al. 2015). PGPR are also involved in plant growth promotion by suppressing harmful pathogens through induced systemic resistance (ISR) and competitive exclusion (Tripathi et al. 2012; Thakur et al. 2020).

PGPR can be classified as extracellular PGPR (ePGPR) and intracellular (iPGPR) (Martínez-Viveros et al. 2010). ePGPR reside in the rhizosphere or in the spaces between the root cortex, whereas iPGPR reach the nodular structures of roots. The bacteria known to be ePGPR are *Azotobacter*, *Micrococcus*, *Serratia*, *Azospirillum*, *Bacillus*, *Arthrobacter*, *Caulobacter*, *Erwinia*, *Flavobacterium,* and *Pseudomonas*. Some endophytic iPGPR, for example, Rhizobia (*Allorhizobium, Bradyrhizobium, and Mesorhizobium*) and *Frankia,* are known to fix atmospheric nitrogen (Bhattacharyya and Jha 2012; Mondal et al. 2020; Rai et al. 2020).

9.2 Applications of PGPR in Agriculture

The green revolution was possible due to two major reasons: use of chemicals like pesticides, herbicides, and chemical fertilizers; and development of improved crop varieties through breeding and genetic manipulations. Intensive use of fertilizer and water and other sources increase the crop yield in the last decades, but these inputs damage the soil health, affect water quality, cause imbalances in trophic level, and lead to environmental degradation. The chemical fungicides are not only creating resistant development in pathogens but also causing environmental pollution to other trophic level as well. In spite of the green revolution in agriculture, new options in agricultural are required to fulfill the requirement of food for growing global population. The use of PGPR as fertilizer could be good for agriculture and related sector both in economically and environmentally sustainable productivity (Rana et al. 2020b; Yadav et al. 2020a). PGPRs can help in security of food by enhancing crop productivity in sustainable manner (Table 9.1). PGPRs are proved as environmentally safe for plant growth compared to synthetic biofertilizers and fungicides. Considering the uses of PGPRs, it will be good for agriculture production.

Studies have also demonstrated the use of PGPR in crop production, either by synthesizing compounds with plant growth-promoting properties (Glick 1995), making availability of essential nutrients such as phosphorus, nitrogen, calcium, and magnesium (Çakmakçi et al. 2006; Belimov and Dietz 2000), or averting plant diseases (Lugtenberg and Kamilova 2009). Improved growth and productivity of many commercial crops have been achieved in maize (Sandhya et al. 2010), cotton (Anjum et al. 2007), rice (Ashrafuzzaman et al. 2009), black pepper (Dastager et al. 2011), wheat (Çakmakçi et al. 2007), and cucumber (Maleki et al. 2010).

PGPR promotes plant growth by different mechanisms such as nitrogen fixation and nodulation. PGPR can be exploited for enhanced crop production for sustainable agriculture (Gonzalez et al. 2015). PGPR are reported to enhance plant growth by different mechanisms like biofertilization (nitrogen fixation, phosphate solubilization) (Ahemad and Khan 2012; Glick 2012); induction of phytohormones production like indole-3-acetic acid (IAA) (Tiwari et al. 2020) and 1-amino-cyclopropane-1-carboxylate (ACC) deaminase, production of siderophores (Jahanian et al. 2012) and hydrogen cyanide (Liu et al. 2016); production of antibiotics or protective enzymes (Kour et al. 2019a; Yadav et al. 2016) and xenobiotic degradation (Sharaff et al. 2020).

Table 9.1 Applications of various PGPRs and their mechanisms in promoting plant growth

S. No.	PGPRs name	Plant name	Improved character	Hypothesized causes/ mechanism	References
Biofertilizers					
1	*Rhizobium*, *Bradyrhizobium*	Soybean	Better growth	Convert N_2 into ammonia	Spaink (1995)
2	*B. subtilis* GB03	*A. thaliana*	Increment in photosynthesis	Increment of chlorophyll	Zhang et al. (2008)
3	*Azospirillum*	Wheat	Yield increment	Root development	Lugtenberg and Kamilova (2009)
4	*Pesudomonas chlororaphis* SZY6	*Brassica napus*	Copper resistant	Root length promotion	He et al. (2010)
5	*Burkholderia phytofirmans* PsJN	Grapevine	Reduce chilling-induced damage	Modification of carbohydrate metabolism	Fernandez et al. (2012)
6	*Bacillus licheniformis* NCCP-59	*Oryza sativa*	Improved seed germination under Ni stress and safeguarded against toxicity	Increase in macro- and micro-nutrient content and total protein and organic nitrogen	Jamil et al. (2014)
7	*Paenibacillus lentimorbus* NRRL B-30488	Chickpea	Growth under nutrient stress condition	Mineral nutrient uptake for better growth and development	Bisht et al. (2019)
8	*Bacillus amyloliquefaciens* IARI-HHS2–30	Wheat	Plant growth promotion	K-solubilization	Verma et al. (2015)
9	*Halolamina pelagica* CDK2	*Suaeda nudiflora*	Plant growth promotion	P-solubilization	Gaba et al. (2017)
10	*Acinetobacter guillouiae* EUB2RT.R1	Wheat	Plant growth promotion and nutrient uptake	P-solubilization	Rana et al. (2020c)
11	*Pseudomonas libanensis* EU-LWNA-33	Wheat, maize, rice, sorghum, and finger millet	Alleviation of drought stress and plant growth promotion	P-solubilization	Kour et al. (2020e)

12	*Streptomyces laurentii* EU-LWT$_3$–69, *Penicillium* sp. EU-DSF-10	*Amaranthus*, buck wheat, millets, and maize	Microbe-mediated alleviation of drought stress and acquisition of phosphorus in great millet (*Sorghum bicolor* L.)	P-solubilization	Kour et al. (2020d)
13	*Acinetobacter calcoaceticus* EU- LRNA-72	Wheat, maize, foxtail millet, and finger millet	Amelioration of drought stress in foxtail millet (Setaria italica L.)	P-solubilization	Kour et al. (2020f)
14	*Penicillium* sp. EU-FTF-6	Wheat, maize, foxtail millet, and finger millet	Amelioration of drought stress in foxtail millet (Setaria italica L.)	P-solubilization	Kour et al. (2020f)
Stress management					
1	*Bradyrhizobium* sp.	Mungbean wilt Fusarium wilt	*R. solani* *F. solani*	Mungbean Sunflower	Ehteshamul-Haque and Ghaffar (1993)
2	*P. Putida*	Fusarium wilt	*Fusarium sp.*	Cucumber Radish	Tari and Anderson (1988), de Boer et al. (2003)
3	*P. putida* PCL1444	–	Increment in germination	Degrades naphthalene	Kuiper et al. (2001)
4	*Rhizobium japonicum*	Soya bean wilt	*Macrophomina phaseolina*	Soya bean	Al-Ani et al. (2012)
5	*P. fluorescens*	Dirty panicle disease of rice	*Sarocladium oryzae*	Rice	Sutruedee et al. (2013)
6	*Bacillus amyloliquefaciens* Bk7 and *Brevibacillus laterosporus* B4	Rice	Cold and drought stress tolerance	Elicitors salicylic acid and β-aminobutyric acid (SB) and their mixture (BBSB)	Kakar et al. (2016)
7	*Dietzia natronolimnaea* STR1	Wheat	Protecting plants from salt stress salinity	Modulating the transcriptional machinery for salt tolerance	Bharti et al. (2016)

(continued)

Table 9.1 (continued)

S. No.	PGPRs name	Plant name	Improved character	Hypothesized causes/ mechanism	References
8	*P. agglomerans* RSO6 and RSO7 and B. aryabhattai RSO25	*Spartina densiflora*	Salinity	Oxidative stress index (OSI) decreased	Paredes-Páliz et al. (2018)
9	*P. putida*	Fusarium wilt	*F. solani*	Bean	Kalantari et al. (2018)
Biocontrol					
1	*Pseudomonas* sp. strain WCS417r	Dianthus	Better growth	Resistant against Fusarium wilt of carnation	Van Peer et al. (1991)
2	*Pseudomonas* sp. strain WCS417r	Cucumber	Better growth	Resistant against fungus	Wei et al. (1991)
3	*B. subtilis*, *B. amyloliquefaciens*	Arabidopsis	Promote plant growth	Volatiles	Ryu et al. (2003)
4	*Bacillus subtilis* (BSCBE4), *Pseudomonas chlororaphis* (PA23), *P. fluorescens* (ENPF1)	*Phyllanthus amarus*	Resistant against stem blight Corynespora cassiicola	Increased the defense-related enzymes such as peroxidase, polyphenol oxidase, chitinase, and β-1,3 glucanase	Mathiyazhagan et al. (2004)
5	*Bacillus subtilis* strain BEB-DN (BsDN)	*Solanum lycopersicum*	Resistant against whitefly *Bemisia tabaci*	–	Valenzuela-Soto et al. (2010)
6	*Burkholderia sp.* RX232	*Salix caprea*	Increased plant growth	Increases in ACCD activity, siderophore production, and Cd-immobilization	Kuffner et al. (2010)
7	*Flavimonas oryzihabitans*	*Tomato*	Resistance against whitefly	Tomato whitefly resistant	Reddy (2014)
8	*Bacillus amyloliquefaciens* subsp. *plantarum* FZB42	Beet	Growth-promoting	Biofertilizers/biocontrol agents	Fan et al. (2019)
9	*Cupriavidus taiwanensis* and *P. aeruginosa*	Rice	Increased Cd tolerance Thai jasmine rice	Decreased the accumulation of Cd	Siripornadulsil and Siripornadulsil (2013)
10	*Pseudomonas aeruginosa*	–	Growth promotion	Siderophore production	Sujatha and Ammani (2013)

11	*Alcaligenes faecalis* RZS2 and *Pseudomonas aeruginosa* RZS3	Wheat and peanut	Promoted growth of sown seeds in heavy metal contaminated soil	Chelated various heavy metal ions	Patel et al. (2016)
12	*Planomicrobium chinense* strain P1 *Bacillus cereus* strain P2	*Helianthus annuus*	Enhancing tolerance of plants to drought and heavy metals	Enhance the translocation and accumulation of micronutrients and heavy metals	Khan et al. (2018)
Phytohormones					
1	*P. fluorescens* WCS365	Radish	Increment of root weight	Auxin-generating	Kamilova et al. (2006)
2	*Pantoea dispersa* strain 1A and *Serratia marcescens* strain SRM	Wheat	Biomass production and nutrient uptake	IAA production	Mishra et al. (2012)
3	*Bacillus licheniformis* K11	Pepper	Drought resistant	Expression of stress proteins, *Cadhn*, *VA*, *sHSP*, *CaPR-10*, auxin, and ACC deaminase	Lim and Kim (2013)
4	*Bacillus subtilis*	*Platycladus orientalis*	Cytokinin production by PGPR elevated ABA levels in shoots	And increased the stomatal conductance	Liu et al. (2013)
5	*Pseudomonas* and *Rhizobium*	Legumes	Alleviation of salt stress	Phytohormonal changes	Egamberdieva et al. (2013)
6	*P. putida* H-2–3	Soybean	Improved plant growth	Secretion of gibberellins hormone and	Kang et al. (2014)
7	*Bacillus pumilus Pseudomonas pseudoalcaligenes*	Rice	Resistant towards the abiotic stress like salinity	PGPRs induces the accumulation of osmoprotectants and sugar	Jha and Subramanian (2018)

9.3 Mechanisms of Plant Growth Promotion by PGPR

Rhizospheric soil is rich in several nutrients like amino acids and sugars useful for rhizobacterial growth. Along with these, several other compounds like Strigolactones are exudated by plant root system to attract the infection of friendly microbes (Dobbelaere et al. 2003; Gray and Smith 2005; Aly et al. 2014). 1–2% of microbes inhabiting in these regions can promote the plant growth called as PGPR colonizing this rhizospheric region (Antoun and Kloepper 2001; Schroth and Hancock 1982). Further, PGPRs also act against pathogens by inducing the ISR by jasmonic acid and ethylene-mediated perception similar to salicylic acid-dependent SAR (systemic acquired resistance) pathway (Beneduzi et al. 2012). Antagonistic nature of PGPRs used in the development of several biocontrol agents. Further, these PGPRs produce siderophores and antibiotics which help in improvement of plant fitness, growth, and yield (Beneduzi et al. 2012).

Several reports have shown that PGPR help in growth promotion of plant by increased yield, solubilization of phosphorus (Gaba et al. 2017; Yadav et al. 2015), potassium (Kour et al. 2020c; Verma et al. 2017a), and nitrogen uptake and other elements availability (Rana et al. 2020a). Further, PGPR also enhance root growth and hairs with lateral branches. Several PGPRs modulate the phytohormone level of indole-3-acetic acid (IAA), zeatin, ethylene, gibberellic acid (GA3), and abscisic acid (ABA) thus helping in architecture maintaining of root system (Table 9.1). PGPRs secrete organic acids which lowers rhizopsheric pH, that makes release of phosphate compounds and becomes available to the plants. PGPRs indirectly enhance the plant growth by producing antibacterial, antifungal elements and competing with phytopathogens for niche and nutrients (Table 9.1). Applications of PGPR lead to the development of resistance against several viruses attacking plants. PGPR produce several antipathogenic enzymes, such as glucanases, chitinases, and proteases, damaging the pathogenic cell walls (Beneduzi et al. 2012; Neeraja et al. 2010; Maksimov et al. 2011). Further, authors also suggested that production of antibiotics (e.g., phenazines, phloroglucinols, pyoluteorin, pyrrolnitrin, cyclic lipopeptides, hydrogen cyanide (HCN), siderophores, and bacteriocins) also help in inhibiting the phytopathogenic proliferation (Haas and Défago 2005; Beneduzi et al. 2012). Along with fertilization activity PGPRs also help in managing the plants during different stresses (Lugtenberg and Kamilova 2009; Kumar et al. 2019d; Verma et al. 2017b; Yadav et al. 2018).

9.3.1 Biofertilization

9.3.1.1 Nitrogen Fixation

Atmospheric nitrogen fixation is carried out by symbiotic or non-symbiotic microbes in association with plants (Shridhar 2012). Some examples of atmospheric N_2-fixing symbiotic PGPR are *Rhizobium* spp. (Ahemad and Kibret 2014) *Azoarcus* sp. (Egener et al. 1999), *Herbaspirillum seropedicae* and *Burkholderia* spp. (Baldani et al. 2000), *Serratia marcescens* (Gyaneshwar et al. 2001), and *Rhizobia* (Chaintreuil et al. 2000). However, such processes are limited to legumes but developing symbiotic or non-symbiotic association in other non-leguminous plants can be applied to increase human food supply and soil fertility as well. However, combination of rhizobacteria inoculums can improve soil quality by enhancing N_2 fixation mediated by *nif* gene along with other structural genes. Inoculation of N_2-fixing PGPR to the crops in the field activates growth and disease management of plants and maintains higher nitrogen level in agricultural soil (Damam et al. 2016).

9.3.1.2 Phosphate Solubilization

Rhizobium and phosphorus (P)-solubilizing bacteria like *Pseudomonas* and *Bacillus* sp. are important to plant nutrition. They play a role as PGPR by biofertilization of soil for better growth of crops. Organic acids (e.g., carboxylic acid) are secreted by these bacteria which help in releasing the bound forms of phosphates from calcareous soils by lowering the pH in the rhizosphere. Use of these PGPRs is an environment-friendly biofertilizer which helps to reduce the use of expensive phosphate fertilizers by increasing the availability of free phosphate (by solubilization) and thus, increasing the efficiency of biological nitrogen fixation and also increase the availability of Fe and Zn in rhizosphere (Kaur et al. 2020; Subrahmanyam et al. 2020).

9.3.1.3 Potassium Solubilization

Potassium is the major macronutrient crucial for the plant growth. More than 90% of potassium exists in the form of insoluble rock and silicate minerals so very small fraction of potassium is available for plants as soluble form in soil (Parmar and Sindhu 2013). The low availability of potassium for plants leads to slow growth, poor root growth, and less seed production. This leads to reduced crops yield (Verma et al. 2016). PGPR have been reported to solubilize this insoluble potassium present in rock by secreting organic acids which lower pH and help in phosphate solubilization (Satyaprakash et al. 2017). Potassium-solubilizing PGPR, such as *Acidithiobacillus* sp., *Paenibacillus* spp., *Bacillus edaphicus*, *Ferrooxidans* sp., *Bacillus mucilaginosus*, *Pseudomonas* sp., and *Burkholderia* have the capacity to solubilize K from the soil minerals (Etesami et al. 2017; Kour et al. 2020b; Rajawat et al. 2020).

9.3.1.4 Exopolysaccharide Production

Exopolysaccharides (EPSs) are heterogeneous mixture composed of: polysaccharides, proteins, nucleic acids, and lipids (Sutherland 1972; Wingender et al. 1999). EPSs help plants to maintain water potential, aggregate soil particles, and help plant roots to make contact with rhizobacteria, and help host plant to sustain during abiotic stress conditions (like salinity drought or water stress) or pathogens and enhance plant growth and crop production (Pawar et al. 2013). PGPR, for example, *Azotobacter vinelandii, Enterobacter cloacae,* and *Rhizobium* sp. are reported to produce EPS and enhance soil fertility and help plants' growth for sustainable agriculture (Mahmood et al. 2016).

9.3.2 Stress Management

Plants face various abiotic and biotic stresses in the field which affect their growth and productivity though plants respond to these stresses on their own, but PGPR that help plants against stress management for plants have also been studied (Yadav 2017; Yadav et al. 2019; Yadav and Yadav 2018). PGPR help plants to grow under abiotic stresses or from pathogens (Akhgar et al. 2014). PGPR produce neutralizing substances against phytopathogens to increase resistance in the host plants (Singh and Jha 2015). PGPR mediate stress management in plants by means of production of ROS-scavenging enzymes, siderophores, various antibiotics and protective enzymes (chitinases, cellulases, proteases, etc.) against plant pathogen or disease resistance, and induction of systematic resistance against various pathogens and pests (Nivya 2015; Gupta et al. 2014; Kour et al. 2019b; Kumar et al. 2019b).

9.3.2.1 Abiotic Stress

Environmental cues like salinity, drought, heat, and cold adversely affect survival and production of crops and are responsible for reduction of food supply to the population worldwide. Abiotic stress tolerance is multigenic trait and includes accumulation of osmoprotectants, eeactive oxygen-scavenging enzymes like ascorbate peroxidase (APX), catalase (CAT), superoxide dismutase (SOD), glutathione reductase, antioxidants like ascorbic acid, α-tocopherol, and glutathione (Agami et al. 2016). Apart from crops' abiotic stress tolerance, it can also be achieved by application of PGPR; it is also proved to mitigate the deleterious effect of stress that has been discussed with respect to the effect of PGPR association with plants (Goswami and Deka 2020). The improvement in drought tolerance in association with PGPR has been observed in crops like maize, soybean, chickpea, and wheat (Ngumbi and Kloepper 2016). Application of PGPR *Brachybacterium saurashtrense* strain JG-06, *Brevibacterium casei* strain JG-08, and *Haererohalobacter* sp. strain JG-11 promoted growth of peanut plants in salt stress (Shukla et al. 2012). PGPR application to okra resulted in salinity tolerance in terms of improved water-use efficiency (WUE) mediated by ROS-scavenging enzymes (Habib et al. 2016).

Arthrobacter protophormiae and *Dietzia natronolimnaea* application to wheat enhanced IAA content and reduced abscisic acid (ABA) and/1-aminocyclopropane-1-carboxylate (ACC) content which modulated expression of genes like *CTR1* as well as *DREB2* transcription factor (Barnawal et al. 2017) conferred salinity tolerance in wheat. Maize plants inoculated with the PGPRs *Azospirillum brasilense* and *Herbaspirillum seropedicae* improved drought tolerance (Curá et al. 2017).

9.3.2.2 Biotic Stress

Biotic stress imposed by phytopathogens like bacteria fungi, protists, nematodes, and viruses results in loss of agricultural yield (Haggag et al. 2015). Biotic stress is a major reason for loss in crop yield, and hence it is required to develop resistant crops against biotic stress. Application of PGPR like *Paenibacillus polymyxa*, *Bacillus amyloliquefaciens*, *B. licheniformis*, *B. thuringiensis*, and *B. subtilis* can solve the problem of biotic stress. PGPR strains belonging to Pseudomonas and

Bacillus sp. have been described to induce and respond against phytopathogens like virus, fungus, and bacteria (Bhattacharyya and Jha 2012). Plants inoculated with PGPR showed enhanced disease resistance (Ngumbi and Kloepper 2016). Several crops including rice, wheat, maize, pepper, tomato, chickpea, etc. have shown development of biotic stress-resistant trait against attacking pathogen during post-treatment of PGPRs (Jha and Subramanian 2018; Valenzuela-Soto et al. 2010; Mathiyazhagan et al. 2004). For example, tomato plant shows the resistant against whitefly *Bemisia tabaci* during post-treatment of *Bacillus subtilis* strain BEB-DN (BsDN) (Valenzuela-Soto et al. 2010). Further, *Bacillus subtilis* (BSCBE4), *Pseudomonas chlororaphis* (PA23), endophytic *P. fluorescens* (ENPF1) like PGPR treated *Phyllanthus amarus* plant showed resistant against stem blight *Corynespora cassiicola* due to increased production of defense-related enzymes such as peroxidase, polyphenol oxidase, chitinase, and β-1,3 glucanase (Mathiyazhagan et al. 2004). Further, *Pseudomonas fluorescens*-mediated resistant against various bacterial and oomycete pathogens in *Arabidopsis* have been reported (Ton et al. 2002).

9.3.2.3 Rhizoremediation

The increase in soil and water pollution causes problem in ecosystem and becomes a threat to the life of organisms throughout the world. Pollution can be alleviated by bioremediation; being time consuming, it can be a better alternative to remediate soil and water pollution. Techniques like phytoremediation, bio-slurry, bio-augmentation, bio-pile, land farming, and bio-venting can be applied to remove pollutants from contaminated sites. However, a combined approach of using phytoremediation and bio-augmentation known as rhizoremediation can be applied for better results (Kumar et al. 2019a, c; Malyan et al. 2019).

PGPR are best suited candidates for rhizoremediation which show symbiotic and non-symbiotic relationships with plants. Rhizospheric plant–microbe interaction is essential for remediation of hazardous pollutants (Chaudhry et al. 2005). Studies on rhizoremediation are limited to few microbes like *Pseudomonas aeruginosa*, genetically engineered *Pseudomonas fluorescens*, and certain *Bacillus* species (Kuiper et al. 2004). So identification of more PGPRs for rhizoremediation is the need for removal of specific and large-scale pollutants from soil and water. The use of PGPR like *Pseudomonas putida* and *Pseudomonas fluorescens* has also been studied to impart protecting barley plants from cadmium toxic effect in soil (Baharlouei et al. 2011).

9.3.3 Biocontrol

PGPR produce substances that also protect them against various diseases. PGPR also show the biocontrol properties against a wide range of soil-borne plant pathogens to protect plants by mediating antagonistic interactions between the biocontrol agent and the pathogen to develop host resistance. The PGPRs are known to inhibit the growth of harmful bacteria and fungi by siderophore, antibiotics, and HCN production (Kour et al. 2020a; Saxena et al. 2020; Singh et al. 2020).

9.3.3.1 Siderophores Production

PGPR are reported to use secretion of extracellular metabolites called siderophores. Siderophores are microbial low molecular weight organic compounds which chelate Fe and are produced under Fe-stressed condition. Their primary function is to sequester iron (III) from rhizosphere in higher plants and direct accumulation in micro-biota from rhizospheric soil solution. The resulting iron-chelates are recognized by specific receptor proteins (iron-regulated outer membrane proteins—IROMPs) and transported into the cell by their respective permeases. Typically, these are categorized into the following three classes—catecholates, hydroxymates, and alpha-carboxylates depending upon the coordination site for iron in the ligand and side-chain chemistry (Ali and Vidhale 2013). Yersiniabactin, a siderophore isolated from *Yersinia pestis* and several strains of *Enterobacter* comes under phenolates (Haag et al. 1993). The other category is "mixed" for having characters of both catecholates and hydroxymates, for example, Pyoverdine (Meyer and Abdallah 1978). Under limiting iron condition, siderophores chelate iron from the soil solution and create hindrance in the nutrition of pathogens. The PGPRs are also known for their role in induced systemic resistance via siderophore production (Rastegari et al. 2020a, 2020b).

This approach is being explored in biocontrol of pathogenic forms (Lemanceau and Albouvette 1993). Nanogram amount if present in the rhizospheric soil solution is sufficient for induction of systemic resistance (Pieterse et al. 2001). PGPRs are being implied in versatile ways for increasing the growth and yield of crop plants (Yadav et al. 2020b, 2020c). Their iron-chelating attribute is studied for enhanced rhizospheric iron concentration mechanism and thus making Fe more bio-available in the soil for crop plants.

9.3.3.2 Disease Resistance by Antibiotics

The production of antibiotics is considered to be one of the most powerful and studied biocontrol mechanisms of plant growth-promoting rhizobacteria against phytopathogens (Shilev 2013; Ulloa-Ogaz et al. 2015). A wide variety of antifungal antibiotic compounds like amphisin, 2,4-diacetylphloroglucinol (DAPG), oomycin A, phenazine, pyoluteorin, pyrrolnitrin, tensin, tropolone, and cyclic lipopeptides are produced by Pseudomonads (Loper and Gross 2007) and oligomycin A, kanosamine, zwittermicin A, and xanthobaccin produced by *Bacillus*, *Streptomyces*, and *Stenotrophomonas* sp. to suppress growth of plant pathogens (Compant et al. 2005). A wide variety of antibacterial antibiotics are also produced by *Bacillus* sp.. *Bacillus amyloliquifaciens* that produces lipopeptide antibiotics like surfactin, iturins, and bacillomycin (Wang et al. 2015).

9.3.3.3 Induced Systemic Resistance

ISR is the first line of defense that provides protection against invasion. PGPR promoted ISR confers broad spectrum pathogen resistance (Van der Ent et al. 2009). PGPRs produce signals which activate defense mechanism during pathogenic invasion which involve defense enzymes like ROS-scavenging enzymes SOD, CAT, and APX and plant defense-related enzymes like β-1, 3 glucanase, chitinase, polyphenol oxidase, peroxidase, and some proteinase inhibitors. ISR is a nonspecific mechanism used and it helps plant to resist against any pathogens (Kamal et al. 2014).

Pseudomonas strains are known to induce systemic resistance in carnation, radish, and *Arabidopsis* in which "O antigenic side chain" of the bacterial outer membrane lipopolysaccharides act as an inducer of response. Another type of siderophores, pseudomanine produced by strains of *Pseudomonas* induces salicylic acid production in radish which responds for enhancing plant's defense (Van Loon and Bakker 2007). The translocatable signal induced by rhizobacteria in the plant roots spreads systemically within the plant and increases the defensive capacity against pathogens defense (Van Loon and Bakker 2007).

Rhizobacteria-mediated ISR signaling involves JA and ethylene-mediated signaling induces defense responses (Pieterse et al. 2001). A variety of bacterial compounds like acetoin, 2,3 butanediol, acyl homoserine lactones, cyclic lipopeptides, lipopolysaccharides, and siderophores are reported to induce ISR (Ryu et al. 2004; Berendsen et al. 2015). Majority of PGPR have been confirmed for inducing ISR which can be utilized in improving tolerance against pathogen of the crop plants.

9.3.3.4 Protective Enzymes

PGPR-induced plant growth is mediated by secretion of protecting enzymes like chitinase, ACC deaminase, and β-1,3-glucanase against phytopathogenic agents. Protecting enzymes are involved in lysis of cell walls of plant pathogens to reduce plant loss (Goswami et al. 2016). Fungal cell wall is made up of N-acetyl-glucoseamine and chitin; thus can be controlled by the β-1,3 glucanase- and chitinase. Beta-glucanases and chitinase producing *Pseudomonas fluorescens* and *Sinorhizobium fredii* are known to control fusarium wilt caused by *Fusarium oxysporum* and *Fusarium udum* (Ramadan et al. 2016). The major crop pathogens like *Phytophthora capsici* and *Rhizoctonia solani* are also reported to be controlled by PGPRs application (Islam et al. 2016; Devi et al. 2020).

9.3.4 PGPR as Plant Growth Regulators

Plant growth and development is coordination of organized cell division, cell expansion, and cell differentiation. The interactions of plants with PGPRs is responsible for influencing these processes that have long been of interest, as they can be the option for sustainable agricultural applications. PGPRs produce growth regulators which are responsible for plant growth promotion that include indole-3-acetic acid (IAA), cytokinin, and gibberellins. Growth is also promoted by breakdown of ethylene produced by plants through bacterial 1-aminocyclopropane-1-carboxylate deaminase (ACC deaminase). Growth regulators are also applied exogenously as extracted substances or synthetic analogues to plants or plant tissues.

Plant growth regulators are organic molecules, which can promote, inhibit, or modify growth and development of plants at very low concentrations (Bisht et al. 2018). PGPRs association with plants can also influence plants to synthesize plant growth regulators. Microbes like *Rhizobium leguminosarum, Rhodospirillum rubrum*, *Pseudomonas fluorescens*, *P. aeruginosa*, *P. putida*, *Bacillus subtilis*, *Azotobacter chroococcum*, *Paenibacillus polymyxa*, *Mesorhizobium ciceri*, and *Klebsiella oxytoca* are reported to induce the production of phytohormones which are regarded as PGPR (Prathap and Ranjitha 2015). Application of IAA-producing rhizobacteria for long term increased the plant growth (Amara et al. 2015; Kaymak 2011) and highly developed roots that helped plants in uptake of better nutrients for plant growth (Aeron et al. 2011). Combined effect of PGPR and growth regulator has also been investigated in chickpea by 3 PGPR: *Bacillus subtilis*, *Bacillus thuringiensis*, and *Bacillus megaterium* and PGR was observed to enhance chlorophyll, protein, and sugar contents in seedlings (Khan et al. 2018).

9.4 Future Prospects and Perspective

Use of excess herbicides, pesticides, fertilizer, and different biotic/abiotic factors limit the PGPR growth-promoting effect. Application of biosensors and nanofertilizers in Agricultural biotechnology has improved in agriculture. An approach for combined application of biotechnology, nanotechnology, and other disciplines can be used in farming practices to enhance food production and security to fulfill the need of growing world population. Nanoparticles application can be used in targeting plant pathogens for crop protection and their management, and enhancement of shelf-life of fruits and vegetables. Application of PGPR as biofertilizer can be achieved by conjugation of gold, aluminum, and silver nanoparticles, with PGPRs like *Bacillus subtilis*, *Paenibacillus elgii*, *Pseudomonas fluorescens*, and *Pseudomonas putida* and used in eliminating harmful fungal parasite in rhizosphere. Encapsulation of nano biofertilizers will also help in prolonged fertilizer release to target cell. Application of titanium nanoparticles to the roots of oilseed rape-mediated adhesion of beneficial bacteria which protected the plants from fungal pathogens. Use of nano-biofertilizers is eco-friendly and is also a good alternative for harmful chemicals. Precision farming can also be used to minimize input of harmful chemicals to the soil although maximize crop production by focusing on environmental variables. Zeolites like crystalline aluminum silicates are best option to improve soil water retention. Rhizoremediation can be utilized for the removal of metal toxicity by cleaning contaminants/from the environment which will help plant in better establishment and growth and finally increasing the productivity. An alternate way of engineering the PGPRs to improve their growth-promoting traits can also be done by genetic manipulations which will be low-input, sustainable, and environment-friendly. The PGPR-based crop growth and production strategies are farmer-friendly and eco-friendly and can supplement the long-term goals of sustainable development.

Applications of PGPRs in agriculture will help in reducing the use of chemical fertilizers and pesticides. Biocontrol agents derived from applications of PGPRs are safe and no cost-benefit ratio will surpass the chemical pesticides. Biotechnological methods of identification and characterization of genes relating to siderophorogenesis, antibiosis, and other antagonistic aspects would be imperative in developing new strategies for crop improvement.

9.5 Conclusion

Excess exploitation of resources due to increased population and their limited occurrence has led to find alternative sources for fulfillment of needs. PGPR can be used in enhancing plant growth and crop production; removing pollutants from wastelands and water bodies; and pesticides degradation in soil. To fulfill the human needs excessively used chemical fertilizers and pesticides have also interrupted plant–microbe interactions. Utilization of modern tools in farming along with application of PGPR can play game changer role in improving soil fertility, crop productivity, and nutrients availability in soil along with tolerance against pathogens. Further research can be focused on developing a rhizosphere with diverse microbial communities and application of multidisciplinary research approaches to utilize PGPR potential in growth enhancement and stress tolerance. Future goal for understanding PGPR mechanism of growth promotion and ability to colonize in rhizospheric area can be used to develop as key player in the management of sustainable agriculture.

Acknowledgments K.G. and S.G. thank UGC Dr. DS Kothari Postdoc Fellowship for financial support.

References

Aeron A, Kumar S, Pandey P, Maheshwari DK (2011) Emerging role of plant growth promoting rhizobacteria in agrobiology. In: Maheshwari DK (ed) Bacteria in agrobiology: crop ecosystems. Springer, Berlin, Heidelberg, pp 1–36. https://doi.org/10.1007/978-3-642-18357-7

Agami RA, Medani RA, Abd El-Mola IA, Taha RS (2016) Exogenous application with plant growth promoting rhizobacteria (PGPR) or proline induces stress tolerance in basil plants (*Ocimum basilicum* L.) exposed to water stress. Int J Environ Agric Res 2:78–92

Ahemad M, Khan MS (2012) Evaluation of plant-growth-promoting activities of rhizobacterium *Pseudomonas putida* under herbicide stress. Ann Microbiol 62:1531–1540

Ahemad M, Kibret M (2014) Mechanisms and applications of plant growth promoting rhizobacteria: current perspective. J King Saud Univ Sci 26:1–20

Akhgar AR, Arzanlou M, Bakker PA, Hamidpour M (2014) Characterization of 1-aminocyclopro pane-1-carboxylate (ACC) deaminase-containing *Pseudomonas* spp. in the rhizosphere of salt-stressed canola. Pedosphere 24:461–468

Al-Ani RA, Adhab MA, Mahdi MH, Abood HM (2012) *Rhizobium japonicum* as a biocontrol agent of soybean root rot disease caused by *Fusarium solani* and *Macrophomina phaseolina*. Plant Prot Sci 48(4):149–155

Ali SS, Vidhale NN (2013) Bacterial siderophore and their application: a review. Int J Curr Microbiol Appl Sci 2:303–312

Aly R, Dubey NK, Yahyaa M, Abu-Nassar J, Ibdah M (2014) Gene silencing of CCD7 and CCD8 in P*helipanche aegyptiaca* by tobacco rattle virus system retarded the parasite development on the host. Plant Signal Behav 9:e29376

Amara U, Khalid R, Hayat R (2015) Soil bacteria and phytohormones for sustainable crop production. In: Bacterial metabolites in sustainable agroecosystem. Springer, Cham, pp 87–103. https://doi.org/10.1007/978-3-319-24654-3_5

Anjum MA, Sajjad MR, Akhtar N, Qureshi MA, Iqbal A, Rehman A (2007) Response of cotton to plant growth promoting rhizobacteria (PGPR) inoculation under different levels of nitrogen. J Agric Res Pak 45:135–142

Antoun H, Kloepper JW (2001) Plant growth promoting rhizobacteria. In: Brenner S, Miller JH (eds) Encyclopedia of genetics. Academic, New York, pp 1477–1480. https://doi.org/10.1006/rwgn.2001.1636

Ashrafuzzaman M, Hossen FA, Ismail MR, Hoque A, Islam MZ, Shahidullah SM, Meon S (2009) Efficiency of plant growth-promoting rhizobacteria (PGPR) for the enhancement of rice growth. Afr J Biotechnol 8:7

Baharlouei J, Pazira E, Khavazi K, Solhi M (2011) Evaluation of inoculation of plant growth-promoting rhizobacteria on cadmium uptake by canola and barley. 2nd Int Conf Env Sci Tech 2:28–32

Baldani VLD, Baldani JI, Döbereiner J (2000) Inoculation of rice plants with the endophytic diazotrophs *Herbaspirillum seropedicae* and *Burkholderia* spp. Biol Fertil Soils 30:485–491. https://doi.org/10.1007/s003740050027

Barea JM, Pozo MJ, Azcon R, Azcon-Aguilar C (2005) Microbial co-operation in the rhizosphere. J Ex Bot 56:1761–1778

Barnawal D, Bharti N, Pandey SS, Pandey A, Chanotiya CS, Kalra A (2017) Plant growth-promoting rhizobacteria enhance wheat salt and drought stress tolerance by altering endogenous phytohormone levels and TaCTR1/TaDREB2 expression. Physiol Plant 161:502–514

Belimov AA, Dietz K-J (2000) Effect of associative bacteria on element composition of barley seedlings grown in solution culture at toxic cadmium concentrations. Microbiol Res 155(2):113–121

Beneduzi A, Ambrosini A, Passaglia LM (2012) Plant growth-promoting rhizobacteria (PGPR): their potential as antagonists and biocontrol agents. Genet Mol Biol 35:1044–1051

Berendsen RL, Verk MCV, Stringlis IA, Zamioudis C, Tommassen J, Pieterse CMJ, Bakker PAHM (2015) Unearthing the genomes of plant-beneficial *Pseudomonas* model strains WCS358, WCS374 and WCS417. BMC Genomics 16:539

Bharti N, Pandey SS, Barnawal D, Patel VK, Kalra A (2016) Plant growth promoting rhizobacteria *Dietzia natronolimnaea* modulates the expression of stress responsive genes providing protection of wheat from salinity stress. Sci Rep-UK 6:1–16

Bhattacharyya PN, Jha DK (2012) Plant growth-promoting rhizobacteria (PGPR): emergence in agriculture. World J Microbiol Biotechnol 28:1327–1350

Bisht N, Tiwari S, Singh PC, Niranjan A, Chauhan PS (2019) A multifaceted rhizobacterium *Paenibacillus lentimorbus* alleviates nutrient deficiency-induced stress in *Cicer arietinum* L. Microbiol Res 223:110–119

Bisht TS, Rawat L, Chakraborty B, Yadav V (2018) A recent advances in use of Plant Growth Regulators (PGRs) in fruit crops—a review. Int J Curr Microbiol App Sci 7:1307–1336

Çakmakçi R, Dönmez F, Aydın A, Şahin F (2006) Growth promotion of plants by plant growth-promoting rhizobacteria under greenhouse and two different field soil conditions. Soil Biol Biochem 38(6):1482–1487

Çakmakçi R, Erat M, Erdogan U, Dönmez MF (2007) The influence of plant growth–promoting rhizobacteria on growth and enzyme activities in wheat and spinach plants. J Plant Nutr Soil Sci 170:288–295

Chaintreuil C, Giraud E, Prin Y, Lorquin J, Ba A, Gillis M, Lajudie P, Dreyfus B (2000) Photosynthetic bradyrhizobia are natural endophytes of the African wild rice *Oryza breviligulata*. Appl Environ Microbiol 66:5437–5447

Chaudhry Q, Blom-Zandstra M, Gupta SK, Joner E (2005) Utilising the synergy between plants and rhizosphere microorganisms to enhance breakdown of organic pollutants in the environment (15 pp). Environ Sci Pollut Res 12:34–48

Compant S, Reiter B, Sessitsch A, Nowak J, Clément C, Barka C (2005) Endophytic colonization of Vitis vinifera L. by plant growth-promoting bacterium *Burkholderia* sp. strain 45. PsJN. Appl Environ Microbiol 71:1685–1693

Compant S, Saikkonen K, Mitter B, Campisano A, Mercado-Blanco J (2016) Editorial special issue: soil, plants and endophytes. Plant Soil 405:1–11

Curá JA, Franz DR, Filosofía JE, Balestrasse KB, Burgueño LE (2017) Inoculation with *Azospirillum* sp. and *Herbaspirillum* sp. bacteria increases the tolerance of maize to drought stress. Microorganisms 5:41

Damam M, Kaloori K, Gaddam B, Kausar R (2016) Plant growth promoting substances (phytohormones) produced by rhizobacterial strains isolated from the rhizosphere of medicinal plants. Int J Pharm Sci Rev 37:130–136

Dastager SG, Deepa CK, Pandey A (2011) Potential plant growth-promoting activity of *Serratia nematodiphila* NII-0928 on black pepper (*Piper nigrum* L.). World J Microbiol Biotech 27:259–265

de Boer M, de Boer M, Bom P, Frodo Kindt F, Keurentjes JJB, van der Sluis L, Loon LC, Bakker PAHM (2003) Control of *Fusarium* wilt of radish by combining *Pseudomonas putida* strains that have different disease-suppressive mechanisms. Phytopathology 93:626–632

Devi R, Kaur T, Guleria G, Rana K, Kour D, Yadav N et al (2020) Fungal secondary metabolites and their biotechnological application for human health. In: Rastegari AA, Yadav AN, Yadav N (eds) Trends of microbial biotechnology for sustainable agriculture and biomedicine systems: perspectives for human health. Elsevier, Amsterdam, pp 147–161. https://doi.org/10.1016/B978-0-12-820528-0.00010-7

Dobbelaere S, Vanderleyden J, Okon Y (2003) Plant growth promoting effects of diazotrophs in the rhizosphere. CRC Crit Rev Plant Sci 22:107–149

Egamberdieva D, Jabborova D, Wirth S (2013) Alleviation of salt stress in legumes by co-inoculation with *Pseudomonas* and *Rhizobium*. In: Arora NK (ed) Plant microbe symbiosis: fundamentals and advances. Springer, New Delhi, pp 291–303. https://doi.org/10.1007/978-81-322-1287-4_11

Egener T, Hurek T, Reinhold-Hurek B (1999) Endophytic expression of nif genes of *Azoarcus* sp. strain BH72 in rice roots. Mol Plant Microbe Interact 12:813–819

Ehteshamul-Haque S, Ghaffar A (1993) Use of Rhizobia in the control of root rot diseases of sunflower, okra, soybean and mungbean. J Phytopathol 138:157–163

Etesami H, Emami S, Alikhani HA (2017) Potassium solubilizing bacteria (KSB): mechanisms, promotion of plant growth, and future prospects a review. J Soil Sci Plant Nutr 17(4):897–911

Fan B, Wang C, Ding X, Zhu B, Song X, Borriss R (2019) AmyloWiki: an integrated database for *Bacillus velezensis* FZB42, the model strain for plant growth-promoting Bacilli. Database 2019:baz071

Fasciglione G, Casanovas EM, Quillehauquy V, Yommi AK, Goñi MG, Roura SI, Barassi CA (2015) *Azospirillum* inoculation effects on growth, product quality and storage life of lettuce plants grown under salt stress. Scientia Hortic 195:154–162

Fernandez O, Theocharis A, Bordiec S, Feil R, Jacquens L, Clément C, Fontaine F, Barka EA (2012) *Burkholderia phytofirmans* PsJN acclimates grapevine to cold by modulating carbohydrate metabolism. Mol Plant Microbe Interact 25:496–504

Gaba S, Singh RN, Abrol S, Yadav AN, Saxena AK, Kaushik R (2017) Draft genome sequence of *Halolamina pelagica* CDK2 isolated from natural salterns from Rann of Kutch, Gujarat, India. Genome Announc 5:1–2. https://doi.org/10.1128/genomeA.01593-16

Glick BR (1995) The enhancement of plant growth by free-living bacteria. Can J Microbiol 41:109–117

Glick BR (2012) Plant growth-promoting bacteria: mechanisms and applications. Scientifica 2012:963401

Gonzalez AM, Victoria DE, Merino FCG (2015) Efficiency of plant growth promoting rhizobacteria (PGPR) in sugarcane. Terra Latinoam 33:321–330

Goswami D, Thakker JN, Dhandhukia PC (2016) Portraying mechanics of plant growth promoting rhizobacteria (PGPR): a review. Cogent Food Agric 2:1127500

Goswami M, Deka S (2020) Plant growth-promoting rhizobacteria-alleviators of abiotic stresses in soil: a review. Pedosphere 30:40–61

Gray EJ, Smith DL (2005) Intracellular and extracellular PGPR: commonalities and distinctions in the plant-bacterium signaling processes. Soil Biol Biochem 37:395–412

Gupta S, Meena MK, Datta S (2014) Isolation, characterization of plant growth promoting bacteria from the plant *Chlorophytum borivilianum* and in-vitro screening for activity of nitrogen fixation, phosphate solubilization and IAA production. Int J Curr Microbial Appl Sci 3:1082–1090

Gyaneshwar P, James EK, Mathan N, Reddy PM, Reinhold-Hurek B, Ladha JK (2001) Endophytic colonization of rice by a diazotrophic strain of *Serratia marcescens*. J Bacteriol 183:2634–2645

Haag H, Hantke K, Drechsel H, Stojiljkovic I, Jung G, Zähner H (1993) Purification of yersiniabactin: a siderophore and possible virulence factor of *Yersinia enterocolitica*. Microbiol 139:2159–2165

Haas D, Défago G (2005) Biological control of soil-borne pathogens by fluorescent pseudomonads. Nat Rev Microbiol 3:307–319

Habib SH, Kausar H, Saud HM (2016) Plant growth-promoting rhizobacteria enhance salinity stress tolerance in okra through ROS-scavenging enzymes. BioMed Res Int 2016:6284547

Haggag WM, Abouziena HF, Abd-El-Kreem F, El Habbasha S (2015) Agriculture biotechnology for management of multiple biotic and abiotic environmental stress in crops. J Chem Pharm Res 7:882–889

He LY, Zhang YF, Ma HY, Chen ZJ, Wang QY, Qian M, Sheng XF (2010) Characterization of copper-resistant bacteria and assessment of bacterial communities in rhizosphere soils of copper-tolerant plants. Appl Soil Ecol 44:49–55

Islam S, Akanda AM, Prova A, Islam MT, Hossain MM (2016) Isolation and identification of plant growth promoting rhizobacteria from cucumber rhizosphere and their effect on plant growth promotion and disease suppression. Front Microbiol 6:1–12

Jahanian A, Chaichi MR, Rezaei K, Rezayazdi K, Khavazi K (2012) The effect of plant growth promoting rhizobacteria (PGPR) on germination and primary growth of artichoke (*Cynara scolymus*). Int J Agric Crop Sci 4:923–929

Jamil M, Zeb S, Anees M, Roohi A, Ahmed I, Rehman S, Rha ES (2014) Role of *Bacillus licheniformis* in phytoremediation of nickel contaminated soil cultivated with rice. Int J Phytoremediation 16:554–571

Jha Y, Subramanian RB (2018) From interaction to gene induction: an eco-friendly mechanism of PGPR-mediated stress management in the plant. In: Egamberdieva D, Ahmad P. (eds) Plant microbiome: stress response SpringerSingapore, pp 217–232. doi:https://doi.org/10.1007/978-981-10-5514-0_10

Kakar KU, Ren XL, Nawaz Z, Cui ZQ, Li B, Xie GL, Ehteshamul Hassan MA, Ali E, Sun GC (2016) A consortium of rhizobacterial strains and biochemical growth elicitors improve cold and drought stress tolerance in rice (*Oryza sativa* L.). Plant Biol 18:471–483

Kalantari S, Marefat A, Naseri B (2018) Improvement of bean yield and *Fusarium* root rot biocontrol using mixtures of *Bacillus, Pseudomonas* and *Rhizobium*. Trop Plant Pathol 43:499

Kamal R, Gusain YS, Kumar V (2014) Interaction and symbiosis of AM fungi, *Actinomycetes* and plant growth promoting rhizobacteria with plants: strategies for the improvement of plants health and defense system. Int J Curr Microbial Appl Sci 3:564–585

Kamilova F, Kravchenko LV, Shaposhnikov AI, Makarova N, Lugtenberg B (2006) Effects of the tomato pathogen *Fusarium oxysporum* f. sp. radicis-lycopersici and of the biocontrol bacterium *Pseudomonas fluorescens* WCS365 on the composition of organic acids and sugars in tomato root exudate. Mol Plant Microbe Int 19:1121–1126

Kang SM, Radhakrishnan R, Khan AL, Kim MJ, Park JM, Kim BR, Shin DH, Lee IJ (2014) Gibberellin secreting rhizobacterium, *Pseudomonas putida* H-2-3 modulates the hormonal and stress physiology of soybean to improve the plant growth under saline and drought conditions. Plant Physiol Biochem 84:115–124

Kaur T, Rana KL, Kour D, Sheikh I, Yadav N, Kumar V et al (2020) Microbe-mediated biofortification for micronutrients: present status and future challenges. In: Rastegari AA, Yadav AN, Yadav N (eds) Trends of microbial biotechnology for sustainable agriculture and biomedicine systems: perspectives for human health. Elsevier, Amsterdam, pp 1–17. https://doi.org/10.1016/B978-0-12-820528-0.00002-8

Kaymak HC (2011) Potential of PGPR in agricultural innovations. In: Maheshwari DK (ed) Plant growth and health promoting bacteria. Springer, Berlin, Heidelberg, pp 45–79. https://doi.org/10.1007/978-3-642-13612-2_3

Khan N, Zandi P, Ali S, Mehmood A, Shahid MA (2018) Impact of Salicylic acid and PGPR on the Drought tolerance and phytoremediation potential of Helianthus annuus. Front Microbiol 9:2507

Khan Z, Rho H, Firrincieli A, Hung SH, Luna V, Masciarelli O, Kim SH, Doty SL (2016) Growth enhancement and drought tolerance of hybrid poplar upon inoculation with endophyte consortia. Curr Plant Biol 6:38–47

Kour D, Kaur T, Devi R, Rana KL, Yadav N, Rastegari AA et al (2020a) Biotechnological applications of beneficial microbiomes for evergreen agriculture and human health. In: Rastegari AA, Yadav AN, Yadav N (eds) Trends of microbial biotechnology for sustainable agriculture and biomedicine systems: perspectives for human health. Elsevier, Amsterdam, pp 255–279. https://doi.org/10.1016/B978-0-12-820528-0.00019-3

Kour D, Kaur T, Yadav N, Rastegari AA, Singh B, Kumar V et al (2020b) Phytases from microbes in phosphorus acquisition for plant growth promotion and soil health. In: Rastegari AA, Yadav AN, Yadav N (eds) Trends of microbial biotechnology for sustainable agriculture and biomedicine systems: diversity and functional perspectives. Elsevier, Amsterdam, pp 157–176. https://doi.org/10.1016/B978-0-12-820526-6.00011-7

Kour D, Rana KL, Kaur T, Devi R, Yadav N, Halder SK et al (2020c) Potassium solubilizing and mobilizing microbes: biodiversity, mechanisms of solubilization and biotechnological implication for alleviations of abiotic stress. In: Rastegari AA, Yadav AN, Yadav N (eds) Trends of microbial biotechnology for sustainable agriculture and biomedicine systems: diversity and functional perspective. Elsevier, Amsterdam, pp 177–202. https://doi.org/10.1016/B978-0-12-820526-6.00012-9

Kour D, Rana KL, Kaur T, Sheikh I, Yadav AN, Kumar V et al (2020d) Microbe-mediated alleviation of drought stress and acquisition of phosphorus in great millet (*Sorghum bicolor* L.) by drought-adaptive and phosphorus-solubilizing microbes. Biocatal Agric Biotechnol 23:101501. https://doi.org/10.1016/j.bcab.2020.101501

Kour D, Rana KL, Kaur T, Singh B, Chauhan VS, Kumar A et al (2019a) Extremophiles for hydrolytic enzymes productions: biodiversity and potential biotechnological applications. In: Molina G, Gupta VK, Singh B, Gathergood N (eds) Bioprocessing for biomolecules production, John Wiley & Sons, Hoboken, pp 321–372. https://doi.org/10.1002/9781119434436.ch16

Kour D, Rana KL, Sheikh I, Kumar V, Yadav AN, Dhaliwal HS et al (2020e) Alleviation of drought stress and plant growth promotion by *Pseudomonas libanensis* EU-LWNA-33, a drought-adaptive phosphorus-solubilizing bacterium. Proc Natl Acad Sci India B. https://doi.org/10.1007/s40011-019-01151-4

Kour D, Rana KL, Yadav AN, Sheikh I, Kumar V, Dhaliwal HS et al (2020f) Amelioration of drought stress in foxtail millet (*Setaria italica* L.) by P-solubilizing drought-tolerant microbes with multifarious plant growth promoting attributes. Environ Sustain 3:23–34. https://doi.org/10.1007/s42398-020-00094-1

Kour D, Rana KL, Yadav AN, Yadav N, Kumar M, Kumar V et al (2020g) Microbial biofertilizers: bioresources and eco-friendly technologies for agricultural and environmental sustainability. Biocatal Agric Biotechnol 23:101487. https://doi.org/10.1016/j.bcab.2019.101487

Kour D, Rana KL, Yadav AN, Yadav N, Kumar V, Kumar A et al (2019b) Drought-tolerant phosphorus-solubilizing microbes: biodiversity and biotechnological applications for alleviation of drought stress in plants. In: Sayyed RZ, Arora NK, Reddy MS (eds) Plant growth promoting Rhizobacteria for sustainable stress management, Rhizobacteria in abiotic stress management, vol 1. Springer, Singapore, pp 255–308. https://doi.org/10.1007/978-981-13-6536-2_13

Kuffner M, De Maria S, Puschenreiter M, Fallmann K, Wieshammer G, Gorfer M, Strauss J, Rivelli AR, Sessitsch A (2010) Culturable bacteria from Zn-and Cd-accumulating *Salix caprea* with differential effects on plant growth and heavy metal availability. J Appl Microbiol 108:1471–1484

Kuiper I, Bloemberg GV, Lugtenberg BJ (2001) Selection of a plant-bacterium pair as a novel tool for rhizostimulation of polycyclic aromatic hydrocarbon-degrading bacteria. Mol Plant Microbe Int 14:1197–1205

Kuiper I, Lagendijk EL, Bloemberg GV, Lugtenberg BJJ (2004) Rhizoremediation: a beneficial plant-microbe interaction. Mol Plant Microbe Int 7:6–15

Kumar A, Chaturvedi AK, Yadav K, Arunkumar KP, Malyan SK, Raja P et al (2019a) Fungal phytoremediation of heavy metal-contaminated resources: current scenario and future prospects. In: Yadav AN, Singh S, Mishra S, Gupta A (eds) Recent advancement in white biotechnology through fungi, Perspective for sustainable environments, vol 3. Springer, Cham, pp 437–461. https://doi.org/10.1007/978-3-030-25506-0_18

Kumar M, Kour D, Yadav AN, Saxena R, Rai PK, Jyoti A et al (2019b) Biodiversity of methylotrophic microbial communities and their potential role in mitigation of abiotic stresses in plants. Biologia 74:287–308. https://doi.org/10.2478/s11756-019-00190-6

Kumar M, Saxena R, Rai PK, Tomar RS, Yadav N, Rana KL et al (2019c) Genetic diversity of methylotrophic yeast and their impact on environments. In: Yadav AN, Singh S, Mishra S, Gupta A (eds) Recent advancement in white biotechnology through fungi, Perspective for sustainable environments, vol 3. Springer International Publishing, Cham, pp 53–71. https://doi.org/10.1007/978-3-030-25506-0_3

Kumar V, Joshi S, Pant NC, Sangwan P, Yadav AN, Saxena A et al (2019d) Molecular approaches for combating multiple abiotic stresses in crops of arid and semi-arid region. In: Singh SP, Upadhyay SK, Pandey A, Kumar S (eds) Molecular approaches in plant biology and environmental challenges. Springer, Singapore, pp 149–170. https://doi.org/10.1007/978-981-15-0690-1_8

Kundan R, Pant G, Jado N, Agrawal PK (2015) Plant growth promoting rhizobacteria: mechanism and current prospective. J Fertilizers Pesticides 6:9

Lemanceau P, Albouvette A (1993) Suppression of Fusarium wilts by fluorescent pseudomonads: mechanism and applications. Biocontrol Sci Tech 3:219–234

Lim JH, Kim SD (2013) Induction of drought stress resistance by multi-functional PGPR *Bacillus licheniformis* K11 in pepper. T Plant Pathol J 29:201

Liu F, Xing S, Ma H, Du Z, Ma B (2013) Cytokinin-producing, plant growth-promoting rhizobacteria that confer resistance to drought stress in *Platycladus orientalis* container seedlings. Appl Microbial Biotech 97:9155–9164

Liu W, Wang Q, Hou J, Tu C, Luo Y, Christie P (2016) Whole genome analysis of halotolerant and alkalotolerant plant growth-promoting rhizobacterium *Klebsiella* sp. D5A. Sci Rep 6:1–10

Loper JE, Gross H (2007) Genomic analysis of antifungal metabolite production by *Pseudomonas fluorescens* Pf-5. Eur J Plant Pathol 119:265–278

Lugtenberg B, Kamilova F (2009) Plant-growth-promoting rhizobacteria. Annu Rev Microbiol 63:541–556

Mahmood S, Daur I, Al-Solaimani SG, Ahmad S, Madkour MH, Yasir M, Hirt H, Ali S, Ali Z (2016) Plant growth promoting rhizobacteria and silicon synergistically enhance salinity tolerance of mung bean. Front Plant Sci 7:876

Maksimov IV, Abizgildina RR, Pusenkova LI (2011) Plant growth promoting rhizobacteria as alternative to chemical crop protectors from pathogens (review). Appl Biochem Microbiol 47:333–345

Maleki M, Mostafaee S, Mokhtarnejad L, Farzaneh M (2010) Characterization of *Pseudomonas fluorescens* strain CV6 isolated from cucumber Rhizosphere in Varamin as a potential biocontrol agent. Aust J Crop Sci 4:676

Malyan SK, Kumar A, Baram S, Kumar J, Singh S, Kumar SS et al (2019) Role of fungi in climate change abatement through carbon sequestration. In: Yadav AN, Singh S, Mishra S, Gupta A (eds) Recent advancement in white biotechnology through fungi, Perspective for sustainable environments, vol 3. Springer, Cham, pp 283–295. https://doi.org/10.1007/978-3-030-25506-0_11

Martínez-Viveros O, Jorquera MA, Crowley DE, Gajardo GM, Mora ML (2010) Mechanisms and practical considerations involved in plant growth promotion by rhizobacteria. J Soil Sci Plant Nutr 10:293–319

Mathiyazhagan S, Kavitha K, Nakkeeran S, Chandrasekar G, Manian K, Renukadevi P, Krishnamoorthy AS, Fernando WG (2004) PGPR mediated management of stem blight of *Phyllanthus amarus* (Schum and Thonn) caused by *Corynespora cassiicola* (Berk and Curt) Wei. Arch Phytopathol Plant Protec 37:183–199

Meyer JA, Abdallah MA (1978) The fluorescent pigment of *Pseudomonas fluorescens*: biosynthesis, purification and physicochemical properties. J Gen Microbiol 107:319–328

Mishra PK, Bisht SC, Bisht JK, Bhatt JC (2012) Cold-tolerant PGPRs as bioinoculants for stress management. In: Bacteria in agrobiology: stress management. Springer, Berlin, Heidelberg, pp 95–118. https://doi.org/10.1007/978-3-662-45795-5_6

Mondal S, Halder SK, Yadav AN, Mondal KC (2020) Microbial consortium with multifunctional plant growth promoting attributes: future perspective in agriculture. In: Yadav AN, Rastegari AA, Yadav N, Kour D (eds) Advances in plant microbiome and sustainable agriculture, Functional annotation and future challenges, vol 2. Springer, Singapore, pp 219–254. https://doi.org/10.1007/978-981-15-3204-7_10

Neeraja C, Anil K, Purushotham P, Suma K, Sarma P, Moerschbacher BM, Podile AR (2010) Biotechnological approaches to develop bacterial chitinases as a bioshield against fungal diseases of plants. Crit Rev Biotechnol 30:231–241

Ngumbi E, Kloepper J (2016) Bacterial-mediated drought tolerance: current and future prospects. Appl Soil Ecol 105:109–125

Nivya RM (2015) A study on plant growth promoting activity of the endophytic bacteria isolated from the root nodules of *Mimosa pudica* plant. Int J Innov Res Sci Er Technol 4:6959–6968

Paredes-Páliz K, Rodríguez-Vázquez R, Duarte B, Caviedes MA, Mateos-Naranjo E, Redondo-Gómez S, Caçador MI, Rodríguez-Llorente ID, Pajuelo E (2018) Investigating the mechanisms underlying phytoprotection by plant growth-promoting rhizobacteria in *Spartina densiflora* under metal stress. Plant Biol 20:497–506

Parmar P, Sindhu SS (2013) Potassium solubilisation by rhizosphere bacteria: influence of nutritional and environmental conditions. J Microbial Res 3:25–31

Patel PR, Shaikh SS, Sayyed RZ (2016) Dynamism of PGPR in bioremediation and plant growth promotion in heavy metal contaminated soil. Indian J Exp Biol 54:286–290

Pawar ST, Bhosale AA, Gawade TB, Nale TR (2013) Isolation, screening and optimization of exopolysaccharide producing bacterium from saline soil. J Microbiol Biotechnol Res 3:24–31

Pieterse CM, Van Pelt JA, Van Wees SC, Ton J, Léon-Kloosterziel KM, Keurentjes JJ, Verhagen BW, Knoester M, Van der Sluis I, Bakker PA, Van Loon LC (2001) Rhizobacteria-mediated induced systemic resistance: triggering, signalling and expression. Eur J Plant Pathol 107:51–61

Prathap M, Ranjitha KBD (2015) A critical review on plant growth promoting rhizobacteria. J Plant Pathol Microbiol 6:1–4

Rai PK, Singh M, Anand K, Saurabhj S, Kaur T, Kour D et al (2020) Role and potential applications of plant growth promotion rhizobacteria for sustainable agriculture. In: Rastegari AA, Yadav AN, Yadav N (eds) Trends of microbial biotechnology for sustainable agriculture and biomedicine systems: diversity and functional perspectives. Elsevier, Amsterdam, pp 49–60. https://doi.org/10.1016/B978-0-12-820526-6.00004-X

Rajawat MVS, Singh R, Singh D, Yadav AN, Singh S, Kumar M et al (2020) Spatial distribution and identification of bacteria in stressed environments capable to weather potassium aluminosilicate mineral. Braz J Microbiol 51:751–764. https://doi.org/10.1007/s42770-019-00210-2

Ramadan EM, Abdel Hafez AA, Hassan EA, Saber FM (2016) Plant growth promoting rhizobacteria and their potential for biocontrol of phytopathogens. Afr J Microbiol Res 10:486–504

Rana KL, Kour D, Kaur T, Devi R, Yadav AN, Yadav N et al (2020a) Endophytic microbes: biodiversity, plant growth-promoting mechanisms and potential applications for agricultural sustainability. Antonie Van Leeuwenhoek. https://doi.org/10.1007/s10482-020-01429-y

Rana KL, Kour D, Kaur T, Devi R, Yadav N, Subrahmanyam G et al (2020b) Biotechnological applications of seed microbiomes for sustainable agriculture and environments. In: Rastegari AA, Yadav AN, Yadav N (eds) Trends of microbial biotechnology for sustainable agriculture and biomedicine systems: diversity and functional perspectives. Elsevier, Amsterdam, pp 127–143. https://doi.org/10.1016/B978-0-12-820526-6.00008-7

Rana KL, Kour D, Kaur T, Sheikh I, Yadav AN, Kumar V et al (2020c) Endophytic microbes from diverse wheat genotypes and their potential biotechnological applications in plant growth promotion and nutrient uptake. Proc Natl Acad Sci India B. https://doi.org/10.1007/s40011-020-01168-0

Rastegari AA, Yadav AN, Yadav N (2020a) New and Future Developments in Microbial Biotechnology and Bioengineering: Trends of Microbial Biotechnology for Sustainable Agriculture and Biomedicine Systems: Diversity and Functional Perspectives. Elsevier, Amsterdam

Rastegari AA, Yadav AN, Yadav N (2020b) New and Future Developments in Microbial Biotechnology and Bioengineering: Trends of Microbial Biotechnology for Sustainable Agriculture and Biomedicine Systems: Perspectives for Human Health. Elsevier, Amsterdam

Reddy PP (2014) Plant growth promoting rhizobacteria for horticultural crop protection, vol 19. Springer, New Delhi, pp 35–54

Ryu CM, Farag MA, Hu CH, Reddy MS, Wei HX, Paré PW, Kloepper JW (2003) Bacterial volatiles promote growth in Arabidopsis. P Natl Acad Sci U S A 100(8):4927–4932

Ryu CM, Farag MA, Hu CH, Reddy MS, Kloepper JW, Paré PW. (2004) Bacterial volatiles induce systemic resistance in Arabidopsis. Plant physiology, 134(3): 1017–1026.

Sandhya V, Ali SZ, Grover M, Reddy G, Venkateswarlu B (2010) Effect of plant growth promoting Pseudomonas spp. on compatible solutes, antioxidant status and plant growth of maize under drought stress. Plant Growth Regul 62(1):21–30

Satyaprakash M, Nikitha T, Reddi EU, Sadhana B, Vani SS (2017) Phosphorous and phosphate solubilising bacteria and their role in plant nutrition. Int J Curr Microbiol App Sci 6:2133–2144

Saxena AK, Padaria JC, Gurjar GT, Yadav AN, Lone SA, Tripathi M et al. (2020) Insecticidal formulation of novel strain of *Bacillus thuringiensis* AK 47. Indian Patent 340541

Schroth MN, Hancock JG (1982) Disease-suppressive soil and root-colonizing bacteria. Science 216:1376–1381

Sharaff MS, Subrahmanyam G, Kumar A, Yadav AN (2020) Mechanistic understanding of root-microbiome interaction for sustainable agriculture in polluted soils. In: Rastegari AA, Yadav AN, Yadav N (eds) Trends of microbial biotechnology for sustainable agriculture and biomedicine systems: diversity and functional perspectives. Elsevier, Amsterdam, pp 61–84. https://doi.org/10.1016/B978-0-12-820526-6.00005-1

Shilev S (2013) Soil Rhizobacteria regulating the uptake of nutrients and undesirable elements by plants. In: Arora NK (ed) Plant microbe symbiosis: fundamentals and advances. Springer, New Delhi, pp 147–167. https://doi.org/10.1007/978-81-322-1287-4_5

Shridhar BS (2012) Review: nitrogen fixing microorganisms. Int J Microbial Res 3:46–52

Shukla PS, Agarwal PK, Jha B (2012) Improved salinity tolerance of *Arachis hypogaea* (L.) by the interaction of halotolerant plant-growth-promoting rhizobacteria. J Plant Growth Regul 31:195–206

Singh A, Kumar R, Yadav AN, Mishra S, Sachan S, Sachan SG (2020) Tiny microbes, big yields: microorganisms for enhancing food crop production sustainable development. In: Rastegari AA, Yadav AN, Yadav N (eds) Trends of microbial biotechnology for sustainable agriculture and biomedicine systems: diversity and functional perspectives. Elsevier, Amsterdam, pp 1–15. https://doi.org/10.1016/B978-0-12-820526-6.00001-4

Singh RP, Jha PN (2015) Molecular identification and characterization of rhizospheric bacteria for plant growth promoting ability. Int J Curr Biotechnol 3:12–18

Siripornadulsil S, Siripornadulsil W (2013) Cadmium-tolerant bacteria reduce the uptake of cadmium in rice: potential for microbial bioremediation. Ecotox Environ Safe 94:94–103

Spaink HP (1995) The molecular basis of infection and nodulation by rhizobia: the ins and outs of sympathogenesis. Annu Rev Phytopathol 33(1):345–368

Subrahmanyam G, Kumar A, Sandilya SP, Chutia M, Yadav AN (2020) Diversity, plant growth promoting attributes, and agricultural applications of rhizospheric microbes. In: Yadav AN, Singh J, Rastegari AA, Yadav N (eds) Plant microbiomes for sustainable agriculture. Springer, Cham, pp 1–52. https://doi.org/10.1007/978-3-030-38453-1_1

Sujatha N, Ammani K (2013) Siderophore production by the isolates of fluorescent pseudomonads. Int J Cur Res Rev 5:1

Sutherland IW (1972) Bacterial exopolysaccharides. Adv Microb Physiol 8:143–213

Sutruedee P, Dusit A, Wilawan C, Tiyakhon C, Natthiya B (2013) Bioformulation *Pseudomonas fluorescens* SP007s against dirty panicle disease of rice. Afr J Microbiol Res 7:5274–5283

Tari PH, Anderson AJ (1988) Fusarium Wilt suppression and agglutinability of *Pseudomonas putida*. Appl Environ Microb 54:2037–2041

Thakur N, Kaur S, Tomar P, Thakur S, Yadav AN (2020) Microbial biopesticides: current status and advancement for sustainable agriculture and environment. In: Rastegari AA, Yadav AN, Yadav N (eds) Trends of microbial biotechnology for sustainable agriculture and biomedicine systems: diversity and functional perspectives. Elsevier, Amsterdam, pp 243–282. https://doi.org/10.1016/B978-0-12-820526-6.00016-6

Tiwari P, Bajpai M, Singh LK, Mishra S, Yadav AN (2020) Phytohormones producing fungal communities: metabolic engineering for abiotic stress tolerance in crops. In: Yadav AN, Mishra S, Kour D, Yadav N, Kumar A (eds) Agriculturally important fungi for sustainable agriculture, volume 1: perspective for diversity and crop productivity. Springer, Cham. https://doi.org/10.1007/978-3-030-45971-0_8

Ton J, Van Pelt JA, Van Loon LC, Pieterse CM (2002) Differential effectiveness of salicylate-dependent and jasmonate/ethylene-dependent induced resistance in *Arabidopsis*. Mol Plant Microbe In 15:27–34

Tripathi DK, Singh VP, Kumar D, Chauhan DK (2012) Impact of exogenous silicon addition on chromium uptake, growth, mineral elements, oxidative stress, antioxidant capacity, and leaf and root structures in rice seedlings exposed to hexavalent chromium. Acta Physiol Plant 34:279–289

Ulloa-Ogaz AL, Munoz-Castellanos LN, Nevarez-Moorillon GV (2015) Biocontrol of phytopathogens: antibiotic production as mechanism of control, the battle against microbial pathogens. In: Mendez Vilas A (ed) Basic science, technological advance and educational programs, vol 1. Formatex Research Center, Badajoz, pp 305–309

Valenzuela-Soto JH, Estrada-Hernández MG, Ibarra-Laclette E, Délano-Frier JP (2010) Inoculation of tomato plants (*Solanum lycopersicum*) with growth-promoting *Bacillus subtilis* retards whitefly *Bemisia tabaci* development. Planta 231:397

Van der Ent S, Van Wees SC, Pieterse CM (2009) Jasmonate signaling in plant interactions with resistance-inducing beneficial microbes. Phytochemistry 70:1581–1588

Van Loon LC, Bakker PAHM (2007) Root-associated bacteria inducing systemic resistance. In: Plant-associated bacteria. Springer, Dordrecht, pp 269–316

Van Peer R, Niemann GJ, Schippers B (1991) Induced resistance and phytoalexin accumulation in biological control of *Fusarium* wilt of carnation by *Pseudomonas* sp. strain WCS 417 r. Phytopathology 81:728–734

Verma P, Yadav AN, Khannam KS, Kumar S, Saxena AK, Suman A (2016) Molecular diversity and multifarious plant growth promoting attributes of Bacilli associated with wheat (*Triticum aestivum* L.) rhizosphere from six diverse agro-ecological zones of India. J Basic Microbiol 56:44–58

Verma P, Yadav AN, Khannam KS, Saxena AK, Suman A (2017a) Potassium-solubilizing microbes: diversity, distribution, and role in plant growth promotion. In: Panpatte DG, Jhala YK, Vyas RV, Shelat HN (eds) Microorganisms for green revolution, Microbes for sustainable crop production, vol 1. Springer Singapore, Singapore, pp 125–149. https://doi.org/10.1007/978-981-10-6241-4_7

Verma P, Yadav AN, Kumar V, Singh DP, Saxena AK (2017b) Beneficial plant-microbes interactions: biodiversity of microbes from diverse extreme environments and its impact for crop improvement. In: Singh DP, Singh HB, Prabha R (eds) Plant-microbe interactions in agro-ecological perspectives, Microbial interactions and agro-ecological impacts, vol 2. Springer, Singapore, pp 543–580. https://doi.org/10.1007/978-981-10-6593-4_22

Verma P, Yadav AN, Shukla L, Saxena AK, Suman A (2015) Alleviation of cold stress in wheat seedlings by *Bacillus amyloliquefaciens* IARI-HHS2-30, an endophytic psychrotolerant K-solubilizing bacterium from NW Indian Himalayas. Natl J Life Sci 12:105–110

Wang X, Mavrodi DV, Ke L, Mavrodi OV, Yang M, Thomashow LS, Zheng N, Weller DM, Zhang J (2015) Biocontrol and plant growth-promoting activity of rhizobacteria from Chinese fields with contaminated soils. Microbial Biotechnol 8:404–418

Wei G, Kloepper JW, Tuzun S (1991) Induction of systemic resistance of cucumber to *Colletotrichum orbiculare* by select strains of plant growth-promoting rhizobacteria. Phytopathology 81:1508–1512

Wingender J, Neu TR, Flemming HC (1999) What are bacterial extracellular polymeric substances? In: Wingender J, Neu TR, Flemming HC (eds) Microbial extracellular polymeric substances. Springer, Berlin, Heidelberg, pp 1–19. https://doi.org/10.1007/978-3-642-60147-7_1

Yadav AN (2017) Beneficial role of extremophilic microbes for plant health and soil fertility. J Agric Sci 1:1–4

Yadav AN, Gulati S, Sharma D, Singh RN, Rajawat MVS, Kumar R et al (2019) Seasonal variations in culturable archaea and their plant growth promoting attributes to predict their role in establishment of vegetation in Rann of Kutch. Biologia 74:1031–1043. https://doi.org/10.2478/s11756-019-00259-2

Yadav AN, Kour D, Kaur T, Devi R, Gukeria G, Rana KL et al (2020a) Microbial biotechnology for sustainable agriculture: current research and future challenges. In: Rastegari AA, Yadav AN, Yadav N (eds) Trends of microbial biotechnology for sustainable agriculture and biomedicine systems: diversity and functional perspectives. Elsevier, Amsterdam, pp 331–343. https://doi.org/10.1016/B978-0-12-820526-6.00020-8

Yadav AN, Kumar V, Dhaliwal HS, Prasad R, Saxena AK (2018) Microbiome in crops: diversity, distribution, and potential role in crop improvement. In: Crop improvement through microbial biotechnology. Elsevier, Amsterdam, pp 305–332. https://doi.org/10.1016/B978-0-444-63987-5.00015-3

Yadav AN, Rastegari AA, Yadav N, Kour D (2020b) Advances in plant microbiome and sustainable agriculture: diversity and biotechnological applications. Springer, Singapore

Yadav AN, Rastegari AA, Yadav N, Kour D (2020c) Advances in plant microbiome and sustainable agriculture: functional annotation and future challenges. Springer, Singapore

Yadav AN, Sachan SG, Verma P, Saxena AK (2016) Bioprospecting of plant growth promoting psychrotrophic Bacilli from cold desert of north western Indian Himalayas. Indian J Exp Biol 54:142–150

Yadav AN, Sharma D, Gulati S, Singh S, Dey R, Pal KK et al (2015) Haloarchaea endowed with phosphorus solubilization attribute implicated in phosphorus cycle. Sci Rep 5:12293

Yadav AN, Singh J, Rastegari AA, Yadav N (2020d) Plant microbiomes for sustainable agriculture. Springer, Cham

Yadav AN, Verma P, Kour D, Rana KL, Kumar V, Singh B et al (2017) Plant microbiomes and its beneficial multifunctional plant growth promoting attributes. Int J Environ Sci Nat Resour 3:1–8. https://doi.org/10.19080/IJESNR.2017.03.555601

Yadav AN, Yadav N (2018) Stress-adaptive microbes for plant growth promotion and alleviation of drought stress in plants. Acta Sci Agric 2:85–88

Zhang H, Xie X, Kim MS, Kornyeyev DA, Holaday S, Paré PW (2008) Soil bacteria augment *Arabidopsis* photosynthesis by decreasing glucose sensing and abscisic acid levels in planta. Plant J 56:264–273

Beneficial Microbiomes for Sustainable Agriculture: An Ecofriendly Approach

10

Hina Upadhyay, Diptanu Banik, Muhammad Aslam, and Jatinder Singh

Abstract

Fertilizer plays an important role in increasing crop productivity though chemical fertilizer demand is increasing, and due to this reduction in soil organic matter, natural chemistry and its health is decreasing day by day. Beneficial microbiomes are economical, organic, and biodegradable than chemical fertilizers. Beneficial microbiomes have the capacity of improving the organic microorganisms in the soil and also build up the quality of the soil. The microorganisms present in the biofertilizers play an important role because they produce the nutrients which benefit the plants. Biofertilizers are also less costly, safe which provide wide scope for the research areas and fields related to organic farming. Overall, the significant role of biofertilizers in plant growth and development is great so it makes them an integral and important tool for the organic and sustainable agriculture. This book chapter describes about the various biofertilizers, their types, their mode of actions, and their benefits.

Keywords

Biofertilizers · Sustainable management · Nitrogen fixation · Soil organic matter

H. Upadhyay · D. Banik · M. Aslam
Department of Agronomy, School of Agriculture, Lovely Professional University, Phagwara, Punjab, India

J. Singh (✉)
Department of Horticulture, School of Agriculture, Lovely Professional University, Phagwara, Punjab, India
e-mail: jatinder.19305@lpu.co.in

A. N. Yadav et al. (eds.), *Current Trends in Microbial Biotechnology for Sustainable Agriculture*, Environmental and Microbial Biotechnology,
https://doi.org/10.1007/978-981-15-6949-4_10

10.1 Introduction

With the past 50 years of Indian history, the chemical pesticides and fertilizers have played an important role in increasing the agricultural productivity. But the large use of chemical fertilizers has negative effects on human health. Indiscriminate use of chemical fertilizers and chemical pesticides contributed in loss of soil productivity along with addition of salts to the soil. In future, the requirement of different types of chemical fertilizers will be less for soil improvement. It will lower the opportunities of water pollution along with unsustainable difficulty on agricultural system. Stewart (1969) described that various microbiologists (Beijerinck 1901; Lipman 1903) were pioneer in the isolation of *Azotobacter* spp., whereas scientist, Winodgradsky isolated the basic strain of *Eubacterium pasteurianum*. After a long time, the invention of the biological development in cyanobacteria was recognized. Subsequently then, analysis efforts in this fields have gradually overstated resulting in the choice of various strains displaying many advantageous options (Podile and Kishore 2007).

These extra adoptions of fertilizers which are chemical in nature in agriculture are costly with negative effect chemical and physical properties of soils (Aggani 2013). Therefore, with the obvious harmful and serious consequences of the chemicals, Khan et al. (2007) also described the introduction of many organic fertilizers that may stimulate the growth process in the plant in a positive way.

Abdul Halim (2009) also submitted that the role of these natural stimulants has an early history that goes from generation to generation on small-scale manure production and generation of farmers. A particular group of such fertilizers includes products in the form of microorganisms that promote bio-growth, called biofertilizers, which are nitrogen-fixing, phosphate-solubilizing, or living or cellulolytic microorganisms. They are used for the application of seeds, soil, or compost areas with the aim of increasing the number of such microorganisms and to increase the availability of nutrients by accelerating some microbial process, which is done by plants (Khosro and Yousef 2012). These biofertilizers play an important role as a major component of integrated nutrient management in soil, which will be helpful for soil productivity and sustainability.

Biofertilizers may contain living microorganisms that, when applied to the seeds (seed treatment) of plants, result in growth of the plant with improvement in hardness property along with increased availability of nutrients to the host plant (Kour et al. 2020c; Rana et al. 2020; Singh et al. 2020b). These biofertilizers are ecologically renewable sources of plant nutrients, which may replace the costing of chemical fertilizers effectively. Three important groups of microbes such as arbuscular mycorrhizal fungi (AMF), plant growth-promoting rhizobacteria (PGPR), and nitrogen-fixing rhizobia act as a biocontrol agents that are extensively used globally (Podile and Kishore 2007).

In 1886, German scientists Hellerl and Wilfarth were responsible for the discovery of nitrogen fixation, which stated that legumes with root nodules can use gaseous (molecular) nitrogen. Dutch microbiologist Beijerinck, in the year 1998, isolated a bacterial strain from root nodules. It was later found to be a

Rhizobium leguminosarum strain. Stewart (1969) reported that in the year 1901, a microbiologist Beijerinck was responsible for isolating *Azotobacter* spp. and subsequently, Lipman in 1903. Apart from this, Winodgradsky in (1901) isolated the first strain of *Clostridium pasteurianum* (Thakur et al. 2014). Stewart 1969 reported that nitrogen fixation was found much later in cyanobacteria (blue-green algae). Podile and Kishore (2007) reviewed that there has been a steady increase in research efforts in these areas, resulting in the selection of several beneficial characteristics to many strains. The uses of fertilizers in fields for crop production showed the adverse effects on soil health. Today, several fertilizers are made available for us which act may as growth enhancer substance. Such kind of information about natural stimulators or microbial inoculums that passes from one generation to another farmer generation includes the use of culture of small-scale production. Plant growth enhancing microbes are known as biofertilizer or "microbial inoculants" (live and dormant microbes) they are economical and feasible in nature.

Biofertilizers can be applied to seed (as seed treatment), soil, and composting areas with a purpose to encourage the number of microbes and to improve the reactions for enhancing the supply of required plant materials which can be easily available to the plant (Khosro and Yousef 2012; Kaur et al. 2020; Kour et al. 2020a, b). Use of such biofertilizers is mandatory part of integrated nutrient management (INM) within soil but its role in productivity and property of soil is significant and indispensable. With passage of time, chemical fertilizers are replaced by these biofertilizers as they are cost-effective because of their eco-friendly approach and recycled supply of plant nutrients. Three primary teams of biocontrol are arbuscular mycorrhizal fungi (AMF), nitrogen-fixing rhizobia (Franche et al. 2009) and plant growth-promoting rhizobacteria (PGPR). Nevertheless, we should focus towards fixing of the chemical elements. *Rhizobium* lives in the nodules of leguminous plants such as peas, beans, or grams. *Rhizobium* has a symbiotic relationship with the plant. These bacteria transform the nitrogen from the air into nitrogen substances that the plant can use.

The rhizobia enter into the plant nodule tissue through a plant-derived infection thread—a tubular structure—to enable the entry of bacterium to deeper layers. These infection-causing threads may grow transcellular and ultimately, rhizobia enveloped into a plant-derived sheath, presently referred to as symbiosome membrane, are transported into plant cells.

10.2 National Scenario

Indian Government and various State Governments are encouraging the usage of biofertilizers through subsidies on sales, grants, and extension programs and with degrees of stress. Over time, farmers are taking practical information related to technology based on the scientific realities of their fields. Thus, farmers are being inspired to adopt the utilization of biofertilizers. The government of Asian countries has been implementing the theme for the promotion of biofertilizers since underneath this theme, one national center-NCOF, and six regional centers—RCOFs are established.

The most frequent operations of those centers include the promotion of biofertilizer through coaching, demonstration, and often providing an economic culture for combining biofertilizers. The theme additionally aims for giving grants up to Rs. 40 large integers per unit of a 150 tons per annum to line up biofertilizer manufacturing units. Since origination, the biofertilizer production capability often, 525 tons, has been envisaged by putting in 83 biofertilizer production units. Out of those units, 9 units are sanctioned by the Department of Fertilizers underneath their theme of providing monetary help for the aim and 74 units are supported; 39 units are created by completely different organizations and personal entrepreneurs with a production capability of seven, 975 tons per annum. The total calculable current demand for biofertilizers in Asian countries is eighteen, 500 tons per annum, whereas calculable production is concerning 10 tons per annum within the country. One of the most economical and pollution-free ways of all energy is to use the flexibility of certain microorganisms such as bacteria, algae, and fungi to modify the chemical element of the area, dissolve phosphorus, destroy organic materials or sulfur within the soil. Once planted within the soil, they increase crop growth and yield, improve soil fertility, and reduce pollution. Therefore, biofertilizers are inhabited by biological or active materials of bacteria, algae, and fungi or microbe inoculants that are prepared to enrich the soil with chemical elements, phosphorus, and organic matter (Figs. 10.1, 10.2, and 10.3).

10.3 Common Nitrogen Fixers

10.3.1 *Azotobacter*

Azotobacter is a free-living bacteria belonging to the subclass of the Proteobacteria, which grows well on a nitrogen-free medium which can be utilizing atmospheric nitrogen. It is an aerobic group of bacteria that has no symbiotic nitrogen fixation activity. These bacteria enhance the development of growth-promoting hormones and further help to increase plant growth and yield (Kumar et al. 2019). A total of seven species are found of genus *Azotobacter*, i.e., *A. chroococcum*, *A. vinelandii*, *A. beijerinckii*, *A. paspali*, *A. nigricans*, *A. salinestri*, and *A. armeniacus*. These groups of microbes are heterotrophic in nature, aerobic and commonly found in soils that have the ability to cure nitrogen non-symbiotic (Doroshenko et al. 2007). *Azotobacter* is very sensitive to salivary and acidic pH. The optimum pH for growth and nitrogen fixation is 7.0–7.5 but fail to grow below the pH of 6. Studies reported that 10–12% of increments in crop production of agricultural crop by the application of *Azotobacter* into the soil also increases wheat crop and improves grain yield.

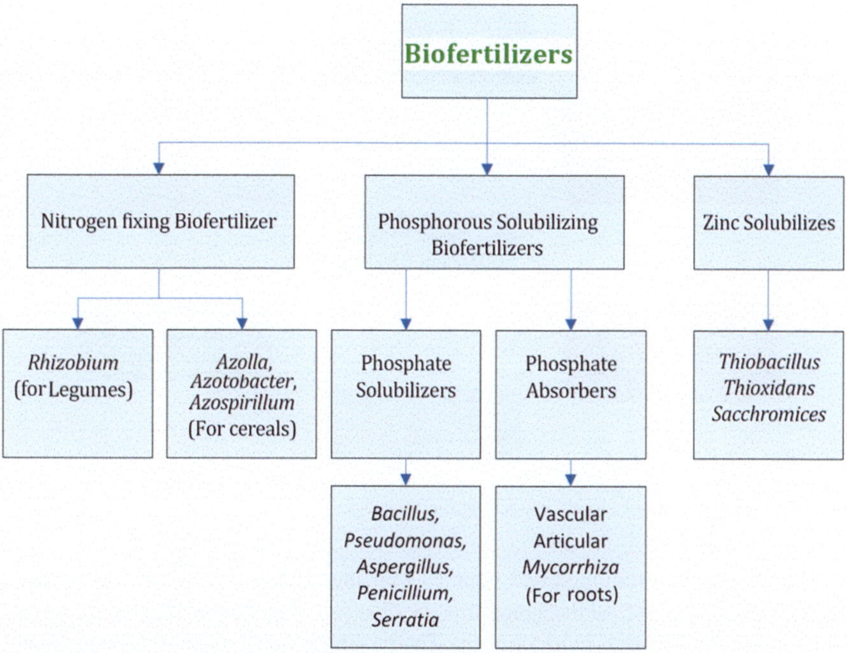

Fig. 10.1 Classification of different biofertilizers

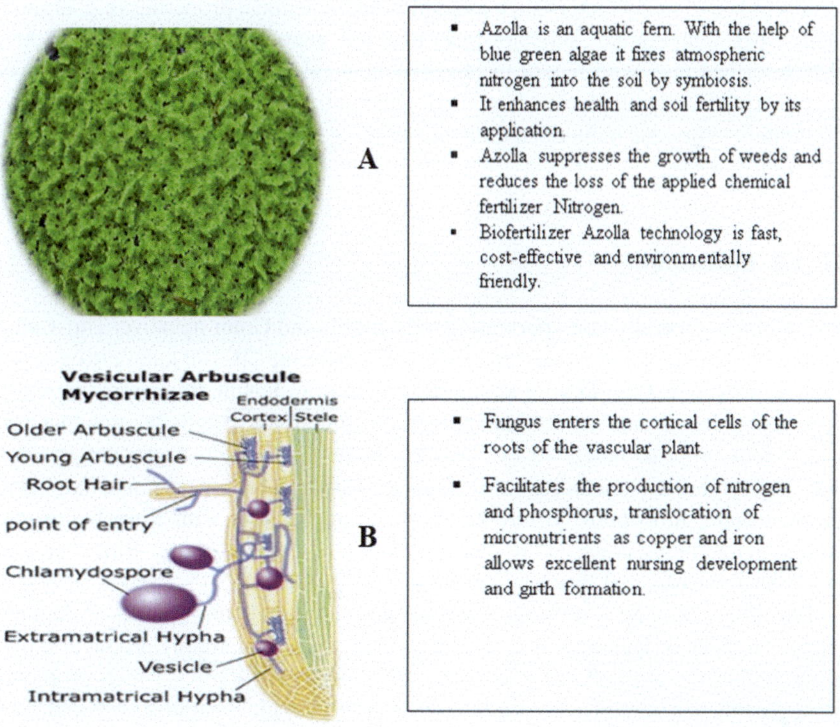

Fig. 10.2 (**a**) Azolla culture (**b**) VAM

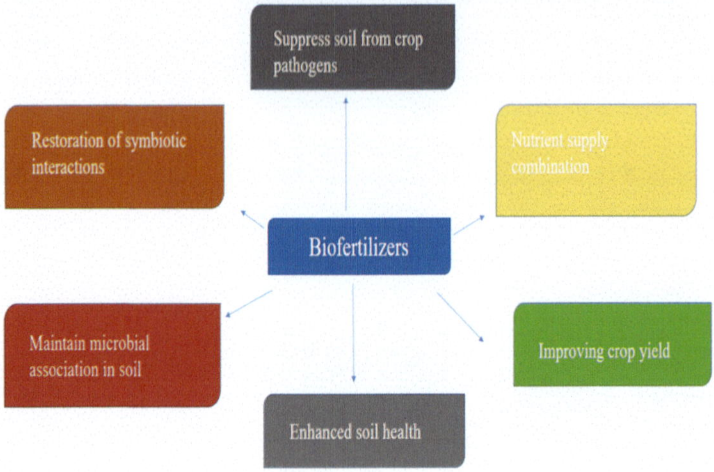

Fig. 10.3 Applications of biofertilizers in agroecosystems

The Dutch microbiologist and botanist Martinus Beijerinck in 1901 introduced the genus *Azotobacter*. At the same time, he discovered and described the important organism of the genus *A. chroococcum*. The proliferation of these soil-borne bacteria is related to several factors, mainly soil pH and fertility. *Azotobacter* activity in soils has beneficial effects on plant metabolism, but the concentration of these bacteria is linked to many factors such as soil Physico-chemical properties (e.g. organic matter, pH, temperature, soil moisture) and soil microorganisms. Ridvan (2009) also described that, with regard to physico-chemical properties of soil, number of studies have mainly emphasized on major plant nutrients like P, K, Ca, etc. and also the content of organic matter and their encouraging influence on the soil populations of *Azotobacter* spp.

10.3.2 *Rhizobium*

Rhizobium is generally non-sporulating rods, motile and Gram-negative. Those who are not non-symbiotic in nature are capable of recovering 50–100 kg of nitrogen per hectare. Effective *Rhizobium* nodulation formation in leguminous crops basically depends on the advancement of an appropriate strain for a specific legume in a particular zone. Approximately 80% of biologically determined nitrogen usually occurs due to symbiosis, including the family Rhizobiaceae, *Sinorhizobium,* and *Azorhizobium, Rhizobium, Bradyrhizobium*, Proteobacteria with Rhizobiales and various leguminous plants. PGP *Rhizobium* belongs to the family Rhizobiaceae, Phyllobacteriaceae, and Bradyrhizobiaceae.

Rhizobium may be a dependent bacterium that forms root nodules in legume plants. Nodules of leguminous crop plants function as miniature N production factories within the agricultural ecosystem. *Rhizobacteria* (PGPR), which promotes plant growth, is potential agent for biological control of plant pathogens. Population of *Rhizobium* is directly affected in soil due to the type of farm crops grown. It is also studied that without leguminous crops, *Rhizobium* populations decrease day by day in agroecosystem (Mahdi et al. 2010). It is also noticed that effective pod formation requires a specific species of *Rhizobium* in their root nodule. Cooper (2004) also studied, in leguminous crops the pod formation is directly influenced by various strains of rhizobia, but growth is increased only when nodules are produced by effective strains of rhizobia.

10.3.3 *Azolla*

As a green manure crop, *Azolla* can be applied to the main field and also as a dual crop. Generally, farmers are growing *Azolla* as a green manure crop. In the flooded fields, *Azolla* can be allowed to grow 2–3 weeks before transplanting. After that excess water is drained and *Azolla* is incorporated by ploughing. When a thick mat is form, it is incorporated. Generally as a dual crop, farmers grow *Azolla* 1000–5000 kg per hectare in the soil 1 week after transplanting. The remaining *Azolla* is regrown and mowed as another crop. Improved growth of azollae, 25–50 kg/ha of superphosphate is applied and 5–10 cm of standing water is maintained continuously in rice fields (Fig. 10.2 and Tables 10.1 and 10.2).

10.4 Need of Biofertilizers for Sustainable Management of Agroecosystem

Non-ethical use of chemical fertilizer doses creating pollutions to our agroecosystem. Side by side they have also reduced soil fertility and increased the soil toxicity. The current research is promoting more development and application of commercial biofertilizers for the food security, human health, and environmental sustainability (Malyan et al. 2020; Sharaff et al. 2020; Thakur et al. 2020). These plant growth-promoting rhizobacteria shows more targeted activity, and their small quantity can be utilized in small amounts; they multiply on their own, but are controlled by the plant as well as indigenous microbial populations present in the soil. These microbial populations can be decomposed more rapidly than commercial chemical pesticides (Rastegari et al. 2020a, b; Tiwari et al. 2020). Reducing feedstock/fossil fuel (energy crisis) rises fertilizer prices. In addition to the above, the long-term use of biofertilizers is inexpensive, environmentally friendly, more effective, competitive, and accessible to marginal and small farmers without chemical fertilizers.

Biofertilizers that play an important role in organic agriculture, by mobilizing fastened different nutrient (macro- and micronutrients) or converting the insoluble P into a plant-accessible form, play an important role in sustaining long-lasting fertility and biodiversity by fixing atmospheric di-nitrogen (N = N). Taking into account the price as well as the environmental impacts of chemical fertilizers, additional dependence on chemical fertilizers is not a viable long-term approach because of the costs of building fertilizer plants and maintaining production in both domestic

Table 10.1 Classification of different types of microorganisms commonly used as biofertilizers

	Type of biofertilizers	Type of microorganism	Role of biofertilizer	References
N_2-fixing bacteria				
1.	Free-living	*Azotobacter, Closteridium, Nostoc*	*Azotobacter* sp. have a full range of enzymes needed to perform the nitrogen fixation: Ferredoxin, hydrogenase, and an important enzyme nitrogenase	Vessey (2003), Shanware et al. (2014), Thammana et al. (2006)
2.	Symbiotic	*Rhizobium, Anabaena azollae*	Bacteria infiltrate the root of the legume and form root nodules that reduce the molecular nitrogen to ammonia used by plants to produce proteins, vitamins as well as other nitrogen compounds *Anabaena azollae* increases the nitrogen levels in the rice paddies	Yoneyama et al. (1987), Ahmed et al. (2007), Zimmer et al. (2016)
3.	Associative symbiotic	*Azospirillum*	In non-leguminous plants, fix a considerable amount of nitrogen in the range of 20–40 kg N/ha	Reynders and Vlassak (1979)
Phosphorous-solubilizing biofertilizer				
1.	Bacteria	*Bacillus subtilis, pseudomonas striata*	Plant growth hormone synthesis (IAA, GA, cytokinins, and spermidines) stimulates the development of plant	Radhakrishnan and Lee (2016), Khan et al. (2009), Linu et al. (2019), Sadiq et al. (2013)
2.	Fungi	*Penicillium* sp., *Aspergillus awamori*	Growing the available P in soil and increase dry matter yield, yield of grain	Vessey and Heisinger (2001)
Phosphorous-mobilizing biofertilizers				
1.	Arbuscular Mycorrhiza	*Glomus* sp., *Scutellospora* sp.	Fungus enters the cortical cells of the roots of the vascular plant	Ryan and Graham (2002), Chalk et al. (2006)
2.	Ectomycorrhiza	*Laccaria* sp., *Oisolithus* sp., *Boletus* sp., *Amanita* sp.		Chang and Yang (2009)
3.	Ericoid Mycorrhiza	*Pezizella ericae*	Helps in the uptake of phosphorus by plants	Bolan (1991)

(continued)

Table 10.1 (continued)

	Type of biofertilizers	Type of microorganism	Role of biofertilizer	References
Biofertilizers for micronutrients				
1	Silicate and zinc solubilizers	*Bacillus* sp.	Bacteria sp. deteriorates silicates and silicates of aluminum. Several organic acids are produced during the metabolism of microbes that have a dual role in silicate weathering	Mahdi et al. (2010), Saravanan et al. (2011)
Plant growth-promoting Rhizobacteria				
1.	*Pseudomonas*	*Pseudomonas fluorescence*	Colonize roots and promote plant growth	Jorquera et al. (2012), Vessey (2003)

Table 10.2 Biofertilizers commonly used in different field crops

Type of crop	Biofertilizers	Nutrient element
Pulses (chickpea, green gram, black gram, pigeon pea, etc.)	*Rhizobium*/PSB	Nitrogen, phosphorous
Cereals (wheat, oat, rice, etc.)	*Azotobacter*/Azospirillum, PSB	Nitrogen, phosphorous
Oil seed (mustard, sesame, sunflower, castor, etc.)	*Azotobacter*/PSB	Nitrogen, phosphorous
Millets (pearl millets, finger millets, kodo millets, etc.)	*Azotobacter*/PSB	Nitrogen, phosphorous

and foreign exchange capital. This can be the viable approach for farmers to improve the productivity per unit area by application of organic manures (biofertilizers) and may even be used in integrated pest management systems.

10.5 Applications of the Biofertilizers

Biofertilizers can be replaced by chemical fertilizers, as excess use of chemical fertilizers is not beneficial to plants as well as agroecosystem.

- Biofertilizers generally helpful to concerned crops to get not only high yield but also improving soil health with nutrients and beneficial microflora, which will be helpful in the maintenance of the sustainability of the agroecosystem.
- Substitution by 25% of chemical nitrogen and phosphorus.
- Promote the growth of plants.
- Biologically activate the soil. Restoring the fertility of soil naturally.
- Protect against drought and some diseases borne by the soil.

These biofertilizers can be applied as seed treatment, seedling dipping, and soil application. These biofertilizers can be utilized as combined application, but biofertilizers cannot be mixed with insecticides, fungicides, and herbicidal application.

Nowadays, scientists are developing genetically engineered microorganisms by the application of many biotechnological researches including a number of uses that are important for present scenario. It includes intergeneric microorganisms. Symbiotic nitrogen-fixers are *Bradyrhizobium japonicum*. These microbial inoculants can also be further classified based on their use, methods, and time of application. The physical form of the biofertilizers can be applied either in slurry form or as dry inoculants can be produced using different kinds of soil materials like coal, peat, soil, etc. as inorganic form, organic and inert materials.

By the excess use of chemical fertilizers and modern agricultural practices, soil fertility and beneficial microflora is degrading day by day. So, we have to promote more and more organic farming and apply the uses of more green chemistry either by using biofertilizers, integrated nutrient management, and integrated pest and weed management programs. For the food security and sustainable management of crop cultivation technology, the role of government, private, and NGOs is very important.

Seed Treatment: 200 g biofertilizer with nitrogen and 200 g of phosphatic is suspended and thoroughly mixed in 300–400 mL water. Ten kilograms of seeds are treated with this paste and dried in shade. These treated seeds must be sown as early as possible.

Seedling root dip: A bed is filled in the field with water for rice cultivation. In this water, recommended biofertilizers are mixed and the seedlings roots are dipped for 8–10 h.

Soil treatment: 4 kg of each of the recommended biofertilizers are mixed and kept overnight in 200 kg of compost. At the time of sowing or planting, this mixture is incorporated in soil.

10.6 Potential Traits of Some Biofertilizers

Rhizobium nitrogen-fixers: They are symbiotic in nature and belong to the family Rhizobiaceae, and may be able to fix 50–100 kg of nitrogen/ha in conjunction with only legumes. It is helpful for pulses such as chickpea, red grams, pea, lentils, black grams, oil-seed legumes like soybeans, groundnut, and forage legumes such as berseem and lucerne. The productive nodulation of *Rhizobium* leguminous crops largely depends on the supply of a companionable strain for a specific legume.

It is able to colonize the roots of various leguminous substances into tumors, which serve as factories for the manufacture of ammonia, such as root nodules. The *Rhizobium* can be used in symbiosis of legumes and some non-legumes such as *Parasponia* to fix atmospheric nitrogen. The population of *Rhizobium* in the soil is dependent on legumes in field. The population decreases in the absence of legumes. The population of active strains of the *Rhizobium* near the rhizosphere needs artificial inoculation of seeds often to speed up N-fixation. For each legume, an effective nodule must be produced by a certain *Rhizobium* species.

10.6.1 *Azospirillum*

Azospirillum is a heterotrophic and associative existence, belonging to Spirilaceae family. They develop growth controlling substances in addition to their nitrogen-fixing capacity of approx. 20–40 kg/ha. Although many species of the genus, such as *A. mazonense*, *A.halopraeferens*, *A. brasilense*, there has been a global distribution and inoculation advantage primarily through *A. lipoferum* and *A. brasilense*. A positive effect of *Azospirillum* on plant growth is likely to involve several effects, such as phytohormone synthesis, N_2 fixation, nitrate reductase activity as well as mineral uptake (Kour et al. 2020c). As a result of their growing and fixing nitrogen on the salts of organic acids, such as malic aspartic acid, *Azospirillum* forms an associative symbiosis with several plants, particularly those with the C4 dicarboxyl pathways of photosynthesis. It is therefore recommended primarily for maize, sugarcane, sorghum, pearl millet, etc. In addition to remaining on the root surface, the *Azotobacter* colonizing the roots also penetrates into radicular tissues and lives with plants in harmony. However, they do not result in the formation of nodules or produce any growth on the root tissue.

10.6.2 *Azotobacter*

These are free-living, aerobic, heterotrophic, and belong to family Azotobacteriaceae. Their existence is frequent in neutral or basic soils while *A. chroococcum* is a specific type of *Azotobacter* which occurs in productive soils. Above all, some other alternative reportable species are *A. beijerinckii, A. vinelandii, A. macrocytogenes,* and *A. insignis*. Their number often exceeds than 10^4–10^5 g^{-1} of soil since lack of organic matter along with existence of other antagonistic microorganisms in soil. The bacterium gives rise to anti-fungal antibiotics that prevent the growth of many pathogenic fungi in the root zone thereby avoiding seedling mortality. But the total number of *Azotobacter* is usually less in the rhizosphere of the crops including uncultivated soils. The presence of this bacterium has been reported from the rhizosphere of various crops such as rice, sugarcane, maize, vegetables, bajra, and plantation crops.

10.6.3 *Azolla* and Blue Green Algae (Cyanobacteria)

They belong to eight different families and phototrophic in nature. They are plentiful in paddy field and sometimes also referred as "paddy species." They are very helpful in the development of plant growth regulators like **Gibberellic** acid, auxin, and may fix 20–30 kg N/ha in submerged rice fields. Contribution of N element cannot be ignored for lowland rice production in large amounts. BNF along with soil N are main sources of N for lowland cultivation of rice. It has been estimated that from mixture of soil N and BNF mineralization rice plant is able to meet the 50–60% N requirement. Fixation of N element should be met by BNF and not by other source from fertilizer industry to achieve food security through sustainable agriculture. BGA is interdependent in nature, but the frequently symbiotic combination can be recorded between a free-floating aquatic fern, *Anabaena azollae* (BGA) and the Azolla. Above all, BGA has the potential to fix N with fungi, flora, and liverworts. *Azolla* contains both types of nitrogen, i.e., 4–5% N dry and 0.2–0.4% wet. The fern forms a green mat-like formation over the water surface, its leaves, stem, and roots intensely bilobed. *Azolla* may be used as green manure. Under Indian conditions, the most common species is Pinnata, and it can be used by vegetative means on a marketable scale. Recently, few species of *Azolla* has been introduced in India like *A. caroliniana* and *Microphylla*, for large biomass production.

10.6.4 Phosphate-Solubilizing Bacteria

Effectiveness of P-fertilizer is very poor particularly in acidic and alkaline-natured soils and, inopportunely, such types of soils are common in India. It is estimated that more than 34% acidic soils are with considerable amount of saline and alkaline soils. In such soils, it in essential to use the inoculation of PSB and other useful microbial inoculants to restore and continue with effective microbial inhabitants in order to harvest good justifiable crops. From reviewed literature, it is clear that some bacterial species have remarkable ability to solubilize insoluble phosphate complexes, such as dicalcium phosphate, rock phosphate, tricalcium phosphate, and hydroxyapatite. This type of potential can be found in certain genera of bacteria such as *Pseudomonas*, *Bacillus*, *Rhizobium*, *Agrobacterium*, *Achromobacter*, *Erwinia*, *Flavobacterium*, *Halococcus*, *Enterobacter*, and *Micrococcus* (Verma et al. 2016; Yadav et al. 2019, 2015). These include both types of strains, namely anaerobic and aerobic, with a resemblance to aerobic strains in the submerged soils. Large volumes of phosphate-soluble bacteria are often available in the rhizosphere than in non-rhizosphere soils. Bacillus, pseudomonas, and fungi are very common examples of soil microflora. Significant populations of phosphate-soluble bacteria are present in soil and in plant rhizosphere.

10.6.5 Mycorrhiza

The term "Mycorrhiza" denotes to "fungus roots." The fungal spp. gets benefits from the process of photosynthesis, carried out by the host plant and in turn, the host is provided with some essential nutrients, particularly zinc, copper, calcium, and phosphates. This is a symbiotic type of union of certain fungal groups but in the root system of the host plants. These nutrients are otherwise unreachable for the plant, but they are made available with the help of the fine absorbing hyphae of the fungus species. Such fungi are accompanied with many crops, except those of the

Amaranthaceae Caryophyllaceae, Chenopodiaceae, Polygonaceae, Commelinaceae, Brassicaceae, and Cyperaceae families (Yadav et al. 2020a, b, c).

10.6.6 Zinc Solubilizers

Azospirillum, BGA, PSB (phosphate-solubilizing bacteria), *Rhizobium*, *Azotobacter*, *Pseudomonas striata*, and *Mycorrhiza* are frequently used biofertilizers. Furthermore, several microorganisms (present in the soil) that are able to process various kinds of micronutrients such as copper, iron, and zinc including others. Various microorganisms like *Thiobacillus*, *Thiooxidans,* and *Saccharomyces sp.* can make this Zn solubilize. Research studies conducted at various places revealed that *Bacillus* sp. and zinc solubilizator bacteria have the potential to be used as zinc biofertilizer, or as native zinc compounds [(zinc oxide (ZnO), zinc sulfide (ZnS), and zinc carbonate ($ZnCO_3$))]. Under temperate conditions, *Rhizobium* inoculation inclines the number of grains and seed weight and thus better yield can be expected. Using *Azospirillum* + BGA in rice fields in low lying area has been proved useful for the development of LAI and yields contributing characters including yield (Mondal et al. 2020; Rai et al. 2020; Singh et al. 2020a).

10.7 Safeguards to Use Biofertilizers

- Biofertilizer packages must not be directly exposed to sunlight or heat, store the packages in a cool and dry place.
- Appropriate biofertilizer combinations are essential.
- Since *Rhizobium* is crop-specific, only the specified crop should be used.
- The biofertilizers should not be mixed with other chemicals.
- When buying one should ensure that each product contains the necessary information, such as the name of the product intended for as manufacturer's name and address, the date of manufacture, the expiry date, the batch number, and instructions.
- The package must only be used for the specified crop and for the required application method before expiry.
- All nitrogen and phosphatic biofertilizers are live products and need storage care to be used for the most effective results.
- Use of biofertilizers alongside chemical fertilizers and organic manures is important.
- Biofertilizers are not fertilizer substitutes but can complement the nutrient requirements of plants.

10.8 Certain Problems Using Biofertilizers

Always store these biofertilizers in cold areas, where there is no strong direct sunlight. These conditions are hard to meet. The accurate and precise combination concerning these biofertilizers is very important, and it is not expected at farmer level. It is true that biofertilizers are very useful for crops but under field conditions their maintenance is difficult. The storage requires low temperature conditions as they are live products. Nevertheless, it is not a substitute to use chemical natured fertilizers in place of biofertilizers. Government officials as well as growers require more awareness and research centers are also expected to encourage INM (Integrated Nutrient Management) approach which should include biofertilizers. Government officials as well as growers require more awareness. The research centers are also expected to encourage INM (Integrated Nutrient Management) approach, for which they should include biofertilizers.

10.9 Conclusion and Future Prospects

Generally, biofertilizers can be helpful in increasing the supply of plant nutrients and soil fertility can also be maintained in agricultural field. Furthermore, with the application of these biofertilizers chemical toxicity in agroecosystems can be prevented. Chemicals utilized during crop production and its residual effect can also be minimized. Additionally, we may compare that they are economical, renewable, and works on ecofriendly process, and hence we can say that there is no alternative of these biofertilizers. However, in current scenario of agriculture we cannot replace them with chemical fertilizer. Application of biofertilizers is very important as an Integrated Nutrient approach as well as sustainable nutrient and management of agricultural field. These practices are most important in current day agriculture with few latest researches and modern agriculture may also involve in present scenario. This will be helpful in the prevention of environmental hazards related to the chemical fertilizers. The demands of biofertilizers are increasing day by day with an additional vital role in present scenario of our agricultural ecosystem.

We have reviewed the influences of biofertilizers towards crop production along with their impact on justifiable management of agroecosystem. It is concluded from available literature that biofertilizers supplement the requirement of nutrient as well as encourage seed germination and plant protection against various soil borne pathogens. In biofertilizers, microbes are frequently used, and these nutrients not only help in INM but also play a significant role in managing soil health. Today due to population growth and urbanization, increasing demand of food supply, all these circumstances compel the farmer to use lethal and harmful chemical fertilizer include pesticides for high and quick production. Therefore, it harms our natural ecosystem along with soil and human health. In this concern, the integrated approach is utmost important and need of the hour. But certain problem of biofertilizers like storage under cool conditions and visible results after long time need to be emphasized. More awareness should be there particularly at farmer and consumer level.

Strong and useful initiative is expected from government and associated agencies to promote more and more steps like organic farming.

Acknowledgments I would like to thank the Dean and Head of the Department Agronomy, School of Agriculture, Lovely Professional University for their constant support for research-related activities in the university.

References

Aggani SL (2013) Development of biofertilizers and its future perspective. Sch Acad J Pharm 2(4):327–332

Ahmed M, Elesheikh EAE, Mahdi AA (2007) The in vitro compatibility of some *Rhizobium* and *Bradyrhizobium* strains with fungicides. Afr Crop Sci Conf Proc 8:1171–1178

Beijerinck MW (1901) Ueber oligonitro-phile Mikroben. Centr Bakt Parasitenk, Abt. II, 7: 561–582

Bolan NS (1991) A critical review on the role of mycorrhizal fungi in the uptake of phosphorus by plants. Plant Soil 134:189–207

Chalk PM, Souza RDF, Urquiaga S, Alves BJR, Boddey RM (2006) The role of arbuscular mycorrhiza in legume symbiotic performance. Soil Biol Biochem 38:2944–2951

Chang CH, Yang SS (2009) Thermo-tolerant phosphate-solubilizing microbes for multi-functional biofertilizer preparation. Bioresour Technol 100(4):1648–1658

Cooper JE (2004) Multiple responses of rhizobia to flavonoids during legume root infection. Adv Bot Res 41:1–62

Doroshenko EV, Boulygina ES, Spiridonova EM, Tourova TP, Kravchenko IK (2007) Isolation and characterization of nitrogen-fixing bacteria of the genus *Azospirillum* from the soil of a *Sphagnum* peat bog. Microbiology 76(1):93–101

Franche C, Lindström K, Elmerich C (2009) Nitrogen-fixing bacteria associated with leguminous and non-leguminous plants. Plant Soil 321(1–2):35–59

Halim NA (2009) Effects of using enhanced biofertilizer containing N-fixer bacteria on patchouli growth (Doctoral dissertation, UMP). Malaysia Pahang. p 145

Jorquera MA, Shaharoona B, Nadeem SM, de la Luz MM, Crowley DE (2012) Plant growth-promoting rhizobacteria associated with ancient clones of creosote bush (*Larrea tridentata*). Microb Ecol 64(4):1008–1017

Kaur T, Rana KL, Kour D, Sheikh I, Yadav N, Kumar V et al (2020) Microbe-mediated biofortification for micronutrients: present status and future challenges. In: Rastegari AA, Yadav AN, Yadav N (eds) Trends of microbial biotechnology for sustainable agriculture and biomedicine systems: perspectives for human health. Elsevier, Amsterdam, pp 1–17. https://doi.org/10.1016/B978-0-12-820528-0.00002-8

Khan MS, Zaidi A, Wani PA (2007) Role of phosphate-solubilizing microorganisms in sustainable agriculture—a review. Agron Sustain Dev 27(1):29–43

Khan W, Rayirath UP, Subramanian S (2009) Seaweed extracts as biostimulants of plant growth and development. J Plant Growth Regul 28:386–399

Khosro M, Yousef S (2012) Bacterial biofertilizers for sustainable crop production: a review. J Agric Biol Sci 7:307–316

Kour D, Kaur T, Devi R, Rana KL, Yadav N, Rastegari AA et al (2020a) Biotechnological applications of beneficial microbiomes for evergreen agriculture and human health. In: Rastegari AA, Yadav AN, Yadav N (eds) Trends of microbial biotechnology for sustainable agriculture and biomedicine systems: perspectives for human health. Elsevier, Amsterdam, pp 255–279. https://doi.org/10.1016/B978-0-12-820528-0.00019-3

Kour D, Kaur T, Yadav N, Rastegari AA, Singh B, Kumar V et al (2020b) Phytases from microbes in phosphorus acquisition for plant growth promotion and soil health. In: Rastegari AA, Yadav

AN, Yadav N (eds) Trends of microbial biotechnology for sustainable agriculture and biomedicine systems: diversity and functional perspectives. Elsevier, Amsterdam, pp 157–176. https://doi.org/10.1016/B978-0-12-820526-6.00011-7

Kour D, Rana KL, Yadav AN, Yadav N, Kumar M, Kumar V et al (2020c) Microbial biofertilizers: bioresources and eco-friendly technologies for agricultural and environmental sustainability. Biocatal Agric Biotechnol 23:101487. https://doi.org/10.1016/j.bcab.2019.101487

Kumar M, Kour D, Yadav AN, Saxena R, Rai PK, Jyoti A et al (2019) Biodiversity of methylotrophic microbial communities and their potential role in mitigation of abiotic stresses in plants. Biologia 74:287–308. https://doi.org/10.2478/s11756-019-00190-6

Linu M, Asok AK, Thampi M, Sreekumar J, Jisha M (2019) Plant growth promoting traits of indigenous phosphate solubilizing *Pseudomonas aeruginosa* isolates from Chilli (*Capsicum annuum* L.) Rhizosphere. Commun Soil Sci Plant Anal 50:444–457

Lipman JG (1903) Nitrogen-fixing bacteria. Doctoral dissertation, Cornell University

Mahdi SS, Dar SA, Ahmad S, Hassan GI (2010) Zinc availability–a major issue in agriculture. Res J Agric Sci 3(3):78–79

Mahdi SS, Hassan G, Samoon S, Rather H, Dar SA, Zehra B (2010) Bio-fertilizers in organic agriculture. J Phytol 2:42–54

Malyan SK, Singh S, Bachheti A, Chahar M, Sah MK, Narender et al (2020) Cyanobacteria: a perspective paradigm for agriculture and environment. In: Rastegari AA, Yadav AN, Yadav N (eds) Trends of microbial biotechnology for sustainable agriculture and biomedicine systems: diversity and functional perspectives. Elsevier, Amsterdam, pp 215–224. https://doi.org/10.1016/B978-0-12-820526-6.00014-2

Mondal S, Halder SK, Yadav AN, Mondal KC (2020) Microbial consortium with multifunctional plant growth promoting attributes: future perspective in agriculture. In: Yadav AN, Rastegari AA, Yadav N, Kour D (eds) Advances in plant microbiome and sustainable agriculture, Functional annotation and future challenges, vol 2. Springer, Singapore, pp 219–254. https://doi.org/10.1007/978-981-15-3204-7_10

Podile AR, Kishore GK (2007) Plant growth-promoting rhizobacteria. In: Gnanamanickam SS (eds) Plant-Associated Bacteria. Springer, Dordrecht, pp 195–230, https://doi.org/10.1007/978-1-4020-4538-7_6

Radhakrishnan R, Lee IJ (2016) Gibberellins producing *Bacillus methylotrophicus* KE2 supports plant growth and enhances nutritional metabolites and food values of lettuce. Plant Physiol Biochem 109:181–189

Rai PK, Singh M, Anand K, Saurabhj S, Kaur T, Kour D et al (2020) Role and potential applications of plant growth promotion rhizobacteria for sustainable agriculture. In: Rastegari AA, Yadav AN, Yadav N (eds) Trends of microbial biotechnology for sustainable agriculture and biomedicine systems: diversity and functional perspectives. Elsevier, Amsterdam, pp 49–60. https://doi.org/10.1016/B978-0-12-820526-6.00004-X

Rana KL, Kour D, Kaur T, Devi R, Yadav AN, Yadav N et al (2020) Endophytic microbes: biodiversity, plant growth-promoting mechanisms and potential applications for agricultural sustainability. Antonie Van Leeuwenhoek. https://doi.org/10.1007/s10482-020-01429-y

Rastegari AA, Yadav AN, Yadav N (2020a) New and Future Developments in Microbial Biotechnology and Bioengineering: Trends of Microbial Biotechnology for Sustainable Agriculture and Biomedicine Systems: Diversity and Functional Perspectives. Elsevier, Amsterdam

Rastegari AA, Yadav AN, Yadav N (2020b) New and Future Developments in Microbial Biotechnology and Bioengineering: Trends of Microbial Biotechnology for Sustainable Agriculture and Biomedicine Systems: Perspectives for Human Health. Elsevier, Amsterdam

Reynders L, Vlassak K (1979) Conversion of tryptophan to indoleacetic-acid by *Azospirillum brasilense*. Soil Biol Biochem 11(5):547–548

Ridvan K (2009) Nitrogen fixation capacity of *Azotobacter* spp. strains isolated from soils in different ecosystems and relationship between them and the microbiological properties of soils. J Environ Biol 30:73–82

Ryan MH, Graham JH (2002) Is there a role for arbuscular mycorrhizal fungi in production agriculture? Plant Soil 244(1–2):263–271

Sadiq HM, Jahangir GZ, Nasir IA, Iqtidar M, Iqbal M (2013) Isolation and characterization of phosphate-solubilizing bacteria from rhizosphere soil. Biotechnol Biotechnol Equip 27(6):4248–4255

Saravanan VS, Kumar MR, Sa TM (2011) Microbial zinc solubilization and their role on plants. In: Bacteria in agrobiology: plant nutrient management. Springer, Berlin, Heidelberg, pp 47–63

Shanware AS, Kalkar SA, Trivedi MM (2014) Potassium solublisers: occurrence, mechanism and their role as competent biofertilizers. Int J Curr Microbiol App Sci 3(9):622–629

Sharaff MS, Subrahmanyam G, Kumar A, Yadav AN (2020) Mechanistic understanding of root-microbiome interaction for sustainable agriculture in polluted soils. In: Rastegari AA, Yadav AN, Yadav N (eds) Trends of microbial biotechnology for sustainable agriculture and biomedicine systems: diversity and functional perspectives. Elsevier, Amsterdam, pp 61–84. https://doi.org/10.1016/B978-0-12-820526-6.00005-1

Singh A, Kumar R, Yadav AN, Mishra S, Sachan S, Sachan SG (2020a) Tiny microbes, big yields: microorganisms for enhancing food crop production sustainable development. In: Rastegari AA, Yadav AN, Yadav N (eds) Trends of microbial biotechnology for sustainable agriculture and biomedicine systems: diversity and functional perspectives. Elsevier, Amsterdam, pp 1–15. https://doi.org/10.1016/B978-0-12-820526-6.00001-4

Singh B, Boukhris I, Pragya KV, Yadav AN, Farhat-Khemakhem A et al (2020b) Contribution of microbial phytases to the improvement of plant growth and nutrition: a review. Pedosphere 30:295–313. https://doi.org/10.1016/S1002-0160(20)60010-8

Stewart WDP (1969) Biological and ecological aspects of nitrogen fixation by free-living microorganisms. Proc Royal Soc London Ser B Biol Sci 172(1029):367–388

Thakur N, Kaur S, Tomar P, Thakur S, Yadav AN (2020) Microbial biopesticides: current status and advancement for sustainable agriculture and environment. In: Rastegari AA, Yadav AN, Yadav N (eds) Trends of microbial biotechnology for sustainable agriculture and biomedicine systems: diversity and functional perspectives. Elsevier, Amsterdam, pp 243–282. https://doi.org/10.1016/B978-0-12-820526-6.00016-6

Thakur S, Bothra AK, Sen A (2014) Exploring the genomes of symbiotic diazotrophs with relevance to biological nitrogen fixation. In: Agricultural bioinformatics. Springer, New Delhi, pp 235–257

Thammana S, Suzuki H, Lobkovsky E, Clardy J, Shimizu Y (2006) Isolation and structure assignment of an iminotetra saccharide from a cultured filamentous cyanobacterium *Anabaena* sp. J Nat Prod 69(3):365–368

Tiwari P, Bajpai M, Singh LK, Mishra S, Yadav AN (2020) Phytohormones producing fungal communities: metabolic engineering for abiotic stress tolerance in crops. In: Yadav AN, Mishra S, Kour D, Yadav N, Kumar A (eds) Agriculturally important fungi for sustainable agriculture, volume 1: perspective for diversity and crop productivity. Springer, Cham. https://doi.org/10.1007/978-3-030-45971-0_8

Verma P, Yadav AN, Khannam KS, Kumar S, Saxena AK, Suman A (2016) Molecular diversity and multifarious plant growth promoting attributes of Bacilli associated with wheat (*Triticum aestivum* L.) rhizosphere from six diverse agro-ecological zones of India. J Basic Microbiol 56:44–58

Vessey JK (2003) Plant growth promoting rhizobacteria as biofertilizers. Plant Soil 255(2):571–586

Vessey JK, Heisinger KG (2001) Effect of *Penicillium bilaii* inoculation and phosphorus fertilisation on root and shoot parameters of field-grown pea. Can J Plant Sci 81(3):361–366

Yadav AN, Gulati S, Sharma D, Singh RN, Rajawat MVS, Kumar R et al (2019) Seasonal variations in culturable archaea and their plant growth promoting attributes to predict their role in establishment of vegetation in Rann of Kutch. Biologia 74:1031–1043. https://doi.org/1010.2478/s11756-019-00259-2

Yadav AN, Rastegari AA, Yadav N, Kour D (2020a) Advances in plant microbiome and sustainable agriculture: diversity and biotechnological applications. Springer, Singapore

Yadav AN, Rastegari AA, Yadav N, Kour D (2020b) Advances in plant microbiome and sustainable agriculture: functional annotation and future challenges. Springer, Singapore

Yadav AN, Sharma D, Gulati S, Singh S, Dey R, Pal KK et al (2015) Haloarchaea endowed with phosphorus solubilization attribute implicated in phosphorus cycle. Sci Rep 5:12293

Yadav AN, Singh J, Rastegari AA, Yadav N (2020c) Plant microbiomes for sustainable agriculture. Springer, Cham

Yoneyama TA, Ladha JK, Watanabe I (1987) Nodule bacteroids and *Anabaena*: natural 15N enrichment in the legume-*Rhizobium* and *Azolla-Anabaena* symbiotic systems. J Plant Physiol 127(3–4):251–259

Zimmer S, Messmer M, Haase T, Piepho HP, Mindermann A, Schulz H et al (2016) Effects of soybean variety and *Bradyrhizobium* strains on yield, protein content and biological nitrogen fixation under cool growing conditions in Germany. Eur J Agron 72:38–46

Endophytic Microbiomes and Their Plant Growth-Promoting Attributes for Plant Health

11

Sougata Ghosh, Tanay Bhagwat, and Thomas J. Webster

Abstract

Endophytes reside within internal tissues of living plants without causing any harm to the host. The influence of these microbial communities on plant growth, yield, stress, and disease resistance, has been identified as potential research priorities in agriculture. In this chapter, we aim to explore the diverse host–endophyte interactions for plant growth promotion and health. Initially, the colonization of endophytes in specific plant tissues is discussed along with their mechanism of entry, habitat selection, response to stimuli, and evasion of the plant immunity. Endophytic microbes promote plant growth through different types of direct and indirect mechanisms. Plant growth-promoting endophytes (PGPE) play a vital role in phytohormone production, nutrient acquisition, nitrogen fixation, and solubilization of minerals. Further, indirect mechanisms (like suppression of plant pathogens by producing volatile organic compounds, antagonizing agents, and quorum quenchers) are also discussed in detail. Siderophores production and the secretion of different hydrolytic enzymes like chitinases, glucanases, and proteases also help in the induction of systemic resistance and protection of the host plants. Bioactive metabolites derived from endophytes serve as excellent therapeutic agents and have potential applications in agriculture, cosmetics, pharmaceutical, and food industries. Hereby, this chapter highlights the scientific rationale behind using endophytic microbiomes as potential biofertilizers, biopesticides, and biocontrol agents.

S. Ghosh (✉)
Department of Microbiology, School of Science, RK University, Rajkot, Gujarat, India

Department of Chemical Engineering, Northeastern University, Boston, USA

T. Bhagwat · T. J. Webster
Department of Chemical Engineering, Northeastern University, Boston, USA

A. N. Yadav et al. (eds.), *Current Trends in Microbial Biotechnology for Sustainable Agriculture*, Environmental and Microbial Biotechnology,
https://doi.org/10.1007/978-981-15-6949-4_11

Keywords

Endophytes · Plant growth promotion · Direct and indirect mechanism · Induced systemic resistance · Bioactive compounds

11.1 Introduction

Increasing crop yield has attracted wide attention in order to meet global demand considering the increase in the world's population. However, conventional farming practices have certain limitations under increasing challenges like shortage of fertile lands, climate change, pests, and other associated abiotic and biotic stress. Thus, various plant growth-promoting microbes are being explored as biofertilizers in agriculture which seems to be a promising innovation to provide viable and environmentally friendly solutions with the potential to ensure food security (Glick 2014). However, this can only be achieved through in-depth knowledge about the underlying plant–microbe interactions. Microbes that reside within the plants without causing any negative impacts are called endophytes. Stimulation of plant defense responses is some inherent properties of endophytes (de Matos et al. 2001).

Plants are significant atmospheric CO_2 fixers on Earth. The solar energy enables plants to utilize CO_2 and reduce it to glucose and further various carbonaceous compounds. Hence, plant-associated heterotrophic microbes derive carbon, nitrogen, and energy from the host plants (Vandenkoornhuyse et al. 2007). On the other hand, plants require the microflora for their growth and stress tolerance. Thus, mutualistic relationships and interdependence exist between microbes and their host plants (Thrall et al. 2007; Sharaff et al. 2020; Suman et al. 2016). Potential uses of plant-associated bacteria as plant growth stimulating agents and management of soil as well as plant health have been portrayed in numerous literatures. Plant growth-promoting bacteria (PGPB) are associated with many, if not all, plant species and are commonly present in many environments (Bashan and Holguin 1998). PGPB are generally plant growth-promoting rhizobacteria (PGPR) that colonize the root surfaces and the rhizosphere (the closely adhering soil interface). Some of these PGPR can also enter the root interior and establish endophytic populations. Prime sites for bacterial colonization are lateral root emergence sites, outer cell layers, root cortex, phloem, and xylem, which may occur both intracellularly and inside the apoplast (Fig. 11.1).

Microbes can evade the endodermis barrier, moving from the root cortex to the vascular system, and eventually colonize as endophytes in roots, shoot, leaves, tubers, flowers, and other organs. Internal tissues of root, internodes, and leaves of grapevine are colonized by the PGPB *Burkholderia* sp. strain PsJN. Similarly, the surface and interior of roots, stems, and needles of lodgepole pine (*Pinus contorta* Dougl. var. *latifolia* Engelm.) harbor the diazotrophic bacterial strain *Paenibacillus polymyxa* P2b-2R (Liu et al. 2017). A facultative intracellular symbiont of *Methylobacterium extorquens* strain DSM13060 was isolated from the Scots pine

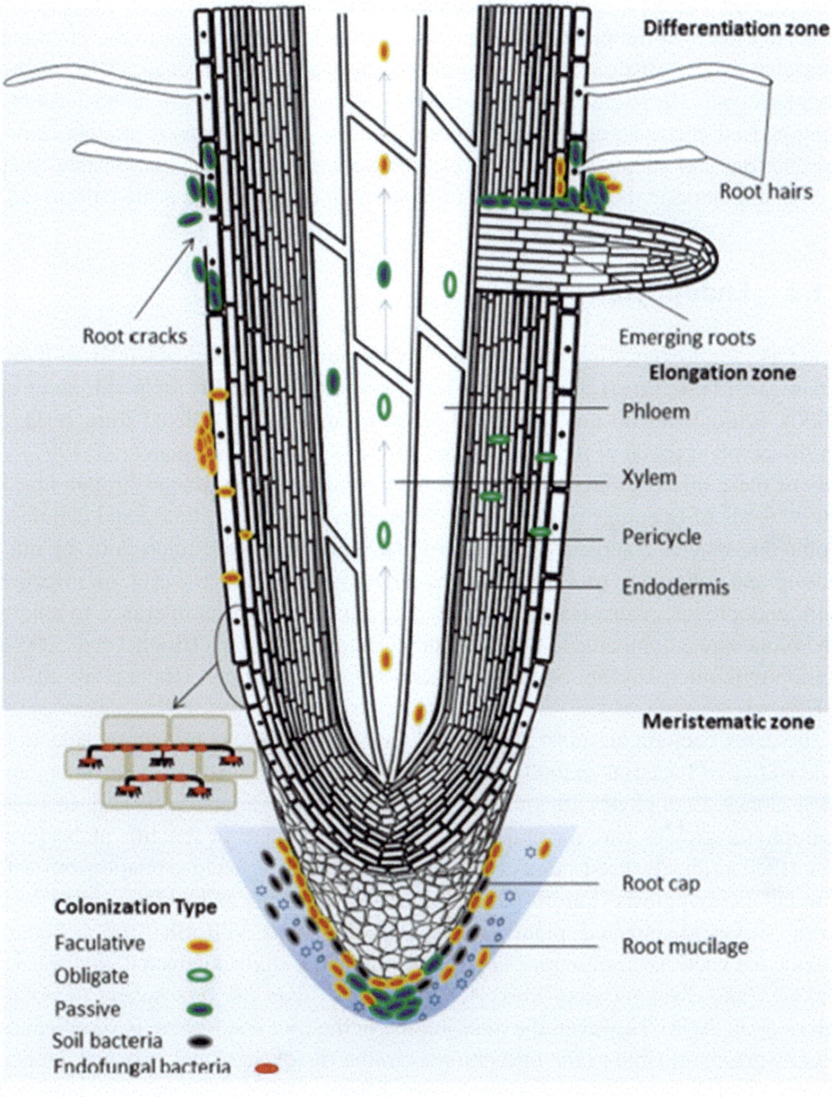

Fig. 11.1 A schematic representation of the bacterial distribution and colonization patterns in the endosphere of a plant root. The emerging sites of lateral roots are among the hotspots of bacterial colonization. Arrows represent the translocation of bacteria inside the xylem and phloem. Endophytic bacteria may engage in different lifestyles as depicted by different colored ovals. Adapted with permission from Liu et al. (2017)

(*Pinus sylvestris* L.) shoot tips where the bacteria aggregated within the living cells surrounding the nucleus (Koskimäki et al. 2015). Microbes adapt to particular internal tissue environment by varying its extent of colonization within host plant organs and tissues (Gray and Smith 2005; Rana et al. 2019b). Consequently, close associations between endophytes and host plants are formed without causing any adverse effects to the plant. These endophytes do not cause harm to the plant and establish a mutualistic association with the host plant (Rana et al. 2019c). This chapter covers diverse aspects of plant growth-promoting endophytic bacteria and fungus. Endophyte-associated distribution patterns; nutrient uptake, phytohormone production, and stress tolerance are elaborated with minute details. Further, their role in augmenting the phytoremediation potential of host plants is also discussed.

11.2 Endophytes

The term "endophyte" is a microbe that asymptomatically colonizes internal living tissues of plants (host) during a particular period of their life span (Stone et al. 2000). Endophytes do not harm the host plant and can be isolated from surface-sterilized plant tissue or the inner tissues of the host plant (Hallman et al. 1997). A few of these microbes are believed to actively infiltrate plant tissues through invading wounds or openings or using hydrolytic enzymes like pectinase and cellulase. Some endophytes emerge from the rhizosphere or phylloplane microflora, by infiltrating and colonizing root tissue as a passage to the xylem. However, on infection with endophytes, plants become healthy and exhibit enhanced tolerance to abiotic and biotic stress compared to their endophyte-free counterparts (Bonnet et al. 2000). Endophytic microbes can be bacteria, actinomycetes, or fungi (Rana et al. 2019a; 2020a, c).

It seems bacteria are most suitable for living inside plants by natural selection. The source of bacterial endophytes is microbial diversity of soil or rhizosphere and their clones. Endophytes are known for >120 years (Hardoim et al. 2009). In 1926, endophytic growth was recognized as a particular stage in the life of bacteria, described as an advanced stage of infection, and as having a close relationship with mutualistic symbiosis (Perotti 1926). Since then, various endophytes are isolated from surface-disinfected plant organs (Henning and Villforth 1940). Potato-associated bacterial communities indicated, in a large study conducted, that species richness and diversity were lower for endophytes than the rhizosphere of potato (Berg et al. 2005). However, the microbiome in the root endosphere is significantly less diverse compared to the microbiomes in the rhizosphere and bulk soil. Hence, roots can work as the most effective habitat filters, restricting community membership resulting in more narrowly defined lineages as the niche from soil to roots. Root endophytic bacterial communities are typically dominated by Proteobacteria (~ 50% in relative abundance), Actinobacteria (~ 10%), Firmicutes (~ 10%), and Bacteroidetes (~ 10%) apart from other bacterial phyla that include Chloroflexi, Cyanobacteria, Armatimonadetes, Verrucomicrobia, Planctomycetes, and Nitrospirae.

11.3 Ubiquity of Endophytes

The presence of endophytes is thought to be ubiquitous in plants as they can be detected in almost all parts including root, shoot, leaves, internodes, and reproductive tissues as well. The differences between the endosphere microbiomes of the root and shoot determine the source of dominant endophytes in them. Root-associated endophytes are primarily derived from soil, which then colonizes internal tissues of stems and leaves through the apoplast in xylem vessels. Therefore, it is common to have microbes of the plant leaf/shoot endosphere significantly overlapping with those in roots at both the taxonomic and functional levels. Recent molecular identification provides a strong evidence of diverse genera and species in endophytes. Kobayashi and Palumbo (2000) reported both Gram-positive and Gram-negative bacterial endophytes from different internal tissues of diverse plant species. Significant variations in populations of both indigenous and infiltrated endophytes were reported which might be attributed to the tissue type, source, plant age, time of sampling, and the environment. Interactions of the internal microflora of plants are needed to be investigated that might lead to beneficial effects due to their combined activities.

There is a deep underlying genetic basis for the differential colonization of various plant tissues by endophytes. Degradation of the cell wall facilitates entry of the bacteria within the interior for translocation to the apoplast. The genome of endophytic bacteria harbors numerous genes encoding cell wall–degrading enzymes (Straub et al. 2013). Genes encoding plant polymer degrading enzymes like cellulases, endoglucanase, xylanases, cellobiohydrolases, and cellulose-binding proteins have been reported in high copy numbers in the metagenome of rice root endophytic bacterial communities (Sessitsch et al. 2012). Bacteria in the phyllosphere may be

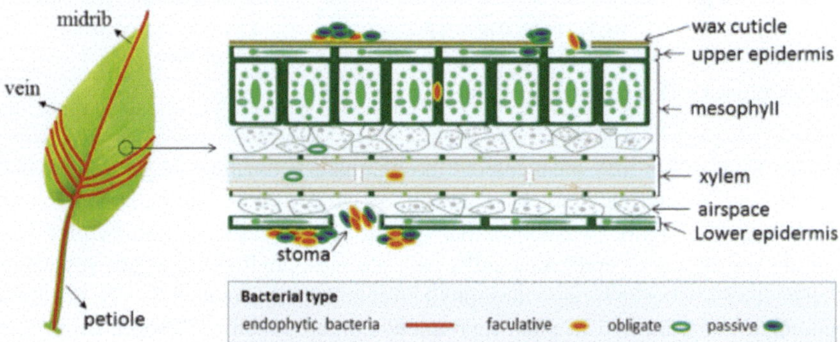

Fig. 11.2 A schematic representation of bacterial colonization patterns in a leaf. The picture shown on the left demonstrates that the presence of bacteria has been detected in the leaf petiole, midrib, and veins. The picture shown on the right is a magnified leaf cross section, which demonstrates that endophytic bacteria may not only colonize the apoplast but are also present intracellularly. Endophytic bacteria are believed to be able to ascend from roots to the leaf via the vascular tissues of the xylem and phloem. Adapted with permission from Liu et al. (2017)

Table 11.1 Complete genomes from bacterial endophytes and their plant-growth promoting traits

Endophytic microbes	Genome size Mb (Replicons)	Host plant	PGP traits
Azoarcus sp. BH72	4.37 (1 chr, 0 pl)	Rice	Nitrogen fixation
Azospirillum lipoferum 4B	6.85 (1 chr, 6 pl)	Rice, maize, wheat	Nitrogen fixation, phytohormone secretion
Azospirillum sp. B510	7.6 (1 chr, 6 pl)	Rice	Nitrogen fixation, phytohormone secretion
Burkholderia phytofirmans PsJN	8.2 (2 chr, 1 pl)	Potato, tomato, maize, barley, onion, canola, grapevine	IAA synthesis, ACC deaminase
Burkholderia spp. KJ006	6.6 (3 chr, 1 pl)	Rice	ACC deaminase, *nif* gene cluster, antifungal action (indirect PGP)
Enterobacter cloacae ENHKU01	4.7 (1 chr, 0 pl)	Pepper	Unknown role in PGP
Enterobacter sp. 638	4.67 (1 chr, 1 pl)	Poplar	Siderophore, IAA, acetoin and 2,3-butanediol synthesis, antifungal action (indirect PGP)
Gluconacetobacter diazotrophicus Pal5	3.9 (1 chr, 2 pl)	Sugarcane, rice, coffee, tea	Nitrogen fixation, auxin synthesis
Klebsiella pneumoniae 342	5.9 (1 chr, 2 pl)	Maize, wheat	Nitrogen fixation
Pseudomonas putida W619	5.77 (1 chr, 0 pl)	Poplar	AA synthesis, ACC deaminase
Pseudomonas stuzeri A1501	4.5 (1 chr, 0 pl)	Rice	Nitrogen fixation
Serratia proteamaculans 568	5.5 (1 chr, 1 pl)	Soybean	IAA synthesis, ACC deaminase, acetoin, and 2,3-butanediol synthesis
Stenotrophomonas maltophilia R551–3	4.57 (1 chr, 0 pl)	Poplar	IAA synthesis, ACC deaminase

Source: Adapted with permission from Santoyo et al. (2016)

derived from soil or may have entered through natural openings (e.g., stomata and hydathodes), wounds, and cracks generated by wind, insects, and pathogen attacks (Vorholt 2012). Figure 11.2 shows that specific sites of bacterial colonization in a leaf are mostly upper epidermis cells, palisade mesophyll cells, xylem vessels as well as spaces between spongy mesophyll layer cells (Olivares et al. 1997). Bacterial endophytes are detected in plant reproductive organs, such as flowers, fruits, and seeds, although in small numbers. Table 11.1 represents various bacterial endophytes from crop plants.

11.4 Role of Endophytes in Plant Growth Promotion

Benefits conferred by endophytes are well recognized but it may not always be clear which population of microorganisms (endophytes or rhizospheric bacteria) promotes plant growth. Differential gene expression might facilitate entry, colonization, and also plant growth promotion. Nitrogen fixation (Iniguez et al. 2004) or the production of phytohormones, by enhancing the availability of minerals (Sessitsch et al. 2004; Sturz et al. 2000), may help to promote plant growth. Further endophytes may lay a critical role in biocontrol of phytopathogens as they colonize the same ecological niche. Various mechanism of biocontrol includes production of antifungal or antibacterial agents, siderophore production, nutrient competition, and induction of systematic-acquired host resistance or immunity (Thakur et al. 2020). Endophytic microorganisms have the capacity to control pathogens, insects, and nematodes (Rana et al. 2020b). In some cases, they also have the capacity to accelerate seedling emergence and promote plant establishment under adverse conditions. Endophytes can confer metal resistance to plants and reduce metal toxicity due to their own metal resistance capability (Ma et al. 2016).

11.5 Mechanisms of Plant Growth Promotion

A deficiency in macro and micronutrients in the soil is detrimental to crop yield and the affected plants become more prone to soil-borne pathogens such as *Fusarium, Pythium,* and *Phytophthora*. Hence chemical fertilizers, herbicides, fungicides, and pesticides are largely used in order to overcome the problems. However, these harmful and toxic chemicals pose a potential threat to human health and the environment as well (Aktar et al. 2009; (Kour et al. 2020b). Endophytes enable the plants to overcome habitat-imposed abiotic and biotic stresses which otherwise result in major losses in plant yield. Endophytic bacteria are capable of promoting plant growth and development through a wide variety of not only direct mechanisms which include nutrient (e.g., phosphorous, nitrogen, and iron) acquisition and production of various phytohormones (Santoyo et al. 2016; Yadav et al. 2020) but also indirect mechanisms for plant growth promotion such as antagonistic effects toward phytopathogens (Compant et al. 2010; Rastegari et al. 2020a, b). It also includes the production of defense-related enzymes like chitinase and β-1,3-glucanase, secreting antimicrobial compounds, lowering endogenous stress-related ethylene (ET), induction of systemic resistance (ISR), quenching the quorum sensing (QS) of phytopathogens, and competition for niche and/or resources (Compant et al. 2010; Glick 2014; Santoyo et al. 2016; Singh et al. 2020a). In the following section, various direct and indirect mechanisms of plant growth promotion by endophytic bacteria are elaborated (Ma et al. 2016).

11.5.1 Direct Mechanisms

Endophytes directly promote plant growth using various mechanisms that include phytohormone production, nutrient acquisition, nitrogen fixation, and solubilization of minerals.

11.5.1.1 Phytohormone Production

Five types of phytohormones, e.g., ethylene, indole-3-acetic acid (IAA), cytokinins, gibberellins, and abscisic acid may play an important role in several stages such as cell elongation, cell division, tissue differentiation, and apical dominance. Both host plants and their endophytes can synthesize these hormones. Hormonal balance of the plant can be altered by plant-associated bacteria as well.

Ethylene is an important example to show that the balance is most important for the effect of hormones. An ubiquitous plant hormone, it plays a vital role in plant growth and survival, to abiotic and biotic stresses including root initiation and nodulation, cell elongation, leaf senescence, abscission, and fruit ripening as well as auxin transport (Ma et al. 2016). While normally considered as an inhibitor of plant growth and known as a senescence hormone, at reduced levels it can stimulate plant growth in *Arabidopsis thaliana* (Pierik et al. 2006). Stress-mediated ethylene production inhibits root elongation, lateral root growth, and root hair formation. It is interesting to note that the endophytes can reduce the ethylene level. The compound 1-aminocyclopropane-1-carboxylate (ACC) is a precursor of ethylene in plants. ACC-deaminase-producing bacteria can degrade ACC into α-ketobutyrate and ammonia, which can be metabolized by the microbes as nitrogen source. Thus, bacteria-mediated reduction of endogenous ACC levels results in root growth (Glick 2005). ACC deaminase-producing bacteria have an additional potential to protect plants against biotic and abiotic stress owing to the fact that ethylene is also a stress hormone (Ma et al. 2016; Saleem et al. 2007).

Indole acetic acid (IAA), one of the most physiologically active auxins, is produced by various plant organs like young leaves and germinating seeds by utilizing the amino acid tryptophan. IAA plays a significant role in plant growth by bringing about apical dominance, promoting root development and proliferation, tropisms (phototropism in the case of shoots and gravitropism in the case of roots), and inducing cell division and differentiation (Tiwari et al. 2020). IAA is a common product of L-tryptophan metabolism by several endophytes leading to plant morphogenic effects. Evidence suggests that endophytes produce IAA while colonizing the internal plant tissues and thereby promoting plant growth. The *Pseudomonas stutzeri* P3 strain was found to produce IAA in *Echinacea* plants and help in the proliferation of these plants even after micropropagation, likewise, a number of bacteria such as *Agrobacterium tumefaciens*, *A. rhizogenes*, *Pseudomonas savastanoi*, *Pseudomonas* spp., *Rhizobium* spp., *Bradyrhizobium* spp., *Azospirillum* spp., and *Acinetobacter* spp. associated with the plants are known to produce IAA (Huddedar et al. 2002; Rao 1986; Baldi et al. 1991; Leinhos 1994).

11.5.1.2 Nutrient Acquisition

Nitrogen

Improved nutrient acquisition helps to promote plant growth directly. Plant-associated microorganisms can supply macronutrients and micronutrients, most significant example being bacterial nitrogen fixation. Nitrogen-fixing bacteria can use root exudates (carbohydrates) and in return provide nitrogen to the plant that can be used for amino acid synthesis. *Azospirillum*, *Burkholderia*, and *Stenotrophomonas* are free-living nitrogen-fixing bacteria (Dobbelare et al. 2003). Brazilian sugarcane requires minimum amounts of fertilizer and shows no N_2 deficiencies due to N_2 fixing endophytes within them. However, level of N_2 fixed by endophytes and amount available to the host plant is still needed to be investigated (Giller and Merckx 2003). Different reports suggest that 30–80 kg N/ha/year are available (Boddey et al. 1995). Under optimal conditions, some plant genotypes seem to obtain part of their N requirements from nitrogen fixation. Kallar grass grows in nitrogen-deficient soils in Pakistan and a diversity of *Azoarcus* spp. was recovered (Reinhold-Hurek et al. 1993). Inside wheat, *Klebsiella* sp. strain Kp342 fixes N_2 that also increases maize yield in the field (Iniguez et al. 2004; Riggs et al. 2001).

Similarly, nitrogen-fixing endophytes seem to relieve N_2 deficiencies of sweet potato (*Ipomoea batatas*) in N_2-poor soils (Reiter et al. 2003). Grasses growing in nutrient-poor sand dunes contain members of genera *Pseudomonas*, *Stenotrophomonas* as well as *Burkholderia*. *Burkholderia* endophytes could contribute nitrogen to the grasses because nitrogenase was detected in roots and cell walls of stems and rhizomes (Dalton et al. 2004). Similarly, the endophytic genera *Burkholderia, Rahnella, Sphingomonas,* and *Acinetobacter* isolated from the stem of *Populus trichocarpa* and *Salix sitchensis* enhanced the growth of plants by providing abundant nitrogen owing to their nitrogen-fixing ability (Doty et al. 2009). Some endophytic bacteria possess both nitrogen fixation (e.g., *nifH*) and denitrification genes. The nitrogen-fixing isolates *P. polymyxa* P2b-2R isolated from lodgepole pine tissue could effectively colonize both rhizosphere and endosphere of maize plants resulting in plant growth promotion (Puri et al. 2016).

Phosphorous

Phosphorous (P) is an essential micronutrient that helps in the proper functioning of metabolic activities, glucose transport, development of roots, and many other physiological processes (Ahemad 2015). Since more than 75% of applied phosphorus forms complexes and are unavailable for plant uptake, endophytes may either solubilize precipitated phosphates by acidification, chelation (i.e., PO_4^{3-}), ion exchange, and release of organic acid or secrete extracellular acid phosphatase to mineralize organic phosphorus resulting in phosphorous availability to plants (van der Hiejden et al. 2008; Kour et al. 2020a; Singh et al. 2020b).

Endophytic bacteria possess the capacity to solubilize phosphates. It was suggested that the endophytic bacteria from soybeans may also participate in phosphate assimilation (Kuklinsky-Sobral et al. 2004). Recently, de Werra et al. (2009) reported that *Pseudomonas fluorescens* CHA0 could reduce the pH of its surrounding environment that helps in solubilization of mineral phosphate. This acidification was strongly dependent gluconic acid-producing ability of the endophyte that can be strongly correlated with antagonistic activity against plant pathogens. Further, Idriss et al. (2002) demonstrated that plants inoculated with a phytase-secreting *Bacillus amyloliquefaciens* FZB45 under P-limitation may result in significant growth enhancement in maize seedlings compared to non-inoculated controls. However, there are no reports of naturally occurring endophytic bacteria with phytase-secreting ability (Ma et al. 2016).

Iron

Iron (Fe) is vital as iron-containing proteins involved in enzymatic reactions are essential for various physiological activities like transpiration (Bothwell 1995). Iron exists in soil in highly insoluble ferric (Fe^{3+}) forms such as oxides, hydroxides, phosphates, and carbonates not available for plant uptake. Microbially secreted chelating agents (e.g., siderophores) help to solubilize Fe under conditions of iron deficiency. Siderophores, low-molecular weight organic compounds (500–1500 Da) having an affinity for Fe^{3+} ions, also bind other bivalent metal ions or Fe^{2+} that can be assimilated by the plant (Rajkumar et al. 2009). The siderophore is discussed in more detail in the indirect mechanisms section.

11.5.2 Indirect Mechanisms of Plant Growth Promotion

Indirect mechanisms mainly include the suppression of the growth or survival of plant pathogens (phytopathogens) and, thus, bring about the promotion of plant growth by microbial antagonism. Endophytes may produce substances like volatile organic compounds, antagonizing agents, and quorum quenchers that may effectively resist phytopathogen-associated disease. Further, siderophore production and secretion of diverse hydrolytic enzymes (such as chitinases, proteases, and glucanases) and induction of systemic resistance also protect the host plants (Sheoran et al. 2015; Mondal et al. 2020).

11.5.2.1 Competition for Colonization Sites

The root surface and internal tissues of plants are significant carbon sinks (Rovira et al. 1965) and nutrient-rich niches that attract diverse groups of microbes including phytopathogens. PGPB protects plants by competing with the phytopathogens over these nutrients and niches (Duffy 2001). Brock et al. (2013) reported that a potent endophyte, *Enterobacter radicincitans* DSM 16656, induced priming in *Arabidopsis* via SA- and JA/ET-dependent pathways. Likewise, endofungal bacterium *R. radiobacter* F4 exhibited nonspecific plant root colonization and enhanced plant resistance against the bacterial leaf pathogens *Xanthomonas translucens* pv. *translucens* and *Pseudomonas syringae* pv. tomato DC3000 (Liu et al. 2017). However, it is yet to be investigated whether endophytes contribute to priming and ISR.

11.5.2.2 Volatile Organic Compounds and Antagonizing Agents

Endophytic bacteria produce volatile organic compounds (VOCs) that can render resistance to the host plants against the phytopathogens (Chung et al. 2016). On inoculation with endophytic *Enterobacter aerogenes* that produce VOC 2,3-butanediol (2,3-BD), maize plants exhibited increased resistance against *Setosphaeria turcica* associated northern corn leaf blight disease (D'Alessandro et al. 2014). The endophytic *Pseudomonas poae* strain RE*1-1-14 isolated from sugar beet roots suppressed the fungal pathogen *Rhizoctonia solani* (Zachow et al. 2015). Further, *P. poae* produced a novel lipopeptide poaeamide that suppressed *R. solani*-associated pathogenesis in sugar beetroots. Similarly, endophytic *B. amyloliquefaciens* was reported to produce a series of isoforms of iturins that can confer protection to its host against pathogens (Han et al. 2015). VOCs produced by *P. fluorescens* and *Serratia plymuthica* inhibited tumorigenic strains of *A. tumefaciens* and *A. vitis* induced crown gall disease in tomatoes. Solid-phase microextraction–gas chromatography–mass spectrometry analysis revealed dimethyl disulfide (DMDS) and 1-Undecene as the major VOCs produced by *S. plymuthica* IC1270 and *P. fluorescens* strains, respectively (Dandurishvili et al. 2011).

11.5.2.3 Quorum Quenching

Quorum sensing is an important phenomenon exhibited by numerous pathogenic microbes in order to survive in a specific ecological niche, communicate between cells, undergo multiplication, control biofilm formation, and induce competence and also adaptation (Miller and Bassler 2001). Certain endophytic bacteria employ QS quenching as an antivirulence strategy to control phytopathogen. Endophytic bacterial strains, *Bacillus* sp. strain B3, *Bacillus megaterium* strain B4, *Brevibacillus borstelensis* strain B8, and *Bacillus* sp. strain B11 from *Cannabis sativa* L. efficiently disrupt cell-to-cell communication in *Chromobacterium violaceum* via quenching its QS signals (Kusari et al. 2014). It is important to note that a diffusible signal factor (DSF) is essential in several *Xanthomonas* species and *Xylella fastidiosa*-associated phytopathogenesis (Newman et al. 2008). *Bacillus* and *Pseudomonas* were reported to complement *carAB*, a gene responsible for fast DSF degradation in the *Pseudomonas* spp. strain G. This mechanism can be exploited as a powerful strategy in the biocontrol of DSF producing pathogens and, thus, can be deployed in agriculture (Liu et al. 2017).

11.5.2.4 Siderophores Production

Iron is a vital metal for growth in all living organisms. There is a great competition for bioavailable iron in soil habitats as well as on plant surfaces. Under iron-limiting conditions, endophytes produce low-molecular-weight compounds called siderophores to competitively acquire ferric ion (Whipps 2001). Although various bacterial siderophores differ in their abilities to sequester iron, in general, they deprive pathogenic fungi of this essential element since the fungal siderophores have lower affinity (Loper and Henkels 1999; O'Sullivan and O'Gara 1992). Some plant growth-promoting endophytes go one step further and draw iron from heterologous siderophores produced by cohabiting microorganisms (Wang et al. 2003; Whipps

2001). Primarily, siderophores help to acquire iron either from iron adsorbed to solid surfaces or from insoluble hydroxides. Siderophores can also extract iron from soluble and insoluble iron compounds, such as ferric-citrate, Fe-transferrin, ferric phosphate, ferritin, or iron bound to sugars, plant flavone pigments, and glycosides or even from artificial chelators like EDTA and nitritriacetate by Fe(III)/ligand-exchange reactions. Hence, although siderophores don't play a direct role in iron solubilization, they can act as carrier for exchange between extracellular iron stores and membrane-located siderophore-transport systems (Winkelmann 2002). Siderophores play a significant role in microbial metabolism because of the following facts:

1. Siderophores mainly consist of hydroxamate, catecholate, or *a*-hydroxycarboxylate ligands that form hexadentate Fe(III) complexes, satisfying the six coordination sites on ferric ions which make them most significant iron-binding ligands.
2. Siderophore biosynthesis is a highly regulated process which is triggered by iron limitation resulting in building up of high local concentrations of siderophores in the vicinity of microbial cells.
3. Siderophores exhibit structural and conformational specificities to fit into membrane receptors and/or transporters besides their ability to solubilize iron and to function as external iron carriers (Stintzi et al. 2000; Huschka et al. 1986; Ecker et al. 1988).

Endophytic isolates of *Phialocephala fortinii* from *P. sylvestris* root, *Carex curvula, Abies alba, Picea abies,* and *P. sylvestris* showed that siderophore production is a function of pH values and iron(III) concentrations; 4.0–4.5 was the range of pH at which maximum siderophore production was found with the optimal ferric iron concentration of 20–40 μg iron (III) L^{-1} (0.36–0.72 μM, respectively). The most predominant siderophores produced by *P. fortinii* is ferricrocin (a hydroxamate siderophore) followed by ferrirubin and ferrichrome C (Bartholdy et al. 2001). An endophytic *Streptomyces* sp. GMKU 3100 isolated from the roots of a Thai jasmine rice plant (*Oryza sativa* L. cv. KDML105) exhibited remarkably high level of siderophore production. Inactivation of desD-like gene that codes a key enzyme responsible for the final step in siderophore biosynthesis resulted in impairment of siderophore production. Rice and mungbean plants inoculated with the wild-type strain-enhanced plant growth and significantly increased root and shoot biomass and lengths unlike siderophore-deficient mutant treatments (Figs. 11.3 and 11.4). Endophytic actinomycetes, therefore, can be applied as a potentially safe and environmentally friendly biofertilizer in agriculture (Rungin et al. 2012).

11.5.2.5 Lytic Enzyme Production

Various extracellular enzymes from microbes perform their function outside the cell which is significant to host–endophyte interdependence. Bacteria and fungi produce various extracellular enzymes that include hydrolases, lyases, oxidoreductases, and transferases (Traving et al. 2015; Kour et al. 2019b). The substrates are mostly

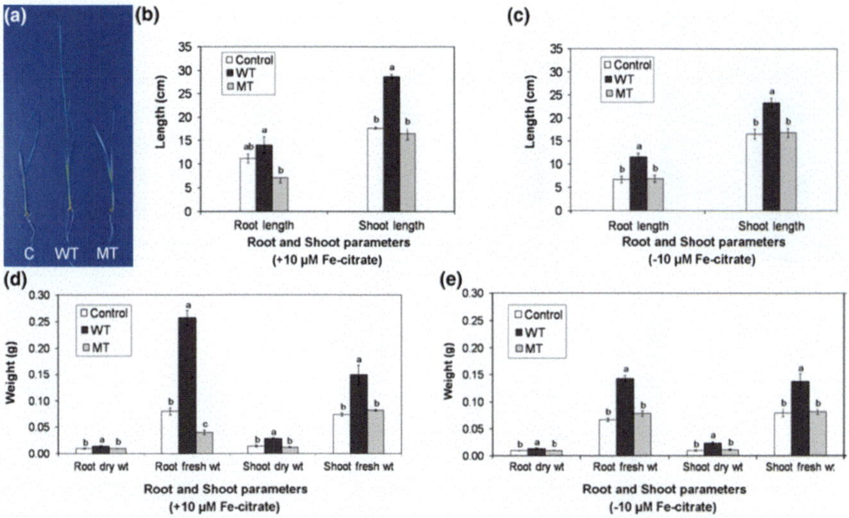

Fig. 11.3 Plant growth parameters of rice plants (Oryza sativa L. cv. KDML105) inoculated with *Streptomyces* sp. GMKU 3100 and the siderophore-deficient mutant after 14 days. (**a**) 14-day rice plants; (**b**) root and shoot lengths with 10 μM Fe-citrate; (**c**) root and shoot lengths without 10 μM Fe-citrate; (**d**) root dry/fresh weights and shoot dry/fresh weights with 10 μM Fe-citrate; (**e**) root dry/fresh weights and shoot dry/fresh weights without 10 μM Fe-citrate; C uninoculated plant (control), WT *Streptomyces* sp. GMKU 3100, MT siderophore-deficient mutant. Data are the mean of ten replicates. Means designated with different letters are significantly different ($p = 0.05$). Error bars show standard deviation ($n = 10$). Adapted with permission from Rungin et al. (2012)

macromolecules such as carbohydrates, proteins, lignin, sugar-based polymers, and organic phosphate which are broken down into simpler forms that can be easily transported, absorbed, and assimilated. Enzymes secreted by endophytes help to initiate the association with the host and symbiosis process. Extracellular hydrolyases counteract plant pathogenic infection (Leo et al. 2016). In fact, certain categories of enzymes namely, cellulases, xylanases, phytases, hemicellulases, asparaginase, proteases, gelatinase, pectinases, tyrosinase, chitinase, amylases, etc., are some of the key enzymes produced by endophytic bacteria and fungi.

Endophytic bacterial strains have been isolated from various plants such as pea (*P. sativum*), tomato (*Lycopersicum esculentum*), corn (*Zea mays*), wheat (*Triticum aesitivum*), oat (*Avena sativa*), canola (*Brassica napus*), barley (*Hordeum vulgare*), radish (*Raphanus sativus*) soybean (*Glycine max*), potato (*Solanum tuberosum*), lettuce (*Lactuca serriola*), and cucumber (*Cucumis sativa*) were identified and characterized that belong to the genus *Arthrobacter, Actinobacter, Aeromonas, Agrobacterium, Alcaligenes, Bacillus, Azospirillium, Enterobacter, Flavobacterium Pseudomonas, Acinetobacter, Azotobacter, Beijerinckia, Burkholderia, Enterobacter, Erwinia, Flavobacterium, Rhizobium,* and *Serratia* (Khan et al. 2017; Gray and Smith 2005). Vijayalakshmi et al. (2016) isolated endophytic bacteria from medicinally important plants, producing α-amylase, protease, and cellulase. Similarly, Leo et al. (2016) reported endophytic bacteria, *Alcaligenes faecalis, Burkholderia*

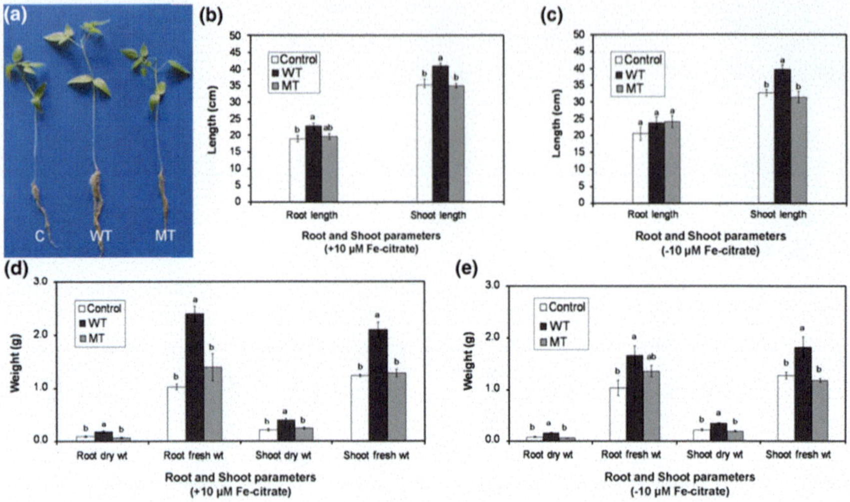

Fig. 11.4 Plant growth parameters of mungbean plants (*Vigna radiata* (L.) Wilczek cv. CN72) inoculated with *Streptomyces* sp. GMKU 3100 and the siderophore-deficient mutant after 28 days. (**a**) 28-day mungbean plants; (**b**) root and shoot lengths with 10 μM Fe-citrate; (**c**) root and shoot lengths without 10 μM Fe-citrate; (**d**) root dry/fresh weights and shoot dry/fresh weights with 10 μM Fe-citrate; (**e**) root dry/fresh weights and shoot dry/fresh weights without 10 μM Fe-citrate; C uninoculated plant (control), WT *Streptomyces* sp. GMKU 3100, MT siderophore-deficient mutant. Data are the mean of ten replicates. Means designated with different letters are significantly different ($p = 0.05$). Error bars show standard deviation ($n = 10$). Adapted with permission from Rungin et al. (2012)

cepacia, and *Enterobacter hormaechei* from perennial grasses that showed the hyper-enzymatic activity of α-amylase, protease, and cellulose (Table 11.2).

A variety of microorganisms also exhibited hyperparasitic activity, attacking pathogens by excreting cell wall hydrolases. Chitinase produced by *S. plymuthica* C48 inhibited spore germination and germ-tube elongation in *Botrytis cinerea* (Frankowski et al. 2001). The ability to produce extracellular chitinases is considered crucial for *Serratia marcescens* to antagonize *Sclerotium rolfsii* (Ordentlich et al. 1988). Using similar mechanisms, *Paenibacillus* sp. and *Streptomyces* sp. suppress *Fusarium oxysporum* while *Pseudomonas* sp. suppresses *Fusarium solani,* the commonly known plant pathogen (Lim et al. 1991). Many endophytic fungi like *Alternaria alternate, Hymenoscyphus ericae,* and *Aspergillus terreus* also produce extracellular enzyme xylanase producers including those found in Table 11.3. Similarly, the endophyte *Periconia* sp. produced β-glucosidase, while *Acremonium* species produced cellulases and hemicellulases.

11.5.2.6 Induced Systemic Resistance

Induced systemic resistance (ISR) is the immunity response mechanism inherent in crop plants that can be triggered by beneficial microbial endophytes during biotic and abiotic stress conditions which may include temperature, salinity, drought,

Table 11.2 Endophytic bacterial strains producing extracellular enzymes

Endophytic Microbes	Enzyme	Detection method
Actinomyces pyogenes, Bacillus circulans, Bacillus coagulans, Bacillus licheniformis, Bacillus megaterium, Corynebacterium renale, Pseudomonas stutzeri, Staphylococcus sp., *Bacillus* sp.	Amylase, esterase Lipase, protease	Agar medium
Pseudomonas oryzihabitans	Asparaginase	Spectrophotometer
Bacillus sp., *Bacillus clausii, Bacillus pumilus, Bacillus licheniformis*	Amylase, protease, cellulose, lipase	Agar medium
Pseudomonas sp.	Exo-β-agarase	Spectrophotometer, NMR
Bacillus sp.	L-asparaginase	Spectrophotometer
Bacillus amyloliquefaciens	Phytase	Spectrophotometer
Paenibacillus polymyxa	Fibrinolytic enzymes	Agar medium, SDS Page
Rhizobium, Massilia, Kosakonia, Pseudorhodoferax, Caulobacter, Pantoea, Sphingomonas, Burkholderia, Methylobacterium, Bacillus, Curtobacterium, Microbacterium, Mucilaginibacter, Chitinophaga	ACC deaminase, endoglucanase, protease	Agar medium
Acinetobacter sp., *Bacillus* sp.	ACC deaminase, cellulase, protease, amylase, pectinase	Agar medium
Bacillus licheniformis, Bacillus pseudomycoides, Paenibacillus senitriformus	L-asparaginase	M9 medium
Pseudomonas hibiscicola, Macrococcus caseolyticus, Enterobacter ludwigii, Bacillus anthracis, Bacillus tequilensis, Pseudomonas entomophila, Chryseobacterium indologenes, Bacillus aerophilus	Cellulase, xylanase, amalyase, pectinase	Agar diffusion method
Bacillus thuringiensis	Anthracene	Spectrophotometer
Bacillus amyloliquefaciens	Exopolysaccharides	Colorimetric method
Bacillus subtilis	YbdN protein	SDS-PAGE, MALD-TOF-MS
Serratia marcescens, Bacillus subtilis, Bacillus methylotrophicus, Bacillus siamensis	L-asparaginase	Spectrophotometer
Paenibacillus polymyxa, Bacillus sp.	Cellulase, xylanase, pectinase	Agar diffusion method
Paenibacillus amylolyticus	Pectin lyase	Spectrophotometer
Alcaligenes faecalis, Burkholderia cepacia, Enterobacter hormaechei	Cellulosic, hemicellulosic, lignin	National renewable energy laboratory methods

Sources: Adapted with permission from Khan et al. (2017)

Table 11.3 Enzyme production from different endophytic fungal species

Microbes	Enzyme produced	Detection method
Penicillium funiuclas, *Trichoderma viride*	Amylase, cellulose, protease, lipase	Agar plate base test
Colletrotrichum, *Fusarium*, *Phoma*, *Penicillium*	L-Asparaginase	Pink zones on agar, nesslerization
Aspergillus sp.	Amylase	Agar medium
Pochonia chlamydosporia	Protease	Spectrophotometer
Colletotrichum gloeosporioides	Protease, chitinase, amylase	
Fusarium sp., *Chaetomium* sp., *Colletotrichum* sp., *Aspergillus flavus*, *Cylindrocephalum* sp., *Coniothyrium* sp., *Phoma* sp., *Aspergillus niger*, *Colletotrichum* sp., *Mycelia sterilia* sp., *Aspergillus fumigates*, *Alternaria* sp., *Colletotrichum gleosporoides*, *Colletotrichum* sp., *Myrotheium* sp., *Fusaruim chlamydosporum*, *Xylaria* sp., *Fusicoccum* sp., *Mycelia sterilia* sp., *Aspergillus* sp., *Pestalotiopsis* sp., *Colletotrichum* sp., *Talaromyces emersonii*, *Pyllosticta* sp., *Pestalotiopsis* sp., *Discosia* sp., *Aspergillus* sp., *Mycelia streilia* sp., *Isaria* sp., *Xylaria* sp., *Phoma* sp., *Pestalotiopsis disseminate*, *Fusarium oxysporum*, *Paecilomyces variotii*, *Fusarium chlamydosporum*, *Acremonium implicatum*, *Nigrospora sphaerica*, *Fusarium solani*, *Penicillium* sp., *Mycelia sterilia* sp., *Phoma* sp., *Basidiomycetes* sp., *Colletotrichum falcatum*, *Phomopsis longicolla* *Fusarium oxysporum*, *Colletotrichum gleosporoides*, *Colletotrichum truncatum*, *Drechsclera* sp., *Cladosporium* sp., *Myrothecium* sp.	Amylase, cellulase, laccase, lipase, pectinase, protease	Agar medium
Cladosporium sp., *Rhizoctonia* sp., *Aspergillus* sp., *Chaetomium* sp., *Biosporus* sp., *Fuzarium* sp., *Curvularia* sp., *Cladosporium* sp., *Colletotrichum* sp.	Amylase, protease, cellulose, lipase	Agar medium, spectrophotometer
Cladosporium cladosporioides, *Curvularia brachyspira*, *C. verruciformis*, *Drechslera awaiiensis*, *Colletotrichum carssipes*, *Colletotrichum falcatum*, *Colletotrichum gloeosporioides*, *Lasiodiplodia theobromae*, *Nigrospora sphaerica*, *Phyllosticta* sp. *Xylariales*	Amylase, cellulase, laccase, lipase, protease	Agar medium
Cladosporium cladosporioides, *C. sphaerospermum*, *Acremonium terricola*, *Monodictys castaneae*, *Penicillium glandicola*, *Phoma tropica*, *Tetraploa aristata*	Pectinases, cellulases, xylanases, proteases	Agar medium

(continued)

Table 11.3 (continued)

Microbes	Enzyme produced	Detection method
Amanita muscaria, A. muscaria, A. spissa, Boletus luridus, Cenococcum geophilum, Cortinarius glaucopus, C. purpurascens, Hydnum rufescens, Hymenoscyphus ericae, Laccaria cf., *Lactarius acerrimus, L. auriolla, L.chrysorrheus, L. controversus, L. deliciosus, L. deterrimus, L. evosmus, L. pubescens, L. quieticolor, L. quietus, L. rufus, L. semisanguifluus, L. subdulcis, L. subumbonatus, L. zonarius, Piceirhiza bicolorata, Piloderma fallax, Piloderma byssinum, Russula chloroides, R. sanguinea, Suillus luteus, S. luteus, Tricholoma* cf. *equestre, S. variegatus, T. fulvum, T. scalpturatum*	Protease	Agar medium
Eurotiales, Chaelomiaceae, Incertae sadis, Aureobasiduaceae, Nectriaceae, Sporomiaceae	Celluloses, phosphatases, glucosidases	Spectrophotometer
Colletotrichum sp., *Macrophomina phaseolina, Nigrospora sphaerica, Fusarium solani*	Cellulase, protease, amylase	Agar medium
Cochliobolus lunatus, C. australiensis, Gibberella baccata, Myrmecridium schulzeri, Penicillium commune, Phoma putaminum, Acremonium curvulum, Aspergillus Niger, A. ochraceus, P. glabrum, C. lunatus, G. fujikuroi, Myrothecium verrucaria, Nodulisporium, Trichoderma piluliferum, A. chartarum, A. ochraceus, P. glabrum, Pithomyces atro-olivaceus	Cellulase, protease, xylanase, lipase	Agar medium
Penicillium chrysogenum, Alternaria alternate, Sterile hyphae	Amylase, pectinase, cellulase, gelatinase, xylanase, tyrosinase	Agar medium
Aspergillus terreus	L-asparaginase	Agar medium, spectrophotometer
Phialocephala fortinii s.l., Meliniomyces variabilis, Umbelopsis isabellina, Hebeloma incarnatulum, Laccaria bicolor	Protease	
Hormonema sp., *Pringsheimia smilacis, Ulocladium* sp., *Neofusicoccum luteum, Neofusicoccum australe*	Laccase	Agar medium, spectrophotometer
Acremonium sp., *Alternaria* sp., *Aspergillus* sp., *Fusarium* sp., *Pestalotiopsis* sp.	Amylase, cellulase, lipase, protease	Agar medium

(continued)

Table 11.3 (continued)

Microbes	Enzyme produced	Detection method
Chaetomium sp., *Preussia* sp., *Penicillium citrinum, Thielavia arenaria, Phoma medicaginis, Aureobasidium* sp., *Preussia* sp., *Dothideomycetes* sp., *Aureobasidium pullulans, Phoma* sp., *Penicillium citrinum, Aureobasidium pullulans, Aureobasidium pullulans, Thielavia arenaria, Sordariomycetes* sp., *Fusarium proliferatum, Preussia* sp.	Glucosidase, phosphatases, cellulases	Fluorescence spectrophotometer

Source: Adapted with permission from Khan et al. (2017)

heavy metal, and phytopathogenic infections. A diverse group of metabolites produced by the endophytes can impart the host plant to overcome the stress (Khan et al. 2017). Immunized through ISR plays a vital role in the protection from pathogenic invasions, exhibition of varied resistance methods, efficient utilization of energy, and exploitation of genetic ability to induce resistance in the plants which are vulnerable for diseases (Latha et al. 2019). Plants are also protected from the parasitic nematodes due to ISR. Bacterial endophytes like *B. amyloliquefaciens, Bacillus pumilus, Bacillus subtilis, P. fluorescens, P. syringae,* and *S. marcescens* can induce ISR (Latha et al. 2019). The following section gives an elaborate account of the endophyte-associated ISR in plants.

Detoxification and Degradation of Virulence Factors

Detoxification of pathogen virulence factors is another mechanism of biological control. For example, certain biocontrol agents are able to detoxify albicidin toxin produced by *Xanthomonas albilineans* (Basnayake and Birch 1995; Zhang and Birch 1997). Endophytic bacterial strains of *B. cepacia* and *Ralstonia solanacearum* were reported to suppress the activity of fusaric acid, a toxin secreted by *Fusarium* species, a major wilt-causing pathogen (Toyoda and Utsumi 1991). The autoinducer-mediated quorum-sensing of endophytes can impair the virulence of pathogens to inflict diseases, which is of paramount importance (Latha et al. 2019).

Insect and Pest Tolerance

Endophytes also play a critical role in insect and pest-induced biotic stress in plants. Entomopathogenic microorganisms inhibit/antagonize other pathogenic microbes that not only help to protect plants but also reduce use of chemical pesticides. Since being established due to their capacity to protect their hosts against insects–pests, pathogens and even herbivores endophytic microorganisms have received considerable attention in the last 20 years. Webber (1981) first reported that endophytic fungus *Phomopsis oblonga* protected elm trees from the beetle *Physocnemum brevilineum. P. oblonga* controlled the beetle *P. brevilineum* which is the vector for *Ceratocystis ulmi*, responsible for the elm Dutch disease. Another endophytic fungi belonging to the *Xylariacea* family synthesized secondary metabolites in hosts of the genus *Fagus* that affected the beetle larvae. Owing to toxin production, endophytic fungus repels insects, induces weight loss, inhibits growth and development, and even increases pest mortality. Another mode of action might be rendering the plant unpalatable to several types of pests like aphids, grasshoppers, beetles, etc. due to metabolites secreted by the endophytes. Endophytic isolates of *Neotyphodium* sp. produced N-formilonine and a paxiline in the host *Echinopogum ovatus* that exhibited insecticidal activity against *L. bonariensis* and other insects (Azevedo et al. 2000).

White spruce *Picea glauca*, death rate in the Homoptera *Adelges abietis* increased when galls were infected with the endophytic fungus *Cladosporium sphaerosperum* while weight gain and survival of the insect–pest, *Spodoptera frugiperda,* were severely compromised when their hosts were colonized by endophytic fungi like *Balansia cyperi.* It is important to note that larvae from the bluegrass webworm *Parapediasia teterrella* preferred endophyte-free plants of *L. perenne* and *F. arundinacea*, to a point that the larvae would starve to death if only plants infected with *Acremonium* were available. Field studies revealed that endophyte-free species were severely attacked by insects, whereas those infected with *Acremonium* stayed almost free of insect larvae (Azevedo et al. 2000).

Cold and Drought Stress Tolerance

Endophytic microbes render the plant its ability to tolerate abiotic stress during severe temperatures and water scarcity. Tomato plants inoculated with psychrotolerant endophytic bacteria *Pseudomonas vancouverensis* OB155 and *P. frederiksbergensis* OS261 were able to overcome cold stress (10–12 °C). Lesser membrane damage with increased antioxidant activity was observed in endophyte-colonized plants compared to endophyte-free control plants. Further, cold acclimation genes (*LeCBF1* and *LeCBF3*) were induced in bacteria-inoculated plants (Subramanian et al. 2015). Similarly, the bacterial endophyte *Burkholderia phytofirmans* strain PsJN resulted in enhancement of *Arabidopsis* growth and strengthened its cell wall, and thereby increased cold stress resistance (Su et al. 2015). Increased plant tolerance to drought was also seen due to endophytic bacteria. *B. phytofirmans* PsJN modulated transcriptional regulation, cellular homeostasis, and ROS detoxification in a drought stress-affected potato (Sheibani-Tezerji et al. 2015). These facts strongly rationalize that endophytes can be potential protective agents in crops under extreme climatic environments as they can influence plant physiological responses to stresses (Liu et al. 2017).

Metal Stress Tolerance

Endophytes can mitigate metal toxicity in plants through their own metal resistance system and encourage plant growth under metal stress. Endophytes improve plant growth in metal-polluted soils either directly or indirectly by metal detoxification, accumulation, or translocation in plants. They can even alter metal accumulation capacity in plants by excreting metal immobilizing extracellular polymeric substances as well as metal mobilizing organic acids and biosurfactants. The metal stress can be circumvented by various mechanisms, which include efflux of metal ions exterior to the cell, transformation of metal ions to less toxic forms, sequestration of metals on the cell surface or in intracellular polymers, and precipitation, adsorption/desorption, or biomethylation (Rajkumar et al. 2013). Inoculation of seeds or seedlings of hyperaccumulator plants with metal resistant endophytes results in accelerated phytoremediation in naturally and/or artificially metal-contaminated soil and improved plant growth.

The endophytic bacterial strain *Bacillus* sp. MN3-4 exhibited metal-resistance owing to active export via a P-type ATPase efflux pump that can transport metal ions across biological membranes against the concentration gradient using energy released by ATP hydrolysis (Shin et al. 2012). Further, endophytic bacteria can modulate the activity of plant antioxidant enzymes (such as POS, CAT, SOD, glutathione peroxidase, and ascorbate peroxidase) as well as lipid peroxidation (malondialdehyde formation) that collectively enable the host plant to overcome heavy metal-induced oxidative stress. Methylation is another significant way to gain metal resistance or detoxification. Endophytic bacteria with mercury-resistant (Mer) operons express MerB gene-encoding organomercurial lyase, which cleaves organomercurials into mercuric ion (Hg^{2+}) (Brown et al. 2003). MerA gene encodes mercuric reductase that converts highly toxic ionic Hg^{2+} into less toxic volatile Hg^0 (Cursino et al. 2000), thus alleviating metal toxicity and improving the efficiency of phytovolatilization. Lead-resistant endophytic bacteria *Bacillus* sp. MN3-4 isolated from the roots of the metal hyperaccumulator plant *Alnus firma* enhanced reduced metal phytotoxicity by extracellular sequestration and intracellular accumulation (Shin et al. 2012).

Similarly, cadmium-resistant endophytic bacterium *Serratia* sp. LRE07 reduced metal stress by absorbing over 65% of Cd and 35% of Zn in bacterial cells from a single metal solution. Endophytes can also alter phytoavailability of heavy metals through the release of metal chelating agents (e.g., siderophores, biosurfactants, and organic acid), acidification of soils, redox activity, and phosphate solubilization. Extracellular polymeric substances (EPS) secreted by endophytes are composed of polysaccharides, proteins, nucleic acids, and lipids that are significantly responsible in metal complexation thereby reducing their bioaccessibility and bioavailability (Ma et al. 2016).

Nickel (Ni)-resistant endophytic bacterium *Pseudomonas* sp. A3R3 increased plant biomass (nonhost *Brassica juncea*) and Ni accumulation in plants (host *A. serpyllifolium*) grown in artificially Ni-contaminated soil (Ma et al. 2011). These effects can be attributed to the ability of endophytes to produce plant growth-promoting substances (ACC deaminase, siderophores, IAA, and P solubilization) and plant polymer-hydrolyzing enzymes like cellulase and pectinase (Table 11.4).

11.6 Bioactive Compounds from Endophytes

Gouda et al. (2016) have summarized the discovery of a number of bioactive metabolites from endophytes that serve as an excellent source of drugs for the treatment against various diseases and with potential applications in agriculture, medicine, food, and the cosmetic industries (Table 11.5). Ezra et al. (2004) reported that

Table 11.4 Endophytic bacterial enhanced phytoremediation of metal contaminated soil

Endophytic bacteria	Host plant	Metal stress	Plant growth-promoting traits	Mechanisms
Bacillus thuringiensis GDB-1	*Alnus firma*	As, Cu, Cd, Ni, Pb, and Zn	Production of IAA, siderophores, ACCD, and solubilization of P	Bioremoval of Pb, Zn, As, Cd, Cu, and Ni in metal-amended and mine tailing extract medium; increased biomass, chlorophyll content, nodule number and metal (As, Cu, Pb, Ni, and Zn) accumulation in *A. firma*
Pseudomonas koreensis AGB-1	*Miscanthus sinensis*	As, Cd, Cu, Pb, and Zn	nd	Increased plant biomass, chlorophyll, protein content, superoxide dismutase and catalase activities, and metal uptake; however, decreased malondialdehyde content in plants
Staphylococcus, Curtobacterium, Bacillus, Pseudomonas, Microbacterium, Arthrobater, Leifsonia, Paenibacillus	*Alyssum bertolonii*	Ni, Co, Cr, Cu, and Zn	Production of siderophores	Had an ability to colonize plant tissues
Serratia nematodiphila LRE07, *Enterobacter aerogenes* LRE17, *Enterobacter* sp. LSE04 *Acinetobacter* sp. LSE06	*Solanum nigrum* L.	Cd	Production of IAA, siderophores, ACCD, and solubilization of P	Increased Cd mobilization in soils; stimulated plant growth and influenced Cd accumulation in plant tissues; colonized the rhizosphere soil and some colonized plant interior tissues
Pseudomonas sp. Lk9	*Solanum nigrum*	Cd, Zn, and Cu	nd	Improved soil Fe, P, and heavy metal availability, shoot dry biomass, and uptake of Cd, Zn, and Cu
P. monteilii PsF84, *P. plecoglossicida* PsF610	*Pelargonium graveolens*	Cr	Production of IAA and siderophores, solubilization of P	Increased plant dry biomass, essential oil yield, and chlorophyll helped Cr(VI) sequester in roots

(continued)

Table 11.4 (continued)

Endophytic bacteria	Host plant	Metal stress	Plant growth-promoting traits	Mechanisms
Rahnella sp. JN6	*Polygonum pubescens*	Cd, Pb, and Zn	Production of IAA, siderophores, ACCD, and solubilization of P	Showed high Cd, Pb, Zn tolerance and mobilization; promoted plant growth and Cd, Pb, Zn uptake by rapes; high level of colonization in tissue interior of rapes
Actinobacterium	*Salix caprea*	Cd and Zn	Production of siderophores and ACCD	Enhanced plant growth and metal accumulation in leaves
Burkholderia cepacia L.S.2.4, *Herbaspirillum seropedicae* LMG2284	*Lupinus luteus* L	Cu, Cd, Co, Ni, Pb, and Zn	nd	Bioremoval of Ni, thus reduced metal toxicity;*B. cepacia* L.S.2.4 increased Ni concentration in roots, while *H. seropedicae* LMG2284 decreased Ni concentration in roots and shoots of *Lolium perenne*
Pseudomonas fluorescens VI8L1, *Bacillus pumilus* VI8L2, *P. fluorescens* II8L4, *P. fluorescens* VI8R2, *Acinetobacter calcoaceticus* II2R3	*Sedum alfredii*	Zn and Cd	Production of IAA, siderophores, fixation of nitrogen, solubilization of $ZnCO_3$, and $Zn_3(PO_4)_2$	Mobilized Zn in soil, thus increased soil Zn bioavailability; improved growth and Zn accumulation by *S. alfredii*
Serratia marcescens LKR01, *Arthrobacter* sp. LKS02, *Flavobacterium* sp. LKS03, *Chryseobacterium* sp. LKS04	*Solanum nigrum* L.	Zn, Cd, Pb, and Cu	Production of IAA, siderophores, ACCD, and solubilization of P	Decreased Cd phytotoxicity; improved plant growth and total Cd accumulation in host plants
Serratia sp. LRE07	*S. nigrum* L.	Cd, Cr, Pb, Cu, and Zn	Production of IAA, siderophores, and solubilization of P	Bioaccumulation or removal of metals (Cd, Zn) in both single-ion and multi-ions systems
Bacillus sp. SLS18	*Sorghum bicolor* L.	Cd and Mn	Production of IAA, siderophores, and ACCD	Improved plant biomass production and its total metal uptake

(continued)

Table 11.4 (continued)

Endophytic bacteria	Host plant	Metal stress	Plant growth-promoting traits	Mechanisms
Pseudomonas sp. A3R3	*Alyssum serpyllifolium*	Ni	Production of IAA, siderophores, ACCD, and solubilization of P; excreted cellulase and pectinase	Increased the biomass of *B. juncea* and Ni content in*A. serpyllifolium*; showed high level of colonization in tissue interior of both plant species
B. pumilus E2S2, *Bacillus* sp. E1S2, *Bacillus* sp. E4S1, *Achromobacter* sp. E4L5, and *Stenotrophomonas* sp. E1L	*Sedum plumbizincicola*	Cd, Pb, and Zn	Production of IAA, siderophores, ACCD, and solubilization of P	Bacterial inoculation increased water-extractable Cd and Zn contents in soil; improved plant growth and metal uptake
Methylobacterium oryzae CBMB20, *Burkholderia* sp. CBMB40	*Lycopersicon esculentum*	Ni and Cd	nd	Biosorption considerable amount of Ni and Cd, thus reduced the metal toxicity; promoted plant growth and reduced accumulation of Ni and Cd in roots and shoots of tomato plants
P. fluorescens G10, *Microbacterium* G16	*Brassica napus*	Pb, Cd, Zn, Cu, and Ni	Production of IAA, siderophores, ACCD	Increased water-soluble Pb in solution and Pb-added soil; increased biomass production and total Pb uptake
Bacillus sp. MN3-4	*Alnus firma* and *B. napus*	Pb, Cd, Zn, Ni, and Cu	Production of IAA and siderophores	Exhibited bioremoval of Pb; increased root elongation of *B. napus* seedlings; reduced metal phytotoxicity and increase Pb accumulation in *A. firma*
Endophytes belonged to Firmicutes, Actinobacteria, Proteobacteria	*Elsholtzia splendens*, *Commelina communis*	Cu	Production of IAA, siderophores, ACCD, and arginine decarboxylase	Increased plant dry weights and Cu content in aboveground tissue of rapes

(continued)

Table 11.4 (continued)

Endophytic bacteria	Host plant	Metal stress	Plant growth-promoting traits	Mechanisms
Microbacterium sp. NCr-8, *Arthrobacter* sp. NCr-1, *Bacillus* sp. NCr-5, *Bacillus* sp. NCr-9, and *Kocuria* sp. NCr-3	*Noccaea caerulescens*, *Thlaspi perfoliatum*	Ni	Production of IAA, siderophores, and ACCD	Enhanced growth and Ni translocation in plants
Serratia nematodiphila LRE07	*Solanum nigrum* L.	Cd	nd	Promoted biomass production; increased higher photosynthetic pigments content of leaves
Rahnella sp. JN27	*Amaranthus hypochondriacus* and *A. mangostanus*	Cd	Production of IAA, siderophores, ACCD, and solubilization of P	Enhanced plant growth and Cd uptake by both plant species
Acinetobacter sp. Q2BJ2, *Bacillus* sp. Q2BG1	*Commelina communis*	Pb, Cu, Cd, and Ni	Production of IAA, siderophores, and ACCD	Increased plant dry weights; increased Pb contents in aboveground tissue of rapes
Ralstonia sp. J1–22-2, *Pantoea agglomerans* Jp3–3, *Pseudomonas thivervalensis* Y1–3-9	*B. napus*	Cu, Pb, Cd, and Ni	Production of IAA, siderophores, ACCD, and solubilization of P	Increased the biomass of rapes and increased Cu content in above-ground tissues
Burkholderia sp. SaZR4, *Burkholderia* sp. SaMR10, *Sphingomonas* sp. SaMR12 and *Variovorax* sp. SaNR1	*Sedum alfredii* Hance	Cd and Zn	nd	SaMR10 had little effect on phytoextraction, while SaMR12 and SaNR1 promoted plant growth and phytoextraction of Zn and Cd; SaZR4 only promoted Zn extraction
Endophytes belonged to Firmicutes, Proteobacteria, and Actinobacteria	*Pteris vittata* and *P. multifida*	As	Production of IAA	Possessed ability of both AsV reduction and AsIII oxidation.

IAA, indole-3-acetic acid; *ACCD*, 1-aminocyclopropane-1-carboxylate deaminase; *P*, phosphorus; *nd*, not determined

Sources: Adapted with permission from Ma et al. (2016)

Table 11.5 Source of bioactive compounds from endophytes and their use against pathogenic microorganisms

Source of endophytes	Bioactive compounds from endophytes	Cure against pathogen	Mode of pathogen transmission
Boesenbergia rotunda, Streptomyces coelicolo	Munumbicins	*Escherichia coli*	Ground meats, raw or under pasteurized milk
Chloridium sp.	Javanicin	*Pseudomonas* sp.	Contaminated water or surgical instruments
Allamanda cathartica	Munumbicins Phomopsilactone		
Cladosporium sp.	Cardiac glycosides, phenolic compounds	*Klebsiella pneumoniae*	Contaminated water and aerosols
Cladosporium sp.	Cardiac glycosides, phenolic compounds	*Proteus* sp.	Canned food products
Cryptosporiopsis quercina	Saadamycin	*Campylobacter jejuni*	Raw or uncooked poultry and milk
Cytonaema sp.	Cytonic acids A and B	*Human cytomegalovirus. Hepatitis virus*	Shellfish, berries, or contaminated water
Diaporthe helianthi	Fabatin, tyrosol	*Enterococcus hirae*	Nosocomial infection through hospitalized patients
Fusarium proliferatum	Beauvericin	*Clostridium botulinum*	Improperly processed, canned food
Fusarium proliferatum	Kakadumycin, beauvericin	*Listeria monocytogenes*	Raw or under pasteurized milk, smoked fish
Fusarium sp., *Cryptosporiopsis quercina*	Xularosides, munumbicins, Saadamycin, cryptocandin	*Candida albicans*	Contaminated sweet fruits and milk products
Ganoderma boninense	Rapamycin, cyclododecane, petalostemumol	*Bacillus subtilis*	Rice, pastas, raw milk, and meat products
Hypericum perforatum, Diaporthe helianthi	Hypericin, emodin, tyrosol	*Salmonella* sp.	Meat, eggs, and untreated tree nuts
Nigrospora sp.	Saadamycin	*Fusarium oxysporum*	Maize, cereals, groundnuts, and tree nuts
Phomopsis sp., *Cinnamomum mollissimum*	Munumbicins, Saadamycin	*Aspergillus niger*	Maize, cereals, groundnuts, and tree nuts
Saccharothrix mutabilis, *Streptomyces* sp.	Capreomycin Munumbicins	*Mycoplasm* (TB)	Uncooked meat, eggs, or poultry

(continued)

Table 11.5 (continued)

Source of endophytes	Bioactive compounds from endophytes	Cure against pathogen	Mode of pathogen transmission
Streptomyces hygroscopicus	Clethramycin	*Cryptococcus Neoformans*	Lettuce harvested from tropical regions
Streptomyces lygroscopicus	Coronamycin, rapamycin	*Saccharomyces cerevisiae*	Bakery and fermented products
Streptomyces sp.	Kakadumycin A, hypericin	*Shigella* sp.	Contaminated food, water, and fecal waste
Streptomyces sp., *Achyranthes bidentata*, *Phoma* sp., *Saurauia scaberrinae*	Terephthalic acid Phomodione	*Staphylococcus aureus*	Meat, eggs, and dairy products
Streptomyces sp., *Kennedia nigricans*	Munumbicins	*Vibrio cholerae*	Raw or undercooked shellfish, particularly oysters
Streptomyces tsusimaensis	Valinomycin	*Corona virus*	Food or water contaminated with infected fecal matter
Thottea grandiflora, *Xylaria* sp.	Streptomyces dihydroxynaphthol, glucopyranoside	*Bacillus cereus* *Herpes virus*	Uncooked meat and raw milk Contaminated body fluid or saliva
Xylaria sp.	Phenolic compounds	*Streptococcus pyogenes*	Contaminated water, raw milk, salads, and eggs
Xylaria sp., *Ginkgo biloba, Fusarium proliferatum*	Sordaricin 7 amino-4-methylcoumarin, Beauvericin	*Yersinia enterocolitica*	Swine meat and meat products, milk, and dairy products

Sources: Adapted with permission from Gouda et al. (2016)

coronamycin, a complex of novel peptide antibiotics with activity against pythiaceous fungi and the human fungal pathogen *Cryptococcus neoformans*, was produced by a verticillate *Streptomyces* sp. isolated as an endophyte from an epiphytic vine *Monstera* sp. It was also active against the malarial parasite, *Plasmodium falciparum.*

Undoubtedly, one of the most revolutionary findings of endophyte studies was the isolation of taxol-producing endophyte *Taxomyces andreanae* (Stierle et al. 1993). The diterpenoid taxol was approved by the FDA as one of the most potent anticancer drugs, but the supply of this drug was limited for the destructive collection of yew tree, the main source of taxol. Later taxol (paclitaxol) was also reported to be produced by the endophyte *Metarhizium anisopliae* found in the bark of a Taxus tree and is one of the most promising anticancer agents (Gouda et al. 2016).

As a selectively cytotoxic quinone dimmer, torreyanic acid is another important anticancer agent. Lee et al. (1996) reported the isolation of an endophyte strain *P. microspore* from *T. taxifolia* (*Florida torreya*) and the extraction of torreyanic acid from cultures of this endophyte. Camptothecin and its derivatives show strong antineoplastic activity. The fungus, which belongs to the family Phycomycetes, isolated from the inner bark of the plant *Nothapodytes foetida*, produced the anticancer drug lead compound camptothecin (Puri et al. 2005).

Endophytes are a potential source of novel secondary metabolites with antiarthritic, antimicrobial, anticancer, antidiabetic, anti-insect, and immunosuppressant activities (Devi et al. 2020; Kour et al. 2019a; Yadav et al. 2019). Bioactive compounds, such as camptothecin, diosgenin, hypericin, paclitaxel, podophyllotoxin, and vinblastine, are commercially produced by different endophytes colonizing respective plants which are agriculturally and pharmaceutically significant (Gouda et al. 2016; Godstime et al. 2014; Joseph and Priya 2011).

11.7 Conclusions and Future Perspectives

Promising plant growth-promoting activity and an ability to induce stress tolerance to host plants have drawn wide attention for developing not only culture dependent but also independent characterization of endophytic diversity. However, reports on successful application of endophytes in plants under field conditions are extremely scarce. Future studies should aim to explore the interrelationship between plant immunity and function of the microbial population of endosphere. Similarly, breeding of endophyte-colonized crops, genetic engineering of endophytes, maintenance, and adaptation to benefit plants at various growth stages of plants should be investigated. Further, endophytes impart resistance to hosts against pests, insects, nematodes, and plant pathogenic fungi and bacteria.

Similarly, host plants obtain tolerance to abiotic stress induced by drought, salinity, and toxic metals. Diverse bioactive compounds have been synthesized by microbial endophytes that may include antimicrobials (vanillin, essential oils), antifungals, antivirals (alkaloids), antioxidants (eugenol), anti-inflammatories (cineole), etc. Therefore, commercial processes can be developed to exploit the rich source of endophytic biodiversity to produce natural products for use in pharmaceutics, food, and cosmetics. Activity-based rapid screening technologies should be developed that may help in the selective isolation of beneficial endophytes. Establishing a target endophytic library for plant breeding may help to protect endangered medicinal plants from overexploitation. Endophytes can be envisioned to be the future of biofertilizers and biocontrol agents that can be promising alternatives to environmentally hazardous chemical fertilizers and pesticides resulting in a paradigm shift in agricultural best practices.

Acknowledgments Dr. Sougata Ghosh acknowledges the Department of Science and Technology (DST), Ministry of Science and Technology, Government of India and Jawaharlal Nehru Centre for an Advanced Scientific Research, India for funding under the Post-doctoral Overseas Fellowship in Nano Science and Technology (Ref. JNC/AO/A.0610.1(4) 2019-2260 dated August 19, 2019).

References

Ahemad M (2015) Phosphate-solubilizing bacteria-assisted phytoremediation of metalliferous soils: a review. 3 Biotech 5:111–121

Aktar MW, Sengupta D, Chowdhury A (2009) Impact of pesticides use in agriculture: their benefits and hazards. Interdiscip Toxicol 2:1–12

Azevedo JL, Maccheroni W Jr, Pereira JO, de Araujo WL (2000) Endophytic microorganisms: a review on insect control and recent advances on tropical plants. Electron J Biotechnol 3(1)

Baldi BG, Maher BR, Slovin JP, Choen JD (1991) Stable isotope labeling, in vivo, of d- and l-tryptophan pools in lemna gibba and the low incorporation of label into indole-3-acetic acid. Plant Physiol 95:1203–1223

Bartholdy BA, Berreck M, Haselwandter K (2001) Hydroxamate siderophore synthesis by *Phialocephala fortinii*, a typical dark septate fungal root endophyte. Biometals 14:33–42

Bashan Y, Holguin G (1998) Proposal for the division of plant growth promoting rhizobacteria into two classifications: biocontrol-PGPB (plant growth-promoting bacteria) and PGPB. Soil Biol Biochem 30:1225–1228

Basnayake WVS, Birch RG (1995) A gene from *Alcaligenes denitrificans* that confers albicidin resistance by reversible antibiotic binding. Microbiology 141:551–560

Berg G, Krechel A, Ditz M, Sikora RA, Ulrich A, Hallmann J (2005) Endophytic and ectophytic potato-associated bacterial communities differ in structure and antagonistic function against plant pathogenic fungi. FEMS Microbiol Ecol 51:215–229

Boddey RM, de Oliveira OC, Urquiaga S, Reis VM, Olivares FL, Baldani VLD, Döbereiner J (1995) Biological nitrogen fixation associated with sugar cane and rice: contributions and prospects for improvement. Plant Soil 174:195–209

Bonnet M, Camares O, Veisseire P (2000) Effects of zinc and influence of *Acremonium lolii* on growth parameters, chlorophyll *a* fluorescence and antioxidant enzyme activities of ryegrass (*Lolium perenne* L. cv Apollo). J Exp Bot 51:945–953

Bothwell TH (1995) Overview and mechanisms of iron regulation. Nutr Rev 53:237–245

Brock AK, Berger B, Mewis I, Ruppel S (2013) Impact of the PGPB *Enterobacter radicincitans* DSM 16656 on growth, glucosinolate profile, and immune responses of *Arabidopsis thaliana*. Microb Ecol 65:661–670

Brown NL, Stoyanov JV, Kidd SP, Hobman JL (2003) The MerR family of transcriptional regulators. FEMS Microbiol Rev 27:145–163

Chung JH, Song GC, Ryu CM (2016) Sweet scents from good bacteria: case studies on bacterial volatile compounds for plant growth and immunity. Plant Mol Biol 90:677–687

Compant S, Clément C, Sessitsch A (2010) Plant growth-promoting bacteria in the rhizo-and endosphere of plants: their role, colonization, mechanisms involved and prospects for utilization. Soil Biol Biochem 42:669–678

Cursino L, Mattos SV, Azevedo V, Galarza F, Bucker DH, Chartone-Souza E, Nascimento A (2000) Capacity of mercury volatilization by *mer* (from *Escherichia coli*) and *glutathione S-transferase* (from *Schistosoma mansoni*) genes cloned in *Escherichia coli*. Sci Total Environ 261:109–113

D'Alessandro M, Erb M, Ton J, Brandenburg A, Karlen D, Zopfi J, Turlings TCJ (2014) Volatiles produced by soil-borne endophytic bacteria increase plant pathogen resistance and affect tritrophic interactions. Plant Cell Environ 37:813–826

Dalton DA, Kramer S, Azios N, Fusaro S, Cahill E, Kennedy C (2004) Endophytic nitrogen fixation in dune grasses (*Ammophila arenaria* and *Elymus mollis*) from Oregon. FEMS Microbiol Ecol 49:469–479

Dandurishvili N, Toklikishvili N, Ovadis M, Eliashvili P, Giorgobiani N, Keshelava R, Tediashvili M, Vainstein A, Khmel I, Szegedi E, Chernin L (2011) Broad-range antagonistic rhizobacteria *Pseudomonas fluorescens* and *Serratia plymuthica* suppress *Agrobacterium* crown-gall tumors on tomato plants. J Appl Microbiol 110:341–352

Devi R, Kaur T, Guleria G, Rana K, Kour D, Yadav N et al (2020) Fungal secondary metabolites and their biotechnological application for human health. In: Rastegari AA, Yadav AN, Yadav N (eds) Trends of microbial biotechnology for sustainable agriculture and biomedicine systems: perspectives for human health. Elsevier, Amsterdam, pp 147–161. https://doi.org/10.1016/B978-0-12-820528-0.00010-7

Dobbelare S, Vanderleydern J, Okon Y (2003) Plant-growth promoting effects of diazotrophs in the rhizosphere. Crit Rev Plant Sci 22:107–149

Doty SL, Oakley B, Xin G, Kang JW, Singleton G, Khan Z, Vajzovic A, Staley JT (2009) Diazotrophic endophytes of native black cottonwood and willow. Symbiosis 47:23–33

Duffy BK (2001) Competition. In: Maloy OC, Murray TD (eds) Encyclopedia of plant pathology. John Wiley & Sons, Inc., New York, N.Y, pp 243–244

Ecker DJ, Loomis LD, Cass ME, Raymond KN (1988) Substituted complexes of enterobactin and synthetic analogues as probes the ferric-enterobactin receptor in *Escheichia coli*. J Am Chem Soc 110:2457–2464

Ezra D, Hess WM, Strobel GA (2004) New endophytic isolates of *Muscodor albus*, a volatile-antibiotic-producing fungus. Microbiology 150:4023–4031

Frankowski J, Lorito M, Scala F, Schmidt R, Berg G, Bahl H (2001) Purification and properties of two chitinolytic enzymes of *Serratia plymuthica* HRO-C48. Arch Microbiol 176:421–426

Giller KE, Merckx R (2003) Exploring the boundaries of N2-fixation in cereals and grasses: an hypothetical and experimental framework. Symbiosis 35:3–17

Glick BR (2005) Modulation of plant ethylene levels by the bacterial enzyme ACC deaminase. FEMS Microbiol Lett 251:1–7

Glick BR (2014) Bacteria with ACC deaminase can promote plant growth and help to feed the world. Microbiol Res J 169:30–39

Godstime OC, Enwa FO, Augustina JO, Christopher EO (2014) Mechanisms of antimicrobial actions of phytochemicals against enteric pathogens– a review. J Pharm Chem Biol Sci 2:77–85

Gouda S, Das G, Sen SK, Shin H-S, Patra JK (2016) Endophytes: a treasure house of bioactive compounds of medicinal importance. Front Microbiol 7:1538

Gray EJ, Smith DL (2005) Intracellular and extracellular PGPR: commonalities and distinctions in the plant-bacterium signaling processes. Soil Biol Biochem 37:395–412

Hallman J, Quadt-Hallman A, Mahafee WF, Kloepper JW (1997) Bacterial endophytes in agricultural crops. Can J Microbiol 43:895–914

Han Q, Wu F, Wang X, Qi H, Shi L, Ren A, Liu Q, Zhao M, Tang C (2015) The bacterial lipopeptide iturins induce *Verticillium dahliae* cell death by affecting fungal signalling pathways and mediate plant defence responses involved in pathogen associated molecular pattern-triggered immunity. Environ Microbiol 17:1166–1188

Hardoim PR, van Overbeek LS, van Elsas JD (2009) Properties of bacterial endophytes and their proposed role in plant growth. Trends Microbiol 16:463–472

Henning K, Villforth F (1940) Experimentelle untersuchungen zur frage der bacteriesymbiose in ho¨heren pflanzen und ihre beeinflussung durch 'Leitemente'. Biochem Z 305:299–309

van der Hiejden MGA, Bardgett RD, van Straalen NM (2008) The unseen majority: soil microbes as drivers of plant diversity and productivity in terrestrial ecosystems. Ecol Lett 11:296–310

Huddedar SB, Shete AM, Tilekar JN, Gore SD, Dhavale DD, Chopade BA (2002) Isolation, characterization and plasmid pUPI126-mediated indole-3-acetic acid production in *Acinetobacter* strains from rhizosphere of wheat. Appl Biochem Biotechnol 102–103:21–39

Huschka H, Jalal MAF, van der Helm D, Winkelmann G (1986) Molecular recognition of siderophores in fungi: role of iron-surrounding *N*-acyl residues and the peptide backbone during membrane transport in *Neurospora crassa*. J Bacteriol 167:1020–1024

Idriss EE, Makarewicz O, Farouk A, Rosner K, Greiner R, Bochow H, Richter T, Borriss R (2002) Extracellular phytase activity of *Bacillus amyloliquefaciens* FZB45 contributes to its plant-growth-promoting effect. Microbiology 148:2097–2109

Iniguez AL, Dong Y, Triplett EW (2004) Nitrogen fixation in wheat provided by *Klebsiella pneumoniae* 342. Mol Plant-Microbe Interact 17:1078–1085

Joseph B, Priya RM (2011) Bioactive compounds from endophytes and their potential in pharmaceutical effect: a review. Am J Biochem Mol Bio 1:291–309

Khan AL, Shahzad R, Al-Harrasi A, Lee IJ (2017) Endophytic microbes: a resource for producing extracellular enzymes. In: Maheshwari D, Annapurna K (eds) Endophytes: crop productivity and protection. Sustainable development and biodiversity, vol 16. Springer, Cham. Copyright © 2017 Springer International Publishing AG

Kobayashi DY, Palumbo JD (2000) Bacterial endophytes and their effects on plants and uses in agriculture. In: Bacon CW, White JF (eds) Microbial endophytes. Marcel Dekker, NY, pp 199–233

Koskimäki JJ, Pirttilä AM, Ihantola EL, Halonen O, Frank AC (2015) The intracellular scots pine shoot symbiont *Methylobacterium extorquens* DSM13060 aggregates around the host nucleus and encodes eukaryote-like proteins. mBio 6(2):e00039–e00015

Kour D, Rana KL, Kaur T, Singh B, Chauhan VS, Kumar A et al (2019a) Extremophiles for hydrolytic enzymes productions: biodiversity and potential biotechnological applications. In: Molina G, Gupta VK, Singh B, Gathergood N (eds) Bioprocessing for biomolecules production, pp 321–372. https://doi.org/10.1002/9781119434436.ch16

Kour D, Rana KL, Yadav N, Yadav AN, Singh J, Rastegari AA et al (2019b) Agriculturally and industrially important fungi: current developments and potential biotechnological applications. In: Yadav AN, Singh S, Mishra S, Gupta A (eds) Recent advancement in White biotechnology through Fungi, Perspective for value-added products and environments, vol 2. Springer, Cham, pp 1–64. https://doi.org/10.1007/978-3-030-14846-1_1

Kour D, Kaur T, Yadav N, Rastegari AA, Singh B, Kumar V et al (2020a) Phytases from microbes in phosphorus acquisition for plant growth promotion and soil health. In: Rastegari AA, Yadav AN, Yadav N (eds) Trends of microbial biotechnology for sustainable agriculture and biomedicine systems: diversity and functional perspectives. Elsevier, Amsterdam, pp 157–176. https://doi.org/10.1016/B978-0-12-820526-6.00011-7

Kour D, Rana KL, Yadav AN, Yadav N, Kumar M, Kumar V et al (2020b) Microbial biofertilizers: bioresources and eco-friendly technologies for agricultural and environmental sustainability. Biocatal Agric Biotechnol 23:101487. https://doi.org/10.1016/j.bcab.2019.101487

Kuklinsky-Sobral J, Araujo WL, Mendes R, Geraldi IO, Pizzirani-Kleiner AA, Azevedo JL (2004) Isolation and characterization of soybean-associated bacteria and their potential for plant growth promotion. Environ Microbiol 6:1244–1251

Kusari P, Kusari S, Lamshöft M, Sezgin S, Spiteller M, Kayser O (2014) Quorum quenching is an antivirulence strategy employed by endophytic bacteria. Appl Microbiol Biotechnol 98:7173–7183

Latha P, Karthikeyan M, Rajeswari E (2019) Endophytic bacteria: prospects and applications for the plant disease management. In: Ansari R, Mahmood I (eds) Plant health under biotic stress. Springer, Singapore

Lee JC, Strobel GA, Lobkovsky E, Clardy J (1996) Torreyanic acid: a selectively cytotoxic quinone dimer from the endophytic fungus *Pestalotiopsis microspora*. J Org Chem 61:3232–3233

Leinhos V (1994) Effects of pH and glucose on auxin production by phosphate-solubilizing rhizobacteria *in vitro*. Microbiol Res 194:135–138

Leo VV, Passari AK, Joshi JB, Mishra VK, Uthandi S, Ramesh N, Gupta VK, Saikia R, Sonawane VC, Singh BP (2016) A novel triculture system (CC3) for simultaneous enzyme production and hydrolysis of common grasses through submerged fermentation. Front Microbiol 7. https://doi.org/10.3389/fmicb.2016.00447

Lim HS, Kim YS, Kim SD (1991) *Pseudomonas stutzeri* YPL-1 genetic transformation and antifungal mechanism against *Fusarium solani*, an agent of plant root rot. Appl Environ Microbiol 57:510–516

Liu H, Carvalhais LC, Crawford M, Singh E, Dennis PG, Pieterse CMJ, Schenk PM (2017) Inner plant values: diversity, colonization and benefits from endophytic bacteria. Front Microbiol 8:2552

Loper JE, Henkels MD (1999) Utilization of heterologous siderophores enhances levels of iron available to *Pseudomonas putida* in the rhizosphere. Appl Environ Microbiol 65:5357–5363

Ma Y, Rajkumar M, Luo YM, Freitas H (2011) Inoculation of endophytic bacteria on host and non-host plants e effects on plant growth and Ni uptake. J Hazard Mater 195:230–237

Ma Y, Rajkumar M, Zhang C, Freitas H (2016) Beneficial role of bacterial endophytes in heavy metal phytoremediation. J Environ Manag 174:14–25

de Matos NE, Vinagre F, Masuda HP, Vargas C, de Pádua VLM, da Silva FR, dos Santos RV, Baldani JI, Gomes Ferreira PC, Hemerley AS (2001) Expression of sugarcane genes induced by inoculation with *Gluconacetobacter diazotrophicus* and *Herbaspirillum rubrisubalbicans*. Genet Mol Biol 24:199–206

Miller MB, Bassler BL (2001) Quorum sensing in bacteria. Annu Rev Microbiol 55:165–199

Mondal S, Halder SK, Yadav AN, Mondal KC (2020) Microbial consortium with multifunctional plant growth promoting attributes: future perspective in agriculture. In: Yadav AN, Rastegari AA, Yadav N, Kour D (eds) Advances in plant microbiome and sustainable agriculture, Functional annotation and future challenges, vol 2. Springer, Singapore, pp 219–254. https://doi.org/10.1007/978-981-15-3204-7_10

Newman KL, Chatterjee S, Ho KA, Lindow SE (2008) Virulence of plant pathogenic bacteria attenuated by degradation of fatty acid cell-to-cell signaling factors. Mol Plant-Microbe Interact 21:326–334

O'Sullivan DJ, O'Gara F (1992) Traits of fluorescent *Pseudomonas* spp. involved in suppression of plant root pathogens. Microbiol Rev 56:662–676

Olivares F, James E, Baldani JI, Döbereiner J (1997) Infection of mottled stripe disease-susceptible and resistant sugar cane varieties by the endophytic diazotroph *Herbaspirillum*. New Phytol 135:723–737

Ordentlich A, Elad Y, Chet I (1988) The role of chitinase of *Serratia marcescens* in biocontrol of *Sclerotium rolfsii*. Phytopathology 78:84–88

Perotti R (1926) On the limits of biological inquiry on soil science. Proc Int Soc Soil Sci 2:146–161

Pierik R, Tholen D, Poorter H, Visser EJW, Voesenek LACJ (2006) The Janus factor of ethylene: growth inhibition and stimulation. Trends Plant Sci 11:176–183

Puri SC, Verma V, Amna T, Qazi GN, Spiteller MJ (2005) An endophytic fungus from *Nothapodytes foetida* that produces camptothecin. J Nat Prod 68:1717–1719

Puri A, Padda KP, Chanway CP (2016) Seedling growth promotion and nitrogen fixation by a bacterial endophyte *Paenibacillus polymyxa* P2b-2R and its GFP derivative in corn in a long-term trial. Symbiosis 69:123–129

Rajkumar M, Ae N, Freitas H (2009) Endophytic bacteria and their potential to enhance heavy metal phytoextraction. Chemosphere 77:153–160

Rajkumar M, Prasad MNV, Sandhya S, Freitas H (2013) Climate change driven plant-metal-microbe interactions. Environ Int 53:74–86

Rana KL, Kour D, Sheikh I, Dhiman A, Yadav N, Yadav AN et al (2019a) Endophytic fungi: biodiversity, ecological significance and potential industrial applications. In: Yadav AN, Mishra S, Singh S, Gupta A (eds) Recent advancement in White biotechnology through Fungi, Diversity and enzymes perspectives, vol 1. Springer, Switzerland, pp 1–62

Rana KL, Kour D, Sheikh I, Yadav N, Yadav AN, Kumar V et al (2019b) Biodiversity of endophytic fungi from diverse niches and their biotechnological applications. In: Singh BP (ed) Advances in endophytic fungal research: present status and future challenges. Springer, Cham, pp 105–144. https://doi.org/10.1007/978-3-030-03589-1_6

Rana KL, Kour D, Yadav AN (2019c) Endophytic microbiomes: biodiversity, ecological significance and biotechnological applications. Res J Biotechnol 14:142–162

Rana KL, Kour D, Kaur T, Devi R, Negi C, Yadav AN et al (2020a) Endophytic fungi from medicinal plants: biodiversity and biotechnological applications. In: Kumar A, EKR (eds) Microbial endophytes. Woodhead Publishing, Cambridge, MA, pp 273–305. https://doi.org/10.1016/B978-0-12-819654-0.00011-9

Rana KL, Kour D, Kaur T, Sheikh I, Yadav AN, Kumar V et al (2020b) Endophytic microbes from diverse wheat genotypes and their potential biotechnological applications in plant growth promotion and nutrient uptake. Proc Natl Acad Sci India B. https://doi.org/10.1007/s40011-020-01168-0

Rana KL, Kour D, Yadav N, Yadav AN (2020c) Endophytic microbes in nanotechnology: current development, and potential biotechnology applications. In: Kumar A, Singh VK (eds) Microbial endophytes. Woodhead Publishing, USA, pp 231–262. https://doi.org/10.1016/B978-0-12-818734-0.00010-3
Rao NS (1986) The rhizosphere. Oxford and IBH, New Delhi, pp 50–80
Rastegari AA, Yadav AN, Yadav N (2020a) New and future developments in microbial biotechnology and bioengineering: Trends of microbial biotechnology for sustainable agriculture and biomedicine systems: diversity and functional perspectives. Elsevier, Amsterdam
Rastegari AA, Yadav AN, Yadav N (2020b) New and future developments in microbial biotechnology and bioengineering: Trends of microbial biotechnology for sustainable agriculture and biomedicine systems: perspectives for human health. Elsevier, Amsterdam
Reinhold-Hurek B, Hurek T, Gillis M, Hoste B, Vancanneyt M, Kersters K, De-Ley J (1993) *Azoarcus* gen. Nov., nitrogen-fixing proteobacteria associated with roots of Kallar grass (*Leptochloa fusca* (L.) Kunth), and description of two species, *Azoarcus indigens* sp. nov. and *Azoarcus communis* sp. nov. Int J Syst Evol Microbiol 43:574–584
Reiter B, Bürgmann H, Burg K, Sessitsch A (2003) Endophytic nifH gene diversity in African sweet potato. Can J Microbiol 49:549–555
Riggs PJ, Chelius MK, Iniguez AL, Kaeppler SM, Triplett EW (2001) Enhanced maize productivity by inoculation with diazotrophic bacteria. Aust J Plant Physiol 28:829–836
Rungin S, Indananda C, Suttiviriya P, Kruasuwan W, Jaemsaeng R, Thamchaipenet A (2012) Plant growth enhancing effects by a siderophore-producing endophytic streptomycete isolated from a Thai jasmine rice plant (*Oryza sativa* L. cv. KDML105). Antonie Van Leeuwenhoek 102:463–472
Saleem M, Arshad M, Hussain S, Bhatti AS (2007) Perspective of plant growth promoting rhizobacteria (PGPR) containing ACC deaminase in stress agriculture. J Ind Microbiol Biotechnol 34:635–648
Santoyo G, Moreno-Hagelsieb G, Orozco-Mosqueda Ma del C, Glick BR (2016) Plant growth-promoting bacterial endophytes. Microbiol Res J 183:92–99
Sessitsch A, Reiter B, Berg G (2004) Endophytic bacterial communities of field grown potato plants and their plant-growth-promoting and antagonistic abilities. Can J Microbiol 50:239–249
Sessitsch A, Hardoim P, Döring J, Weilharter A, Krause A, Woyke T, Mitter B, Hauberg-Lotte L, Friedrich F, Rahalkar M, Hurek T, Sarkar A, Bodrossy L, van Overbeek L, Brar D, van Elsas JD, Reinhold-Hurek B (2012) Functional characteristics of an endophyte community colonizing rice roots as revealed by metagenomic analysis. Mol Plant-Microbe Interact 25:28–36
Sharaff MS, Subrahmanyam G, Kumar A, Yadav AN (2020) Mechanistic understanding of root-microbiome interaction for sustainable agriculture in polluted soils. In: Rastegari AA, Yadav AN, Yadav N (eds) Trends of microbial biotechnology for sustainable agriculture and biomedicine systems: diversity and functional perspectives. Elsevier, Amsterdam, pp 61–84. https://doi.org/10.1016/B978-0-12-820526-6.00005-1
Sheibani-Tezerji R, Rattei T, Sessitsch A, Trognitz F, Mitter B (2015) Transcriptome profiling of the endophyte *Burkholderia phytofirmans* PsJN indicates sensing of the plant environment and drought stress. mBio 6:e00621–e00615
Sheoran N, Valiya Nadakkakath A, Munjal V, Kundu A, Subaharan K, Venugopal V, Rajamma S, Eapen SJ, Kumar A (2015) Genetic analysis of plant endophytic *Pseudomonas putida* BP25 and chemo-profiling of its antimicrobial volatile organic compounds. Microbiol Res 173:66–78
Shin M, Shim J, You Y, Myung H, Bang KS, Cho M, Kamala-Kannan S, Oh BT (2012) Characterization of lead resistant endophytic *Bacillus* sp. MN3-4 and its potential for promoting lead accumulation in metal hyperaccumulator *Alnus firma*. J Hazard Mater 199-200:314–320
Singh A, Kumar R, Yadav AN, Mishra S, Sachan S, Sachan SG (2020a) Tiny microbes, big yields: microorganisms for enhancing food crop production sustainable development. In: Rastegari AA, Yadav AN, Yadav N (eds) Trends of microbial biotechnology for sustainable agriculture and biomedicine systems: diversity and functional perspectives. Elsevier, Amsterdam, pp 1–15. https://doi.org/10.1016/B978-0-12-820526-6.00001-4

Singh B, Boukhris I, Pragya KV, Yadav AN, Farhat-Khemakhem A et al (2020b) Contribution of microbial phytases to the improvement of plant growth and nutrition: a review. Pedosphere 30:295–313. https://doi.org/10.1016/S1002-0160(20)60010-8
Stierle A, Strobel GA, Stierle D (1993) Taxol and taxane production by *Taxomyces andreanae*, an endophytic fungus of Pacific yew. Science 260:214–216
Stintzi A, Barnes C, Xu J, Raymond KN (2000) Microbial iron transport via a siderophore shuttle: a membrane ion transport paradigm. Proc Natl Acad of Sci USA 97:10691–10696
Stone JK, Bacon CW, White JF (2000) In: Bacon CW, White JF Jr (eds) An overview of endophytic microbes: Endophytism defined. M. Dekker, Inc, New York, pp 3–5
Straub D, Rothballer M, Hartmann A, Ludewig U (2013) The genome of the endophytic bacterium *H. frisingense* GSF30(T) identifies diverse strategies in the *Herbaspirillum* genus to interact with plants. Front Microbiol 4:168
Sturz AV, Christie BR, Nowak J (2000) Bacterial endophytes: potential role in developing sustainable systems of crop production. Crit Rev Plant Sci 19:1–30
Su F, Jacquard C, Villaume S, Michel J, Rabenoelina F, Clément C, Barka EA, Dhondt-Cordelier S, Vaillant-Gaveau N (2015) *Burkholderia phytofirmans* PsJN reduces impact of freezing temperatures on photosynthesis in *Arabidopsis thaliana*. Front Plant Sci 6:810
Subramanian P, Mageswari A, Kim K, Lee Y, Sa T (2015) Psychrotolerant endophytic *Pseudomonas* sp. strains OB155 and OS261 induced chilling resistance in tomato plants (*Solanum lycopersicum* mill.) by activation of their antioxidant capacity. Mol Plant-Microbe Interact 28:1073–1081
Suman A, Yadav AN, Verma P (2016) Endophytic microbes in crops: diversity and beneficial impact for sustainable agriculture. In: Singh DP, Singh HB, Prabha R (eds) Microbial inoculants in sustainable agricultural productivity, Research perspectives, vol 1. Springer, New Delhi, pp 117–143. https://doi.org/10.1007/978-81-322-2647-5_7
Thakur N, Kaur S, Tomar P, Thakur S, Yadav AN (2020) Microbial biopesticides: current status and advancement for sustainable agriculture and environment. In: Rastegari AA, Yadav AN, Yadav N (eds) Trends of microbial biotechnology for sustainable agriculture and biomedicine systems: diversity and functional perspectives. Elsevier, Amsterdam, pp 243–282. https://doi.org/10.1016/B978-0-12-820526-6.00016-6
Thrall PH, Hochberg ME, Burdon JJ, Bever JD (2007) Coevolution of symbiotic mutualists and parasites in a community context. Trends Ecol and Evol 22:120–126
Tiwari P, Bajpai M, Singh LK, Mishra S, Yadav AN (2020) Phytohormones producing fungal communities: metabolic engineering for abiotic stress tolerance in crops. In: Yadav AN, Mishra S, Kour D, Yadav N, Kumar A (eds) Agriculturally important Fungi for sustainable agriculture, Perspective for diversity and crop productivity, vol 1. Springer, Cham, pp 1–25. https://doi.org/10.1007/978-3-030-45971-0_8
Toyoda H, Utsumi R (1991) Method for the prevention of *Fusarium* diseases and microorganisms used for the same. U.S. patent 4988586
Traving SJ, Thygesen UH, Riemann L, Stedmon CA (2015) A model of extracellular enzymes in free-living microbes: which strategy pays off? Appl Environ Microbiol 81:7385–7393
Vandenkoornhuyse P, Mahe S, Ineson P, Staddon P, Ostle N, Cliquet J-B, Francez A-J, Fitter AH, Young JPW (2007) Active root-inhabiting microbes identified by rapid incorporation of plant-derived carbon into RNA. Proc Natl Acad of Sci USA 104:16970–16975
Vijayalakshmi R, Kairunnisa K, Sivvaswamy SN, Dharan SS, Natarajan S (2016) Enzyme production and antimicrobial activity of endophytic bacteria isolated from medicinal plants. Indian J Sci Technol 9(14)
Vorholt JA (2012) Microbial life in the phyllosphere. Nat Rev Microbiol 10:828–840
Wang WX, Barak T, Vinocur B, Shoseyov O, Altman A (2003) Abiotic resistance and chaperones: possible physiological role of SP1, a stable and stabilizing protein from *Populus*. In: Vasil IK (ed) Plant Biotechnology 2000 and Beyond. Dordrecht, Kluwer, pp 439–443
Webber J (1981) A natural control of Dutch elm disease. Nature (London) 292:449–451
de Werra P, Pechy-Tarr M, Keel C, Maurhofer M (2009) Role of gluconic acid production in the regulation of biocontrol traits of *Pseudomonas fluorescens* CHA0. Appl Environ Microb 75:4162–4174

Whipps JM (2001) Microbial interactions and biocontrol in the rhizosphere. J Exp Bot 52:487–511
Winkelmann G (2002) Microbial siderophore-mediated transport. Biochem Soc Trans 30:691–696
Yadav AN, Kour D, Rana KL, Yadav N, Singh B, Chauhan VS et al (2019) Metabolic engineering to synthetic biology of secondary metabolites production. In: Gupta VK, Pandey A (eds) New and future developments in microbial biotechnology and bioengineering. Elsevier, Amsterdam, pp 279–320. https://doi.org/10.1016/B978-0-444-63504-4.00020-7
Yadav AN, Singh J, Rastegari AA, Yadav N (2020) Plant microbiomes for sustainable agriculture. Springer, Cham
Zachow C, Jahanshah G, de Bruijn I, Song C, Ianni F, Pataj Z, Gerhardt H, Pianet I, Lammerhofer M, Berg G, Gross H, Raaijmakers JM (2015) The novel lipopeptide poaeamide of the endophyte *Pseudomonas poae* RE*1-1-14 is involved in pathogen suppression and root colonization. Mol Plant-Microbe Interact 28:800–810
Zhang L, Birch RG (1997) The gene for albicidin detoxification from *Pantoea dispersa* encodes an esterase and attenuates pathogenicity of *Xanthomonas albilineans* to sugarcane. Proc Natl Acad of Sci USA 94:9984–9989

Mycorrhiza: A Sustainable Option for Better Crop Production

12

Sanjeev Kumar, Saurabh Saxena, and Samiksha

Abstract

Agriculturists and farmers have been under pressure to fulfill the demands of the increasing population. Although the use of inorganic fertilizers benefited farmers by providing good and fast yields, but become progressively worse on the quality of soil by decreasing biomass and microbial activity. Therefore, now pressure to increase yield along with making sustainable progress has led to the usage of mycorrhizal fungi as biofertilizers. These fungi have been found to provide numerous and diverse benefits to soil, plants, and ecosystem by improving soil quality, concentrating nutrients in plants, providing resistance against drought and diseases, and helping in nutrient cycling. The AMF is known to be a very difficult fungus because its culture is difficult in vitro conditions and therefore proper procedures are needed to be followed and precautions are needed to be taken to get the desired pure yield of the fungus. The fungus if studied and researched properly can open many doors to new developments in the field of science and agriculture.

Keywords

Arbuscular mycorrhiza · Bioremediation · Sustainable agriculture · Zygomycetes

S. Kumar
Department of Genetics and Plant Breeding, School of Agriculture, Lovely Professional University, Jalandhar, India

S. Saxena
Department of Medical Laboratory Sciences, Lovely Faculty of Applied Medical Sciences (LFAMS), Lovely Professional University, Jalandhar, Punjab, India

Samiksha (✉)
Department of Biotechnology, School of Biosciences, Lovely Professional University, Jalandhar, India

A. N. Yadav et al. (eds.), *Current Trends in Microbial Biotechnology for Sustainable Agriculture*, Environmental and Microbial Biotechnology,
https://doi.org/10.1007/978-981-15-6949-4_12

12.1 Introduction

The arbuscular mycorrhizal (AM) symbiosis is known over more than 400 million years ago. It is association between higher plants and soil fungi belonging to phylum Glomeromycota. Since both partners coevolved at the same time, with more than 80% of all terrestrial plants belonging to diverse ecological niche associated with AMF. The main economic importance of these fungal symbionts is estimated for adding phosphorus in phosphorus in P-deficient soil to sustain the productivity of soil and also the same time reduced the impact of chemical fertilizers. A global land area of 9.2 billion hectares encompassing six biomes with terrestrial plants, relative proportion of plants likely to form AM symbioses in each biome. Native AM fungi drastically reduced the burden of inorganic phosphoric fertilizers. About $549 billion of P input would be needed to substitute for native AM fungi (Morton 1988). This cost does not include the account of the mortality of many trees and other plant species. Therefore, many times continuous use of inorganic fertilizer would be ineffective in regard to plant growth and development (Yadav et al. 2019a; Yadav et al. 2020c). Moreover, apart from P-uptake, these fungi are also play important role in many key processes of nutrient cycling, soil conservation, and plant health.

The benefits imparted to the plant by mycorrhizal association range from adsorbing toxic elements from the soil such as heavy metals, tolerance for biotic and abiotic stress soil, bioremediation, soil restoration, establishment of green cover, disease resistance, etc. These microorganisms have a role in sustaining the ecosystem by enriching the soil and providing nutrition to the plant and in turn getting carbon from the plant. This, in turn, reduces the dependence if not eliminates external chemical inputs and makes the utilization of the soil nutrient highly efficient. Thus, AM fungi are considered as promising biofertilizers for sustainable crop productivity and mitigation of problems by marginal farmers of developing countries (Kour et al. 2020b; Rastegari et al. 2020). These fungi have emerged out as potential tools for agriculture, forestry, and bioremediation and wasteland reclamation. However, to exploit them for various purposes, the functional characterization of these fungi is mandatory.

12.2 Role and Limitations of Inorganic Chemicals in Environmental Sustainability

To meet the demand of feeding ever-increasing population and to save the expense of over costly manure practices, farmers switched to the use of fertilizers which were easily available and less expensive (Hera 1996). These inorganic fertilizers have Nitrogen, Phosphorus, and Potassium as their main components and their usage can lead to modest yet immediate increases in the yield of crops if used in an appropriate amount (Larson and Frisvold 1996; Yadav et al. 2019b). Although farmyard manure (FYM) provides all the essential nutrients like N, P, K, Ca, Mg, S and helps in improvement of physical, chemical, and biological properties of soil, the dependency of agriculture on inorganic chemical fertilizers and pesticides increased (Tadesse et al. 2013).

The usage of chemical fertilizers made agricultural systems efficient and able to produce quality products (Savci 2012a, b). Inorganic chemical fertilizers benefit agriculture in many ways which include agricultural fields when deficient in various inorganic nutrients, are provided with the chemical fertilizers to fulfill the nutritional demand of the soil which further provides the nutrients to the plants. The most commonly used fertilizers are N and P fertilizers, which are applied to fields by different methods like fertilizer placement to ensure nutrient availability to the plants. Studies have suggested that fertilizers when combined with organic manures help in increasing the yield in agricultural fields. The subsurface placement of fertilizers has yielded many benefits to farmers, which include root growth stimulation, less loss of nutrients to the environment, availability of high levels of nutrients to the plants as they are placed close to the plant roots, increase in yield of crops (Nkebiwe et al. 2016). Due to increasing global population, the worldwide food production needs to be increased by 2050 but at the same time, the dependency on conventional fertilizer practices needs to be reduced as it has been found that continuous and long-term use of these chemicals leads to degradation of land, deterioration in soil health, and has become a threat to human and animal health. There are many limitations to using these inorganic chemicals in the agricultural fields (Igiehon and Babalola 2017).

The nitrates contained in fertilizers, which are not absorbed by the plants leach into the soil through rainwater and thus reach groundwater. Groundwater, therefore, when used for consumption by animals and humans, causes deteriorating health effects. Long-term use of chemicals affects soil quality drastically. These chemicals deteriorate soil quality as some of the inorganic fertilizers contain sodium and potassium which impact soil negatively by changing the pH. In addition, the fertilizers containing heavy metals lead to accumulation of these heavy metals into the plants and fruits which affect when fed to animals and humans. Chemical fertilizers when applied in large amounts cause air pollution by emissions of various oxides nitrogen like NO, N_2O, NO_2. These oxides not only cause air pollution but also act as greenhouse gases, thus affecting the environment. The application of chemicals containing urea leads to evaporation of ammonia which after getting oxidized turns into nitric acid and pours down with rain as acid rain (Savci 2012a, b). Thus, intensive use of chemicals undoubtedly provides good yield but at the same time, it deteriorates the environment and human health, whereas, the use of organic manures can improve characteristics of soil as well as crop production. Therefore, to enhance the recovery of nutrients and promote good plant growth and yield, the focus is now on usage of both organic and inorganic fertilizers in an appropriate combination (Mahmood et al. 2017).

12.3 Types and Functions of AM Fungal Biodiversity in Rhizospheric Soil

Rhizosphere is the zone of soil surrounding root where the microbial population is stimulated by root activities. Out of many soil organisms present in soil microbiota, fungi are a very important component of the soil and play a very beneficial role in soil by decomposing organic matter and promoting element release by mineralization (Chandrashekar et al. 2014), however, fungal diversity in the rhizosphere also essential for biogeochemical cycles, plant growth, and disease development and control (Yadav et al. 2020a, b). The rhizosphere fungal communities are determined by many factors which include the type of plant, root exudate, and organic carbon content (Wang et al. 2017; Kour et al. 2020a). The studies have suggested that on an overall, millions of species of fungi make a rich diversity but, the diversity found in soil is much lesser (Hawksworth and Lücking 2017). Many studies suggested a deal with diverse roles of AMF fungi in different ecosystems; however, very little is known about how agricultural practices create selection pressure to change microbial diversity and its function (Shennan 2007; Toljander et al. 2008). It is also very little knowledge about the induction of microbial diversity and its role and function in a diverse sustainable ecosystem (Martini et al. 2004; Saxena et al. 2016; Verma et al. 2019).

The fungal diversity can be found out by many methods including checking for the fruiting bodies or the culture obtained from soil samples. But these methods are not fully reliable because organisms existing in the mycelial form are not easily detectable and therefore cannot give the true measure of diversity. To get an accurate and reliable measurement and calculation, various molecular techniques like polymerase chain reaction (PCR) detection, coupled with single-strand conformation polymorphisms (SSCP) or denaturing gradient gel electrophoresis (DGGE) have been developed (Bridge and Spooner 2001). The diversity of fungal species using molecular techniques when estimated was found to be 3000 fungal species from a 400 Ha site, thus giving a much reliable and accurate data (Fierer et al. 2007). Studies done by Wang et al. (2017) reported the abundance of various fungal phyla across the soil samples which was Ascomycota (average 68.7%), Zygomycota (average 13.3%), and Basidiomycota (average 4.1%) (Wang et al. 2017).

12.4 Types of Mycorrhiza and its Role in Functional Diversity

Life is considered to have emerged on land during the pre-Cambrian period and is supposed to have been it colonized by microorganisms, which are phototrophic in nature and were probably prokaryotic. Although the recent evidences have suggested that land plants might have emerged during Ordovician period but it was earlier believed that the establishment of land plants was in the late Silurian period. It is believed that the vascular plants have arisen from the green aquatic algae which during the course of evolution became semiaquatic and on further evolution, fully terrestrial to become first land plants. After the emergence of semiaquatic algae, it began to invade land around 490 million years ago (mya) and faced very harsh environmental conditions.

It was suggested by Pirozynski and Malloch (1975) that the mutualistic relationship between fungi and the plants is of great significance and important not only in land colonization by plants and for improvement of nutritional status but also to help plants to sustain themselves in harsh environmental conditions. The whole plant diversity that is found in the present in various ecosystems and environments, in tropical rainforests or temperate habitats is all due to these associations between plants and fungi. The oldest fossil evidence of mycorrhiza was recorded in the early Devonian period, around 400 million years ago (Mya) in the form of fossil arbuscules. It was observed by Simon et al. (1993) that the origin of arbuscular mycorrhizal fungi between 462 and 363 Mya, and that too around the Ordovician, Silurian, and Devonian period. These dates easily place them at the time of land plant emergence.

The evidences revealed by fossil and molecular studies have suggested that AM fungi have been forming symbiotic colonies in the terrestrial habitats since ancient times. Redecker et al. (2000) discovered some spores and hypha which belonged to the glomalean fungi. The rocks from which those spores and hypha had been found are about 460 million years old and are of the Ordovician period which is believed to have liverwort-like plants in dominance. This suggested that the fungi have been in symbiosis with vascular land plants. There are hundreds of different types of mycorrhizal associations which involve different plants and fungal species. It has been found that around 80% of plant species and about 92% families of plants (Wang and Qiu 2006) form symbiotic association with Arbuscular mycorrhizal fungus belonging to the phylum Glomeromycota.

On the basis of fungal hyphae location when the hyphae enter the root tissues of the plant, the mycorrhiza has been classified into two types: ectotrophic mycorrhiza and endotrophic mycorrhiza, where the word *ecto* means outside the root and the word *endo* means inside the root. Ectomycorrhiza is known to be the most advanced symbiotic association found between higher plants and fungi, including about 3% of seed plants and including the majority of forest trees.

12.4.1 Endomycorrhizas

Endomycorrhizas are the type of associations in which the fungal structure penetrates the host root. The root comprising three major and two minor groups are listed in Table 12.1. The only plant families which are found to be non-mycorrhizal are Brassicaceae (e.g., cabbage, *Arabidopsis*), Caryophyllaceae (e.g., carnation), and Chenopodiaceae (e.g., spinach). The only plant species unable to grow at all, in the absence of mycorrhiza, are orchids and mycoheterotrophic. The mycorrhizal symbiosis enhances availability of soluble phosphorus from through network of mycelium and ultimately overall nutritional status of plants (Smith and Gianinazzi-Pearson 1988). Their importance and significance have been studied mainly in low phosphorus and marginal areas but these fungi also play a significant role in soil containing

Table 12.1 Summary of main characters of seven types of mycorrhiza

Eeature	Ectomycorrhiza		Endomycorrhiza				
Mycorrhiza Taxa	Ecto	Ectoendo	Arbuscular mycorrhiza	Ericoid Mycorrhiza	Orchid	Arbutoid	Mono-tropoid
Plant involved	Perennial, typically woody tree		Taxonomically diverse, many species	Ericales	Orchidales	Gyminoangio	Gyminoangio
Fungus involved	Basidomycota, Ascomycota	Basido-Asco (Glomeromycota)	Glomeromycota	Ascomycota	Basidomycota	Basidomycota	Basidomycota
Morphology							
Hartig net	+	+	–	–	–	+	+
Intracellular hypha in epidermis	–	–	–	+	–	–	–
Coil hypha in cortical cells	–	–	–	–	+	–	–
Sheath	+	±	–	–	–	±	+
Fungi septate	+	+	–	+	+	+	+
Fungi aseptate	–	–	+	–	–	–	–
Functionally							
P acquisition	Important	Important	Important	Important	Dominant	Important	Important
Pathogen protection	Important	Important	Important	Unknown	Unknown	Unknown	Unknown

Source: Modified from Smith and Read (2008)

high phosphorus conditions. In high phosphorus conditions of soil, the availability of soluble P is very low.

Therefore, very little amount of available P available for growth and development of plants. Moreover, the major amount of P-fertilizer applied in soil not readily available for plant growth and development on the other hand fixed in the soil as a nonavailable form. However, AM fungi have the ability to solubilize phosphorus and are one of the solutions to this problem. In specific circumstances when the amount on inorganically phosphorous is very high, much phosphate has been fixated and not easily available for plants. Fertilization with inorganic in nature is of no use in those circumstances. Also, rock phosphate amendments are not wise in this case. This because there is enough of this phosphate in the soil already, but it won't move. AM fungi can free this fixed phosphate and bring it into the roots of the plants. AM fungi decrease the need for fertilizers by contributing to the satisfaction of a crop plant P demand at non-inhibitive levels of P supply (Koide 1991). Thus, the potential for AM utilization in P nutrition is present but requires testing and evaluation of the supply and demand relationships of the symbiosis.

Although it is a common belief/misbelief that AM fungi do not perform in high phosphorus conditions, the interaction between AM fungi and P fertilization within the context of sustainable agriculture is complex and needs more understanding. On the one hand, high levels of soil P, although transient (Bolan 1991), maybe deleterious to some AM fungi (Abbott et al. 1984). On the other hand, some species of AM fungi are able to colonize roots under high P regimes or in fertile soils (Young et al. 1985), fertilization may eventually result in the selection of AM fungal tolerant of high P content, while others may be eliminated. These fungal isolates are of importance in conditions especially temperate where high P and high organic matter are prevailing. Currently, many AM species/isolate conservation bank has been routinely and extensively collected of AM fungal isolates including from different agroclimatic zone. These further apply in phosphorus-deficient soil. Many AM fungi collected from high P conditions and thus can provide a good source for selecting AM fungal isolates tolerant of high P content.

The association of mycorrhiza with plants is ancient and its occurrence can be observed in almost all terrestrial ecosystems (Johnson et al. 2006). During the formation of such associations, the host root gets modified due to the infection caused by mycorrhizal fungi and thus there is an establishment of the intimate relationship between the host root and fungus (Gerdemann 1968). These fungi can live around root epidermal cells, on the root surface or inside the plant root cortex. There are four major types of Mycorrhizal fungi: arbuscular mycorrhizal fungi (AMF), ectomycorrhiza (EM), and ericoid mycorrhiza. Mycorrhizal fungi have the potential to improve the growth of plants, provide resistance to pests, influence soil stability, and carbon storage and nutrient cycling (Johnson et al. 2006).

12.4.2 Arbuscular Mycorrhizal Fungi

AMF is an inhabitant of belowground systems and is distributed globally in abundance (Munkvold et al. 2004). It belongs to phylum Glomeromycota forms a symbiosis with approximately 80% of plant species. AMF is known to be an important biotic component associated with plants as it is considered that their absence can lead to a reduction in the working efficiency of the ecosystem (Berruti et al. 2016).

12.4.3 Ectomycorrhiza

Ectomycorrhizal fungi are biotrophs that are known to play an important role in the process of organic nitrogen mobilization. In terrestrial ecosystems, these play a significant role in the cycling of nutrients like N and C. The rapid turnover of this fungus helps in good functioning of the ecosystem as the turnover provides good ectomycorrhizal biomass which adds up to soil organic matter (SOM) (Fernandez et al. 2016). These fungi using oxidative mechanisms can convert organic matter into SOM. The ECM fungi have genes that encode for enzymes responsible for the degradation of lignocellulose and the production of hydrogen peroxide (Shah et al. 2016).

12.4.4 Ericoid Mycorrhiza

The ericoid mycorrhizal fungi have the potential to retrieve N and P from the litter of plants found in forests and are considered good decomposers (Lindahl and Tunlid 2015). The habitats of this fungus mainly include soils which are acidic in nature and which are having high amounts of recalcitrant phenolic compounds. These fungi show their main contribution in mobilizing the nutrients from complex organic matter to host plants (Martino et al. 2018). The increasing awareness among people has led scientists and farmers to find out new ways to fertilize agricultural fields and thus Mycorrhizal Fungi can emerge as new biofertilizers. Usage and inoculation of Mycorrhizal fungi with plants can help in achieving the overall fitness of plants as it helps plants by providing many benefits. The mycorrhizal fungi are diverse in their functions and services that they provide to the plants, soils, and ecosystem. These functions include.

12.5 Effect of Organic and Inorganic Fertilizer and its Role in AM Diversity

Although application of fertilizers and manure to the agricultural soil influences the nutrition, pH, amount of humic acid and organic substances, soil aggregation, microbial diversity, and many other important aspects of the soil, the response of AMF community to the fertilization depends upon the amount or the doze of fertilizers provided to the soils. Mycorrhizal fungi being symbiotic remain in association with plants and play a vital role in mineral mobilization. But the presence of various sources of nutrients in soil also affects the communities and the diversity of mycorrhizal fungi (Qin et al. 2015). The experiments performed by Hassan et al. (2013) were focused on observing the effect of organic and inorganic fertilizer applications on the mycorrhizal communities. Sunflower plants were grown for 12 years and the field was fertilized with organic or inorganic N fertilizers. They observed a significant difference in the mycorrhizal diversity after these treatments. The soil having organic inputs of farmyard manure, sewage sludge, etc. was observed to have species such as *Rhizophagus intraradices* whereas soils treated with inorganic mineral fertilizers were found to have *Claroideoglomus* in more amount. Not only this, other studies done, Chen et al., revealed that a long-term application of inorganic N fertilizers, resulted in decrease in the *Glomeromycota* abundance due to which the species richness and diversity of AMF was influenced significantly. However, on the other hand, application of P fertilizers did not result in significant changes in the structure of mycorrhizal community but a significant reduction was seen in rate of colonization of mycorrhiza, arbuscule colonization, and density of hyphal length (Chen et al. 2016).

12.6 AMF in Sustainable Crop Production

The AMF is known to improve the composition and yield of the plants by supplying nutrients to the plants and providing photosynthate products by increasing the photosynthetic activity of the plants. Experiments and studies were performed by inoculating mycorrhiza with tomato plants and it was found out that the nutritional quality of tomatoes was increased. The effects of these fungi over plants are considered ameliorative as the nutrients tend to get more concentrated in the fruits formed on plants which are in symbiosis with these fungi. Not only this, the fungi are being considered as part of the "Second Green Revolution" as these can play their role in supplementing nutritive food and thus alleviating malnutrition (Hart et al. 2015). Drought and salinity are two of the major constraints in the field of agriculture which lead to huge losses to farmers each year (Kour et al. 2019; Rana et al. 2019). The mycorrhiza has the ability to maintain the hydration status of a plant and thus can avoid drought stress. Role and application of AM fungi in stress condition are mentioned in Fig. 12.1. Mycorrhiza helps in improving the osmotic system of the plant which in turn helps in maintaining the hydration level and turgor pressure of leaves even when the water potentials of leaf are low (Rapparini and Peñuelas 2014). Due to salinity plants, experience changes at osmotic level and thus growth of plants is hampered.

Furthermore, the reactive oxygen species produced due to stress response pose detrimental effects on the plants like oxidation of chlorophyll and other important plant cell components. But plants inoculated with AMF showed enhancement in chlorophyll levels and it was found that the negative effect of salinity stress was mitigated. Besides this, experiments showed that AMF helps in enhancement of the

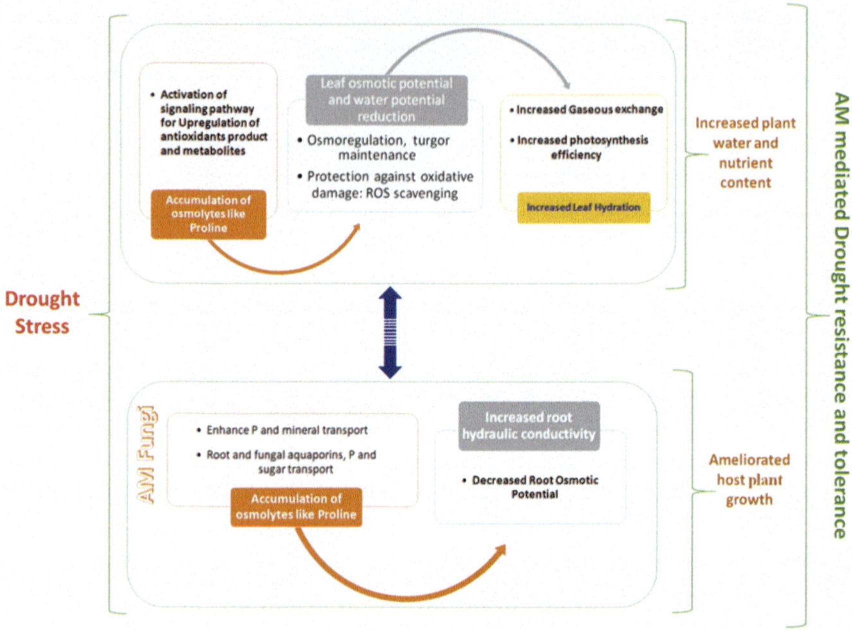

Fig. 12.1 Schematic diagram of role of AM fungi under abiotic (drought) stress condition modified from Rapparini and Peñuelas (2014)

defense responses which a plant shows under stress and thus protects the plants under stress conditions (Hashem et al. 2015). Plants respond to biotic and abiotic stresses in a way that they tend to protect themselves under such conditions by using certain sophisticated mechanisms. One of those mechanisms is the ability of plants to form associations with mycorrhiza in the roots. Mycorrhizal fungi have been known to provide resistance among various plant diseases.

There have been many evidences which suggest the same. Colonization of mycorrhiza *Funneliformis mosseae* in onion plants significantly helped in alleviation of pink rot disease caused by *Pyrenochaetaterrestris*. Further experiments on tomato plants showed that inoculation of tomato plants with mycorrhizal fungi helped in reducing the early blight disease incidence (Song et al. 2015). The loss of nutrients from the soil due to leaching has been a major concern in the field of agriculture as the soil gets deprived of nutrients and this, in turn, affects plant growth and development. Mobile nutrients like nitrates and sulfates are highly susceptible to get lost by the leaching process. The AMF has been found to significantly reduce the loss of inorganic nutrients like N and P. These fungi have developed the mechanism to enhance the nutrient interception zone around them and thus save nutrients from getting leached by rainwater or any other agricultural activity (Cavagnaro et al. 2015). Soil aggregates are very important component of soil structure as soil aggregation affects the water-holding capacity and infiltration rate. The mycorrhizal fungi produce hyphae in which the soil particles get entangled and form aggregates (Leifheit et al. 2014).

12.7 Diversity of AMF for Sustainable Agriculture: Methods and Constrain

Arbuscular mycorrhizal fungi are important symbionts of plants that improve plants' nutrient uptake and in turn helps in plant growth promotion. Genetic analysis of arbuscular mycorrhizal fungi for community study has been a complicated task because of its difficulties in isolation and cultivation from contaminate-free system and heterogeneity of rDNA sequence within single arbuscular mycorrhizal spore. This diversity allows us to find reliable methods for genotyping of arbuscular mycorrhiza fungi. Cluster analysis has been performed by many authors on FAME profiles using unweighted pair group method with arithmetic mean and result were compared to a neighbor joining of rDNA sequence within same species (Kumar and Adholeya 2018). Many authors suggested combination of the morphological, biochemical, and molecular (sequencing of highly variable D1–D2 region of LSU and ITS rRNA gene) method could be employed for phylogenetic analysis and species-level resolution of *Glomeromycota* (Ryberg et al. 2009; Stockinger et al. 2010; Walker et al. 2007). D1–D2 of LSU and ITS region of rRNA gene to evaluate the quality of arbuscular mycorrhiza produced on a large scale and to track the selected arbuscular mycorrhiza after inoculation into the field. Many studies on the basis of molecular and morphological data sets revealed that the species *Scutellospora* and *Gigaspora*were present in low or negligible amounts in field soil samples.

This supports the hypothesis proposed earlier by Jansa et al. (2003), who also found that there is a lower number of *Gigasporacaea sp.* inland which is managed chemically. Moreover, the morphological analysis shows that such significant differences in number of *Glomeraceae* spores between trap cultures are due to conventional tillage and zero tillage fields. More recent investigations done by Mirás-Avalos et al. (2011) were based on denaturing gradient gel electrophoresis (DGGE) sequencing and found that there is an increase in the presence of *Glomus* fungi in the soil which is receiving the conventional tillage practice. Morphological data revealed that there is presence of *Gigasporaceae* spores in large amounts in trap culture set up field soil of raised bed plantation under zero tillage practices (Kumar and Adholeya 2016; Kumar and Adholeya 2018).

Another morphological study has reported that the fields which have received chemical fertilizers for a long time have *Rhizophagus* and *Funneliformis sp*in dominance and *F. mosseae* with frequent occurrence (Oehl et al. 2003). Similar studies on the basis of molecular analysis by Mathimaranhe et al. (2005) found that the fields in which the conventional farming practices are followed have *R. intraradices* as the dominant species of AMF and also suggested that, frequent inputs of chemical fertilizers can decrease the availability of the AMF propagules in the tropical soils. The presence of fatty acids also indicates presence of fungal species as observed by Madan et al. (2002a, b) that the soils which have originated from intensive farming have *Gigasporacae* spores in them. These data show that the diversity of arbuscular mycorrhizal fungi is not always found to be low in cultivated lands (Hijri et al. 2006). The AMF monoaxenic cultures can be encapsulated in alginate beads for use as inoculums to diversify the culturing process (Saito and Marumoto 2002).

12.8 Methods of Isolation and Propagation of Mycorrhizal Species

Spores are collected from rhizosphere soil and kept in a watch glass or a small petri dish. These spores are then sorted into morphotypes (SMT) or groups in which the spores within one group or morphotypes appear similar in morphology, based on the external features of the spore like size, shape, color, the contents which are visible and shape of the subtending hypha. The external morphology and color of the spores from the trap culture can be identified by using prepared slide of the collected spores and visualizing that slide with the help of a dissecting microscope with reflected light. For color description color chart of Glomalean fungi (INVAM website) is used. The abundance of AM spores in each group is estimated on the following semiquantitative scale 1, upto 5 spores; 2.6–20 spores; 3.21–50 spores; 4.50–200 spores; 5, more than 200 spores. Spore sample are kept at 4 °C until the analysis of AMF spores for abundance and species richness and molecular analysis. Rhizospheric soil from each trap culture is examined for AM fungal spores. Once the data are obtained, the following are calculated for AM diversity analysis (1) Spore density (Total number of spores in 100 g of soil sample). (2) AM fungal species richness (the total number of AM fungal species in each site), (3) relative abundance (the ratio between the number of sores of particular fungal species to the total number of AM spores), (4) Shannon–Weiner index (H′) is calculated for each sites using Eq. (1), where $Pi = ni/N$, ni is the number of individuals of species i, and N is the total number of individuals in all species).

$H' = -\sum (Pi) \ln (Pi)$ (1). Spore density: The air-dried, weighed (100 g) soil samples are mixed in a substantial amount of water, and suspension is decant through a series of sieves. For the quantification of spores, the measured volume of sieving is transferred onto gridded petri-plate and observed under a stereomicroscope. The number of spores in petri-plate is counted and expressed as spores/mL of soil suspension. Finally, spore density per gram of soil is calculated by dividing the total number of spores present in total suspension by the quantity of soil sieved. Species richness and relative spore abundance are calculated according to modified methods described by Oehl et al. (2003). Species richness is also required to calculate from the trap culture; the number of spores belonging to different AM species is also calculated. Species richness can be defined as the number of AM species occurred per soil sample. Total number of AMF species identified per site (total amount of soil explored, 100 g). Relative spore abundance: The relative abundance of the spores present is identified for each and every species of AMF and site (the total amount of soil 100 g).

12.9 Monosporal Culture of AMF: Source of Pure Mycorrhizal Species

Trap culture contains mixture of diverse mycorrhizal species which create condition sporulation of cryptic AMF species. However, monosporal culture contains single species culture of AMF originated from trap culture with host plants. For this, one to ten AMF spores of the from single morphotypes isolated from trap pots are placed on germinating roots of maize, which is then grown for 6 months in a pot filled with sterile substrate. Successive pot culture of trap cultures isolate can cause an unexpected outbreak of the contaminant. The micropipette tips are filled with the substrate (Terragreen and sand; 1:1) and 3–5 host seeds are placed over the substrate (*Allium porrum*). A healthy spore is placed over the host seed to ensure the colonization of germinated spore on the host (*Sorghum bicolor*). The cultures are kept in a tray containing water for 1–2 days so that seeds and seedling in micropipette tips get enough moisture and incubated in the controlled condition. When the seedling emerges out from tray, the tips are taken out of the tray containing water. The tray is left out to dry for 1–2 days and the cycle is repeated three times so as to get vigorous root production. The roots are chopped off from the tips and the whole seedling is transferred to big sized pots. More seeds are placed in the pot and regular and proper watering is done to allow the plant to complete its life cycle (3–4 months). After completion of the cycle, the aboveground portions of plants are chopped off to initiate a new cycle using different host plants. This cycle is repeated three times. Life cycle of AMF collected from field soil and further characterized for morphological, molecular, and biochemical analysis is showing in Fig. 12.2.

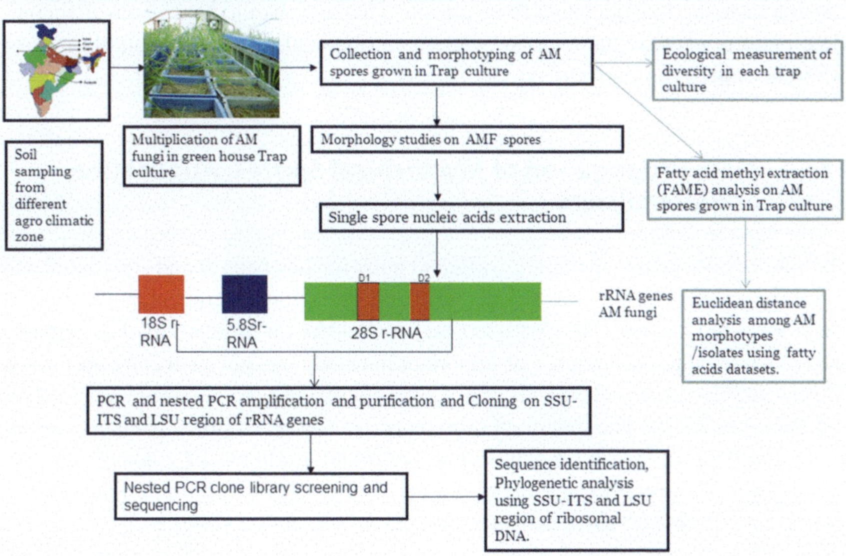

Fig. 12.2 The schematic representation of methods of isolation, multiplication, and identification of AMF species collected from field soil

12.10 Root Organ Culture of AMF: Benefit in Biofertilizers Production

As it is quite difficult to define species concept of AMF using any one method, building up a comprehensive total profile of an isolate using a combination of data sets achieved by molecular, biochemical, and microscopy studies can classify the uniqueness of an isolate. Use of a combined approach is not new in the field of mycorrhiza and other fungi as several species were characterized using more than one feature. Mycelia of ectomycorrhizal fungi (Tuber sp.) from pure culture were characterized by combining morphological and molecular tools by Lotti et al. (2002). Some similarly characterized fungi are *Emmonsia pasteuriana* (Drouhet et al. 1998) and *Pseudotomentella ochracea* (Koljalg and Larsson 1998). Recent reports by (Morton and Msiska 2010) and (Krüger et al. 2012) led to the revision of classification features of family *Gigasporaceae* (Glomeromycota) and *Acaulospora brasileinsis* respectively based on combined morphological and molecular characters.

Declerck et al. (2000) described an AMF species raised in a root organ culture using for the first time the data based on FAME profile, ultrastructure studies of spores, and n-rDNA. They concluded that long-term maintenance of AM fungi under strict controlled conditions, without contamination is a suitable platform for comparative analysis using morphological, biochemical, and molecular tools of that isolate. As a larger number of AMF isolates are brought into root organ cultures, the availability of consistent material will increase and developing the complete comprehensive profile, which will be the isolate's signature and unique profile, will be feasible. This has been demonstrated by our study comparing ten isolates from ROC and emphasizing that dependence on any one character would have given an incomplete picture whereas the overlap of the data obtained resolves the similarities or differences between each isolate.

12.11 Mass Propagation of Mycorrhizal Spores: Application as Biofertilizers

The role of mycorrhiza in plant growth and nutrition has been found significant and therefore it is now being used as a biofertilizer. The microbial inoculants of mycorrhizal fungi are known as biofertilizers and are helpful for production of sustainable food as these help in utilization of important nutrients like nitrogen and phosphorus without the use of any chemicals and without damaging the environment (Igiehon and Babalola 2017). The inoculum of mycorrhiza can be obtained from any soil as the mycorrhizal fungi are ubiquitous and are found in almost all kinds of soil. For the purpose of mass propagation of mycorrhizal fungi, mainly three methods are used in vitro, substrate-free, and substrate-based production systems.

12.11.1 Substrate-Based Production System

In this the symbionts of AMF associated with respective plant species are cultivated in soil-based substrate, After the AMF species are obtained and identified, these species are further propagated in plastic or clay pots. This kind of production system has an advantage that single AMF species can be propagated in mass amounts.

12.11.2 Substrate-Free Production System

In this type of system, the plant roots and the AMF are provided with a nutrient media required for the growth of roots and propagation of the AMF. This is the most widely used method as it provides fungi an environment and medium which mimics the field conditions. In this, the availability of oxygen and nutrients is ensured by switching the pump on from time to time. Another such way is aeroponics in which the nutrition-rich fog is sprayed on the roots which also allows the exchange of gases. The advantage of this system is that the production of mycorrhiza is done without using any substrate.

12.11.3 In Vitro Production System

This system uses t-DNA modified root of *Daucus carota* and large-scale propagation of fungi is achieved in bioreactors containing perlite or any other solid medium as substrate. This system poses a great advantage to researchers that the interference of other unwanted microorganisms is very less due to which pure cultures can be obtained (Akhtar and Abdullah 2014; Selvakumar et al. 2018).

The use of AMF strains for re-enriching soil with nutrients and to act as an alternative to conventional fertilizing practices. Due to the potential of AMF to provide nutritional benefits to the soil and plants both infield and in vitro, it can be of significant use to the farmers to be used as a biofertilizer. But unfortunately, large-scale production of AMF is a bit challenging task because it can only be grown with the host plant, therefore, to achieve the goal of sustainable crop production, the AMF containing soil can be used as inoculum for mass propagation. Studies have shown that the inoculation of AMF in the field is also as effective as the inoculation done in a greenhouse (Berruti et al. 2016).

12.12 Quality Production of AMF Fungi: Limitation and Prospects

The main limitation that is faced during the in vitro culture of AMF is due to the fact that AMF is an obligate biotroph. Another noticeable limitation is that the AMF spores lose their infectivity if they are subcultured successively in vitro. Moreover, proper inoculation and its maintenance require very skilled people and time and are very expensive. The prospects of AMF culture can bring many benefits and discoveries to the light. The potential benefits of inoculating the Arbuscular Mycorrhizal Fungi not only include sustainable food production but also conservation of environment. The AMF monoaxenic cultures can be encapsulated in alginate beads for use as inoculum to diversify the culturing process (Diop 2003; Saito and Marumoto 2002).

12.13 Growth and Propagation of Arbuscular Mycorrhizal Fungi

Propagation of pure mycorrhizal cultures on a large scale is very important for use in the agriculture but it is limited due to the obligate biotrophic nature of the AMF. Due to this nature, the AMF is dependent on a host plant for their survival and thus it is difficult to grow and propagate AMF in vitro conditions. Although many strategies and methods like aeroponics etc. are used for large-scale production of the spores but all these methods have their own advantages and disadvantages out of which the major disadvantage is lack of contamination-free culture. Therefore, depending on the requirements, different methods are used.

12.13.1 Trap Culture

Trap culture is done in cases when either the roots of plants contain mycorrhiza but the sporulation is negligible or in soils where the spores have undergone so many structural changes that the identification of species has become almost impossible. In trap culture, the rhizosphere soil is dug up and a root ball is collected. The shoots are removed from it at the crown and roots are chopped into small fragments and mixed thoroughly with the help of an axe. This chopped blend is then mixed with autoclaved soil in ration 1:1 in zip-loc bags and massaged thoroughly to break the lumps of soil and roots and mix properly so that more homogenous product can be obtained. After this, a 15-cm plastic pot is taken and the material is transferred to the plastic pot. For reasons like minimizing bare surface and forcing plant to grow slowly and at similar height, the pot soil is overseeded and the plants which will grow out of seeds will be used as hosts. These are then cultured for 4 months in greenhouse and fertilization is kept minimal or is done only in cases where P and N deficiencies are observed. Pots are then left in a shaded room to dry slowly at stable temperature and spores are extracted from them before these pots get too dry. If sporulation is low then the pots are retained and reseeded. The trap cultures can be stored in ziploc bags for a period of at least 30 days. Although some species require a dormant period before becoming infective but the time period may change according to the change in habitat (Selvakumar et al. 2018; https://invam.wvu.edu/).

12.14 Conclusion and Future Prospects

The role of AMF in the field of agriculture is very important as its proper inoculation and propagation can bring many benefits to the farmers and help in maintaining the nutrient quality of the crops and protecting crops from diseases without damaging the field soil and the soil microflora. Although many methods have been designed for the production of AMF fungi but the large-scale production of fungus is still a challenge and requires a lot of work in the same field so that the fungus is easily available to farmers at low costs. The prospects of AMF culture can bring many benefits and discoveries to the light. The potential benefits of inoculating the arbuscular mycorrhizal fungi not only include sustainable food production but also conservation of environment. In future, quality production of in vitro grown AMF inoculum for biofertilizers production would add sustainable growth in agriculture.

Acknowledgments Authors thank the University Grants Commission (UGC), the Government of India for the award of a doctoral fellowship to carry out research work at The Energy and Resources Institute (TERI), New Delhi. This research was supported by funds provided to TERI by the Department of Biotechnology, Government of India.

References

Abbott LK, Robson AD, De Boer G (1984) The effect of phosphorus on the formation of hyphae in soil by the vesicular-arbuscular mycorrhizal fungus, *Glomus fasciculatum*. New Phytol 97:437–446

Akhtar MS, Abdullah SNA (2014) Mass production techniques of arbuscular mycorrhizal fungi: major advantages and disadvantages: a review. Biosci Biotechnol Res Asia 11:1199–1204

Berruti A, Lumini E, Balestrini R, Bianciotto V (2016) Arbuscular mycorrhizal fungi as natural biofertilizers: let's benefit from past successes. Front Microbiol 6:1559

Bolan N (1991) A critical review on the role of mycorrhizal fungi in the uptake of phosphorus by plants. Plant Soil 134:189–207

Bridge P, Spooner B (2001) Soil fungi: diversity and detection. Plant Soil 232:147–154

Cavagnaro TR, Bender SF, Asghari HR, van der Heijden MG (2015) The role of arbuscular mycorrhizas in reducing soil nutrient loss. Trends Plant Sci 20:283–290

Chandrashekar M, Soumya P, Raju N (2014) Fungal diversity of rhizosphere soils in different agricultural fields of Nanjangud Taluk of Mysore District, Karnataka, India. Int J Curr Microbiol App Sci 3:559–566

Chen C, Zhang J, Lu M, Qin C, Chen Y, Yang L, Shen Q (2016) Microbial communities of an arable soil treated for 8 years with organic and inorganic fertilizers. Biol Fertil Soil 52:455–467

Declerck S, Cranenbrouck S, Dalpé Y, Séguin S, Grandmougin-Ferjani A, Fontaine J, Sancholle M (2000) *Glomus proliferum* sp. nov.: a description based on morphological, biochemical, molecular and monoxenic cultivation data. Mycologia 92:1178–1187

Diop TA (2003) In vitro culture of arbuscular mycorrhizal fungi: advances and future prospects. Afr J Biotechnol 2:692–697

Drouhet E, Gueho E, Gori S, Huerre M, Provost F, Borgers M, Dupont B (1998) Mycological, ultrastructural and experimental aspects of a new dimorphic fungus *Emmonsia pasteuriana* sp. nov. isolated from a cutaneous disseminated mycosis in AIDS. J Mycol Med 8:64–77

Fernandez CW, Langley JA, Chapman S, McCormack ML, Koide RT (2016) The decomposition of ectomycorrhizal fungal necromass. Soil Biol Biochem 93:38–49

Fierer N, Breitbart M, Nulton J, Salamon P, Lozupone C, Jones R et al (2007) Metagenomic and small-subunit rRNA analyses reveal the genetic diversity of bacteria, archaea, fungi, and viruses in soil. Appl Environ Microb 73:7059–7066

Gerdemann JW (1968) Vesicular-arbuscular mycorrhiza and plant growth. Annu Rev Phyto Pathol 6:397–418

Hart M, Ehret DL, Krumbein A, Leung C, Murch S, Turi C, Franken P (2015) Inoculation with arbuscular mycorrhizal fungi improves the nutritional value of tomatoes. Mycorrhiza 25:359–376

Hashem A, Abd Allah EF, Alqarawi AA, Aldubise A, Egamberdieva D (2015) Arbuscular mycorrhizal fungi enhance salinity tolerance of *Panicum turgidum* for ssk by altering photosynthetic and antioxidant pathways. J Plant Interact 10:230–242

Hassan SED, Liu A, Bittman S, Forge TA, Hunt DE, Hijri M, St-Arnaud M (2013) Impact of 12-year field treatments with organic and inorganic fertilizers on crop productivity and mycorrhizal community structure. Biol Fert Soils 49:1109–1121

Hawksworth D L, and Lücking R (2017) Fungal diversity revisited: 2.2 to 3.8 million species. The fungal kingdom 79–95

Hera C (1996) The role of inorganic fertilizers and their management practices. Fertil Environ:131–149

Hijri I, Sýkorová Z, Oehl F, Ineichen K, Mäder P, Wiemken A, Redecker D (2006) Communities of arbuscular mycorrhizal fungi in arable soils are not necessarily low in diversity. Mol Ecol 15:2277–2289

Igiehon NO, Babalola O (2017) Biofertilizers and sustainable agriculture: exploring arbuscular mycorrhizal fungi. Appl Micro boil Biot 101:4871–4881

Jansa J, Mozafar A, Kuhn G, Anken T, Ruh R, Sanders IR, Frossard E (2003) Soil tillage affects the community structure of mycorrhizal fungi in maize roots. Ecol Appl 13:1164–1176

Johnson NC, Hoeksema JD, Bever JD, Chaudhary VB, Gehring C, Klironomos J et al (2006) From Lilliput to Brobdingnag: extending models of mycorrhizal function across scales. Bioscience 56:889–900

Koide RT (1991) Nutrient supply, nutrient demand and plant response to mycorrhizal infection. New Phytol 117:365–386

Koljalg U, Larsson E (1998) *Pseudotomentella ochracea* sp. Nov based on morphological and molecular data. Folia Cryptogamica Estonica 33:53–56

Kour D, Rana KL, Yadav N, Yadav AN, Singh J, Rastegari AA et al (2019) Agriculturally and industrially important fungi: current developments and potential biotechnological applications. In: Yadav AN, Singh S, Mishra S, Gupta A (eds) Recent advancement in white biotechnology through Fungi, Perspective for value-added products and environments, vol 2. Springer International Publishing, Cham, pp 1–64. https://doi.org/10.1007/978-3-030-14846-1_1

Kour D, Kaur T, Devi R, Rana KL, Yadav N, Rastegari AA et al (2020a) Biotechnological applications of beneficial microbiomes for evergreen agriculture and human health. In: Rastegari AA, Yadav AN, Yadav N (eds) Trends of microbial biotechnology for sustainable agriculture and biomedicine systems: perspectives for human health. Elsevier, Amsterdam, pp 255–279. https://doi.org/10.1016/B978-0-12-820528-0.00019-3

Kour D, Rana KL, Yadav AN, Yadav N, Kumar M, Kumar V et al (2020b) Microbial biofertilizers: bioresources and eco-friendly technologies for agricultural and environmental sustainability. Biocatal Agric Biotechnol 23:101487. https://doi.org/10.1016/j.bcab.2019.101487

Krüger M, Krüger C, Walker C, Stockinger H, Schüßler A (2012) Phylogenetic reference data for systematics and phylotaxonomy of arbuscular mycorrhizal fungi from phylum to species level. New Phytol 193:970–984

Kumar S, Adholeya A (2016) Impact of land use and soil types on arbuscular mycorrhizal fungal diversity in tropical soil of India. Afr J Microbiol Res 10:1595–1606

Kumar S, Adholeya A (2018) Congruence of morphology and fatty acid methyl ester profile (FAME profile) revealed low Mycorrhizal diversity in soil contaminated with tannery sludge. Pollut Res 37:S71–S81

Larson BA, Frisvold GB (1996) Fertilizers to support agricultural development in sub-Saharan Africa: what is needed and why. Food Policy 21:509–525

Leifheit EF, Veresoglou SD, Lehmann A, Morris EK, Rillig MC (2014) Multiple factors influence the role of arbuscular mycorrhizal fungi in soil aggregation—a meta-analysis. Plant Soil 374:523–537

Lindahl BD, Tunlid A (2015) Ectomycorrhizal fungi–potential organic matter decomposers, yet not saprotrophs. New Phytol 205:1443–1447

Lotti M, Amicucci A, Stocchi V, Zambonelli A (2002) Morphological and molecular characterization of mycelia of some tuber species in pure culture. New Phytol 155:499–505

Madan R, Pankhurst C, Hawke B, Smith S (2002a) Use of fatty acid for identification of AM fungi and estimation of the biomass of AM spores in soil. Soil Biol Biochem 34:125–128

Madan R, Pankhurst C, Hawke B, Smith S (2002b) Use of fatty acid for identification of AM fungi and estimation of the biomass of AM spores in soil. Soil Biol Biochem 34:125–128

Mahmood F, Khan I, Ashraf U, Shahzad T, Hussain S, Shahid M et al (2017) Effects of organic and inorganic manures on maize and their residual impact on soil physico-chemical properties. J Soil Sci Plant Nut 17:22–32

Martini EA, Buyer JS, Bryant DC, Hartz TK, Denison RF (2004) Yield increases during the organic transition: improving soil quality or increasing experience? Field Crops Res 86:255–266

Martino E, Morin E, Grelet GA, Kuo A, Kohler A, Daghino S et al (2018) Comparative genomics and transcriptomics depict ericoid mycorrhizal fungi as versatile saprotrophs and plant mutualists. New Phytol 217:1213–1229

Mathimaranhe N, Ruh R, Vullioud P, Frossard E, Jansa J (2005) *Glomus intraradices* dominates arbuscular mycorrhizal communities in a heavy textured agricultural soil. Mycorrhiza 16:61–66

Mirás-Avalos J, Antunes Pedro M, Koch A, Khosla K, Klironomos John N, Dunfield (2011) The influence of tillage on the structure of rhizosphere and root-associated arbuscular mycorrhizal fungal communities. Pedobiologia 54:235–241

Morton J (1988) Taxonomy of VA mycorrhizal fungi: classification, nomenclature, and identification. Mycotaxon 32:267–324

Morton JB, Msiska Z (2010) Phylogenies from genetic and morphological characters do not support a revision of Gigasporaceae (Glomeromycota) into four families and five genera. Mycorrhiza 20:483–496

Munkvold L, Kjøller R, Vestberg M, Rosendahl S, Jakobsen I (2004) High functional diversity within species of arbuscular mycorrhizal fungi. New Phytol 164:357–364

Nkebiwe PM, Weinmann M, Bar-Tal A, Müller T (2016) Fertilizer placement to improve crop nutrient acquisition and yield: a review and meta-analysis. Field Crop Res 196:389–401

Oehl F, Sieverding E, Ineichen K, Mäder P, Boller T, Wiemken A (2003) Impact of land use intensity on the species diversity of arbuscular mycorrhizal fungi in agroecosystems of Central Europe. Appl Environ Microb 69:2816–2824

Pirozynski KA, Malloch DW (1975) The origin of land plants: a matter of mycotropism. Biosystems 6:153–164

Qin H, Lu K, Strong PJ, Xu Q, Wu Q, Xu Z et al (2015) Long-term fertilizer application effects on the soil, root arbuscular mycorrhizal fungi and community composition in rotation agriculture. Appl Soil Ecol 89:35–43

Rana KL, Kour D, Sheikh I, Dhiman A, Yadav N, Yadav AN et al (2019) Endophytic fungi: biodiversity, ecological significance and potential industrial applications. In: Yadav AN, Mishra S, Singh S, Gupta A (eds) Recent advancement in white biotechnology through Fungi, Diversity and enzymes perspectives, vol 1. Springer, Switzerland, pp 1–62

Rapparini F, Peñuelas J (2014) Mycorrhizal fungi to alleviate drought stress on plant growth. Use of microbes for the alleviation of soil stresses 1. Springer, New York, NY, pp 21–42

Rastegari AA, Yadav AN, Yadav N (2020) New and Future Developments in Microbial Biotechnology and Bioengineering: Trends of Microbial Biotechnology for Sustainable Agriculture and Biomedicine Systems: Diversity and Functional Perspectives. Elsevier, Amsterdam

Redecker D, Kodner R, Graham LE (2000) Glomalean fungi from the Ordovician. Science 289:1920–1921

Ryberg M, Kristiansson E, Sjökvist E, Nilsson RH (2009) An outlook on fungal internal transcribed spacer sequences in Gene Bank and the introduction of a web-based tool for the exploration of fungal diversity. New Phytol 181:471–477

Saito M, Marumoto T (2002) Inoculation with arbuscular mycorrhizal fungi: the status quo in Japan and the future prospects. Diversity and integration in Mycorrhizas. Springer, Dordrecht, pp 273–279

Savci S (2012a) An agricultural pollutant: chemical fertilizer. APCBEE Proc 3:73

Savci S (2012b) Investigation of effect of chemical fertilizers on environment. APCBEE Proc 1:287–292

Saxena AK, Yadav AN, Rajawat M, Kaushik R, Kumar R, Kumar M et al (2016) Microbial diversity of extreme regions: an unseen heritage and wealth. Indian J Plant Genet Resour 29:246–248

Selvakumar G, Shagol CC, Kang Y, Chung BN, Han SG, Sa TM (2018) Arbuscular mycorrhizal fungi spore propagation using single spore as starter inoculum and a plant host. J Appl Microbiol 124:1556–1565

Shah F, Nicolás C, Bentzer J, Ellström M, Smits M, Rineau F et al (2016) Ectomycorrhizal fungi decompose soil organic matter using oxidative mechanisms adapted from saprotrophic ancestors. New Phytol 209:1705–1719

Shennan C (2007) Biotic interactions, ecological knowledge and agriculture. Philos T Roy Soc B 363:717–739

Simon L, Bousquet J, Levesque RC, Lalonde M (1993) Origin and diversification of endomycorrhizal fungi and coincidence with vascular land plants. Cah Rev The 362:67–69

Smith SE, Gianinazzi-Pearson V (1988) Physiological interactions between symbionts in vesicular-arbuscular mycorrhizal plants. Annu Rev Plant Phys 39:221–244

Smith SE, Read DJ (2008) Mycorrhizal symbiosis. Academic Press, New York, USA

Song Y, Chen D, Lu K, Sun Z, Zeng R (2015) Enhanced tomato disease resistance primed by arbuscular mycorrhizal fungus. Front Plant Sci 6:786

Stockinger H, Krüger M, Schüßler A (2010) DNA barcoding of arbuscular mycorrhizal fungi. New Phytol 187:461–474

Tadesse T, Dechassa N, Bayu W, Gebeyehu S (2013) Effects of farmyard manure and inorganic fertilizer application on soil physico-chemical properties and nutrient balance in rain-fed lowland rice ecosystem. Am J Plant Sc 4:309

Toljander JF, Santos-González JC, Tehler A, Finlay RD (2008) Community analysis of arbuscular mycorrhizal fungi and bacteria in the maize mycorrhizosphere in a long-term fertilization trial. FEMS Micro biol Ecol 65:323–338

Verma P, Yadav AN, Khannam KS, Mishra S, Kumar S, Saxena AK et al (2019) Appraisal of diversity and functional attributes of thermotolerant wheat associated bacteria from the peninsular zone of India. Saudi J Biol Sci 26:1882–1895. https://doi.org/10.1016/j.sjbs.2016.01.042

Walker C, Vestberg M, Demircik F, Stockinger H, Saito M, Sawaki H, NishmuraI (2007) Molecular phylogeny and new taxa in the Archaeosporales (Glomeromycota): *Ambispora fennica* gen. Sp. nov., Ambisporaceae fam. Nov., and emendation of Archaeospora and Archaeosporaceae. Mycol Res 111:137–153

Wang B, Qiu YL (2006) Phylogenetic distribution and evolution of mycorrhizas in land plants. Mycorrhiza 16:299–363

Wang Z, Li T, Wen X, Liu Y, Han J, Liao Y, De Bruyn JM (2017) Fungal communities in rhizosphere soil under conservation tillage shift in response to plant growth. Front Microbiol 8:1301

Yadav AN, Mishra S, Singh S, Gupta A (2019a) Recent advancement in white biotechnology through Fungi. Volume 1: diversity and enzymes perspectives. Springer International Publishing, Cham

Yadav AN, Singh S, Mishra S, Gupta A (2019b) Recent advancement in white biotechnology through Fungi. Volume 3: perspective for sustainable environments. Springer International Publishing, Cham

Yadav AN, Mishra S, Kour D, Yadav N, Kumar A (2020a) Agriculturally important Fungi for sustainable agriculture, volume 1: perspective for diversity and crop productivity. Springer International Publishing, Cham

Yadav AN, Mishra S, Kour D, Yadav N, Kumar A (2020b) Agriculturally important Fungi for sustainable agriculture, volume 2: functional annotation for crop protection. Springer International Publishing, Cham

Yadav AN, Singh J, Rastegari AA, Yadav N (2020c) Plant microbiomes for sustainable agriculture. Springer, Cham

Young AW, Hay DC, McWeeny KH, Flude BM, Ellis AW (1985) Matching familiar and unfamiliar faces on internal and external features. Perception 14:737–746

Phyllospheric Microbes: Diversity, Functions, Interaction, and Applications in Agriculture

13

Rupal Gupta and Ravishankar Patil

Abstract

Phyllospheric microbes refer to the microbes resides on the above-ground portion of the different part of the plants such as stems, leaves, flowers, and fruits. It includes bacteria, fungi, archaea, protist, algae, etc., and maybe either advantageous or deleterious to the host plant. The valuable microbial community plays a role in plant disease management, plant growth, nutrient acquisition, protection of plant from external environment, and also provides resistance to stress. Phyllospheric microbes are being explored as a biocontrol agent and biofertilizers for sustainable development of agriculture. In the present chapter, we thoroughly discussed about diverse pathogenic and non-pathogenic phyllospheric microorganisms, their ecosystem interaction, and important function in host plant. Furthermore, microbial food safety and application of phyllospheric microflora in sustainable agriculture growth have also been discussed.

Keywords

Ecosystem dynamics · Microbial interaction · Phyllospheric microbes · Plant–microbe ecology · Sustainable agriculture

13.1 Introduction

Phytomicrobiome refers to the diverse microbes associated with plants and its contiguous environment. This microbial population maybe parasitic, commensal, or mutualistic (Jones et al. 2019; Leveau 2019) and thus may have beneficial, harmful,

R. Gupta · R. Patil (✉)
Amity Institute of Biotechnology, Amity University, Mumbai, Maharashtra, India

A. N. Yadav et al. (eds.), *Current Trends in Microbial Biotechnology for Sustainable Agriculture*, Environmental and Microbial Biotechnology,
https://doi.org/10.1007/978-981-15-6949-4_13

or no detectable effect on growth, development, and function of the plant (Steven et al. 2018; Uroz et al. 2019). They may provide protection to the host plant from pathogenic attack and in turn deal resource exchange affecting plant growth (Stone et al. 2018). The plant environment provides an ecological niche to microbiota in their phyllosphere, rhizosphere, and endosphere (Dong et al. 2019; Yadav et al. 2020b). Phyllosphere is aboveground plant organs that are habitat for various microbes (Farré-Armengol et al. 2016). The volume of soil in the vicinity of a living plant that is influenced by root activity is known as rhizosphere whereas the inner root tissues are inhabited by certain microorganisms which are referred to as endosphere (Bulgarelli et al. 2013). Notably, the microbial community of individual plant hosts is similar within a given environment and it may change according to respective environmental factor changes (Espenshade et al. 2019) (Fig. 13.1).

Phyllosphere is the largest and most prevalent habitats for microbes (Dong et al. 2019) influencing host plants in terms of growth and resistance to stress due to biotic and abiotic factors (Glick 2005; Saleem et al. 2017; Compant et al. 2019). They are sufficiently rich enough to impact the global carbon and nitrogen cycle (Delmotte et al. 2009; Sivakumar et al. 2020). These phyllospheric microflora plays an important role in the plant's development by secreting auxins, cytokinins, gibberellin, and biosurfactants (Whipps et al. 2008; Fu et al. 2016; Steven et al. 2018). Phyllosphere microbes possess the capacity to affect ecosystem function and biogeography of host plants (Kembel et al. 2014; Verma et al. 2017). The phyllospheric microbiome provides resistance against the pathogen (Monteiro et al. 2012), increases the metabolic capacity of the plant, and enhances nutrient uptake (Parasuraman et al. 2019; Yadav et al. 2018).

The endosphere microbes are involved in nitrogen fixation, production of phytohormones, phosphate solubilization, enhance nutrient uptake, provide protection against pathogen and the high-stress environment of the soil (Pacifico et al. 2019; Rana et al. 2020a), whereas the rhizospheric microbiomes support nutrient supply (Mwajita et al. 2013; Kour et al. 2019), suppress pathogens (Nihorimbere et al. 2011), stimulate plant growth hormone (PGH), solubilize nutrients (Marschner 2007; Azcón-Aguilar and Barea 2015), and help in the decomposition of minerals

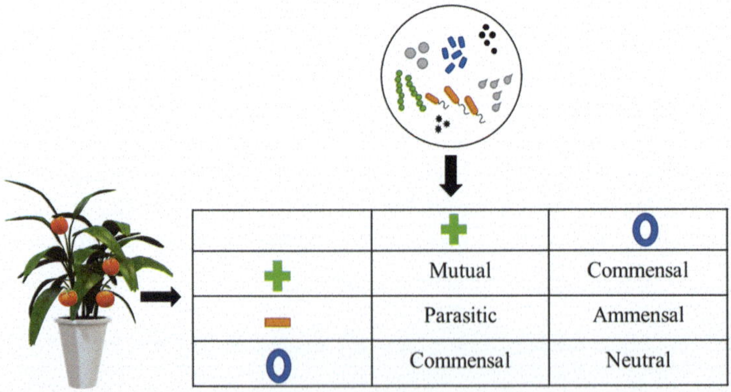

Fig. 13.1 Relation between the host plant and associated microbiota

Table 13.1 Various functions of phytomicrobiome

Microbiome	Roles	References
Phyllospheric	Promote plant growth by secreting auxins, cytokinins, gibberellin, or secreting biosurfactants	Fu et al. (2016), Steven et al. (2018)
	Provide resistance against pathogens	Monteiro et al. (2012)
	Contribute to carbon, phosphate cycle	Delmotte et al. (2009), Sivakumar et al. (2020)
	Nitrogen fixation	Fürnkranz et al. (2008), Lambais et al. (2017)
	Expand the metabolic capacity of the plant	Knief et al. (2012), Parasuraman et al. (2019)
	Enhances resource uptake efficiency	Parasuraman et al. (2019)
	Emission of volatile compound	Farré-Armengol et al. (2016)
	Metabolic dynamics	Knief et al. (2012)
	Provide tolerance to stress conditions	Glick (2005), Saleem et al. (2017), Compant et al. (2019)
Endophytic	Protects from the high-stress soil environment	Compant et al. (2005)
	Nitrogen fixation	Rilling et al. (2018)
	Phytohormone production	Pacifico et al. (2019)
	Phosphate solubilization	
	Disease suppression	Carrión et al. (2019)
	Increased enzymatic catalysis, enhanced water, and nutrient uptake	Pacifico et al. (2019)
	Plant growth, mitigation of cold	Verma et al. (2015b)
	Plant growth promotion and nutrient uptake	Rana et al. (2020b)
Rhizospheric	Ability to decompose	Katznelson et al. (1948)
	Nutrient solubilization a cycling	Marschner (2007), Azcón-Aguilar and Barea (2015)
	Stimulate PGH secretion	Fu et al. (2016)
	Nitrogen fixation	Xu et al. (2020)
	Suppression of plant pathogen	Nihorimbere et al. (2011)
	Support nutrient supply to the plant	Mwajita et al. (2013)
	Alleviation of drought stress and plant growth promotion	Kour et al. (2020d)
	Microbe-mediated alleviation of drought stress and acquisition of phosphorus	Kour et al. (2020c)
	Amelioration of drought stress and plant growth promotion	Kour et al. (2020e)
	Hydrogel-based bio-inoculant formulations for plant growth	Suman et al. (2016)

(Katznelson et al. 1948; Azcón-Fu et al. 2016). Table 13.1 gives an outline of the diverse functions of phytomicrobiome (phyllospheric, endospheric, and rhizospheric microbes). The present chapter deliberates phyllospheric microbiota, its subdivisions, diversity, role in ecosystem dynamic along with their interaction with host and environment. Role of phyllospheric microbes in the plant growth and defense mechanism and maintenance of the ecosystem is exhaustively discussed.

13.2 Phyllospheric Subdivision and Dominant Microbes

The microbial communities distributed in various regions of the phyllosphere are known as subdivisions of the phyllosphere. The surface of the stems refers to caulosphere and the portion in the vicinity of leaves is known as phylloplane. The surface of the flowers, fruits, and seeds is known as the asthenosphere, carposhere, and spermosphere, respectively (Fig. 13.2) (Hardoim et al. 2015; Farré-Armengol et al. 2016). However, if inhabiting microbes are found within leaf, stem, fruit, or flower then that microbiota are known as endophytes (Reddy and Saravanan 2013).

The phyllosphere regions represented in Fig. 13.2 show association with a variety of microbes. It differs from species to species as well as individuals of the same species. Here, we have given some noticeable examples of microbial genera found in the different regions of the phyllosphere reported in diverse plant species. Anthosphere contains *Aureobasidium, Bacillus, Cladosporium,* and *Pseudomonas* (Van Toor et al. 2005; Steven et al. 2018). Caulosphere comprises *Alternaria, Aureobasidium, Candida, Cladosporium, Coniothyrium, Fusarium, Geotrichum, Hanseniaspora, Lachancea, Meyerozyma, Nodulosporium, Penicillium, Phoma, Pichia, Rhinocladiella, Saccharomyces,* and *Trichoderma* (Cotter and Blanchard 1982; Koricha et al. 2019). Carposhere may involve *Actinobacteria, Bacillus, Citrobacter, Curtobacterium, Enterobacter, Erwinia, Firmicutes Frigobacterium, Gluconobacter, Massilia, Methylobacterium, Pantoea, Pseudomonas,* and *Sphingomonas* (Pascazio et al. 2015; Kecskeméti et al. 2016). Phylloplane may include *Ascomycetes, Aspergillus, Chaetomium, Cladosporium, Curvularia, Penicillium, Phycomycetes, Pseudomonas, Rhodotorula, Rhizopus, Mucormycetes,* and *Trichoderma* spp. (Southwell et al. 1999; Tanti et al. 2016), whereas

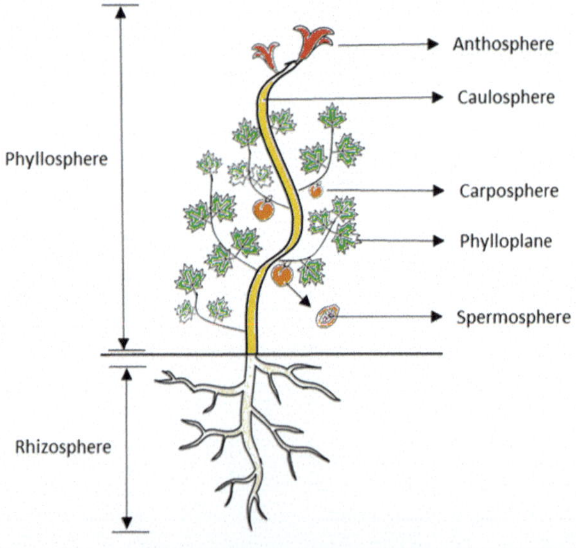

Fig. 13.2 Subdivision of phyllosphere on the basis of plant organ

Spermosphere may have *Aspergillus, Bacillus, Bacteroidetes, Cloacimonetes, Penicillium, Fusarium, Spirochaetae, Enterobacteriaceae, Staphylococcaceae, Pseudomonadaceae, Bacillaceae,* and *Rhizobiaceae* spp., (Lopez-Velasco et al. 2013; Saritha and Sreeramulu 2013; Tanti et al. 2016; Dong et al. 2019). Examples of microbes associated with the phyllospheric region of various plant hosts have been listed in Table 13.2.

13.3 Diversity of Phyllospheric Microbiome

The microbiota or community of microorganisms in the phyllosphere includes bacteria, fungi, yeast, protists, archaea, viruses, algae, nematode, and other microeukaryotes (Andrews and Harris 2000; Turner et al. 2013; Compant et al. 2019; Sivakumar et al. 2020). Bacteria are abundantly found in the phyllosphere as compared to other microbial communities. Around more than 10^{26} bacterial cells share the same habitat, without including other taxonomic groups (Leveau 2007; Carvalho and Castillo 2018). In a healthy environment, it is found that around 10^5–10^7 microbes reside on a cm^2 area of host plant tissue (Parasuraman et al. 2019). Microbes of phylum *Actinobacteria, Bacteroidetes, Chloroflexi, Elusimicrobia, Firmicutes, Proteobacteria (α-proteobacteria, β-proteobacteria γ-proteobacteria, δ-proteobacteria, and ε-proteobacteria), Sphingobacteria, Solibacteres,* and genus *Arthrobacter, Bacillus, Massilia, Pantoea, Pseudomonas,* and *Thermomicrobia* are abundantly found in the phyllosphere. Table 13.2 describes an outline of microbial diversity commonly found in plants.

13.4 Structure and Function of Phyllosphere-Associated Microbiome

13.4.1 Structure of Phyllospheric Microbes

Phyllosphere structure refers to the composition of taxa and their relative abundance whereas function refers to the global phenotype (role) observed, which results from the sum of all the functions of the members of the complex assembly (Uroz et al. 2019). The host plant physiology, morphology, and nutrients availability impact bacterial growth on a leaf surface. The rate of volatile organic compounds (VOC) production and antimicrobial compounds synthesized by the host plant also governs plant growth. These common VOCs are single carbon atom compounds such as formaldehyde (CH_2O), methanol (CH_3OH), and chloromethane (CH_3Cl) (Bringel and Couée 2015). It is observed that leaves of different host plant of same species show similar bacterial community profiles, e.g., fungi genus *Sporobolomyces* is found on phylloplane of barley and wheat plant (Aleklett et al. 2014) indicating dependence for a particular type of nutrients and other characteristics for growth and development.

Table 13.2 Microbes found in the different phyllospheric region of plants

Phyllospheric region	Plant species	Microbes associated	References
Anthosphere	*Malus domestica*	*Entrobacteriaceae* and *Pseudomonas*	Steven et al. (2018)
	Camellia japonica	*Aureobasidium* sp. Ch3F3, *Bacillus* sp. LU1006, *Bacillus subtilis* LU1007, *Cladosporium* sp. LU172, *Cladosporium cladosporioides* LU174, *Pseudomonas* sp. A3/B1, *Pseudomonas fluorescens* LU1004, and *Pseudomonas marginalis* LU1009	Van Toor et al. (2005)
Caulosphere	*Harbuu and Qilxuu*	*Candida blattae, Candida humilis, Hanseniaspora uvarum, Lachancea thermotolerans, Meyerozyma guilliermondii, Pichia kudriavzevii*, and *Saccharomyces cerevisiae,*	Koricha et al. (2019)
	Fagus grandifolia	*Alternaria, Aureobasidium, Cladosporium, Coniothyrium, Fusarium, Geotrichum, Nodulosporium, Penicillium, Phoma, Rhinocladiella,* and *Trichoderma*	Cotter and Blanchard (1982)
Carposphere	*Vitis vinifera*	*Bacillus, Citrobacter, Curtobacterium, Enterobacter, Erwinia, Frigobacterium, Gluconobacter, Massilia, Methylobacterium, Pantoea, Pseudomonas,* and *Sphingomonas*	Kecskeméti et al. (2016)
	Olive fruit	*Actinobacteria, Firmicutes* and *Proteobacteria*	Pascazio et al. (2015)
Phylloplane	*Camellia sinensis*	*Aspergillus flavus,* and *Penicillium chrysogenum* *Chaetomium, Curvularia, Rhizopus, Trichoderma*	Tanti et al., (2016)
	Triticum aestivum and *Hordeum vulgare*	*Cladosporium cladosporioides, Pseudomonas fluorescens,* and *Rhodotorula glutinis*	Southwell et al. (1999)
	Triticum aestivum	*Methylobacterium* sp., *Methylobacterium phyllosphaerae, Bacillus atrophaeus*	(Verma et al. 2013)
	Triticum aestivum	*Corynebacterium callunae, Arthrobacter humicola, Paenibacillus amylolyticus, Bacillus aryabhattai, Methylobacterium extorquens, Methylobacterium mesophilicum, Methylobacterium radiotolerans, Psychrobacter fozii, Pseudomonas fuscovaginae,*	(Verma et al. 2014)
	Triticum aestivum	*Methylobacterium mesophilicum, M. radiotolerans, M. phyllosphaerae,*	(Verma et al. 2015a)
	Triticum aestivum	*Achromobacter spanius, M. mesophilicum, Methylobacterium* sp.	(Verma et al. 2019)

(continued)

Table 13.2 (continued)

Phyllospheric region	Plant species	Microbes associated	References
Spermosphere	*Celosia argentea*	*Aspergillus niger, Bacillus subtilis, Fusarium solani,* and *Penicillium notatum*	Saritha and Sreeramulu (2013)
	Lycopersicon esculentum	*Bacteroidetes, Cloacimonetes* and *Spirochaetae*	Dong et al. (2019)
	Spinacia oleracea	Phyla: *Actinobacteria, Firmicutes, Gammaproteobacteria,* and *Proteobacteria* Genera: *Enterobacteriaceae* is abundance present followed by *Staphylococcaceae, Pseudomonadaceae, Bacillaceae* and *Rhizobiaceae.*	Lopez-Velasco et al. (2013)

13.4.2 Functions of Phyllospheric Microbes

Phyllosphere bacterial communities considerably influence the fitness and function of the host plant (Chaudhary et al. 2017; Thapa et al. 2017). The phytobiome influences plant growth pattern, evolutionary dynamics of phytomicrobiome, and physiological functions (Hawkes and Connor 2017; Baltrus 2017; Kour et al. 2020f; Singh et al. 2020). Although it is a nutrient poor region, the phyllosphere shows high species richness when compared with the rhizosphere (Parasuraman et al. 2019). Phyllosphere-associated microbes helps in maintaining plant growth by synthesizing phytohormones (as discussed in the introduction), growth-promoting nutrients, and also works as a biofertilizer, biopesticide (biocontrol agent), and phytostimulator (Mitter et al. 2013; Mondal et al. 2020; Thakur et al. 2020). Some examples of phyllosphere-associated microbes in different plants and their functions are mentioned in Table 13.3.

13.4.2.1 Recycling

Some phyllospheric microbes such as *Carludovica drudei*, *Costus laevis,* and *Grias cauliflora* are capable of fixing nitrogen and make it available to plants and animals. In the rainforest ecosystem, cyanobacteria and proteobacteria facilitate the intake of nitrogen (Fürnkranz et al. 2008). The phyllosphere and their inhabiting microbiota play a crucial role in the remediation of pesticide residues and recalcitrant atmospheric pollutants, recycling of essential elements (Morris and Riffaud 2004).

13.4.2.2 Biocontrol Agents

Some plant microbes act as a biocontrol agent, for example, *Cladasporium* spp., *Penicillum* spp., *Aspergillus flavus, Bacillus subtilis* provide resistance to *Helminthosporium oryzae* which causes brown spot on rice. *Bacillus pumilus* and *Trichoderma* are commercial biocontrol agents against the blast and blight diseases in *Oryza sativa* (De Costa et al. 2006; Harish et al. 2007). In *Camellia japonica,* microbes such as *Aureobasidium* sp. Ch3F3, *Bacillus subtilis* LU1007, *Bacillus* sp. LU1006, *Cladosporium cladosporioides* LU174, *Cladosporium* sp. LU172, *Pseudomonas*

Table 13.3 Function of phyllospheric microbes in specific host plant

Host plant	Microbes	Function	References
Saccharum officinarum L.	*M. extorquens* PPFM-So78	Improve plant growth, plant height, number of internodes, and cane yield	Madhaiyan et al. (2005)
Oryza sativa	*Aspergillus flavus*, *Bacillus subtilis*, *Cladasporium* sp., and *Penicillum* sp.	Biocontrol agent against *Helminthosporium oryzae* which causes brown spot on rice	Harish et al. (2007)
	B. pumilus and *Trichoderma*	Commercial biocontrol agents that are resistant to blast and blight diseases.	De Costa et al. (2006)
Camellia japonica	*Aureobasidium* sp. Ch3F3, *Bacillus* sp. LU1006, *Bacillus subtilis* LU1007, *Cladosporium cladosporioides* LU174, *Cladosporium* sp. LU172, *Pseudomonas* sp. A3/B1, *Pseudomonas fluorescens* LU1004, and *Pseudomonas marginalis* LU1009	They protect flowers from ascospore infection by *Ciborinia camelliae*.	Van Toor et al. (2005)
Lycopersicon esculentum	*Azospirillum brasilense*	Nitrogen-fixing bacterium in the phyllosphere	Pascazio et al. (2015)
	Methylobacterium spp.	Increases plant height, produce PGH like IAA, cytokinin, kinetin, and benzyl adenine purine (BAP). It stimulated cell division, cell enlargement, increase fruit weight by improving plant nutrient uptake, and hence increases the yield of agriculture.	Senthilkumar and Krishnamoorthy (2017)
	Acinetobacter was abundant. *Weissella cibaria* (OTU120) present around seeds, *Bacteroides thetaiotaomicron* (OTU628) present inside and outside the tomato seeds.	Helps in seed development and ripening of fruits by facilitating carbohydrate degradation.	Dong et al. (2019)

(continued)

Table 13.3 (continued)

Host plant	Microbes	Function	References
Triticum aestivum	*Acinetobacter, Bacillus, Kineococcus, Microbacterium, Proteus, Pseudomonas, Psychrobacter,* and *Streptomyces* sp.	Production of plant growth-promoting hormones such as auxin in phylloplane region. Nitrogen fixation, phosphate solubilization, HCN production.	Batool et al. (2016)
Zea mays	*Buttiauxella* sp. and *Raoultella* sp. belong to genera *Enterobacteriaceae* are the beneficial microbes found. Other genera such as *Alternaria, Cryptococcus, Penicillium, Pyrenochaeta,* and *Rhizopus* are also found.	Capable of degrading herbicides atrazine and 2,4,5-trichlorophenoxyacetic acid as well as promote plant growth.	Kong et al. (2019)
Arabidopsis thaliana	Phylum -*Proteobacteria, Actinobacteria,* and *Bacteroidetes* Genus—*Massilia,* and *Flavobacterium*	Protects against fluctuating temperature, UV radiation, and humidity.	Bodenhausen et al. (2013)
Vitis vinifera	Genera *Bacillus* and *Staphylococcus* are found in the endosphere whereas *Cupriavidus, Microbacterium, Methylobacterium,* and *Sphingomonas* inhabitants in the phyllosphere.	Protection against solar irradiation, drastic humidity, or temperature changes.	Vionnet et al. (2018)

fluorescens LU1004, *Pseudomonas* sp. A3/B1, and *Pseudomonas marginalis* LU1009 protects from ascospore infection produced by *Ciborinia camelliae* (Van Toor et al. 2005). Phyllospheric microbes may diminish the effect of pathogens by triggering plant immune response and producing antimicrobial compounds or by competing for limiting food and space on the host surface (Leveau 2019).

13.4.2.3 Growth Promoters

Microbes such as *Raoultella* sp., and *Buttiauxella* sp. promotes plant growth in maize. The genera *Rhizopus, Cryptococcus, Penicillium, Pyrenochaeta,* and *Alternaria* also act as plant growth promoters (Kong et al. 2019). In tomato plant, *Methylobacterium* spp., and *Bacteroides thetaiotaomicron* increases plant height by producing several plant growth hormone and facilitate ripening of tomato fruits, respectively (Senthilkumar and Krishnamoorthy 2017; Dong et al. 2019). In sugarcane, *Methylobacterium extorquens* PPFM-So78 improves plant growth and yield by increasing the internode number and plant height (Madhaiyan et al. 2005). The phyllospheric bacterial species *Acinetobacter, Bacillus, Kineococcus, Microbacterium, Pseudomonas, Psychrobacter, Proteus,* and *Streptomyces* sp. in wheat (Batool et al. 2016) and *Azospirillum brasilense* in tomato have been reported for nitrogen fixation (Pascazio et al. 2015).

13.4.2.4 Stress Tolerance

According to the study of Vionnet et al. (2018), *Microbacterium, Cupriavidus, Methylobacterium,* and *Sphingomonas* populaces in grapevine phyllosphere to combat against environmental stress conditions including solar irradiation, drastic humidity, temperature change, and desiccation. There are many reports on stress tolerance and mitigation of different abiotic stresses by plant microbiomes (Kour et al. 2020b; Rai et al. 2020; Sharaff et al. 2020; Yadav et al. 2020a).

13.4.2.5 Pathogenic Phyllospheric Microbes

There are many phyllospheric microbes that cause disease to plants (Parasitic nature). For example, *Burkholderia cepacia* causes sour skin in onion and are also pathogenic to human (Parke 2000), *Erwinia amylovora* inhibits the growth of apple and pear trees by causing pustules or mosaic patterns and spots on fruits and leaves, or smelly tuber rots to plant death. In many plants, *Agrobacterium tumefaciens* can induce crown gall infection. In potato, *Clavibacter michiganensis subsp. sepedonicus* infection results in ring rot disease (Vidaver and Lambrecht 2004; Uroz et al. 2019). *Phakopsora pachyrhizi* in soybean causes soybean rust (Young et al. 2012), *Podosphaera pannosa* in roses induce powdery mildew (Suthaparan et al. 2012), *Podosphaera xanthii, Corynespora cassiicola, and Sphaerotheca fuliginea* are pathogenic to cucumber (Suthaparan et al. 2014), *Bipolaris oryzae*, and *Magnaporthe oryzae* are harmful to rice (Shirasawa et al. 2012; Parada et al. 2015), *Botrytis cinereal* causes disease in broad bean (Khanam et al. 2005), grapevine (Ahn et al. 2015), and tomato (Xu et al. 2017). Mansfield et al. (2012) reported pathogenic bacteria ten plants from genus such as *Pseudomonas, Agrobacterium, Xanthomonas, Erwinia,* and *Xylella* have been reported.

13.5 Factors Effecting Structure and Function of Phytomicrobiome

Plant growth, health, and productivity are governed by the microbial community associated with it (Chaudhary et al. 2017). However, the microbiota found in the phyllosphere also depends on the physicochemical traits of the host, microbial ecological niche, and plant species in which that microbes reside. The structure and function of phytomicrobiome are regulated by biotic (intrinsic) and abiotic (extrinsic) factors (Steven et al. 2018; Kour et al. 2020a; Kumar et al. 2019; Subrahmanyam et al. 2020). Furthermore, the environment can alter the structure and size of the microbial community (Bittar et al. 2018).

Effect of macro and micronutrients such as sodium, potassium, phosphorous, and iron governs the microbial population profile (Kaur et al. 2020; Rastegari et al. 2020a, b). Even if phyllospheric region promotes stable microbial population, the major variation occurs due to heterogeneous nutrient distribution (e.g., glucose, fructose, sucrose) in the different regions of phyllosphere (Thapa et al. 2017). In phylloplane, carbon sources facilitate the modification of diverse microbiota (Mercier and Lindow 2000). According to Carvalho and Castillo (2018), continuous change in the environmental conditions such as availability of air, water, seeds, soil, and migration of microbes, animal-borne sources, wind, carbon dioxide, light, temperature, nitrogen nutrition, and environment exchange, phyllospheric microbiome is not stable and cannot be estimated precisely.

Other environmental conditions or abiotic factors include pH, temperature, geographic locations, water, and nutrition availability (Berg and Koskella 2018). Exposure to gamma irradiation, ozone, fertilizers, chemical pollutants, or biostimulants also affects considerably (Leveau 2019). Additionally, some important biotic factors such as plant organs, genotype, phenotype, age, generation, health status (immune system), and development stage of the host plant also define phyllospheric microbial community (Steven et al. 2018). Different parts of the same plant represent the same genotype but possess different phenotypes, which result in microbial communities' variation (Singh et al. 2018). The characteristic of phyllospheric microflora could be determined by plant volatile organic compound (VOC) emission such as methanol, terpenoids, and aldehydes (Farré-Armengol et al. 2016). Anthropogenic activities such as urbanization, fertilizer, pesticides, insecticides, herbicides, fungicides, and industrialization (Southwell et al. 1999; Hartmann et al. 2015) also have a great impact on phyllosphere microbiome (Fig. 13.3).

13.6 Phyllospheric Interaction and Ecosystem Dynamic

13.6.1 Microbes Interaction

The interactions of phyllospheric microbiome with respective host plants are very complex and dynamic in nature. Production of secondary metabolites supports the growth and survival of plant microbiota. Generally, glucosinolates are produced during the interaction of plant and microbes and works as a defensive molecule against plant pathogen. The interaction of microbiota with phyllosphere, increases plant fitness, and productivity of economically important crops (Chaudhary et al. 2017). Hence, this knowledge could be utilized to increase the yield of crops and plant resistance to diseases for the betterment of the agricultural sector.

Due to a wide range of carbon sources (sugars and organic acids), phyllospheric microflora undergo genomic alteration in order to get acclimatize nutrient composition for survival (Lindow and Brandl 2003). Terpenoids, aromatic compounds, fatty acid derivatives, nitrogen-containing compounds, and volatile sulfur compounds

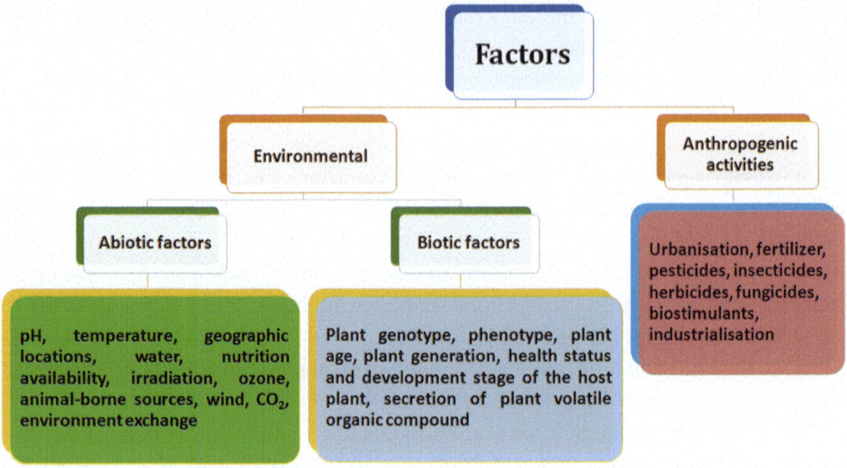

Fig. 13.3 Factors affecting phyllosphere microbiome

which are synthesized by plants acts as a carbon source for microbes (Bedia et al. 2018). Carbon source determines whether the microbial growth would be promoted or hindered. Methanol supports the growth of bacteria and fungi, whereas aldehydes and terpenoids inhibit. In return, the microbial VOCs act as an antimicrobial agent for plant pathogens and enhance plant growth and stress resistance (Farré-Armengol et al. 2016).

In adverse conditions, they can proliferate on one-carbon compounds; for example, methylotrophs can proliferate in the presence of methanol or organic acid (Vorholt 2012). Therefore, phyllospheric bacteria are capable to modify their local environment and production of extracellular polysaccharides to resist harsh environmental conditions suffered by the phyllospheric region such as variation in nutrient availability and other physical condition like temperature, light, and desiccation (Vorholt 2012). Plant alters its surface properties to shape its microbiome and ecosystem by interaction with keystone microbial species. The keystones play an important role in stabilizing species diversity. Future research focusing on beneficial microbial species recruitment would contribute significantly to sustainable agriculture development (Jones et al. 2019).

13.6.2 Chemical Exchange

Phytohormones also contribute to the degradation of chemical molecules which are harmful to plants and the atmosphere. Phylum *Actinobacteria* and genus *Arthrobacter* degrade detrimental organic compounds like pesticides and aromatic hydrocarbons produced by car exhausts that affect ozone by depollution process. Phyllosphere microbes also metabolizes some organic compounds emitted by plants such as chloromethane and isoprene. These microbes have ability to degrade aromatic hydrocarbons and therefore maybe used for the removal of atmospheric pollutants. Sometimes, they act as probiotic agents to the plant (Bringel and Couée 2015).

13.6.3 Climate Interaction

Phyllospheric microorganisms contribute geochemical changes in the environment, such as trace gas composition of an ecosystem, nitrogen fixation, and climate dynamics. The growth of climate regulating microbes in the phyllosphere is facilitated by volatile compounds (VOC). These microbes help in maintaining important global climate regulating organic gases in the atmosphere including methane, isoprene, chloromethane, and volatile dimethyl sulfide (DMS) (Bringel and Couée 2015). Methane and isoprene are the precursors for photochemical ozone production. Certain bacteria like *M. extorquens* CM4 and other methylotrophs are adapted to the phyllosphere condition and could survive on a single atom carbon source (Bringel and Couée 2015). The phyllosphere microbiota also plays an important role in carbon, nitrogen, phosphorus, and sulfur biogeochemical cycles, ecosystemic signaling, and climate regulation (Batool et al. 2016).

13.6.4 Environment Interaction

Processes such as immigration, multiplication, dispersal, emigration, death trigger the phyllospheric microbial community. Microbes arrive on the leaf surface, find suitable conditions to survive, produce a large number of offspring, move to a different compartment of the plants, departure to other plant, or die. These processes are influenced by the plant–microbe–environment triad (Leveau 2019).

13.7 Phyllospheric Microbes and Food Safety

Phyllosphere microbiota of edible plant contains a human pathogen, antibiotic-resistant bacteria such as *Escherichia coli* and *Salmonella* (Lindow and Leveau 2002; Chen et al. 2019). Worldwide distribution and consumption of unprocessed vegetables and fruits may exert a higher risk to the consumer. If extensive information is available on the behavior of plant pathogens and other indigenous plants bacteria, protocol to minimize contamination can be effectively developed (Lindow and Leveau 2002). Plant microbiome acts as a connecting channel between human and natural microbiota. Consumption of fruits and raw vegetables becomes a route for the spreading of antibiotic resistance in the human body. However, self-transmissible plasmids having resistance to antibiotics such as tetracycline have been studied in vegetable supermarkets (Chen et al. 2019). Therefore, food must be properly processed and preserve to obtain harmless products, extended shelf life, and maximum returns. Such practices decrease the chance of food-borne bacterial contamination and infections.

A number of bacterial genera of phyllosphere such as *Raoultella, Klebsiella, Serratia, Sphingomonas, Kluyvera, Chryseobacterium,* and *Hafnia* are capable of causing human diseases (Kong et al. 2019). Some of the strains of *B. cepacia* cause fatal lung infections with cystic fibrosis (Parke 2000). Epidemics of *Salmonella poona* infection reported in the United States are linked with melons consumption imported from Mexico. It was found that melons were produced and packaged in unhygienic conditions. The gastroenteritis outbreak that happened in the United States and Canada caused Guatemalan raspberries contamination with the protozoan *Cyclospora* (Chen et al. 2019). These pathogens may enter and create resistome in the human body by direct consumption of fresh fruits and vegetables without proper sterilization using safe methods as outlined in following Fig. 13.4 (Chen et al. 2019).

13.8 Applications of Phyllospheric Microbiota in Agriculture

Increasing human population has led to a rise in food demand. Unfortunately, increasing environmental pollution due to civilization and industrialization influences plant health by disturbing plant–microbes association and results in low agriculture yield. It is important to understand the concept of phytomicrobiome to achieve higher food production through sustainable agriculture practices. The colonization of the microbial community in every host plant impacts their health and disease management (Andrews 1992). Phyllospheric microbiota could be utilized for the protection and growth promotion of plants. They are also efficient in phytoremediation of toxic recalcitrant pollutants and the determination of pathogens. However, it is not possible to isolate and identify the phyllospheric microbes by

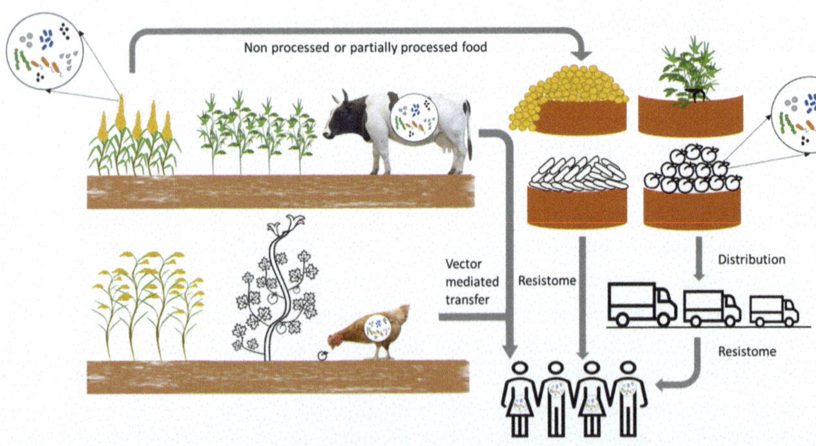

Fig. 13.4 Phyllospheric microbes (Pathogenic), food safety, and resistome development

culture-based method, since they failed to grow in the in-vitro conditions. Thus, molecular and analytical techniques, environmental genomics and metagenomics (Remus-Emsermann and Schlechter 2018), high-throughput sequencing techniques (Bodenhausen et al. 2013) can be employed to understand the plant physiology, function, and ecological properties of host, phyllosphere microbiomes, their interaction and identification at phylum, genus, and species level (Arya and Harel 2019). These advanced biotechnological techniques may support agro-industrial practice through understanding the plant–microbe relationship and further helps in breeding programs for improving crop yield.

Phyllospheric microorganisms could be utilized for disease control and management without harming the host plant. This phyllospheric microbe-mediated host plant defense leads to the activation of the antioxidants by reprogramming defense-related enzymes and activation of pathways for lignin deposition and phenolic production.

As discussed earlier, phytomicrobiota acts as biofertilizers, plant growth stimulators, biocontrol agents, and plays a vital role in nutrient cycling and VOC emission. The biocontrol agents can tolerate environmental stress like direct radiation, water scarcity, or moisture in rainy season, etc. They also increase crop resistance to biotic–abiotic stress and improves nutrient mobilization.

Depending on various techniques, the role of plant-associated microbe has been studied by many researchers. *Serratia marcescens* GPS 5 and *Bacillus circulans* GRS 243 are found in the groundnuts as biocontrol agents against *P. personata* which causes foliar disease (Kishore et al. 2005). Similarly, pyrosequencing studies showed that *Lysobacter capsica* AZ78 is a biocontrol agent of grapevine (Puopolo et al. 2014; Perazzolli et al. 2014). Many phyllospheric microbes synthesise phytohormones like IAA, auxin, cytokinins, abscisic acid, gibberellins, etc. and help in the plant growth by promoting cell division, elongation, and apical dominance (Egamberdieva et al. 2017).

Many *Methylobacterium* species are employed in numerous functions of phyllosphere including carbon, nitrogen, and phosphate cycle, maintain abiotic stress tolerance, seed vigor index, seedling length, and act as a plant probiotic. For example, *Methylobacterium oryzae* protects plants against oxidative stress, dryness, radiation and helps in defense mechanisms, and signaling. *Methylobacterium* is found in different types of crop plants such as *Saccharum officinarum, Cajanus cajan, Brassica campestris, Arachis hypogaea, Solanum tuberosum,* and *Raphanus sativus* (Parasuraman et al. 2019).

Southwell et al. (1999) showed that the three applications of mancozeb prevented recolonization of dominant microbes for 24 days in wheat and barley plants whereas the application of triadimefon did not reduce fungal populations. *Xanthomonas oryzae* cause Bacterial Leaf Blight (BLB) which decreases the rice yield. Ilsan et al. (2016) isolated *Streptomyces, Actinomadura, Nonomuraea,* and *Micromonospora* nonpathogenic actinomycetes from *O. sativa* capable of managing BLB disease caused by *Xanthomonas oryzae*. The study showed that the most efficient actinomycetes are *Nonomuraea* spp. Saleem et al. (2017) tried a mixture of bacterial species to act against the pathogens. However, it is proven that the mixture performed poorer than monocultures in host plant growth promotion. This may happen due to negative interspecific interactions among the bacteria.

Thus, data suggest that phyllospheric microbes play a very diverse and important aspect in host plant physiology, growth, and development, and hence can be utilized for the improvement of current farming practices.

13.9 Future Prospective

Modern agriculture technologies aim to fulfill the need for sufficient food for ever-increasing population of the world. However, current agriculture practices prefer overuse of chemical-based fertilizers and pesticides, which adversely affects the natural harmony of the plants and plant-associated microbes making the agriculture land sterile. Moreover, such chemicals are very harmful to human health posing diverse types of deadly diseases. According to the study conducted by Mishra et al. (2020), very less scientific literature is available on the phyllosphere as compared to soil microbial diversity from the past two decades. Therefore, future studies can be dedicated to the investigation of phyllospheric microbiota, its interaction, and benefits to the ecosystem and mankind.

Research data reveals the potential of phyllospheric microbes in the growth and development of plants through mutual benefits. These microbes help plants in developing biotic–abiotic stress and balance other physiological functions. Hence, the time has come to explore phyllospheric microflora in the agriculture sector for the purpose of higher and healthy crop production. So far, farmers have knowledge about the role of rhizobium in agriculture. Hence, awareness can be created about the phyllospheric microorganisms as an eco-friendly and cost-effective approach. Future, plant breeding programs can be designed by considering the benefits of phyllosphere microbiome. It can be done through the development of smart microbial consortia, suitable formulation, and delivery approach (Compant et al. 2019). Extensive studies shall be carried out for "Plant Probiotics" development that might prove economically feasible for boosting agricultural crops under biotic and abiotic stress conditions.

References

Ahn S, Kim S, Yun H (2015) Inhibition of *Botrytis cinerea* and accumulation of stilbene compounds by light-emitting diodes of grapevine leaves and differential expression of defense-related genes. Eur J Plant Pathol 143:753–765

Aleklett K, Hart M, Shade A (2014) The microbial ecology of flowers: an emerging frontier in phyllosphere research. Botany 92:253–266. https://doi.org/10.1139/cjb-2013-0166

Andrews JH (1992) Biological control in the phyllosphere. Annu Rev Phytopathol 30:603–635. https://doi.org/10.1146/annurev.py.30.090192.003131

Andrews JH, Harris RF (2000) The ecology and biogeography of microorganisms on plant surfaces. Annu Rev Phytopathol 38:145–180. https://doi.org/10.1146/annurev.py.30.090192.003131

Arya GC, Harel A (2019) Phyllosphere and its potential role in sustainable agriculture. In: Tripathi V, Kumar P, Tripathi P, Kishore A (eds) Microbial genomics in sustainable Agroecosystems: volume 1. Springer, Singapore, pp 39–65. https://doi.org/10.1007/978-981-13-8739-5_3

Azcón-Aguilar C, Barea JM (2015) Nutrient cycling in the mycorrhizosphere. J Soil Sci Plant Nutr 15:372–396. https://doi.org/10.4067/s0718-95162015005000035

Baltrus DA (2017) Adaptation, specialization, and coevolution within phytobiomes. Curr Opin Plant Biol. https://doi.org/10.1016/j.pbi.2017.04.023

Batool F, Rehman Y, Hasnain S (2016) Phylloplane associated plant bacteria of commercially superior wheat varieties exhibit superior plant growth promoting abilities. Front Life Sci 9:313–322. https://doi.org/10.1080/21553769.2016.1256842

Bedia C, Cardoso P, Dalmau N, Garreta-Lara E, Gómez-Canela C, Gorrochategui E et al (2018) Applications of metabolomics analysis in environmental research. Compr Anal Chem 82:533–582. https://doi.org/10.1016/bs.coac.2018.07.006

Berg M, Koskella B (2018) Nutrient- and dose-dependent microbiome-mediated protection against a plant pathogen. Curr Biol 28(15):2487–2492.e3. https://doi.org/10.1016/j.cub.2018.05.085

Bittar TB, Pound P, Whitetree A, Moore LD, Van Stan JT (2018) Estimation of Throughfall and Stemflow bacterial flux in a subtropical oak-cedar Forest. Geophys Res Lett 45(3):1410–1418. https://doi.org/10.1002/2017GL075827

Bodenhausen N, Horton MW, Bergelson J (2013) Bacterial communities associated with the leaves and the roots of *Arabidopsis thaliana*. PLoS One 8(2):e56329. https://doi.org/10.1371/journal.pone.0056329

Bringel F, Couée I (2015) Pivotal roles of phyllosphere microorganisms at the interface between plant functioning and atmospheric trace gas dynamics. Front Microbiol 6:486. https://doi.org/10.3389/fmicb.2015.00486

Bulgarelli D, Schlaeppi K, Spaepen S, Van Themaat EVL, Schulze-Lefert P (2013) Structure and functions of the bacterial microbiota of plants. Annu Rev Plant Biol 64:807–838. https://doi.org/10.1146/annurev-arplant-050312-120106

Carrión VJ, Perez-Jaramillo J, Cordovez V, Tracanna V, De Hollander M, Ruiz-Buck D et al (2019) Pathogen-induced activation of disease-suppressive functions in the endophytic root microbiome. Science 366(6465):606–612. https://doi.org/10.1126/science.aaw9285

Carvalho SD, Castillo JA (2018) Influence of light on plant–Phyllosphere interaction. Front Plant Sci 9:1482. https://doi.org/10.3389/fpls.2018.01482

Chaudhary D, Kumar R, Sihag K, Kumari A (2017) Phyllospheric microflora and its impact on plant growth: a review. Agric Rev 38(1):51–59. https://doi.org/10.18805/ag.v0iof.7308

Chen QL, Cui HL, Su JQ, Penuelas J, Zhu YG (2019) Antibiotic Resistomes in plant microbiomes. Trends Plant Sci 24(6):530–541. https://doi.org/10.1016/j.tplants.2019.02.010

Compant S, Reiter B, Sessitsch A, Nowak J, Clément C, Barka EA (2005) Endophytic colonization of *Vitis vinifera* L. by plant growth-promoting bacterium *Burkholderia* sp. strain PsJN. Appl Environ Microbiol 71(4):1685–1693. https://doi.org/10.1128/AEM.71.4.1685-1693.2005

Compant S, Samad A, Faist H, Sessitsch A (2019) A review on the plant microbiome: ecology, functions, and emerging trends in microbial application. J Adv Res 19:29–37. https://doi.org/10.1016/j.jare.2019.03.004

Cotter HVT, Blanchard RO (1982) The fungal Flora of bark of *Fagus grandifolia*. Mycologia 74(5):836–843. https://doi.org/10.1080/00275514.1982.12021593

De Costa DM, Pinto MTC, Geethanjalee HDN, Dissanayake N (2006) Suppression of rice pathogens by phyllosphere associated microflora of different rice varieties in Sri Lanka. Trop Sci 46(2):97–104. https://doi.org/10.1002/ts.63

Delmotte N, Knief C, Chaffron S, Innerebner G, Roschitzki B, Schlapbach R et al (2009) Community proteogenomics reveals insights into the physiology of phyllosphere bacteria. Proc Natl Acad Sci U S A 106(38):16428–16433. https://doi.org/10.1073/pnas.0905240106

Dong CJ, Wang LL, Li Q, Shang QM (2019) Bacterial communities in the rhizosphere, phyllosphere and endosphere of tomato plants. PLoS One 14(11):e0223847. https://doi.org/10.1371/journal.pone.0223847

Egamberdieva D, Wirth SJ, Alqarawi AA, Abd-Allah EF, Hashem A (2017) Phytohormones and beneficial microbes: essential components for plants to balance stress and fitness. Front Microbiol 8:2104. https://doi.org/10.3389/fmicb.2017.02104

Espenshade J, Thijs S, Gawronski S, Bové H, Weyens N, Vangronsveld J (2019) Influence of urbanization on epiphytic bacterial communities of the *Platanus x hispanica* tree leaves in a biennial study. Front Microbiol 10:675. https://doi.org/10.3389/fmicb.2019.00675

Farré-Armengol G, Filella I, Llusia J, Peñuelas J (2016) Bidirectional interaction between phyllospheric microbiotas and plant volatile emissions. Trends Plant Sci 21(10):854–860. https://doi.org/10.1016/j.tplants.2016.06.005

Fu SF, Sun PF, Lu HY, Wei JY, Xiao HS, Fang WT et al (2016) Plant growth-promoting traits of yeasts isolated from the phyllosphere and rhizosphere of *Drosera spatulata* lab. Fungal Biol 120(3):433–448. https://doi.org/10.1016/j.funbio.2015.12.006

Fürnkranz M, Wanek W, Richter A, Abell G, Rasche F, Sessitsch A (2008) Nitrogen fixation by phyllosphere bacteria associated with higher plants and their colonizing epiphytes of a tropical lowland rainforest of Costa Rica. ISME J 2(5):561–570. https://doi.org/10.1038/ismej.2008.14

Glick BR (2005) Modulation of plant ethylene levels by the bacterial enzyme ACC deaminase. FEMS Microbiol Lett 251(1):1–7. https://doi.org/10.1016/j.femsle.2005.07.030

Hardoim PR, Van Overbeek LS, Berg G, Pirttilä AM, Compant S, Campisano A et al (2015) The hidden world within plants: ecological and evolutionary considerations for defining functioning of microbial Endophytes. Microbiol Mol Biol Rev 79(3):293–320. https://doi.org/10.1128/mmbr.00050-14

Harish S, Saravanakumar D, Kamalakannan A, Vivekananthan R, Ebenezar EG, Seetharaman K (2007) Phylloplane microorganisms as a potential biocontrol agent against *Helminthosporium oryzae* Breda de Hann, the incitant of rice brown spot. Arch Phytopathol Plant Prot 40(2):148–157. https://doi.org/10.1080/03235400500383651

Hartmann M, Frey B, Mayer J, Mäder P, Widmer F (2015) Distinct soil microbial diversity under long-term organic and conventional farming. ISME J 9:1177–1194. https://doi.org/10.1038/ismej.2014.210

Hawkes CV, Connor EW (2017) Translating phytobiomes from theory to practice: ecological and evolutionary considerations. Phytobiomes J 1(2):57–69. https://doi.org/10.1094/PBIOMES-05-17-0019-RVW

Ilsan NA, Nawangsih AA, Wahyudi AT (2016) Rice phyllosphere actinomycetes as biocontrol agent of bacterial leaf blight disease on rice. Asian J Plant Pathol 10(2):1–8. https://doi.org/10.3923/ajppaj.2016.1.8

Jones P, Garcia BJ, Furches A, Tuskan GA, Jacobson D (2019) Plant host-associated mechanisms for microbial selection. Front Plant Sci 10:862. https://doi.org/10.3389/fpls.2019.00862

Katznelson H, Lochhead AG, Timonin MI (1948) Soil microorganisms and the rhizosphere. Bot Rev 14:543–586. https://doi.org/10.1007/BF02861843

Kaur T, Rana KL, Kour D, Sheikh I, Yadav N, Kumar V et al (2020) Microbe-mediated biofortification for micronutrients: present status and future challenges. In: Rastegari AA, Yadav AN, Yadav N (eds) Trends of microbial biotechnology for sustainable agriculture and biomedicine systems: perspectives for human health. Elsevier, Amsterdam, pp 1–17. https://doi.org/10.1016/B978-0-12-820528-0.00002-8

Kecskeméti E, Berkelmann-Löhnertz B, Reineke A (2016) Are epiphytic microbial communities in the carposphere of ripening grape clusters (*Vitis vinifera L.*) different between conventional, organic, and biodynamic grapes? PLoS One 11(8):e0160852. https://doi.org/10.1371/journal.pone.0160852

Kembel SW, O'Connor TK, Arnold HK, Hubbell SP, Wright SJ, Green JL (2014) Relationships between phyllosphere bacterial communities and plant functional traits in a neotropical forest. Proc Natl Acad Sci U S A 111(38):13715–13720. https://doi.org/10.1073/pnas.1216057111

Khanam NN, Ueno M, Kihara J, Honda Y, Arase S (2005) Suppression of red light-induced resistance in broad beans to *Botrytis cinerea* by salicylic acid. Physiol Mol Plant Pathol 66:20–29. https://doi.org/10.1016/j.pmpp.2005.03.006

Kishore GK, Pande S, Podile AR (2005) Biological control of late leaf spot of peanut (*Arachis hypogaea*) with chitinolytic bacteria. Phytopathology 95(10):1157–1165. https://doi.org/10.1094/PHYTO-95-1157

Knief C, Delmotte N, Chaffron S, Stark M, Innerebner G, Wassmann R et al (2012) Metaproteogenomic analysis of microbial communities in the phyllosphere and rhizosphere of rice. ISME J 6:1378–1390. https://doi.org/10.1038/ismej.2011.192

Kong X, Han Z, Tai X, Jin D, Ai S, Zheng X et al (2019) Maize (*Zea mays* L. Sp.) varieties significantly influence bacterial and fungal community in bulk soil, rhizosphere soil and phyllosphere. FEMS Microbiol Ecol 96(3). https://doi.org/10.1093/femsec/fiaa020

Koricha AD, Han DY, Bacha K, Bai FY (2019) Occurrence and molecular identification of wild yeasts from Jimma zone, south West Ethiopia. Microorganisms 7(12):633. https://doi.org/10.3390/microorganisms7120633

Kour D, Rana KL, Yadav N, Yadav AN, Kumar A, Meena VS et al (2019) Rhizospheric microbiomes: biodiversity, mechanisms of plant growth promotion, and biotechnological applications for sustainable agriculture. In: Kumar A, Meena VS (eds) Plant growth promoting Rhizobacteria for agricultural sustainability: from theory to practices. Springer, Singapore, pp 19–65. https://doi.org/10.1007/978-981-13-7553-8_2

Kour D, Kaur T, Devi R, Rana KL, Yadav N, Rastegari AA et al (2020a) Biotechnological applications of beneficial microbiomes for evergreen agriculture and human health. In: Rastegari AA, Yadav AN, Yadav N (eds) Trends of microbial biotechnology for sustainable agriculture and biomedicine systems: perspectives for human health. Elsevier, Amsterdam, pp 255–279. https://doi.org/10.1016/B978-0-12-820528-0.00019-3

Kour D, Rana KL, Kaur T, Devi R, Yadav N, Halder SK et al (2020b) Potassium solubilizing and mobilizing microbes: biodiversity, mechanisms of solubilization and biotechnological implication for alleviations of abiotic stress. In: Rastegari AA, Yadav AN, Yadav N (eds) Trends of microbial biotechnology for sustainable agriculture and biomedicine systems: diversity and functional perspective. Elsevier, Amsterdam, pp 177–202. https://doi.org/10.1016/B978-0-12-820526-6.00012-9

Kour D, Rana KL, Kaur T, Sheikh I, Yadav AN, Kumar V et al (2020c) Microbe-mediated alleviation of drought stress and acquisition of phosphorus in great millet (*Sorghum bicolour* L.) by drought-adaptive and phosphorus-solubilizing microbes. Biocatal Agric Biotechnol 23:101501. https://doi.org/10.1016/j.bcab.2020.101501

Kour D, Rana KL, Sheikh I, Kumar V, Yadav AN, Dhaliwal HS et al (2020d) Alleviation of drought stress and plant growth promotion by *Pseudomonas libanensis* EU-LWNA-33, a drought-adaptive phosphorus-solubilizing bacterium. Proc Natl Acad Sci India B. https://doi.org/10.1007/s40011-019-01151-4

Kour D, Rana KL, Yadav AN, Sheikh I, Kumar V, Dhaliwal HS et al (2020e) Amelioration of drought stress in foxtail millet (*Setaria italica* L.) by P-solubilizing drought-tolerant microbes with multifarious plant growth promoting attributes. Environ Sustain 3:23–34. https://doi.org/10.1007/s42398-020-00094-1

Kour D, Rana KL, Yadav AN, Yadav N, Kumar M, Kumar V et al (2020f) Microbial biofertilizers: bioresources and eco-friendly technologies for agricultural and environmental sustainability. Biocatal Agric Biotechnol 23:101487. https://doi.org/10.1016/j.bcab.2019.101487

Kumar V, Joshi S, Pant NC, Sangwan P, Yadav AN, Saxena A et al (2019) Molecular approaches for combating multiple abiotic stresses in crops of arid and semi-arid region. In: Singh SP, Upadhyay SK, Pandey A, Kumar S (eds) Molecular approaches in plant biology and environmental challenges. Springer, Singapore, pp 149–170. https://doi.org/10.1007/978-981-15-0690-1_8

Lambais MR, Barrera SE, Santos EC, Crowley DE, Jumpponen A (2017) Phyllosphere Metaproteomes of trees from the Brazilian Atlantic Forest show high levels of functional redundancy. Microb Ecol 73(1):123–134. https://doi.org/10.1007/s00248-016-0878-6

Leveau JHJ (2007) Microbia communities in the Phyllosphere. In: Riederer M, Müller C (eds) Biology of the plant cuticle. Annual Plant Reviews, pp 334–367. https://doi.org/10.1002/9780470988718.ch11

Leveau JH (2019) A brief from the leaf: latest research to inform our understanding of the phyllosphere microbiome. Curr Opin Microbiol 49:41–49. https://doi.org/10.1016/j.mib.2019.10.002

Lindow SE, Brandl MT (2003) Microbiology of the phyllosphere. Appl Environ Microbiol 69(4):1875–1883. https://doi.org/10.1128/AEM.69.4.1875-1883.2003

Lindow SE, Leveau JHJ (2002) Phyllosphere microbiology. Curr Opin Biotechnol 13(3):238–243. https://doi.org/10.1016/S0958-1669(02)00313-0

Lopez-Velasco G, Carder PA, Welbaum GE, Ponder MA (2013) Diversity of the spinach (*Spinacia oleracea*) spermosphere and phyllosphere bacterial communities. FEMS Microbiol Lett 346(2):146–154. https://doi.org/10.1111/1574-6968.12216

Madhaiyan M, Poonguzhali S, Lee HS, Hari K, Sundaram SP, Sa TM (2005) Pink-pigmented facultative methylotrophic bacteria accelerate germination, growth and yield of sugarcane clone Co86032 (*Saccharum officinarum* L.). Biol Fertil Soils 41:350–358. https://doi.org/10.1007/s00374-005-0838-7

Mansfield J, Genin S, Magori S, Citovsky V, Sriariyanum M, Ronald P et al (2012) Top 10 plant pathogenic bacteria in molecular plant pathology. Mol Plant Pathol 13(6):614–629. https://doi.org/10.1111/j.1364-3703.2012.00804.x

Marschner P (2007) Plant-microbe interactions in the Rhizosphere and nutrient cycling. In: Marschner P, Rengel Z (eds) Nutrient cycling in terrestrial ecosystems vol 10. Springer-Verlag, Berlin Heidelberg, pp 159–182. https://doi.org/10.1007/978-3-540-68027-7_6

Mercier J, Lindow SE (2000) Role of leaf surface sugars in colonization of plants by bacterial epiphytes. Appl Environ Microbiol 66(1):369–374. https://doi.org/10.1128/AEM.66.1.369-374.2000

Mishra S, Hättenschwiler S, Yang X (2020) The plant microbiome: a missing link for the understanding of community dynamics and multifunctionality in forest ecosystems. Appl Soil Ecol 145:103345. https://doi.org/10.1016/j.apsoil.2019.08.007

Mitter B, Brader G, Afzal M, Compant S, Naveed M, Trognitz F et al (2013) Advances in elucidating beneficial interactions between plants, soil, and bacteria. In: Advances in agronomy, vol 121. Elsevier Inc, pp 381–445. https://doi.org/10.1016/B978-0-12-407685-3.00007-4

Mondal S, Halder SK, Yadav AN, Mondal KC (2020) Microbial consortium with multifunctional plant growth promoting attributes: future perspective in agriculture. In: Yadav AN, Rastegari AA, Yadav N, Kour D (eds) Advances in plant microbiome and sustainable agriculture, Functional annotation and future challenges, vol 2. Springer, Singapore, pp 219–254. https://doi.org/10.1007/978-981-15-3204-7_10

Monteiro RA, Balsanelli E, Wassem R, Marin AM, Brusamarello-Santos LCC, Schmidt MA et al (2012) Herbaspirillum-plant interactions: Microscopical, histological and molecular aspects. Plant Soil 356:175–196. https://doi.org/10.1007/s11104-012-1125-7

Morris CE, Riffaud CMH (2004) Bacterial Survival Strategies. In: Encyclopedia of plant and crop science. https://doi.org/10.1081/e-epcs-120010611

Mwajita MR, Murage H, Tani A, Kahangi EM (2013) Evaluation of rhizosphere, rhizoplane and phyllosphere bacteria and fungi isolated from rice in Kenya for plant growth promoters. Springerplus 2:606. https://doi.org/10.1186/2193-1801-2-606

Nihorimbere V, Ongena M, Smargiassi M, Thonart P (2011) Beneficial effect of the rhizosphere microbial community for plant growth and health. Biotechnol Agron Soc Environ 15(2)

Pacifico D, Squartini A, Crucitti D, Barizza E, Schiavo FL, Muresu R et al (2019) The role of the Endophytic microbiome in the grapevine response to environmental triggers. Front Plant Sci 10:1256. https://doi.org/10.3389/fpls.2019.01256

Parada RY, Mon-nai W, Ueno M, Kihara J, Arase S (2015) Red-light-induced resistance to Brown spot disease caused by *Bipolaris oryzae* in Rice. J Phytopathol 163(2). https://doi.org/10.1111/jph.12288

Parasuraman P, Pattnaik S, Busi S (2019) Phyllosphere microbiome: functional importance in sustainable agriculture. In: Singh JS, Singh DP (eds) New and future developments in microbial biotechnology and bioengineering: microbial biotechnology in agro-environmental sustainability, pp 135–148. https://doi.org/10.1016/B978-0-444-64191-5.00010-9

Parke JL (2000) *Burkholderia cepacia*: friend or foe? Plant Heal Instr. https://doi.org/10.1094/phi-i-2000-0926-01

Pascazio S, Crecchio C, Ricciuti P, Palese AM, Xiloyannis C, Sofo A (2015) Phyllosphere and carposphere bacterial communities in olive plants subjected to different cultural practices. Int J Plant Biol 6(1). https://doi.org/10.4081/pb.2015.6011

Perazzolli M, Antonielli L, Storari M, Puopolo G, Pancher M, Giovannini O et al (2014) Resilience of the natural phyllosphere microbiota of the grapevine to chemical and biological pesticides. Appl Environ Microbiol 80(12):3585–3596. https://doi.org/10.1128/AEM.00415-14

Puopolo G, Giovannini O, Pertot I (2014) *Lysobacter capsici* AZ78 can be combined with copper to effectively control *Plasmopara viticola* on grapevine. Microbiol Res 169(7-8):633–642. https://doi.org/10.1016/j.micres.2013.09.013

Rai PK, Singh M, Anand K, Saurabhj S, Kaur T, Kour D et al (2020) Role and potential applications of plant growth promotion rhizobacteria for sustainable agriculture. In: Rastegari AA, Yadav AN, Yadav N (eds) Trends of microbial biotechnology for sustainable agriculture and biomedicine systems: diversity and functional perspectives. Elsevier, Amsterdam, pp 49–60. https://doi.org/10.1016/B978-0-12-820526-6.00004-X

Rana KL, Kour D, Kaur T, Devi R, Yadav AN, Yadav N et al (2020a) Endophytic microbes: biodiversity, plant growth-promoting mechanisms and potential applications for agricultural sustainability. Antonie Van Leeuwenhoek. https://doi.org/10.1007/s10482-020-01429-y

Rana KL, Kour D, Kaur T, Sheikh I, Yadav AN, Kumar V et al (2020b) Endophytic microbes from diverse wheat genotypes and their potential biotechnological applications in plant growth promotion and nutrient uptake. Proc Natl Acad Sci India B. https://doi.org/10.1007/s40011-020-01168-0

Rastegari AA, Yadav AN, Yadav N (2020a) New and future developments in microbial biotechnology and bioengineering: Trends of microbial biotechnology for sustainable agriculture and biomedicine systems: diversity and functional perspectives. Elsevier, Amsterdam

Rastegari AA, Yadav AN, Yadav N (2020b) New and future developments in microbial biotechnology and bioengineering: Trends of microbial biotechnology for sustainable agriculture and biomedicine systems: perspectives for human health. Elsevier, Amsterdam

Reddy CA, Saravanan RS (2013) Polymicrobial multi-functional approach for enhancement of crop productivity. In: Advances in applied microbiology. Volume 82, pp 53–113. https://doi.org/10.1016/B978-0-12-407679-2.00003-X

Remus-Emsermann MNP, Schlechter RO (2018) Phyllosphere microbiology: at the interface between microbial individuals and the plant host. New Phytol 218(4):1327–1332. https://doi.org/10.1111/nph.15054

Rilling JI, Acuña JJ, Sadowsky MJ, Jorquera MA (2018) Putative nitrogen-fixing bacteria associated with the rhizosphere and root endosphere of wheat plants grown in an andisol from southern Chile. Front Microbiol 9:2710. https://doi.org/10.3389/fmicb.2018.02710

Saleem M, Meckes N, Pervaiz ZH, Traw MB (2017) Microbial interactions in the phyllosphere increase plant performance under herbivore biotic stress. Front Microbiol 8:41. https://doi.org/10.3389/fmicb.2017.00041

Saritha P, Sreeramulu A (2013) Spermosphere micro organisms of *Celosia argentea* L. and it's relationship with germination studies. Int J LifeSc Bt Pharm Res 2(1):114–118

Senthilkumar M, Krishnamoorthy R (2017) Isolation and characterization of tomato leaf phyllosphere *Methylobacterium* and their effect on plant growth. Int J Curr Microbiol Appl Sci 6(11):2121–2136. https://doi.org/10.20546/ijcmas.2017.611.250

Sharaff MS, Subrahmanyam G, Kumar A, Yadav AN (2020) Mechanistic understanding of root-microbiome interaction for sustainable agriculture in polluted soils. In: Rastegari AA, Yadav AN, Yadav N (eds) Trends of microbial biotechnology for sustainable agriculture and biomedicine systems: diversity and functional perspectives. Elsevier, Amsterdam, pp 61–84. https://doi.org/10.1016/B978-0-12-820526-6.00005-1

Shirasawa H, Ueno M, Kihara J, Arase S (2012) Protective effect of red light against blast disease caused by *Magnaporthe oryzae* in rice. Crop Prot 39:41–44. https://doi.org/10.1016/j.cropro.2012.03.026

Singh P, Gobbi A, Santoni S, Hansen LH, This P, Péros JP (2018) Assessing the impact of plant genetic diversity in shaping the microbial community structure of vitis vinifera phyllosphere in the mediterranean. Front Life Sci 11(1):35–46. https://doi.org/10.1080/21553769.2018.1552628

Singh B, Boukhris I, Pragya KV, Yadav AN, Farhat-Khemakhem A et al (2020) Contribution of microbial phytases to the improvement of plant growth and nutrition: a review. Pedosphere 30:295–313. https://doi.org/10.1016/S1002-0160(20)60010-8

Sivakumar N, Sathishkumar R, Selvakumar G, Shyamkumar R, Arjunekumar K (2020) Phyllospheric microbiomes: diversity, ecological significance, and biotechnological applications. Plant Microb Sust Agricult:113–172. https://doi.org/10.1007/978-3-030-38453-1_5

Southwell RJ, Brown JF, Welsby SM (1999) Microbial interactions on the phylloplane of wheat and barley after applications of mancozeb and triadimefon. Australas Plant Pathol 28:139–148. https://doi.org/10.1071/AP99024

Steven B, Huntley RB, Zeng Q (2018) The influence of flower anatomy and apple cultivar on the apple flower phytobiome. Phytobiomes J 2(3):171–179. https://doi.org/10.1094/PBIOMES-03-18-0015-R

Stone BWG, Weingarten EA, Jackson CR (2018) The role of the phyllosphere microbiome in plant health and function. Annual Plant Reviews Online 1:1–24. https://doi.org/10.1002/9781119312994.apr0614

Subrahmanyam G, Kumar A, Sandilya SP, Chutia M, Yadav AN (2020) Diversity, plant growth promoting attributes, and agricultural applications of rhizospheric microbes. In: Yadav AN, Singh J, Rastegari AA, Yadav N (eds) Plant microbiomes for sustainable agriculture. Springer, Cham, pp 1–52. https://doi.org/10.1007/978-3-030-38453-1_1

Suman A, Verma P, Yadav AN, Srinivasamurthy R, Singh A, Prasanna R (2016) Development of hydrogel based bio-inoculant formulations and their impact on plant biometric parameters of wheat (*Triticum aestivum* L.). Int J Curr Microbiol Appl Sci 5:890–901

Suthaparan A, Stensvand A, Solhaug KA, Torre S, Mortensen LM, Gadoury DM et al (2012) Suppression of powdery mildew (*Podosphaera pannosa*) in greenhouse roses by brief exposure to supplemental UV-B radiation. Plant Dis 96(11):1653–1660. https://doi.org/10.1094/PDIS-01-12-0094-RE

Suthaparan A, Stensvand A, Solhaug KA, Torre S, Telfer KH, Ruud AK et al (2014) Suppression of cucumber powdery mildew by supplemental UV-B radiation in greenhouses can be augmented or reduced by background radiation quality. Plant Dis 98(10):1349–1357. https://doi.org/10.1094/PDIS-03-13-0222-RE

Tanti A, Bhattacharyya PN, Dutta P, Sarmah SR, Madhab M, Saikia D et al (2016) Diversity of phylloplane microflora in certain tea cultivars of Assam, north-East India. Eur J Biol Res 6(4):287–292

Thakur N, Kaur S, Tomar P, Thakur S, Yadav AN (2020) Microbial biopesticides: current status and advancement for sustainable agriculture and environment. In: Rastegari AA, Yadav AN, Yadav N (eds) Trends of microbial biotechnology for sustainable agriculture and biomedicine systems: diversity and functional perspectives. Elsevier, Amsterdam, pp 243–282. https://doi.org/10.1016/B978-0-12-820526-6.00016-6

Thapa S, Prasanna R, Ranjan K, Velmourougane K, Ramakrishnan B (2017) Nutrients and host attributes modulate the abundance and functional traits of phyllosphere microbiome in rice. Microbiol Res 204:55–64. https://doi.org/10.1016/j.micres.2017.07.007

Turner TR, James EK, Poole PS (2013) The plant microbiome. Genome Biol 14(6):209. https://doi.org/10.1186/gb-2013-14-6-209

Uroz S, Courty PE, Oger P (2019) Plant Symbionts are engineers of the plant-associated microbiome. Trends Plant Sci 24(10):905–916. https://doi.org/10.1016/j.tplants.2019.06.008

Van Toor RF, Pay JM, Jaspers MV, Stewart A (2005) Evaluation of phylloplane microorganisms for biological control of camellia flower blight. Australas Plant Pathol 34:525–531. https://doi.org/10.1071/AP05063

Verma P, Yadav AN, Kazy SK, Saxena AK, Suman A (2013) Elucidating the diversity and plant growth promoting attributes of wheat (*Triticum aestivum*) associated acidotolerant bacteria from southern hills zone of India. Natl J Life Sci 10:219–227

Verma P, Yadav AN, Kazy SK, Saxena AK, Suman A (2014) Evaluating the diversity and phylogeny of plant growth promoting bacteria associated with wheat (*Triticum aestivum*) growing in central zone of India. Int J Curr Microbiol Appl Sci 3:432–447

Verma P, Yadav AN, Khannam KS, Panjiar N, Kumar S, Saxena AK et al (2015a) Assessment of genetic diversity and plant growth promoting attributes of psychrotolerant bacteria allied with wheat (*Triticum aestivum*) from the northern hills zone of India. Ann Microbiol 65:1885–1899

Verma P, Yadav AN, Shukla L, Saxena AK, Suman A (2015b) Alleviation of cold stress in wheat seedlings by *Bacillus amyloliquefaciens* IARI-HHS2-30, an endophytic psychrotolerant K-solubilizing bacterium from NW Indian Himalayas. Natl J Life Sci 12:105–110

Verma P, Yadav AN, Kumar V, Singh DP, Saxena AK (2017) Beneficial plant-microbes interactions: biodiversity of microbes from diverse extreme environments and its impact for crop improvement. In: Singh DP, Singh HB, Prabha R (eds) Plant-microbe interactions in agro-ecological perspectives, Microbial interactions and agro-ecological impacts, vol 2. Springer, Singapore, pp 543–580. https://doi.org/10.1007/978-981-10-6593-4_22

Verma P, Yadav AN, Khannam KS, Mishra S, Kumar S, Saxena AK et al (2019) Appraisal of diversity and functional attributes of thermotolerant wheat associated bacteria from the peninsular zone of India. Saudi J Biol Sci 26:1882–1895. https://doi.org/10.1016/j.sjbs.2016.01.042

Vidaver AK, Lambrecht PA (2004) Bacteria as plant pathogens. Plant Heal Instr. https://doi.org/10.1094/phi-i-2004-0809-01

Vionnet L, De Vrieze M, Dutartre A, Gfeller A, Lüthi A, L'Haridon F, Weisskopf L (2018) Microbial life in the grapevine: what can we expect from the leaf microbiome? Oeno One 52(3):205–210. https://doi.org/10.20870/oeno-one.2018.52.3.2120

Vorholt JA (2012) Microbial life in the phyllosphere. Nat Rev Microbiol 10:828–840. https://doi.org/10.1038/nrmicro2910

Whipps JM, Hand P, Pink D, Bending GD (2008) Phyllosphere microbiology with special reference to diversity and plant genotype. J Appl Microbiol 105(6). https://doi.org/10.1111/j.1365-2672.2008.03906.x

Xu H, Nan FY, Lai LT, Wang R (2017) Effects of different LED light wavelengths on the resistance of tomato against Botrytis cinerea and the corresponding physiological mechanisms. J Integr Agric 16(1):106–114. https://doi.org/10.1016/S2095-3119(16)61435-1

Xu H, Yang Y, Tian Y, Xu R, Zhong Y, Liao H (2020) *Rhizobium* inoculation drives the shifting of rhizosphere fungal community in a Host Genotype Dependent Manner. Front Microbiol 10:3135. https://doi.org/10.3389/fmicb.2019.03135

Yadav AN, Kumar V, Dhaliwal HS, Prasad R, Saxena AK (2018) Microbiome in crops: diversity, distribution, and potential role in crop improvement. In: Crop improvement through microbial biotechnology. Elsevier, Amsterdam, pp 305–332. https://doi.org/10.1016/B978-0-444-63987-5.00015-3

Yadav AN, Kaur T, Kour D, Rana KL, Yadav N, Rastegari AA et al (2020a) Saline microbiome: biodiversity, ecological significance and potential role in amelioration of salt stress in plants. In: Rastegari AA, Yadav AN, Yadav N (eds) Trends of microbial biotechnology for sustainable agriculture and biomedicine systems: diversity and functional perspectives. Elsevier, Amsterdam, pp 283–309. https://doi.org/10.1016/B978-0-12-820526-6.00018-X

Yadav AN, Singh J, Rastegari AA, Yadav N (2020b) Plant microbiomes for sustainable agriculture. Springer, Cham

Young HM, George S, Narváez DF, Srivastava P, Schuerger AC, Wright DL et al (2012) Effect of solar radiation on severity of soybean rust. Phytopathology 102(8):794–803. https://doi.org/10.1094/PHYTO-10-11-0294

Mitigation Strategies for Abiotic Stress Tolerance in Plants Through Stress-Tolerant Plant Growth-Promoting Microbes

14

Salma Mukhtar, Kauser Abdulla Malik, Samina Mehnaz, and Zabish Khaliq

Abstract

Crop production is adversely affected by a number of abiotic stresses that arise due to anthropogenic activities and inherent edaphic factors. Several agronomic strategies have been used to mitigate the abiotic stresses to increase crop yield. Recently, researchers have been intrigued by the rhizosphere associated microorganisms from the plants growing in extreme environments. Bacterial strains belonging to the phyla Proteobacteria, Actinobacteria, Firmicutes, and archaeal strains related to the phyla Crenarchaeota and Euryarchaeota were abundantly found in the rhizosphere of plants growing under abiotic stress conditions. The well-known PGP strains include *Bacillus, Rhizobium, Frankia, Azotobacter, Azospirillum, Paenibacillus, Serratia, Pseudomonas,* and *Klebsiella*. Plant associated microbial communities promote plant growth under extreme conditions by mineral solubilization, phytohormones production, nitrogen fixation, siderophore, and HCN production. A number of rhizobacterial and archaeal strains have the ability to enhance plant defense mechanisms against different bacterial and fungal pathogens by the production of different antibacterial and antifungal compounds. Meta-omics approaches including metagenomics, metatranscriptomics, and metaproteomics are commonly used for microbial diversity analysis and microbe-mediated stress alleviation in different crops growing under extreme conditions. This chapter gives an overview of the archaeal and bacterial diversity

S. Mukhtar (✉)
School of Life Sciences, Forman Christian College (A Chartered University), Lahore, Pakistan

School of Biological Sciences, University of the Punjab, Lahore, Pakistan

K. A. Malik · S. Mehnaz · Z. Khaliq
School of Life Sciences, Forman Christian College (A Chartered University), Lahore, Pakistan

A. N. Yadav et al. (eds.), *Current Trends in Microbial Biotechnology for Sustainable Agriculture*, Environmental and Microbial Biotechnology,
https://doi.org/10.1007/978-981-15-6949-4_14

residing in the rhizosphere and root endosphere of plants growing under extreme environments and also explained different microbe-mediated mitigation strategies in plants under various abiotic stresses.

Keywords

Biofertilizers · Extreme environments · Meta-omics · Plant–microbe interactions · Rhizosphere microbiome

14.1 Introduction

Agricultural land is adversely affected due to various abiotic stresses such as drought, salinity, acidity, alkalinity, low/high temperatures, and nutrient starvation and this ultimately affects the crop production (Onaga and Wydra 2016; Pareek et al. 2009). More than 60% of the area is affected by drought globally, about 6% of the global land has been affected by salinity, 15% by acidic soils, 9% by minerals deficiency, and 57% by cold environments (Bui 2013; Cramer et al. 2011; Mittler 2006). In different regions of the world, about 30–70% plant growth is affected by abiotic or biotic stresses. Water uptake, biochemical, and physiological processes of plants were affected and production of major crops such as wheat, rice, maize, and sugarcane is reduced and ultimately a threat to global food security is potentially increasing (El-Beltagy and Madkour 2012; Mahalingam 2015; Tigchelaar et al. 2018).

Plants growing in extreme environments have adapted different protective, physiological, and genetic strategies to deal with adverse environmental conditions (Yolcu et al. 2016; Verma et al. 2019). A number of chemical compounds known as plant growth regulators produced by plants are usually used to modulate plant growth under different abiotic and biotic stresses (Vineeth et al. 2016; Wakchaure et al. 2018; Zhao et al. 2009). Plant hormones such as auxins, cytokinins, gibberellins, abscisic acid, and salicylic acid are considered as important growth regulators that control plant growth by playing an important role in plant metabolism and ultimately mitigation of abiotic stresses (Hu et al. 2013; Kazan 2013; Teale et al. 2006; Sharaff et al. 2020). The level of phytohormone production may be changed with the increase in abiotic stresses that adversely affect plant growth (Debez et al. 2001; Khan et al. 2014). Some synthetic compounds, for example, thiourea can be used as a plant growth regulator which promotes growth and productivity, particularly under extreme environments (Garg et al. 2006; Iqbal and Ashraf 2013; Islam et al. 2016).

Microbial communities associated with the plants growing under extreme conditions play a vital role in plant growth by increasing the nutrients available to the plants, help to tolerate abiotic stresses and provide resistance against different plant pathogens (Bulgarelli et al. 2012; Liljeqvist et al. 2015; Sessitsch et al. 2012; Turner et al. 2013; Yadav 2017). Extremophilic microorganisms including xerophiles, halophiles, acidophiles, alkaliphiles, and thermophiles have a genetic and physiological modification to survive under extreme conditions (Mukhtar et al. 2018a; Souza et al. 2015). Plant growth-promoting microbes enhance plant growth by increasing the nutrient availability to the plants such as nitrogen (N), potassium (K), phosphorus (P), and zinc (Zn), nitrogen fixation, production of phytohormones, including auxins, cytokinins, gibberellins, abscisic acid, and salicylic acid, production of siderophores and hydrogen cyanide (HCN) (Mukhtar et al. 2017; Yadav et al. 2017a; Yadav et al. 2020e, f). Root-associated bacteria and archaea also produce a variety of antifungal and antibacterial compounds that can be used to control various fungal and bacterial plant diseases (Jaisingh et al. 2016; Kumar et al. 2011; Subrahmanyam et al. 2020). Plant microbiome also improves plant health by suppressing bacterial and fungal pathogens such as *Xanthomonas* sp., *Fusarium* sp., *Aspergillus flavus,* and *Alternaria* sp. (Mehnaz et al. 2010; Khan et al. 2018; Singh et al. 2020a).

With the progress in the next sequencing approaches, interest in the microbial diversity analysis from the rhizosphere of plants growing under extreme environments has been increased (Mukhtar et al. 2018c, 2019a, b; Naik et al. 2009). Meta-omics approaches such as metagenomics, metatranscriptomics, and metaproteomics help us to understand the functional characterization of plant-associated microbial communities from extreme environments (Venter et al. 2004; Wilmes and Bond 2006; Zeyaullah et al. 2009; Zhou et al. 2015). These techniques can also be used to study the potential of plant growth-promoting bacteria and their role in the mitigation of abiotic stresses under various extreme environments (Castro et al. 2013; Liu et al. 2015; Wang et al. 2016). In this chapter, we have discussed the plant-associated microbial communities from various extreme environments and their role in growth promotion of economically important crops grown in areas that are affected by abiotic stresses.

14.2 Microbial Diversity of Microbes of Plants Growing Under Extreme Environments

The plant microbiome can be classified according to plant parts, such as rhizosphere, phyllosphere, and endosphere microbiomes (Fig. 14.1). The plant microbiome plays an important role in plant health and productivity. Rhizosphere and root endospheric bacteria, archaea, and fungi enable host plants to survive under extreme conditions (Hashem et al. 2016; Mukhtar et al. 2018b, c; Verma et al. 2014). Rhizosphere associated microbial communities have the ability to carry out metabolic processes that improve the soil health and promote the plant growth under abiotic stresses (Egamberdieva 2009; Khan et al. 2014; Biswas et al. 2018). Plant growth-promoting microorganisms can directly enhance plant health and productivity through mineral solubilization, fixation of atmospheric nitrogen, and production of phytohormones (Browne et al. 2009; Mehnaz et al. 2010; Mukhtar et al. 2019e). Some PGP microorganisms produce antibacterial and antifungal compounds, such as siderophores, HCN, and triazole to protect plants against different bacterial and fungal pathogens under extreme conditions. These microbes also trigger plant

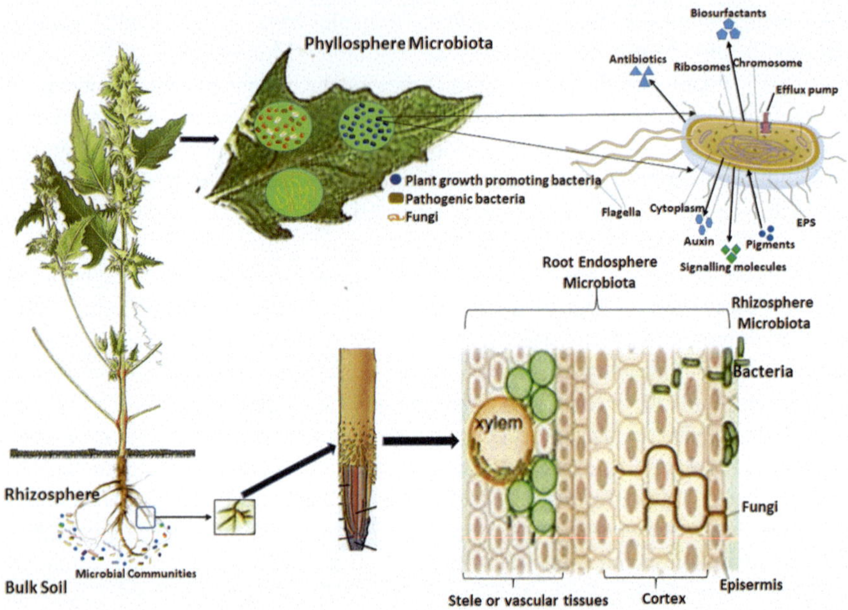

Fig. 14.1 Overview of the halophilic microbiome, their functions, and impact of microbial communities in the rhizosphere, endosphere, and phyllosphere of halophytes, Adapted from Mukhtar et al. (2019b)

immunity and increase resistance against pathogens (Khan et al. 2017; Mehnaz et al. 2010; Mukhtar et al. 2019e).

14.2.1 Saline Environments

Abiotic factors, including soil salinity and drought, are affecting the plant's growth and decreases crop yield by more than 40% and it increases day by day (Pitman and Lauchl 2002). At least 0.2 M NaCl is required for the growth of halophilic microorganisms from the hypersaline environments. Based on different salt concentrations, halophiles are classified as slight, moderate, and extreme halophiles. About 0.2–0.9 M NaCl concentrations are required for slight halophiles growth, 0.9–3.4 M NaCl concentrations are required for moderate halophiles growth, and 3.4–5.2 M NaCl concentrations are required for the optimal growth of extremophilic halophiles (DasSarma and DasSarma 2015; Mukhtar et al. 2018a). Halophiles have tolerance for different salt concentrations and can grow in various saline environments (Yadav et al. 2020a). Different parameters, such as pH, salt concentration, nutrients, and temperature variations affect the physiology of halophiles (Ruppel and FrankenP 2013). Halophilic bacteria and archaea use two main strategies to tolerate high osmotic stress. Mostly halophilic archaea and methanogenic bacteria use "Salt in" strategy. They acquire high KCl ions concentration copes with the high salt stress environment. Halotolerant and halophilic bacteria have the ability to grow in salt-affected environments by using small organic molecules, such as betaine, proline, ectoine, glutamine, and trehalose (DasSarma and DasSarma 2015; Oren 2015). Plant growth-promoting halophilic bacteria and archaea have also the ability to increase plant salt tolerance (Yadav et al. 2019; Yadav et al. 2017b). Halotolerant and halophilic bacterial genera including *Pseudomonas, Halomonas, Micrococcus, Planococcus, Marinococcus, Halobacillus, Virgibacillus, Arthrobacter, Nesterenkonia, Brachybacterium, Brevibacillus,* and *Pantoea* have been isolated from the rhizosphere of different halophytes as shown in Fig. 14.2 and Table 14.1 (Meng et al. 2018; Rueda-Puente et al. 2010; Zhao et al. 2016; Yadav et al. 2015d). Growth of barley and oat was increased in salinity environment by inoculation of *Pseudomonas* and *Bacillus* strains (Chang et al. 2014; Orhan 2016; Roy et al. 2014). *Burkholderia* strain PsN also positively affects the salt stress and increase maize growth (Naveed et al. 2014). *Halobacillus* and *Halomonas* were reported to increase of wheat growth and *Streptomyces* strain for tomato growth under salinity-affected environments (Palaniyandi et al. 2014). Soil and roots of halophytes, such as *Sporobolus, Dichanthium, Suaeda, and Cenchrus* have been used for the isolation and characterization of halophilic archaeal strains. Haloarchaeal strains such as *Halococcus, Halobacterium, Haloarcula,* and *Haloferax* have been studied for their plant growth-promoting abilities under hypersaline conditions (Wang et al. 2009; Yadav et al. 2015d) (Fig. 14.3 and Table 14.1).

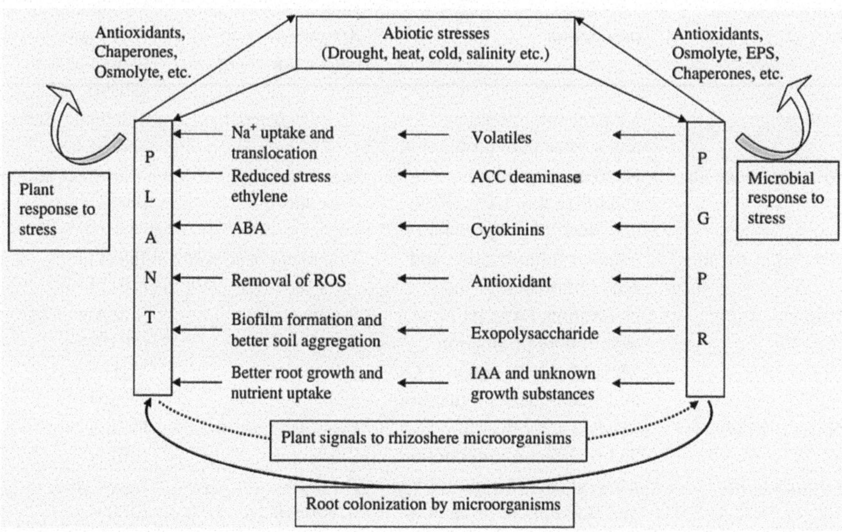

Fig. 14.2 A Conceptual diagram on the plant–microbe interactions under abiotic stress. Adapted from Grover et al. (2011)

Table 14.1 Plant growth-promoting microorganisms from different extreme environments

Extreme habitats/ microbe	PGP attributes	Host–plants	Reference
Salinity			
Virgibacillus	P-solubilization and siderophore production	*Acacia* spp.	Yadav et al. (2015e)
Halomonas	IAA production and ACC deaminase activity	*Salicornia bigelovii*	Rueda-Puente et al. (2010)
Marinococcus	P-solubilization, IAA production, and nitrogen fixation	*Salicornia* spp.	Mapelli et al. (2013)
Halobacillus	P-solubilization, IAA production, and biocontrol activity	*Salicornia europaea*	Zhao et al. (2016)
Micrococcus	P-solubilization and siderophore production	*Urochloa mutica*	Mukhtar et al. (2016)
Oceanobacillus	Mineral solubilization, IAA and siderophore production	*Atriplex amnicola*	Mukhtar et al. (2019a); (Mukhtar et al. 2019d)
Planococcus	P-solubilization and IAA production	*Triticum aestivum*	Rajput et al. (2013)
Pseudomonas	P-solubilization, nitrogen fixation, and siderophore production	*Hordeum vulgare*	Chang et al. (2014)
Salinivibrio	IAA and siderophore production	*Salsola stocksii* and *Atriplex Atriplex leucoclada amnicola*	Mukhtar et al. (2019a); (Mukhtar et al. 2019d)
Arthrobacter	Mineral solubilization, IAA, and siderophore production	*Atriplex leucoclada*	Ullah and Bano (2015)
Nesterenkonia	N_2 fixation, mineral solubilization, IAA, HCN, and siderophore production	*Salicornia strobilacea*	Mapelli et al. (2013)
Brachybacterium	Mineral solubilization and IAA production	*Salicornia brachiata*	Jha et al. (2012)
Pantoea	N_2 fixation, IAA, HCN, and siderophore production	*Suaeda salsa*	Siddikee et al. (2010)
Brevibacillus	Mineral solubilization, IAA, and siderophore production	*Wheat*	Yadav et al. (2018)
Haererohalobacter	Mineral solubilization, IAA, and siderophore production	*Salicornia brachiate*	Gontia et al. (2011)
Lysinibacillus	Mineral solubilization, IAA, and siderophore production	*Prosopis strombulifera*	Sgroy et al. (2009)
Halobacterium	P-solubilization and Nitrogen fixation	*Oryza sativa*	Wang et al. (2009)
Haloferax	IAA production and biocontrol activity	*Suaeda nudiflora*	Saxena et al. (2015)

(continued)

Table 14.1 (continued)

Extreme habitats/ microbe	PGP attributes	Host–plants	Reference
Halococcus	P-solubilization and siderophore production	*Sporobolus indicus*	Yadav et al. (2015d)
Drought			
Bacillus	P-solubilization, ACC deaminase activity, and IAA production	*Cupressus dupreziana*	Jorquera et al. (2012)
Kocuria	P-solubilization, ACC deaminase activity, and nitrogen fixation	*Zygophyllum dumosum*	Steinberger et al. (1995)
Frankia	P-solubilization and nitrogen fixation	*Aristida plumosa*	Bhatnagar and Bhatnagar (2009)
Virgibacillus	P-solubilization, IAA, HCN, and siderophore production	*Triticum aestivum*	Verma et al. (2016)
Azotobacter	P-solubilization, IAA production, and nitrogen fixation	*Artemesia* sp.	Hamdi and Yousef (1979)
Rhizobium	N_2 fixation, IAA and siderophore production	*Psoralea corylifolia*	Sorty et al. (2016)
Enterobacter	P-solubilization, nitrogen fixation, IAA, HCN, and siderophore production	*Phoenix dactylifera*	Ferjani et al. (2015)
Chryseobacterium	Nitrogen fixation, HCN, and siderophore production	*Glycine max*	Dardanelli et al. (2010)
Azoarcus	Nitrogen fixation, IAA, and siderophore production	*Leptochloa fusca*	Malik et al. (1997)
Pantoea	N_2 fixation, IAA, HCN, and siderophore production	*Suaeda salsa*	Siddikee et al. (2010)
Halobacterium	P-solubilization and nitrogen fixation	*Oryza sativa*	Wang et al. (2009)
Halococcus	P-solubilization and siderophore production	*Sporobolus indicus*	Yadav et al. (2015d)
Pseudomonas libanensis	Alleviation of drought stress and plant growth promotion	Wheat, maize, rice, sorghum, and finger millet	Kour et al. (2020b)
Streptomyces laurentii	Microbe-mediated alleviation of drought stress and acquisition of phosphorus in great millet (*Sorghum bicolour* L.)	*Amaranthus*, buckwheat, millets, and maize	Kour et al. (2020a)
Acinetobacter calcoaceticus	Amelioration of drought stress in foxtail millet (Setaria italica L.)	Wheat, maize, foxtail millet, and finger millet	Kour et al. (2020c)
Acidity			
Acidithiobacillus	P-solubilization, IAA, HCN, and siderophore production	*Pinus rigida*	Dang et al. (2017)

(continued)

Table 14.1 (continued)

Extreme habitats/ microbe	PGP attributes	Host–plants	Reference
Methylobacterium	P-solubilization, ACC deaminase activity, IAA, HCN, and siderophore production	*Triticum aestivum*	Wellner et al. (2011)
Lysinibacillus	P-solubilization, IAA, HCN, and siderophore production	*Triticum aestivum*	Verma et al. (2013)
Flavobacterium	P and K solubilization and biocontrol activity	*Hordeum vulgare*	Verma et al. (2014)
Azotobacter	P-solubilization, IAA production, and nitrogen fixation	*Artemesia* sp.	Upadhyay et al. (2009)
Pseudomonas	P-solubilization, IAA, HCN, and siderophore production	*Triticum aestivum*	Verma et al. (2013)
Pyrococcus	P and K solubilization and biocontrol activity	Thermal marine sediments	Gao et al. (2003)
Alkalinity			
Pseudorhodoplanes	IAA production, P-solubilization, and nitrogen fixation	*Photinia fraseri*	Seker et al. (2017)
Sphingomonas	P-solubilization and IAA production	*Smallanthus sonchifolius*	Moraes et al. (2012)
Curtobacterium	IAA production and P-solubilization	*Chrysanthemum morifolium*	Zawadzka et al. (2014)
Kocuria	P-solubilization, IAA production, and nitrogen fixation	*Dichanthium annulatum*	Mukhtar et al. (2018b)
Burkholderia	IAA and ACC deaminase production and nitrogen fixation	*Vitis vinifera*	Barka et al. (2006)
Paenibacillus	IAA production, P-solubilization, and nitrogen fixation	*Photinia fraseri*	Seker et al. (2017)
Heat			
Bacillus	P-solubilization, IAA, and siderophore production	*Triticum aestivum*	(Verma et al. 2018)
Arthrobacter	P-solubilization, IAA, and biocontrol activity	*Triticum aestivum*	Kumar et al. (2011)
Pseudomonas	P and Zn solubilization, IAA, HCN, and siderophore production	*Triticum aestivum*	Vyas et al. (2009)
Providencia	P and Zn solubilization, IAA production, and nitrogen fixation	*Amaranthus viridis*	Forchetti et al. (2007)
Staphylococcus	P-solubilization, HCN, and siderophore production	*Cupressus dupreziana*	Jorquera et al. (2012)

(continued)

Table 14.1 (continued)

Extreme habitats/ microbe	PGP attributes	Host–plants	Reference
Streptomyces	P-solubilization and biocontrol activity	*Vigna unguiculata*	Dimkpa et al. (2008)
Geobacillus	P-solubilization, IAA, and siderophore production and biocontrol activity	Petroleum contaminated Kuwait soil	Zeigler (2014) Al-Hassan et al. (2011)
Halococcus	P-solubilization and siderophore production	*Sporobolus indicus*	Yadav et al. (2015d)
Cold			
Kocuria	P-solubilization, IAA production, and nitrogen fixation	*Triticum aestivum*	Yadav et al. (2015a)
Bacillus	P-solubilization, IAA, and siderophore production	*Capsicum annuum*	Barka et al. (2006)
Arthrobacter	P-solubilization, IAA, and biocontrol activity	*Pinus roxburghii*	Singh et al. (2016)
Klebsiella	P-solubilization, IAA, and siderophore production	*Zea mays*	Rana et al. (2017)
Lysinibacillus	Mineral solubilization, IAA, and siderophore production	*Prosopis strombulifera*	Sgroy et al. (2009)
Pseudomonas	P-solubilization, nitrogen fixation, IAA, HCN, and siderophore production	*Solanum tuberosum*	Sati et al. (2013)
Methanosarcina	P-solubilization, IAA, and siderophore production and biocontrol activity	Siberian permafrost	Morozova and Wagner (2007)
Methylobacterium	P-solubilization, ACC deaminase activity, IAA, HCN, and siderophore production	*Triticum aestivum*	Saxena et al. (2016)

14.2.2 Arid and Semi-Arid Environments

Moisture content of the soil also affects the microbial communities associated with plants growing under arid and semi-arid environments. Moisture content is the main abiotic factor that affects microbial diversity associated with xerophytes, such as *Leptochloafusca, Aristida plumose, Zygophyllum dumosum, Artemesia* sp. and *Cupressus dupreziana* (Bhatnagar and Bhatnagar 2009; Buyanovsky et al. 1982). These microorganisms use small organic solutes, such as sugars, amino acids, and some other organic molecules including glutamine, ectoine, betaine, and trehalose to maintain their internal environment. The rhizosphere microbiome of xerophytes is getting more attention than other soil microbiomes since the last decade, due to its effectiveness (Jorquera et al. 2012). Microbiome of xerophytes has about 54% microbial diversity of Gram-positive bacteria especially Actinomycetes, such as *Kocuria, Streptomyces, Frankia,* and *Micrococcus* (Eppard et al. 1996; Steinberger et al. 1995). Some other genera such as *Azoarcus, Azotobacter, Bacillus, Enterobacter,* and

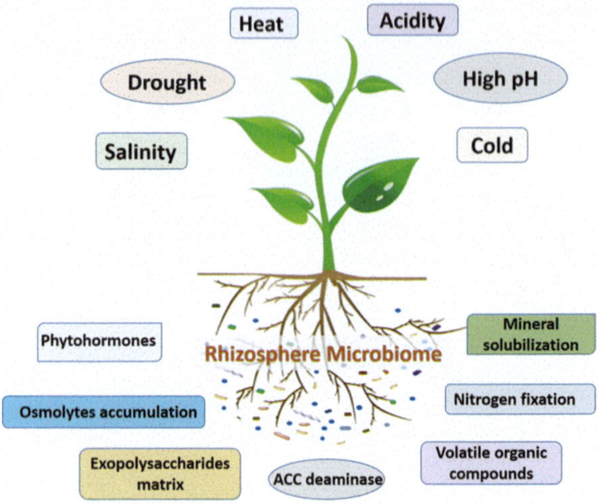

Fig. 14.3 Overview of microbe-mediated mitigation of abiotic stresses by plants. Adapted from Mukhtar et al. (2019c)

Virgibacillus have also been identified from the rhizosphere of xerophytes (Bhatnagar and Bhatnagar 2009; Kour et al. 2017; Malik et al. 1997). *Bacillus licheniformis* strain K11 has been reported to increase the growth of pepper plants in drought stress conditions (Figs. 14.2 and 14.3; Table 14.1). *Kocuria, Bacillus,* and *Pseudomonas* being drought-tolerant bacterial genera also have plant growth-promoting abilities, such as nitrogen fixation, HCN, P-solubilization, IAA, and siderophore production. These bacteria can also be used as bioformulation and biocontrol agents for different crops growing in arid and semi-arid environments (Jorquera et al. 2012; Kour et al. 2017; Lim and Kem 2013; Saxena et al. 2020; Thakur et al. 2020).

14.2.3 Acidic Environments

Soil pH plays an important role in shaping the composition of microbial communities associated with plants growing in acidic or alkaline environments (Feliatra et al. 2016; Wellner et al. 2011). Rhizosphere is the most active site for microbial diversity analysis from acidic environments. Many acidophilic and acidotolerant bacteria and archaea including *Pseudomonas, Azotobacter, Lysinibacillus, Acidithiobacillus, Serratia, Flavobacterium,* and *Pyrococcus* have been isolated and characterized from the various acidic environments (Dang et al. 2017; Feliatra et al. 2016; Upadhyay et al. 2009; Wellner et al. 2011). These microorganisms stimulate plants to withstand extremely acidic conditions and maintain their internal pH (Figs. 14.2 and 14.3; Table 14.1). Many PGP bacterial strains identified from the acidophilic environments promote plant growth of various crops such as rice, wheat, maize, and sugarcane to grow under acidic conditions (Verma et al. 2013; Wellner et al. 2011). Acidophilic microorganisms produce siderophores that are important for their survival under acidic conditions. These microbes have the ability to convert Fe^{3+} to Fe^{2+} in an acidic environment (Sorty et al. 2016; Vansuyt et al. 2007). Acid-tolerant microorganisms have been used as bio-inoculants for crops growing under acid-affected soil.

14.2.4 Alkaline Environments

Microbial diversity of different soda lakes around the world have been studied extensively during the last decade. The pH range of soda lake water is usually from 8 to 10 and even sometimes more than 12 (Antony et al. 2013; Grant and Sorokin 2011). The rhizosphere of plants such as *Dichanthium annulatum, Chrysanthemum morifolium, Photinia fraseri,* and *Smallanthus sonchifolius* present in the alkaline environment has unique microbial diversity as compared to soils with neutral pH because alkaline soils have less carbon and more methane and hydrogen content (Pikuta et al. 2003; Tiago et al. 2004). These microorganisms maintain their functional and structural integrity of cytoplasmic proteins by using specific proteins and enzymes (Jones et al. 1998; Zawadzka et al. 2014). Many alkaliphilic bacterial and archaeal strains such as *Sphingomonas, Pseudorhodoplanes, Paenibacillus, Arthrobacter, Burkholderia,* and *Curtobacterium* have been characterized by alkaline environments (Figs. 14.2 and 14.3; Table 14.1). A huge number of microbes identified from alkaline environments showed phytohormones production and P-solubilization ability (Rastegari et al. 2020; Yadav 2020). Rhizosphere microbiome of crops such as wheat, rice, maize, and barley are considered as important sources for maintaining the production and yield of these crops. These alkaliphilic bacteria having multi PGP abilities can be used for the improvement of plant growth in alkaline environments (Mukhtar et al. 2018b; Nautiyal et al. 2000).

14.2.5 Hot Environments

Temperature is one of the important abiotic factors which has effects on seed germination, photosynthesis rate, and membrane permeability of plants (Xu et al. 2014). Various plants growing in hot environments such as *Triticum aestivum, Vigna unguiculata, C. dupreziana,* and *Sporobolus indicus* have special enzymes and proteins to survive under hot environments. Rhizosphere and root-associated microbial communities from these environments have the ability to promote plant growth by increasing phytohormones production, nitrogen fixation, HCN and siderophores production, and P-solubilization as shown in Fig. 14.2 and Table 14.1 (Mukhtar et al. 2017; Vyas et al. 2009; Verma et al. 2018). Many bacteria have the ability to solubilize different minerals such as P, Zn, Al, and K by producing different organic acids, gluconic acid, formic acid, and citric acid in high temperature (Verma et al. 2014, 2016). A huge number of microbial genera such as *Staphylococcus, Arthrobacter, Streptomyces, Pseudomonas, Providencia,* and *Geobacillus* could be used as biofertilizers for plants growth under hot environments (Dimkpa et al. 2008; Gao et al. 2003; Zeigler 2014).

14.2.6 Cold Environments

Microbial diversity from cold environments is of particular importance in global ecology. A number of lakes and other aquatic ecosystems have very low temperatures permanently or seasonally (Singh 2014; Yadav et al. 2015b; Yadav et al. 2015c). Some plant species such as *Pinus roxburghii, Zea mays, Capsicum annuum,* and *T. aestivum* can grow under cold conditions by freezing tolerance or avoiding cooling of the tissue water (Thomashow 2010). Psychrophilic microorganisms have maximum functional activities at low temperatures as compared to mesophiles. Cold-tolerant plants have different microbial diversity and ability to tolerate cold and drought stress by solubilization of minerals, activation of defense-related and cold-active enzymes, production of phytohormones and exopolysaccharides (Ait Bakra et al. 2006; Kaushal and Wani 2016; Yadav et al. 2016) (Figs. 14.2 and 14.3; Table 14.1).

Many cold-tolerant bacterial strains including *Bacillus, Kocuria, Arthrobacter, Janthinobacterium, Klebsiella, Lysinibacillus, Paenibacillus, Providencia, Methylobacterium,* and *Methanosarcina* were characterized from cold-tolerant plants (Selvakumar et al. 2011; Shukla et al. 2016; Singh 2014; Singh et al. 2016; Yadav et al. 2015a). A number of endophytic cold-tolerant bacterial strains were isolated from crops growing under the low-temperature conditions (Rana et al. 2020). These bacterial strains showed the ability to solubilize minerals, produce phytohormones, siderophores, and HCN (Rana et al. 2017). Psychrophilic plant growth microorganisms can be used as biofertilizers for improvement of crops such as wheat, rice, and sugarcane growing under cold environments (Kour et al. 2020a; Kour et al. 2020b; Kour et al. 2020c; Kour et al. 2020d).

14.3 Mitigation Strategies for Abiotic Stress Tolerance in Plants

14.3.1 Phytohormones Production

Among the production of many plant beneficial chemicals, the production of phytohormones, such as auxins, cytokinins, gibberellins, ethylene, and abscisic acid, is key striking aspects of extremophilic bacteria imparting plant growth promotion under the unsuitable salt-affected area (Dodd and Perez-Alfocea 2012). The cellular mechanisms of plant growth promotion along with increased root length, due to IAA producing PGPR are direct stimulation of cell differentiation and division (Desale et al. 2014; Gonzalez et al. 2015; Shakirova 2007; Trindade et al. 2010; Tiwari et al. 2020). The genera of halophilic/tolerant bacteria described as PGPRs are *Bacillus, Enterobacter, Micrococcus, Pseudomonas, and Serratia.* These bacteria, when used as inoculants for the host plants, showed improved growth of wheat, sugarcane, and corn, improved catalase and peroxidase activity along with the increased level of TSS (total soluble sugar) content, some amino acids and K+/Na + ratio under salt stress (Gontia et al. 2011; Mukhtar et al. 2017a, b; Mukhtar et al. 2019d). Cytokinins, the plant growth-stimulating phytohormone, are revealed to be produced by hypersaline soil isolated *Halobacillus* strain which increased shoot biomass under salt stress (Figs. 14.3 and 14.4; Table 14.1). The cytokinins signaling is not one-way signaling mechanism as shown by many studies, cytokinins producing *Bacilli* increased shoot biomass but reduced root length which may be due to the presence of abscisic acid in the roots (Arkhipova et al. 2007; Ilangumaran and Smith 2017). Some plant-associated methylotrophs, such as *Methylobacterium* and *Methylovorusmays,* synthesize and excrete indole acetic acid and cytokinins (Ivanova et al. 2001).

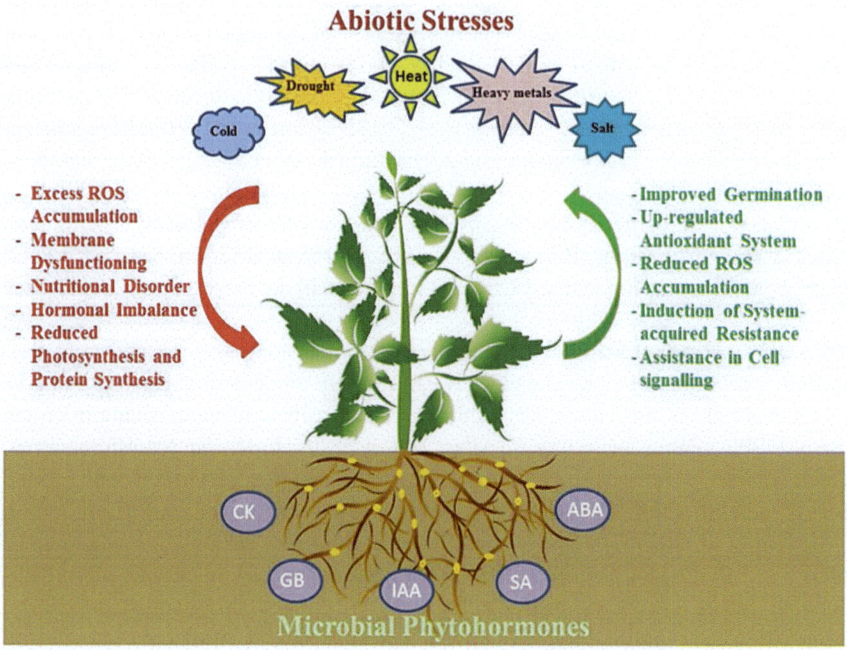

Fig. 14.4 An overview of mechanisms in microbial phytohormone-mediated plant stress tolerance. Rhizosphere-associated microorganisms produce indole-3-acetic acid (IAA), cytokinin (CK), gibberellin (GB), abscisic acid (ABA), and salicylic acid (SA) that help plants to withstand stress by enhancing its antioxidant potential, by upregulation of the antioxidant system and by the accumulation of compatible osmolytes thus reducing oxidative stress-induced damage; improving photosynthetic capacity and membrane stability; promoting cell division and stomatal regulation; stimulating the growth of root system, and acquisition of water and nutrients. Adapted from Egamberdieva et al. (2017)

14.3.2 Nitrogen Fixation

Nitrogen fixation by microbes is considered as one of the major methods for plant growth promotion because these microbes have the ability to fix atmospheric nitrogen and change it to nitrate that requires for the healthy and enhanced plant growth (Glick 2012; Kour et al. 2020d; Kaur et al. 2020). Frequently documented bacterial nitrogen-fixing genera include *Azotobacter, Azospirillum, Frankia, Bacillus, Klebsiella, Paenibacillus, Pantoea, Pseudomonas, Rhizobium, Salinibacter, and Serratia* (Ahmad and Kebret 2014; Jaisingh et al. 2016; Kuan et al. 2016). Apart from atmospheric nitrogen fixation, most of the plant growth-promoting rhizobacteria, root endophytic bacteria, as well as archaea, can produce phytopathogen (bacterial or fungal) limiting compounds to be used for biocontrol (Jaisingh et al. 2016; Kumar et al. 2011; Mondal et al. 2020). Rhizosphere microbiome was recognized as a source of suppressing fungal phytopathogens like, *Alternaria* sp., *Aspergillus flavus*, and *Fusarium* sp. making plants resistant to tested pathogens (Mehnaz et al. 2010). Plants from extreme environments have been explored to exploit associated microbiomes and several studies reported successful isolation and use of these isolates for the plant growth promotion. Such N_2-fixing reported genera are *Azospirillum* (Omar et al. 2009), *Bacillus* (Mukhtar et al. 2018a; Sorty and Shaikh 2015; Sorty et al. 2016), *Bradyrhizobium* (Panlada et al. 2013; Swaine et al. 2007), *Burkholderia* (Barka et al. 2006), *Enterobacter* and *Klebsiella* (Sorty et al. 2016; Mukhtar et al. 2017a, b), *Frankia* (Tani et al. 2003), *Micrococcus* (Dastager et al. 2010; Oliveira et al. 2009; Steinberger et al. 1995), *Pseudomonas* (Ali et al. 2009; Grichko and Glick 2001), *Rhizobium* (Remans et al. 2008; Sorty et al. 2016) with successful plant growth promotion (Figs. 14.2 and 14.3; Table 14.1).

14.3.2.1 Mineral Solubilization

Extremophilic microbes used as PGPR can directly enhance plant nutrient uptake by the roots (Figs. 14.2 and 14.3; Table 14.1). Apart from nitrogen-fixing microbe, many PGPR genera, including *Bacillus, Halobacillus, Enterobacter, Micrococcus, Pseudomonas, Virgibacillus, Pantoea, Rhizobium,* and *Serratia* have been reported for the solubilization of minerals (P, K, Zn) along with plant growth promotion (Mukhtar et al. 2017a, b; Sgroy et al. 2009; Yadav et al. 2020b, c). In the case of phosphate, PGPR converts its inorganic form into bioavailable organic phosphates and they can be used as a biofertilizer for the cultivation of barley, sugarcane, maize, rice, and wheat (Farrar et al. 2014; Jaisingh et al. 2016; Mukhtar et al. 2019d; Siddikee et al. 2010). The underlying mechanism for phosphate solubilization by microbes is their ability to produce organic acids; acetic acid, oxalic acid, lactic acid, and citric acid, responsible for phosphate conversion and the reported genera of phosphate solubilizing bacteria are *Bacillus, Enterobacter,* and *Pseudomonas* (Berendsen et al. 2012; Kumar et al. 2011; Ramaekers et al. 2010). The mineral-solubilizing and mobilizing microbes play important role in plant growth promotion, nutrient uptake, and soil health for sustainable agriculture (Kumar et al. 2019; Kumar et al. 2017; Singh et al. 2020b).

Potassium is the third most essential nutrient for plant growth; therefore, potassium solubilizing bacteria are used as biofertilizers in potassium limiting soils for agriculture. The reported PGPR genera for potassium solubilization are *Bacillus, Acidothiobacillus, Paenibacillus, Azospirillum, Marinococcus, Serratia, Streptomyces,* and *Azotobacter* (Zhao et al. 2016; Rana et al. 2019; Verma et al. 2017a, b). Several studies have reported potassium-solubilizing bacteria as biofertilizers for the cultivation of wheat, rice, maize, and sugarcane, to reduce the use of potassium fertilizer (Badar et al. 2006; Etesami et al. 2017). Zinc solubilizing bacteria, isolated from extreme saline environments, showed the ability to convert its inorganic form to organic form for plant uptake and utilization. The reported genera of zinc solubilization from various extreme environments include *Bacillus, Pseudomonas, Burkholderia, Brevibacillus,* and *Gluconacetobacter* (Figs. 14.2 and 14.3; Table 14.1) (Desai et al. 2012). These strains possess potentials to be used as chemo-attractants for the plant roots as well as PGPR for enhanced growth (Singh et al. 2020a; Singh and Yadav 2020; Yadav et al. 2020d).

14.3.2.2 ACC Deaminase Production

ACC deaminases, a viral compound for helping plants grow in unsupportive environmental conditions. Many rhizobacteria including *Oceanobacillus, Bacillus, Achromobacter, Halobacillus, Micrococcus, Virgibacillus,* and *Planococcus* can produce ACC deaminase for lowering the amount of ethylene (Figs. 14.2 and 14.3; Table 14.1). Ethylene is a two-step production and enzymatic conversion system; ACC synthase converts AdoMet (*S-adenosylmethionine*) to ACC (1-aminocyclopro pane-1-carboxylic acid), and ACC is converted to ethylene with the help of ACC oxidase (Etesami et al. 2015; Glick, 2014; Nadeem et al. 2007). The ACC deaminase producing plant-associated microbes protect against many abiotic stresses such as salinity, drought, heavy metal, water-logging, and petroleum exposure. ACC deaminase-producing rhizobacteria act as bioprotectant for maintaining ACC levels inside the host plant and its surroundings by hydrolyzing ACC through deaminase. It is indirectly involved in root elongation by lowering the inhibitory effects of ethylene on plant roots (Lima et al. 2011; Nikolic et al. 2011; Yadav et al. 2020g).

14.3.2.3 Exopolysaccharides Matrix

The production of EPS (exo-polysaccharides) by extremophilic rhizobacteria includes *Halobacillus, Pseudomonas, Corynebacterium, Nesterenkonia, Acinetobacter*, and *Planococcus*, works by creating a matrix for attachment of soil particles to plant roots and associated microbes thereafter creating a complex network in the soil within the plant root vicinity. The formation of such complex plant microbe-associated meshwork around the roots helps in establishing successful plant-microbe interactions and imparting bioprotection against phytopathogens such as protest, fungal, and bacterial (Mapelli et al. 2013; Sorty et al. 2016). Apart from providing biological benefits, the production of EPS supports beneficial physical properties of soil, such as water-holding capacity along with stabilizing the soil structure (Figs. 14.2 and 14.3; Table 14.1). Halotolerant PGPR with the ability of

EPS production has been successfully used under arid and saline conditions for chickpea, maize, sugarcane, and wheat (Mukhtar et al. 2019d; Oren 2015).

14.3.2.4 Siderophores Production and Biocontrol

Iron is considered one of the most crucial elements for the plant's growth. It is involved in many plant growth essential mechanisms such as nitrogen fixation, respiration, and photosynthesis (Figs. 14.2 and 14.3; Table 14.1). Iron availability for plant decreased in sodic, saline, arid, and acidic soils hindering healthy plant growth (Abbas et al. 2015). Many PGPR has the ability to produce siderophores which help in iron chelation thus, helping in iron availability for plants (Kour et al. 2019a, b). Production of siderophores by PGPR indirectly provides biocontrol to host plants, many PGPRs such as *Halobacillus, Bacillus, Pseudomonas, Halovibrio, Klebsiella,* and *Rhizobium* isolated from the arid and saline environments have the ability to produce siderophores (Singh et al. 2015).

The most fascinating aspect of PGPRs is the production of antifungal and antibacterial compounds; HCN (hydrogen cyanide), 2,4-diacetylphloroglucinol, pyoluteorin, gliotoxin, pyrrol-nitrin, and tensin. The reported extremotolerant PGPRs genera for antipathogenic compounds include *Aeromonas, Rhizobium, Bacillus, Halomonas, Acinetobacter, Pseudomonas,* and *Enterobacter* (Bhattacharyya and Jha 2012; Singh et al. 2015). The application of these bacteria has successfully protected the plants against tested fungal and bacterial pathogens. Hydrogen cyanide (HCN) is one of the most frequently reported antifungal compounds and has been reported in a number of PGPRs isolated from diversified environments (Barea et al. 2005). Apart from imparting antifungal protection, HCN-producing PGPRs have been reported for mineral (Zn, P, K) mobilization in soils (Frey et al. 2010; Rai et al. 2020; Suman et al. 2016). Some studies have shown that HCN-producing PGPRs in acidic soils play a vital role in iron sequestration, phosphate mobilization, thus increasing the bioavailability of phosphate for the host plants (Ström et al. 2002).

14.4 Conclusion and Future Prospects

Food production has increased as the world population doubled during the last few decades. Plants growing under harsh environments have special genetic and physiological modifications. Microbe-mediated stress alleviations have been extensively studied during the last few years. PGP microorganisms isolated and characterized from the rhizosphere and roots of plants growing under extreme environments can be used as bio-inoculants for increasing crop production under various abiotic stresses. A number of bacterial, archaeal, and fungal strains have the potential to be used as biocontrol agents against different bacterial and fungal diseases. Microbe-mediated abiotic stresses alleviation in crops may also be involved in the production of different organic compounds, especially extracellular enzymes, and can be used to improve soil properties, promote plant growth, and provide as signaling molecules to the plants. By using meta-omics approaches, plant growth-promoting microorganisms can be studied and utilized in a better way for crop improvement and production under abiotic stresses. New information from metagenomics, metatranscriptomics, and metaproteomics will help us to find out new roles of plant-associated microorganisms under extreme environments. Different microbial osmoregulatory and other stress-tolerant genes identified from a number of extreme environments may be used for the development of stress-tolerant transgenic crops in the future.

Acknowledgments We are highly thankful to the Higher Education Commission of Pakistan [Project # HEC (FD/2012/1843)] for the research grant.

References

Abbas G, Saqib M, Akhtar J (2015) Interactive effects of salinity and iron deficiency on different rice genotypes. J Plant Nutr Soil Sci 178:306–311

Ahmad M, Kebret M (2014) Mechanisms and applications of plant growth promoting rhizobacteria: current perspective. J King Saud Univ Sci 26:1–20

Ait Bakra E, Nowak J, Clement C (2006) Enhancement of chilling resistance of inoculated grapevine plantlets with a plant growth promoting rhizobacterium, *Burkholderia phytofirmans* strain PsJN. Appl Environ Microbiol 72:7246–7252

Al-Hassan JM, Al-Awadi S, Oommen S, Alkhamis A, Afzal M (2011) Tryptophan oxidative metabolism catalyzed by *Geobacillus stearothermophilus*: a thermophile isolated from Kuwait soil contaminated with petroleum hydrocarbons. Int J Tryptophan Res 4:1–6

Ali SZ, Sandhya V, Grover M, Kishore N, Rao LV, Venkateswarlu B (2009) *Pseudomonas* sp. strain AKM-P6 enhances tolerance of sorghum seedlings to elevated temperatures. Biol Fertil Soil 46:45–55

Antony CP, Kumaresan D, Hunger S, Drake HL, Murrell JC, Shouche YS (2013) Microbiology of Lonar Lake and other soda lakes. ISME J 7:468–476

Arkhipova TN, Prinsen E, Veselov SU, Martinenko EV et al (2007) Cytokinin producing bacteria enhances plant growth in drying soil. Plant Soil 292:305–315

Badar MA, Shafei AM, Sharaf El-Deen SH (2006) The dissolution of K and P-bearing minerals by silicate dissolving bacteria and their effect on sorghum growth. Res J Agri Biol Sci 2:5–11

Barea JM, Pozo MJ, Azcon R, Aguilar CA (2005) Microbial co-operation in the rhizosphere. J Exp Bot 56:1761–1778

Barka EA, Nowak J, Clément C (2006) Enhancement of chilling resistance of inoculated grapevine plantlets with a plant growth-promoting rhizobacterium *Burkholderia phytofirmans* strain PsJN. Appl Environ Microbiol 72:7246–7252

Berendsen RL, Pieterse CMJ, Bakker PAHM (2012) The rhizosphere microbiome and plant health. Tren Plant Sci 17:478–486

Bhatnagar A, Bhatnagar M (2009) Microbial diversity in desert ecosystems. Curr Sci 89:91–100

Bhattacharyya PN, Jha DK (2012) Plant growth-promoting rhizobacteria (PGPR): emergence in agriculture. World J Microbiol Biotechnol 28:1327–1350

Biswas S, Kundu D, Mazumdar S, Saha A, Majumdar B, Ghorai A et al (2018) Study on the activity and diversity of bacteria in a new Gangetic alluvial soil (Eutrocrept) under rice-wheat-jute cropping system. J Environ Biol 39:379–386. https://doi.org/10.22438/jeb/39/3/MRN-523

Browne P, Rice O, Miller SH et al (2009) Superior inorganic phosphate solubilization is linked to phylogeny within the *Pseudomonas fluorescens* complex. Appl Soil Ecol 43:131–138

Bui EN (2013) Soil salinity: a neglected factor in plant ecology and biogeography. J Arid Environ 92:14–25

Bulgarelli D, Rott M, Schlaeppi K, Ver Loren van Themaat E, Ahmadinejad N, Assenza F (2012) Revealing structure and assembly cues for *Arabidopsis* root-inhabiting bacterial microbiota. Nature 488:91–95

Buyanovsky G, Dicke M, Berwick P (1982) Soil environment and activity of soil microflora in the Negev desert. J Arid Environ 5:13–28

Castro AP, Sartori A, Silva MR, Quirino BF, Kruger RH (2013) Combining "omics" strategies to analyze the biotechnological potential of complex microbial environments. Curr Protein Pept Sci 14:447–458

Chang P, Gerhardt KE, Huang XD, Yu XM, Glick BR et al (2014) Plant growth promoting bacteria facilitate the growth of barley and oats in salt impacted soil: implications for phytoremediation of saline soils. Int J Phytoremediation 16:1133–1147

Cramer GR, Urano K, Delrot S, Pezzotti M, Shinozaki K (2011) Effects of abiotic stress on plants: a systems biology perspective. BMC Plant Biol 11:163

Dang P, Yu X, Le H, Liu J, Shen Z, Zhao Z (2017) Effects of stand age and soil properties on soil bacterial and fungal community composition in Chinese pine plantations on the loess plateau. PLoS One 12:e0186501

Dardanelli MS, Manyani H, Gonzalez-Barroso S, Rodriguez-Carvajal MA, Gil-Serrano AM, Espuny MR et al (2010) Effect of the presence of the plant growth promoting rhizobacterium (PGPR) *Chryseobacterium balustinum* Aur9 and salt stress in the pattern of flavonoids exuded by soybean roots. Plant Soil 328:483–493

DasSarma S, DasSarma P (2015) Halophiles and their enzymes: negativity put to good use. Curr Opin Microbiol 25:120–126

Dastager SG, Deepa CK, Pandey A (2010) Isolation and characterization of novel plant growth promoting *Micrococcus* sp. NII-0909 and its interaction with cowpea. Plant Physiol Biochem 48:987–992

Debez A, Chaibi W, Bouzid S (2001) Effect du NaCl et de regulatoeurs de croissance sur la germination d' *Atriplex halimus* L. Cah Agric 10:135–138

Desai S, Kumar PG, Sultana U, Pinisetty S, Ahmed MHSK, Amalraj LDE, Reddy G (2012) Potential microbial candidate strains for management of nutrient requirements of crops. Afr J Microbiol Res 6:3924–3931

Desale P, Patel B, Singh S, Malhotra A, Nawani N (2014) Plant growth promoting properties of *Halobacillus* sp. and *Halomonas* sp. in presence of salinity and heavy metals. J Basic Microbiol 54:781–791

Dimkpa C, Svatoš A, Merten D, Büchel G, Kothe E (2008) Hydroxamate siderophores produced by *Streptomyces acidiscabies* E13 bind nickel and promote growth in cowpea (*Vigna unguiculata* L.) under nickel stress. Can J Microbiol 54:163–172

Dodd IC, Perez-Alfocea F (2012) Microbial amelioration of crop salinity stress. J Exper Bot 63:3415–3428

Egamberdieva D (2009) Alleviation of salt stress by plant growth regulators and IAA producing bacteria in wheat. Acta Physiol Plant 31:861–864

El-Beltagy A, Madkour M (2012) Impact of climate change on arid lands agriculture. Agric Food Sec 1:3

Eppard M, Krumbein WE, Koch C, Rhiel E, Staley JT, Stackebrandt E (1996) Morphological, physiological, and molecular characterization of actinomycetes isolated from dry soil, rocks, and monument surfaces. Arch Microbiol 166:12–22

Etesami H, Emami S, Alikhani HA (2017) Potassium solubilizing bacteria (KSB): mechanisms, promotion of plant growth, and future prospects - a review. J Soil Sci Plant Nutr 7:897–911

Farrar K, Bryant D, Cope-Selby N (2014) Understanding and engineering beneficial plant-microbe interactions: plant growth promotion in energy crops. Plant Biotechnol J 12:1193–1206

Feliatra F, Lukistyowati I, Yoswaty D, Rerian H, Melina D, Hasyim W, Nugroho TT, Fauzi AR, Yolanda R (2016) Phylogenetic analysis to compare populations of acid tolerant bacteria isolated from the gastrointestinal tract of two different prawn species *Macrobrachium rosenbergii* and *Penaeus monodon*. AACL Bioflux 9:360–368

Ferjani R, Marasco R, Rolli E, Cherife H et al (2015) The date palm tree rhizosphere is a niche for plant growth promoting bacteria in the oasis ecosystem. Biomed Res Int 2015:153851

Forchetti G, Masciarelli O, Alemano S, Alvarez D, Abdala G (2007) Endophytic bacteria in sunflower (*Helianthus annuus* L.): isolation, characterization, and production of jasmonates and abscisic acid in culture medium. Appl Microbiol Biotechnol 76:1145–1152

Frey B, Rieder SR, Brunner I, Plötze M, Koetzsch S (2010) Weathering-associated bacteria from the Damma glacier forefield: physiological capabilities and impact on granite dissolution. Appl Environ Microbiol 76:4788–4796

Gao J, Bauer MW, Shockley KR, Pysz MA, Kelly RM (2003) Growth of hyperthermophilic archaeon *Pyrococcus furiosus* on chitin involves two family 18 chitinases. Appl Environ Microbiol 69:3119–3128

Garg BK, Burman U, Kathju S (2006) Influence of thiourea on photosynthesis, nitrogen metabolism and yield of cluster bean (*Cyamopsis tetragonoloba* (L.) Taub.) under rainfed conditions of Indian arid zone. Plant Growth Regul 48:237–245

Glick BR (2012) Plant growth-promoting Bacteria: mechanisms and applications. Hindawi Pub Corpor Sci 2012:23–30

Gontia I, Kavita K, Schmid M, Hartmann A, Jha B (2011) *Brachybacterium saurashtrense* sp. nov., a halotolerant root-associated bacterium with plant growth-promoting potential. Int J Syst Evol Microbiol 61:2799–2804

Grant WD, Sorokin DY (2011) Distribution and diversity of soda lake Alkaliphiles. In: Horikoshi K (ed) Extremophiles handbook. Springer, Japan, pp 27–54

Grichko VP, Glick BR (2001) Amelioration of flooding stress by ACC deaminase containing plant growth promoting bacteria. Can J Microbiol 47:77–80

Grover M, Ali SZ, Sandhya V, Rasul A, Venkateswarlu B (2011) Role of microorganisms in adaptation of agriculture crops to abiotic stresses. World J Microbiol Biotechnol 27:1231–1240

Hamdi Y, Yousef AN (1979) Nitrogen fixers in the rhizosphere of certain desert plants. Zentralbl Bakteriol Naturwiss 134:19–24

Hashem A, Abd Allah EF, Alqarawi A, Al-Huqail AA, Wirth S, Egamberdieva D (2016) The interaction between arbuscular mycorrhizal fungi and endophytic bacteria enhances plant growth of *Acacia gerrardii* under salt stress. Front Plant Sci 7:1089

Ilangumaran G, Smith DL (2017) Plant growth promoting rhizobacteria in amelioration of salinity stress: a systems biology perspective. Front Plant Sci 8:1768

Iqbal M, Ashraf M (2013) Gibberellic acid mediated induction of salt tolerance in wheat plants: growth, ionic partitioning, photosynthesis, yield and hormonal homeostasis. Environ Exp Bot 86:76–85

Islam F, Yasmeen T, Arif MS, Riaz M, Shahzad SM, Imran Q et al (2016) Combined ability of chromium (Cr) tolerant plant growth promoting bacteria (PGPB) and salicylic acid (SA) in attenuation of chromium stress in maize plants. Plant Physiol Biochem 108:456–467

Ivanova EG, Doronina NV, Trotsenko YA (2001) Aerobic methylobacteria are capable of synthesizing auxins. Microbiology 70:392–397

Jaisingh R, Kumar A, Dhiman M (2016) Isolation and characterization of PGPR from rhizosphere of *Sesame indicum* L. Int J Adv Res Biol Sci 3:238–244

Jones BE, Grant WD, Duckworth AW, Owenson GG (1998) Microbial diversity of soda lakes. Extremophiles 2:191–200

Jorquera MA, Shaharoona B, Nadeem SM, de la Luz MM, Crowley DE (2012) Plant growth-promoting rhizobacteria associated with ancient clones of creosote bush (*Larrea tridentata*). Microb Ecol 64:1008–1017

Kaur T, Rana KL, Kour D, Sheikh I, Yadav N, Kumar V et al (2020) Microbe-mediated biofortification for micronutrients: present status and future challenges. In: Rastegari AA, Yadav AN, Yadav N (eds) Trends of microbial biotechnology for sustainable agriculture and biomedicine systems: perspectives for human health. Elsevier, Amsterdam, pp 1–17. https://doi.org/10.1016/B978-0-12-820528-0.00002-8

Kaushal M, Wani SP (2016) Plant-growth-promoting rhizobacteria: drought stress alleviators to ameliorate crop production in drylands. Ann Microbiol 66:35–42

Kazan K (2013) Auxin and the integration of environmental signals into plant root development. Ann Bot 112:1655–1665

Khan MIR, Asgher M, Khan NA (2014) Alleviation of salt-induced photosynthesis and growth inhibition by salicylic acid involves glycine betaine and ethylene in mungbean (*Vigna radiata* L.). Plant Physiol Biochem 80:67–74

Khan N, Martínez-Hidalgo P, Ice TA, Maymon M, Humm EA, Nejat N, Sanders ER, Kaplan D, Hirsch AM (2018) Antifungal activity of Bacillus species against Fusarium and analysis of the potential mechanisms used in biocontrol. Front Microbiol 9:2363

Khan N, Maymon M, Hirsch AM (2017) Combating *Fusarium* infection using *Bacillus*-based antimicrobials. Microorganisms 5:E75

Kour D, Rana KL, Verma P, Yadav AN, Kumar V, Dhaliwal HS (2017) Drought tolerant phosphorus solubilizing microbes: diversity and biotechnological applications for crops growing under rainfed conditions. In: Proceeding of national conference on advances in food science and technology, pp 166–167

Kour D, Rana KL, Yadav N, Yadav AN, Kumar A, Meena VS et al (2019a) Rhizospheric microbiomes: biodiversity, mechanisms of plant growth promotion, and biotechnological applications for sustainable agriculture. In: Kumar A, Meena VS (eds) Plant growth promoting Rhizobacteria for agricultural sustainability : from theory to practices. Springer Singapore, Singapore, pp 19–65. https://doi.org/10.1007/978-981-13-7553-8_2

Kour D, Rana KL, Yadav N, Yadav AN, Singh J, Rastegari AA et al (2019b) Agriculturally and industrially important fungi: current developments and potential biotechnological applications. In: Yadav AN, Singh S, Mishra S, Gupta A (eds) Recent advancement in white biotechnology through Fungi, Perspective for value-added products and environments, vol 2. Springer International Publishing, Cham, pp 1–64. https://doi.org/10.1007/978-3-030-14846-1_1

Kour D, Rana KL, Kaur T, Sheikh I, Yadav AN, Kumar V et al (2020a) Microbe-mediated alleviation of drought stress and acquisition of phosphorus in great millet (*Sorghum bicolour* L.) by drought-adaptive and phosphorus-solubilizing microbes. Biocatal Agric Biotechnol 23:101501. https://doi.org/10.1016/j.bcab.2020.101501

Kour D, Rana KL, Sheikh I, Kumar V, Yadav AN, Dhaliwal HS et al (2020b) Alleviation of drought stress and plant growth promotion by *Pseudomonas libanensis* EU-LWNA-33, a drought-adaptive phosphorus-solubilizing bacterium. Proc Natl Acad Sci India B. https://doi.org/10.1007/s40011-019-01151-4

Kour D, Rana KL, Yadav AN, Sheikh I, Kumar V, Dhaliwal HS et al (2020c) Amelioration of drought stress in foxtail millet (*Setaria italica* L.) by P-solubilizing drought-tolerant microbes with multifarious plant growth promoting attributes. Environ Sustain 3:23–34. https://doi.org/10.1007/s42398-020-00094-1

Kour D, Rana KL, Yadav AN, Yadav N, Kumar M, Kumar V et al (2020d) Microbial biofertilizers: bioresources and eco-friendly technologies for agricultural and environmental sustainability. Biocatal Agric Biotechnol 23:101487. https://doi.org/10.1016/j.bcab.2019.101487

Kuan KB, Othman R, Abdul Rahim K, Shamsuddin ZH (2016) Plant growth-promoting rhizobacteria inoculation to enhance vegetative growth, nitrogen fixation and nitrogen remobilisation of maize under greenhouse conditions. PLoS One 11:e0152478

Kumar A, Prakash A, Johri B (2011) *Bacillus* as PGPR in crop ecosystem. In: Maheshwari (ed) Bacteria in agrobiology: crop ecosystems, vol 201. Springer-Verlag, Berlin, Heidelberg, pp 37–59

Kumar V, Yadav AN, Verema P, Sangwan P, Abhishake S, Singh B (2017) β-Propeller phytases: diversity, catalytic attributes, current developments and potential biotechnological applications. Int J Biol Macromolec 98:595–609. https://doi.org/10.1016/j.ijbiomac.2017.01.134

Kumar M, Kour D, Yadav AN, Saxena R, Rai PK, Jyoti A et al (2019) Biodiversity of methylotrophic microbial communities and their potential role in mitigation of abiotic stresses in plants. Biologia 74:287–308. https://doi.org/10.2478/s11756-019-00190-6

Liljeqvist M, Ossandon FJ, Gonzalez C, Rajan S, Stell A, Valdes J et al (2015) Metagenomic analysis reveals adaptations to a cold-adapted lifestyle in a low-temperature acid mine drainage stream. FEMS Microbiol Ecol 91:fiv011

Lim JH, Kem SD (2013) Induction of drought stress resistance by multi-functional PGPR *Bacillus licheniformis* K11 in pepper. Plant Pathol J 29:201–208

Lima JC, Arenhart RA, Margis-Pinheiro M, Margis R (2011) Aluminum triggers broad changes in microRNA expression in rice roots. Genet Mol Res 10:2817–2832

Liu Z, Li Y, Cao H, Ren D (2015) Comparative phospho-proteomics analysis of salt-responsive phosphoproteins regulated by the MKK9-MPK6 cascade in *Arabidopsis*. Plant Sci 241:138–150
Mahalingam R (2015) Consideration of combined stress: a crucial paradigm for improving multiple stress tolerance in plants. In: Mahalingam R (ed) Combined stresses in plants. Springer, Berlin, pp 1–25
Malik KA, Bilal R, Mehnaz S, Rasool G, Mirza MS, Ali S (1997) Association of nitrogen-fixing, plant growth promoting rhizobacteria (PGPR) with kallar grass and rice. Plant Soil 194:37–44
Mapelli F, Marasco R, Rolli E, Barbato M et al (2013) Potential for plant growth promotion of rhizobacteria associated with *Salicornia* growing in Tunisian hypersaline soils. Biomed Res Int 2013:248078
Mehnaz S, Baig DN, Lazarovits G (2010) Genetic and phenotypic diversity of plant growth promoting rhizobacteria isolated from sugarcane plants growing in Pakistan. J Microbiol Biotechnol 20:1614–1623
Meng X, Zhou J, Sui N (2018) Mechanisms of salt tolerance in halophytes: current understanding and recent advances. Open Life Sci 13(1):149–154
Mittler R (2006) Abiotic stress, the field environment and stress combination. Trends Plant Sci 11:15–19
Mondal S, Halder SK, Yadav AN, Mondal KC (2020) Microbial consortium with multifunctional plant growth promoting attributes: future perspective in agriculture. In: Yadav AN, Rastegari AA, Yadav N, Kour D (eds) Advances in plant microbiome and sustainable agriculture, Functional annotation and future challenges, vol 2. Springer, Singapore, pp 219–254. https://doi.org/10.1007/978-981-15-3204-7_10
Moraes RM, Melo IS, Samyanto J, Chandra S, Joshi V (2012) Bacterial community associated with autotrophic and heterotrophic cultures of medicinal plant *Smallanthus sonchifolius* (Yacon). Am J Plant Sci 3:1382–1389
Morozova D, Wagner D (2007) Stress response of methanogenic archaea from Siberian permafrost compared with methanogens from nonpermafrost habitats. FEMS Microbiol Ecol 61:16–25
Mukhtar S, Mirza MS, Awan HA, Maqbool A, Mehnaz S, Malik KA (2016) Microbial diversity and metagenomic analysis of the rhizosphere of Para grass (*Urochloa mutica*) growing under saline conditions. Pak J Bot 48:779–791
Mukhtar S, Ishaq A, Hassan S, Mehnaz S, Mirza MS, Malik KA (2017a) Comparison of microbial communities associated with halophyte (*Salsola stocksii*) and non-halophyte (*Triticum aestivum*) using culture-independent approaches. Pol J Microbiol 66:375–386
Mukhtar S, Shahid I, Mehnaz S, Malik KA (2017b) Assessment of two carrier materials for phosphate solubilizing biofertilizers and their effect on growth of wheat (*Triticum aestivum*). Microbiol Res 205:107–117
Mukhtar S, Mirza MS, Mehnaz S, Mirza BS, Malik KA (2018a) Diversity of *Bacillus*-like bacterial community in the rhizospheric and nonrhizospheric soil of halophytes (*Salsola stocksii* and *Atriplex amnicola*) and characterization of osmoregulatory genes in halophilic bacilli. Can J Microbiol 64:567–579
Mukhtar S, Kauser AM, Samina M (2018b) Isolation and characterization of haloalkaliphilic bacteria isolated from the rhizosphere of *Dichanthium annulatum*. J Adv Res Biotech 3:1–9
Mukhtar S, Mirza BS, Mehnaz S, Mirza MS, Mclean J, Kauser AM (2018c) Impact of soil salinity on the structure and composition of rhizosphere microbiome. World J Microbiol Biotechnol 34:136
Mukhtar S, Mehnaz S, Mirza MS, Malik KA (2019a) Isolation and characterization of halophilic bacteria from the rhizosphere of halophytes and non-rhizospheric soil samples. Braz J Microbiol 50:85–97
Mukhtar S, Mehnaz S, Malik KA (2019b) Microbiome of halophyte: diversity and importance for plant health and productivity. Microbiol Biotechnol Lett 47(1):1–10
Mukhtar S, Mehnaz S, Malik KA (2019c) Microbial diversity in the rhizosphere of plants growing under extreme environments and its impact on crops improvement. Environ Sustain. https://doi.org/10.1007/s42398-019-00061-5

Mukhtar S, Ahmad S, Bashir A, Mehnaz S, Malik KA (2019d) Identification of plasmid encoded osmoregulatory genes from halophilic bacteria isolated from the rhizosphere of halophytes. Microbiol Res 228:126307

Mukhtar S, Zareen M, Khaliq Z, Mehnaz S, Malik KA (2019e) Phylogenetic analysis of halophyte-associated rhizobacteria and effect of halotolerant and halophilic phosphate-solubilizing biofertilizers on maize growth under salinity stress conditions. Appl Microbiol. https://doi.org/10.1111/jam.14497

Nadeem SM, Zahir ZA, Naveed M, Arshad M (2007) Preliminary investigations on inducing salt tolerance in maize through inoculation with rhizobacteria containing ACC deaminase activity. Can J Microbiol 53:1141–1149

Naik BS, Shashikala J, Krishnamurthy Y (2009) Study on the diversity of endophytic communities from rice (*Oryza sativa* L.) and their antagonistic activities in vitro. Microbiol Res 164:290–296

Nautiyal CS, Bhadauria S, Kumar P, Lal H, Mondal R, Verma D (2000) Stress induced phosphate solubilization in bacteria isolated from alkaline soils. FEMS Microbiol Lett 182:291–296

Naveed M, Hussain MB, Zahir ZA, Mitter B, Sessitsch A (2014) Drought stress amelioration in wheat through inoculation with *Burkholderia phytofirmans* strain PsJN. Plant Growth Regul 73:121–131

Nikolic B, Schwab H, Sessitsch A (2011) Metagenomic analysis of the 1-aminocyclopropane-1-carboxylate deaminase gene (acdS) operon of an uncultured bacterial endophyte colonizing *Solanum tuberosum* L. Arch Microbial 193:665–676

Oliveira CA, Alves VMC, Marriel IE, Gomes EA et al (2009) Phosphate solubilizing microorganisms isolated from rhizosphere of maize cultivated in an oxisol of the Brazilian Cerrado biome. Soil Biol Biochem 41:1782–1787

Omar MNA, Osman MEH, Kasim WA, Abd El-Daim IA (2009) Improvement of salt tolerance mechanisms of barley cultivated under salt stress using *Azospirillum brasiliense*. Tasks Veg Sci 44:133–147

Onaga G, Wydra K (2016) Advances in plant tolerance to abiotic stresses. In: Abdurakhmonov IY (ed) Plant genomics. InTech, Rijeka

Oren A (2015) Halophilic microbial communities and their environments. Curr Opin Microbiol 33:119–124

Orhan F (2016) Alleviation of salt stress by halotolerant and halophilic plant growth-promoting bacteria in wheat (*Triticum aestivum*). Braz J Microbiol 47:621–627

Palaniyandi SA, Damodharan K, Yang SH, Suh JW (2014) *Streptomyces* sp. strain PGPA39 alleviates salt stress and promotes growth of 'micro tom' tomato plants. J Appl Microbiol 117:766–773

Panlada T, Pongdet P, Aphakorn L, Rujirek NN, Nantakorn B, Neung T (2013) Alleviation of the effect of environmental stresses using co-inoculation of mungbean by *Bradyrhizobium* and rhizobacteria containing stress-induced ACC deaminase enzyme. Soil Sci Plant Nut 59:559–571

Pareek A, Sopory SK, Bohnert HJ, Govindjee (2009) Abiotic Stress Adaptation in Plants: Physiological, Molecular and Genomic Foundation, Springer, the Netherlands

Pikuta EV, Hoover RB, Bej AK, Marsic D, Whitman WB, Cleland D et al (2003) *Desulfonatronum thiodismutans* sp. nov., a novel alkaliphilic, sulfate-reducing bacterium capable of lithoautotrophic growth. Inter J Syst Evol Microbiol 53:1327–1332

Pitman MG, Lauchl A (2002) In: Lauchli A, Luttage U (eds) Global impact of salinity and agricultural ecosystems in salinity: environment - plants - molecules. Kluwer Academic Publishers, Amsterdam, pp 3–20

Rai PK, Singh M, Anand K, Saurabhj S, Kaur T, Kour D et al (2020) Role and potential applications of plant growth promotion rhizobacteria for sustainable agriculture. In: Rastegari AA, Yadav AN, Yadav N (eds) Trends of microbial biotechnology for sustainable agriculture and biomedicine systems: diversity and functional perspectives. Elsevier, Amsterdam, pp 49–60. https://doi.org/10.1016/B978-0-12-820526-6.00004-X

Rajput L, Imran A, Mubeen F, Hafeez FY (2013) Salt-tolerant PGPR strain *Planococcus rifietoensis* promotes the growth and yield of wheat (*Triticum aestivum* L.) cultivated in saline soil. Pak J Bot 45:1955–1962

Rana KL, Kour D, Verma P, Yadav AN, Kumar V, Singh DH (2017) Diversity and biotechnological applications of endophytic microbes associated with maize (*Zea mays* L.) growing in Indian Himalayan regions. In: Proceeding of national conference on advances in food science and technology, pp 41–42

Rana KL, Kour D, Yadav AN (2019) Endophytic microbiomes: biodiversity, ecological significance and biotechnological applications. Res J Biotechnol 14:142–162

Rana KL, Kour D, Kaur T, Sheikh I, Yadav AN, Kumar V et al (2020) Endophytic microbes from diverse wheat genotypes and their potential biotechnological applications in plant growth promotion and nutrient uptake. Proc Natl Acad Sci India B. https://doi.org/10.1007/s40011-020-01168-0

Rastegari AA, Yadav AN, Yadav N (2020) New and future developments in microbial biotechnology and bioengineering: Trends of microbial biotechnology for sustainable agriculture and biomedicine systems: diversity and functional perspectives. Elsevier, Amsterdam

Remans R, Ramaekers L, Shelkens S, Hernandez G et al (2008) Effect of *Rhizobium, Azospirillum* co-inoculation on nitrogen fixation and yield of two contrasting *Phaseolus vulgaris* L. genotypes cultivated across different environments in Cuba. Plant Soil 312:25–37

Roy SJ, Negrão S, Tester M (2014) Salt resistant crop plants. Curr Opin Biotech 26:115–124

Rueda-Puente E, Castellanos-Cervantes T, Diaz de Leon-Alvarez J, Preciado-Rangel P, Almaguer-Vargas G (2010) Bacterial community of rhizosphere associated to the annual halophyte *Salicornia bigelovii* (Torr.). Terra Latinoamericana 28:345–353

Ruppel S, FrankenP WK (2013) Properties of the halophyte microbiome and their implications for plant salt tolerance. Func Plan Biol 40:940–951

Sati P, Dhakar K, Pandey A (2013) Microbial diversity in soil under potato cultivation from Cold Desert Himalaya, India. Hindawi Publishing Corporation, ISRN Biodiversity, p 767453

Saxena AK, Kaushik R, Yadav AN, Gulati S, Sharma D (2015) Role of Archaea in sustenance of plants in extreme saline environments. In: Proceeding of 56th AMI-2015 and international symposium on "Emerging Discoveries in Microbiology". https://doi.org/10.13140/RG.2.1.2073.9925

Saxena AK, Yadav AN, Rajawat M, Kaushik R, Kumar R, Kumar M, Prasanna R, Shukla L (2016) Microbial diversity of extreme regions: an unseen heritage and wealth. Indian J Plant Genet Resour 29:246–248

Saxena AK, Padaria JC, Gurjar GT, Yadav AN, Lone SA, Tripathi M et al (2020) Insecticidal formulation of novel strain of *Bacillus thuringiensis* AK 47. Indian Patent 340541

Seker MG, Sah I, Kırdök E, Ekinci H, Çiftçi YO, Akkaya O (2017) A hidden plant growth promoting bacterium isolated from in vitro cultures of Fraser Photinia (*Photinia fraseri*). Int J Agric Biol 19:1511–1519

Selvakumar G, Joshi P, Suyal P, Mishra PK, Joshi GK, Bisht JK, Bhatt JC, Gupta HS (2011) *Pseudomonas lurida* M2RH3 (MTCC 9245), a psychrotolerant bacterium from the Uttarakhand Himalayas, solubilizes phosphate and promotes wheat seedling growth. World J Microbiol Biotechnol 27:1129–1135

Sessitsch A, Hardoim P, Döring J, Weilharter A, Krause A, Woyke T et al (2012) Functional characteristics of an endophyte community colonizing roots as revealed by metagenomic analysis. Mol Plant-Microbe Interact 25:28–36

Shakirova FM (2007) Role of hormonal system in manifestation of growth promoting and anti-stress action of salicylic acid. In: Hayat S, Ahmad A (eds) Salicylic acid a plant hormone. Springer, Dordrecht, pp 69–89

Sharaff MS, Subrahmanyam G, Kumar A, Yadav AN (2020) Mechanistic understanding of root-microbiome interaction for sustainable agriculture in polluted soils. In: Rastegari AA, Yadav AN, Yadav N (eds) Trends of microbial biotechnology for sustainable agriculture and biomedicine systems: diversity and functional perspectives. Elsevier, Amsterdam, pp 61–84. https://doi.org/10.1016/B978-0-12-820526-6.00005-1

Shukla L, Suman A, Yadav AN, Verma P, Saxena AK (2016) Syntrophic microbial system for ex situ degradation of paddy straw at low temperature under controlled and natural environment. J Appl Biol Biotechnol 4:30–37

Singh S (2014) A review on possible elicitor molecules of cyanobacteria: their role in improving plant growth and providing tolerance against biotic or abiotic stress. J Appl Microbiol 117:1221–1244

Singh J, Yadav AN (2020) Natural bioactive products in sustainable agriculture. Springer, Singapore

Singh RN, Gaba S, Yadav AN, Gaur P, Gulati S, Kaushik R, Saxena AK (2016) First, high quality draft genome sequence of a plant growth promoting and cold active enzymes producing psychrotrophic *Arthrobacter agilis* strain L77. Stand Genom Sci 11:54

Singh A, Kumar R, Yadav AN, Mishra S, Sachan S, Sachan SG (2020a) Tiny microbes, big yields: microorganisms for enhancing food crop production sustainable development. In: Rastegari AA, Yadav AN, Yadav N (eds) Trends of microbial biotechnology for sustainable agriculture and biomedicine systems: diversity and functional perspectives. Elsevier, Amsterdam, pp 1–15. https://doi.org/10.1016/B978-0-12-820526-6.00001-4

Singh B, Boukhris I, Pragya KV, Yadav AN, Farhat-Khemakhem A et al (2020b) Contribution of microbial phytases in improving plants growth and nutrition: a review. Pedosphere 30:295–313. https://doi.org/10.1016/S1002-0160(20)60010-8

Sorty AM, Shaikh NR (2015) Novel co-enrichment method for isolation of magnetotactic bacteria. J Basic Microbiol 55:520–526

Sorty AM, Meena KK, Choudhary K, Bitla UM, Minhas PS, Krishnani KK (2016) Effect of plant growth promoting bacteria associated with halophytic weed (*Psoralea corylifolia* L.) on germination and seedling growth of wheat under saline conditions. Appl Biochem Biotechnol 180:872–882

Souza RD, Ambrosini A, Passaglia LMP (2015) Plant growth-promoting bacteria as inoculants in agricultural soils. Genet. Mol Biol 38:401–419

Steinberger Y, Degani R, Barnen G (1995) Decomposition of root litter and related microbial population dynamics of a Negev desert shrub, *Zygophyllum dumosum*. J Arid Environ 31:383–389

Ström L, Owen AG, Godbold DL, Jones DL (2002) Organic acid mediated P mobilization in the rhizosphere and uptake by maize roots. Soil Biol Biochem 34:703–710

Subrahmanyam G, Kumar A, Sandilya SP, Chutia M, Yadav AN (2020) Diversity, plant growth promoting attributes, and agricultural applications of rhizospheric microbes. In: Yadav AN, Singh J, Rastegari AA, Yadav N (eds) Plant microbiomes for sustainable agriculture. Springer, Cham, pp 1–52. https://doi.org/10.1007/978-3-030-38453-1_1

Suman A, Yadav AN, Verma P (2016) Endophytic microbes in crops: diversity and beneficial impact for sustainable agriculture. In: Singh D, Abhilash P, Prabha R (eds) Microbial inoculants in sustainable agricultural productivity, research perspectives. Springer-Verlag, India, pp 117–143. https://doi.org/10.1007/978-81-322-2647-5_7

Swaine EK, Swaine MD, Killham K (2007) Effects of drought on isolates of *Bradyrhizobium elkanii* cultured from *Albizia adianthifolia* seedlings on different provenances. Agrofor Syst 69:135–145

Tani C, Sasakawa H, Takenouchi K (2003) Isolation of endophytic *Frankia* from root nodules of *Casuarina equisetifolia* and infectivity of the isolate to host plants. Soil Sci Plant Nutr 49:137–142

Teale WD, Paponov IA, Palme K (2006) Auxin in action: signalling, transport and the control of plant growth and development. Nat Rev Mol Cell Biol 7:847–859

Thakur N, Kaur S, Tomar P, Thakur S, Yadav AN (2020) Microbial biopesticides: current status and advancement for sustainable agriculture and environment. In: Rastegari AA, Yadav AN, Yadav N (eds) Trends of microbial biotechnology for sustainable agriculture and biomedicine systems: diversity and functional perspectives. Elsevier, Amsterdam, pp 243–282. https://doi.org/10.1016/B978-0-12-820526-6.00016-6

Thomashow MF (2010) Molecular basis of plant cold acclimation: insights gained from studying the CBF cold response pathway. Plant Physiol 154:571–577

Tiago I, Chung AP, Verissimo A (2004) Bacterial diversity in a nonsaline alkaline environment: heterotrophic aerobic populations. Appl Environ Microbiol 70:7378–7387

Tigchelaar M, Battisti DS, Naylor RL, Ray DK (2018) Future warming increases probability of globally synchronized maize production shocks. Proc Natl Acad Sci 115:6644–6649

Tiwari P, Bajpai M, Singh LK, Mishra S, Yadav AN (2020) Phytohormones producing fungal communities: metabolic engineering for abiotic stress tolerance in crops. In: Yadav AN, Mishra S, Kour D, Yadav N, Kumar A (eds) Agriculturally important Fungi for sustainable agriculture, Perspective for diversity and crop productivity, vol 1. Springer, Cham, pp 1–25. https://doi.org/10.1007/978-3-030-45971-0_8

Trindade I, Capitao C, Dalmay T, Fevereiro MP, Santos DM (2010) miR398 and miR408 are up-regulated in response to water deficit in *Medicago truncatula*. Planta 231:705–716

Turner TR, James EK, Poole PS (2013) The plant microbiome. Genome Biol 14:209

Upadhyay SK, Singh DP, Saikia R (2009) Genetic diversity of plant growth promoting rhizobacteria isolated from rhizospheric soil of wheat under saline condition. Curr Microbiol 59(5):489–496

Vansuyt G, Robin A, Briat JF, Curie C, Lemanceau P (2007) Iron acquisition from Fe-pyoverdine by *Arabidopsis thaliana*. Mol Plant-Microbe Interact 20:441–447

Venter JC, Remington K, Heidelberg JF, Halpern AL, Rusch D, Eisen JA (2004) Environmental genome shotgun sequencing of the Sargasso Sea. Science 304:66–74

Verma P, Yadav AN, Kazy SK, Saxena AK, Suman A (2013) Elucidating the diversity and plant growth promoting attributes of wheat (*Triticum aestivum*) associated acido-tolerant bacteria from southern hills zone of India. Natl J Life Sci 10:219–226

Verma P, Yadav AN, Kazy SK, Saxena AK, Suman A (2014) Evaluating the diversity and phylogeny of plant growth promoting bacteria associated with wheat (*Triticum aestivum*) growing in central zone of India. Int J Curr Microbiol Appl Sci 3:432–447

Verma P, Yadav AN, Khannam KS, Kumar S, Saxena AK, Suman A (2016) Molecular diversity and multifarious plant growth promoting attributes of *Bacilli* associated with wheat (*Triticum aestivum* L.) rhizosphere from six diverse agro-ecological zones of India. J Basic Microbiol 56:44–58

Verma P, Yadav AN, Khannam KS, Saxena AK, Suman A (2017a) Potassium-solubilizing microbes: diversity, distribution, and role in plant growth promotion. In: Panpatte DG, Jhala YK, Vyas RV, Shelat HN (eds) Microorganisms for green revolution, Microbes for sustainable crop production, vol 1. Springer Singapore, Singapore, pp 125–149. https://doi.org/10.1007/978-981-10-6241-4_7

Verma P, Yadav AN, Kumar V, Singh DP, Saxena AK (2017b) Beneficial plant-microbes interactions: biodiversity of microbes from diverse extreme environments and its impact for crop improvement. In: Singh DP, Singh HB, Prabha R (eds) Plant-microbe interactions in agro-ecological perspectives, Microbial interactions and agro-ecological impacts, vol 2. Springer Singapore, Singapore, pp 543–580. https://doi.org/10.1007/978-981-10-6593-4_22

Verma JP, Jaiswal DK, Krishna R, Prakash S, Yadav J, Singh V (2018) Characterization and screening of thermophilic *Bacillus* strains for developing plant growth promoting consortium from hot spring of Leh and Ladakh region of India. Front Microbiol 9:1293

Verma P, Yadav AN, Khannam KS, Mishra S, Kumar S, Saxena AK et al (2019) Appraisal of diversity and functional attributes of thermotolerant wheat associated bacteria from the peninsular zone of India. Saudi J Biol Sci 26:1882–1895. https://doi.org/10.1016/j.sjbs.2016.01.042

Vineeth TV, Kumar P, Krishna GK (2016) Bioregulators protected photosynthetic machinery by inducing expression of photorespiratory genes under water stress in chickpea. Photosynthetica 54:234–242

Vyas P, Rahi P, Gulati A (2009) Stress tolerance and genetic variability of phosphate-solubilizing fluorescent *Pseudomonas* from the cold deserts of the trans-Himalayas. Microb Ecol 58:425–434

Wakchaure GC, Minhas PS, Meena KK, Singh NP, Hegade PM, Sorty AM (2018) Growth, bulb yield, water productivity and quality of onion (*Allium cepa* L.) as affected by deficit irrigation regimes and exogenous application of plant bio-regulators. Agric Water Manag 199:1–10

Wang Y, Ke X, Wu L, Lu Y (2009) Community composition of ammonia-oxidizing bacteria and archaea in rice field soil as affected by nitrogen fertilization. Syst Appl Microbiol 32:27–36

Wang Y, Hu B, Du S, Gao S, Chen X, Chen D (2016) Proteomic analyses reveal the mechanism of *Dunaliella salina* ds-26-16 gene enhancing salt tolerance in *Escherichia coli*. PLoS One 11:e0153640

Wellner S, Lodders N, Kämpfer P (2011) Diversity and biogeography of selected phyllosphere bacteria with special emphasis on *Methylobacterium* spp. Syst Appl Microbiol 34:621–630

Wilmes P, Bond PL (2006) Metaproteomics: studying functional gene expression in microbial ecosystems. Trend microbial 14:92–97

Xu Z, Shimizu H, Ito S, Yagasaki Y, Zou C, Zhou G et al (2014) Effects of elevated CO2, warming and precipitation change on plant growth, photosynthesis and peroxidation in dominant species from North China grassland. Planta 239:421–435

Yadav AN (2017) Agriculturally important microbiomes: biodiversity and multifarious PGP attributes for amelioration of diverse abiotic stresses in crops for sustainable agriculture. Biomed J Sci Tech Res 1:1–4. https://doi.org/10.26717/BJSTR.2017.01.000321

Yadav AN (2020) Plant microbiomes for sustainable agriculture: current research and future challenges. In: Yadav AN, Singh J, Rastegari AA, Yadav N (eds) Plant microbiomes for sustainable agriculture. Springer International Publishing, Cham, pp 475–482. https://doi.org/10.1007/978-3-030-38453-1_16

Yadav A, Verma P, Sachan S, Kaushik R, Saxena A (2015a) Microbes mediated alleviation of cold stress for growth and yield of wheat (*Triticum aestivum* L.). In: Proceeding of international conference on "Low Temperature Science and Biotechnological Advances", p 179. https://doi.org/10.13140/RG.2.1.2374.2883

Yadav AN, Sachan SG, Verma P, Saxena AK (2015b) Prospecting cold deserts of north western Himalayas for microbial diversity and plant growth promoting attributes. J Biosci Bioeng 119:683–693. https://doi.org/10.1016/j.jbiosc.2014.11.006

Yadav AN, Sachan SG, Verma P, Tyagi SP, Kaushik R, Saxena AK (2015c) Culturable diversity and functional annotation of psychrotrophic bacteria from cold desert of Leh Ladakh (India). World J Microbiol Biotechnol 31:95–108. https://doi.org/10.1007/s11274-014-1768-z

Yadav AN, Sharma D, Gulati S, Singh S, Dey R, Pal KK et al (2015d) Haloarchaea endowed with phosphorus solubilization attribute implicated in phosphorus cycle. Sci Rep 5:12293. https://doi.org/10.1038/srep12293

Yadav AN, Verma P, Kumar M, Pal KK, Dey R, Gupta A et al (2015e) Diversity and phylogenetic profiling of niche-specific bacilli from extreme environments of India. Ann Microbiol 65:611–629. https://doi.org/10.1007/s13213-014-0897-9

Yadav AN, Sachan SG, Verma P, Saxena AK (2016) Bioprospecting of plant growth promoting psychrotrophic bacilli from cold desert of north western Indian Himalayas. Indian J Exp Biol 54:142–150

Yadav AN, Kumar R, Kumar S, Kumar V, Sugitha T, Singh B et al (2017a) Beneficial microbiomes: biodiversity and potential biotechnological applications for sustainable agriculture and human health. J Appl Biol Biotechnol 5:45–57. https://doi.org/10.7324/JABB.2017.50607

Yadav AN, Verma P, Kaushik R, Dhaliwal HS, Saxena AK (2017b) Archaea endowed with plant growth promoting attributes. EC Microbiol 8:294–298

Yadav AN, Kumar V, Prasad R, Saxena AK, Dhaliwal HS (2018) Microbiome in crops: diversity, distribution and potential role in crops improvements. In: Prasad R, Gill SS, Tuteja N (eds) Crop improvement through microbial biotechnology. Elsevier, USA, pp 305–332. https://doi.org/10.1016/B978-0-444-63987-5.00015-3

Yadav AN, Gulati S, Sharma D, Singh RN, Rajawat MVS, Kumar R et al (2019) Seasonal variations in culturable archaea and their plant growth promoting attributes to predict their role in establishment of vegetation in Rann of Kutch. Biologia 74:1031–1043. https://doi.org/10.2478/s11756-019-00259-2

Yadav AN, Kaur T, Kour D, Rana KL, Yadav N, Rastegari AA et al (2020a) Saline microbiome: biodiversity, ecological significance and potential role in amelioration of salt stress in plants. In: Rastegari AA, Yadav AN, Yadav N (eds) Trends of microbial biotechnology for sustainable agriculture and biomedicine systems: diversity and functional perspectives. Elsevier, Amsterdam, pp 283–309. https://doi.org/10.1016/B978-0-12-820526-6.00018-X

Yadav AN, Mishra S, Kour D, Yadav N, Kumar A (2020b) Agriculturally important Fungi for sustainable agriculture, volume 1: perspective for diversity and crop productivity. Springer International Publishing, Cham

Yadav AN, Mishra S, Kour D, Yadav N, Kumar A (2020c) Agriculturally important Fungi for sustainable agriculture, volume 2: functional annotation for crop protection. Springer International Publishing, Cham

Yadav AN, Rastegari AA, Yadav N (2020d) Microbiomes of extreme environments: biodiversity and biotechnological applications. CRC Press, Taylor & Francis, Boca Raton

Yadav AN, Rastegari AA, Yadav N, Kour D (2020e) Advances in plant microbiome and sustainable agriculture: diversity and biotechnological applications. Springer, Singapore

Yadav AN, Rastegari AA, Yadav N, Kour D (2020f) Advances in plant microbiome and sustainable agriculture: functional annotation and future challenges. Springer, Singapore

Yadav AN, Singh J, Rastegari AA, Yadav N (2020g) Plant microbiomes for sustainable agriculture. Springer, Cham

Yolcu S, Ozdemir F, Güler A, Bor M (2016) Histone acetylation influences the transcriptional activation of POX in *Beta vulgaris* L. and *Beta maritima* L. under salt stress. Plant Physiol Biochem 100:37–46

Zawadzka M, Trzciński P, Nowak K, Orlikowska T (2014) The impact of three bacteria isolated from contaminated plant cultures on in vitro multiplication and rooting of microshoots of four ornamental plants. J Hort Res 21:41–51

Zeigler DR (2014) The *Geobacillus* paradox: why is a thermophilic bacterial genus so prevalent on a mesophilic planet? Microbiology 160:1–11

Zeyaullah M, Kamli MR, Islam B, Atif M, Benkhayal FA, Nehal M, Rizvi MA, Ali A (2009) Metagenomics - an advanced approach for non-cultivable microorganisms. Biotechnol Mol Biol Rev 4:49–54

Zhao MG, Chen L, Zhang LL, Zhang WH (2009) Nitric reductase-dependent nitric oxide production is involved in cold acclimation and freezing tolerance in *Arabidopsis*. Plant Physiol 151:755–767

Zhao S, Zhou N, Zhao ZY, Zhang K, Wu GH, Tian CY (2016) Isolation of endophytic plant growth-promoting bacteria associated with the halophyte *Salicornia europaea* and evaluation of their promoting activity under salt stress. Curr Microbiol 73:574–581

Zhou J, He Z, Yang Y, Deng Y, Tringe SG, Alvarez-Cohen L (2015) High-throughput metagenomic technologies for complex microbial community analysis: open and closed formats. MBio 6:2288–2214

Plant- and Microbes-Mediated Secondary Metabolites: Remunerative Venture for Discovery and Development

15

Prasann Kumar, Shipa Rani Dey, and Padmanabh Dwivedi

Abstract

Secondary metabolites derived from plants and microbes are natural substances that can be used for potential drug production. Simultaneously, there is very little concern in the current implementation of research and drug development. None other than combinational chemistry that shows the prospects for new low molecular weight lead substances provides continuing unique structural diversity through these natural products. Biodiversity, in the world of essential biological activities, only less than 10% has been reported and many more are expected to be found through determined research and classification of biochemical miscellaneous natural products. There have been a few instances about the predominant wellspring of data of regular substances' use from therapeutic herbs that are a result of human testing by preliminaries and disappointment for a considerable length of time through heavenliness of preliminaries or disastrous passings, and discovering the accessible nourishments for sickness medicines. Moreover, secondary (minor) metabolites are manufactured by consequences of the mother creature taking on to its near environment or by defense means in opposition of predators to aid the continued existence of the organisms. These metabolites are in general not required for the growth, reproduction, and development of an organism. The essential progressions of glycolysis, photosynthesis, and the

P. Kumar (✉)
Divisions of Research and Development, Lovely Professional University, Jalandhar, Punjab, India

S. R. Dey
Department of Agronomy, School of Agriculture, Lovely Professional University, Jalandhar, Punjab, India

P. Dwivedi
Department of Plant Physiology, I.Ag.Sci. BHU, Varanasi, Uttar Pradesh, India

A. N. Yadav et al. (eds.), *Current Trends in Microbial Biotechnology for Sustainable Agriculture*, Environmental and Microbial Biotechnology,
https://doi.org/10.1007/978-981-15-6949-4_15

Krebs cycle are the sources of biosynthesis of optional metabolites that can manage the cost of biosynthetic intermediates.

Keywords

Abiotic · Biotic · Crop · Density · Energy · Forage · Gap

15.1 Introduction

Plant-derived secondary metabolites are natural substances that can be used for the manufacturing of potential drugs. At the same time, current execution in the discovery and development of drugs has a very less extent of interest. Nonetheless, continued unique structural diversity is provided by these natural products in contrast with combinatorial chemistry, which shows the prospects for finding out the new low molecular weight lead substances. From the world's biodiversity, only less than 10% has been estimated for important biological activity, and still, many more are awaited to be found through the dare, which tells how to obtain and identify the natural products of biochemical miscellany (Cragg and Newman 2005). The most primitive accounts of natural substances were illustrated on the isolation of oils commencing *Cupressus sempervirens* (Cypress) as well as *Commiphora* species (myrrh) and utilized in the production of Mesopotamia (2600 B.C.) clay tablets in cuneiform, which are still used in the treatment of colds, coughs, and inflammation (Cragg and Newman 2005).

An Egyptian pharmaceutical account of *Ebers Papyrus* in 2900 B.C. has documented more than 700 plant-based medicines such as pills, gargles, infusions, and ointments. In 1100 B.C. the Chinese medical matter like Shennong herbal (~100 B.C.) contained 365 medicines, 52 medicines were contained by Wu Shi Er Bing Fang, however, 659 medicines were contained in Herbal Tang, and A.D. containing around 850 medicines are accounts for the utilization of natural substances (Cragg and Newman 2005). Dioscorides (100 A.D.), the Greek physician, accounted the utilization, storage, and collection of medicinal plants, whereas Theophrastus, a Greek rationalist and characteristic researcher (~300 B.C.) addressed using pharmaceutical plants. All through the obscure and mid eternities the nunneries, Arabs conserved the Greco-Roman information and stretched out the utilization of their relationship with Chinese and Indian therapeutic herbs that were obscure for the Greco-Roman world while England, Germany, France, and Ireland monitored this Western information (Cragg and Newman 2005). The Arabs were the first to represent secretly as the proprietor of drug stores in the eighth century with a Persian drug specialist, savant, doctor, and artist, who contributed an incredible work in the study of the drug store and prescription (Cragg and Newman 2005).

15.2 Medicinal or Therapeutic Plants

All through the history, the utilization of common substances has been clarified, for example, customary drugs, tonics, cures, and oils, yet a significant number of bioactive normal items are still unidentified. There are a few instances of the predominant wellspring of data of regular substances use from therapeutic herbs as a result of human testing by preliminaries and disappointment for a considerable length of time through heavenliness of preliminaries or disastrous passings and discovering the accessible nourishments for sickness medicines (Hicks 2014; Kinghorn et al. 2011). The plant class Salvia is a case of restorative plant that is found all through the southwestern district of the United States in addition to northwestern Mexico and was used by Indian farmers (Hicks 2014).

Tribes used to cook food for male newborn babies in the ashes of hot Salvia as they believed that by doing this, these infants were raised to be fit and solid and be in the good physical shape as per their particular community and they seemed to have better immunity against all respiratory complaints (Hicks 2014). During the hot summer, a shrub named Alhagi maurorum Medik also known as Camel's thorn liberates a sweetish, gluey substance (sap) commencing its stems and leaves (Duke et al. 2008). The gummy substance is known as "manna" and contains sucrose, melezitose, and invert sugar and it is used for the treatment of constipation, anorexia, fever, dermatitis, epitasis, leprosy, and obesity by the Ayurvedic doctors (Duke et al. 2008). It was also accounted by the Israelis for the treatment of bloody diarrhoea by drinking the boiled extract of the roots of camel's thorn. The Romans used this plant for the treatment of nasal polyps while the smoke of this plant is used by Konkani people for the treatment of asthma (Duke et al. 2008). The people of Northern Europe and Eastern North America believed that eating the plant (*Ligusticum scoticum Linnaeus*) raw in an empty stomach in the morning protects them from daily infection (Dillenius 1724). The roots of this plant were even used in the treatment of flatulence (Martin 1934; Lightfoot 1977; Beith 1999) and used in the Faeroe Islands as an aphrodisiac in the form of a sedative (Svabo 1959). In many countries like Asia, Europe, New Zealand, Africa, and America a toxic plant discovered named Atropa belladonna Linnaeus (destructive nightshade), which three berries can execute the new brought into the world (Allen and Hatfield 2004).

15.2.1 The Other Sources of Medicinal Natural Substances

On birches, a fungus named *Piptoporus betulinus* grows that was used as an antiseptic and disinfectant when steamed to produce charcoal (Swanton 1915). *P. betulinus's* strips were used to stop the bleeding and making of comfy pads of corn (Swanton 1932). The fungal growth *Agaricus campestris* Linnaeux ex Fries (field mushroom), which by and large is found in the southern and northern mild zones and the Caribbean had accounted to be milk stewed to mollify the throat malignant growth (Hatfield 2005). In the early 17th to eighteenth century, lichens accounted for a significant job in colors showcase and esteemed more than the oriental flavors.

To date, no drug acquired from lichens got authorization, although their importance in people writing has been revealed (Müller 2001). In assembling of scents, beauty care products, and drugs lichens have been utilized from the antiquated time of Chinese and Egyptian advancements (Purvis 2000). *Usnea dillenius ex Adanson* conventionally had been utilized to cure dry scalp diseases, plus at a standstill was sold in pharmaceutical industries as an important ingredient for anti-dandruff shampoos, while in Ireland it was used to resolve the issue of sore eyes (Purvis 2000). *Usnea subfloridana Stirton* is lichen that was used to make a lotion by mixing it with butter and boiled then cooled (Allen and Hatfield 2004). In the British Isles, *Parmelia omphalodes (Linnaeus) Acharius* was found in abundance and had been used to prepare brown dyes. It was also sprinkled on stockings traditionally in the highlands to avoid inflammation of the feet (Macfarlane 1929; Cameron 1900). It was also found useful in curing bad abscesses underneath the chin and blisters and slashes in Ireland (Allen and Hatfield 2004). On the other hand, the marine atmosphere got fewer accounts in the application of traditional medicine. *P. omphalodes (Linnaeus) Acharius* are the red algae that were used in the beverage for the reduction or abolition of colds, tender throats, and chest contagions laidback with tuberculosis. The alga helps to cure kidney trouble and burns when consumed with boiled milk and boiled water (Vickery 1995; Moloney 1919). Moreover, drinking three spoons of the juice of *Porphyra umbilicalis* (Linnaeus) Kützing (a red alga)every morning, and accompanying it with 3 weeks of fasting was effective against cancers, especially breast cancer. In the Aran Islands, *P. umbilicalis* had been used to cure indigestion and was also given to the cows for relaxation in springtime constipation (ÓhEithir 1983).

15.2.2 Plants Metabolites

The compounds obtained from the breakdown and biosynthesis of fats, proteins, carbohydrates, and nucleic acids are "major (primary) metabolites" involved in primary metabolism essential in all living organisms (Dewick 2002). Secondary metabolites are compounds that are biosynthesized by an organism and are frequently found to be distinct to an organism or is a characteristic of individual species and stated as "secondary (minor) metabolism" (Dewick 2002; Maplestone et al. 1992).

Secondary (minor) metabolites are manufactured, moreover, by consequences of the mother creature taking on to its near environment or by defense means in opposition of predators to aid in the continued existence of the organism and these metabolites are in general not required for growth, reproduction, and development of an organism (Dewick 2002; Colegate and Molyneux 2008). The essential progressions of glycolysis, photosynthesis, and the Krebs cycle are the sources of biosynthesis of optional metabolites that can manage the cost of biosynthetic intermediates and their outcomes in the generation of normal substances (Dewick 2002). It tends to be discovered that the structure squares are less, yet the improvement of novel optional metabolites is endless. Most of the head building hinders being used for the formation of optional products that are individuals acquired after the intermediary: acetyl-CoA, corrosive of shikimic, corrosive of mevalonic, and 1-deoxy-D-xylulose-5-phosphate. Numerous system responses like alkylation, classes, Schiff base development, and decarboxylation happened by uncountable pathways of biosynthesis, which are included by optional metabolites (Dewick 2002). As per some theory, it is expected that auxiliary digestion makes use of amino acids and the acetic acid derivation and shikimic acid pathways to create "shunt metabolites" that pursued a substituting biosynthetic way (Sarker et al. 2006).

The adaptations in the pathways of biosynthesis because of regular reasons (infections) or abnormal reasons (e.g., compound or emission) are a push to supply solidness for the life form (Sarker et al. 2006). It is the restrictive biosynthesis of these common harvests, created by the incalculable number of earthbound and seagoing life forms, which make accessible the trademark substance designs that have a game plan of organic activities.

15.3 Natural Substances

The natural plant products have been the reason for most of the ancient medicines sought after by a resulting medical, pharmaceutical, and substance concoction think about (Butler 2004). Probably the most mainstream and comprehended manual for a date would be the amalgamation of the calming specialist, acetylsalicylic corrosive (1) (ibuprofen) got normally from salicin (2) Salix Alba L. (the willow tree) bark (DerMarderosian and Beutler 2002). Assessment of opium poppy (*Papaver somniferum* L.) achieved the division of a couple of alkaloids including morphine (3), a monetarily noteworthy drug that came to light in the year 1803 as shown in Fig. 15.1. It was during the 1870s that *P. somniferum* used to obtain morphine drug was bubbled in acidic anhydride to yield diacetylmorphine (heroin) and was quickly changed over to a painkiller. It is recorded that the earliest Greeks, as well as Sumerians people used the extricates of poppy therapeutically, while opium was to found to be addictive by the Arabians (DerMarderosian and Beutler 2002). During the tenth century in Europe, Foxglove botanically called *Digitalis purpurea* L. had been pursued back, yet it was not until the 1700s that the dynamic component called digitoxin (4) was used to overhaul heart transmission, as such refining the quality of contractibility of cardiovascular.

Drug digitoxin showed in Fig. 15.1 in fourth number, its analogs have for a long while been used in the organization of congestive cardiovascular breakdown and contain likely long stretch negative effects along with its removal by various prescriptions at the time of "heart insufficiency" (DerMarderosian and Beutler 2002). Counter prescription for malaria is quinine (Fig. 15.1(5)) embraced by the U.S. Food Drug Administration, withdrew the bark product *Cinchona succirubra Pav.* exKlotsch, for an impressive period for the treatment of high temperature, acid reflux, mouth and throat ailments, and dangerous development. Ancient use of the bark to care for intestinal ailment was set up in mid-1800 when the British people began the general improvement of related plants (DerMarderosian and Beutler 2002). *Pilocarpus jaborandi* obtains Pilocarpine shown in Fig. 15.1 (6), is L-histidine-decided alkaloid, which has medicinal properties in the curing of perpetual open-edge glaucoma and serious edge end glaucoma since the last 100 years. An oral itemizing in the year 1994, pilocarpine (Fig. 15.1(6)) was supported by the Food Drug Administration to take care of dryness in the mouth at the time of radiation with head and neck and besides used to vitalize sweating organs to determine the combinations of Na and Cl as shown in Fig. 15.1 (Aniszewski 2007). In the previous era, oral arranging was insisted on the organization of Sjogren's issue, resistant framework contamination that indemnified the salivary and lacrimal organs.

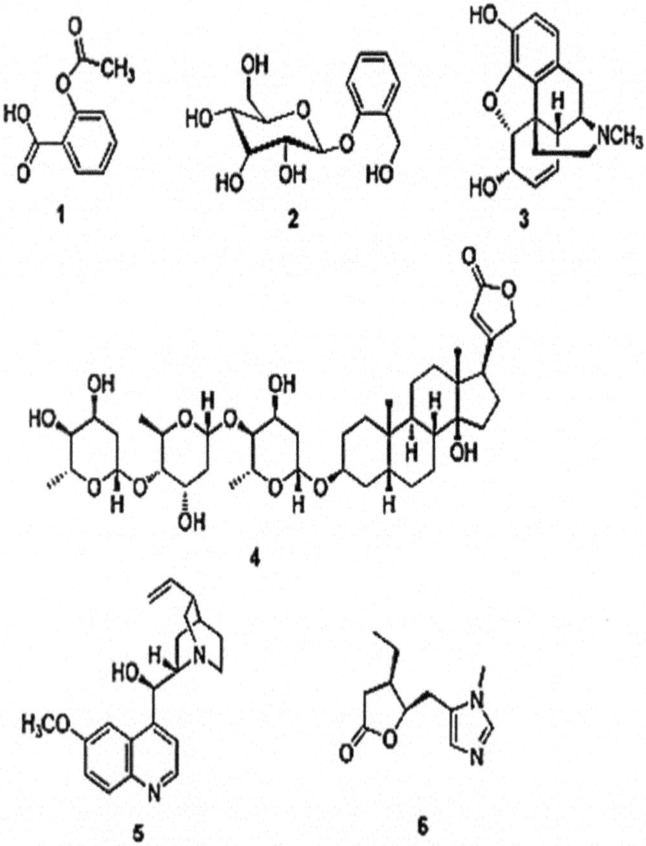

Fig. 15.1 (1) Acetylsalicylic acid, (2) Salicin, (3) Morphine, (4) Digitoxin, (5) Quinine, (6) Pilocarpine

15.3.1 Natural Substances from Fungi

For many years, micro and macrofungi have been parts of a life of human beings. These have been utilized as edibles like mushrooms, in the arrangement of mixed refreshments (Saccharomyces cerevisiae yeast), and conventional medication just as social reasons, with propels in the science of micro, their uses have been applied to proteins, natural guideline, anti-infection agents, and other therapeutic dynamic substances (Mann 1994). Apparently, one of the most celebrated discoveries of regular items separated from microorganisms is penicillin (Fig. 15.2 (7)) from a parasite, *Penicillium notatum* found by scientist Fleming in the year 1929 (Abraham et al. 1941). Extractive detachment by countercurrent method, which yielded in significant returns, in vivo research, was required, which at last spared endless lives (Fig. 15.2) (together with Fleming) (Alder 1970). The clinical investigations of Chain, Florey, and their associates in the middle of the 1940s revealed what drove the business of manufactured penicillin, which inevitably altered medication (Lax 2004; Wainwright 1990; Mann 1999; Buss and Waigh 1995).

Subsequently, in the years 1942–1944, a worldwide exertion was made to find new antibodies structure from microorganisms and bioactive regular substances for penicillin G (7) (Fabbretti et al. 2011; Zjawiony 2004). Until 1968, the distinguishing proof strategy for β-lactams was being utilized and was inferred that β-lactams (natural) had been found in the old technique (Fabbretti et al. 2011). Typically, this isn't the situation, just like the case with the execution of new selection strategies during the 1970s, the improvement of β-lactam supersensitivity bacterial strains, hindrance trial of β-lactamases, and the particularity of sulfur-containing metabolites, which finished in the revelation of novel anti-infection basic (norcardicins, carbapenems, and monobactams) including antimicrobial assurance, norcardicin, imipenem, and aztreonam 8,9,and 10, respectively, shown in Fig. 15.2 (Fabbretti et al. 2011; Zjawiony 2004; Stamets 2002). There are at present nine β-lactams, which are distributed as two cephalosporins, six carbapenems, and one penem in

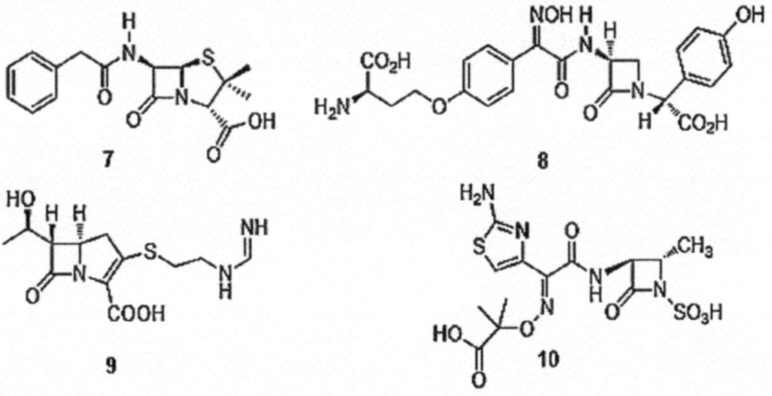

Fig. 15.2 (7)Penicillin, (8)Norcardicin, (9) Imipenem, (10) Aztreonam

medical preliminaries or experiencing medicines enlistment, going to a novel wide range of antimicrobial class called glycylcyclines (Stamets 2002).

Polypores, for example, full-scale organisms are an enormous gathering of wood-rotting growths of the phylum Basidiomycete species, which are around 25,000 in number, of which around 500 are individuals from the Aphyllophorales (Gwinn et al. 1992). Roughly 75% of the polypore organisms tried demonstrated solid antimicrobial movement and might be a reasonable hotspot for the assembling of novel antitoxins. Most substances have antiviral, cytotoxic, against neoplastic, neurological, mitigating, safe animating, and anticancer action (Gwinn et al. 1992; Tan and Zou 2001). All the more regular microorganisms are organisms, some of which may spend a base piece of lifespan into tissue with no grounds of any mistaken indications of contamination or infection (Petrini 1986; Newman and Cragg 2007). Trees, grasses, green growth, and herbaceous plants have been found to possess and live in intercellular zones of a plant without affecting the host organism (Martin et al. 2007). All things considered; these parasites are alluded as endophytes. New bioactive auxiliary metabolites that have evolved from contagious resources have created the absolute most significant regular assembling items for the pharmaceutical business (Cragg and Newman 2005). Edmund Kornfeld, in 1953, originally segregated vancomycin (11) an antitoxin glycopeptide delivered in *Amycolatopsis Orientalis* societies that is dynamic beside a broad scope of gram +ve living beings, for example, Staphylococcus and Streptococcus and against gram −ve microscopic organisms, mycobacterium, and parasites and was acknowledged by the Food &Drug Administration in the year 1958 (Fig. 15.3). It is utilized for the finding of serious contamination and inclined

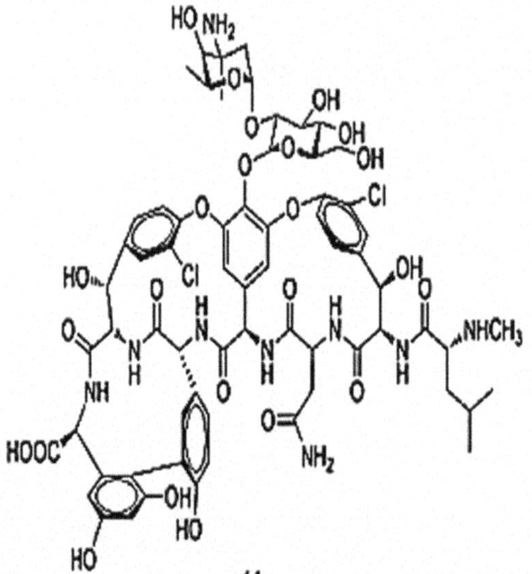

11

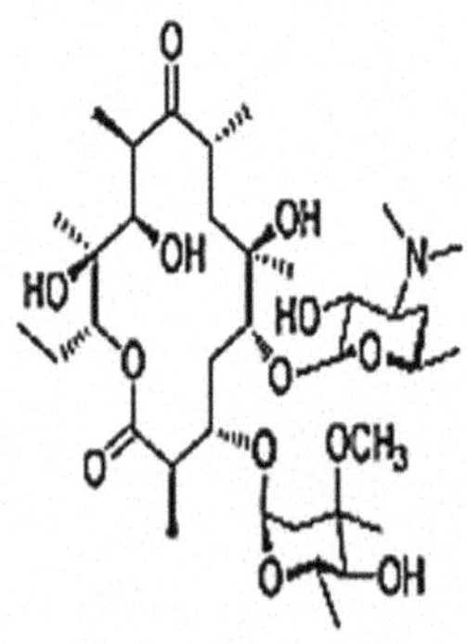

12

Fig. 15.3 (11) Vancomycin, (12) Erythromycin

species inpatient (7) (Butler 2004). Fig. 15.3(12), obtained from a fungi *Saccharopolyspora erythraea,* is an antibacterial medicine including a 14-part macrocycle comprising completely of propionate (ester of propionic acid) units (Fig. 15.3). Activity in opposition to gram +ve coccus and bacillus and is utilized for gentle to direct, respiratory tract contaminations are wide range against erythromycin (12) (Butler 2004; Dewick 2002). There are currently three semi-artificial ketolide subordinates of erythromycin (12), cethromycin (ABT-773, RestanzaTM), EP-420 (Enanta Pharmaceuticals), and BAL-19403 (Basilea) in medical advancement.

Cell infections are diminutive types of life in nature that ascent to an activity fever, flu, and Ebola. A set number of regular antiviral items or misleadingly got analogs from parasites was observed (Kashiwada et al. 1996). Betulinic corrosive shown in Fig. 15.4 (13) is a bark-based triterpenoid. At first, Betula pubescens was known as a frail HIV replication inhibitor (Yogeeswari and Sriram 2005). Betulinic corrosive can impede topoisomerase I and is being tried as a chemopreventive malignancy operator in Phase I think about (Fig. 15.4) (Min et al. 1998). Bevirimat (PA-457) (14), from *Syzygium claviform,* is in stage II b clinical case and is thought to abridge the last advances in the generation of HIV GAG protein (Lee et al. 1996). Ganoderic corrosive β (15), available from *Ganoderma lucidum,* indicated eminent enemy of HIV-1 protease interest with an IC50 estimation of 20 μM (Fig. 15.4) (Li et al. 2003).

In Fig. 15.5 (Butler 2004; Dewick 2002), compounds 16, 17, and 18 were isolated to treat various diseases. Torreyanic corrosive (18) was acquired from the endophyte of the jeopardized plant, Torreyataxifolia (McRae et al. 2007) and was tried in a few cell lines of malignant growth and saw five to ten times increasingly strong/cytotoxic in cell lines verification against protein kinase C causing apoptosis cell passing (Fig. 15.5) (Fellows and Scofield 1995).

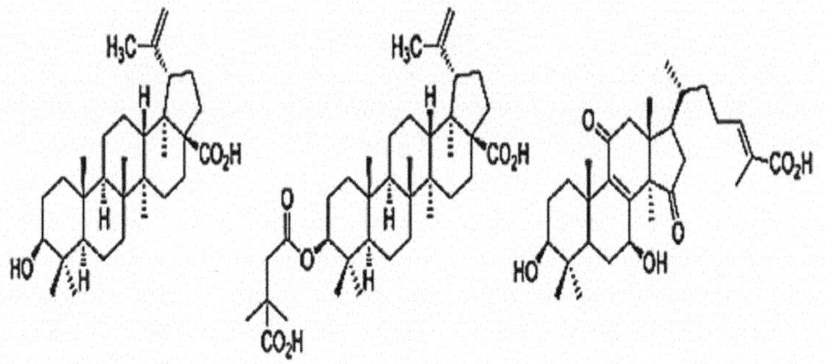

13 14 15

Fig. 15.4 (13) Betulinic acid, (14) Bevirimat (PA-457), (15) Ganoderic acid

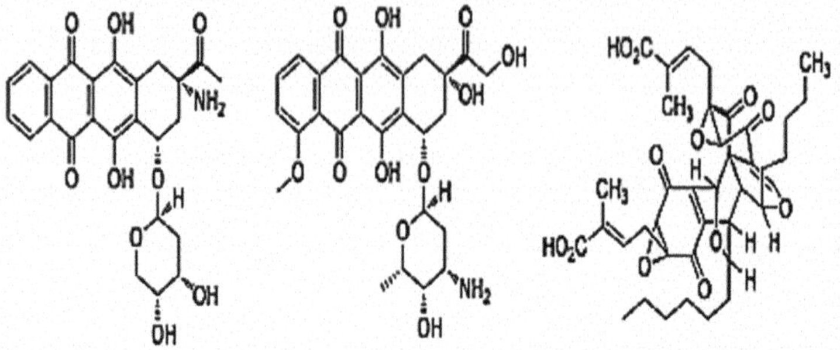

16 17 18

Fig. 15.5 (16) Amrubicin hydrochloride, (17) Doxorubicin, (18) Torreyanic acid

15.3.2 Plants as Ordinary Substances

For 1000 years, normal plant substances were notable for their pharmaceutical (therapeutic) employments. Ethnopharmacological properties were utilized for early medication angling as an essential wellspring of medicines for early distinguishing proof of restorative medications (Farnsworth et al. 1985; Fabricant and Farnsworth 2001). World Health Organization (WHO), announced that 80% of individuals still depended on plant-based outside essential human services medicines (Cragg 1998) and 80% of 122 plant-inferred restorative medications were related to their unique ethnopharmacological reason (Nicolaou et al. 1994). The mindfulness related to conventional medication (integral and elective homegrown items) has supported further work on therapeutic plants as conceivable restorative items and has added to the rejection of numerous characteristic items that have become understood pharmaceutical items. Paclitaxel (Taxol®) (19) is the most normally utilized in bosom malignant growth and this medication is made from the *Taxus brevifolia* bark. In 1962, the bark was used by the National Cancer Institute (NCI) as a major aspect of its exploratory plant screening program as shown in Fig. 15.6 (Nicolaou et al. 1994). Around three developed trees' (100 years of age) bark is utilized to give 1 g of 19 as a treatment plan that gives 2 g of the prescription. The current interest for 19 is around 100–200 kg each year (50,000 medicines per year) and is presently misleadingly created (Dewick 2002). In the year 1992, Taxol was presented the numerous acknowledgment (Kedei et al. 2004). Taxol (19) stayed available in restricted amounts from characteristic sources, yet its amalgamation has been effectively completed (Ogbourne et al. 2004). A particularly lower and energetically open from T needles is Baccatin III (20). Brevifolia and other related subordinates are a few instances of fundamentally comparable intensify that can be changed productively into 19 (Fig. 15.6) (Dewick 2002).

Instances of antitumor mixes that are as of now experiencing clinical preliminaries incorporate ingenious 3-O-angelate shown in Fig. 15.7 number 21, it is a polyhydroxy diterpenoid ingenol subordinate got from *Euphorbia peplus* sap, a

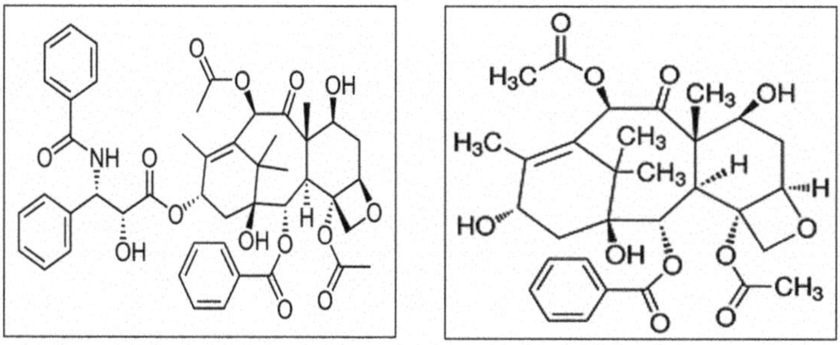

(19) (20)

Fig. 15.6 (19) Paclitaxel (Taxol), (20) Baccatin III

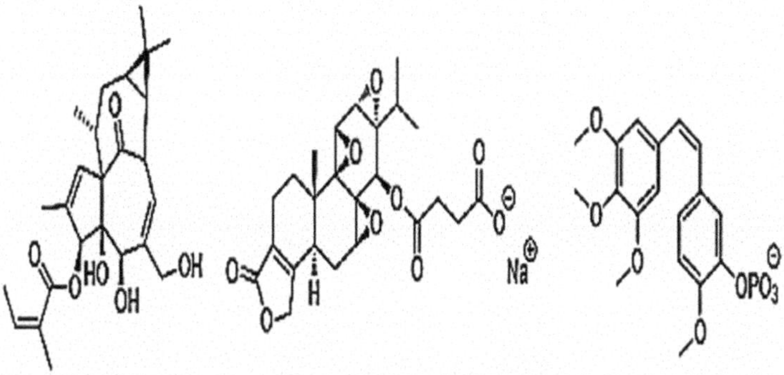

21 22 23

Fig. 15.7 (21)ingenol 3-O-angelate, (22) PG490–88, (23) Combretastation A-4 phosphate

prospective chemotherapeutic operator for skin malignancy is presently under investigation in the healing of cancers by biotechnologists shown in Fig. 15.7 (Kiviharju et al. 2002). The compound 22, a semi engineered comparative compound that is utilized by the Chinese for the immune system and incendiary syndromes (Newman and Cragg 2005). Combretastatin A-4 phosphate (23), which has a job as an enemy of the angiogenic operator, that actuates vascular shutdown in tumors (rot) and at present experiencing clinical preliminaries in Phase II research(Fig. 15.7) (Holwell et al. 2002; Kashman et al. 1992; Gustafson et al. 1992).

In the 1980s the pandemic of AIDS, which incited the National Cancer Institute (NCI) and different offices for the examination of characteristics items for wellsprings of a potential medication. The most noteworthy consequences of the examinations are the compound class called calanolides. The segregation of calanolide A& B respectively shown in Fig. 15.8 number 24, 25 from the Calonphyllum gathering, together with prostratin (26) from Homalanthusnutans, has advanced toward different clinical and preclinical improvement in like manner. (Fig. 15.8) (Cox 2001; Davidson et al. 2009; Carroll et al. 2005). CalanolideA (24) was assessed and

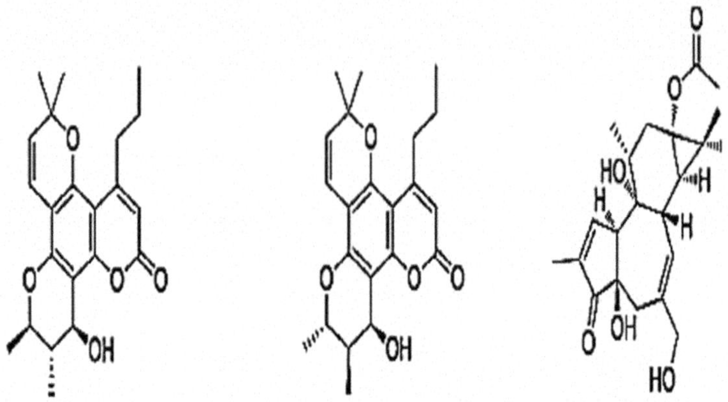

24 25 26

Fig. 15.8 (24)Calanolide A, (25) Calanolide B, (26) Prostratin

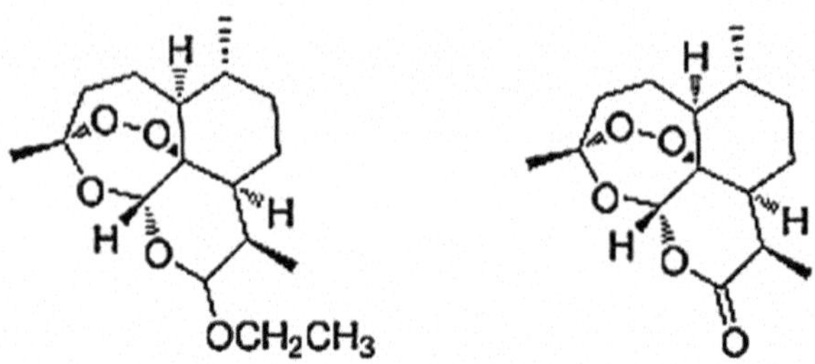

27 28

Fig. 15.9 (27) Arteether, (28) Artemisinin

affirmed by Sarawak Medichem Pharmaceuticals for Phase II clinical preliminaries, yet further medication advancement declaration was not made. In 2010 the AIDS research coalition conduct in Los Angeles, California (Fig. 15.8) led some human clinical preliminaries of prostratin (26).

In 2000, Arteether as shown in Fig. 15.9 (27), introduced Artemotil derived artemisinin shown in Fig. 15.9 (28) introduced as Artemisinin 1987, which was initially obtained from the *Artemisia annua* plant and they are both anti-malaria drugs as shown in Fig. 15.9 (Kashiwada et al. 1996). Initially, the plant was used for the healing of chills and fever in some conventional Chinese medical therapy. Other artemisinin derivatives (28) are also being used as anti-malaria drugs in Europe at various stages of development (Cragg and Newman 2005; Dewick 2002). To date, efforts are being made to cure malaria, based on the 28 pharmacophores evaluated in an arrangement through piperaquine (artificial anti-malarial bisquino line drug) (Howes et al. 2003).

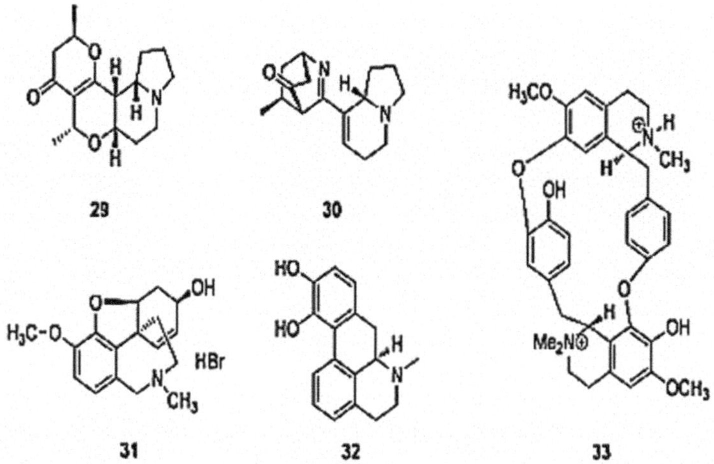

Fig. 15.10 (29) Grandisine A, (30) Grandisine B, (31) Morphine, (32) apomorphine, (33) Tubocaurarine

The two indole alkaloids confined from the Australian rainforest tree leaves are Elaeocarpusgrandis are Grandisines A (29) and B (30) (Fig. 15.10). The compound 29, has an unconventional structure, while compound 30 has an irregular structure. The mixes have a coupling like the human receptor and are potential leads for some pain-relieving operators (Heinrich and Teoh 2004). The compound (31) extracted from the plant Galanthusnivalis, is truly utilized in neurological conditions in Turkey and Bulgaria and used to treat Alzheimer's illness (Deleu et al. 2004; Marris 2006). Apomorphine (32) a morphine subordinate (31) a strong dopamine receptor agonist utilized in the healing of illness of Parkinson's as shown in Fig. 15.10 (Haefner 2003)."Cuwhich an uncommon" is the South American Indigenous bolt poison and is likewise arranged in the Amazon and Orinoco rainforests. Chondrodendrontomentosum (Menispermaceae), tubocurarine showed in number 33, which is secluded from the plant with climbing nature. Tubocurarin's restricted accessibility (33) has prompted the advancement of a progression of manufactured analogs that are currently liked to the normal item (Fig. 15.10) (Dewick 2002).

15.3.3 Marine Environment and Products

Although a portion of the plants has demonstrated hotspot regular items, the marine condition had an unmistakable reputation likewise in offering novel auxiliary substances. Fenical says, "We are not marine living beings," "so until around 1970. It was left as a profound mystery. It appeared to be crazy to me that the sea with such a tremendous territory had gotten away from anybody's notice. In any case, there are valid justifications for doing as such. Individuals fear the sea; it was seen as a perilous, ungracious place"(Alejandro et al. 2010). Seventy percent of the outside of the planet earth is secured by seas, and some pharmaceutical organizations have begun to understand that the sea would have surprising biodiversity and could be a potential wellspring of potential medication candidates (Rinehart and Lithgow-Bertelloni 1991). Such historic advancements in the previous 40 years of the marine conditions have created a huge number of various bioactive marine characteristic items being extricated from the framework. Until now, the worldwide marine pharmaceutical pipeline comprises of three medicines affirmed by the Food and Drug Administration, one EU-enrolled sedate, 13 characteristic items (or their subordinates) in dissimilar periods of the restorative pipeline and an enormous number of marine synthetic concoctions in the pre-medical channel (Urdiales et al. 1996). For model, Ziconotide, a peptide initially found in a humid cone snail, endorsed for torment treatment in December 2004.

Plitidepsin (34), a depsipeptide secluded from Aplidium albicans from the Mediterranean Sea tunicate a marine invertebrate (Henríquez et al. 2005, 80). Plitidepsin (34) is a functioning compound in the treatment of different malignant growths, including melanoma, little and non-little cell lung, prostate, and is at present experiencing clinical preliminaries in stage II research (Fig. 15.11) (Urdiales et al. 1996; Rinehart et al. 1990; Wright et al. 1990). Ecteinascidin 743 (35) has been disconnected from the ascidian *Ecteinascidia turbinata* in extremely low yields (Manzanares et al. 2001; Cuevas and Francesch 2009). By actualizing exceptionally huge scale aquaculture of E, the amounts of ET743 (35) and is required for cutting edge pre-clinical and clinical examinations that have been accomplished. All things considered, the semi-amalgamation of ET743 (35) was outstanding in open lakes shown in Fig. 15.11 (Wright et al. 1990; Alvarez-Miranda et al. 2003; Cuadros et al. 2000). In October 2007, the main marine enemy of malignant growth medication affirmed was Trabectedin (35). Trabectedin (35) was endorsed by the European Medicinal Product Assessment Agency (EMEA) and finished key stage III investigations for endorsement in the United States (Urdiales et al. 1996). Spisulosine (36), isolated from the Spisula polynomial marine mollusk, demonstrated huge specific conduct against tumor cells in correlation with sound cells (Salcedo et al. 2003a). This continued in clinical stage I preliminaries against strong tumors, however, was expelled toward the finish of 2006 (Salcedo et al. 2003b; Trimurtulu et al. 1994; Faulkner 2002). A drug Cryptophycin shown in number 37 was picked for

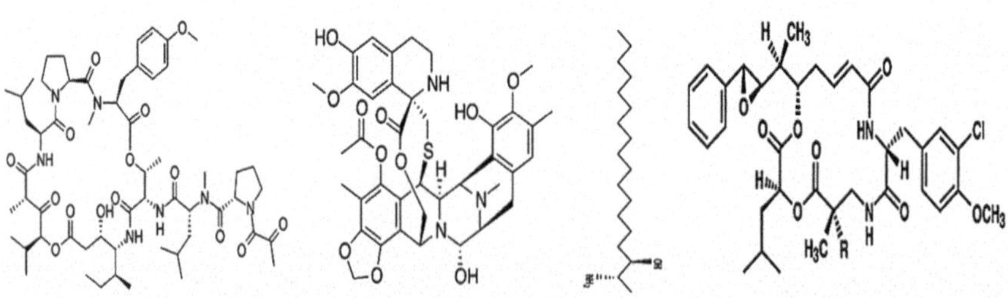

(34) (35) (36) (37)

Fig. 15.11 (34) Plitidepsin, (36) ET743, Spisulosine, (37) Cryptophycin

preliminaries and afterward, shifted to stage II test, yet was expelled because of poisonous quality and absence of viability as shown in Fig. 15.11 (Bhakuni and Rawat 2005).

15.3.4 Algae and its Products

To the biosphere at any rate 30,000 types of green growth (large scale green growth, ocean growth) are providing oxygen, nourishment for fish and man, drug and manures, and go about as a significant wellspring of fundamentally one of a kind common item (Baslow 1969). From the marine green growth in the 1970s–1980s to the terpenoids a class of mixes was secluded. These terpenoid type configurations have prompted the confinement of numerous modules subsidiaries, subordinates from the substance examinations (Yotsu-Yamashita et al. 1993). The green, dark-colored, and red-green growth has been seriously estimated for antimicrobial exercises for organic action (White et al. 2001). For ocean depths harming polycavernoside-A (38) that is detached from *Polycaverosa tsudai* (red algae) be the dangerous compound is mindful. This condition turned 13 individuals to be sick and three kicked the bucket in Japan (Fig. 15.12) (Paquette et al. 2000; Ishitsuka and Kusumi 1988; Faulkner 1988). The antitumor exercises were shown respectively in Fig. 15.12 (39–42) (Tringali et al. 1988; San-Martin et al. 1997). Another is crenuladial (43), which is severed from the darker algae *Dilophus ligatus,* likewise shows antimicrobial movement against as shown in Fig. 15.12 (San-Martin et al. 1997; Elsworth 1989).

Red algae growth, specifically the variety Laurencia (Rhodophyta), is going about as a more prominent wellspring of halogenated sesquiterpenes. From the 1970s the compound examinations concerning the variety Laurencia for optional metabolites have been dynamic. Halogenated sesquiterpenes and diterpenes are the most ordinarily happening auxiliary metabolites. Additionally, this variety is exceptional in delivering C15-acetogenins, similar to those components, which had a terminal enyne, for example, compound 44 (Elsworth 1989). Different models may incorporate the class of mixes called the chamigrenes, which are halogenated terpenes having various chemical structures like 45 and 46 shown in Fig. 15.13. From the class Laurencia numerous chamigrenes have been disengaged to date, which is developed in a wide range of geological zones (Dias et al. 2009; Duke et al. 1993; Georghiou 1990; El Sayed et al. 1997).

Because of manufactured synthetic pesticides (SCPS) efficiency in farming in the last 50 years has gotten progressed in the nuisance control (San-Martin et al. 1991). Be that as it may, because of ascending in the protection from ebb and flow, control operators, the quest for new pesticides are required. During 1984-–1990 protection from manufactured compound pesticides by creepy crawlies and vermin expanded by 13% (Fukuzawa and Masamune 1981; Watanabe et al. 1989). A lot of research has begun on the confinement of insecticidal leads from marine green growth. This has prompted the disconnection of more than 40 dynamic constituents

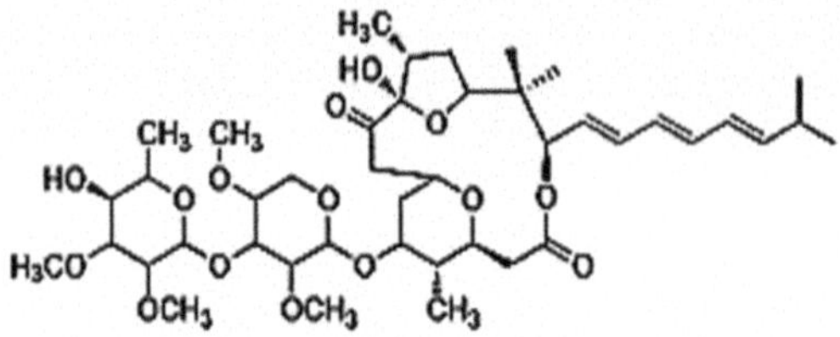

38

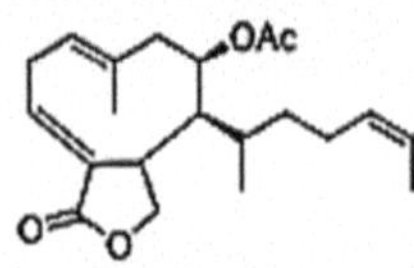

39

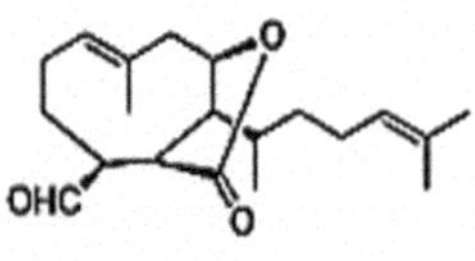

40

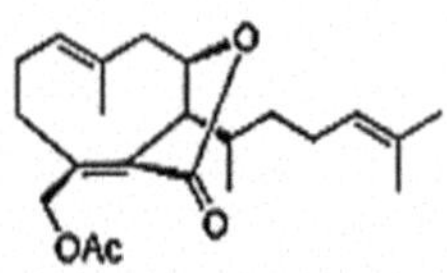

41

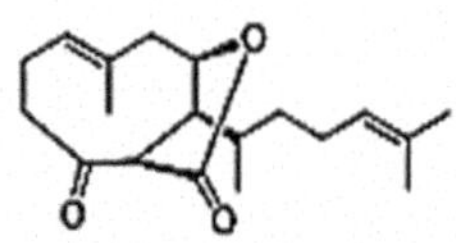

42

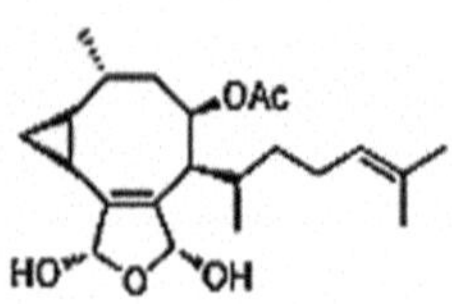

43

Fig. 15.12 (38) Polycavernoside-A, (39) 4-acetoxydictylolactone, (40) Dictyolide A, (41) Dictyolide B, (42) Nordictyolide, (43) Crenuladial

(Watanabe et al. 1989). A few instances of regular bug sprays incorporate the seclusion as shown in Fig. 15.13. The Aster leafhopper, *Macrosteles fascifrons* will get insecticidal action impact from these mixes (McConnell et al. 1994). Different models incorporate laurepinnacin (49), a mix that appeared to show dominant movement hatchling (Fig. 15.13) (Chin et al. 2006).

15.3.5 Porifera and Products Derived

The absence of methods for self-headway living beings, which come up short on an apprehensive, stomach related and circulatory framework and keep up steady water move all the way acquire nourishment, O_2, and to evacuate squanders. Completely wipes remain "current" or "channel" not many bodily methods for the barrier that can be utilized against predators. In around 500 years these are considered to be the principal multicellular creatures that have changed practically nothing. Bergmann reported on the disengagement and recognizable proof of C-nucleosides, spongouridine, and spongothymidine as shown in Fig. 15.14

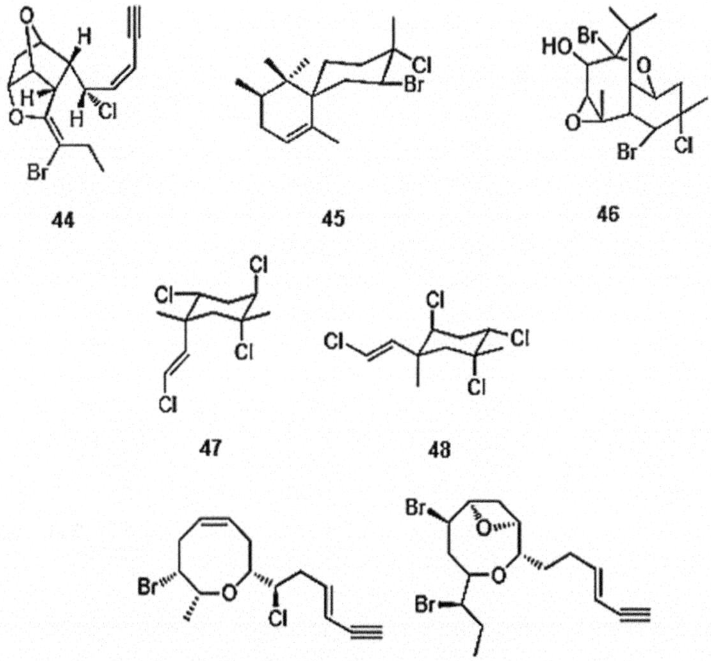

Fig. 15.13 (44) Compound, (45) Compound, (46) Compound, (47) 1ª-dimethylcyclohexane, (48) trichloro-1ª, 5β-dimethylcyclohexane, (49) Laurepinnacin, (50) Z-laureatin

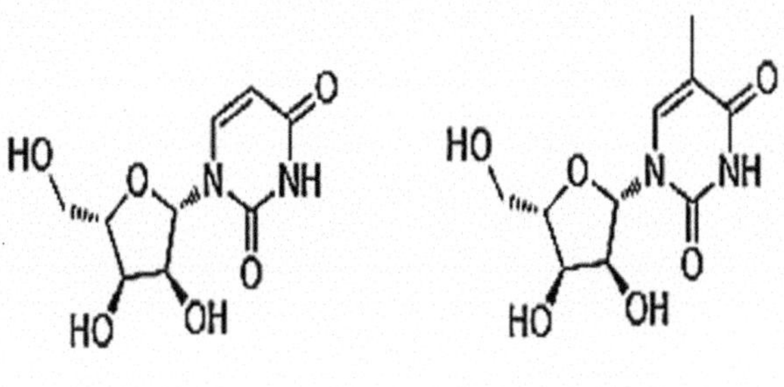

Fig. 15.14 (51) Spongouridine, (52) Spongothymidine

(51–52) from the Caribbean wipe, *Cryptotheca crypta* in the middle 1950s has demonstrated the principal striking disclosure of naturally dynamic mixes from marine sources shown in Fig. 15.14 (Uemura et al. 1985; Pettit et al. 1991). The incredible enemy of the leukemic operator is distinguishing as an (Ara-c) because of these examinations (Pettit et al. 1993).

15.3.6 Marine Sources of Natural Substances

The gathering of false compounds known as bryologists, for example, 53, is obtained from bryostatin 1 (54), an antineoplastic specialist got from *Bulgula neritina* (Butler 2004; Pettit et al. 1991). Bryostatin 1 (54) has been segregated in adequate adds up to take into account more than 80 clinical preliminaries, with 20 finished in both stages I and stage II levels shown in Fig. 15.15 (Urdiales et al. 1996). It has been demonstrated that positive reactions to results running from aggregate to fragmentary abatement (Colegate and Molyneux 2008). Four (stage I) and eight (stage II) clinical preliminaries have been performed since 2007, all mix thinks about expectations various carcinomas with biologics or cytotoxins. In reality, 54 are clinical preliminaries in two-stage I and are being tried as an enemy of Alzheimer's treatment (affirmed Phase I preliminary) (Urdiales et al. 1996). HalichondrinB (55) was disengaged from various wipes, together with *Halichondria okadai* from Japan(Litaudon et al. 1994); *Axinella sp.* from the West Pacific (Aicher et al. 1992), *Phakellia carteri* from the East Indian Ocean(Baker et al. 2007) and *Lissodendoryx* sp. from the East Coast of New Zealand's South Island shown in Fig. 15.15 (Ojima 2008). Halichondrin B (55) has been effectively combined (Nussbaum et al. 2006) alongside an assortment of basic analogs, including halichondrin E-7389 (56), chose for further examination and right now experiencing stage III clinical preliminaries for the treatment of bosom carcinoma (Fig. 15.15) [Pettit et al. 1993].

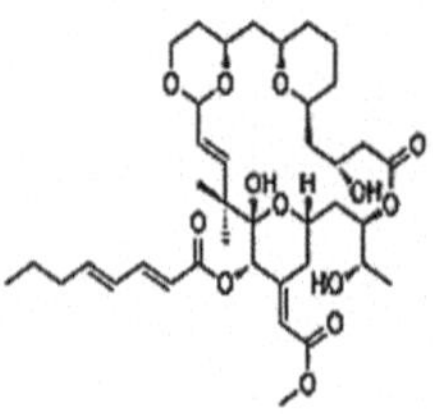

53

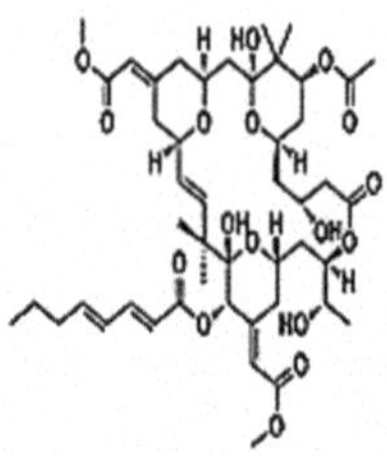

54

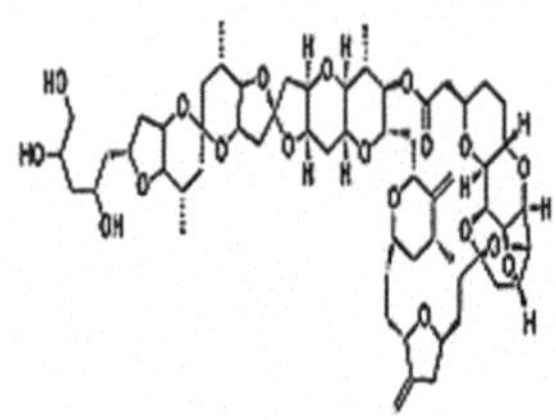

55

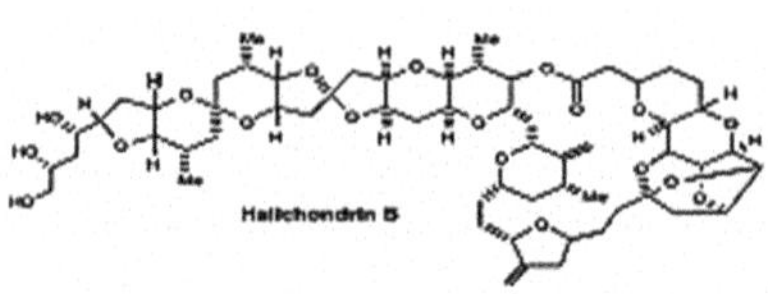

56

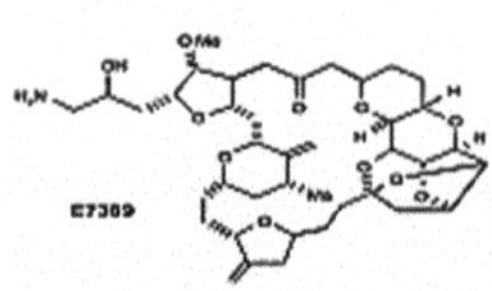

Fig. 15.15 (53) Compound, (54) Bryostatin 1, (55) HalichondriaB, (56) Halichondrin E-7389

15.4 Drug Innovation: Natural Substance

Support toward finding spanking new anti-infection agents, a few significant pharmaceutical ventures presented activities that focused on antibacterial and antifungal objectives, as well as on irresistible illnesses. Such administrations offered lead mixes for the treatment of organ transplantation, bacterial diseases, and hypercholesteremia and tissue dismissal (Luzhetskyy et al. 2007; Newman 2008). All things considered, during the 1990s and mid-2000s, a large number of the significant pharmaceutical organizations decommissioned their NPD frameworks. It was the appearance of robotized elite broadcasting which expanded the prevalence of organic testing and began to be advanced as a superior way to deal with the improvement of "medicate like" HTS mixes. Therefore, a considerable lot of the pharmaceutical organizations separated or sold their broadcasting assortments (Ramakrishna et al. 1993; Sashidhara and Rosaiah 2007), accepted customary concentrate, brought about the constant reevaluation of recently segregated mixes, and that the auxiliary idea of characteristic items requested total monetarily and artificially risky blend and derivatization.

The time required to build up a characteristic item from a concentrated hit to a pharmaceutical was considered too long because of supply issues; HTS advancements rely upon combinatorial science to deliver enormous compound libraries. "Traditional regular substances science" has been to a great extent substituted by restorative medication disclosure concentrated on sub-atomic objectives in the course of recent decades, utilizing huge accomplished powerful "hits" (Newman 2008). Regardless, progresses, innovation, responsive quick recognition of the original regular items, and the explanation of structures keep on improving the procedure of characteristic item revelation (Luzhetskyy et al. 2007). Combinatorial science was believed to be the potential starting point of numerous new carbon skeletons and medication leads or new combinatorial elements (NCEs) from the 1980s onward. This was not the situation since there was just a single combinatorial new synthetic substances affirmed by the U.S. Nourishment, Food and Drug Administration for renal carcinoma, the kinase inhibitor sorafenib shown in structure 57 in Fig. 15.16 (Cordell and Shin 1999).

Truly, combinatorial science has changed the advancement of new dynamic substance drives prompting the amalgamation of basic analogs (Cordell and Shin 1999). However, in the late 1990s, physicists found a goupe of compounds that comprised of

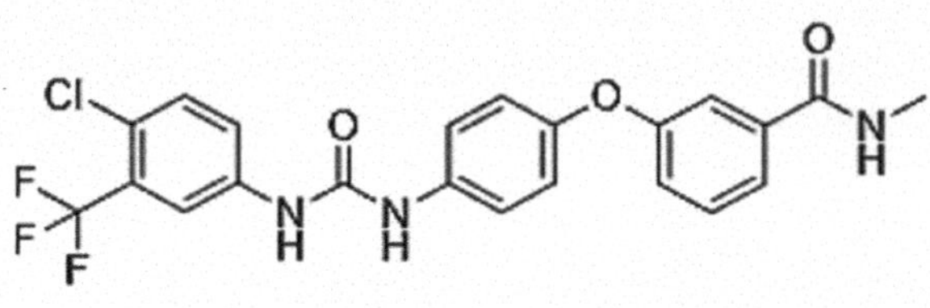

57

Fig. 15.16 (57) Sorafendib

100–1000 new mixes with different compositions, (Cordell and Shin 1999). The possibility of a decent variety situated union (DOS) was presented in which manufactured scientific experts can blend to intensify that look common or appear items (mirrors) or depend on topologies of regular items. Such mixes are right now being tried in a wide scope of organic screens to decide their role(s) as prompting new medication substances (Cordell and Shin 1999). Examination of the NCE endorsement rate uncovers that characteristic items keep on adding to or take an interest in about half of all little particle preliminaries somewhere in the range of 2000 and 2006 (Kashiwada et al. 1996). Albeit a lot of assets were contributed by the pharmaceutical business on science (Kashiwada et al. 1996), generally speaking of the 1184 NCEs covering all maladies/nations/sources somewhere in the range of 1981 and 2006, 30% were seen as fake. It ought to likewise be noticed that 52% (aggregate) of these mixes are a characteristic item, a clone, or a synthetic adjustment of a current pharmacophore regular item.

15.4.1 Dereplication

The medication revelation package tries to look for (a) new bioactive invention(s) that are natural, which have approximately the same type of intense organic action. The separation of known and unfortunate characteristic items with no concoction or pharmacological intrigue is unavoidable. The way toward distinguishing known compounds liable for the exercise of concentrate before bioassay-guided detachment stands alluded (Colegate and Molyneux 2008; Sarker et al. 2006). Many propelled strategies, conventions that recognize original substances regular compounds at the beginning time of a medication revelation program or in a characteristic item seclusion system (Sarker et al. 2006). Necessary to realize that the disengagement of novel characteristic items was an undeniably increasing visit during the 1970s and is consistently decreasing, albeit common sources (e.g., plants, organisms, marine, and microbial sources) are as yet viewed as unlimited hotspots for novel synthetic substances. Actions and charges to determine new compound substances revelation unquestionably rare progressively (Brkljaca and Urban 2011). Besides, the determination of different kinds of life forms (earthly or potentially marine) will in general lead to the rediscovery of recently announced common items as they are frequently found (Blunt and Munro 2009). In this manner, it is highly essential to perceive from the get-go, for setting aside time and cost, and yet to assign assets to progressively productive concentrates. It is obvious that regular items and tirelessness for the identification of sufficient lead mixes than programs are carefully dependent on manufactured synthetic concoctions. This is additionally subject to accessibility of bioassay-guided fractionation, in-house screening, availability to higher field NMR, and mass spectrometers, which are all important to productively run such a program. Lead compounds emerging from normal item revelation programs are fundamentally one of a kind because of their co-development with target destinations in organic frameworks. In any case, the speed at which lead compounds can be created and continuously progressed is slower than relating manufactured medication disclosure approaches (Luzhetskyy et al. 2007). With the approach of new hyphenated spectroscopy advances, further methods for quick complex recognition are presently conceivable (Marinlit 2011; Blunt et al. 2012).

15.4.1.1 Methods of Dereplication

Dereplication methodologies, for the most part, include a blend of bioassay, science of separation, spectroscopic strategies, and searching of database and can be viewed as concoction or organic screening forms. There are various manners by which regular item programs approach dereplication, which depends on the accessibility of screening techniques and the expense recognize conceivable "natural leads or novel mixtures" an unrefined concentrate.

15.4.2 Searching of Database

MarinLit-The Marine Natural Products Database collecting cutting-edge bibliographic information on marine living beings with the number of references from 1200 diaries/books and information for ~21,000 commixture (Scifinder 2011; Blunt et al. 2012; Lang et al. 2008); Anti Marin, a recent database, is where the number of methyl gatherings, the quantity of sp3-hybridized methylene or methine protons, alkene, acetal, ether and formyl gatherings can be looked (Scopus 2011; Napralert 2011). SciFinder Scholar and SCOPUS is an exploration disclosure apparatus (Chemical Abstracts on-line) (Urban and Separovic 2005; Wolfender et al. 2003) and NAPRALERTTM is a database of every regular item, including ethnomedical data, pharmacological/biochemical data of concentrates of living beings in vitro, in situ, in vivo, in people (case reports, non-clinical preliminaries) and clinical investigations (Wolfender et al. 2003). Admittance to scientific databases aforementioned is a major, necessary, and vital step in a well-governed ordinary product program. Detailed and widespread literature explorations are needed once the subsequent queries need to be addressed:

- Was there a history of writing covering the objective life forms (terrestrial or marine?) • Is the potential for limiting novel mixtures (geological or regular varieties?) What kind of compound classes are confined to a species and, if not species, to a species, to a species or a family?
- Is the NMR spectroscopic information inadequate or destitute for previously unspecified innate items?
- Are there new organic exercises that were ignored for known exacerbates?

15.5 Hyphenated Instrumentation "Classical Versus Hyphenated (on-line) Approaches"

Common item extracts frequently contain an enormous number of components including those, which are difficult to isolate. The combination of old-style technologies such as UV intake, UR, Mass spectrometry (MS), and MRS allows for unambiguous functional confirmation of unadulterated materials regularly. In the same way that the spectroscopy of the element can be done periodically, and where it is not possible to resolve the supreme arrangement, a single crystal or blended X-ray method will be used. The immediate hyphenation of an effective divisive procedure with unbelievable spectroscopic systems can contribute to the dereplication procedure because traditional separation methods are dreary and difficult (Schroeder and Gronquist 2006). These hyphened systems integrate HPLC-FTIR, which is useful in recognizing important samples in major components of mixtures (although they are not widely used). The developers of HPLC-FTIR were used, but owing to the confines of similitude they have not encountered a thorough use, which implies preferably separated and properly identified (Schroeder and Gronquist 2006). FIA-NMR consists of an example that is infused into the liquid stream as an attachment and then cleared into the NMR curl indicator. For the transmission from the injector port to the NMR cell of the infused test, FIA-NMR utilizes a versatile stage as a water-powered push. The spectrometer, after the siphon stops, determines the area of the dissolvable pinnacles and then gains a dissolving stifled range. Upon processing, the old specimen from the NMR flow cell is being flushed by a signal to the dissolving pump (Schroeder and Gronquist 2006). HPLC-NMR-MS is a spectrometric progression-hyphenated approach for the removal of common objects (Lewis et al. 2005). In addition to the skills, it's unmistakable coordination of MS information to the NMR range is the major leeway of HPLC-NMR-MS.

Moreover, given that the HPLC-NMR does not provide data on quiet functional groups (e.g., hydroxyl and amino moieties), the D2O swap can promptly recognize these functionalities using MS. The workout, disconnection, filtrating, and structure explanation (standard natural product separation methodologies) for chemical and organic analysis in the search for new bioactive natural products may be challenging and increasingly repetitive. The extraction of marine as well as earthbound forms is typically the initial step. The decision of the dissolvable extraction by the solvable distribution or by shredding can lead to many problems, including the development of ancient rarities. Homogenization and freeze-drinking with natural solvents will affect the structure and relative sizes of the available isolated metabolites.

HPLC-NMR was seen as an amazing spectroscopic instrument that has advanced over the last decade, especially with higher field magnets and cryo-tests in its use in unrefined concentrates (PDA HPLC NMR and UV profile). Recently, in addition to capillary NMR (Cap NMR), the advances made in micro coils HPLC-NMR have taken into account smaller amounts of tests needed for the 40–120 μL method, and this has remarkably improved the affectability in profiling and standard dereplication elements (Dias and Urban 2008; Lin et al. 2008) following high-field magnets. The HPLC-NMR microcoil is usually suitable for Online HPLC-NMR where parts present in higher fixes in concentrate are isolated and examined using either the on-stream or off-stream analysis (Sun Lin et al. 2007, Clarkson et al. 2006, Cogne et al. 2006) in an ordinary HPLC-NMR framework.

Capillary NMR takes into consideration the utilization of non-deuterated solvents in the disconnected HPLC division giving a more extensive scope of solvents to be utilized and low expenses. Separated compounds are re-resolved in deuterated solvents and infused with the CapNMR strip test with the capacity of distinguishing novel auxiliary metabolites of low-level by volumes of approximately 6 μL and 1H-NMR spectrums acquired on request for 2–30 icons. Also, the data obtained both from 1D and 2D NMR spectrums may be sufficient to differentiate between the compound groups, providing data that will provide for reasonable choices as to the best fractional approach or whether you should try to disengage. The findings will, therefore, be assumed to represent compound classes. This technique was used in many late developments (Clarkson et al. 2006). Although the process and use of the HPLC-NMR in regular items can be recognized, the compound profiling of plants was mainly handled by use (Roessner and Beckles 2009; Roessner et al. 2011; Beckles and Roessner 2011). The collapsing frequency of chromatography interfaced with the basic knowledge provided by NMR is followed by different methods of HPLC-NMR (overwhelmingly onstream and stop-stream modes).

The biological system is a growing field covering instruments such as transcriptomics, proteomics, glycomics, and fluxomics that are built to most thoroughly explain both genes and products in cells, including mRNA, protein, glycan structures, and metabolites. To find links across all accessible information, the philosophy of metabolomics is to create unbiased expectations using extremely reproductive explanatory methods used for knowledge analysis. A solitary systemic strategy to profile the metabolites of all low atomic weight of a given being does not exist in the increasing field of metabolomics. This field is consolidating explanatory science, natural chemistry, and complex information technology, which enables the analysis in all organic contexts of thousands of small particles (metabolites). The key forensic processes are the mass spectrometry of nuclear reverberations hyphenated by gas chromatography (CG), liquid chromatography (LC), and slender electrophoresis (CE).

Essential and auxiliary metabolites are segregated through an unbiased rugged extraction process in tissues and biofluids, which ensures that every or most of the metabolites in their characteristic form is extracted successfully before analyzing the solvents used. Because of the incredibly complex artificial decent metabolite variety with which they interact, the metabolite extricates are not prepared for a specific diagnosis stage and solution to split all of the metabolites down at all times. Various division sciences should be used to carry out the best possible work comprehension (Rochfort 2005). Due to the increased sensitivity of instrumentation, resolution, and progress, several mixes can simultaneously be broken up with subsequently refined information technology tools to separate information from the information, sift calculations to remove foundation concussion, location, and mix of peaks through huge information indexes, standardization, and changes in the coming. The greatest metabolomics limitation is the ability to identify the symbol for its complex type. Nonetheless, about 60–80% of each distinct compound is still dark (Cortina et al. 2012; Guoxiang et al. 2008) and the metabolomics network has modified a range of activities to address this problem through the development of large mass libraries of NMRs worldwide. A considerable lot of these obscure structures (i.e., optional metabolites) recognized might be unfamiliar common item assets. Fingerprinting, footprinting, profiling, or target examinations are regular terms utilized in this field. Fingerprinting means to take a "preview" of the living being the place the sign can't really be utilized to distinguish/recognize explicit metabolites and depends unequivocally on the method utilized. Metabolite profiling systems necessitate that signs can be appointed to a particular metabolite whether it is of a known or novel nature. The term target investigation means to decide and evaluate a particular metabolite of premium (Politi et al. 2008). There are not many reports in the logical writing, which talk about the harmony of old-style regular item science which slants by metabolomics to distinguish innovative bioactive common items. These have commonly centered on the investigation of plants (Politi et al. 2008). The distinguishing proof of bioactive characteristic items commencing plants stays a multifaceted errand on account of their high synthetic assorted variety and intricacy. By estimating the metabolome of various concentrates or divisions of a plant and consolidating this information with its comparing natural movement, signals identified with the mixes identified with the showed action can conceivably be resolved. In one model, myxoprincomide (58), a novel NRPS/PKS common item from *Myxococcus Xanthus* DK1622 was recognized by consolidating strategies for focused mutagenesis, liquid chromatography coupled to high goals mass spectrometry (LC-HRMS), and a factual information assessment (Fig. 15.17). Mutant and wild-type strains were developed in little scale maturation in quadruplicate, imitate separates were examined by LCHRMS, and information was pretreated by utilizing a compound discovering calculation, bringing about the meaning of >1000 subatomic highlights per test. Subatomic highlights explicitly missing in culture removes from DK1622 mutant strains were recognized utilizing PCA to the preprocessed LC–MS datasets (Verpoorte et al. 2005).

Five therapeutic Panax herbs (ginseng species) were exposed to a metabolomic examination utilizing Ultra-Performance Liquid Chromatography-Quadrupole Time of Flight Mass Spectrometry (UPLCQTOFMS) and multivariate factual investigation systems. PCA of the diagnostic information clarified that the five Panax herbs could be isolated into five unique gatherings of phytochemicals (Wang et al. 2005). PCA recognized, ginsenoside Rf (59), 20(S)—pseudoginsenoside F11 (60), malonyl gisenoside Rb1 (61), and gisenoside Rb2 (62), which represented the change that was distinguished through the loadings plot of the PCA, and likely by the exact mass of TOFMS (Fig. 15.18) (Wang et al. 2005). The outcomes and philosophy exhibited this technique to be solid for the fast investigation of a gathering of metabolites present in normal items extricate (Wang et al. 2005). On account of

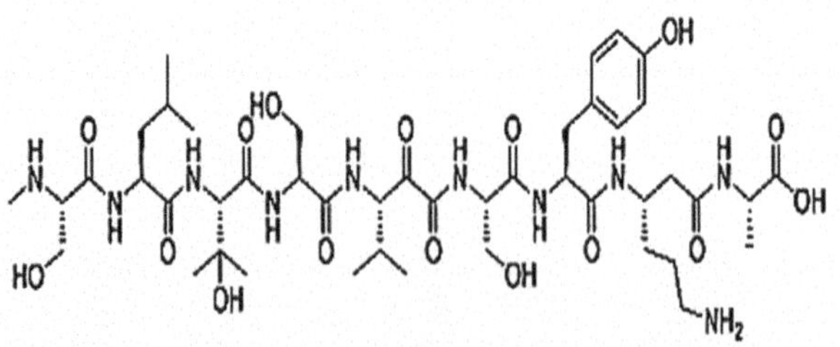

58

Fig. 15.17 (58) Myxoprincomide

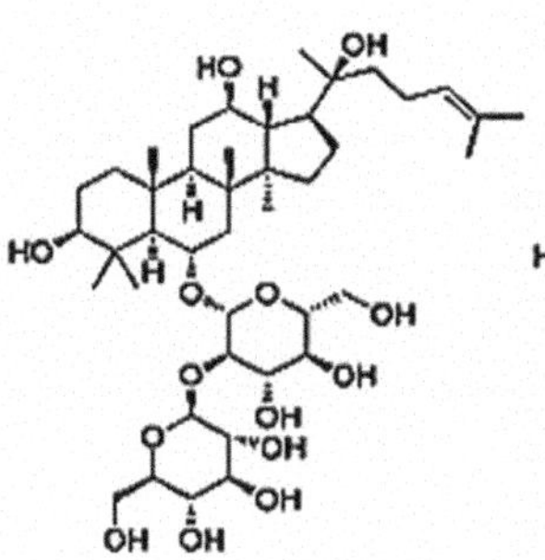

59

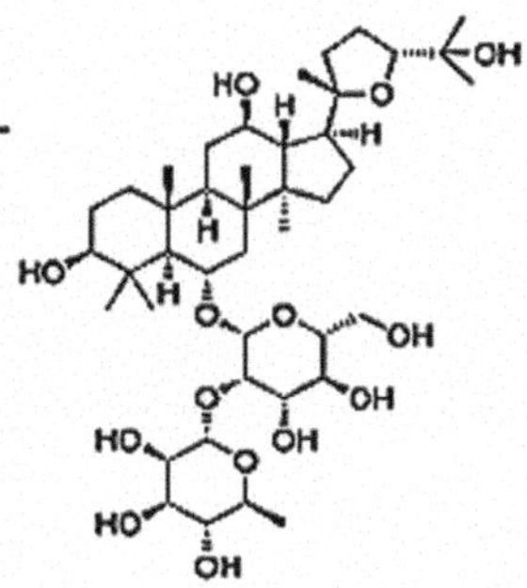

60

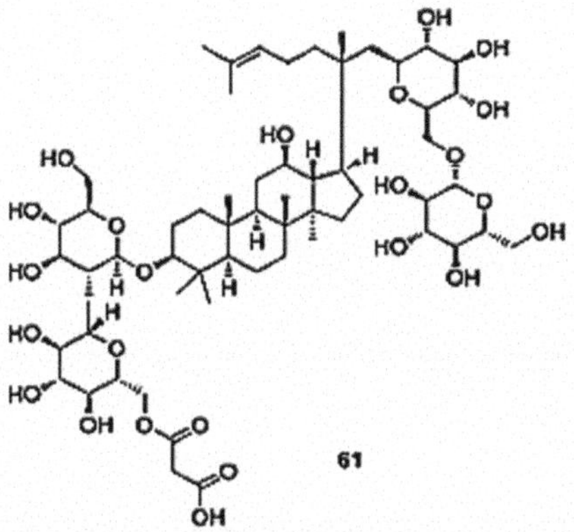

61

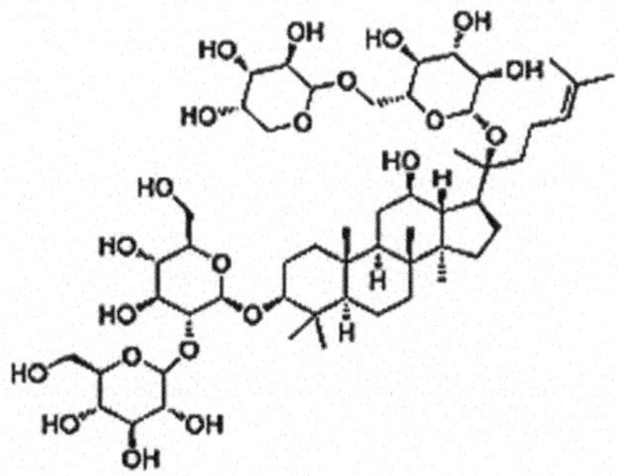

62

Fig. 15.18 (59) GinsenosideRF, (60) 20(S)-pseudoginsenoside F11, (61) Malonyl gisenoside Rb1, (62) Gisenoside Rb2

NMR of rough concentrates, examples can be imagined and translated which is commonly joined with multivariate information investigation. This can be done in a near way recognizing contrasts between moderately comparable concentrates or it tends to be connected with a particular (for the most part in vitro) natural movement. Eventually, this empowers the development of an intricate database of the metabolome (Cardoso-Taketa et al. 2008). An NMR based metabolomics approach has as of late been used in the investigation of *Galphimia glauca*, a Mexican plant that has been utilized in conventional medication for the treatment of focal anxious disarranges (Biao-Yi et al. 2008; Kim et al. 2010; Deyrupa et al. 2011). Six assortments from the Mexican territory showed calming and anxiolytic exercises, with just two assortments of G. glauca demonstrating huge action.1H-NMR metabolomic profiling was directed on every one of the six concentrates and was examined by halfway least square-discriminant examination (PLS-DA) utilizing past data on their bioactivities (Deyrupa et al. 2011). The PLS-DA loadings plot exhibited a sign emphatically associated with calming and anxiolytic exercises was seen as galphimine (63) (Fig. 15.19). A focused HPLC metabolomic approach was likewise directed which gave proof that the two assortments having strong sedative and

Fig. 15.19 (63)
Galphimine

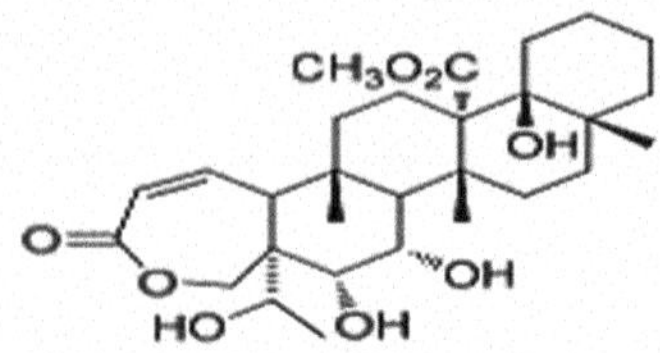

63

Fig. 15.20 (64)
Dihydrocucurbitacin
F-25-O-acetate

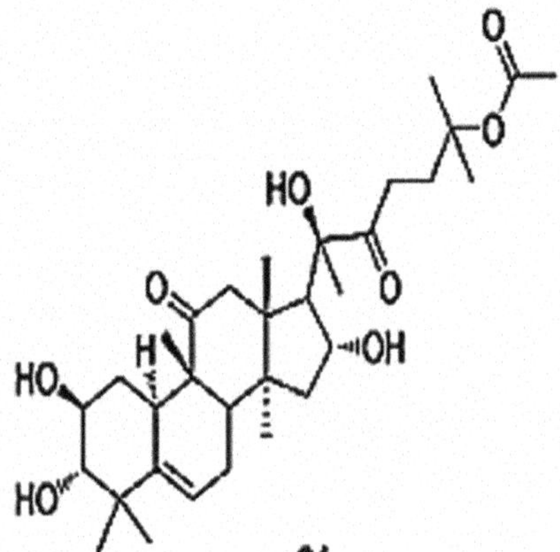

64

anxiolytic exercises contained high measures of galphimine while the other two (less dynamic) examples did not (Deyrupa et al. 2011).

Zhi and associates examined the impact of various anti-infection agents with various methods of activity on different organisms (Deyrupa et al. 2011). The results contemplated that dihydrocucurbitacin F-25-O-acidic corrosive inference (64), a noteworthy constituent of the Chinese plant Hemsleya pengxianensis demonstrated antimicrobial activity (Deyrupa et al. 2011). The metabolome of a Staphylococcus aureus culture treated with a plant remove 64 and a couple of acknowledged enemy of disease specialists was contemplated. PCA examination uncovered that 64 was the part liable for the principle antimicrobial action on *S. aureus* in *H. pengxianensis* through its capacity to repress cell divider amalgamation, as on account of vancomycin (Fig. 15.20).

NMR based metabolomics has numerous applications in plant science and can be utilized in practical genomics to separate plants from the various cause, or after various medicines. Kim and associates depict the benefit of an NMR metabolomics investigation and the probability of recognizing metabolites by contrasting NMR information and references or by structure explanation utilizing 2D-NMR (Deyrupa et al. 2011). Drupa and colleagues additionally exhibited the utilization of 2D-NMR spectroscopy to screen a library of rationally chosen creepy-crawly metabolite tests for incomplete structures. This examination empowered the identification of novel mixes in complex metabolite blends without earlier fractionation or confinement. This prompted the revelation and detachment of two groups of tricyclic pyrones in

Delphastus catalinae, a modest ladybird bug that is utilized monetarily as an organic irritation control specialist (Deyrupa et al. 2011). The D. catalinae pyrones speak to ring frameworks not recently found in nature (Deyrupa et al. 2011).

15.6 Conclusion and Prospects

Concluded mix advancements through normal item exposure procedures will be helpful on different levels. Right off the bat, by increasing the quantity of distinguishing pieces of proof information give fresh constructions to be tried illness under scrutiny. Examination utilizing innovations likewise upgrade build compound portrayal procedures of a wide range of animal groups from normal assets. Furthermore, as referenced earlier, characteristic item scientific experts have gathered a lifetime of compound libraries of dynamic and dormant unadulterated mixes that would now be able to be examined to build mass phantom and NMR ghastly libraries and along these lines improve organic elucidations of metabolomics information. The progressions in investigative instrumentation and complex hyphenation of partition procedures with high touchy indicators have taken into account more noteworthy recognition of little particle mixes quantifiable in organic frameworks (i.e., essential and optional metabolites) and without a doubt will presently be utilized to propel the disclosure of common item science to distinguish potential novel medications applicants that will help in supporting wellbeing and fighting the battle against malady and sickness.

Acknowledgments P.K. and S.R.D. gratefully acknowledge the support provided by Lovely Professional University. The P.D gratefully acknowledged the support provided by Banaras Hindu University, Varanasi.

References

Abraham EP, Chain E, Fletcher CM (1941) Further observations on penicillin. Lancet 16:177–189

Aicher TD, Buszek KR, Fang FG, Forsyth CJ, Jung SH, Kishi Y, Matelich MC, Scola PM, Spero DM, Yoon SK (1992) Total synthesis of halichondrin B and norhalichondrin B. J Am Chem Soc 114:3162–3164

Alder AL (1970) The history of penicillin production. American Institute of Chemical Engineers, New York, NY

Alejandro M, Glaser KB, Cuevas C, Jacobs RS, Kem W, Little RD, McIntosh JM, Newman DJ, Potts BC, Shuster DE (2010) The odyssey of marine pharmaceuticals: a current pipeline perspective. Trends PharmSci 31:255–265

Allen DE, Hatfield G (2004) Medicinal plants in folk tradition: an Ethnobotany of Britain and Ireland. Timber Press, Cambridge, UK, p 431

Alvarez-Miranda M, Rodriguez-Gonzalez A, Ptero G, Lacal JC (2003) Characterization of the mechanism of action of ES-285, a novel antitumor drug from Mactomeris polynyma. Clin Cancer Res 9:Abstract C17

Aniszewski T (2007) Alkaloids—secrets of life. In: Alkaloid chemistry, biological significance, applications and ecological role. Elsevier Science, Amsterdam, p 334

Baker DD, Chu M, Oza U, Rajgarhia V (2007) The value of natural products to future pharmaceutical discovery. Nat Prod Rep 24:1225–1244

Baslow MN (1969) Marine pharmacology; a study of toxins and other biologically active substances of marine origin, vol 286. Williams & Wilkins Co, Baltimore, MD

Beckles DM, Roessner U (2011) Plant metabolomics—applications and opportunities for agricultural biotechnology. In: Altmann A, Hasegawa PM (eds) Plant biotechnology and agriculture: prospects for the 21st century. Elsevier/Academic Press, Boston, MA, USA

Beith M (1999) Healing threads: traditional medicines of the highlands and islands Edinburgh. Edinburgh, UK, Polygon

Bhakuni DS, Rawat DS (2005) Bioactive marine natural products, 1st edn. Springer-Verlag, New Delhi, p 404

Biao-Yi Z, Yan Yu Y, Zeng-Liang Y (2008) Investigation of antimicrobial model of Hemsleya pengxianensis W.J.Chang and its main active component by metabolomics technique. J Ethnopharmacol 116:89–95

Blunt JW, Munro MHG (2009) Dictionary of Marine Natural Products. http://dmnp.chemnetbase.com/. Accessed 21 July 2009

Blunt JW, Munro MHG, Laatsch H (2012) AntiMarin Database. http://www.chem.canterbury.ac.nz/marinlit/marinlit.shtml. Accessed 10 April 2012

Brkljaca R, Urban S (2011) Recent advances in HPLC-NMR and applications for natural product profiling and identification. J Liq Chroma Rel Technol 34:1063–1076

Buss AD, Waigh RD (1995) Antipearasitic drugs. In: Wolff ME (ed) Burger's medicinal chemistry and drug discovery, vol 1, 5th edn. Wiley-Interscience, New York, NY, pp 1021–1028

Butler MS (2004) The role of natural product in chemistry in drug discovery. J Nat Prod 67:2141–2153

Cameron J (1900) The Gaelic names of plants (Scottish, Irish and Mnax) collected and Artanged in scientific order, with notes on their etymology, uses, plant superstitions, etc., among the Celts, with copious Gaelic, English and scientific indices. John Mackay, Glasgow, Scotland, p 160

Cardoso-Taketa AT, Pereda-Miranda R, Choi YH, Verpoorte R, Villarreal ML (2008) Metabolic profiling of the Mexican anxiolytic and sedative plant Galphimia glauca using nuclear magnetic resonance spectroscopy and multivariate data analysis. Planta Med 74:1295–1301

Carroll AR, Arumugan G, Quinn RJ, Redburn J, Guymer G, Grimshaw P (2005) GrandisineA and B, novel indolizidine alkaloids with -opioid receptor binding affinity from the leaves of the human australian rainforest tree Elaeocarpus grandis. J Org Chem 70:1889–1892

Chin YW, Balunas MJ, Chai HB, Kinghorn AD (2006) Drug discovery from natural sources. AAPSJ 8:239–253

Clarkson C, Stærk D, Hansen SH, Smith PJ, Jaroszewski JW (2006) Discovering new natural products directly from crude extracts by HPLC-SPE-NMR: Chinane diterpenes Harpagophytum procumbens. J Nat Prod 69:527–530

Cogne AL, Queiroz EF, Marston A, Wolfender JL, Mavi S, Hostettmann K (2006) On-line identification of unstable iridoids from Jamesbrittenia fodina by HPLC-MS and HPLC-NMR. Phytochem Anal 16:429–439

Colegate SM, Molyneux RJ (2008) Bioactive natural products: detection, isolation and structure determination. CRC Press, Boca Raton, FL, USA, pp 421–437

Cordell GA, Shin YG (1999) Finding the needle in the haystack. The dereplication of natural product extracts. Pure ApplChem 71:1089–1094

Cortina NS, Krug D, Plaza A, Revermann O, Muller R (2012) Myxoprincomide: a natural product from Myxococcus xanthus discovered by comprehensive analysis of the secondary metabolome. Angew Chem Int Ed 51:811–816

Cox PA (2001) Saving the Ethnopharmacological heritage of Samoa. Pharm Biol 39:33–40

Cragg GM (1998) Paclitaxel (Taxol): a success story with valuable lessons for natural product drug discovery and development. Med Res Rev 18:315–331

Cragg GM, Newman DJ (2005) Biodiversity: a continuing source of novel drug leads. Pure Appl Chem 77:7–24

Cuadros R, Montejo de Garcini E, Wandosell F, Faircloth G, Fernandez-Sousa JM, Avila J (2000) The marine compound spisulosine, an inhibitor of cell proliferation, promotes the disassembly of actin stress fibers. Cancer Lett 152:23

Cuevas C, Francesch A (2009) Development of Yondelis® (trabectedin, ET-743). A semisynthetic process solves the supply problem. Nat Prod Rep 26:322–337

Davidson RN, den Boer M, Ritmeijer K (2009) Paromomycin. Trans R Soc Trop Med Hyg 103:653–660

Deleu D, Hanssens Y, Northway MG (2004) Subcutaneous apomorphine: an evidence-based review of its use in Parkinson's disease. Drugs Aging 21:687–709

DerMarderosian A, Beutler JA (2002) The review of natural products, 2nd edn. Facts and Comparisons, Seattle, WA, pp 13–43

Dewick PM (2002) Medicinal natural products: a Biosynthentic approach, 2nd edn. John Wiley and Son, West Sussex, p 520

Deyrupa ST, Eckman LE, McCarthy PH, Smedley SR, Meinwald J, Schroeder FC (2011) 2D NMR-spectroscopic screening reveals polyketides in ladybugs. Proc Natl Acad Soc USA 108:9753–9758

Dias D, Urban S (2008) Phytochemical analysis of the southern australian marine alga, *Plocamium mertensii* using HPLC-NMR. Phytochem Anal 19:453–470

Dias DA, White JM, Urban S (2009) *Laurencia filiformis*: phytochemical profiling by conventional and HPLC-NMR approaches. Nat Prod Commun 4:157–172

Dillenius JJ (1724) Synopsis Methodica Stirpium Britannicarum, 3rd edn. G. and J.Innys, London, UK, p 482

Duke SO, Menn JJ, Plimmer JR (1993) In: Duke SO, Menn JJ, Plimmer JR (eds) Pest control with enhanced environmental safety; ACS symposium series no. 514. American Chemical Society, Washington, DC

Duke JA, Duke PAK, du Cellier JL (2008) Duke's handbook of medicinal plants of the bible. CRC Press Taylor and Francis Group, Boca Raton, FL, USA, p 552

El Sayed KA, Dunbar DC, Perry TL, Wilkins SP, Hamann MT, Greenplate JT, Wideman MA (1997) Marine natural products as prototype insecticidal agents. J Agric Food Chem 45:2735–2739

Elsworth JM (1989) A new chamigrane from *Laurencia glomerata*. J Nat Prod 52:893–895

Fabbretti A, Gualerzi CO, Brandi L (2011) How to cope with the quest for new antibiotics. FEBS Lett 585:1673–1681

Fabricant DS, Farnsworth NR (2001) The value of plants used in traditional medicine for drug discovery. Environ Health Perspect 109:69–75

Farnsworth NR, Akerele RO, Bingel AS, Soejarto DD, Guo Z (1985) Medicinal plants in therapy. Bull WHO 63:965–981

Faulkner DJ (1988) Marine natural products. Nat Prod Rep 20:269–309

Faulkner DJ (2002) Marine natural products. J Nat Prod Rep 19:1–48

Fellows L, Scofield A (1995) Chemical diversity in plants. In: Intellectual property rights and biodiversity conservation—an interdisciplinary analysis of the values of medicinal plants. University Press, Cambridge, UK

Fukuzawa A, Masamune T (1981) Laurepinnacin and isolaurepinnacin, new acetylenic cyclic ethers from the marine red alga Laurencia pinnata Yamada. Tetrahedron Lett 22:4081–4084

Georghiou GP (1990) Overview of insecticide resistance. In: Green MB, Le Baron HM, Moberg WK (eds) Managing resistance to agrochemicals: from fundamental research to practical strategies. ACSSymp. Ser. 421. Am ChemSoc, Washington, DC, pp 18–41

Guoxiang X, Robert P, Mingming S, Zhaohui X, Aihua Z, Mingfeng Q, Xiangbao L, Zhong L, Wei J (2008) Ultra-performance LC/TOF MS analysis of medicinal Panax herbs for metabolomic research. J Sep Sci 31:1015–1026

Gustafson KR, Cardellin JH, McMahon JB, Gulakowski RJ, Ishitoya J, Szallasi Z, Lewin NE, Blumberg PM, Weislow OS, Beutler JA (1992) A nonpromoting phorbol from the Samoan medicinal plant Homalanthus nutans inhibits cell killing by HIV-1. J MedChem 35:1978–1986

Gwinn KD, Schardl CL, Friburg A (1992) Southern regional information exchange group (SRIEG-37) on the tall fescue endophyte. J Prod Agric 5:189–190

Haefner B (2003) Drugs from the deep: marine natural products as drug candidates. Drug Discov Today 8:536–544

Hatfield G (2005) Country remedies: traditional east Anglian plant remedies in the twentieth century. Boydell Press, Woodbridge, UK

Heinrich M, Teoh HL (2004) Galanthamine from snowdrop-the development of a modern drug against Alzheimer's disease from local Caucasian knowledge. J Ethnopharmacol 92:147–162

Henríquez R, Faircloth G, Cuevas C (2005) In Ecteinascidin 743 (ET-743, Yondelis), aplidin, and kahalalide F. In: Cragg GM, Kingston DGI, Newman DJ (eds) Anticancer agents from natural products. Taylor and Francis, Boca Raton, FL, p 215

Hicks S (2014) Desert plants and people, 1st edn. Naylor Co, San Antonio, TX, p 75

Holwell SE, Cooper PA, Grosios JW, Lippert JW, Pettit GR, Snyder SD, Bibby MC (2002) Combretastatin A-1 phosphate, a novel tubulin-binding agent with in-vivo anti-vascular effects in experimental tumors. Anticancer Res 22:707–712

Howes MJR, Perry NSL, Houghton PJ (2003) Plants with traditional uses and activitities, relevant to the management of Alzheimer's disease and other cognitive disorders. Phytother Res 17:1–18

Ishitsuka MO, Kusumi TKH (1988) Antitumor xenicane and norxenicane lactones from the brown alga *Dictyota dichotoma*. J OrgChem 53:5010–5013

Kashiwada Y, Hashimoto F, Cosentino LM (1996) Betulinic acid and dihydrobetulinic acid derivatives as potent HIV agents. J Med Chem 39:1016–1017

Kashman Y, Gustafson KR, Fuller RW, Cardellin JH, McMahon JB, Currens MJ, et al. (1992) The Calanolides, a novel HIVinhibitory class of coumarin derivatives from the tropical rainforest tree, *Calophyllum lanigerum*. J Med Chem 35:2735–2743

Kedei N, Lundberg DJ, Toth A, Welburn P, Garfield SH, Blumberg PM (2004) Characterization of the interaction of ingenol 3-angelate with protein kinase C. Cancer Res 64:3243–3255

Kim HK, Choi YH, Verpoorte R (2010) NMR-based metabolomic analysis of plants. Nat Protoc 5:536–549

Kinghorn AD, Pan L, Fletcher JN, Chai H (2011) The relevance of higher plants in lead compound discovery programs. J Nat Prod 74:1539–1555

Kiviharju TM, Lecane PS, Sellers RG, Peehl DM (2002) Antiproliferative and proapoptic of triptolide (PG490), a natural product entering clinical trials, on primary cultures of human prostatic epithelial cells. Clin Cancer Res 8:2666–2674

Lang G, Mayhudin NA, Mitova MI, Sun L, van der Sar S, Blunt JW, et al. (2008) Evolving trends in the dereplication of natural product extracts: new methodology for rapid, small-scale investigation of natural product extracts. J Nat Prod 19:1595–1599

Lax E (2004) The Mold in Dr.Florey's coat: the story of the *Penicillian miracle*. John Macrae/ Henry Hol, New York, NY, p 308

Lee JC, Strobel GA, Lobkovsky E, Clardy JC (1996) Torreyanic acid: a selectively cytotoxic quinone dimer from the endophytic fungus *Pestalotiopsis microspora*. J Org Chem 61:3232–3233

Lewis RJ, Bernstein MA, DuncanSJ SCJA (2005) Comparison of capillary-scale LC-NMR with alternative techniques: spectroscopic and practical considerations. Magn Reson Chem 43:783–789

Li C, Johnson RP, Porco JA (2003) Total synthesis of the quinine epoxide dimer (+)-torreyanic acid: application of a biomimetic oxidation/electrocyclization/Diels-Alder dimerization cascade. J Am ChemSoc 125:5059–5106

Lightfoot J (1977) FloraScotica, vol 2. BenjaminWhite, London

Lin Y, Schiavo S, Orjala J, Vouros P, Kautz R (2008) Microscale LC-MS-NMR platform applied to the identification of active cyanobacterial metabolites. Anal Chem 80:8045–8054

Litaudon M, Hart JB, Blunt JW, Lake RJ, Munro MHG (1994) Isohomohalichondrin B, a new antitumour polyether macrolide from the New Zealand deep-water sponge, Lyssodendoryx sp. Tetrahedron Lett 35:9435

Luzhetskyy A, Pelzer S, Bechthold A (2007) The future of natural products as a source of new antibiotics. Curr Opin Investig Drugs 8:608–613

MacFarlane AM (1929) Gaelic names of plants: study of their uses and lore. Trans Gaelic Soc Inverness 32:1–48

Mann J (1994) Murder, magic, and medicine. OxfordUniversity Press, New York, NY, pp 164–170
Mann J (1999) The elusive magic bullet: the search for the perfect drug. Oxford University Press, New York, NY, USA
Manzanares I, Cuevas C, Garcia-Nieto R, Marco E, Gago F (2001) Advances in the chemistry and pharmacology of ecteinascidins, a promising new class of anticancer agents. Curr Med Chem Antican Agents 1:257
Maplestone RA, Stone MJ, Williams DH (1992) The evolutionary role of secondary metabolites—a review. Gene 115:151–157
Marin Lit. 2011. http://www.chem.canterbury.ac.nz/marinlit/marinlit.shtml. Accessed 23 June 2011
Marris E (2006) Marine natural products: drugs from the deep. Nature 443:904–905
Martin MA (1934) In: Macleod DJ (ed) Description of the Western isles of Scotland, 4th edn. Stirling: EneasMackay, Cornhill, UK
Martin DE, Blum R, Wilton J (2007) Safety and pharmacokinetics of beririmat (PA-457) a novel inhibitor of human immunodeficiency virus maturation, in healthy volunteers. Antimicrob. Agents Chem 51:3063–3066
McConnell O, Longley RE, Koehn FE (1994) In: Gullo VP (ed) The discovery of natural products with therapeutic potential. Butterworth-Heinemann, Boston, MA, pp 109–174
McRae J, Yang Q, Crawford R, Palombo W (2007) Review of the methods used for isolating pharmaceutical lead compounds from traditional medicinal plants. Environment 27:165–174
Min BS, Nakamura N, Miyashiro H, Bae KW, Hattori M (1998) Triterpenes from the spores of *Ganoderma lucidum* and their inhibitory activity against HIV-1 protease. Chem Pharm Bull 46:1607–1612
Moloney MF (1919) In: Dublin MH (ed) Irish ethno-botany and the evolution of medicine in Ireland. Gill and Son, Dublin, Ireland
Müller K (2001) Pharmaceutically relevant metabolites from lichens. ApplMicrobiolBiotechnol 56:9
Napralert. 2011. http://www.napralert.org/. Accessed 3 May 2011
Newman DJ (2008) Natural products as leads to potential drugs: an old process or the new hope for drug discovery? J Med Chem 51:2589–2599
Newman DJ, Cragg GM (2005) In: Zhang L, Fleming A, Demain AL (eds) In drug discovery, therapeutics, and preventive medicine. Humana Press, Totowa, NJ, p 74
Newman DJ, Cragg GM (2007) Natural products as sources of new drugs over the last 25 years. J Nat Prod 70:461–477
Nicolaou KC, Yang Z, Liu JJ, Ueno H, Nantermet PG, Guy RK, et al. (1994) Total synthesis of Taxol. Nature 367:630–634
Nussbaum FV, Brands M, Hinzen B, Weigand S, Habich D (2006) Antibacterial natural products in medicinal chemistry - exodus or revival? Angew Chem Int Ed 45:5072–5129
Ogbourne SM, Suhrbier A, Jones B (2004) Antitumour activity of ingenol 3-angelate: plasma membrane and mitochondrial disruption and necrotic cell death. Cancer Res 64:2833–2839
ÓhEithir R (1983) Folk medical beliefs and practices in the Aran Islands. Master's thesis. National University of Ireland, Galway, Ireland
Ojima I (2008) Modern natural products chemistry and drug discovery. J Med Chem 51:2587–2588
Paquette LA, Barriault L, Pissarnitski D, Johnston JN (2000) Stereocontrolled elaboration of natural (−)-Polycavernoside a, a powerfully toxic metabolite of the red alga Polycavernosa tsudai. J Am ChemSoc 122:619–631
Petrini O (1986) Taxonomy of endophytic fungi of aerial plant tissues. In: Fokkema NJ, van den Heuvel J (eds) Microbiology of Phyllosphere. Cambridge University Press, Cambridge, pp 175–187
Pettit GR, Herald CL, Boyd MR, Leet JE, Dufresne C, Doubek DL, et al. (1991) Antineoplastic agents. 219. Isolation and structure of the cell growth inhibitory constituents from the western Pacific marine sponge *Axinella* sp. J Med Chem 34:3339–3340
Pettit GR, Srirangam JK, Herald DL, Erickson KL, Doubek DL, Schmidt JM, et al. (1993) Antineoplastic agents. 251. Isolation and structure of stylostatin 1 from the Papua New Guinea marine sponge *Stylotella* sp. J Org Chem 58:3222

Politi M, Peschel W, Wilson N, Zloh M, Prieto JM, Heinrich M (2008) Cannabis water extracts and tinctures analysed by NMR spectroscopy; different strategies to reduce the content of D9-THC. Phytochemistry 69:562–570

Purvis W (2000) Lichens. Natural History Museum, London/Smithsonian Institution, WashingtonD.C., USA, p 112

Ramakrishna NVS, Nadkarni SR, Bhat RG, Naker SD, Kumar EKSV, Lal B (1993) Screening of natural product extracts for antibacterial activity: early identification and elimination of known compounds by dereplication. Ind J Chem 38:1384–1387

Rinehart KL, Lithgow-Bertelloni AM (1991) Novel antiviral and cytotoxic agent, dehydrodidemnin B. PCT Int Pat Appl 15:248086q

Rinehart KL, Holt TG, Fregeau NL, Stroh JG, Keifer PA, Sun F, Li LH, Martin DG (1990) Ecteinascidins 729, 743, 745, 759A, 759B, and 770: potent antitumor agents from the Caribbean tunicate *Ecteinascidia turbinata*. J OrgChem 55:4512–4515

Rochfort S (2005) Metabolomics reviewed: a new "Omics" platform technology for systems biology and implications for natural products research. J Nat Prod 68:1813–1820

Roessner U, Beckles DM (2009) Metabolite measurements. In: Junker B, Schwender J (eds) Plant metabolic networks. Springer, Heidelberg, Germany

Roessner U, Nahid A, Hunter A, Bellgard M (2011) Metabolomics—the combination of analytical chemistry, biology and informatics. In: Moo-Young M (ed) Comprehensive biotechnology, vol 1, 2nd edn. Springer, Heidelberg, pp 447–459

Salcedo M, Cuevas C, Otero G, Sanchez-Puelles JM, Fernandez-Sousa JM, Avila J, Wandosell F (2003a) The marine antitmor compound ES 285 activates EGD receptors. Clin Cancer Res 9:Abstract C24

Salcedo M, Cuevas C, Sanchez-Puelles JM, Otero G, Sousa JMF, Avila J, Wandosell F (2003b) ES-285, a novel antitumoral compound, interacts with EDG receptors. Proceedings of American Association for Cancer Research 94th Meeting, Washington, 11–14 July, 2003, Abstract 3649

San-Martin A, Negrete R, Rovirosa J (1991) Insecticide and acaricide activities of polyhalogenated monoterpenes from Chilean Plocamium cartilagineum. Phytochemistry 30:2165–2169

San-Martin A, Darias J, Soto H, Contreras C, Herrera JS, Rovirosa J (1997) A new C15 acetogenin from the marine alga *Laurencia claviformis*. Nat Prod Lett 10:303–311

Sarker SD, Latif Z, Gray AI (2006) In: Satyajit D (ed) Methods in biotechnology: natural product isolation. Human Press Inc, Totowa, NJ, USA, p 528

Sashidhara KV, Rosaiah JN (2007) Various dereplication strategies using LC-MS for rapid natural product lead identification and drug discovery. Nat Prod Commun 2:193–202

Schroeder FC, Gronquist M (2006) Extending the scope of NMR spectroscopy with microcoil probes. Angew Chem Int Ed 245:7122–7131

Sci Finder Scholar. 2011. http://www.cas.org/SCIFINDER/SCHOLAR/. Accessed 2 July 2011

SCOPUS. 2011. http://www.scopus.com/home.url. Accessed 25 August 2011

Stamets P (2002) Novel antimicrobials from mushrooms. Herbal Gram 54:28–33

Sun Lin J, Mahyudin NA, Chamyuang S, Blunt JW, Cole T, Lang G, et al. (2007) Less is more: Dereplication and discovery using CapNMR techniques. In Proceedings of ManaproXII: 12th International Symposium of Marine Natural Products, Queenstown, New Zealand, Feb 4th–9th, 2007

Svabo JC (1959) Indberetninger fra en Reise I Færøe 1781 og 1782. Selskabet til Udgivelse af FærøskeKildeskrifter og Studier, Copenhagen, Denmark, p 497

Swanton EW (1915) Economic and folklore notes. Trans Br MycolSoc 5:408–409

Swanton EW (1932) Sussex County magazine, vol 6. T.R. Beckett, Eastbourne, UK, p 709

Tan RX, Zou WX (2001) Endophytes: a rich source of functional metabolites (1987 to 2000). Nat Prod Rep 18:448–459

Trimurtulu G, Ohtani I, Patterson GML, Moore RE, Corbett TH, Valeriote FA, Demchik L (1994) Total structures of cryptophycins, potent antitumor depsipeptides from the bluegreen alga *Nostoc* sp. strain GSV 224. J Am ChemSoc 116:4729–4737

Tringali C, Oriente G, Piattelli M, Geraci C, Nicolosi G, Breitmaier E (1988) Crenuladial, an antimicrobial diterpenoid from the brown alga *Dilophus ligulatus*. Can J Chem 66:2799–2802

Uemura D, Takahashi K, Yamamoto T, Katayama C, Tanaka J, Okumura Y, Hirata Y (1985) Norhalichondrin a: an antitumor polyether macrolide from a marine sponge. J Am ChemSoc 107:4796–4798

Urban S, Separovic F (2005) Developments in hyphenated spectroscopic methods in natural product profiling. Front Drug Des Discov 1:113–166

Urdiales JL, Morata P, De Castro IN, Sanchez-Jimenez F (1996) Anti-proliferative effect of dehydrodidemnin B (DDB), a depsipeptide isolated from *Mediterranean tunicates*. Cancer Lett 102:31–37

Verpoorte R, Choi YH, Kim HK (2005) Ethnopharmacology and systems biology: a perfect holistic match. J Ethnopharmacol 100:53–56

Vickery R (1995) A dictionary of plant-Lore. Oxford University Press, Oxford, UK

Wainwright M (1990) Miracle cure: the story of penicillin and the Golden age of antibiotics. Blackwell Scientific, Oxford, p 57

Wang M, Lamers RJAN, Korthout HA, Van Nesselrooij JHJ, Witkamp RF, Van der Heijden R (2005) Metabolomics in the context of systems biology: bridging traditional Chinese medicine and molecular pharmacology. Phytother Res 3:173–182

Watanabe K, Umeda K, Miyakado M (1989) Isolation and identification of three insecticidal principles from the red alga *Laurencia nipponica* Yamada. Agric Biol Chem 53:2513–2515

White JD, Blakemore PR, Browder CC, HongJ LCM, Nagornyy PA, Robarge LA, Wardrop DJ (2001) Total synthesis of the marine toxin polycavernoside a via selective macrolactonization of a trihydroxy carboxylic acid. J Am Chem Soc 123:8593–8595

Wolfender JL, Ndjoko K, Hostettman K (2003) Liquid chromatography with ultraviolet absorbance-mass spectrometric detection and with nuclear magnetic resonance spectroscopy: a powerful combination for the on-line structural investigation of plant metabolites. J Chromatogr A1000:437–455

Wright AE, Forleo DA, Gunawardana GP, Gunasekera SP, Koehn FE, McConnell OJ (1990) Antitumor tetrahydroisoquinoline alkaloids from the colonial ascidian *Ecteinascidia turbinata*. J Org Chem 55:4508–4512

Yogeeswari P, Sriram D (2005) Betulinic acid and its derivatives: a review on their biological properties. Curr Med Chem 12:763–771

Yotsu-Yamashita M, Haddock RL, Yasumoto T (1993) PolycavernosideA: a novel glycosidic macrolide from the red alga *Polycavernosa tsudai* (*Gracilaria edulis*). J Am Chem Soc 115:1147–1148

Zjawiony JK (2004) Biologically active compounds from aphyllophorales (polypore) fungi. J Nat Prod 67:300–310

Potential Strategies for Control of Agricultural Occupational Health Hazards

16

Vinod Pravin Sharma, Simranjeet Singh, Daljeet Singh Dhanjal, Joginder Singh, and Ajar Nath Yadav

Abstract

Although the development and progression in agriculture have increased crop production, in return, it has resulted in various health problems. International Labour Organization has considered the agricultural sector as one of the most hazardous to health worldwide. Many hazardous toxic non-biodegradable chemical compounds like, persistent organic pollutants, xenoestrogens are persisting in the agricultural environment, and have entered the food chain as well as started accumulating in agriculture workers. Farmers are highly exposed to biological, chemical and environmental hazards but can sustain their life by taking home remedies as well as breathing fresh air released by plants, but the risk associated with chemicals, heat, musculoskeletal injuries, noise, poisonous insects, reptiles, grain bins and silos still prevail. We need to take utmost care of the agriculture sector people and educate them with new concepts of farming and simultaneously understand the experiences gained by them. We need to increase the level of health education, safety literacy and subsequently improve the farmers' quality of life for improved living standards. This book chapter provides an overview

V. P. Sharma (✉)
Academy of Scientific and Innovative Research, CSIR-Indian Institute of Toxicology Research, Lucknow, Uttar Pradesh, India

S. Singh · J. Singh
Department of Microbiology, Division of Research and Development, Climate Mitigation and Sustainable Agriculture Research lab (CMaSAR), Lovely Professional University, Phagwara, Punjab, India

D. S. Dhanjal
Department of Biotechnology, Lovely Professional University, Phagwara, Punjab, India

A. N. Yadav
Department of Biotechnology, Dr. Khem Singh Gill Akal College of Agriculture, Eternal University, Baru Sahib, Sirmour, Himachal Pradesh, India

A. N. Yadav et al. (eds.), *Current Trends in Microbial Biotechnology for Sustainable Agriculture*, Environmental and Microbial Biotechnology,
https://doi.org/10.1007/978-981-15-6949-4_16

of the health hazards associated with modern agricultural practices and suitable alternatives to cope up with them.

Keywords

Agriculture · Practice · Health · Pesticides · Risk · Factors

16.1 Introduction

For improving the living standards, there is a need for spreading awareness among the farmers for a quality life. The policy and strategies are required to be farmed with the help of both researchers and policymakers to inculcate new strategies with prior policies and programmes to improve the health of farmers (De Boer et al. 2017). Nowadays, farmers are highly exposed to biological, chemical and environmental hazards but are able to sustain their life by taking home remedies as well as breathing fresh air released by plants. But the risk associated with chemicals, heat, musculoskeletal injuries, noise, poisonous insects, reptiles, grain bins and silos still prevails (Tulchinsky and Varavikova 2014). Previously comprehended literature discussing the control as well as solution to curb these hazards needs to spread among the common and illiterate people with the help of short movies, nukkadnataks and plays (Coman 2020). The indirect environmental hazards have largely affected the health and well-being of the agricultural workers (Damalas and Eleftherohorinos 2011). Therefore, it has become important to make them conscious about the initial symptoms of pulmonary diseases, chemical toxicities, cancers, heat-induced complications, noise-induced hearing loss and skin disorders (Kumar et al. 2018b, c, d; Kaur et al. 2018; Koul et al. 2019; Singh et al. 2019b). Moreover, there is a need for the establishment of precautions to curtail and eliminate the potential risk without creating chaos (Sharma et al. 2020). Besides this, affected farmers may also affect their animals as well as family members and can spread diseases.

The clinical zoonotic abnormalities like Anthrax, Brucellosis, Cryptosporidiosis, Giardiasis, Leptospirosis, Psittacosis, Rabies and Tuberculosis are the ones which get transmitted from animals to humans. These diseases get triggered by bacteria, fungi, parasites, protozoans and viruses, which are part of their normal microflora. These diseases get spread due to touching of the products (infected wounds, manure, placenta) or animal itself (Esch and Petersen 2013). Other than this, direct transmission takes place due to ingestion of raw animal products or through an animal bite. This has made humans indirectly highly vulnerable to get infected by contaminated food, water and soil (Kumar et al. 2014a). Farmers, veterinarians and other workers are at higher risk of getting infected by zoonoses due to their close interaction with animals. The agriculturalist is at higher risk of acquiring animal-borne or zoonotic infection due to nature of work they perform or the kind of animals they interact (Kumar et al. 2019a, b; Singh et al. 2020k). Hence, there is a need for maintaining adequate resources for workers operating in the area where they are prone to get exposed to both animals and animal-borne diseases. This chapter intends to highlight about chemical hazards of toxic compounds, protective measures taken to curb it and discuss their effect on human health.

16.2 Chemical Hazards of Toxic Compounds

16.2.1 Persistent Organic Pollutants (POPs)

These are non-biodegradable chemical compounds persisting in the environment and have entered our food chain as well as started accumulating in our body (Pramanik et al. 2015). The accumulation of the chemical compounds has affected not only the environment but also the microflora and human health (Kumar et al. 2016). Moreover, the organic nature of these compounds makes them resistant to biological, chemical and photolytic degradation (Singh et al. 2019f; Kumar et al. 2020a). These compounds have low as well as high water solubility, which allows their accumulation in the fatty tissues (Datta et al. 2020). Additionally, these compounds are also semi-volatile in nature (Girdhar et al. 2014; Kumar et al. 2019e). Furthermore, various amended forms of this organic pollutant exist because of anthropogenic and natural activities, which are determined via bioaccumulative and high persistence characteristics (Bhatia et al. 2014). This POPs group involves chemical-like industrial chemicals (like polychlorinated biphenyls), residues of numerous industrial processes (like dioxins and furans) and pesticides (like DDT) (Kumar et al. 2013b, 2014b; Singh et al. 2016, 2018; Sidhu et al. 2019).

A large number of compounds have an adverse effect because of their high persistence in the environment, although they possess the ability to bioaccumulate as well as biomagnified. Polychlorinated benzene is the one of the POPs, which has been comprehended to persist in the environment for a long period of time (Kumar et al. 2013a, 2015a, 2016; Sharma et al. 2015; Makkar et al. 2016).

POPs releases into the environment in various ways, directly or indirectly. It releases as a by-product in combustion emissions or direct introduction into the environment (Singh et al. 2020a). Pesticides were used, as a result, to protect plant protection from various insects, weeds, etc. (Gill et al. 2015). The residues of these pesticides are highly persistent in nature and have a negative impact on microflora and soil fertility (Singh et al. 2020j, k). Pesticides released into the environment by their use, during their transport and storage PCBs are used in capacitors as cooling fluids and are released into the environment via evaporations (Kumar et al. 2019c, g; Bhati et al. 2019). These compounds are semi-volatile and have accumulative characteristics. Certain compounds such as dioxins, furans polycyclic aromatic compounds are directly emitted into the air and are synthesized by chlorinated substances (Kumar et al. 2019f; Singh et al. 2019e; Kapoor et al. 2019). The dioxin and furan are emitted in the environment by the combustion of waste or stationary fuel combustion or by thermal processes in iron industry and fire forests (Dhanjal et al. 2018; Kumar et al. 2018b; Datta et al. 2018; Singh et al. 2019c). The chlorinated compounds are leached into the environment by use of these compounds in bleaching paper or use of these compounds in the production of aluminium and chlorinated solvents (Bhale et al. 2018; Kumar et al. 2018d; Kaur et al. 2018). The sources of PCBs emission in the environment consist of the cooling fluids used in capacitors and transformers (Kumar et al. 2018a).

16.3 Chemical Hazards Due to Pesticides Usage

Agrochemicals in general or pesticides, in particular, may pose risks of variable durations, i.e. short or long term in target groups and their families. The workers who perform the various functions viz. mixing, loading or application of different types of pesticides expose them to the chemicals due to direct contact, accidental spills or inadequate protective equipment (Kumar and Singh 2018). Employs those who conduct these works contractually works with their bare hands and are directly exposed to direct spray or with pesticide residues remaining in the soil and crops (Kumar et al. 2019d; Singh et al. 2019d). Several reports indicate about accidents in villages or unintentional exposure to children (Singh et al. 2020b). Fig. 16.1 illustrates the consequences of pesticides on farmers.

The applicators of agrochemicals or harvesters, family members, are at increased risks to contamination and toxic residues present in air, food and water (Hussain et al. 2020; Singh et al. 2020c, e, f). The sprayers also get exposed to agrochemicals in different ways. These may include situations of working in a farm where chemicals have been stored or admixed, inhalation in quantities above than the permissible/admissible levels or persons associated with other toxic features (Damalas and

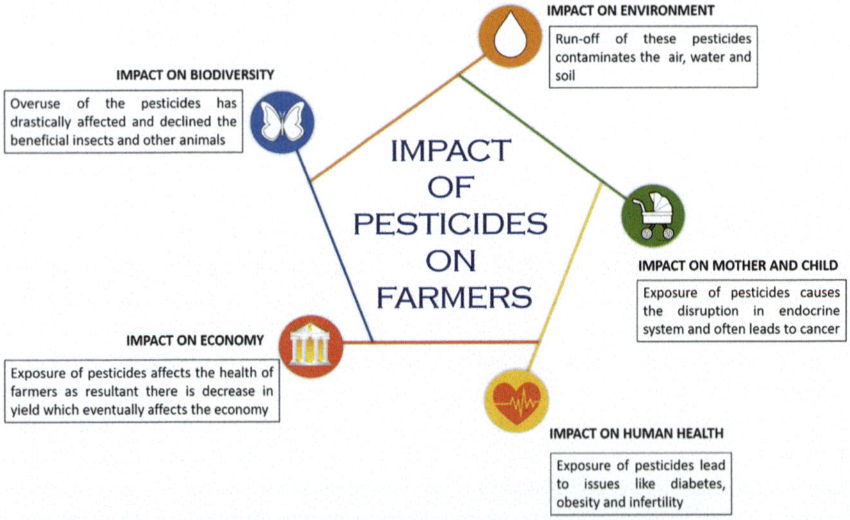

Fig. 16.1 Impact of pesticides on farmer's health and wealth

Koutroubas 2016). Sometimes, the workers do not have adequate personal protective equipment [PPEs] or they are not able to use it properly. Few incidences are related to persons consuming food with contaminated hand or food grown in the contaminated fields. Other than this, workers are at high risk to these toxic chemicals, if they consume the water from the sources near these toxic chemical manufacturing units (Protano et al. 2009; Vitali et al. 2009).

16.4 Use of Xenoestrogens in Day-to-Day Life and Health Hazards

Most of the perfumes, PVC pipes, cosmetics, liquid varnishes, lacquers and drugs used in pharmaceutical industry contain Phthalates which is a common environmental xenoestrogen and exposure to these xenoestrogens have an adverse effect on human health (Fig. 16.2). Children and men may have low risk as compared to females due to their increased cosmetic use (Datta et al. 2020; Singh et al. 2020j). Phthalates are omnipresent in personal care products and cosmetics. A study by López-Carrillo et al. 2010 found a connotation between endocrine disruption and exposure of phthalates resulting in the development of breast cancer (López-Carrillo et al. 2010). In rodents, Phthalates usually suppresses steroidogenesis. Extreme doses result in hormonal imbalance and cause defects in birth. Various diseases and disorders related to xenoestrogens include premature ovarian failure, prostate cancer, disruption in thyroid hormones, infertility in males and females, cancer in

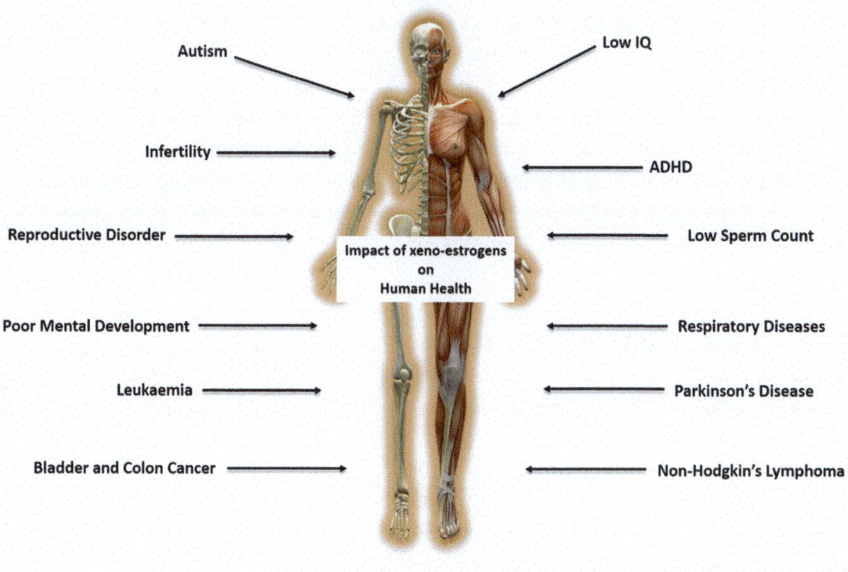

Fig. 16.2 Consequences of xenoestrogens on human health

reproductive sites (ovarian, breasts and uterine) and obesity (Diamanti-Kandarakis et al. 2009).

In our daily life, xenoestrogens absorbs by inhalation, direct skin contacts and ingestion. Usually, drinks and beverages in Styrofoam cup contain xenoestrogens which are ingested or absorbed by the skin (Basini et al. 2017). When our water supply contains a chemical used to kills bacteria, it accumulates in our body. Even thermal paper, food preserving cans, shampoos, hair spray, nail paints, sodas, chlorinated pools, tampons perfumes etc. contains traces of xenoestrogens found to have cell proliferation and endocrine disruption mechanism (Gonsioroski et al. 2020).

On a daily basis, many workers working in agricultural fields get exposed to toxic chemical multiple times in a season. If they do not follow the safety measures, then it may lead to infection and sometimes even death (Chakraborty et al. 2009). The Environmental Protection Agency [EPA] and various other organization keep the record of mortality and make them available online for public access to keep them updated and aware (Singh et al. 2014; Pramanik et al. 2015; Kumar et al. 2015c). Most of the poisoning cases recorded are from home. For which, it is considered that exposure to pesticides may have varied route and can have a different impact on the person depending on their age and sex (Kumar et al. 2020b, d; Singh et al. 2020i). For agricultural use, one should be well aware of how to use the agrochemicals correctly. The personal protective equipment should be made compulsory to keep workers from safe from dermal topical contact and inhalation of these toxic compounds (Kumar et al. 2013a, 2015b; Kaur et al. 2015; Wani et al. 2017; Singh et al. 2017; Dhanjal et al. 2018; Mukherjee et al. 2018). Additionally, when the concentration of these toxic compounds is diluted, it is suggested to wear protective coveralls like a face mask, rubber gloves as well as shoes and full-length apron. However, the multiple usage aprons and shoes can be cleaned daily with water and soap to remove residues of pesticides (Karnwal et al. 2019). All the cleaning should be done separately from other clothes. Other than this, it is recommended to wear a respirator while working with a specific chemical which demands it (Garrigou et al. 2020). Selection of PPEs requires considerable evaluation because of the specificity of pesticide to be used in agricultural land. Fitting all the equipments should be well assured to prevent undesirable accidents. As accidents happen due to negligence, and the family has to bear an irreparable loss of life and also, the economy is adversely affected (Thouvenin et al. 2017).

16.5 Protection from Pesticides

The EPA endorses that employers showed follow the Worker Protection Standard (WPS) while using pesticides. Because of these standards aid in reducing the chances of pesticide poisonings and injuries among handlers and workers. This assist in protecting the farmers, employer working in greenhouses, farmers and nurseries (Werts et al. 2017). There are two categories of workers according to the stated regulations.

- Pesticide handlers: these workers blend, loads and spray the pesticides, repair the equipment's used for handling pesticides.
- Agricultural workers: these workers perform the function like cultivation and harvesting of plants either in farms, nurseries or greenhouse.

The standard is well documented and imbibes prerequisite for safety measures, use of PPE kit, pesticide application schedule, labelling, sanitization of supplies. We need to take all preventive approaches as per the law of the land and also taking into consideration all imaginable emergencies. Thus, the backup plans may be boundless and must be best suitable for the workplace in remotest situations (Yarpuz-Bozdogan 2018).

16.6 Communication of Risks and Potential Hazards

There is a need for the proper labelling of the chemicals for farmworkers, so they are aware of the toxic and hazards of chemicals they are being exposed to. The international regulations promote and suggest that workers should be aware of the hazard associated with the chemicals they are using for performing agricultural activities (Singh et al. 2020g). Additionally, it also addresses the standard norms and requirements for risk assessment and interrelated communications. The communication should be simple and understandable (Singh et al. 2019a, 2020d). The regulatory compliances are vital and need to be implemented in true spirits from safety perspectives.

16.7 Respiratory Hazards and Protection

The prevailing situation in the agriculture sector leads to various chronic respiratory diseases among farmers. The exposure of these toxic compounds leads to undue coughing as well as congestion in farmers. Signs of chronic bronchitis are usually observed in grain handlers as they don't follow the safety measures effectively and already have obscured medical history. The organic dust shows the symptoms of a pulmonary illness similar to influenza-like illnesses like muscular pain, fever and headache (Kumar et al. 2020b, c; Singh et al. 2020d; Sisodia et al. 2020). Rice husks, grain and silage dust also exhibit the same symptoms. Suspended particles of 2.5–10 PM, devoid of mouldy spores, are stated as nuisance dust. Recurring exposure of the toxic compounds hardens the lungs and affects the non-functioning tissue leading to the development of chronic bronchitis as well as occupational asthma. Various gases like carbon dioxide (CO_2), ammonia (NH_3), hydrogen sulphide (H_2S), methane (CH_4) and nitrogen dioxide (NO_2) are also produced on the regular use of these toxic compounds (Singh et al. 2019a). The mild exposure of NH_3, H_2S or NO_2 also shows symptoms like dizziness, headache and eye irritation (Weiss and Lakshminarayan 1994).

The most prominent way to prevent the use of this respiratory disease is to employ respirator approved by the National Institute of Occupational Safety and Health (NIOSH) and accredited as purifiers to remove toxic chemicals from the air. Even standard policies have been framed to use air respirators under oxygen-limited conditions or in environments with acute toxic gas levels (Casey and Mazurek 2017). Further, medical assistance should be taken in an earlier stage to circumvent complications. Barns, manure pits and silos are considered as the respiratory hazards and can range from acute to chronic air contamination (Kuhn and Ghannoum 2003). Most communal respiratory hazards for workers involve aerosols, endotoxins, microorganisms, organic dust and xenobiotics synthesized by the catabolizing of animal and grain waste. Moreover, the silicates dust from harvesting and tilling activities are also prevalent but are not up to a significant level (Badirdast et al. 2017). Presently, farming practices have changed and improved due to which workers are now less prone to respiratory hazards. In order to regulate the aerosols, certain preventive measures are taken like applying moisture to friable material, enclosure as well as ventilation of tractors and installation of respirators (Janssen et al. 2013).

Other than this, the shortage of drinking water, poor handwashing and sanitation facilities also lead to several health issues. The workers working in the agricultural fields may suffer from heatstroke due to inadequate intake of water or others due to other infection. Other reason for this may be poor sanitation facilities, poisoning by agrichemical also deprives the handwashing facilities, which makes them more prone to microbial and parasitic exposures leading to infectious and communicable diseases (Mendoza-Espinosa and Daesslé 2018). As per standards, the workers working in the field must be provided with good quality drinking water, adequate handwashing and sanitation facilities and also educate them about the importance of good hygiene practices (Ravindra et al. 2019).

16.8 Skin Disorders and Infections

The most common skin disorder prevalent in agricultural workers is contact dermatitis. It may fall under the categories of allergic and irritant. These irritants act on the skin surface and elicit the immune response (Anderson and Meade 2014). The phototoxic implications may occur when sunlight acts in combination with certain substances. Sometimes rashes may be visibly seen due to insect and plant irritants (Singh et al. 2015). Several factors viz. age, sex, temperature, humidity, personal hygiene, sanitation conditions and race play a vital role during this infection. The associated occupational diseases are often detected easily, but the diagnosis is difficult. It is vital for the clinician to know about the chemical agent to which an individual is exposed or sensitive. Therefore, the proper cleaning and washing of protective clothing with disinfectants or putting up appropriate medicated creams are an effective approach to prevent skin infection (Arcury et al. 2012). It is advised to consult the expert skin specialists rather than trying at a personal level and reaching for the medical suggestions at only complicated situations.

16.9 Musculoskeletal Injuries

The musculoskeletal injuries may occur in workers during agricultural operations or repetitive motions in awkward positions or postures. The ergonomic issues are prevalent during prolonged exertions on shoulders, head due to pushing, pulling and lifting of hefty weights and sometimes due to inappropriate postures (Jain et al. 2018). The environmental conditions in a region may vary from season to season. The winds, temperature and humidity may intensify these conditions. The use of sophisticated technology may sometimes appear excellent but may not be economically feasible for the workers (D'Amato et al. 2015). Moreover, in a few cases reduce the chances of natural injuries but increase the chances of other injuries. For instance, dairy farmers are conventionally are highly prone to developing knee or hands osteoarthritis as they have started using modern milking system in the state-of-the-art cooperatives (Xiao et al. 2013).

16.10 Ergonomic Protections

The musculoskeletal injuries may be reduced with the use of techniques or tools, padding for minimized vibration and minimal activities with high repetition. The NIOSH directions on ergonomics may be beneficial for preventing muscular wounds among the planters, safety experts, managers or human resources personnel (Kumaraveloo and Lunner Kolstrup 2018). The workers exposed to the hot and moist environment are at high risk of heat-related abnormalities while performing tenacious work using the heavy protective clothing as well as equipment. Fresh entrants are also at higher risk in contrast to other workers due to incompetency or inappropriate acclimatization (Taffere et al. 2019).

16.11 Heat-Related Stress and Prevention

The heat-related illnesses are also common in areas of deserts and geographical areas with scorching summers. While working in extremely hot conditions, it is suggested to remember the concept of periodic water intake, rest and self-monitoring. Each should care for the next and prevent dehydration of body (Langley et al. 2017). The light cotton clothes with fabric to cover the head, faces, cap/hat. We must take care of fellow workers and know their locations so as to call for assistance as per requirements (Parker et al. 2020). We must not hesitate to take help if there are any signs of illness and contact the nearest medical aids. Usually, heat-related stress prevails when the body temperature is more than the bearable level, and elevated humidity, temperature, sunlight and workload also contribute to heat-related stresses. The hand-operated fans, ventilation systems and shade may be preferred along with ORS or water intake with glucose (Mutic et al. 2018).

Healthy organisms have a healthy mind, and without a doubt, they can contribute to society in better ways. Thus, one needs to sustain the health claimed to be the key element of life (Mishra et al. 2016). We need to protect the health of agricultural workers from having a modernized product which is needed for the sustenance of population. The due considerations must be rendered to a wide range of equipment, animals and manufactured products, which can be used in both outdoor and indoor environments in different geographical regions with varied climatic conditions (Kumar et al. 2017). The innovative entrepreneurs are continuously upgrading tools and mechanisms for producing new varieties in limited spaces with minimum pesticides usage. The era of organic farming is progressing with mechanized tools and products (Ryan et al. 2018).

In India, ICAR institutes are working hand to hand with partners of developed nations. We need to adopt comprehensive and holistic international legislation with quality control from seed to final produce and marketing (Ferroni and Zhou 2017). Hence, there is the need of motivation among the employees working in the agricultural field irrespective of employment status and gender biases. The policies may include the promotion of preventive Occupational Safety and health culture with positive attitudes and behavior (Gasperini 2017). We need to recognize the biological differences between women and men. We must manage occupational accidents efficiently with the prevention of potential diseases. Subsidies may be given, but they are not the ultimate solutions, and the views may differ (Jadhav et al. 2016).

The manufacturers of chemicals, equipment and machinery should also ensure safety and risk associated with the fabricated and developed product for use in the agricultural sector. The instructions inadequate words, preferably bilingual related to first aid and Material Safety Data Sheets [MSDS] be available at the site of installation, storage and maintenance with marking and warning labels on it (Tinc et al. 2018). Moreover, chemicals need to be properly labelled, and its safety datasheets should also be attached to it. The regular evaluation and monitoring of the performance should also strengthen our commitment towards prevention of accident and promote this culture in the ISO/IEC 9001 system (Fouilleux and Loconto 2017). Training programmes may be for all level of workers at the agriculture site. This may be for owners of land, managers and supervisors, migrant and temporary workers. This culture may be difficult initially but beneficial in longer terms for generating competent persons (Spector et al. 2016).

16.12 Conclusion

Agriculture is the oldest occupations of the human civilizations but still needs continual improvements. Although the development and progression in this field have increased crop production, in return, it has resulted in various health problems. Therefore, we need to take utmost care of farmers and educate them with new concepts of farming and simultaneously understand the experiences gained by them. The agricultural workers suffer due to climatic changes and pest infestations. The economic policies are to be decided to take the welfare of agriculturists who are important members of our community. Internationally acceptable models may be attempted to satisfy the stakeholders.

References

Anderson SE, Meade BJ (2014) Potential health effects associated with dermal exposure to occupational chemicals. Environ Health Insig 8s1:EHI.S15258. https://doi.org/10.4137/EHI.S15258

Arcury TA, Weir MM, Summers P et al (2012) Safety, security, hygiene and privacy in migrant farmworker housing. New Solut 22:153–173. https://doi.org/10.2190/NS.22.2.d

Badirdast P, Azari MR, Salehpour S et al (2017) The effect of wood aerosols and bioaerosols on the respiratory Systems of Wood Manufacturing Industry Workers in Golestan Province. Tanaffos 16:53

Basini G, Bussolati S, Grolli S et al (2017) Bisphenol a interferes with swine vascular endothelial cell functions. Can J Physiol Pharmacol 95:365–371. https://doi.org/10.1139/cjpp-2016-0180

Bhale UN, Bansode SA, Singh S (2018) Multifactorial role of arbuscular mycorrhizae in agroecosystem. In: Fungi and their role in sustainable development: current perspective. Springer, Singapore, pp 205–220

Bhati S, Kumar V, Singh S, Singh J (2019) Synthesis, biological activities and docking studies of piperazine incorporated 1, 3, 4-oxadiazole derivatives. J Mol Struct 1191:197–205. https://doi.org/10.1016/j.molstruc.2019.04.106

Bhatia D, Singh S, Vyas A et al (2014) Studies on fungal strains of selected regions of Ludhiana and their biochemical characterization. Curr World Environ 9:192–202. https://doi.org/10.12944/CWE.9.1.27

Casey ML, Mazurek JM (2017) Respirator use among US farm operators with asthma: results from the 2011 farm and ranch safety survey. J Agromedicine 22:78–88. https://doi.org/10.1080/1059924X.2017.1282904

Chakraborty S, Mukherjee S, Roychoudhury S et al (2009) Chronic exposures to cholinesterase-inhibiting pesticides adversely affect respiratory health of agricultural Workers in India. J Occup Health 51:488–497. https://doi.org/10.1539/joh.L9070

Coman, Marcu, Chereches, et al (2020) Educational interventions to improve safety and health literacy among agricultural workers: a systematic review. Int J Environ Res Public Health 17:1114. https://doi.org/10.3390/ijerph17031114

D'Amato G, Holgate ST, Pawankar R et al (2015) Meteorological conditions, climate change, new emerging factors, and asthma and related allergic disorders. A statement of the world allergy organization. World Aller Organ J 8:1

Damalas CA, Eleftherohorinos IG (2011) Pesticide exposure, safety issues, and risk assessment indicators. Int J Environ Res Public Health 8:1402–1419

Damalas CA, Koutroubas SD (2016) Farmers' exposure to pesticides: toxicity types and ways of prevention. Toxics 4

Datta S, Singh J, Singh J et al (2018) Assessment of genotoxic effects of pesticide and vermicompost treated soil with *Allium cepa* test. Sust Environ Res 28:171–178. https://doi.org/10.1016/j.serj.2018.01.005

Datta S, Singh S, Kumar V et al (2020) Endophytic bacteria in xenobiotic degradation. In: Microbial Endophytes. Elsevier, pp 125–156

De Boer B, Hamers JPH, Zwakhalen SMG et al (2017) Quality of care and quality of life of people with dementia living at green care farms: a cross-sectional study. BMC Geriatr 17:1–10. https://doi.org/10.1186/s12877-017-0550-0

Dhanjal DS, Singh S, Bhatia D et al (2018) Pre-treatment of the municipal wastewater with chemical coagulants. Pollut Res 37:S32–S38

Diamanti-Kandarakis E, Bourguignon JP, Giudice LC et al (2009) Endocrine-disrupting chemicals: an Endocrine Society scientific statement. Endocr Rev 30:293–342

Esch KJ, Petersen CA (2013) Transmission and epidemiology of zoonotic protozoal diseases of companion animals. Clin Microbiol Rev 26:58–85. https://doi.org/10.1128/CMR.00067-12

Ferroni M, Zhou Y (2017) The private sector and India's agricultural transformation. Glob J Emerg Mark Econ 9:28–37. https://doi.org/10.1177/0974910117716406

Fouilleux E, Loconto A (2017) Voluntary standards, certification, and accreditation in the global organic agriculture field: a tripartite model of techno-politics. Agric Hum Values 34:1–14. https://doi.org/10.1007/s10460-016-9686-3

Garrigou A, Laurent C, Berthet A et al (2020) Critical review of the role of PPE in the prevention of risks related to agricultural pesticide use. Saf Sci 123:104527

Gasperini FA (2017) Agricultural leaders' influence on the safety culture of workers. J Agromedicine 22:309–311

Gill JPK, Singh S, Sethi N (2015) Synthesis, characterization and antimicrobial activity of protected dipeptides and their deprotected analogs. Orient J Chem 31:417–421. https://doi.org/10.13005/ojc/310149

Girdhar M, Singh S, Rasool HI et al (2014) Evaluating different weeds for phytoremediation potential available in tannery polluted area by conducting pot and hydroponic experiments. Curr World Environ 9:156–167. https://doi.org/10.12944/CWE.9.1.22

Gonsioroski A, Mourikes VE, Flaws JA (2020) Endocrine disruptors in water and their effects on the reproductive system. Int J Mol Sci:21

Hussain T, Singh S, Mohd D et al (2020) Natural metabolites: an eco-friendly approach to manage plant diseases and for better agriculture farming. In: Natural bioactive products in sustainable agriculture. Springer Singapore, Singapore, pp 1–13

Jadhav R, Achutan C, Haynatzki G et al (2016) Review and meta-analysis of emerging risk factors for agricultural injury. J Agromedicine 21:284–297

Jain R, Meena ML, Dangayach GS (2018) Ergonomic intervention for manual harvesting in agriculture: a review. In: Ergonomics in caring for people. Springer, Singapore, pp 183–191

Janssen L, Ettinger H, Graham S et al (2013) Commentary: the use of respirators to reduce inhalation of airborne biological agents. J Occup Environ Hyg 10:D97

Kapoor D, Singh S, Kumar V et al (2019) Antioxidant enzymes regulation in plants in reference to reactive oxygen species (ROS) and reactive nitrogen species (RNS). Plant Gene 19:100182. https://doi.org/10.1016/j.plgene.2019.100182

Karnwal A, Singh S, Kumar V et al (2019) Fungal enzymes for the textile industry. In: Recent advancement in white biotechnology through Fungi. Springer, Cham, pp 459–482

Kaur A, Kumar V, Singh S et al (2015) Toll-like receptor-associated keratitis and strategies for its management. 3 Biotech 5:611–619

Kaur P, Singh S, Kumar V et al (2018) Effect of rhizobacteria on arsenic uptake by macrophyte Eichhornia crassipes (Mart.) Solms. Int J Phytoremediation 20:114–120. https://doi.org/10.1080/15226514.2017.1337071

Koul B, Singh S, Dhanjal DS, Singh J (2019) Plant growth-promoting Rhizobacteria (PGPRs): a fruitful resource. In: Microbial interventions in agriculture and environment. Springer, Singapore, pp 83–127

Kuhn DM, Ghannoum MA (2003) Indoor mold, toxigenic fungi, and *Stachybotrys chartarum*: infectious disease perspective. Clin Microbiol Rev 16:144–172

Kumar V, Singh S (2018) Interactions of Acephate, glyphosate, Monocrotophos and Phorate with bovine serum albumin. Indian J Pharm Sci 80:1151–1155. https://doi.org/10.4172/pharmaceutical-sciences.1000467

Kumar V, Upadhyay N, Singh S et al (2013a) Thin-layer chromatography: comparative estimation of Soil's atrazine. Curr World Environ 8:469–472. https://doi.org/10.12944/CWE.8.3.17

Kumar V, Upadhyay N, Wasit AB et al (2013b) Spectroscopic methods for the detection of organophosphate pesticides-a preview. Curr World Environ 8:313–318. https://doi.org/10.12944/CWE.8.2.19

Kumar V, Singh S, Manhas A et al (2014a) Bioremediation of petroleum hydrocarbon by using *Pseudomonas* species isolated from petroleum contaminated soil. Orient J Chem 30:1771–1776

Kumar V, Upadhyay N, Kumar V et al (2014b) Environmental exposure and health risks of the insecticide monocrotophos—a review. J Biodivers Environ Sci 5:111–120

Kumar G, Sharma PK, Sharma S, Singh S (2015a) Synthesis and antimicrobial studies of fused heterocycles 'pyrimidobenzothiazoles'. J Chem Pharm Res 7:710–714

Kumar V, Singh S, Bhadrecha P et al (2015b) Bioremediation of heavy metals by employing resistant microbial isolates from agricultural soil irrigated with industrial waste water. Orient J Chem 31:357–361. https://doi.org/10.13005/ojc/310142

Kumar V, Singh S, Singh J, Upadhyay N (2015c) Potential of plant growth promoting traits by bacteria isolated from heavy metal contaminated soils. Bull Environ Contam Toxicol 94:807–814. https://doi.org/10.1007/s00128-015-1523-7

Kumar V, Kaur S, Singh S, Upadhyay N (2016) Unexpected formation of N′-phenyl-thiophosphorohydrazidic acid O,S-dimethyl ester from acephate: chemical, biotechnical and computational study. 3 Biotech 6:1–11. https://doi.org/10.1007/s13205-015-0313-6

Kumar V, Singh S, Singh R et al (2017) Design, synthesis, and characterization of 2,2-bis(2,4-dinitrophenyl)-2-(phosphonatomethylamino)acetate as a herbicidal and biological active agent. J Chem Biol 10:179–190. https://doi.org/10.1007/s12154-017-0174-z

Kumar V, Shahi SK, Singh S (2018a) Bioremediation: an eco-sustainable approach for restoration of contaminated sites. In: Microbial bioprospecting for sustainable development. Springer, Singapore, pp 115–136

Kumar V, Singh S, Singh A et al (2018b) Determination of phytochemical, antioxidant, antimicrobial, and protein binding qualities of hydroethanolic extract of Celastrus paniculatus. J Appl Biol Biotechnol 6:11–17. https://doi.org/10.7324/JABB.2018.60602

Kumar V, Singh S, Singh A et al (2018c) Phytochemical, antioxidant, antimicrobial, and protein binding qualities of hydro-ethanolic extract of Tinospora cordifolia. J Biol Act Prod Nat 8:192–200. https://doi.org/10.1080/22311866.2018.1485513

Kumar V, Singh S, Singh R et al (2018d) Spectral, structural and energetic study of acephate, glyphosate, monocrotophos and phorate: an experimental and computational approach. J Taibah Univ Sci 12:69–78. https://doi.org/10.1080/16583655.2018.1451109

Kumar V, Singh S, Bhadouria R et al (2019a) Phytochemical, analytical and medicinal studies of *Holoptelea integrifolia* Roxb. Planch - a review. Curr Trad Med 5:270–277. https://doi.org/10.2174/2215083805666190521103308

Kumar V, Singh S, Kaur S et al (2019b) Spectrophotometric interaction of Carbofuran on food grains. J Gujarat Res Soc 21:841–850

Kumar V, Singh S, Kondalkar SA et al (2019c) High resolution GC/MS analysis of the *Holoptelea integrifoli*'s leaves and their medicinal qualities. Biocatal Agric Biotechnol 22:101405. https://doi.org/10.1016/j.bcab.2019.101405

Kumar V, Singh S, Singh A et al (2019d) Assessment of heavy metal ions, essential metal ions, and antioxidant properties of the most common herbal drugs in Indian Ayurvedic hospital: for ensuring quality assurance of certain Ayurvedic drugs. Biocatal Agric Biotechnol 18:101018. https://doi.org/10.1016/j.bcab.2019.01.056

Kumar V, Singh S, Srivastava B et al (2019e) Volatile and semi-volatile compounds of *Tephrosia purpurea* and its medicinal activities: experimental and computational studies. Biocatal Agric Biotechnol 20:101222. https://doi.org/10.1016/j.bcab.2019.101222

Kumar V, Singh S, Srivastava B et al (2019f) Green synthesis of silver nanoparticles using leaf extract of *Holoptelea integrifolia* and preliminary investigation of its antioxidant, anti-inflammatory, antidiabetic and antibacterial activities. J Environ Chem Eng 7:103094. https://doi.org/10.1016/j.jece.2019.103094

Kumar V, Singh S, Upadhyay N (2019g) Effects of organophosphate pesticides on siderophore producing soils microorganisms. Biocatal Agric Biotechnol 21:101359. https://doi.org/10.1016/j.bcab.2019.101359

Kumar A, Devi S, Agrawal H et al (2020a) Rhizoremediation: a unique plant microbiome association of biodegradation. In: Plant microbe Symbiosis. Springer International Publishing, Berlin, pp 203–220

Kumar S, Singh S, Kumar V et al (2020b) Pathogenesis and antibiotic resistance of *Staphylococcus aureus*. In: Model organisms for microbial pathogenesis, biofilm formation and antimicrobial drug discovery. Springer, Singapore, pp 99–115

Kumar V, Singh S, Singh D et al (2020c) Current trends in *Mycobacterium tuberculosis* pathogenesis and drug resistance. In: Model organisms for microbial pathogenesis, biofilm formation and antimicrobial drug discovery. Springer, Singapore, pp 301–322

Kumar V, Singh S, Singh S et al (2020d) Methods and techniques for the chemical profiling and quality control of natural products and natural product-derived drugs. In: Bioactive natural products in drug discovery. Springer, Singapore, pp 585–598

Kumaraveloo KS, Lunner Kolstrup C (2018) Agriculture and musculoskeletal disorders in low- and middle-income countries. J Agromedicine 23:227–248. https://doi.org/10.1080/1059924X.2018.1458671

Langley R, Hirsch A, Cullen R et al (2017) North Carolina state agencies working to prevent agricultural injuries and illnesses. J Agromedicine 22:358–363. https://doi.org/10.1080/1059924X.2017.1353468

López-Carrillo L, Hernández-Ramírez RU, Calafat AM et al (2010) Exposure to phthalates and breast Cancer risk in northern Mexico. Environ Health Perspect 118:539–544. https://doi.org/10.1289/ehp.0901091

Makkar R, Singh S, Sharma PK (2016) Antibacterial, antifungal and antioxidant activities of substituted 4H-1, 4-benzothiazines. Der Pharma Chemica 8:156

Mendoza-Espinosa LG, Daesslé LW (2018) Consolidating the use of reclaimed water for irrigation and infiltration in a semi-arid agricultural valley in Mexico: water management experiences and results. J Water Sanit Hyg Develop 8:679–687. https://doi.org/10.2166/washdev.2018.021

Mishra V, Gupta A, Kaur P et al (2016) Synergistic effects of Arbuscular mycorrhizal fungi and plant growth promoting rhizobacteria in bioremediation of iron contaminated soils. Int J Phytoremediation 18:697–703. https://doi.org/10.1080/15226514.2015.1131231

Mukherjee D, Singh S, Kumar M et al (2018) Fungal biotechnology: role and aspects. In: Fungi and their role in sustainable development: current perspective. Springer, Singapore, pp 91–103

Mutic AD, Mix JM, Elon L et al (2018) Classification of heat-related illness symptoms among Florida farmworkers. J Nurs Scholarsh 50:74–82. https://doi.org/10.1111/jnu.12355

Parker LE, McElrone AJ, Ostoja SM, Forrestel EJ (2020) Extreme heat effects on perennial crops and strategies for sustaining future production. Plant Sci 110397

Pramanik T, Pathan AH, Gupta R et al (2015) Dihydropyrimidinone derivatives: green synthesis and effect of electronic factor on their antimicrobial properties. Res J Pharma Biol Chem Sci 6:1152–1157

Protano C, Guidotti M, Vitali M (2009) Performance of different work clothing types for reducing skin exposure to pesticides during open field treatment. Bull Environ Contam Toxicol 83:115–119. https://doi.org/10.1007/s00128-009-9753-1

Ravindra K, Mor S, Pinnaka VL (2019) Water uses, treatment, and sanitation practices in rural areas of Chandigarh and its relation with waterborne diseases. Environ Sci Pollut Res 26:19512–19522. https://doi.org/10.1007/s11356-019-04964-y

Ryan SF, Adamson NL, Aktipis A et al (2018) The role of citizen science in addressing grand challenges in food and agriculture research. Proc R Soc B Biol Sci 285:20181977. https://doi.org/10.1098/rspb.2018.1977

Sharma PK, Dar JM, Kumar G, Singh S (2015) Synthesis and antimicrobial activities of substituted phenylthioureas. J Chem Pharm Res 7:133–139

Sharma K, Singh S, Kumar V et al (2020) *Saccharomyces cerevisiae* as model organism to study biological activities of nanoparticles. In: Model organisms to study biological activities and toxicity of nanoparticles. Springer, Singapore, pp 101–115

Sidhu GK, Singh S, Kumar V et al (2019) Toxicity, monitoring and biodegradation of organophosphate pesticides: a review. Crit Rev Environ Sci Technol 49:1135–1187. https://doi.org/10.1080/10643389.2019.1565554

Singh J, Singh S, Kaur P et al (2014) Biochemical characterization of *Tylophora indica* and its immunomodulatary effect. South As J Exp Biol 4:61–67

Singh J, Singh S, Datta S et al (2015) Toxicological effects of lambda-cyhalothrin on liver, kidney and testis of Indian catfish *Clarias batrachus*. Toxicol Int 22:128–136. https://doi.org/10.22506/ti/2015/v22/i3/137637

Singh S, Singh N, Kumar V et al (2016) Toxicity, monitoring and biodegradation of the fungicide carbendazim. Environ Chem Lett 14:317–329

Singh S, Kumar V, Upadhyay N et al (2017) Efficient biodegradation of acephate by *Pseudomonas pseudoalcaligenes* PS-5 in the presence and absence of heavy metal ions [cu(II) and Fe(III)], and humic acid. 3 Biotech 7:1–10. https://doi.org/10.1007/s13205-017-0900-9

Singh S, Kumar V, Chauhan A et al (2018) Toxicity, degradation and analysis of the herbicide atrazine. Environ Chem Lett 16:211–237

Singh S, Kumar V, Kapoor D et al (2019a) Revealing on hydrogen sulfide and nitric oxide signals co-ordination for plant growth under stress conditions. Physiol Plant 168:ppl.13002. https://doi.org/10.1111/ppl.13002

Singh S, Kumar V, Romero R et al (2019b) Applications of nanoparticles in wastewater treatment. In: Nanobiotechnology in bioformulations. Springer, Cham, pp 395–418

Singh S, Kumar V, Sidhu GK et al (2019c) Plant growth promoting rhizobacteria from heavy metal contaminated soil promote growth attributes of Pisum sativum L. Biocatal Agric Biotechnol 17:665–671. https://doi.org/10.1016/j.bcab.2019.01.035

Singh S, Kumar V, Singh J (2019d) Kinetic study of the biodegradation of glyphosate by indigenous soil bacterial isolates in presence of humic acid, Fe(III) and cu(II) ions. J Environ Chem Eng 7:103098. https://doi.org/10.1016/j.jece.2019.103098

Singh S, Kumar V, Singh S, Singh J (2019e) Influence of humic acid, iron and copper on microbial degradation of fungicide Carbendazim. Biocatal Agric Biotechnol 20:101196. https://doi.org/10.1016/j.bcab.2019.101196

Singh S, Sidhu GK, Kumar V et al (2019f) Fungal Xylanases: sources, types, and biotechnological applications. In: Recent advancement in white biotechnology through Fungi. Springer, Cham, pp 405–428

Singh S, Kumar V, Datta S et al (2020a) Glyphosate uptake, translocation, resistance emergence in crops, analytical monitoring, toxicity and degradation: a review. Environ Chem Lett 18:663–702

Singh S, Kumar V, Datta S et al (2020b) Current advancement and future prospect of biosorbents for bioremediation. Sci Total Environ 709:135895

Singh S, Kumar V, Datta S et al (2020c) Plant disease management by bioactive natural products. In: Natural bioactive products in sustainable agriculture. Springer Singapore, Singapore, pp 15–29

Singh S, Kumar V, Datta S et al (2020d) Challenges and future perspectives of Nanotoxicology. In: Model organisms to study biological activities and toxicity of nanoparticles. Springer, Singapore, pp 451–466

Singh S, Kumar V, Dhanjal DS, Singh J (2020e) Juvenoids and its application in crop management. In: Natural bioactive products in sustainable agriculture. Springer Singapore, Singapore, pp 101–112

Singh S, Kumar V, Dhanjal DS, Singh J (2020f) Biological control agents: diversity, ecological significances, and biotechnological applications. In: Natural bioactive products in sustainable agriculture. Springer Singapore, Singapore, pp 31–44

Singh S, Kumar V, Dhanjal DS, Singh J (2020g) Herbicides and plant growth regulators: current developments and future challenges. In: Natural bioactive products in sustainable agriculture. Springer Singapore, Singapore, pp 67–81

Singh S, Kumar V, Singh S et al (2020i) Global scenario of plant–microbiome for sustainable agriculture: current advancements and future challenges. In: Plant Microbiomes for Sustainable Agriculture. Springer, Cham, pp 425–443

Singh S, Kumar V, Singla S et al (2020j) Kinetic study of the biodegradation of Acephate by indigenous soil bacterial isolates in the presence of humic acid and metal ions. Biomol Ther 10:433. https://doi.org/10.3390/biom10030433

Singh S, Kumar V, Upadhyay N, Singh J (2020k) The effects of Fe(II), cu(II) and humic acid on biodegradation of atrazine. J Environ Chem Eng 8:103539. https://doi.org/10.1016/j.jece.2019.103539

Sisodia BS, Kumar V, Singh S et al (2020) Zebra fish infection model: from pathogenesis to therapeutics. In: Model organisms for microbial pathogenesis, biofilm formation and antimicrobial drug discovery. Springer, Singapore, pp 429–440

Spector JT, Bonauto DK, Sheppard L et al (2016) A case-crossover study of heat exposure and injury risk in outdoor agricultural workers. PLoS One 11. https://doi.org/10.1371/journal.pone.0164498

Taffere GR, Bonsa M, Assefa M (2019) Magnitude of occupational exposure to noise, heat and associated factors among sugarcane factory workers in Ethiopia, 2017. J Pub Health:1–7. https://doi.org/10.1007/s10389-019-01070-8

Thouvenin I, Bouneb F, Mercier T (2017) Operator dermal exposure and protection provided by personal protective equipment and working coveralls during mixing/loading, application and sprayer cleaning in vineyards. Int J Occup Saf Ergon 23:229–239. https://doi.org/10.1080/10803548.2016.1195130

Tinc PJ, Gadomski A, Sorensen JA et al (2018) Applying the consolidated framework for implementation research to agricultural safety and health: barriers, facilitators, and evaluation opportunities. Saf Sci 107:99–108. https://doi.org/10.1016/j.ssci.2018.04.008

Tulchinsky TH, Varavikova EA (2014) Environmental and occupational health learning objectives. The New Public Health 471. https://doi.org/10.1016/B978-0-12-415766-8.00009-4

Vitali M, Protano C, Del Monte A et al (2009) Operative modalities and exposure to pesticides during open field treatments among a group of agricultural subcontractors. Arch Environ Contam Toxicol 57:193–202. https://doi.org/10.1007/s00244-008-9225-3

Wani AB, Chadar H, Wani AH et al (2017) Salicylic acid to decrease plant stress. Environ Chem Lett 15:101–123

Weiss SM, Lakshminarayan S (1994) Acute inhalation injury. Clin Chest Med 15:103–116

Werts P, Bernard T, Green T (2017) Crop advisers impacted by requirements in new USEPA worker protection standard. Crops & Soils 50:34–38. https://doi.org/10.2134/cs2017.50.0313

Xiao H, McCurdy SA, Stoecklin-Marois MT et al (2013) Agricultural work and chronic musculoskeletal pain among latino farm workers: the MICASA study. Am J Ind Med 56:216–225. https://doi.org/10.1002/ajim.22118

Yarpuz-Bozdogan N (2018) The importance of personal protective equipment in pesticide applications in agriculture. Current Opinion in Environmental Science & Health 4:1–4. https://doi.org/10.1016/j.coesh.2018.02.001

Insecticides Derived from Natural Products: Diversity and Potential Applications

17

Johnson Wahengbam, Laxman Sonawane Bhushan, Jyoti B. Patil, and Jayakumar Pathma

Abstract

Entomopathogenic microbes viz., bacteria, actinomycetes, fungi, virus, protozoans; microbial metabolites; phytochemicals from plants viz., neem, chrysanthemum, tobacco, derris, basil, citrus, etc., and compounds of animal origins viz., nereistoxin and insect pheromones have evidenced excellent protection against crop pests and are commercially available. Most of these microbes and bioactive molecules of natural origins are target specific, biodegradable, and eco-friendly. Insecticides from natural products will act as an effective alternative to pollution-causing synthetic agrochemicals with high health hazard to nontarget organism in the ecosystem. Extensive research to screen and identify environmentally safe biomolecules of natural origin with high efficacy against target organisms and intensive studies on their biological activity, mode of action, and ways to enhance their bio-efficacy using biotechnological tools may help in improving their bioactivity and target specificity. This chapter discusses about diverse group of entomopathogens, microbial metabolites, botanicals, insecticidal toxins of animal origin, and semiochemicals conferring plant protection against herbivorous insect pests and their potential application for crop protection.

J. Wahengbam · L. S. Bhushan · J. B. Patil
Department of Entomology, School of Agriculture, Lovely Professional University, Phagwara, Punjab, India

J. Pathma (✉)
Department of Entomology, School of Agriculture, Lovely Professional University, Phagwara, Punjab, India

Department of Trans-disciplinary Research, Division of Research and Development, Lovely Professional University, Phagwara, Punjab, India

A. N. Yadav et al. (eds.), *Current Trends in Microbial Biotechnology for Sustainable Agriculture*, Environmental and Microbial Biotechnology,
https://doi.org/10.1007/978-981-15-6949-4_17

Keywords

Biological control · Botanicals · Entomopathogens · Secondary metabolites · Semiochemicals

17.1 Introduction

Insecticides from natural products are year-old method of pest control without upsetting environment. Plants produce numerous secondary metabolites with insecticidal or insect-repelling properties to defend themselves from insect herbivore attack but their use has been abridged with the birth of first synthetic organic insecticide, DDT in 1939. By now thousands of compounds have been synthesized and tested for their insecticidal properties all over the world (Atwal and Dhaliwal 2005). However extensive and imprudent use of pesticides resulted in adverse effects on biotic and abiotic components of the ecosystem ultimately disrupting ecological balance and threatening life on earth. The 3R's factor, i.e. resistance, resurgence, and replacement, has made pest management more complex (Wahengbam et al. 2018). Extensive usage of DDT for insect control triggered resistance development and first case of resistance was documented in housefly in 1946. Initially only environmentalists protested against pesticide usage and it was Rachel Carson's pioneering work the "Silent spring" published in 1962 enlightened common people about the ill effects of organochlorines, their bioaccumulation, and biomagnification in organisms of higher trophic levels via food chain.

Several species of microbes, such as bacteria, fungi, virus, protozoa, and nematode, infects insects and other arthropods under natural conditions and cause diseases. Few among them could be mass cultured. Living entities as such or their by-products have been formulated and commercially available for pest control. This includes bacterium *Bacillus thuringiensis* (*Bt* endotoxins), actinomycetes *Streptomyces avermitilis* (Avermectins and its derivatives—Emamectin), *Streptomyces hygroscopicus* (Milbemectin), and *Saccharopolyspora spinosa* (Spinosad) etc., nuclear polyhedrosis virus (NPV), granulosis virus (GV), muscardine and halo fungi, entomopathogenic nematodes, and nereistoxin from *Lumbriconereis heteropoda* (Kour et al. 2020b; Yadav et al. 2020a, b). Phytochemicals from a wide range of plants viz., neem, chrysanthemum, tobacco, derris, and many more showed insecticidal activity and were used for pest control. In addition, insect pheromones were also relied as an important resource of insect pest management. All these living entities and their products play an integral part of IPM strategy (Singh and Yadav 2020). Among insecticides of natural origin, microbial pesticides lead the products available in biopesticide market and with a global market value of 1944.3 million USD in 2018 which has been estimated to raise around USD 4753.1 million in 2024 (Anonymous 2019).

Microbial insecticides excel chemical and botanical insecticides in aspects of environmental safety and socio-economical benefits. They are safe to nontarget organism due to high species specificity, with no reports of phytotoxicity and are not harmful even if utilized in harvesting stage of the crop. However, their sensitivity and shelf life are major constraints which affect their success rate. Innovations to produce low cost, unique, stable formulations with extended shelf life, as well as training and demonstrations to improve farmers knowledge on their method of application decides the success of the microbial insecticide in farm conditions and this in turn will improve farmers faith and preference for utilizing these eco-friendly insecticide formulations. This chapter gives a concise picture of the diversity and potential applications of naturally occurring entomopathogens and insecticides derived from natural products and their potential in IPM.

17.2 Botanicals for Pest Management

Plants produce a wide array of structurally diverse secondary metabolites such as peptides, amines, antibiotics, alkaloids, cyanogenic glucosides, glucosinolates, organic acids, nonprotein amino acids, phenolics, polyphenols, polyacetylenes, terpenoids, lipophilic terpenes, sesquiterpene lactones, quinones and defensive proteins viz., chitinases, lectins, beta-1,3-glucanases, vicilins, systemins, arcelins, and enzyme inhibitors. These compounds exhibit insect-repelling or insecticidal properties as well as antimicrobial potential against phytopathogens and act as a first line of defense to protect the plants from biotic stress including insect attack. Nearly 6000 plant species have been identified to possess insecticidal properties (Nawaz et al. 2016). Copious research on botanical insecticides has been conducted globally. In 1980, only 1.43% of publications on insecticides dealt with botanical insecticides but the number has been increased up to 21.38% in 2012 (Isman and Grieneisen 2014).

Use of botanicals for pest control dates back to two millennium as evidenced from historical remains, hieroglyphs, literatures, vedas, and scriptures of ancient civilizations of Egypt, Greece, Rome, China, and India. History also documents the usage of phytochemicals for pest management in all parts of world including Europe and North America nearly one and a half-century ago until the discovery of synthetic compounds for pest control (Ware 1983; Thacker 2002; El-Wakeil 2013). Secondary metabolites are compounds produced by plants to cope up with stress and environmental changes and have no role in the primary functioning of the plants. Plant allelochemicals are products of coevolution that activate interspecific communication resulting in plant defense (Regnault-Roger and Philogene 2008). These metabolites are highly bioactive with many antagonistic and pharmaceutical properties and can be efficiently utilized for plant protection and human therapeutics. Pytochemicals commonly used for insect control from the time immemorial includes azadiractin (neem), pyrethrum (chrysanthemum), rotenone (Derris), and essential oils (neem, eucalyptus, etc.). Also, compounds viz., ryanodine (Ryania), nicotine (tobacco), and sabadilla (Schoenocaulon) are used in a comparatively lesser frequency for pest control (Table 17.1).

Table 17.1 Botanicals used in pest management

Plant source	Family	Plant parts	Insecticidal compounds	Active against	Mode of action	References
Azadiracta indica (Neem)	Meliaceae	Leaves, bark, seed, oil	Azadirachtin, Meliantriol, Nimbidnin, Nimbin, Salannin	Lepidoptera, Coleoptera, Hemiptera, Orthoptera	Acetylcholinesterase (AChE) inhibition, molting inhibition, antigonadotropic effects, feeding and ovipositional deterrent	Pradhan et al. (1962); Copping and Duke (2007); Morgan (2009)
Melia spp. (chinaberry)	Meliaceae	Leaves, bark, fruits	Nimbolin, Meliatoxin	Lepidoptera, Coleoptera	Acetylcholinesterase (AChE) inhibition, disruption of hormonal balance, feeding deterrent, ovipositional deterrent	Senthil-Nathan (2006); Schmidt et al. (1998), Chen et al. (1995)
Pongamia glabra (Pongam)	Leguminaceae	Leaves, flower, roots, seeds, oil	Pongamol, Glabrin, Pongapin, Pongone, Pinnatin, Karanjin Pongaglabrone,	Lepidoptera, Coleoptera, Hemiptera	Repellent, antifeedant, nonlethal chronic toxin-affecting insect fertility	Hoa et al. (2016); Kumar et al. (2006)
Derris spp., *Lonchocarpus* spp., *Tephrosia* spp.,	Leguminaceae	Roots	Rotenone	Hemiptera, Coleoptera, Hymenoptera, Acarina	Inhibits cellular respiration by blocking mitochondrial complex I electron transport	Cabras et al. (2002); Isman (2006); Usha Rani et al. (2013)
Gliricidia sepium (Gliricidia)	Leguminaceae	Leaves, bark	Glericidin, Sepinol, Gliricidol	Hemiptera, Diptera	Feeding deterrent	Nukmal et al. (2017); Krishnaveni et al. (2015)
Annona squamosa (custard apple)	Annonaceae	Leaves, bark	Annonin, Squamocin	Lepidoptera, Ixodida, Hemiptera	Repellent, anti ovipositional, decline in free aminoacid for protein synthesis inhibiting metamorphosis	Leatemia and Isman (2004); Bhagawan et al. (1992)

Eucalyptus globulus (Eucalyptus)	Myrtaceae	Leaves, oil	Camphene, limonene, linalool	Diptera, Lepidoptera	Meddles with octopaminergic receptors	Erler and Yalcinkaya (2006); Enan (2001)
Euphorbia tirucalli (Euphorbia)	Euphorbiaceae	Stem, leaves	Euphorbosterol, Euporbol	Dictyoptera, Diptera	Antifeedant and adversely regulates growth	Uma et al. (2009); Okonkwo and Ohaeri (2018)
Datura stramonium (Datura)	Solanaceae	Leaves, roots, fruits, seeds	Atropine, Hyoscyamine	Coleoptera, Lepidoptera, Hemiptera, Acarina	Repellent, antifeedant	Abbasipour et al. (2011)
Nicotinana spp. (tobacco)	Solanaceae	Whole plant	Nicotine, Nornicotine	Hemiptera, Diptera, Acarina	Acts as agonist of cholinergic acetylcholine nicotinic receptor	Pittarelli et al. (1993; Guthrie (2013); Sarker and Lim (2018)
Ryania spp.	Flacourtiaceae	Stem, roots	Ryanodine	Lepidoptera, Hemiptera, Thysanoptera	Disrupts calcium channels	Copping and Menn (2000); Isman (2006)
Lanthana camera (Lantana)	Verbenaceae	Leaves, whole plants	Lantanolic acids, lantic acid	Coleoptera	Respiratory and dermal toxins	Rajashekar et al. (2014)
Vitex negundo (Notchi/Nirgundi)	Verbenaceae	Flowers, leaves, roots	Vitexin, Nirgundoside	Coleoptera	Repellent and enzyme inhibition	David et al. (1988); Chowdhury et al. (2011)

(continued)

Table 17.1 (continued)

Plant source	Family	Plant parts	Insecticidal compounds	Active against	Mode of action	References
Chrysanthemum cinerariaefolium (Chrysanthemum)	Asteraceae	Flowers, leaves, roots	Pyrethrin, Cinerin	Coleoptera	Disrupts exchange of sodium and potassium ions	Casida (1973)
Allium sativum (garlic)	Alliaceae	Whole plant	Allicin, diallyl sulfide, essential oil	Lepidoptera, Hemiptera, Coleoptera, Diptera	Inhibition of acetylcholine esterase, repellent, ovipositional inhibition	Chaubey (2017); Plata-Rueda et al. (2017)
Cymbopogon spp. (lemongrass)	Poaceae	Leaves, roots	Essential oil (citronelal, eugenol, geraniol, limonene)	Lepidoptera, Diptera, Coleoptera	Repellent	Caballero-Gallardo et al. (2012); Hernández-Lambraño et al. (2014)

17.3 Phytochemicals with Insecticidal Properties

Use of Neem (*Azadirachta indica*) for pest control was practiced 4000 years ago and was cited in the Indian Vedic literatures. Pytochemicals from "Meliaceae" were important source of insecticidal compounds. Neem products were effective against a broad range of insects viz., defoliating caterpillars, stemborers, bollworms, armyworms, cutworms, leaf miners, whiteflies, aphids, leafhoppers, thrips, scales, psyllids, and mites (Schmutterer 2002; Dimetry 2012; El-Wakeil 2013). Almost all plant parts are source of insecticidal compounds with seeds harboring bioactive compounds such as azadirachtin and limonoids (nimbolide, nimbin, and salannin). Azadirachtin, the principle compound with insecticidal activity was found to prevent the synthesis and release of ecdysteroids resulting in incomplete molting of immature insects. In addition, "Azadirachtin" acts as a "feeding deterrent," "ovipositional repellant," and "sterility inducer." Kraus (2002a, b) reported that *Melia azedarach*, a relative of neem, produced "meliacarpins" regulating insect growth. "Toosendanin" from *Melia toosendan* acted as stomach poisons controlling chewing insects (Chiu 1988). Nicotine, a naturally occurring alkaloid obtained from tobacco plants (*Nicotiana tabacum, Nicotiana rustica*) showed excellent insecticidal potential against herbivorous insects (Thakur et al. 2020).

As early as 1690s, extracts of tobacco plant were used for managing sucking pests of cereals. However, the active compound "Nicotine" was extracted in 1828 and later it was synthesized in 1904. Nicotine possessed neurotoxicity to insects, mammals, and birds, which limited their usage as insecticide. It mimics the neurotransmitter acetylcholine and acts as an agonist of the cholinergic acetylcholine nicotinic receptor and effectively controls caterpillars, aphids, and thrips (Casanova et al. 2002). Neonicotinoids were synthetic insecticides synthesized based on archetype of this natural compound nicotine. "Acetogenins" from "Annonaceae" were reported to have insecticidal activity. "Acetogenins" viz., annonin I, or squamocin from seeds of "*Anona squamosa*" and "asimicin" from bark of "*Asimina triloba*" with insecticidal properties against chewing insects were identified and patented. These compounds acted as stomach and respiratory poisons (Moeschler et al. 1987; Mikolajczak et al. 1988; McLaughlin et al. 1997; Johnson et al. 2000).

Lewis et al. (1993) reported that "Asimicin" from *A. squamosa* acted as respiratory poisons and reduced the respiration rate of *Ostrinia nubilalis*. Crude ethanolic extracts of *A. squamosa* seeds were found to have insecticidal properties against diamondback moth (*Plutella xylostella*) (Isman 2006). "Rotenone" from leguminaceous plants *Derris elliptica* and *Lonchocarpus nicou* is another natural insecticide. This inhibits the cellular respiration in insects leading to their mortality and is effective against aphids, plant bugs, potato beetles, carpenter ant, and spider mites (Yamamoto and Kurokawa 1970; Khambay et al. 2003; Cabras et al. 2002; Cabizza et al. 2004). Though rotenone is considered as safe for controlling agriculture and veterinary pests concerns raise for its usage near water bodies as they are highly toxic to aquatic life. As rotenone is photodegradable it has not been reported to contaminate underground water (Robertson and Smith-Vaniz 2008). "Pyrethrin" from *Crysanthemum cinerariaefolium* disrupts the sodium and potassium ion exchange in insects and kills them (Casida 1973). Pyrethrum was effective against beetles, leaf-eating caterpillars, cabbage worms, plant bugs, aphids, leafhoppers, and spider mites (Casida 1973; Glynne-Jones 2001). Ryania is another naturally occurring insecticide obtained from the stem of *Ryania speciosa* and was found to be effective against codling moths, corn earworms, onion thrips, and potato aphids (Copping and Menn 2000; Isman 2006).

17.4 Plant Proteins with Anti-Nutritional Effects on Insects

Lectins are a complex group of proteins found in plants, animals, and microbes with high structural and functional diversity and affinity for several carbohydrates resulting in various biological activities including insecticidal properties. The primary function of lectins is to defend the parent organism from foreign pathogens. "Lectin" is derived from a Latin word and means "to choose or select." Few lectins are "classical lectins" that are constitutively expressed while others are "inducible lectins." Among various sources, lectins from leguminous plants were found to possess insecticidal properties owing to their glycosylation reactions in insect's midgut and reduction of intestinal absorption of nutrients in insects. These plant lectins which showed insecticidal or insect-repelling properties especially against sucking pests were mannose-specific and had no effect on human metabolism showing its safety for use as a pest management strategy. Snowdrop "'*Galanthus nivalis* agglutinin" was the first lectin to be identified with insecticidal property (Van Damme et al. 1987). Production of transgenic crop carrying the snowdrop lectin gene was attempted as a possible means of pest control (Stoger et al. 1999). Lectins were also effective against lepidopterans and coleopterans. Wheat germ agglutinin (WGA) was effective against *O. nubilalis* (European corn borer). Lectins from garlic and snowdrop showed moderate toxicity to *Spodoptera litoralis* (tobacco hornworm) and *Callosobruchus maculatus* (cowpea weevil). Chitin-binding lectins from *Oryza sativa* (paddy) and *Urtica dioica* (stinging nettle) also exhibited toxicity toward cowpea weevil. Lectins present in seeds of *Phaseolus vulgaris* (PHA), *Dioclea grandiflora* (DGL), *Dioclea rostrata* (DRL), *Cratylia floribunda* (CFL), and *Canavalia brasiliensis* (concanavalin A) protected them from *C. maculatus* (Peumans and VanDamme 1995).

Zhang et al. (2000) recorded a first inducible lectin, Orysata, an agglutinin from *O. sativa*. This mannose-specific lectin was salt and drought inducible. Transgenic tobacco carrying the Orysata gene was deleterious to *Spodoptera exigua* (beet armyworm), *Myzus persicae* (green peach aphid), and *Acyrthosiphon pisum* (pea aphid) (Atalah et al. 2014; Macedo et al. 2015). Williams et al. (2002) showed that infestation of wheat plant by *Mayetiola destructor* (Hessian fly) larvae resulted in expression of three lectin-like genes (*Hfr*—Hessian fly responsive) leading to plant defense. *N. tabacum* agglutinin (NICTABA), an another inducible lectin from tobacco plant, was the first nucleocytoplasmic lectin identified from plants and was induced by chewing insects (*Manduca sexta* and *Spodoptera littoralis*) and spider mites but not by phloem-feeding guilds (*Trialeurodes vaporarorum* and *Myzus nicotianae*) (Lannoo and Van Damme 2010; Macedo et al. 2015).

Experiments involving lectins in artificial diets or transgenic plants prove their effectiveness in controlling insect belonging to lepidoptera, coleoptera, and hemiptera. Transgenic plants containing lectin genes effectively controlled sucking pests which were not affected by Bt transgenes. *Allium sativum* leaf agglutinin (ASAL) has been tailored into many plant species and the transgenes are effective against hemipterans. Bharathi et al. (2011) reported that rice pyramided with GNA and ASAL conferred resistance to *Nilaparvatha lugens*, *Nephotettix virescens,* and *Sogatella furcifera*. Other plant compounds which were found to be effective targets for utilization in crop protection include protease inhibitors (PI) majority of them belonging to families leguminosae, solanaceae (tomato, potato), and gramineae (Richardson 1991). These compounds are rich in amino acids namely cysteine and lysine and have no deleterious effects on humans.

Lipke et al. (1954) showed that soybean trypsin inhibitor (SBTI) exhibited toxicity to *Tribolium confusum*. Similarly, cowpea trypsin inhibitor (CpTI) from mature cowpea seeds was reported to have insecticidal, nematicidal, and fungicidal activity. Transgenic tobacco plants carrying CpTI genes conferred resistance to lepidopterans (*Spodoptera* sp. and *Heliothis* sp.), coleopterans (*Anthonomnous* sp. and *Diabrotica* sp.), and orthopteras (Locusts) (Hilder et al. 1987; Lawrence and Koundal 2002).

Botanical pesticides lured focus of researchers due to their eco-friendly traits such as volatile nature, low environmental risk, minimal residual activity, and safe to nontarget organisms (Xu et al. 2011). However, several constraints arose in commercialization of botanical pesticides including quality control and product standardization (Nawaz et al. 2016). In addition, injudicious use of synthetic botanical insecticides may also lead to development of pest resistance and phytotoxicity. Extracts of *Acotinum* spp. and *Ricinus communis* are toxic to human; *Tephrosia vogelii* has adverse effect against fish; nicotine extracts are harmful to humans and pollute environment and at high concentrations also leads to phytotoxicity; while neem oil-based insecticides at high concentration is also phytotoxic to tomato, brinjal, and few ornamental plants (Stevenson et al. 2012; Nawaz et al. 2016). All natural insecticides are not safe and cannot be synonymous as organic as some natural insecticides and active ingredients of certain essential oils are categorized as highly toxic (Mossa et al. 2018).

17.5 Microbial Insecticides

Nearly 3000 microbial species have been reported to cause epizootics in insects in natural conditions among which only a few have been identified and documented (Table 17.2). The list of identified entomopathogens includes around 100 bacterial species, over 800 fungal species, 1000 protozoan species, and nearly 1000 viruses with nuclear polyhedrosis virus (NPV) affecting nearly 525 insect species worldwide. Additionally two major genera of entomopathogenic nematodes viz.,

Table 17.2 Important Entomopathogens used for pest management

Entomopathogens	Insecticidal toxins	Host range	Reference
Entomopathogenic bacteria			
Bacillus thuringiensis ssp. *kurstaki*	CryIIA, Cry1Aa, Cry1Ab, Cry1Ac, Cry2Aa	Lepidoptera	Schunemann et al. (2014)
B. thuringiensis ssp. *aizawai*	Cry1Aa, Cry1Ab, Cry1C, Cry1D, Cry1B, Cry1Ca, Cry1Da	Lepidoptera	Roh et al. (2007) Schunemann et al. (2014)
B. thuringiensis ssp. *japonensis*		Coleoptera: Scarabaeidae	Mashtoly et al. (2010)
B. thuringiensis ssp. *israelensis*	Cry4Aa, Cry4Ba, Cry4A, Cry4BCyt1Aa	Diptera	Schunemann et al. (2014) Soares-da-Silva et al. (2015)
B. thuringiensis ssp. *tenebrionis*	Cry3Aa	Coleoptera	Roh et al. (2007)
Paenibacillus popilliae	Toxic metabolites	Coleoptera: Scarabaeidae, *Popillia japonica*	Koppenhofer et al. (2012)
Serratia entomophila	Toxic metabolites	Coleoptera: Scarabaeidae: *Costelytra zealandica*	Jackson et al. (1992, 2001); Jackson (2003), Jackson and Klein (2006)
Entomopathogenic actinomycetes			
Streptomyces avermitilis	Avemectins	Wide host range	Lasota and Dybas (1991); Pitterna et al. (2009)
Saccharopolyspora species	Spinosyns	Wide host range	Snyder et al. 2007; Huang et al. (2009)
Entomopathogenic fungi			
Metarhizium anisopliae	Dextruxins	Subterranean insects	Robles-Acosta et al. (2019)
Beauveria bassiana	Beauvericin, Beuverolides, Bassinolide	Wide host range	Robles-Acosta et al. (2019)
Beauveria brongniartii	Isarolides A,B,C,D	Wide host range	Ruiu (2018)
Verticillium lecanii	Bassinolide	Sucking pests	Aguirre and Krugg (2014)
Isaria spp.	Isarin	Hemiptera, Thysanoptera	Robles-Acosta et al. (2019)
Entomopathogenic virus			
Nuclear Polyhydrosis virus	–	Lepidoptera	Kalha et al. (2014)
Granulosis virus	–	Lepidoptera	Kalha et al. (2014)
Entomopathogenic protozoan			

(continued)

Table 17.2 (continued)

Entomopathogens	Insecticidal toxins	Host range	Reference
Nosema locustae	–	European cornborer caterpillars, grasshoppers, and Mormon crickets	Copping and Menn (2000)
Paranosema locustae	–	*Oedaleus senegalensis*, *Pyrgomorpha cognata*, and *Acrotylus blondeli*	Tounou et al. (2008)
Nosema lymantriae	–	*Lymantria dispar*	Ironside (2013)
Vairimorpha disparis	–	*Lymantria dispar*	Ironside (2013)
Amblyospora connecticus	–	*Aedes cantator*	Johny et al. (2006)

Steinernema (55 species) and *Heterorhabtidis* (12 species) also play an important role under natural conditions to keep the insect population under control (Nawaz et al. 2016; Koul 2011). Soil harbors diverse micro and macro flora and fauna which have coevolved (Rai et al. 2020; Singh et al. 2020). Microbes exhibit a variety of interactions with insects, of which many are mutualistic and few are pathogenic (Rastegari et al. 2020a, b). Insect pathogens have evolved a multiple strategy to invade inside host, overcome the host's immune response, and finally to kill it. As the interaction between host and their pathogens is a long coevolutionary process in-depth studies on the molecular mechanisms are essential to develop a novel insect control strategy with precise microbial formulations with improved stability under storage; quick action, target specificity, and environment-friendly nature on field application.

17.5.1 Entomopathogenic Bacteria

Entomopathogenic bacteria are unicellular prokaryotic organisms, measuring from <1 μm to several microns. Majority of entomopathogenic bacteria belong to family Bacillaceae, Enterobacteriaceae, Micrococcaceae, Pseudomonadaceae, and Streptococcaceae where many of them behave as weak pathogens or epiphytes (Kalha et al. 2014; Singh et al. 2020). Among them Bacillaceae is the major group of entomopathogenic bacteria commercially exploited for pest management (Verma et al. 2016; Yadav et al. 2016). *B. thuringiensis* is a lethal pathogen of major insect orders (Lepidoptera, Coleoptera, Diptera, Orthoptera, etc.,) which act as crop and urban pests *Lysinibacillus sphaericus* is highly virulent to mosquitoes, *Paenibacillus popilliae* cause "milky diseases" to scarabaieds (Suman et al. 2016). Other important bacteria include *Pseudomonas entomophila, Yersinia entomophaga, Serratia* sp., and endosymbionts *Photorhabdus* spp. and *Xenorhabdus* spp., of entomopathogenic nematodes belonging to genera *Steinernema* and *Heterorhabditis*, respectively (Ruiu 2015). Metabolites of *Chromobacterium subtsugae* have broad-spectrum insecticidal activity against Lepidoptera, Hemiptera, Coleoptera, and Diptera (Martin et al. 2007; Verma et al. 2018). A strain of *Burkholderia rinojensis* was identified to possess biocontrol property against various chewing and sucking pests including mites (Cordova-Kreylos et al. 2013).

B. thuringiensis (Bt) is the widely used and intensively researched species in last few decades (Vega 2008). The pathogenic action initiates on ingestion of crystalline inclusions and spores possessing δ-endotoxins (Saxena et al. 2020). They interact with specific receptors in the epithelial cell of the insect midgut. The toxin is activated by the alkaline pH of the insect saliva and acts on midgut cell and disrupts the permeability cell membrane causing cell lysis followed by gut paralysis and death. The insecticidal activity and host specificity rely on biosynthesis of crystal toxins associated with parasporal bodies such as crystal (Cry) and cytolytic (Cyt) proteins that are produced during sporulation phase and also by vegetative insecticidal proteins (VIP) released during vegetative phase of bacterial growth (Bravo et al. 2007; Ruiu 2015). Different bacilli species and strains harbor variation in toxin gene sequences which decide the binding specificity of toxic proteins with insect midgut receptors thereby constituting to diverse strain-specific insecticidal properties (Pigott and Ellar 2007). Studies on Cry 1Ab in *M. sexta* depicted that Cry toxin binds with the receptor cadherin and N-aminopeptidase in surface membrane (Dorsch et al. 2002; Hua et al. 2004).

Such explicit molecular affinities between the receptors and toxins result in proteolytic cleavage of Cry toxin leading to structural changes forming oligomers which function as "pre-pores." These changes affect the integrity of membrane and cause cell death due to osmotic imbalance (Melo et al. 2016). The model of sequential binding views the changes in cell and formation of pore after interacting between toxins and receptors in susceptible host. Similar model is implied in resistance but the receptor delays the cytotoxicity. However, tolerance takes place without any changes in structure of receptor (Vachon et al. 2012; Melo et al. 2016). Nonresponse of receptor in resistance is due to silencing of RNA cadherin and N-aminopeptidases which noticeably showed the importance of receptor in the biological activity of Bt (Schwartz and Laprade 2000).

The sensitivity of midgut cell to various type of Cry toxin regain with the expression of only one of these heterologous receptors (cadherin or N-aminopeptidase), indicating that there may be other mode to resistances and toxicity (Melo et al. 2016). Cry 1Ac-producing transgenic cotton is lethal to many lepidopteran pests. In pink bollworm, Cry 1 Ac binds to recombinant peptides corresponding to extracellular regions of cadherin protein (BtR) and pink bollworm resistance to Cry 1Ac is related with mutation in BtR gene, revealed that BtR is receptor for Cry 1 Ac in pink bollworm (Fabrick and Tabashnik 2007). Similar studies on pink bollworm resistance against Cry 2Ab specifies that mutation in resistant larva (loss of exon 6 caused by alternate splicing of pre-mRNA) disrupts the ATP- binding transporter genes (*PgABCA2*) in laboratory-selected strain from Arizona, USA and field selected populations from India (Mathew et al. 2018).

In recent years, numerous researches on omics of Bt have been conducted for genetic improvement of crops and to delay the development of resistance against

Cry toxins. The studies on pink bollworm resistance to Cry 1Ac and Cry 2Ab showed that focusing on gene response to toxin in receptor site may help to speed up the study on monitoring and managing Bt resistance. Gene pyramiding and non Bt-refugee crops are other strategies to combat against Bt-resistance. In United States, genetically engineered Bt cotton and maize are widely planted as a part of IPM. Though *Helicoverpa zea* developed resistance against Cry toxins (Cry 1A and Cry 2A), to overcome this researchers are proposing to deploy cultivars expressing both Cry and VIP3Aa and non-Bt-refuse crops to favor the susceptibility *H. zea* to Bt toxins (Reisig and Kurtz 2018). As a complementary approach release of transgenic male insects with a female-specific self-limiting gene can reduce the evolution of resistance by introgression of susceptible gene through male.

Release of transgenic males of *P. xylostella* for introgressing susceptible alleles into target population reduced fecundity and male-mating competitiveness thereby suppressed the population size and delayed development of resistance (Zhou et al. 2018). Therefore genetic improvement through gene pyramiding, research on mechanism involved in resistance development, incorporation of non-Bt refugia, and release of transgenic males could assist in development of a reliable pest management strategy minimizing resistance development.

L. sphaericus, formerly known as *Bacillus sphaericus*, has been documented as an important bioagent for mosquito control and also finds its application in bioremediation of toxic metals (Geng et al. 2018). This bacterium produces spherical endospores, closely associated with parasporal crystals containing an equimolar ratio of binary protein toxins (BinA and BinB) and mosquitocidal toxins (Mtx protein). Vegetative cells of few strains are also reported to produce S-layer protein (Baumann et al. 1991; Lozano et al. 2011; Ruiu 2015). The insecticidal action is similar to *B. thuringiensis* attributing to the damage of microvillar epithelial cells in the midgut (Charles et al. 2013). The receptor site of *B. thuringiensis* var. *israelensis* (Bti) is different from receptor site of *L. sphaericus* thus the combined use of *L. sphaericus*/Bti biolarvicide exhibited lower frequency of the resistance alleles and can be used to delay the onset of *L. sphaericus* resistance (de Mendonca Santos et al. 2019) and the toxin Cry48Aa/Cry49Aa from *L. sphaericus* also killed Bin-resistant larvae (Rezende et al. 2019). In addition to larvicidal activity, they also released plant growth hormones and antifungal compounds. *L. sphaericus* ZA9 produced appreciable quantities of IAA, siderophores, HCN, and hydrolytic enzymes capable of solubilizing phosphates, potassium, and silicates and thus their inoculation enhances seed germination and seedling vigor. The cell-free culture supernatant of *L. sphaericus* ZA9 showed antagonistic property against *Alternariaalternata*, *Aspergillus* sp., *Bipolarisspicifera*, *Curvularialunata*, *Sclerotinia* sp., and *Trichophyton* sp. (Naureen et al. 2017). Hence, bacterium *L. sphaericus* has multiple advantages in management of household insect vectors carrying human pathogens as well as in agriculture by production of plant growth regulators and in combating phytopathogens.

Paenibacillus spp. is another bacterium gaining importance in agriculture and industrial biotechnology. This genus *Paenibacillus* is ubiquitous and majorities of their species have been reported to promote plant growth through nitrogen fixation, phosphate solubilization, indole-3-acetic acid (IAA), and siderophores production (Verma et al. 2016; Verma et al. 2019; Verma et al. 2015). They also offer defense against phytophagous insect and phytopathogenic microbes by triggering a hypersensitive defensive response causing induced systemic resistance (ISR) in plants and by inducing production of defensive secondary metabolites by infested plants (Grady et al. 2016). *Paenibacillus japonica,* the causative agent of milky disease in Japanese beetle was first microbial insect control agent registered in the United States though their utilization was not widespread due to their obligatory nature (Klein 1988). Potential of *Paenibacillus* species in combating insect pests was attributed to the production of chitinase and crystal protein (Cry). Chitinase enzymes released by *Paenibacillus* hydrolyze chitin leading to reduced feeding rates and mortality of insects. Chitinase is highly stable at field at 40 °C indicating their potential to be used as insecticides in field (Singh et al. 2009). Genes encoding Cry toxins have also been identified from many species and strains of genus *Paenibacillus* including *P. popilliae*, *P. lentimorbus* strain Semadara, and *Paenibacillus* sp. Kh3 (Zhang et al. 1997; Yokoyama et al. 2004; Gorashi et al. 2014). *Paenibacillus polymyxa* NMO10 genetically engineered with Cry 1C from *B. thuringiensis* for combined activity of insecticidal and growth promotion showed increased toxicity against lepidopteran insects apart from promoting plant growth (Hussie et al. 2011; Ibrahim et al. 2014). Contrastingly, high population of *P. polymyxa* in rhizosphere was reported to increase susceptibility of plant to aphids via increasing levels of IAA (Kim et al. 2016). Presence of *Paenibacillus nematophilus* in rhizosphere was found to restrict the dispersal of entomopathogenic nematode *Heterorhabditis megidis* (Enright and Griffin 2005). Certain species of *Paenibacillus* were evidenced to cause disease on honey bees and wild bees. However, few species as in case of *Paenibacillus* MBD-MB06, a symbiont reported from adult body surface, guts, and nests of wild bees exhibited antimicrobial, chitin-binding and biofilm-forming properties, which prevent penetration of fungal hyphae into bee integument under humid and nutrient-rich environment of wild bee nests which favors fungal growth (Keller et al. 2018).

Pseudomonas is an important group of agriculturally important Gram-negative bacterium with many of their species aiding plant growth and crop protection. They colonize various environmental niches ranging from oil spilled seawater (Viggor et al. 2013) to soil (Weller et al. 2002), plant rhizosphere, and other plant tissues (Hirano and Upper 2000; Loper et al. 2012; Pathma et al. 2019a, b), insect guts (Vodovar et al. 2005), and from different extreme habitats such as low temperature (Verma et al. 2015; Yadav et al. 2015a; Yadav et al. 2015b), high temperature (Sahay et al. 2017; Saxena et al. 2016), salinity (Yadav et al. 2019; Yadav et al. 2015c), acidic soil (Verma et al. 2013), and drought conditions (Kour et al. 2020a). Those colonizing soil rhizosphere protected plants from phytopathogens and few strains also evidenced to cause diseases to insect pests. The list includes *Pseudomonas fluorescens*, *Pseudomonas chlororaphis,* and *Pseudomonas protegens.* Among them, *P. fluorescens* apart from biofertilizing potential also possess biocontrol properties and hence can be used for controlling notorious insect pests and plant pathogens. *Pseudomonas* spp. are considered as PGPRs (plant growth-promoting rhizobacteria) as they produce IAA, ACC-deaminase, gibberellins, and cytokinins and solubilize nutrients which promote plant growth (Yadav et al. 2017; Yadav et al. 2020c). They inhibit phytopathogens by production of siderophores which limits iron availability, production of lytic enzymes viz., chitinases, and β-1,3-glucanases that degrade chitin and glucan present in the cell wall of phytopathogenic fungi and also degrades toxin produced by pathogen (Pathma et al. 2010a, b; Pathma et al. 2011; Pathma and Sakthivel 2013; David et al. 2018; Pathma et al. 2019a, 2020).

PGPR-mediated induced systemic resistance (ISR) combats insect pest by influencing normal growth and development of insect. The growth rate, feeding efficiency, and feed digestibility of *Helicoverpa armigera* got affected when larvae fed on *Pseudomonas gladioli*-treated cotton due to an increase in their polyphenol and terpenoid content (Zhang et al. 1998). Penetration of *Heterodera schactii*, cysts nematode in sugar beet was reported to be inhibited by ISR induced by *P. fluorescens* (Oostendorp and Sikora 1989, 1990). Similarly, the infestation level of *Meloidogyne incognita* on tomato was reduced by root dip with *P. fluorescens* Pf1 (Santhi and Sivakumar 1995). *P. fluorescens* genetically engineered with Cry toxins from *B. thuringiensis* effectively controlled lepidopteran pests. Transgenic *Pseudomonas cepacia* strain 526 containing crystal protein gene from *Bt* effectively controlled tobacco hornworm (Stock et al. 1990; David et al. 2018). Hence PGP Pseudomonas provides a significant scope and potential for use in sustainable organic agriculture.

17.5.2 Entomopathogenic Actinomycetes

Actinomycetes are Gram-positive actinobacteria previously referred to as ray fungi. They are ubiquitous and exhibit a wide taxonomical and functional diversity. They are reservoirs of many antibiotics of therapeutic value in human medicine and biocontrol potential in agriculture (Yadav et al. 2018). They also produce extracellular enzymes such as chitinase that can degrade insect cell wall. Among actinomycetes, *Streptomyces* and *Saccharopolyspora* species gained commercial interests in biopesticides market (Ruiu 2018). Avemectins are macrocyclic lactones derivatives obtained from soil actinomycetes *S. avermitilis* and possess insecticidal, nematicidal, and acaricidal properties (Lasota and Dybas 1991; Pitterna et al. 2009). The mode of action includes blockage of electrical transmission by enhancing the effects of glutamate at the invertebrate-specific glutamate-gated chloride channel with minor effect on gamma-aminobutyric acid receptors causing influx of chloride ions into the cell. This lead to hyperpolarization and paralysis of invertebrate neuromuscular system in low doses that are nontoxic to mammals (Cully et al. 1994; Bloomquist 1993). Avemectin B1 (abamectin) is the major product of fermentation possessing both insecticidal and acaricidal activity and is widely used in agriculture and horticulture. It also showed least toxicity to nontarget organisms. Abamectin is easily degraded by soil microorganisms and hence possesses least environmental persistence (Lasota and Dybas 1990).

Abamectin is also photodegradable and their stability to light can be improved by encapsulation with protein zein (Demchak and Dybas 1997). Compounds such as emamectin benzoate (mixture of avemectin B1a and B1b) are also widely used in control of lepidopterous pest. Thus avermectins are good candidates for inclusion in IPM programs. *S. spinosa* is another soil actinomycetes, producing compounds with insecticidal property based on which spinosad has been formulated. The compound contains two components spinosyn A (major) and spinosyn D (minor) roughly in 17:3 ratio (Mertz and Yao 1990). Mode of action includes binding to the site of nicotinic acetylcholines receptors leading to disruption of acetylcholine neurotransmission. It kills the insect by hyperexcitation (Snyder et al. 2007; Huang et al. 2009). It also can be used in management of storage pests.

17.5.3 Entomopathogenic Fungi

Majority of Entomopathogenic fungi belong to divisions Zygomycota, Ascomycota, and Deuteromycota (Samson et al. 1988). The most studied class of entomopathogenic fungi are Entomophthorales in the Zygomycota and Hyphomycetes in the Deuteromycota and they differ in mode of action. Lymphocytes members are generally considered as opportunistic pathogens with wide insect host range and bring mortality through toxin production (Burges 1981; Samson et al. 1988). While Entomophthorales kills the host with tissue colonization and host utilization with less or no evidence of toxin-induced death (Humber 1984). Unlike other entomopathogens, entomopathogenic fungi infect by penetrating through cuticle and need no oral ingestion. The infection initiates as soon as the spores contact the insect cuticle. The spores from entomopathogenic fungi are of two forms (dry powder and wet slime).

The dry spore is evidenced to adhere to the insect cuticle through a combination of electrostatic force and chemical binding which facilitates fungal spore attachment to hydrophobic, lipophilic epicuticle of insect (Samson et al. 1988) and in some group appressorium may be formed while the wet spores utilize the mucilaginous matrix surrounding them (Evans and Hywel-Jones 1997). After germination of spores, the fungal spores penetrate the insect body through sequential process and led to mummification of the insect cadaver. They also produce toxins that act as immunosuppressive compounds and facilitate fungal infection (Pedrini 2018; Mannino et al. 2019). The recent finding on whole-genome sequencing of *Beuveria bassiana* and *Metarhizium* spp. identified some candidates for oral infection (Xiao et al. 2012; Gao et al. 2011; Hu et al. 2014). The metabolic pathway of *B. bassiana* in oral infection was revealed in the context of insect–pathogen coevolutionary studies (Mannino et al. 2019). Therefore insect has a harder challenge to defend themselves against fungal infection from both routes.

B. bassiana, the white muscardine fungi infects a wide range of arthropod host and is a potent bioagent for control of insect pests of agricultural and medical importance. The genus *Beauveria* also comprise of another species *brongniartii* with potent insecticidal properties. Host range varies with species and strains. Few strains have broad host range while few having narrower host range such as Bba 5653 which is highly virulent but has restricted host range including caterpillar of diamondback moth and few others (EPA 2006). The infection initiates after conidia attach itself into cuticle with physical force followed by germination and penetration through cuticular layers with the help of hydrolytic enzymes, such as protease, chitinases, lipases, etc., and mechanical forces (Mascarin and Jaronski 2016).

The growing fungal hyphae colonize the hemolymph and forms specialized budding single-celled structure, the yeast-like blastospores (or hyphal bodies) that proliferate and exploit nutrients, colonize internal tissue, and weaken the host immune system (Humber 2008). In addition, they also produce toxic metabolites that are capable of suppression of host immune, nutrient depletion, and destruction of host tissues resulting in insect mortality (Ortiz-Urquiza et al. 2010; Gibson et al. 2014). Abiotic factors viz., humidity, rainfall, and solar radiation (McCoy et al. 2000; Jaronski 2010; Fernandes et al. 2015) and biotic factors such host age, behavior, disease susceptibility affects the virulence and pathogenicity of the fungi in natural conditions (Mascarin and Jaronski 2016).

Humidity plays a crucial role in the effectiveness, high temperature, and UV radiation reduces their efficacy and survival. Several formulations which increase the stability, self-life, and effectiveness of microbial pesticides are available in insecticides market. Investigations on insecticidal activity of different bioformulations namely water dispersible granule (WG) and oil-based formulation (OD) of *B. bassiana* conidia produced by solid-state fermentation against *Hypothenemus hampei* evidenced that type of bioformulation and moisture percent strongly impacted the biocontrol potential as well as shelf life of the entomopathogens (Lopes and Faria 2019). Addition of UV protectants such as 10% humic acid increased the spore survival up to 87% at seven days after field application and sesame and colza oil also acted as UV protects (Kaiser et al. 2019). Transgenic *B. bassiana* strain *Bb*-Cyt2Ba carrying *Bt* toxin gene *Cyt2Ba* showed significant reduction in fecundity and survival of mosquitoes (*Aedes* spp.) as compared to wild *B. bassiana* strain (Deng et al. 2019).

Metarhizium anisopiliae, the green muscardine fungus is a promising biocontrol agent of agriculturally important insect pests. The mode of infection is penetration of insect cuticle and involves the release of serine protease that initiates protein degradation during invasion. After the activity of proteases enzymes, action of exopeptidases (carboxypeptidases) initiates break down of host's amino acid for provision of nutrient to the invading fungus (Woessner 2013). The most prevalent and extensively studied toxin produced from *M. anisopliae* in combating insect pest is destruxin which are cyclic depsipeptides reported to have insecticidal, antiviral, and cytotoxic properties (Schrank and Vainstein 2010). Dextruxin is divided chemically into five groups and is labeled as A–E where Dextruxin A (DA), Dextruxin E (DE), and Dextruxin B (DB) showed insecticidal property (Thomsen and Eilenberg 2000). DB and desmethyl-DB were reported to be phytotoxic to Brassica plants (Pedras et al. 2002). The genus *Metarhizium* occurs in soil ecosystem with abundance in species diversity and population. The species complex can be identified with molecular markers such as translation elongation factor 1-α (TEF1-α), nuclear intergenic loci MzIGS3, and few microsatellite markers (Bischoff et al. 2006; Bischoff et al. 2009; Oulevey et al. 2009; Kepler and Rehner 2013; Juliya 2019). Exploration on genetic diversity of *Metarhizium* and enzymes and toxins they produce could offer crucial information on successful utilization of these microbial bioagents in integrated pest management.

Lecanicillium lecani earlier referred to as *Verticillium* lecani is another important member of entomopathogenic fungi. *Lecanicillium* spp. is commonly referred to as white halo fungi as their mycelium are white in color and highly virulent against sucking pests (Horn 1915; Ekbom 1979; Ahmad et al. 2019). An infection starts when the insect comes into contact with sticky spores. As other entomopathogenic fungi utilizes hydrolytic enzymes and mechanical force to penetrate into cuticle of host. The fungi invade the insect body for their nourishment leading to death of the host within 42–78 hours of infection. Many extracellular enzymes such as n-acetyl glucosamine, chitinase, protease, endoprotease, carboxypeptidase A, esterase, PR 1-chymoelastase, and serine protease are accountable for piercing the integument where all enzymes work synergistically to breach the cuticle wall but PR 1 contribute more in breaking cuticle wall through increasing concentration at the point of penetration (Ahmad et al. 2019; Kour et al. 2019; Yadav 2019). The chemicals produced during infection are bassianolide (Kanaoka et al. 1978), beauvericin, dipicolinic acid (Claydon and Grove 1982), vertilecanin-A1, 10-hydroxy 8-decenoic acid, and decenedioic acid (Soman et al. 2001). The hyphae protrude out from body to form a mycelial mat on body surface after exhausting the nutrients inside host body and kill the host (Yeo 2000).

Hirsutella thompsoni is another entomopathogenic fungus that is commercially exploited since 1970s in the United States especially for the control eriophyid mite though it can cause disease in lepidopterans and hemipterans. The genus *Hirsutella* has many species of which three species viz., *H. thompsonii*, *H. citriformis*, and *H. gigantean* are effective entomopathogens. Among various compounds produced by *H. thompsoni,* hirsutellins is an important toxin. Among 162 strains of *H. thompsoni* have been isolated worldwide of which nearly100 isolates showed positive for protein exotoxin hirsutellin A (HtA) gene in PCR amplification (Maimala et al. 2002). The toxin HtA possesses ribonucleolytic activity and acts as ribotoxins (Liu et al. 1995). Apart from ribosomal inhibition, *H. thompsoni* generates oxidative stress in insects and the action of enzymes viz., CAT, SOD, and GPx causes cellular damage (Fornazier et al. 2002; Ahmad et al. 2019). The effectiveness of *H. thompsoni* can be enhanced through integration with other entomopathogenic microbes, arthropod bioagents, and compatible insecticides.

Isaria spp. formerly referred to as *Paecilomyces* spp. caused "yellow muscardine" disease in insects (Prado et al. 2008). Under humid conditions, the fungus grows extensively over leaf surface and easily spreads to the host population (Wraight et al. 2000). *Paecilomyces furiosus* effectively controlled *Culex pipiens* (Sandhu and Mishra 1994). *Paecilomyces fumosorseus* has been well known to cause epizootics and is extensively utilized against whitefly both in greenhouse and field conditions. It produces mycelial threads that cover the host body and stick them on undersides of the leaves (Prado et al. 2008; Ahmad et al. 2019). Amatuzzi et al. (2018) documented that the isolates of *Paecilomyces* cause high mortality of European pepper moth, *Duponchelia fovealis*, greenhouse pests of vegetables and cut flowers as well as pests of aquatic plants in Canada and northern Europe. *Isaria javanica* pf185 was reported to produce an active metabolite dibunyl succinate with antifungal property against *Colletotrichum acutatum* and insecticidal property against aphid, *M. persicae* (Lee et al. 2019). The insecticidal activity of *Isaria fumosorseus* was comparatively higher than *B. bassiana* against *Eublemma ammabilis* (Das et al. 2019).

Isaria isolates ME-33 and ILT-01 were treated against adults of *Sitophilus oryzae* in laboratory using immersion and food mix method in which immersion method showed higher mortality (100%) at LT_{50} 3.63 with highest conidial concentration ($1 \times 107\ mL^{-1}$) indicating the potential of *Isaria* to substitute chemical insecticides in stored pest management (Ahmad et al. 2019). *I. javanica* pf185 a potential entomopathogen and also served as plant growth regulator. A significant change was documented in tomato seedling exposed to volatiles of *I. javanica* pf185 in which the plants and fungus were grown separately and connected through air space. Treated tomato seedlings showed increase in length, weight, chlorophyll content, and number of root branches as compared to the check. Solid-phase micro-extraction and gas chromatography–mass spectrophotometry of the volatiles produced detected the presence of 3-hexanone, 2,4-dimethylhexane, 2-nonanone, and heptane (Lee and Kim 2019). Therefore further research is necessary to elucidate the epizootic caused by *Isaria*, mechanism of plant growth promotion, and their genetic improvement to tailor suitable candidates for organic agriculture.

17.5.4 Entomopathogenic Virus

Baculoviruses viz., nucleopolyhydrosis virus (NPV) and granulosis virus (GV) have found their potential in pest management programs. Both of them possess different types of inclusion bodies in which virions are embedded. The virus particle invades the midgut nucleus, fat body, or other tissues and disintegrate and liquefying tissues of the dead insect cadavers. The mode of entry of entomopathogenic virus is similar to bacteria where infection initiates after ingestion by host therefore they are ideal for controlling insects with biting and chewing mouthparts. Baculoviruses evolve different strategies to overcome host defense, which includes apoptosis, melanization, and RNA interference. The infected larvae climb to higher plant canopy before death and hang upside down due to paralysis of their thoracic legs and the condition is referred to as "tree-top disease" or "wipfelkrankeit" in German. The body liquefies and releases virions into the environment. The larval behavior of reaching tree top before death was found to be advantageous for the viral spread in the canopy below infecting other members of the population (Hoover et al. 2011; Wang and Hu 2019).

Their target specificity ensures its safety to nontarget organisms with no much report of resistance development and persistence adding advantage to environmental safety (Erayya et al. 2013). The disadvantage of using baculoviruses for pest management as compared to entomopathogenic bacteria is the time taken to bring about the kill of the insect pest and its environmental stability. The virulence and pathogenicity of entomopathogens are an aftermath of gene–enzyme compatibility or action. Application of transgene technology to improve the virulence and speed of kill of entomopathogenic viruses by insertion of some neurotoxin genes from spiders, scorpions, or predatory mites or genes coding for insect hormones have been attempted. Kao et al. (1974) reported that insertion of genes-encoding diuretic hormone from *M. sexta* into *Bombyx mori* NPV (BmNPV) enhanced the susceptibility of *B. mori* to the transgenic BmNPV by 20-fold. The hybrid gene from broad-spectrum NPV, AcNPV (*Autographa californica* nucleo polyhedro virus), and BmNPV increased the host range of AcNPV to *B. mori* (Croizier et al. 1994). The deletion of gene-encoding enzyme ecdysosteroid UDP glucosyltransferase (EGT) with recombinant strains of AcNPV reduced time to kill *S. exigua* (Flipsen et al. 1995). The expression of insect hormone-related genes through baculoviruses increased the efficacy as well as speed of action. The recombinant baculoviruses of *H. armigera* (HaNPV) expressing dsRNA of *H. armigera* juvenile hormone acid methyltransferase gene (*HaJHAMT*) and *H. armigera* juvenile hormone acid-binding protein gene (*HaJHBP*) accelerated the larval death where LT_{50} of recombinant baculovirus expressing dsRNA of *HaJHBP* was 54.2% above the control value and also altered the transcription of key gene responsible for hormone signaling pathway such as ecdysone receptor gene (HaEcR) (Liu et al. 2019). Thus in-depth knowledge on molecular biology of baculoviruses as well as genes and proteins of insecticidal properties would help us tailor an effective baculovirus for combating pests.

17.5.5 Entomopathogenic Protozoa

Nearly 1000 species of protozoans have been documented to attack invertebrates including insects (Brooks 1988). In general, protozoan attack does not kill the host insect immediately but produces chronic poisons and debilitates the hosts. The infection starts after ingestion of spores by insects which germinate in their midgut and invade to target cells resulting in reduced feeding, fecundity, longevity, and vigor of insect host (Kachhawa 2017). In addition, EPP-infected host is more susceptible to biotic and abiotic stress. The pathogen–insect relationship between *Nosema pyrausta* and European corn borer showed the transmission of pathogen vertically from parents to offspring and as well as horizontally across population, drastically reducing adult vitality and fecundity (Henry 1981).

17.5.6 Entomopathogenic Nematode

Entomopathogenic nematodes (EPN) most widely used in pest management belong to two families viz., Heterorhabditidae and Steinernematidae (Grewal et al. 2005). They are naturally present in soil and use different cues to locate hosts such as vibration, carbon dioxide, and other chemicals (Kaya and Gaugler 1993). The EPN follows two strategies to attack the host based on which they are classified as "ambushers" and "cruisers" (Grewal et al. 1994a). *Steinernema carpocapsae* follows ambusher's strategy where they conserve energy and wait for mobile host insects in the upper soil, while *Steinernema glaseri* and *Heterorhabditis bacteriophora* are cruisers, mostly subterranean and highly active where they cover long distance to hunt the underground for suitable insect hosts using volatiles cues. Therefore they are highly effective against subterranean pest. Some species of nematode *Steinernema feltiae* and *Steinernema riobrave* utilize the combination of both strategies to locate their host (Grewal et al. 1994a). The nematode penetrates the host via natural openings such as anus, spiracles, mouth, and some species enter through intersegmental membrane of cuticle to reach hemocoel (Bedding and Molyneux 1982). *Heterorhabditis* harbors symbiotic bacterium *Photorhabdus* sp. while *Steinernema* harbors *Xenorhabdus* sp. (Ferreira and Malan 2014). After invading the host body, the juveniles release the bacteria into the host's hemoceol where they multiply and thereby kills the infected host within 24–48 hours. The nematode continues to feed on cadavers, develops through four juvenile stages, and becomes adult.

One or more generation may be completed in the cadavers depending on nutritional resources and large numbers of juveniles are released to environment which attacks new hosts and continues their life cycle (Kaya and Gaugler 1993). The efficacy of EPN in field application is limited by both biotic and abiotic factors. Higher efficacy is obtained under sandy soil with pH 4 and 8. EPN is susceptible to desiccation, UV light, higher temperatures, and freezing. *S. glaseri, Steinernema ribrave,* and *Heterorhabditis indica* are heat tolerant while *S. feltiae*, *Heterorhabditis marelatus,* and *H. megidis* are cold tolerant (Grewal et al. 1994b). Their efficacy can be enhanced by using improved and standard formulations, proper dosage, and method of application, etc. In addition, the orientation and attractiveness of nematode toward host depend on interaction between arthropod herbivores, plants, nematodes, and bacterial colonization on rhizosphere.

The nematodes *H. bacteriophora* were attracted toward the maize roots from seeds coated with both living and dead bacterium *B. pumilus* strain INR-7 in the absence of herbivore *Diabrotica virgifera*. With the presence of *D. virgifera,* nematode attraction was more toward plants treated with living bacteria than other plants showing that the bacterium shapes up tri-trophic interactions belowground and emphasizes the need for further investigations (Disi et al. 2019). Investigations showed that nematode dispersal and efficacy increased when seeds were treated with extracts of EPN-infected hosts cadavers that possess active compounds "ascarocides" which act as nematode pheromones. Pheromone treatment increased the numbers of IJs and attractiveness of EPN *S. carpocapsae* and *S. feltiae* in bait trap with *Tenebriomolitor* larvae (Oliveira-Hofman et al. 2019).

17.6 Semiochemicals

Semiochemicals are compounds which induce intra or interspecific interactions between organisms. Pheromone is semiochemicals that induce intraspecific communication and can be classified into different classes depending on purpose such as trail marking, oviposition, alarm, marking, aggregation, and sex pheromones. Allelochemicals are semiochemicals which induce interspecific interactions and are classified as allomones (benefits the emitter), kairomones (benefits the receiver), apneumone (of synthetic origin), and synomones (benefiting both emitter and receiver). Synthetic analogs mimicking semiochemicals of natural origin had played an important role in pest management. Two different strategies are used for field application. The first strategy is to act directly on insect behavior using slow-release dispensers. Compounds mimicking sex pheromones of many insects including *Spodoptera, Helicoverpa*, pink bollworm, rhinoceros beetle, etc. have been synthesized (e.g., Spodolure, Helilure, Gossyplure, Rhinolure). These play an important role in eco-friendly pest management programs. These pheromones installed in suitable traps in appropriate numbers in field helps in pest monitoring and pest management (by mass trapping and causing mating disturbance). Second strategy induces plant's defense mechanism and also attracts natural enemies and repels herbivores through spray application of synthetic compounds or phytohormones (Simpson et al. 2011a; Colazza et al. 2013; Michereff et al. 2016; Blassioli-Moraes et al. 2019).

17.7 Attract and Reward Strategy

The attractiveness of natural enemies would be meager in pesticide-treated areas as the natural enemies do not find their insect hosts. Therefore this strategy includes presence of additional untreated plant resource that hosts insect pests which act as food for natural enemies (Landis et al. 2000; Gurr et al. 2004). Availability of sufficient food supply increases the action of natural enemies positively by influencing their fecundity, longevity, activity, and host-searching capabilities. The combination of herbivore-induced plant volatiles (HIPVs) and buckwheat as rewards in maize field increased the abundance of Eulophidae and Encyrtidae parasitoids and predators and subsequently reduced the population of *Helicoverpa* spp. (Simpson et al. 2011b; Blassioli-Moraes et al. 2019). Volatile semiochemicals from the hosts, such as sex pheromones, increase the attractiveness and efficiency of natural enemies (Borges et al. 1999). In addition to use of their own semiochemicals (sex pheromones and aggregation pheromones) and their combination with HIPVs was found to enhance the action of natural enemies (Steiner et al. 2006; Aldrich et al. 1997; Jones et al. 2011).

17.8 Push–Pull Strategy

In this strategy, two semiochemicals are used, of which one repels (push away) herbivore insects from crops and another attracts (pulls) natural enemies of insect pests. These strategies involve the use of different chemical stimuli for manipulating the behavior of pests and natural enemies. In an experiment, maize was intercropped with *Desmodium uncinatum*, a repellant plant to stem borer, whereas Napier grass was grown as trap crop around maize plant to attract stem borer. An activity of natural enemies was higher in this scenario as compared to monocultured maize (Midega et al. 2016; Blassioli-Moraes et al. 2019).

17.9 Miscellaneous Compounds of Natural Origin

Nereistoxin, a natural compound isolated from marine annelids *Lumbriconereis heterodopa*, was found to have insecticidal property. The chemical structure and insecticidal property were first discovered by Hashimoto and coworkers. Mode of action includes binding of acetylcholine receptors and inhibition of neurotransmission in insects. Cartap is a chemical analog based on nereistoxin which is widely used as insecticide (Nishiyama et al. 2016). This compound has been used worldwide and considered to be relatively safe for incorporation in IPM module though a few cases of cartap poisoning on human was reported from Japan and China (Kumar et al. 2011).

17.10 Conclusion and Future Perspective

Nature offers a wide array of compounds with insecticidal properties which are products of coevolution of plants, microbes, and insects. Screening of organisms and identification of potent living entities with entomopathogenic traits or the bioactive molecules with insecticidal or insect repellant or attractant properties have proved their potential for integration as a component for integrated pest management programs (IPM), which reduces the usage and ill effects of synthetic crop protection agents which will cause environmental pollution. However, there are many demerits in formulating, commercialization, and application of living entities and their products for pest control due to their sensitivity to introduced environment, reduced shelf-life, slow speed of kill, etc. This reduces their popularity among farmers who expect economical and fast-acting pest control solutions. Intense research to improve the environmental stability of entomopathogens, microbial metabolites, and botanicals and ways to enhance their efficacy, including their speed of kill, reduced dosage, etc., would open up new avenues and gain farmers faith in utility of these eco-friendly natural products for crop protection. Elucidation of the chemistry of natural compounds can provide us information about their structure, biological activity, environmental stability, chances of resistance development, etc., enabling us to choose and mass produce or synthesize efficient, risk-free eco-friendly microbials, or insecticidal compounds for use in pest management. In addition, appropriate use of biotechnological tools could help us devise improved, risk-free transgenic microbial entomopathogens or crops, which would supplement eco-friendly crop protection and enhance agricultural productivity.

References

Abbasipour H, Rastegar F, Mahmoudvand M (2011) Insecticidal activity of extract from *Datura stramonium* against *Callosobruchs maculatus*. Integrat Protect Stored Prod 69:251–256

Aguirre EPA, Krugg JHW (2014) Effect of *Lecanicillium lecanii* and *Beauveria bassiana* on the *Panonychus citri* mite under laboratory conditions. REBIOL 34:42–50

Ahmad T, Rasool A, Gull S et al (2019) Ascomycota and integrated pest management. In: Khan M, Ahmad W (eds) Microbes for sustainable insect Pest management. Sustainability in plant and crop protection. Springer, Champ, pp 151–183

Aldrich JR, Zanuncio JC, Vilela EF et al (1997) Field tests of predaceous pentatomid pheromones and semiochemistry of Podisus and Supputius species (Heteroptera: Pentatomidae: Asopinae). An Soc Entomol Bras 26:1–14

Amatuzzi RF, Cardoso N, Poltronieri AS et al (2018) Potential of endophytic fungi as biocontrol agents of *Duponchelia fovealis* (Zeller) (Lepidoptera: Crambidae). Braz J Biol 78:429–435

Anonymous (2019) Microbial pesticides market - growth, trends and forecasts (2019–2024). https://www.mordorintelligence.com/industry-reports/microbial-pesticides-market

Atalah AB, Smagghe G, Orysata VDEJ (2014) A jacalin-related lectin from rice, could protect plants against biting-chewing and piercing-sucking insects. Plant Sci 21:221–222

Atwal AS, Dhaliwal GS (2005) Agricultural pests of South Asia and their management, vol 280. Kalyani Publishers, New Delhi

Baumann PAUL, Clark MA, Baumann LINDA et al (1991) *Bacillus sphaericus* as a mosquito pathogen: properties of the organism and its toxins. Microbiol Mol Biol Rev 55:425–436

Bedding R, Molyneux A (1982) Penetration of insect cuticle by infective juveniles of *Heterorhabditis* spp. (Heterorhabditidae: Nematoda). Nematol 28:354–359

Bhagawan CN, Reddy KD, Sukumar K (1992) Annona-induced growth anomalies and protein depletion in red cotton bug *Dysdercus koenigii*. Indian J Exp Biol 30:908–912

Bharathi Y, Vijaya Kumar S, Pasalu IC et al (2011) Pyramided rice lines harbouring *Allium sativum* (asal) and *Galanthus nivalis* (gna) lectin genes impart enhanced resistance against major sap-sucking pests. J Biotechnol 152:63–71

Bischoff JF, Rehner SA, Humber RA (2006) *Metarhizium frigidum* sp. nov.: a cryptic species of *M. anisopliae* and a member of the *M. flavoviride* complex. Mycol 98:737–745

Bischoff JF, Rehner SA, Humber RA (2009) A multilocus phylogeny of the *Metarhizium anisopliae* lineage. Mycol 101:512–530

Blassioli-Moraes MC, Laumann RA, Michereff MF et al (2019) Semiochemicals for integrated Pest management. In: Vaz S Jr (ed) Sustainable Agrochemistry. Springer, Cham, pp 85–112

Bloomquist JR (1993) Toxicology, mode of action and target site-mediated resistance to insecticides acting on chloride channels. Comp Biochem Physiol Part C: Pharmacol Toxicol Endocrine 106:301–314

Borges M, Costa MLM, Sujii ER et al (1999) Semiochemical and physical stimuli involved in host recognition by *Telenomus podisi* (Hymenoptera: Scelionidae) toward *Euschistus heros* (Heteroptera: Pentatomidae). Physiol Entomol 24:227–233

Bravo A, Gill SS, Soberon M (2007) Mode of action of *Bacillus thuringiensis* cry and Cyt toxins and their potential for insect control. Toxicon 49:423–435

Brooks FM (1988) Entomogenous protozoa. In: Ignoffo CM, Mandava MB (eds) Handbook of natural pesti-cides, vol V, Microbial insecticides, part A, entomogenous protozoa and fungi. CRC Press Inc., Boca Raton, pp 1–149

Burges HD (1981) Microbial control of pests and plant diseases 1970–1980. Academic Press, London, pp 1–949

Caballero-Gallardo K, Olivero-Verbel J, Stashenko EE (2012) Repellency and toxicity of essential oils from *Cymbopogon martinii*, *Cymbopogon flexuosus* and *Lippia origanoides* cultivated in Colombia against *Tribolium castaneum*. J Stored Prod Res 50:62–65

Cabizza M, Angioni A, Melis M, Cabras M, Tuberoso CV, Cabras P (2004) Rotenone and rotenoids in cubé resins, formulations, and residues on olives. J Agric Food Chem 52:288–293

Cabras P, Caboni P, Cabras M, Angioni A, Russo M (2002) Rotenone residues on olives and in olive oil. J Agric Food Chem 50:2576–2580

Casanova H, Ortiz C, Pel'aez C, Vallejo A, Moreno ME, Acevedo M (2002) Insecticide formulations based on nicotine oleate stabilized by sodium caseinate. J Agric Food Chem 50:6389–6394

Casida JE (1973) Pyrethrum the natural insecticide. Academic Press, New York, p 329

Charles JF, Delecluse A, Nielsen-Le Roux C (eds) (2013) Entomopathogenic bacteria: from laboratory to field application. Springer Science & Business Media, p 532

Chaubey MK (2017) Study of insecticidal properties of garlic, *Allium sativum* (Alliaceae) and Bel, *Aegle marmelos* (Rutaceae) essential oils against *Sitophilus zeamais* (Coleoptera: Curculionidae). J Entomol 14:191–198

Chen W, Isman MB, Chiu SF (1995) Antifeedant and growth inhibitory effects of the limonoid toosendanin and *Melia toosendan* extracts on the variegated cutworm, *Peridroma saucia* (Lep., Noctuidae). J Appl Entomol 119:367–370

Chiu SF (1988) Recent advances in research on botanical insecticides in China. In: Arnason AT, Philogène BJR, Morand P (eds) Insecticides of plant origin. American Chemical Society, Washington, DC, pp 69–77

Chowdhury NY, Islam W, Khalequzzaman M (2011) Insecticidal activity of compounds from the leaves of *Vitex negundo* (Verbenaceae) against *Tribolium castaneum* (Coleoptera: Tenebrionidae). Int J Tropic Insect Sci 31:174–181

Claydon N, Grove JF (1982) Insecticidal secondary metabolic products from the entomogenous fungus *Verticillium lecanii*. J Invertebr Pathol 40:413–418

Colazza S, Peri E, Cusumano A (2013) Application of chemical cues in arthropod pest management for orchards and vineyards. Chem Ecol Insect Parasitoids:245–265

Copping LG, Duke SO (2007) Natural products used commercially as crop protection agents. Pest Manag Sci 63:524–554

Copping LG, Menn JJ (2000) Biopesticides: a review of their action, applications and efficacy. Pest Manag Sci 56:651–676

Cordova-Kreylos AL, Fernandez LE, Koivunen M et al (2013) Isolation and characterization of *Burkholderia rinojensis* sp. nov., a non-Burkholderiacepacia complex soil bacterium with insecticidal and miticidal activities. Appl Environ Microbiol 79:7669–7678

Croizier G, Croizier L, Argaud O et al (1994) Extension of *Autographa californica* nuclear polyhedrosis virus host range by interspecific replacement of a short DNA sequence in the p143 helicase gene. Proc Natl Acad Sci 91:48–52

Cully DF, Vassilatis DK, Liu KK et al (1994) Cloning of an avermectin-sensitive glutamate-gated chloride channel from Caenorhabditise legans. Nature 371:707

Das P, Borah B, Saikia P et al (2019) Efficacy of *Beauveria bassiana* and *Isaria fumosorosea* against *Eublemma amabilis* (Noctuidae: Lepidoptera): a predator of lac insect, Kerria lacca (Kerr). J Entomol Zool Studies 7:1239–1241

David BV, Sukumaran D, Kandasamy C (1988) The Indian privet *Vitex negundo* Linn – a plant possessing promising pesticidal activity. Pesticides (Bombey) 22:27–29

David BV, Chandrasehar G, Selvam PN (2018) Pseudomonas fluorescens: a plant-growth-promoting Rhizobacterium (PGPR) with potential role in biocontrol of pests of crops. In: Prasad R, Gill SS, Tuteja N (eds) Crop improvement through microbial biotechnology. Elsevier, pp 221–243

Demchak RJ, Dybas RA (1997) Photostability of abamectin/zein microspheres. J Agric Food Chem 45:260–262

Deng SQ, Zou WH, Li DL et al (2019) Expression of *Bacillus thuringiensis* toxin Cyt2Ba in the entomopathogenic fungus *Beauveria bassiana* increases its virulence towards Aedes mosquitoes. PLoS Negl Trop Dis 13:e0007590

Dimetry NZ (2012) Prospects of botanical pesticides for the future in integrated pest management programme (IPM) with special reference to neem uses in Egypt. Arch Phytopathol & Plant Prot 45:1138–1161

Disi JO, Mohammad HK, Lawrence K et al (2019) A soil bacterium can shape belowground interactions between maize, herbivores and entomopathogenic nematodes. Plant Soil 437:83–92

Dorsch JA, Candas M, Griko NB et al (2002) Cry1A toxins of *Bacillus thuringiensis* bind specifically to a region adjacent to the membrane-proximal extracellular domain of BT-R1 in *Manduca sexta*: involvement of a cadherin in the entomopathogenicity of *Bacillus thuringiensis*. Insect Biochem Mol Biol 32:1025–1036

Ekbom BS (1979) Investigations on the potential of a parasitic fungus (*Verticillium lecanii*) for biological control of the greenhouse whitefly (*Trialeurodes vaporariorum*). Swedish J Agric Res 9:129–138

El-Wakeil NE (2013) Botanical pesticides and their mode of action. GesundePflanzen 65:125–149

Enan E (2001) Insecticidal activity of essential oils: octopaminergic sites of action. Comp Biochem Physiol 130C:325–337

Enright MR, Griffin CT (2005) Effects of *Paenibacillus nematophilus* on the entomopathogenic nematode *Heterorhabditis megidis*. J Invertebr Pathol 88:40–48

EPA (2006) EPA Factsheet. Accessed 14/12/2006. http://www.iaszoology.com/corcyra-cephalonica/. http://www.uniprot.org/taxonomy/176275

Erayya JJ, Sajeesh PK, Vinod U (2013) Nuclear Polyhedrosis virus (NPV), a potential biopesticide: a review. Res J Agric Forestry Sci 1:30–33

Erler F, Yalcinkaya B (2006) Repellent activity of five essential oils against *Culex pipiens*. Science Direct 77:491–494

Evans HC, Hywel-Jones NL (1997) Entomopathogenic Fungi. World Crop Pests:3–27. https://doi.org/10.1016/s1572-4379(97)80075-9

Fabrick JA, Tabashnik BE (2007) Binding of *Bacillus thuringiensis* toxin Cry1Ac to multiple sites of cadherin in pink bollworm. Insect Biochem Mol Biol 37:97–106

Fernandes EK, Rangel DE, Braga GU et al (2015) Tolerance of entomopathogenic fungi to ultraviolet radiation: a review on screening of strains and their formulation. Curr Genet 61:427–440

Ferreira T, Malan AP (2014) *Xenorhabdus* and *Photorhabdus*, bacterial symbionts of the entomopathogenic nematodes *Steinernema* and *Heterorhabditis* and their in vitro liquid mass culture: a review. Afr Entomol 22:1–14

Flipsen JT, Mans RM, Kleefsman AW et al (1995) Deletion of the baculovirus ecdysteroid UDP-glucosyl transferase gene induces early degeneration of Malpighian tubules in infected insects. J Virol 69:4529–4532

Fornazier RF, Ferreira RR, Vitoria AP et al (2002) Effects of cadmium on antioxidant enzyme activities in sugar cane. Biol Plant 45:91–97

Gao Q, Jin K, Ying SH et al (2011) Genome sequencing and comparative transcriptomics of the model entomopathogenic fungi *Metarhizium anisopliae* and *M. acridum*. PLoS Genet 7:e1001264

Geng P, Wan X, Cheng J et al (2018) vB_LspM-01: a novel myovirus displaying pseudo lysogeny in *Lysinibacillus sphaericus* C3-41. Appl Microbiol Biotech 102:10691–10702

Gibson DM, Donzelli BG, Krasnoff SB et al (2014) Discovering the secondary metabolite potential encoded within entomopathogenic fungi. Natural product rep 31:1287–1305

Glynne-Jones A (2001) *Pyrethrum*. Pestic Outlook 12:195–198

Gorashi NE, Tripathi M, Kalia V et al (2014) Identification and characterization of the Sudanese *Bacillus thuringiensis* and related bacterial strains for their efficacy against Helicoverpa armigera and *Tribolium castaneum*. 52:637–649

Grady EN, MacDonald J, Liu L et al (2016) Current knowledge and perspectives of *Paenibacillus*: a review. Microb Cell Factories 15:203

Grewal P, Lewis E, Gaugler R et al (1994a) Host finding behaviour as a predictor of foraging strategy in entomopathogenic nematodes. Parasitology 108:207–215

Grewal PS, Selvan S, Gaugler R (1994b) Thermal adaptation of entomopathogenic nematodes-niche breadth for infection, establishment and reproduction. J Therm Biol 19:245–253

Grewal PS, Ehlers RU, Shapiro-Ilan DI (2005) Nematodes as Biocontrol Agents. CABI, New York, NY

Gurr GM, Wratten SD, Altieri MA (2004) Ecological engineering: advances in habitat manipulation for arthropods. CSIRO Publishing, Mel-bourne, Vic., Australia

Guthrie FE (2013) The nature and significance of pesticide residues on tobacco and in tobacco smoke. Beitrage zur Tabakforschung. https://doi.org/10.2478/cttr-2013-0190

Henry JE (1981) Natural and applied control of insects by protozoa. Annu Rev Entomol 26:49–73

Hernández-Lambraño R, Caballero-Gallardo K, Olivero-Verbel K (2014) Toxicity and anti-feedant activity of essential oils from three aromatic plants grown in Colombia against *Euprosterna elaeasa* and *Acharia fusca* (Lepidoptera: Limacodidae). Asian Pacific J Tropic Biomed 4:695–700

Hilder VA, Gatehouse AMR, Sheerman SE et al (1987) A novel mechanism of insect resistance engineered into tobacco. Nature 300:160–163

Hirano SS, Upper CD (2000) Bacteria in the leaf ecosystem with emphasis on Pseudomonas syringae–a pathogen, ice nucleus, and epiphyte. Microbiol Mol Biol Rev 64:624–653

Hoa TD, Khac P, Uena T et al (2016) Control efficacy of pongam leaf extract against the turnip aphid. J Fac Agr Kyushu Univ 61:141–145

Hoover K, Grove M, Gardner M (2011) Gene for an extended phenotype. Sci 333:1401

Horn AS (1915) The occurrence of fungi on *Aleurodes vaporariorum* in Great Britain. Ann Appl Biol 2:109–111

Hu X, Xiao G, Zheng P et al (2014) Trajectory and genomic determinants of fungal-pathogen speciation and host adaptation. Proc Natl Acad Sci 111:16796–16801

Hua G, Jurat-Fuentes JL, Adang MJ (2004) Fluorescent-based assays establish *Manduca sexta* Bt-R1a cadherin as a receptor for multiple *Bacillus thuringiensis* Cry1A toxins in Drosophila S2 cells. Insect Biochem Mol Biol 34:193–202

Huang K-x, Xia L et al (2009) Recent advances in the biochemistry of spinosyns. Appl Microbiol Biotechnol 82:13–23

Humber RA (1984) Foundations for an evolutionary classification of the Entomophthorales (Zygomycetes). In: Wheeler Q, Blackwell M (eds) Fungus-insect relationships. Columbia University Press, New York, pp 167–183

Humber RA (2008) Evolution of entomopathogenicity in fungi. J Invertebr Pathol 98:262–266

Hussie AI, Nahed Ibrahim A, Hatem AES et al (2011) Activida dinsecticida y fijadora de nitrogeno de la bacteria transformada *Paenibacillus polymyxa* que expresa Cry1C. Rev Col Entomol 37:192–197

Ibrahim NAGA, Hussien AI, Hatem AES et al (2014) Persistence of the transformed *Paenibacillus polymyxa* expressing CRY1C in the plant leaves and its effect on chlorophyll and carotenoid. Life Sci J 11:433–442

Ironside JE (2013) Diversity and recombination of dispersed ribosomal DNA and protein coding genes in microsporidia. PLoS One 8:e55878

Isman MB (2006) Botanical insecticides, deterrents, and repellents in modern agriculture and an increasingly regulated world. Annu Rev Entomol 51:45–66

Isman MB, Grieneisen ML (2014) Botanical insecticide research: many publications, limited useful data. Trends Plant Sci 19:140–145

Jackson TA (2003) Environmental safety of inundative application of a naturally occurring biocontrol agent, *Serratia entomophila*. In: Environmental impacts of microbial insecticides. Springer, Dordrecht, pp 169–176

Jackson TA, Klein MG (2006) Scarabs as pests: a continuing problem. Coleopt Bull 60:102–119

Jackson TA, Pearson JF, Ocallaghan M et al (1992) Pathogen to product-development of *Serratia entomophila* (Enterobacteriaceae) as a commercial biological control agent for the New Zealand grass grub (*Costelytra zealandica*). In: Jackson TA, Glare TR (eds) Use of pathogens in scarab pest management. Intercept, Andover, UK, pp 191–198

Jackson TA, Boucias DG, Thaler JO (2001) Pathobiology of amber disease caused by *Serratia* spp., in the New Zealand grass grub, *Costelytra zealandica*. J Invertebr Pathol 78:232–243

Jaronski ST (2010) Ecological factors in the inundative use of fungal entomopathogens. Biol Control 55:159–185

Johnson HA, Oberlies NH, Alali FQ, McLaughlin JE (2000) Thwarting resistance: annonaceous acetogenins as new pesticidal and antitumor agents. In: Cutler SJ, Cutler JG (eds) Biological active natural products: pharmaceuticals. CRC Press, Boca Raton, pp 173–183

Johny S, Kanginakudru S, Muralirangan MC et al (2006) Morphological and molecular characterization of a new microsporidian (Protozoa: Microsporidia) isolated from *Spodoptera litura* (Fabricius) (Lepidoptera: Noctuidae). Parasitology 132:803–814

Jones VP, Steffan SA, Wiman NG et al (2011) Evaluation of herbivore-induced plant volatiles for monitoring green lacewings in Washington apple orchards. Biol Control 56:98–105

Juliya RF (2019) Biocontrol potential and genetic diversity of *Metarhizium anisopliae* lineage in agricultural habitats. J Appl Microbiol 127:556–564

Kachhawa D (2017) Microorganisms as a biopesticides. J Entomol Zool Studies 5:468–473

Kaiser D, Bacher S, Mene-Saffrane L et al (2019) Efficiency of natural substances to protect *Beauveria bassiana* conidia from UV radiation. Pest Manag Sci 75:556–563

Kalha CS, Singh PP, Kang SS et al (2014) Entomopathogenic viruses and bacteria for insect-pest control. In: Integrated Pest management. Academic Press, pp 225–244

Kanaoka M, Isogai A, Murakoshi S et al (1978) Bassianolide, a new insecticidal cyclodepsi peptide from *Beauveria bassiana* and *Verticillium lecanii*. Agri Biol Chem 42:629–635

Kao KN, Constabel F, Michayluk MR et al (1974) Plant protoplast fusion and growth of intergeneric hybrid cells. Planta 120:215–227

Kaya HK, Gaugler R (1993) Entomopathogenic nematodes. Annu Rev Entomol 38:181–206

Keller A, Brandel A, Becker MC et al (2018) Wild bees and their nests host Paenibacillus bacteria with functional potential of avail. Microbiome 6:229

Kepler RM, Rehner SA (2013) Genome-assisted development of nuclear intergenic sequence markers for entomopathogenic fungi of the *Metarhizium anisopliae* species complex. Mol Ecol Resour 13:210–217

Khambay BP, Batty D, Jewess PJ, Bateman GL, Hollomon DW (2003) Mode of action and pesticidal activity of the natural product dunnione and of some analogues. Pest Manag Sci 59:174–182

Kim B, Song GC, Ryu CM (2016) Root exudation by aphid leaf infestation recruits root-associated *Paenibacillus* spp. to lead plant insect susceptibility. J Microbiol Biotechnol 26:549–557

Klein MG (1988) Pest management of soil-inhabiting insects with microorganisms. Agric Eco Environ 24:337–349

Koppenhofer AM, Jackson TA, Klein MG (2012) Bacteria for use against soil-inhabiting insects. Manual of Techniques in Invertebrate Pathology pp 129–149

Koul O (2011) Microbial biopesticides: opportunities and challenges. CAB Rev 6:1–26

Kour D, Rana KL, Yadav N, Yadav AN, Singh J, Rastegari AA et al (2019) Agriculturally and industrially important fungi: current developments and potential biotechnological applications. In: Yadav AN, Singh S, Mishra S, Gupta A (eds) Recent advancement in white biotechnology through Fungi, Perspective for value-added products and environments, vol 2. Springer International Publishing, Cham, pp 1–64. https://doi.org/10.1007/978-3-030-14846-1_1

Kour D, Rana KL, Sheikh I, Kumar V, Yadav AN, Dhaliwal HS et al (2020a) Alleviation of drought stress and plant growth promotion by *Pseudomonas libanensis* EU-LWNA-33, a drought-adaptive phosphorus-solubilizing bacterium. Proc Natl Acad Sci India B. https://doi.org/10.1007/s40011-019-01151-4

Kour D, Rana KL, Yadav AN, Yadav N, Kumar M, Kumar V et al (2020b) Microbial biofertilizers: bioresources and eco-friendly technologies for agricultural and environmental sustainability. Biocatal Agric Biotechnol 23:101487. https://doi.org/10.1016/j.bcab.2019.101487

Kraus W (2002a) Biologically active ingredients – azadirachtin and other triterpenoids (part I). In: Schmutterer H (ed) . loc cit, pp 39–78

Kraus W (2002b) Azadirachtin and other triterpenoids - part II. In: Schmutterer H (ed) . loc cit, pp 80–111

Krishnaveni KV, Nayaki RT, Balasubramanian M (2015) Effect of *Gliricidia sepium* leaves extracts on *Aedes aegypti*: Larvicidal activity. J Phytology 7:26–31

Kumar V, Chandrashekar K, Sidhu OM (2006) Efficacy of karanjin and different extracts of *Pongamia pinnata* against selected insect pests. J Ent Res 30:103–108

Kumar AP, Amalnath D, Dutta TK (2011) Cartap poisoning: a rare case report. Indian Journal Critic Care Med 15:233

Landis DA, Wratten SD, Gurr GM (2000) Habitat management to conserve natural enemies of arthropod pests in agriculture. Annu Rev Entomol 45:175–201
Lannoo N, Van Damme EJ (2010) Nucleocytoplasmic plant lectins. Biochem Biophys Acta 1800:190–201
Lasota JA, Dybas RA (1990) Abamectin as a pesticide for agricultural use. Acta Leiden 59:217–225
Lasota JA, Dybas RA (1991) Avermectins, a novel class of compounds: implications for use in arthropod pest control. Annu Rev Entomol 36:91–117
Lawrence PK, Koundal KR (2002) Plant protease inhibitors in control of phytophagous insects. Electron J Biotechnol 5:1. https://doi.org/10.2225/vol5-issue1-fulltext-3
Leatemia JA, Isman MB (2004) Efficacy of crude seed extracts of *Annona squamosa* against *Plutella xylostella* L. in the greenhouse. Int J Pest Manag 50:129–133
Lee YS, Kim YC (2019) Tobacco growth promotion by the entomopathogenic fungus, Isaria javanica pf185. Mycobiol 47:126–133
Lee YS, Han JH, Kang BR et al (2019) Dibutyl succinate, produced by an insect-pathogenic fungus, *Isaria javanica* pf185, is a metabolite that controls of aphids and a fungal disease, anthracnose. Pest Manag Sci 75:852–858
Lewis MA, Arnason J et al (1993) Inhibition of respiration at site I by asimicin, an insecticidal acetogenin of the pawpaw, Asimina triloba (Annonaceae). Pestic Biochem Physiol 45:15–23
Lipke H, Fraenkel GS, Liener IE (1954) Effects of soybean inhibitors on growth of *Tribolium confusum*. J Sci Food Agric 2:410–415
Liu WZ, Boucias DG, McCoy CW (1995) Extraction and characterization of the insecticidal toxin hirsutellin a produced by *Hirsutella thompsonii* var. *thompsonii*. Exp Mycol 19:254–262
Liu Z, Wang X, Dai Y et al (2019) Expressing double-stranded RNAs of insect hormone-related genes enhances baculovirus insecticidal activity. Int J Mol Sci 20:419
Loper JE, Hassan KA, Mavrodi DV et al (2012) Comparative genomics of plant-associated *Pseudomonas* spp.: insights into diversity and inheritance of traits involved in multitrophic interactions. PLoS Genet 8:e1002784
Lopes RB, Faria M (2019) Influence of two formulation types and moisture levels on the storage stability and insecticidal activity of *Beauveria bassiana*. Biocontrol Sci Tech 29:437–450
Lozano LC, Ayala JA, Dussan J (2011) *Lysinibacillus sphaericus* S-layer protein toxicity against Culex quinquefasciatus. Biotechnol Lett 33:2037–2041
Macedo ML, Oliveira CF, Oliveira CT (2015) Insecticidal activity of plant lectins and potential application in crop protection. Molecules 20:2014–2033
Maimala S, Tartar A et al (2002) Detection of the toxin Hirsutellin A from *Hirsutella thompsonii*. J Invertebr Pathol 80:112–126
Mannino MC, Huarte-Bonnet C, Davyt-Colo B et al (2019) Is the insect cuticle the only entry gate for fungal infection? Insights into alternative modes of action of entomopathogenic fungi. J Fungi 5:33
Martin PA, Gundersen-Rindal D, Blackburn M et al (2007) Chromo bacterium subtsugae sp. nov., a betaproteo bacterium toxic to Colorado potato beetle and other insect pests. Int J Syst Evol Microbiol 57:993–999
Mascarin GM, Jaronski ST (2016) The production and uses of *Beauveria bassiana* as a microbial insecticide. World J Microbiol Biotech 32:177
Mashtoly TA, Abolmaaty A, Thompson N et al (2010) Enhanced toxicity of *Bacillus thuringiensis japonensis* strain Buibui toxin to oriental beetle and northern masked chafer (Coleoptera: Scarabaeidae) larvae with *Bacillus* sp. NFD2. J Econ Entomol 103:1547–1554
Mathew LG, Ponnuraj J, Mallappa B et al (2018) ABC transporter mis-splicing associated with resistance to Bt toxin Cry2Ab in laboratory-and field-selected pink bollworm. Sci Rep 8:13531
McCoy C, Quintela ED, De Faria M (2000) Environmental persistence of entomopathogenic fungi. In: Baur ME, Fuxa JR (eds) Factors affecting the survival of entomopathogens. Lousiana State University Agricultural Center, Southern Cooperative Series Bulletin, p 400
McLaughlin JL, Zeng L, Oberlies NH et al (1997) Annonaceous acetogenins as new natural pesticides: recent progress. In: Hedin PA, Hollingworth RM, Masler EP et al (eds) Phytochemicals for pest control. American Chemical Society, Washington, DC, pp 117–133

Melo ALDA, Soccol VT, Soccol CR (2016) *Bacillus thuringiensis*: mechanism of action, resistance, and new applications: a review. Crit Rev Biotechnol 36:317–326

de Mendonca Santos EM, de Melo Chalegre KD, de Albuquerque AL et al (2019) Frequency of resistance alleles to *Lysinibacillus sphaericus* in a *Culex quinquefasciatus* population treated with a *L. sphaericus*/Bti bio larvicide. Biol Control 132:95–101

Mertz FP, Yao RC (1990) *Saccharopolyspora spinosa* sp. nov. isolated from soil collected in a sugar mill rum still. Int J Systematic Bacteriol 40:34–39. https://doi.org/10.1099/00207713-40-1-34

Michereff MFF, Borges M, Aquino MFS et al (2016) The influence of volatile semiochemicals from stink bug eggs and oviposition-damaged plants on the foraging behaviour of the egg parasitoid *Telenomus podisi*. Bulletin Entomol Res 106:663–671

Midega CA, Murage AW, Pittchar JO et al (2016) Managing storage pests of maize: Farmers' knowledge, perceptions and practices in western Kenya. Crop Prot 90:142–149

Mikolajczak KL, McLaughlin JL, Rupprecht JK (1988) Control of Pests with Annonaceous Acetogenins. (divisional patent on asimicin) U.S. Patent No. 4,855,319

Moeschler HF, Pfuger W, Wendlisch D (1987) Pure annonin and a process for the preparation thereof. US Patent No 4:689,323

Morgan ED (2009) Azadirachtin, a scientific gold mine. Bioorg Med Chem 17:4096–4105

Mossa ATH, Mohafrash SM, Chandrasekaran N (2018) Safety of natural insecticides: toxic effects on experimental animals. Biomed Res Int 2018:4308054

Naureen Z, Rehman NU, Hussain H et al (2017) Exploring the potentials of *Lysinibacillus sphaericus* ZA9 for plant growth promotion and biocontrol activities against phytopathogenic fungi. Front Microbiol 8:1477

Nawaz M, Mabubu JI, Hua H (2016) Current status and advancement of biopesticides microbial and botanical pesticides. J Entomol Zool Stud 4:241–246

Nishiyama K, Tsuruta K, Ikeda M et al (2016) Sensitive electrochemical detection of Nereistoxin by reductive desorption from au (111) and au (100). Electrochemistry 84:349–353

Nukmal N, Rosa E, Apriliyani et al (2017) Insecticidal effects of the flavonoid-rich fraction of leaves extract of Gamal (*Gliricidia sepium*) on the coffee Mealybugs (*Planococcus citri* Risso.). ARRB 16:1–9

Okonkwo CO, Ohaeri OC (2018) Essential oils from the leaves of *Euphorbia milieu* exert insecticidal activity through disruption in ionic composition. IOSR J Pharma Biol Sci 13: 46–9

Oliveira-Hofman C, Kaplan F, Stevens G et al (2019) Pheromone extracts act as boosters for entomopathogenic nematodes efficacy. J Invertebr Pathol 164:38–42

Oostendorp M, Sikora RA (1989) Seed treatment with antagonistic rhizobacteria for the suppression of *Heterodera schachtii* early root infection of sugar beet. Rev Nematol 12:77–83

Oostendorp M, Sikora RA (1990) In vitro interrelationships between rhizosphere bacteria and *Heterodera schachtii*. Rev Nematol 13:269–274

Ortiz-Urquiza A, Riveiro-Miranda L, Santiago-Alvarez C et al (2010) Insect-toxic secreted proteins and virulence of the entomopathogenic fungus *Beauveria bassiana*. J Invertebr Pathol 105:270–278

Oulevey C, Widmer F, Kolliker R et al (2009) An optimized microsatellite marker set for detection of *Metarhizium anisopliae* genotype diversity on field and regional scales. Mycol Res 113:1016–1024

Pathma J, Sakthivel N (2013) Molecular and functional characterization of bacteria isolated from straw and goat manure based vermicompost. Appl Soil Ecol 70:33–47

Pathma J, Ayyadurai N, Sakthivel N (2010a) Assessment of genetic and functional relationship of antagonistic fluorescent pseudomonads of rice rhizosphere by repetitive sequence, protein coding sequence and functional gene analyses. J Microbiol 48:715–727

Pathma J, Kamaraj Kennedy R, Sakthivel N (2010b) Mechanisms of fluorescent pseudomonads that mediate biological control of phytopathogens and plant growth promotion of crop plants. In: Maheshwari DK (ed) Bacteria in agro-biology: plant growth. Springer-Verlag, Berlin Heidelberg, pp 77–105

Pathma J, Rahul GR, Kamaraj Kennedy R, Subashri R, Sakthivel N (2011) Secondary metabolite production by bacterial antagonists. J Biol Control 25:165–181

Pathma J, Raman G, Kennedy RK et al (2019a) Recent advances in plant-microbe interaction. In: Sharma SG, Sharma M, Sharma NR (eds) Microcosm: microbial diversity, interventions and scope. Springer International. (in Press)

Pathma J, Raman G, Sakthivel N (2019b) Microbiome of rhizospheric soil and vermicompost and their applications in soil fertility, pest and pathogen management for sustainable agriculture. In: Panpatte DG, Jhala YK (eds) Soil fertility Management for Sustainable Development. Springer, Nature Singapore, pp 189–210

Pathma J, Kennedy RK BLS et al (2020) Microbial biofertilizers and biopesticides: Nature's assets fostering sustainable agriculture. In: Panwar JS, Prasad R (eds) Recent developments in microbial technologies. Springer. (in Press)

Pedras MSC, Zaharia LI, Ward DE (2002) The destruxins: synthesis, biosynthesis, biotransformation, and biological activity. Phytochemistry 59:579–596

Pedrini N (2018) Molecular interactions between entomopathogenic fungi (Hypocreales) and their insect host: perspectives from stressful cuticle and hemolymph battlefields and the potential of dual RNA sequencing for future studies. Fungal Biol 122:538–545

Peumans WJ, VanDamme EJM (1995) Lectins as PlantDefense proteins. Plant Physiol 109:347–352

Pigott CR, Ellar DJ (2007) Role of receptors in Bacillus thuringiensis crystal toxin activity. Microbiol Mol Biol Rev 71:255–281

Pittarelli GW, Buta JG, Neal JW Jr, et al (1993) Biological pesticide derived from *Nicotiana* Plants. U.S. Patent No. 5,260,281

Pitterna TC, Jerome H, Ottmar FJ et al (2009) New ventures in the chemistry of avermectins. Bioorg Med Chem 17:4085–4095

Plata-Rueda A, Martínez LC, Santos MHD et al (2017) Insecticidal activity of garlic essential oil and their constituents against the mealworm beetle, *Tenebrio molitor* Linnaeus (Coleoptera: Tenebrionidae). Sci Rep 7:46406

Pradhan S, Jotwani MG, Rai BK (1962) The neem seed deterrent to locust. Indian Farm 12:7–11

Prado END, Iannacone J, Gomez H (2008) Effect of two entomopathogenic fungi in controlling *Aleurodicus cocois* (Curtis 1846) (Hemiptera: Aleyrodidae). Chil J Agric Res 68:21–30

Rai PK, Singh M, Anand K, Saurabhj S, Kaur T, Kour D et al (2020) Role and potential applications of plant growth promotion rhizobacteria for sustainable agriculture. In: Rastegari AA, Yadav AN, Yadav N (eds) Trends of microbial biotechnology for sustainable agriculture and biomedicine systems: diversity and functional perspectives. Elsevier, Amsterdam, pp 49–60. https://doi.org/10.1016/B978-0-12-820526-6.00004-X

Rajashekar Y, Ravindra KV, Bakthavatsalam N (2014) Leaves of *Lantana camara* Linn. (Verbenaceae) as a potential insecticide for the management of three species of stored grain insect pests. J Food Sci Technol 51:3494–3499

Rastegari AA, Yadav AN, Yadav N (2020a) New and future developments in microbial biotechnology and bioengineering: Trends of microbial biotechnology for sustainable agriculture and biomedicine systems: diversity and functional perspectives. Elsevier, Amsterdam

Rastegari AA, Yadav AN, Yadav N (2020b) New and future developments in microbial biotechnology and bioengineering: Trends of microbial biotechnology for sustainable agriculture and biomedicine systems: perspectives for human health. Elsevier, Amsterdam

Regnault-Roger C, Philogene BJR (2008) Past and current prospects for the use of botanicals and plant allelochemicals in integrated pest management. Pharm Biol 46:41–52

Reisig DD, Kurtz R (2018) Bt resistance implications for *Helicoverpa zea* (Lepidoptera: Noctuidae) insecticide resistance management in the United States. Environ Entomol 47:1357–1364

Rezende TMT, Rezende AM, Wallau GL et al (2019) A differential transcriptional profile by *Culex quinquefasciatus* larvae resistant to *Lysinibacillus sphaericus* IAB59 highlights genes and pathways associated with the resistance phenotype. Parasit Vectors 12(1):407

Richardson MJ (1991) Seed storage proteins: the enzyme inhibitors. In: Richardson MJ (ed) Methods in plant biochemistry. Academic Press, New York, pp 259–305

Robertson DR, Smith-Vaniz WF (2008) Rotenone: an essential but demonized tool for assessing marine fish diversity. Bioscience 58:165–170

Robles-Acosta IN, Chacon-Hernandez JC, Torres-Acosta RI et al (2019) Entomopathogenic fungi as biological control agents of *Phyllocoptruta oleivora* (Prostigmata: Eriophyidae) under greenhouse conditions. Fla Entomol 102:303–308
Roh JY, Choi JY, Li MS (2007) *Bacillus thuringiensis* as a specific, safe, and effective tool for insect pest control. Microb Biotechnol 17:547
Ruiu L (2015) Insect pathogenic bacteria in integrated pest management. Insects 6:352–367
Ruiu L (2018) Microbial biopesticides in agroecosystems. Agronomy 8:235
Sahay H, Yadav AN, Singh AK, Singh S, Kaushik R, Saxena AK (2017) Hot springs of Indian Himalayas: potential sources of microbial diversity and thermostable hydrolytic enzymes. 3. Biotech 7:1–11
Samson RA, Evans HC, Latge JP (eds) (1988) Atlas of Entomopathogenic Fungi. Springer-Verlag, Berlin
Sandhu SS, Mishra M (1994) Larvicidal activity of fungal isolates *Beaveria bassiana, Metarhizium anisopliae* and *Aspergillus flavus* against mosquito sp. Culex pipiens. In: Proc Natl Symp Advan bio control insect pests, pp 145–150
Santhi A, Sivakumar CV (1995) Biocontrol potential of *Pseudomonas fluorescens* (Migula) against root-knot nematode, *Meloidogyne incognita* (Kofoid and white, 1919) Chitwood, 1949 on tomato. J Bio Control 9:113–115
Sarker S, Lim UT (2018) Extract of *Nicotiana tabacum* as a potential control agent of *Grapholita molesta* (Lepidoptera: Tortricidae). PLoS One 13(8):e0198302
Saxena AK, Yadav AN, Rajawat M, Kaushik R, Kumar R, Kumar M et al (2016) Microbial diversity of extreme regions: an unseen heritage and wealth. Indian J Plant Genet Resour 29:246–248
Saxena AK, Padaria JC, Gurjar GT, Yadav AN, Lone SA, Tripathi M et al (2020) Insecticidal formulation of novel strain of *Bacillus thuringiensis* AK 47. Indian Patent 340541
Schmidt GH, Rembold H, Ahmed AAI et al (1998) Effect of *Melia azedarach* fruit extract on juvenile hormone titer and protein content in the hemolymph of two species of noctuid lepidopteran larvae (Insecta: Lepidoptera: Noctuidae). Phytoparasitica 26:283–292
Schmutterer H (ed) (2002) The neem tree. Neem Found, Mumbai
Schrank A, Vainstein MH (2010) Metarhizium anisopliae enzymes and toxins. Toxicon 56:1267–1274
Schunemann R, Knaak N, Fiuza LM (2014) Mode of action and specificity of *Bacillus thuringiensis* toxins in the control of caterpillars and stink bugs in soybean culture. ISRN Microbiol 2014:135675
Schwartz JL, Laprade R (2000) Membrane permeabilisation by *Bacillus thuringiensis* toxins: protein insertion and pore formation. In: Charles JF, Delecluse A, Nielsen-LeRoux C (eds) Entomopathogenic bacteria: from laboratory to field application. Kluwer Associate Publishing, Norwell, MA, pp 199–218
Senthil-Nathan S (2006) Effects of *Melia azedarach* on nutritional physiology and enzyme activities of the rice leaf folder *Cnaphalocrocis medinalis* (Guenée) (Lepidoptera: Pyralidae). Pestic Biochem Physiol 84:98–108
Simpson M, Gurr GM, Simmons AT et al (2011a) Insect attraction to synthetic herbivore-induced plant volatile-treated field crops. Agric For Entomol 13:45–57
Simpson M, Gurr GM, Simmons AT et al (2011b) Field evaluation of the 'attract and reward' biological control approach in vineyards. Ann Appl Biol 159:69–78
Singh J, Yadav AN (2020) Natural bioactive products in sustainable agriculture. Springer, Singapore
Singh AK, Ghodke I, Chhatpar HS (2009) Pesticide tolerance of *Paenibacillus* sp. D1 and its chitinase. J Environ Manag 91:358–362
Singh A, Kumar R, Yadav AN, Mishra S, Sachan S, Sachan SG (2020) Tiny microbes, big yields: microorganisms for enhancing food crop production sustainable development. In: Rastegari AA, Yadav AN, Yadav N (eds) Trends of microbial biotechnology for sustainable agriculture and biomedicine systems: diversity and functional perspectives. Elsevier, Amsterdam, pp 1–15. https://doi.org/10.1016/B978-0-12-820526-6.00001-4

Snyder DE, Meyer J, Zimmermann AG et al (2007) Preliminary studies on the effectiveness of the novel pulicide, spinosad, for the treatment and control of fleas on dogs. Veter Parasitol 150:345–351

Soares-da-Silva J, Pinheiro VCS, Litaiff-Abreu E et al (2015) Isolation of *Bacillus thuringiensis* from the state of Amazonas, in Brazil, and screening against *Aedesa egypti* (Diptera, Culicidae). Revista Brasileira de Entomol 59:1–6

Soman AG, Gloer JB, Angawi RF et al (2001) Vertilecanins: new phenopicolinic acid analogues from *Verticillium lecanii*. J Nat Prod 64:189–192

Steiner S, Hermann N, Ruther J (2006) Characterization of a female-produced courtship pheromone in the parasitoid *Nasonia vitripennis*. J Chem Ecol 32:1687–1702

Stevenson PC, Nyirenda SP, Mvumi BM et al (2012) Pesticidal plants: a viable alternative insect pest management approach for resource-poor farming in Africa. In: Koul O, Dhaliwal GS, Khokhar S et al (eds) Biopesticides in environment and food security: issues and strategies. Scientific Publishers, India, pp 212–238

Stock CA, McLoughlin TJ, Klein JA et al (1990) Expression of *Bacillus thuringiensis* crystal protein gene in *Pseudomonas cepacia* 526. Can J Microbiol 36:879–884

Stoger E, Williams S, Christou P et al (1999) Expression of the insecticidal lectin from snowdrop (*Galanthusnivalis agglutinin*; GNA) in transgenic wheat plants: effects on predation by the grain aphid *Sitobion avenae*. Mol Breed 5:65–73

Suman A, Yadav AN, Verma P (2016) Endophytic microbes in crops: diversity and beneficial impact for sustainable agriculture. In: Singh D, Abhilash P, Prabha R (eds) Microbial inoculants in sustainable agricultural productivity. Research Perspectives. Springer-Verlag, India, pp 117–143. https://doi.org/10.1007/978-81-322-2647-5_7

Thacker JMR (2002) An introduction to arthropod pest control. Cambridge University Press, Cambridge, p 343

Thakur N, Kaur S, Tomar P, Thakur S, Yadav AN (2020) Microbial biopesticides: current status and advancement for sustainable agriculture and environment. In: Rastegari AA, Yadav AN, Yadav N (eds) Trends of microbial biotechnology for sustainable agriculture and biomedicine systems: diversity and functional perspectives. Elsevier, Amsterdam, pp 243–282. https://doi.org/10.1016/B978-0-12-820526-6.00016-6

Thomsen L, Eilenberg J (2000) Time-concentration mortality of *Pieris brassicae* (Lepidoptera: Pieridae) and *Agrotis segetum* (Lepidoptera: Noctuidae) larvae from different destruxins. Environ Entomol 29:1041–1047

Tounou AK, Kooyman C, Douro-Kpindou OK et al (2008) Combined field efficacy of *Paranosema locustae* and *Metarhizium anisopliae* var. *acridum* for the control of Sahelian grasshoppers. BioControl 53:813

Uma MS, Prasanna PM, Manjunathareddy GV et al (2009) Efficacy of some Euphorbiaceae plant extracts against cabbage diamondback moth, *Plutella xylostella* L. Karnataka J Agric Sci 22:688–689

Usha Rani P, Hymavathi A, Suresh Babu K et al (2013) Bioactivity evaluation of prenylated isoflavones derived from *Derris scandens* Benth against two stored pest larvae. J Bio pesticides 6:14–21

Vachon V, Laprade R, Schwartz JL (2012) Current models of the mode of action of *Bacillus thuringiensis* insecticidal crystal proteins: a critical review. J Invertebr Pathol 111:1–12

Van Damme EJM, Allen AK, Peumans WJ (1987) Isolation and characterization of a lectin with exclusive specificity towards mannose from snowdrop (*Galanthus nivalis*) bulbs. FEBS Lett 215:140–114

Vega FE (2008) Insect pathology and fungal endophytes. J Invertebr Pathol 98:277–279

Verma P, Yadav AN, Kazy SK, Saxena AK, Suman A (2013) Elucidating the diversity and plant growth promoting attributes of wheat (*Triticum aestivum*) associated acidotolerant bacteria from southern hills zone of India. Natl J Life Sci 10:219–227

Verma P, Yadav AN, Khannam KS, Panjiar N, Kumar S, Saxena AK et al (2015) Assessment of genetic diversity and plant growth promoting attributes of psychrotolerant bacteria allied with wheat (*Triticum aestivum*) from the northern hills zone of India. Ann Microbiol 65:1885–1899

Verma P, Yadav AN, Khannam KS, Kumar S, Saxena AK, Suman A (2016) Molecular diversity and multifarious plant growth promoting attributes of bacilli associated with wheat (*Triticum aestivum* L.) rhizosphere from six diverse agro-ecological zones of India. J Basic Microbiol 56:44–58

Verma P, Yadav AN, Kumar V, Khan MA, Saxena AK (2018) Microbes in termite management: potential role and strategies. In: Khan MA, Ahmad W (eds) Termites and sustainable management: volume 2 - economic losses and management. Springer International Publishing, Cham, pp 197–217. https://doi.org/10.1007/978-3-319-68726-1_9

Verma P, Yadav AN, Khannam KS, Mishra S, Kumar S, Saxena AK et al (2019) Appraisal of diversity and functional attributes of thermotolerant wheat associated bacteria from the peninsular zone of India. Saudi J Biol Sci 26:1882–1895. https://doi.org/10.1016/j.sjbs.2016.01.042

Viggor S, Juhanson J, Joesaar M et al (2013) Dynamic changes in the structure of microbial communities in Baltic Sea coastal seawater microcosms modified by crude oil, shale oil or diesel fuel. Microbiol Res 168:415–427

Vodovar N, Vinals M, Liehl P et al (2005) *Drosophila* host defense after oral infection by an entomopathogenic *Pseudomonas* species. Proc Natl Acad Sci U S A 102:11414–11419

Wahengbam J, Raut AM, Mandal SK (2018) Efficacy of new generation insecticides against *Trichogramma chilonis* Ishii and *Trichogramma pretiosum* Riley. Mortality 10:100

Wang M, Hu Z (2019) Cross-talking between baculoviruses and host insects towards a successful infection. Philos T R Soc B 374:20180324

Ware GW (1983) Pesticides. Theory and application. Freeman, San Francisco, p 308

Weller DM, Raaijmakers JM, Gardener BBM, Thomashow LS (2002) Microbial populations responsible for specific soil suppressiveness to plant pathogens. Annu Rev Phytopathol 40:309–348

Williams CE, Collier CC, Nemacheck JA et al (2002) A lectin-like wheat gene responds systemically to attempted feeding by avirulent first-instar hessian fly larvae. J Chem Ecol 28:1411–1428

Woessner JF (2013) Carboxy peptidase MeCPA. In: Rawlings ND, Salvesen G (eds) Handbook of Proteolytic enzymes. Academic Press, pp 1329–1331

Wraight SP, Carruthers RI, Jaronski ST et al (2000) Evaluation of the entomopathogenic fungi *Beauveria bassiana* and *Paecilomyces fumosoroseus* for microbial control of the silver leaf whitefly, *Bemisia argentifolii*. Biol Control 17:203–217

Xiao G, Ying SH, Zheng P et al (2012) Genomic perspectives on the evolution of fungal entomopathogenicity in *Beauveria bassiana*. Sci Rep 2:483

Xu XM, Jeffries P, Pautasso M et al (2011) Combined use of biocontrol agents to manage plant diseases in theory and practice. Phytopathology 101:1024–1031

Yadav AN (2019) Fungal white biotechnology: conclusion and future prospects. In: Yadav AN, Singh S, Mishra S, Gupta A (eds) Recent advancement in white biotechnology through Fungi, Perspective for sustainable environments, vol 3. Springer International Publishing, Cham, pp 491–498. https://doi.org/10.1007/978-3-030-25506-0_20

Yadav AN, Sachan SG, Verma P, Saxena AK (2015a) Prospecting cold deserts of north western Himalayas for microbial diversity and plant growth promoting attributes. J Biosci Bioeng 119:683–693. https://doi.org/10.1016/j.jbiosc.2014.11.006

Yadav AN, Sachan SG, Verma P, Tyagi SP, Kaushik R, Saxena AK (2015b) Culturable diversity and functional annotation of psychrotrophic bacteria from cold desert of Leh Ladakh (India). World J Microbiol Biotechnol 31:95–108. https://doi.org/10.1007/s11274-014-1768-z

Yadav AN, Verma P, Kumar M, Pal KK, Dey R, Gupta A et al (2015c) Diversity and phylogenetic profiling of niche-specific bacilli from extreme environments of India. Ann Microbiol 65:611–629

Yadav AN, Sachan SG, Verma P, Saxena AK (2016) Bioprospecting of plant growth promoting psychrotrophic bacilli from cold desert of north western Indian Himalayas. Indian J Exp Biol 54:142–150

Yadav AN, Kumar R, Kumar S, Kumar V, Sugitha T, Singh B et al (2017) Beneficial microbiomes: biodiversity and potential biotechnological applications for sustainable agriculture and human health. J Appl Biol Biotechnol 5:45–57. https://doi.org/10.7324/JABB.2017.50607

Yadav AN, Verma P, Kumar S, Kumar V, Kumar M, Singh BP et al. (2018) Actinobacteria from Rhizosphere: molecular diversity, distributions and potential biotechnological applications. In: Singh B, Gupta V, Passari A (eds) New and future developments in microbial biotechnology and bioengineering. Elsevier, Amsterdam, pp 13–41. https://doi.org/10.1016/B978-0-444-63994-3.00002-3

Yadav AN, Gulati S, Sharma D, Singh RN, Rajawat MVS, Kumar R et al (2019) Seasonal variations in culturable archaea and their plant growth promoting attributes to predict their role in establishment of vegetation in Rann of Kutch. Biologia 74:1031–1043. https://doi.org/10.2478/s11756-019-00259-2

Yadav AN, Mishra S, Kour D, Yadav N, Kumar A (2020a) Agriculturally important Fungi for sustainable agriculture, volume 1: perspective for diversity and crop productivity. Springer International Publishing, Cham

Yadav AN, Mishra S, Kour D, Yadav N, Kumar A (2020b) Agriculturally important Fungi for sustainable agriculture, volume 2: functional annotation for crop protection. Springer International Publishing, Cham

Yadav AN, Singh J, Rastegari AA, Yadav N (2020c) Plant microbiomes for sustainable agriculture. Springer, Cham

Yamamoto C, Kurokawa M (1970) Synaptic potentials recorded in brain slices and their modification by changes in the level of tissue ATP. Exp Brain Res 10:159–170

Yeo H (2000) Mycoinsecticides for aphid management: a biorational approach (doctoral dissertation). University of Nottingham, Nottingham

Yokoyama T, Tanaka M, Hasegawa M (2004) Novel cry gene from *Paenibacillus lentimorbus* strain Semadara inhibits ingestion and promotes insecticidal activity in *Anomala cuprea* larvae. J Invertebr Pathol 85:25–32

Zhang J, Hodgman TC, Krieger L et al (1997) Cloning and analysis of the first cry gene from *Bacillus popilliae*. J Bacteriol 179:4336–4341

Zhang Q, Li P, Wang G et al (1998) On the biochemical mechanism of induced resistance of cotton to cotton bollworm by cutting off young seedling at plumular axis. Acta Phytophylacica Sinica 25:209–212

Zhang W, Peumans WJ, Barre A et al (2000) Isolation and characterization of a jacalin-related mannose-binding lectin from salt-stressed rice (*Oryza sativa*) plants. Planta 210:970–978

Zhou L, Alphey N, Walker AS et al (2018) Combining the high-dose refuge strategy and self-limiting transgenic insects in resistance management - a test in experimental mesocosms. Evol Appl 11:727–738

Bacillus thuringiensis as Potential Biocontrol Agent for Sustainable Agriculture

18

Manoj Baliram Pohare, Sopan Ganpatrao Wagh, and Varatharajalu Udayasuriyan

Abstract

The global agricultural production needs to go up by 70% until 2050, to keep pace with ever-increasing human population and is a major challenge of this century. Green revolution and heavy use of chemical pesticides have helped to achieve feeding objective in the past few decades by developing high-yielding cultivars and reducing yield losses due to weeds, pests, and diseases. The use of biocontrol agents is almost a century-old concept in agriculture, but their applications were limited due to the development of highly effective chemical pesticides in the past half a dozen decades. However, chemical insecticides have prolonged leaching residual effects, leading to environmental damages and adverse effects on other organisms involved in the biogeochemical cycle. Besides, excessive use of chemical pesticides generates tolerance in pests. Therefore, safer pest management options are required to preserve environmental sustainability. The use of biocontrol agents with specificity in their targets is one of the widespread pest control approaches in plant health management. Recently, increased demand for biocontrol agents is mainly the result of the changed perception of human–society preferring a healthy environment and safety over the effectiveness with harmful side effects. Most successful of all the

M. B. Pohare (✉)
Department of Infection, Immunity and Cardiovascular Diseases, University of Sheffield, Sheffield, UK

S. G. Wagh
Department of Bio-Science and Technology, Agri-biotech College, VNMK University, Aurangabad, MS, India

V. Udayasuriyan
Department of Plant Biotechnology, Centre for Plant Molecular Biology & Biotechnology, Tamil Nadu Agricultural University, Coimbatore, India

A. N. Yadav et al. (eds.), *Current Trends in Microbial Biotechnology for Sustainable Agriculture*, Environmental and Microbial Biotechnology,
https://doi.org/10.1007/978-981-15-6949-4_18

biocontrol agents, *Bacillus thuringiensis* and its potential as a biocontrol agent, is discussed in this chapter.

Keywords

Bacillus thuringiensis · Biocontrol agent · *Bt* formulations · *Bt* transgenic

18.1 Introduction

The world's population was estimated to be about one billion in the early 1800s. In the 1920s, it rose to two billion and, at the end of the last century, to six billion. With this alarming speed, the world population will grow from the current 7.7 billion, by the mid-2050s to about 10 billion (Hickey et al. 2019). To keep pace with this ever-increasing human population, the world's agricultural production needs to go up by 70% until 2050. Additionally, all of it must be carried out without affecting the environment and in the presence of depleting resources of land and water. Moreover, agricultural production should be accompanied by preserving natural resources as well as it should take care of the livelihood of the farmers all over the world (Pohare et al. 2017).

Global intensified efforts are now required for agricultural production with clear future objectives in mind. One can achieve this objective either by increasing agricultural production by using high-yielding cultivars or by reducing the losses caused by weeds, pests, and diseases. We are making reasonable efforts to produce the food at the required pace but still losing a bulk of its share, which can be up to 40% due to insect damage (Oerke and Dehne 1997; Anderson 2010). A hefty amount of money is still being spent worldwide on the control of agricultural pests (Pimentel and Burgess 2014). Additionally, the dependence on chemical insecticides may not be a viable option as they come with detrimental environmental hazards (Davies et al. 2012; Liu et al. 2016). Thus, we do need an eco-friendly pest control option, which can maintain the environmental sustainability. The use of biological pesticides, as one of the parts of integrated pest management (IPM) strategies, has been receiving worldwide popularity, as detrimental effects of chemical pesticides both "immediate and long term" are no longer acceptable to the environmentally conscious advanced human–society.

Biocontrol has a history of more than 100 years. Biological substitutes to chemical pesticides mainly include a wide range of biopesticides, like plant extracts, microbial (fungal, bacterial, protozoa, and viral), and bioactive molecules (metabolites) (Kour et al. 2020; Verma et al. 2019). The bacterium *Bacillus thuringiensis* Berliner, commonly known as *Bt*, is one of the most successful biocontrol agents evaluated until now. *B. thuringiensis* have been reported from diverse habitats worldwide such as plant microbiomes (endophytes, rhizospheric, phyllospheric) (Verma et al. 2014; Verma et al. 2016; Verma et al. 2015), bulk soil (Biswas et al. 2018) as well as from different extreme habitats including low temperature (Yadav et al. 2016a; Yadav et al. 2015a, 2016b), high temperature (Sahay et al. 2017), hypersaline environment (Saxena et al. 2016; Yadav and Saxena 2018), acidic soil (Verma et al. 2017), and water-deficient region (Yadav et al. 2015b). *B. thuringiensis* (*Bt*) is a Gram-positive, sporulating, soil bacterium that develops confined pesticide components as inclusion bodies or parasporal crystals (Cry and Cyt proteins) during the sporulation phase as well as vegetative insecticidal proteins (Vips) that are secreted into the culture medium during the vegetative growth phase. Upon sporulation, crystalline (Cry) proteins separated as protoxins mediate unique pathogenicity against insects (Schnepf et al. 1998). *Bt* is the key biopesticides due to its benefits, such as high specificity against the target pests, eco-friendly nature (biodegradable), and safe to nontarget organisms.

Over the past couple of decades, *Bt* has been successfully deployed as *Bt* formulations to check the prevalence of various agricultural insect pests (Schnepf et al. 1998). Recently, new advancements in biotechnology have been used to incorporate *Bt* toxin genes into plants so that each plant part can inherently possess the insecticidal toxins and thus tolerate the lepidopteran pests effectively (Carriere et al. 2015; Liu et al. 2018; Zotti et al. 2018). *Bt* products, however, constituted less than 1% of the global market for insecticides until the Cry toxin gene was incorporated into plants and marketed in 1996. Now it has become a major pest control measure, which was once a minor insecticide, with transgenic *Bt* crops expressing *cry* gene being planted globally on 101 million hectares in 2017 (ISAAA 2017). However, in response to the long-term exposure to the *Bt* toxin, target insects evolved themselves so that they become tolerant to the *Bt* toxicity (Tabashnik et al. 2013). As a solution, scientists are screening new *Bt* isolates and formulating novel tactics for delaying resistance to *Bt* crops (summarized in Tabashnik and Carrière 2017). The most commonly used strategies for novel *Bt* gene screening are based on the PCR method along with high-performance sequencing, which can be very helpful in introspecting the presence of different novel *cry* genes in *Bt* for further biocontrol studies (Sajid et al. 2018).

Recent Cry protein classification was given by Crickmore et al. (1998) and is mainly dependent on the amino acid sequence homology. To the extent that we know more than 993 *Bt* genes, encoding the toxic insecticidal toxins were identified and characterized as a result of an ongoing global effort (Xiao and Wu 2019). Only a few of these toxins, however, are highly insect toxic and have been marketed. Several agriculturally important pests still exist, which are not targeted by the available discovered *Bt* toxins. The control of these pests can be supplemented by either screening new highly toxic wild *Bt* isolates or by making synthetic/chimeric Cry proteins with increased activity (Udayasuriyan et al. 2006). Cry toxin mutagenesis was exploited to a larger extent to produce new recombinant toxins. Such recombinant Cry toxins made by mutations in the known *cry* genes and/or synthetic one could lead to a new protein that has significantly different level of toxicity and specificity to its target group due to sequence variations (Udayasuriyan et al. 1994; Sasaki et al. 1997; Chandra et al. 1999; Abdullah et al. 2003; Karlova et al. 2005; Huang et al. 2012; Chougule et al. 2013). In addition, understanding of Cry toxin, its structure, and receptors will help to explore the better options for IPM (Reinoso et al. 2016). In this chapter, most successful of all the biocontrol agents, *B. thuringiensis*, its diversity and potential as a biocontrol agent are discussed.

18.2 Background

Bt belongs to *Bacillus* genus and can produce endospores that can tolerate heat. Further, *Bt* is similar in shape and morphology to the other members of the genus (Stahly et al. 1991). *Bt* is an anaerobic (facultative though), gram-positive with rod type body. Widthwise the rod-shaped bacterial cells range from 0.5 to 1.0 μm when grown on liquid culture media. *Bt* colonies grown on solid media usually are white, rounded, wavy, or slender edges, with a fried egg-type appearance and are elevated (Astuti et al. 2018) (Fig. 18.1). *Bt* is almost indistinguishable from its most closely related species like *Bacillus cereus, Bacillus anthracis*, etc., except that it can produce the parasporal crystals during the process of endospore formation (Andrews et al. 1985, 1987). They are so similar that Gordon et al. (1973) thought *Bt* is just a crystal producing form of *B. cereus.* Even there are claims reported in the literature that *Bt, B. cereus, and B. anthracis* should be considered as a single species, based on the 16sRNA analysis (Carlson et al. 1994, 1996; Bourque et al. 1995; Helgason et al. 2000).

18.3 Developments of *Bt* Research

Bt was recovered from an ill *Bombyx mori* larvae (silkworm) by Ishiwata for the first time in 1901 and called it as *Bacillus sotto*. After a decade, Berliner had recovered a similar kind of *Bacillus* from *Anagasta kuehniella* larvae (flour moth), and he called it as *B. thuringiensis* (Cannon 1995), which is derived from a word Thuringia, a name of the village where this diseased Mediterranean moth was found (Melo

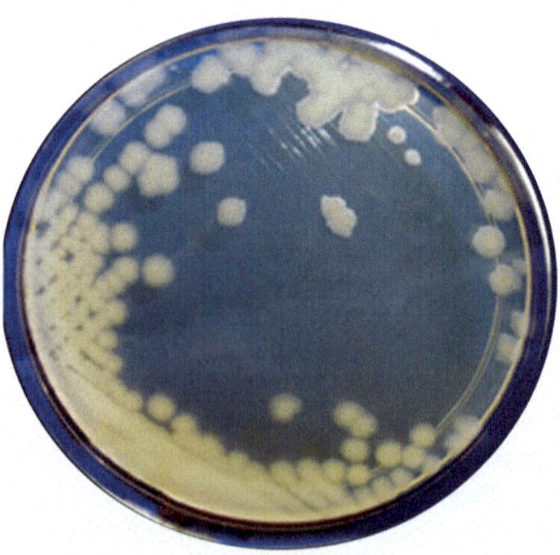

Fig. 18.1 Fried egg-type colony of *Bt* strain, HD1 (Pohare 2010)

et al. 2016). Due to the rise of more potent chemical pesticides, demand for *Bt* formulations in the agricultural sector declined between the 1970s and 1980s. However, *Bt* research was again stimulated by advances in biotechnology in the 1980s. A crystal toxin gene from *Bt* subspecies *kurstaki* was cloned into *Escherichia coli* by Schnepf and Whiteley (1981) for the first time. The first *Bt*-based commercial biopesticide product was Thuricide, derived from *Bt* subspecies *kurstaki* (Beegle and Yamamoto 1992). Dulmage et al. (1971) found even more potent *Bt* subspecies *kurstaki* (HD1) and marketed it as Dipel (Glazer and Nikaido 1995) in the United States. Zhong et al. (2000) described the first *Bt* toxin, which is alone toxic to the pests from three different insect orders (Lepidoptera, Diptera, and Coleoptera). After that, many reports have been cited about the *Bt* and its toxicity against the several other insects' orders (Mendez-Lopez et al. 2003; Asano et al. 2003; Cabreran et al. 2006). Since then, a lot more research has been carried out in *Bt*, varying from finding basic information about *Bt*, its Cry toxins, and its targeted insect groups to applying newly advanced biotechnological tools to express the insecticidal Cry toxins from *Bt* into plants.

18.4 Prevalence and Genetic Diversity of *Bt*

Bt species were reported in different habitats and climatic conditions across the continents, varying from high altitudes to tropics, including Antarctica and aquatic ecosystems (Hastowo et al. 1992; Landen et al. 1994; Forsyth and Logan 2000; Bernhard et al. 1997; Martinez and Caballero 2002). As *Bt* can acclimatize highly diverse ecosystems, they are spread all over the world (De Maagd et al. 2001).

18.4.1 The *Bt* Genome

Bt genome size ranges from 2 to six million base pairs (Carlson et al. 1994). For some *Bt* strains, physical maps were constructed (Carlson and Kolsto 1993). *Bt* chromosomal map comparison to that of *B. cereus* indicates that they have almost the same architecture in the center half of the chromosome around the origin of replication while they vary from each other at the terminal half (Carlson et al. 1996). Most *Bt* strains also possess circular as well as linear plasmids with their size varying from 2 to >200 kb. Large plasmids typically encode the parasporal crystal proteins genes (Reyaz et al. 2019). Sequence hybridization experiments with probes developed from *cry* genes have demonstrated that the bacterial chromosome also possesses *cry* genes (Carlson et al. 1994). Transposable elements (TE) are also reported in *Bt* and their role is believed to be in the duplication of *cry* genes. Another possible role for these TE that could be thought of might be in the horizontal dissemination of the genetic information by mediating in the conjugation process within *Bt* or with other related bacterial species (Mahillon et al. 1994; Schnepf et al. 1998).

18.4.1.1 The *cry* Genes

Insecticidal toxin-coding genes are usually situated on plasmids with molecular weight more than 30 kDa (Gonzales and Carlton 1980). Several *cry* genes have been named (cry and cyt), cloned, expressed, and their toxic proteins characterized. Lepidopteran insect toxic proteins primarily belong to classes of Cry1, Cry2, Cry9, Cry19, Cry20, and Cry59 families. The Cry1, Cry3, Cry8, Cry18, Cry22, Cry34, Cry35, and Cry43 proteins act as toxins for coleopteran insects. Nematicidal proteins mainly belong to Cry5, Cry6, Cry12, Cry13, Cry14, and Cry21 groups. Cry2Aa, a subgroup of Cry2 proteins, is toxic to dipteran insects along with Cry1, Cry4, Cry11, Cry19, Cry21, Cry24, Cry29, Cry30, Cry32, Cry40, Cry52, Cyt1, and Cyt2 (De Maagd et al. 2001). Each of the *Bt* strains may contain one or more crystal toxin genes, thus synthesizing one or more crystal proteins (Table 18.1).

18.4.1.2 Genetic Diversity in *Bt*

A considerable amount of genetic variation among the *Bt* strains is mainly due to the conjugal transfers of plasmids in between them (Thomas et al. 2001; Wilcks et al. 1998). Most of the *Bt* strains possess 2–12 *cry* genes which are situated on large plasmids of usually more than 30 kDa in size. Bacterial chromosomal DNA also harbors some of the *cry* genes in addition to the genes located on low copy high molecular weight plasmid (Carlson and Kolsto 1993). The presence of TEs as flanking sequences of the *cry* gene could be one of the reasons responsible for the diversity among them. Additionally, the presence of conserved blocks in *cry* genes

Table 18.1 Insecticidal activity of *Bt* crystal proteins

Susceptible insect order	Crystal proteins
Lepidoptera	Cry1A, Cry1B, Cry1C, Cry1D, Cry1E, Cry1F, Cry1G, Cry1H, Cry1I, Cry1J, Cry1K, Cry2A, Cry2B, Cry7B, Cry8D, Cry9A, Cry9B, Cry9C, Cry9E, Cry15A, Cry19A, Cry19B, Cry19C, Cry20A, Cry20B, Cry22A, Cry32A, Cry51A, Cry54A, Cry54B, Cry59A, Cry59B, Cyt2B
Coleoptera	Cry1A, Cry1B, Cry1C, Cry1D, Cry1E, Cry1F, Cry1G, Cry1H, Cry1I, Cry2A, Cry3A, Cry3B, Cry3C, Cry7A, Cry8A, Cry8B, Cry8C, Cry8D, Cry8E, Cry8F, Cry8G, Cry9D, Cry14A, Cry18A, Cry18B, Cry18C, Cry22A, Cry22B, Cry23A, Cry26A, Cry28A, Cry34A, Cry34B, Cry35A, Cry35B, Cry36A, Cry37A, Cry38A, Cry43A, Cry43B, Cry43C, Cry55A, Cyt1A, Cyt2C
Diptera	Cry1A, Cry1B, Cry1C, Cry2A, Cry2B, Cry4A, Cry4B, Cry4C, Cry10A, Cry11A, Cry11B, Cry16A, Cry17A, Cry19A, Cry19B, Cry19C, Cry20A, Cry21A, Cry21B, Cry21C, Cry21D, Cry21E, Cry21F, Cry21G, Cry21H, Cry24A, Cry24B, Cry24C, Cry25A, Cry27A, Cry29A, Cry29B, Cry30A, Cry30B, Cry30C, Cry30D, Cry30E, Cry30F, Cry30G, Cry32B, Cry32C, Cry32D, Cry39A, Cry40A, Cry40B, Cry40C, Cry40D, Cry44A, Cry47A, Cry48A, Cry49A, Cry52A, Cry52B, Cyt1A, Cyt1B, Cyt1C, Cyt1D, Cyt2A, Cyt2B, Cyt2C, Cyt2D, Cyt3A
Hymenoptera	Cry2A, Cry3A, Cry11A, Cry22A
Hemiptera	Cry3A, Cry5A, Cry22A, Cry78A
Nematodes	Cry5A, Cry5B, Cry6A, Cry6B, Cry12A, Cry13A, Cry14A, Cry21A

(Gatehouse et al. 2002; De Maagd 2001; Jouzani et al. 2017; Badran et al. 2016; Wang et al. 2018)

located on bacterial chromosomes and large plasmids represents the probability of recombinational events among themselves (Lereclus et al. 1986).

18.4.2 Insecticidal Proteins of *Bt*

Some *Bt* variants are also known to develop few other proteins other than the δ-endotoxins (Cry and Cyt) such as Vips, chitinases, specific proteases, phospholipases, and β-exotoxins, 4A4, during its vegetative life cycle, which might have some role in the insecticidal activity of *Bt* against the insects (Levinson 1990; Lovgren et al. 1990; Zhang et al. 1993; Estruch et al. 1996; Schnepf et al. 1998; Palma et al. 2014; Iatsenko et al. 2014; Chakroun et al. 2016). However, only ICP and Vips have been studied as major insecticidal proteins.

18.4.2.1 Classification of *Bt* Insecticidal Crystal Proteins

There are various criteria by which *Bt* strains can be differentiated from each other such as the presence of enzymes, lectins, antibody raised against them, that is antigenic differentiation, and phage typing (Saxena et al. 2020; Yamamoto and Powell 1993). According to Lecadet et al. (1999), antigenic determinants like flagellar H antigen forms more than 80 serovars of *Bt* that are recognized until now. Other methods that were implemented for *Bt* isolates differentiation are Fourier transform infrared spectroscopy (Beattie et al. 1998) and indirect immunofluorescence (Chitra et al. 1998).

Various *Bt* isolates insecticidal to different groups of insects, nematodes, and mites, are cited in the literature and this can also form the basis for the classification of *Bt* isolates (Marroquin et al. 2000). However, all the criteria as mentioned earlier comes with one or other disadvantages such as many of them don't give a clear information about the specific *cry* gene that *Bt* strain carry, while others are slower and become a significant hurdle in speedy screening of enormous *Bt* isolate collection (Wang et al. 2003; Soufiane and Cote 2009). Hofte and Whiteley (1989) used the presence of conserved blocks among *cry* genes as well as their targeted insect groups as a criterion to differentiate and separated them into four groups such as *cry*I (Lepidoptera), *cry*II (Diptera and Lepidoptera), *cry*III (Coleoptera), and *cry*IV (Diptera). Further, another two groups are added to this classification and extended it to *cry*V and *cry*VI levels (Feitelson et al. 1992). After that, numerous Cry toxins which are toxic to different insect groups have been identified and assigned in the classification table (Table 18.1).

However, the classification put forward by Hofte and Whiteley (1989), does not have room for Cry proteins, which belong to the same group but target different insect orders. To avoid such confusion Crickmore et al. (1998) brought in a new way of classifying *Bt* toxins, which form the basis of homology in the amino acid sequence. In this system of Cry toxin classification, Cry toxins have given a name that consists of symbolic Cry or Cyt and further four hierarchical ranks like Arabic numbers, capital English alphabets, lower case English alphabets, and finally Arabic numbers (e.g., Cry5Ba2). Thus, different primary ranks are allocated to proteins which share less than 45% homology (Cry3, Cry4 etc.), while amino acids sharing of less than 78% and 95% homology belongs to the secondary (Cry3B, Cry3C) and tertiary ranks (Cry3Bb, Cry3Bc), respectively. Finally, the proteins, which share more than 95% homology, are allocated into the quaternary ranks (Cry3Bb1 and Cry3Bb2).

18.4.2.2 Crystal Morphology and Solubility

Crystals produced by *Bt* isolates are of various shapes such as cuboidal, bipyramidal, spherical, square, and different irregular shapes like tiny crystals attached to spores (Fig. 18.2) (Chilcott and Wigley 1994; Ramalakshmi and Udayasuriyan 2010). The shape of *Bt* crystals is generally linked to its insecticidal activity against a target insect group such as crystals that are pyramidal in shape are insecticidal to lepidopteran insects (Attathom et al. 1995), while cuboidal crystals are insecticidal for dipteran and/or lepidopteran insects (Yamamoto and Mclanghlin 1981). Similarly, spherical and irregular shape crystals are lethal for mosquitoes as well as coleopteran group of insects (Krieg et al. 1983). *Bt* crystals are soluble under alkaline conditions but insoluble in water as well as in inorganic solvents. Different Cry toxins have different requirements for their solubility in alkaline conditions like Cry1, Cry4A, Cry5B, and Cyt toxins are generally soluble at pH 9.5, while Cry2 and Cry4D toxins require pH 12 for their solubility. There are the exceptional Cry toxins as well like Cry3A which is soluble at pH below 4 and pH above 9.5 (Koller et al. 1992).

18.4.2.3 Structural Features of Crystal Proteins

Bt isolates produce different types of crystalline inclusions during the sporulation phase of its life cycle, which can be easily observed and differentiated under phase-contrast microscope. These crystalline inclusion bodies are also termed as ICPs or δ-endotoxin or Cry toxins and are highly insecticidal. In contrast, actively dividing cells in its vegetative phase lack the insecticidal crystals and are not as toxic as *Bt* crystals to the insects.

The δ-endotoxins produced by *Bt* are mainly divided into two categories, Cry and Cyt. Cry and Cyt toxins kill the insects in a similar way by forming a pore in the epithelial cells of insect midgut but do not share amino acid sequence homology

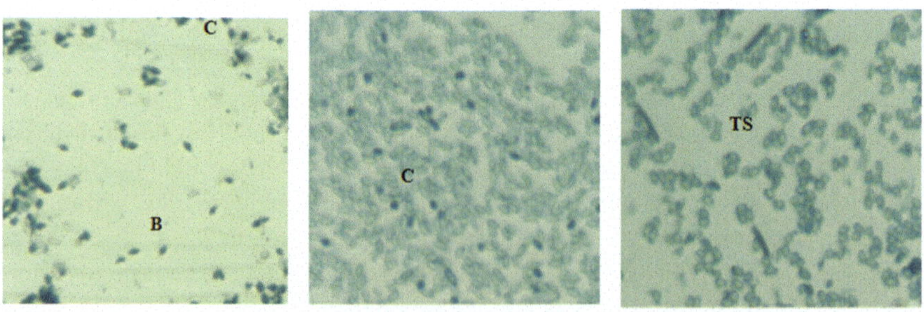

Fig. 18.2 *Bt* crystal morphology; B: Bipyramidal; C: Cuboidal; TS: Tiny attached to Spores (Pohare 2010)

(Gill et al. 1987; Thomas and Ellar 1983; Chang et al. 1993; Guerchicoff et al. 2001). The technique of X-ray crystallography has facilitated a lot to resolve the crystal structure of many Cry proteins (Li et al. 1991; Li et al. 1996; Cody et al. 1992; Grochulski et al. 1995; Derbyshire et al. 2001; Morse et al. 2001; Galitsky et al. 2001; Boonserm et al. 2005, 2006; Guo et al. 2009; Hui et al. 2012), which is useful further in allocating the functions to the different domains of Cry toxins (Thompson et al. 1995). Lopez-pazos and Ceron-salamanca (2007) reviewed the structures of *Bt* δ-endotoxins (Cry1Aa, Cry2Aa, Cry3Aa, Cry3Bb, Cry4, Cry11Bb, Cry45Aa, Cyt1Aa, and Cyt2Aa). (The crystal structures of Cry2Aa δ-endotoxin are shown in Fig. 18.3 (Morse et al. 2001)). *Bt* Cry toxins consist of three domains viz., I, II, and III; located from N to C terminal. The first 250 amino acids of Cry toxins constitute domain I, next approximately 300 amino acids constitute domain II while the last 250 amino acids constitute domain III. Domain I mainly consist of seven α-helices while domain II consists of three β-sheets, and domain III consist of two antiparallel β-sheets (Galitsky et al. 2001; Boonserm et al. 2006).

Participation of domain I in pore development and to that of domain II and III in receptor recognition and interactions with receptors have already been described in the literature (Flores et al. 1997; Burton et al. 1999; Bravo et al. 2005). Site-directed mutagenesis studies as well as biochemical analysis has shown that domain I develop the pore in the insect midgut membrane by penetrating α -helices 4 and 5 into it. α-helix 4 aligns itself along the inside wall of the pore, followed by the oligomerization of either 3 or 4 Cry toxins by intermolecular interactions (Tigue et al. 2001; Soberon et al. 2000). Studies with techniques like single-particle EM and

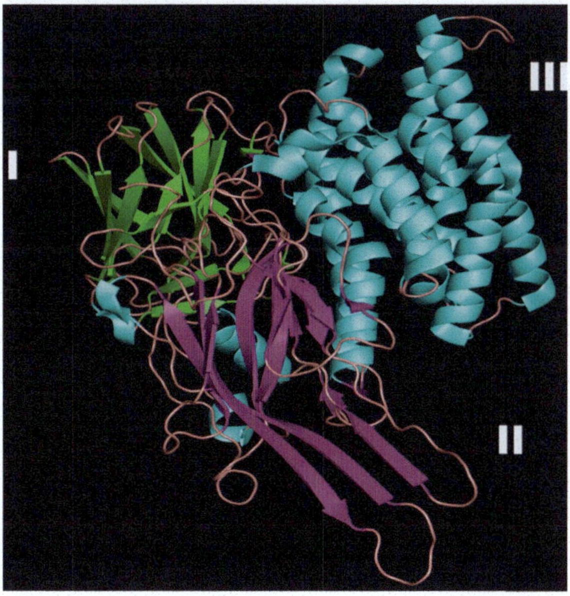

Fig. 18.3 3D structure of Cry2Aa protein (www.rcsb.org/structure/1i5p)

HS-AFM have already proved that pore formed by Cry toxins is an oligomer of either 3 or 4 Cry protein monomers with a lumen of around 0.75 nm radius (Vie et al. 2001; Ounjai et al. 2007; Groulx et al. 2011; Lin et al. 2014).

However, the Cyt2Ba structure is entirely different from that of the three-domain Cry toxin structure. Cyt2Ba structure consists of only one domain instead of three in Cry toxin, as shown in Fig. 18.4, where β-sheets are surrounded by α-helices (Schnepf et al. 1998; Cohen et al. 2008).

Cry toxins after getting activated in the midgut by proteolytic digestion mainly performs two functions; the first one is binding either reversibly or irreversibly to the receptors of susceptible insect midgut and the second one is pore formation (Hofman et al. 1988). In contrast, in the case of Cyt toxins, they can form the pore and are cytolytic (hemolytic) in nature, but they do not recognize any specific receptors on the insect midgut membrane (Butko 2003; De Maagd et al. 2003).

18.4.2.4 Mode of Action (MOA)

It is generally believed that most of the Cry toxins are almost similar in their three-dimensional (domain) structure and they kill their targets by a similar type of mechanism (Li et al. 1991; Crickmore et al. 1998; Boonserm et al. 2005; Angsuthanasombat 2010). The most widely accepted model about the action of Cry toxins on their

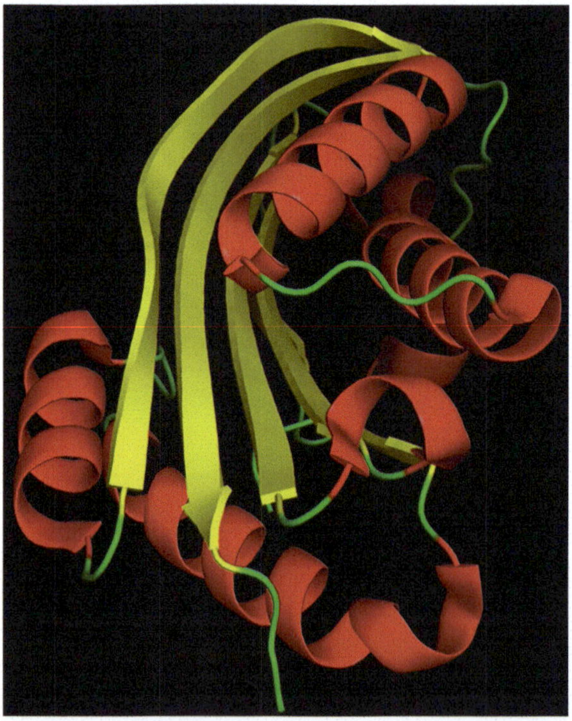

Fig. 18.4 3D structure of Cyt2Ba protein (www.rcsb.org/structure/2rci)

specific targets in the insect midgut is that these toxins are soluble in the basic conditions of insect midgut, where they were activated by proteolytic digestion. These activated toxins then recognize and interact with the receptors in the membrane, which leads to pore development and finally the death of insects (De Maagd et al. 2001; Pardo-Lopez et al. 2013; Adang et al. 2014; Portugal et al. 2014). Although Cry toxins have highly specific insect targets in vivo, many of these Cry toxins can be nonspecific under in vitro conditions, and they can even form the pores in the lipid bilayer after being activated by proteolytic digestion. However, to form the pores in the lipid bilayer, an excessive amount of toxins is required but these pores have different properties to that of formed in the insect midgut in vivo (Peyronnet et al. 2001). Some proteins are reported from the insect midgut, which can interact with the *Bt* Cry toxins; however, interactions do not always interpret as susceptibility.

To kill insect larvae, the three-domain Cry toxin needs to undergo a multiple steps process that then leads to midgut epithelial cells bursting. Two different models have been suggested to explain the MOA of Cry toxins; the first model talks about pore development in the membrane and the second one on signal transduction. The first steps in both the models hypothesized are the same where after ingestion of Cry toxins by larvae, it will be solubilized in the basic conditions of insect midgut. Solubilized toxins are then proteolyzed by the proteases of insect midgut to produce an active 60 kDa-truncated version of Cry toxin (Bravo et al. 2007; Pardo-Lopez et al. 2013; Adang et al. 2014; Portugal et al. 2014). This activated Cry toxin can recognize and interact with the cadherin receptors situated on the microvilli of midgut cells (Vadlamudi et al. 1995). In the next step of the pore-formation model, interaction of activated toxins with receptors leads to the oligomerization of the either three or four toxin monomers (Gomez et al. 2002; Ounjai et al. 2007; Groulx et al. 2011; Lin et al. 2014), which then interact with another receptor attached to the membrane by a glycosylphosphatidylinositol (GPI) anchor, such as aminopeptidase N in *Manduca sexta* or alkaline phosphatase in *Heliothis virescens* (Bravo et al. 2004; Bravo et al. 2007; Jurat-Fuentes and Adang 2006; Zhao et al. 2017). Further, these Cry toxin oligomers are then inserted into insect epithelial midgut membrane, where it forms pores, which ultimately leads to cell bursting and death of insects (Zhuang et al. 2002; Pardo-Lopez et al. 2006, 2013; Bravo et al. 2007; Adang et al. 2014). In contrast, in the signal transduction model, proposed based on studies with insect cell lines, interaction of Cry1A toxin with cadherin receptors is supposed to initiate a series of reactions, which ultimately results in the increased concentration of cAMP that then results in the oncotic cell death by activating protein kinase A (Zhang et al. 2006).

The whole event thus can be summarized in terms specific to their domains I, II, and III, as cited by various authors and which supports the theory of pore formation. The toxicity of Cry proteins in the midgut has been thought to be resites in domain I. It has been postulated that interaction of Cry toxin with membrane receptors leads to conformational change in the domain I structure, where hydrophobic amino acids residues cover the surface of domain I, which will then help the protein to be inserted into the membrane (Knowles 1994; Li et al. 1991; Pardo-Lopez et al. 2013; Adang et al. 2014). Receptor binding is the function of domain II of Cry toxin, and it has been proposed that membrane proteins like vitelline membrane outer layer protein I (VMO I), which share similar 3-D structure with ICPs might have similar carbohydrate-binding site, as N-acetylgalactosamine inhibit binding of Cry toxin to its glycoprotein counterpart and VMO I binds to hexasaccharides of N-acetylglucosamine (Knowles et al. 1991; Kido et al. 1995). Domain III of Cry toxins is thought to be involved in the stability role, where it stabilizes the Cry toxin by protecting it against proteases (Knowles et al. 1991) and it might also play a role in receptor binding (Aronson et al. 1995; Lee et al. 1995). However, only interaction of toxin to the receptor does not lead to the death of the insects, process of pore formation should be followed after binding, that then leads to cell disruption and insect death (Schnepf et al. 1998; De Maagd et al. 1999; Jenkins et al. 2000; De Maagd et al. 2001; Canton et al. 2014).

18.4.3 Other Insecticidal Constituents of *Bt*

Some *Bt* variants are also known to develop few other proteins such as Vips, chitinases, certain proteases, phospholipases, and β-exotoxins during its vegetative life cycle, which might have some role in the insecticidal activity of *Bt* against the insects (Levinson 1990; Lovgren et al. 1990; Zhang et al. 1993; Estruch et al. 1996; Schnepf et al. 1998; Palma et al. 2014; Juarez-Hernández et al. 2015; Dammak et al. 2015). In addition to these proteins, they are also capable of producing some antifungal compounds (Stabb et al. 1994). Vips secreted by certain strains of *Bt* into the extracellular environment are insecticidal and have an MOA different from that of Cry toxins (Lee et al. 2003). One of the members of Vips family, Vip1A, is a 100 kDa protein encoded by *vip1A* gene, but this protein will be processed at its N-terminal to produce an 80 kDa polypeptide and this 80 kDa polypeptide has already proved to be insecticidal (Schnepf et al. 1998; Chakroun et al. 2016). Another class of proteins produced by *Bt* is proteases, commonly known as inhibitor A. These proteases are highly specific in their action and act specifically at the C-terminal hydrophobic region of cecropin and attacins, which are antibacterial and plays a vital role as an antibacterial defense for insects (Dalhambar and Steiner 1984; Dammak et al. 2015). As a result of these proteases actions, antibacterial defense of insects collapses and they become no more immune to the bacterial infections which leads to their death. β-exotoxins, produced by certain strains of *Bt*, are thermostable proteins with a structure like nucleotide and because of its structural similarities to that of nucleotide, it can easily restrict the activity of DNA-dependent RNA polymerase (Glazer and Nikaido 1995; Liu et al. 2014). However, the *Bt* Cry toxins produced as parasporal crystals are the most effective insecticidal toxins compared to the other insecticidal constituents produced by the *Bt*.

18.5 *Bt* as a Biocontrol Agent

18.5.1 *Bt* Formulations

Biopesticides are still considered to be more expensive because of their competitive counterpart (chemical compounds), which are available at a relatively cheaper cost. Although biopesticides are costly but they are better performer as only small quantity is needed for the application (few grams/hectare) compared to almost half a dozen kg of chemical pesticide (Chapa et al. 2019). The biopesticides market has an increasing trend from 2005 onward and increased by more than 10%, while it was decreased by 1.5% for chemical pesticides (Tanwar et al. 2012). In the 1990s, *Bt* products alone have covered 95% of the microbial biocontrol agents, which are decreased to 60% in the last decade due to the development of other biocontrol agents like beneficial insects, viruses, and entomo-pathogenic fungi, etc. (Thakore 2006).

Most of the *Bt* preparations in the market contain spores with endotoxin composed of inclusion bodies. The efficiency of *Bt* products is dependent on the formulation used as well as on environmental factors like UV radiation, water quality, heat, and susceptibility of the insects (Tikar and Prakash 2017). The development of a wide variety of spores and crystal-based formulations has the sole intensions of increasing the contact between the formulations and insects by using matrices, which are made from chemical, vegetable, or animal products (Nava et al. 2012). Additionally, these formulations have further improvements over basic spore and crystal mixture, as they might increase toxic activity, increased palatability to insects, provide longer durability, and can be handled easily for their application in the field.

Different types of formulations used are commonly in the form of granules, emulsions, wettable powders, water-dispersible granules, suspension concentrate, oil dispersions, capsule, or suspension. Each formulation has its own advantages and disadvantages; and the success of the *Bt* formulations depends on our capacity to choose the correct formulations to increase its performance in the field (Brar et al. 2006).

At commercial levels, the *Bt* crystals are produced by fermentation are collected, concentrated, and formulated, according to the need (Baum et al. 1996). Although numerous *Bt* strains have insecticidal activity against insects from various orders (e.g., Lepidoptera, Diptera, Coleoptera) and nematodes, only a few of them have been produced commercially. As commercial providers of *Bt* formulations have the difficult task to fulfill, they have to provide solutions not only to specific crops or against some insect pests but also they have to develop broad spectrum of bioinsecticides against the new high-power *Bt* isolates which are isolated from the agroecosystems where they are to be used (Nava et al. 2012).

18.5.2 Expression of *cry* Genes in Other Microorganisms

Schnepf and Whiteley (1981) were the first to clone and express the *Bt* crystal protein gene in *E. coli.* Lower persistence of *Bt* pesticides due to environmental factors like heat and UV light can be controlled by moving *cry* genes to other microbes that are better persistent.

To counter this, attempts have been made to manipulate DNA, and the cloned cry genes have been expressed in other plant-associated microorganisms. First such incidence was reported when a scientist from Monsanto cloned a *cry1Ab* gene from *Bt* and expressed it into a plant root-associated bacterium (*Pseudomonas fluorescens)* (Watrud et al. 1985). This has increased the foliar persistence of the toxin by twofold, which has provided better protection against lepidopteran and coleopteran pests of cabbage and potato, respectively (Carlton 1996). Since then, many scientists had successfully expressed the *Bt cry* genes into *E. coli* and expressed proteins were found to be biologically active (Lin et al. 2008; Tan et al. 2009; Zheng et al. 2010; Zhu et al. 2010). Nowadays, easier, quicker, and cheaper cloning methods (Pohare and Akita 2017) have made DNA manipulations, plug and play things, and may help to clone and express different *Bt* genes in heterologous hosts to analyze the increased persistence of *Bt* toxins on fields.

18.5.3 Expression of *cry* Genes from *Bt* in Plants through a Transgenic Approach

Well-documented protocols are now available for the introduction of the heterologous genes either into a model plant or into a crop plants. Nonetheless, before these can be expressed in plants, bacterial genes require significant alteration. One such change was the addition of regulators like promotor and signals of polyadenylation that can be then suitable for expression in plants. Merely, the use of strong promotors like CaMV 35S and *rbs*S is not adequate for *cry* gene expression into plants.

Differences like comparatively higher GC content in plants to bacteria and amino acid preferences make this process even more complicated. And this was one of the reasons why the level of *Bt cry* gene expression was very poor in the initial reports (Perlak et al. 1991; Diehn et al. 1998). Additionally, *Bacillus* genes are AT-rich that act as possible signals of polyadenylation, resulting in disrupted transcription of these genes (Diehn et al. 1998). That is why codon optimization is necessary to express *Bacillus* genes into the plants. Perlak et al. (1993) achieved better protections against Colorado beetle by increasing *cry3A* gene GC content from 36 to 49% in transgenic potatoes. One more effective way of expressing *Bt cry* genes in plants is to use their truncated versions, as C terminal stretch is not required for their toxicity. C terminal has its roles in the *Bt* crystal formation in the bacterial cell and not in the plants.

Modifications by Perlak et al. (1991), in *cry1Ab Bt* gene, have achieved almost 100-fold higher expression level compared to the wild-type gene in cotton var., Coker312. These modifications mainly include the use of strong promotors like

CaMV 35S with a couple of enhancers and avoiding secondary structure formation in *mRNA* by doing related gene sequence codon optimization.

Initial efforts of obtaining *Bt* transgenic plants were limited to a dicotyledonous group of plants. The breakthrough in the monocotyledonous group was achieved when a synthetic *cry1Ab* gene under tissue-specific promotor, with 65% GC content, was successfully expressed in maize (Koziel et al. 1993). This has provided better protection of maize against *Ostrinia nubilalis* (European corn borer).

Since then many efforts have been made to produce transgenic plants with various *cry* gene(s) in different crops like cotton, maize, soya bean, rice, potato, tomato, and sugarcane (Khanna and Raina 2002; Kumar and Kumar 2004; Wu et al. 2007; Yu et al. 2013; Ranjithkumar et al. 2013; Kumar and Swamy 2014; Zhang et al. 2017; Liu et al. 2018; Tang et al. 2018) and many of them are commercialized as well. However, only a few specific *cry* genes were exploited in the development of transgenic crops. Therefore, to develop an effective resistance management strategy, it is needed to explore different *Bt cry* genes with diverse MOAs. As a multipronged strategy, not only different genes but also a different combination of multiple genes could also become an effective strategy in combating the development of resistance in the insect population. Shelton et al. (2002) have fused two genes, *cry1Ab* and *cry1Ac,* and introduced to develop GK cotton. Similarly, Dow Agro-Sciences has combined *cry34Ab1* and *cry35Ab1* to develop *Bt* corn (Moellenbeck et al. 2001) and *cry1Ac* and *cry1F* genes to develop Wide Strike cotton.

Although most of today's *Bt* transgenic plants express *cry* genes but *Bt* bacterium also produces an additional insecticidal Vips proteins during vegetative growth stage of its life cycle (Estruch et al. 1996; Chakroun et al. 2016). Currently, more than 152 *vips* genes have been identified in *Bt* that kills insects with an MOA different than that of Cry toxins (Lee et al. 2003; Xiao and Wu 2019) and some of them were already commercialized in crops like maize and cotton (Raybould and Vlachos 2011; Whitehouse et al. 2014). Gene pyramiding with multiple transgenes is the future of biotech crops, which is now being adopted in several countries. The benefit of gene pyramiding is that it can accomplish multiple needs of the end-user (farmer). SmartStax® maize is one such example where eight different genes are stacked together to control three traits, two for pest tolerance against lepidoptera and coleopteran insects and one for herbicide resistance (Head et al. 2017). The use of *Bt* technology is not restricted to the crops such as soybean, maize, and cotton but has recently been deployed in Bangladesh to defend eggplants against eggplant fruit and shoot borer (*Leucinodes orbonalis*) (Hautea et al. 2016). ISAAA (2017) reports show that only 20 farmers have grown transgenic *Bt* eggplants on their field in 2014 but this number has increased dramatically to 27,000 framers in 2018 (Shelton et al. 2018).

18.6 Development of Insect Resistance to *Bt*

The evolution of the insects to develop tolerance to pesticides has long been a major concern within the agricultural community. It has been well documented that without IPM, approximately 645 different species of insects and mites have become tolerant to various chemical insecticides (Geordhiou and Tajeda 1993; Rajamohan 1998).

Transgenic plants can speed up the development of resistance in pests if a transgene is expressed constitutively and all over the plant (Gould 1998). The further wide adoption of the *Bt* technology may enhance the rate at which insects are developing tolerance to *Bt*. One of the examples for this is that now there is a serious threat that *Helicoverpa armigera* has developed the tolerance to Cry1Ac toxin in laboratory as well as on field conditions (Zhang et al. 2011; Wei et al. 2015). This increased adoption of the transgenic *Bt* technology has also raised concerns over the use of the *Bt* formulations as biopesticides. A strain of *Plutella xylostella* has developed the resistance against Cry1Ac and could complete its life cycle on transgenic *Bt* canola plant (Ramachandran et al. 1998).

18.6.1 Mechanism of Insect Resistance to *Bt*

Most of *Bt* crystal proteins kill insects by binding to the insect midgut epithelial membrane, and thus most common mechanism predicted for the development of resistance in insects is by limiting the interactions between toxin and receptors (Bravo et al. 2011). Additionally, most of these Cry toxins share the receptor-binding site that is even more worrying as tolerance development to one toxin makes the insects tolerant to other Cry toxins as well. Reduced binding to the receptor might be the result of structural modifications in it. Therefore, the mechanism of generation of tolerance in insects is proposed to be related to the loss of *glycosyltransferase* genes. These genes are responsible for the assembly of a receptor with glycosyl anchor (Griffitts et al. 2001, 2003; Pardo-Lopez et al. 2013). Additionally, other mechanisms might also exist by which insects generate tolerance to *Bt* toxins (Ballester et al. 1999; Li et al. 2004; Pardo-Lopez et al. 2013).

Reports, which states some other mechanisms by which insects develop resistance, are also reported in the literature (Tabashnik et al. 2011). One such report states the involvement of Cadherin locus in developing tolerance in *Pectinophora gossypiella* (Tabashnik et al. 2005), but not in *P. xylostella* which has evolved tolerance in the field (Baxter et al. 2005). Another report states the involvement of gut proteases, which processes Cry toxins in the generation of tolerance in insects (Oppert et al. 1997). Jurat-Fuentes and Adang (2001) reported modification of receptor, a 170 kDa aminopeptidase N (APN), is a mechanism for *H. virescens* tolerance to Cry1A toxins. For more detailed information about the genetics and biochemistry involved in the process of development of resistance in insects, do refer Ferre and Rie (2002). Sequestering of Cry1Ac toxin by esterase is another mechanism by which insects develop resistance (Gunning et al. 2005).

18.6.2 Strategies for Management of *Bt*-Resistant Insect

Continuous exposure of pests to *Bt* Cry proteins leads to the fast generation of resistance in pests. So, it is required to focus on the identification of novel potential genes to continue the successful journey of *Bt* transgenic crops. Several strategies are proposed for resistant management in *Bt* transgenic crops (Tabashnik et al. 1994; Gelernter 1997; Schnepf et al. 1998). These mainly include the deployment of non-*Bt* refugia crops, mixed seeds, toxin mixtures, and alteration of *Bt* toxins. Denholum and Rowland (1992) and Tabashnik et al. (1994) emphasized the need to develop a resistance management program that integrates into IPM system to take advantage of reduced selection pressure.

18.6.3 Enhancing Toxicity of Cry Proteins

18.6.3.1 Potentiation of Cry Toxin Activity by Additional Proteins

Serine proteases are one such example, which can enhance the Cry protein toxicity by several folds (Deist et al. 2014). MacIntosh et al. (1990) reported a six-fold increase in the toxicity level of Cry1Ab when fused to protease inhibitor in transgenic tobacco plants against *H. virescens,* as compared to transgenic plants expressing Cry1Ab alone. How this works, is not well understood, but it might work by inhibiting the gut proteases, which might inactivate Cry proteins, or they might be involved in limiting the proteolysis of receptor proteins in insect midgut and therefore increases the chances of them binding to the Cry proteins. Similar effects have also been reported in case of endochitinases where it can improve the toxicity of Cry toxins up to ten-fold (Ding et al. 2008). It was also hypothesized that chitinases increase the insecticidal activity by creating holes in the epithelial membrane, which increases the chance of Cry proteins to have a better accessibility to the receptors in the epithelial membrane.

18.6.3.2 Modifications in the Cry Toxin Gene

Site-directed mutagenesis is an important tool to explore the role of individual domains as well as individual amino acid residues in the toxicity of the Cry proteins (Portugal et al. 2014; Deist et al. 2014). Although many of such modifications end in either null effect or even with reduced toxicity levels, but some of these could be effective in enhancing the toxicity of the toxins (Wu and Aronson 1992; Schnepf et al. 1998; Portugal et al. 2014; Deist et al. 2014). Wu and Aronson (1992) reported that replacement of histidine at 168 locations with arginine in helix α-5 of domain I leads to three-fold increase in its toxicity against the *Manduca secta*.

In addition to the site-directed mutagenesis, domain swapping is the other alternative where hybrid proteins can be constructed, and the resulting hybrid toxin could be even more toxic or might have a broader target range against the insects as compared to their parental toxins (Deist et al. 2014; Shao et al. 2016; Torres-Quintero et al. 2018). In this way, several toxins, with low toxicity or negligible toxicity against insects, can be used to swap their domain with some other Cry toxin to make them more toxic (Bideshi et al. 2017). Several such reports are available where Cry toxins with low toxicity to *Spodoptera axigua* like Cry1Ab, Cry1Ac, Cry1Ba, and Cry1Ea become even more toxic when their III domain is replaced by that of Cry1Ca (De Maagd et al. 2000; Deist et al. 2014).

Processing of the toxin in the alkaline insect midgut is an important step in the activation of the Cry protein and it has been extensively studied. Reports indicate that activation of Cry toxin requires removal of 27–29 amino acid residues from N terminal and 500–600 amino acid residues from the C terminal. If somehow one can enhance the cleavage of these toxins at their proteolytic cleavage site, then that will definitely enhance interactions of these toxins with receptors in insect midgut (Bravo et al. 2007; Deist et al. 2014).

According to one of the models explaining MOA of Cry proteins, binding of Cry toxin with cadherin receptors is necessary to remove helix α-1 from toxin, which then oligomerizes with other three or four similar Cry toxin monomers to form prepore structure (Vie et al. 2001; Ounjai et al. 2007; Groulx et al. 2011; Lin et al. 2014). Modifications in the cadherin receptors in the insect midgut are already linked to the development of resistance in insects against Cry toxins (Xu et al. 2005). Therefore, deletion of the helix α-1 from Cry toxin can oligomerize even in the absence of cadherin receptors and still can be toxic to the insects that have developed resistance against the Cry toxins (Soberon et al. 2007; Portugal et al. 2014).

18.7 Conclusion and Future Prospectus

After being in the picture for more than a century, *Bt,* without a doubt, is the most efficient microbial biocontrol agent. As expected, the *Bt* toxins have highly specific targets and rarely have any unintended effects on any other natural enemy insects and most importantly, they don't have any hazardous detrimental effects on the environment as that of their competitive counterparts (chemical pesticides). According to the National Academies of Sciences, Engineering, and Medicine (2016), the adoption of *Bt* crops results in a comparatively higher level of insect biodiversity on the farms than that of similar var. of non-*Bt* crops treated with chemical pesticides. Although different reports indicate the generation of tolerance to the *Bt* toxins in insects, however, various strategies have been implemented in the last two decades to counter this tolerance in insects and this will be continued in the future as well. Discovery of high potency new *Bt* strains with a higher level of toxicity, development of new biopesticides, gaining insights into the plant–microbe interactions (Shelake et al. 2019a), modern transgene-free mutagenesis tools (Shelake et al. 2019b), and molecular analysis of toxin action and insect tolerance using advanced techniques like CRISPR/Cas, RNAi (Garczynski et al. 2017; Karaminejadranjbar et al. 2018; Zotti et al. 2018; Pohare et al. 2019; Wagh and Pohare 2019), will continue the *Bt* era to bloom in the future too. All these kinds of development will help the *Bt* technology to prosper further in the upcoming near future.

Acknowledgments The authors are grateful to Rahul Mahadev Shelake, Division of Applied Life Science (BK21 Plus Program), Plant Molecular Biology and Biotechnology Research Center, Gyeongsang National University, Jinju 52828, Korea for his critical comments and English corrections. Authors are also thankful to Pallavi Kisan Patil, Research Associate, University of Sheffield, UK for her critical English corrections.

References

Abdullah MAF, Alzate O, Mohammad M, McNall RJ, Adang MJ, Dean DH (2003) Introduction of *Culex* toxicity into *B. thuringiensis* Cry4Ba by protein engineering. Appl Environ Microbiol 69:5343–5353

Adang MJ, Crickmore N, Jurat-Fuentes JL (2014) Chapter two-diversity of *Bacillus thuringiensis* crystal toxins and mechanism of action. Adv Insect Physiol 47:39–87

Anderson K (2010) Globalization's effects on world agricultural trade, 1960–2050. Philos Trans R Soc Lond Ser B Biol Sci 365:3007–3021

Andrews RE, Bibilops MM, Bulla LA (1985) Protease activation of the entomocidal protoxin of *Bacillus thuringiensis* subsp. *kurstaki*. Appl Environ Microbiol 50:737–742

Andrews RE, Faust RM, Wabiko H, Raymond KC, Bulla LA (1987) The biotechnology of *Bacillus thuringiensis*. Crit Rev Biotechnol 6:163–232

Angsuthanasombat C (2010) Structural basis of pore formation by mosquito-larvicidal proteins from *Bacillus thuringiensis*. Open Toxinol J 3:119–125

Aronson AI, Wu D, Zhang C (1995) Mutagenesis of specificity and toxicity regions of a *Bacillus thuringiensis* protoxin gene. J Bacteriol 177:4059–4065

Asano SC, Yamashita C, Iizuka T, Takeuchi K, Yamanaka S, Cerf D, Yamamoto T (2003) A strain of *B. thuringiensis* subsp. *galleriae* containing a novel *cry8* gene highly toxic to *Anomala cuprea* (Coleoptera: Scarabaeidae). Biol Control 28:191–196

Astuti DT, Pujiastuti Y, Suparman SHK, Damiri N, Nugraha S, Sembiring ER et al (2018) Exploration of *Bacillus thuringiensis* Berl. From soil and screening test its toxicity on insects of Lepidoptera order. IOP Conf Ser earth. Environ Sci 102:012063

Attathom T, Chongrattanameteekul W, Chanpaisang J, Siriyan R (1995) Morphological diversity and toxicity of delta-endotoxin produced by various strains of *B. thuringiensis*. Bull Ent Res 85:167–173

Badran A, Guzov V, Huai Q, Kemp M, Vishwanath P, Kain W (2016) Continuous evolution of *Bacillus thuringiensis* toxins overcomes insect resistance. Nature 533:58–63

Ballester V, Granoro F, Tabashnik BE, Malvar T, Ferre J (1999) Integrative model for binding of *Bacillus thuringiensis* toxins in susceptible and resistant larvae of the diamondback moth (*Plutella xylostella*). Appl Environ Microbiol 65:1413–1419

Baum JA, Kakefuda M, Gawron-Burke C (1996) Engineering *Bacillus thuringiensis* bioinsecticides with an indigenous site-specific recombination system. Appl Environ Microbiol 62:4367–4373

Baxter SW, Zhao JZ, Gahan LJ, Shelton AM, Tabashnik BE, Heckel DG (2005) Novel genetic basis of field-evolved resistance to *Bt* toxins in *Plutella xylostella*. Insect Mol Biol 14:327–334

Beattie SH, Halt C, Hirst D, Williams AG (1998) Discrimination among *Bacillus cereus, Bacillus mycoides* and *Bacillus thuringiensis* and some other species of the genus *Bacillus* by Fourier transform infrared spectroscopy. FEMS Microbiol Lett 164:201–206

Beegle CC, Yamamoto T (1992) History of *Bacillus thuringiensis* Berliner research and development. Can Entomol 124:587–616

Bernhard K, Jarrett P, Meadows M, Butt J, Ellis DJ, Roberts GM, Pauli S, Rodgers P, Burges HD (1997) Natural isolates of *B. thuringiensis*: worldwide distribution, characterization and activity against insect pests. J Invertebr Pathol 70:59–68

Bideshi DK, Park HW, Hice RH, Wirth MC, Federeci BA (2017) Highly effective broad spectrum chimeric larvicide that targets vector mosquitoes using a lipophilic protein. Sci Rep 7(1):11282

Biswas S, Kundu D, Mazumdar S, Saha A, Majumdar B, Ghorai A et al (2018) Study on the activity and diversity of bacteria in a new Gangetic alluvial soil (Eutrocrept) under rice-wheat-jute cropping system. J Environ Biol 39:379–386

Boonserm P, Davis P, Ellar DJ, Li J (2005) Crystal structure of the mosquito-larvicidal toxin Cry4Ba and its biological implications. J Mol Biol 348:363–382

Boonserm P, Mo M, Angsuthanasombat C, Lescar J (2006) Structure of the functional form of the mosquito larvicidal Cry4Aa toxin from *Bacillus thuringiensis* at a 2.8-angstrom resolution. J Bacteriol 188:3391–3401

Bourque SN, Valero JR, Lavoie MC, Levesque RC (1995) Comparative analysis of the 16S to 23S ribozomal intergenic spacer sequences of *Bacillus thuringiensis* strains and subspecies and of closely related species. Appl Environ Microbiol 61:1623–1626

Brar SK, Verma M, Tyagi RD, Valero JRC (2006) Recent advances in downstream processing and formulations of *Bacillus thuringiensis* based biopesticides. Process Biochem 41:323–342

Bravo A, Gomez I, Conde J, MuÇoz-Garay C, Snchez J, Miranda R, Zhuang M, Gill SS, Soberon M (2004) Oligomerization triggers binding of a *Bacillus thuringiensis* Cry1Ab pore-forming toxin to aminopeptidase N receptor leading to insertion into membrane microdomains. Biochem Biophys Acta 1667:38–46

Bravo A, Gill SS, Sobero NM (2005) Bacillus thuringiensis mechanisms and use. Comprehensive Molecular Insect Science 56:175–206

Bravo A, Gill SS, Soberon NM (2007) Mode of action of *Bacillus thuringiensis* toxins and their potential for insect control. Toxicon 49:423–435

Bravo A, Likitvivatanavong S, Gill SS, Soberón M (2011) *Bacillus thuringiensis*: a story of a successful bioinsecticide. Insect Biochem Mol Biol 41:423–431

Burton SL, Ellar DJ, Li J, Derbyshire DJ (1999) N-acetylgalactosamine on the putative insect receptor aminopeptidase N is recognized by a site on the domain III lectin like fold of a *Bacillus thuringiensis* insecticidal toxin. J Mol Biol 287:1011–1022

Butko P (2003) Cytolytic toxin Cyt1A and its mechanism of membrane damage: data and hypotheses. Appl Environ Microbiol 69:2415–2422

Cabreran JG, Farinós GP, Caccia S, Mendoza MD, Castañera P, Leonardi MG, Giordana B, Ferré J (2006) Toxicity and mode of action of *Bacillus thuringiensis* cry proteins in the Mediterranean corn borer, *Sesamia nonagrioides* (Lefebvre). Appl Environ Microbiol 72:2594–2600

Cannon RJC (1995) *Bacillus thuringiensis* in pest control. In: HMT H, Lynch JM (eds) Biological control: benefits and risks. Cambridge University Press, pp 190–197

Canton PE, López-Días JA, Gill SS, Bravo A, Soberon M (2014) Membrane binding and oligomer membrane insertion are necessary but insufficient for *Bacillus thuringiensis* Cyt1Aa toxicity. Peptides 53:286–291

Carlson CR, Kolsto AB (1993) A complete physical map of a *Bacillus thuringiensis* chromosome. J Bacteriol 175:1053–1060

Carlson CR, Caugant DA, Kolsto AB (1994) Genotypic diversity among *Bacillus cereus* and *Bacillus thuringiensis* strains. Appl Environ Microbiol 60:1719–1725

Carlson CR, Johansen T, Lecadet MM, Kolsto AB (1996) Genomic organization of the entomopathogenic bacterium *Bacillus thuringiensis* subsp. *berliner* 1715. Microbiol 142:1625–1634

Carlton BC (1996) Development and commercialization of new and improved biopesticides. Ann N Y Acad Sci 792:154–163

Carriere Y, Crickmore N, Tabashnik B (2015) Optimizing pyramided transgenic *Bt* crops for sustainable pest management. Nat Biotechnol 33:161–168

Chakroun M, Banyuls N, Bel Y, Escriche B, Ferre J (2016) Bacterial vegetative insecticidal proteins (Vip) from Entomopathogenic Bacteria. Microbiol Mol Biol Rev 80(2):329–350

Chandra A, Ghosh P, Mandoakar AD, Bera AK, Sharma RP, Das S, Kumar PA (1999) Amino acid substitution in α−helix7 Cry1Ac δ endotoxin of *Bacillus thuringiensis* leads to enhanced toxicity to *Helicoverpa armigera Hubner*. FEBS Lett 458:174–179

Chang C, Yu YM, Dai SM, Law SK, Gill SS (1993) High-level *cryIVD* and *cytA* gene expression does not require the 20-kilodalton protein, and the co-expressed gene products are synergistic in their toxicity to mosquitoes. Appl Environ Microbiol 59:815–821

Chapa DF, Villalobos JR, Wong LG (2019) Toxic potential of *Bacillus thuringiensis*: an overview. In: Jia Y (ed) Protecting Rice Grains in the Post-Genomic Era. IntechOpen, pp 1–22

Chilcott CN, Wigley PJ (1994) Isolation and toxicity of *B. thuringiensis* from soil and insect habitats in New Zealand. J Invertebr Pathol 61:244–247

Chitra S, Narayanan R, Balakrishnan A, Jayaraman K (1998) A rapid and specific method for the identification of *Bacillus thuringiensis* strains by indirect immunofluorescence. J Invertebr Pathol 74:263–267

Chougule NP, Li H, Liu S, Linz LB, Narva KE, Meade T, Bonning BC (2013) Retargeting of the *Bacillus thuringiensis* toxin Cyt2Aa against hemipteran insect pests. Proc Natl Acad Sci U S A 110:8465

Cody V, Luft J, Jensen E, Pangborn W, English L (1992) Purification and crystalization of insecticidal delta endotoxins *CryIII* from *Bacillus thuringiensis* proteins. J Struct Funct Gen 14:324–330

Cohen S, Dym O, Albeck S, Ben-Dov E, Cahan R, Firer M, Zaritsky A (2008) High-resolution crystal structure of activated Cyt2Ba monomer from *Bacillus thuringiensis* subsp. *israelensis*. J Mol Biol 380:820–827

Crickmore N, Zeigler DR, Feitelson J, Schnepf E, Van Rie J, Lereclus J, Baum J, Dean DH (1998) Revision of the nomenclature for the *B. thuringiensis* pesticidal crystal proteins. Microbiol Mol Biol Rev 62:807–813

Dalhambar G, Steiner H (1984) Characterization of inhibitor a, a protease from *Bacillus thuringiensis* which degrades attacins and cecropins, two classes of antibacterial proteins in insects. Eur J Biochem 139:247–252

Dammak I, Dammak M, Tounsi S (2015) Histopathological and combinatorial effects of the metalloprotease InhA1 and cry proteins of *Bacillus thuringiensis* against *Spodoptera littoralis*. Int J Biol Macromol 81:759–762

Davies TGE, Field LM, Williamson MS (2012) The re-emergence of the bed bug as a nuisance pest: implications of resistance to the pyrethroid insecticides. Med Vet Entomol 26:241–254

De Maagd RA, Bakkar PL, Masson L, Adang MJ, Sangandala S, Stiekema W, Bosch D (1999) Domain III of the *Bacillus thuringiensis* delta-endotoxin Cry1Ac is involved in binding to *Manduca sexta* brush border membranes and to its purified amino peptidase. Mol Microbiol 31:463–471

De Maagd RA, Weemen-Hendriks M, Stiekema W, Bosch D (2000) *Bacillus thuringiensis* delta-endotoxin Cry1C domain III can function as a specificity determinant for *Spodoptera exigua* in different, but not all, Cry1-Cry1C hybrids. Appl Environ Microbiol 66:1559–1563

De Maagd RA, Bravo A, Crickmore N (2001) How *Bacillus thuringiensis* has evolved specific toxins to colonize the insect world. Trends Genet 17:193–199

De Maagd RA, Bravo A, Berry C, Crickmore N, Schnepf HE (2003) Structure, diversity, and evolution of protein toxins from spore-forming entomopathogenic bacteria. Annu Rev Genet 37:409–433

Deist BR, Rausch MA, Fernandez-Luna MT, Adang MJ, Bonning BC (2014) Bt toxin modification for enhanced efficacy. Toxins 6:3005–3027

Denholum I, Rowland MW (1992) Tactics for management of pesticides resistant in arthropods: theory and practice. Annu Rev Ent 37:91–112

Derbyshire DJ, Ellar DJ, Li J (2001) Crystallization of the *Bacillus thuringiensis* toxin Cry1Ac and its complex with the receptor ligand N-acetyl-D-galactosamine. Acta Cryst D57:1938–1944

Diehn SH, Chiy WL, De Rocher EJ, Green PJ (1998) Premature polyadenylation at multiple sites with a *Bacillus thuringiensis* toxin gene-coding region. Plant Physiol 117:1433–1443

Ding X, Luo Z, Xia L, Gao B, Sun Y, Zhang Y (2008) Improving the insecticidal activity by expression of a recombinant *cry1Ac* Gene with *chitinase*-encoding gene in acrystalliferous *Bacillus thuringiensis*. Curr Microbiol 56:442–447

Dulmage HT, Boening OP, Rehnborg CS, Habsen GD (1971) A proposed standardized bioassay for formulations of *Bacillus thuringiensis* based on the international unit. J Invertebr Pathol 18:240–245

Estruch JJ, Warren GW, Mullins MA, Nye GJ, Craig JA, Koziel MG (1996) Vip3A, a novel *B. thuringiensis* vegetative insecticidal protein with a wide spectrum of activities against lepidopteran insects. Proc Natl Acad Sci U S A 93:5389–5394
Feitelson JS, Payne J, Kim L (1992) *B. thuringiensis*: insects and beyond. Biotechnology 10:271–275
Ferre J, Rie JV (2002) Biochemistry and genetics of insect resistance to *Bacillus thuringiensis*. Annu Rev Entomol 47:501–533
Flores H, Soberon X, Sanchez J, Bravo A (1997) Isolated domain II and III from the *Bacillus thuringiensis* Cry1Ab delta endotoxins binds to lepidopteran midgut membranes. FEBS Lett 414:313–318
Forsyth G, Logan NA (2000) Isolation of *Bacillus thuringiensis* from northern Victoria land, Antarctica. Lett Appl Microbiol 30:263–266
Galitsky N, Cody V, Wojtczak A, Ghosh D, Luft JR, Pangborn W, English L (2001) Structure of the insecticidal bacterial delta-endotoxin Cry3Bb1 of *Bacillus thuringiensis*. Acta Cryst D57:1101–1109
Garczynski SF, Martin JA, Griset M, Willett LS, Cooper WR, Swisher KD, Unruh TR (2017) CRISPR/Cas9 editing of the codling moth (Lepidoptera: Tortricidae) *cpomOR1* gene affects egg production and viability. J Econ Entomol 110:1847–1855
Gatehouse AMR, Ferry N, Raemaekers RJM (2002) The case of the monarch butterfly: a verdict is returned. Trends Genet 18:249–251
Gelernter WD (1997) Resistant to microbial insecticides in agriculture. In: microbial insecticides: novelty or necessity. Proc Br Crop Prot Council Symp 68:243–253
Geordhiou GP, Tajeda LA (1993) The occurrence of resistant to pesticides in arthropods. FAO, Rome pp 123
Gill SS, Sing GJP, Hornung JM (1987) Cell membrane activaiton of *Bacillus thuringiensis* subsp. *israelensis* cytolytic toxins. Infect Immun 55:1300–1308
Glazer AN, Nikaido H (1995) Microbial insecticides. In: Freeman WH (ed) Microbial biotechnology fundamentals of applied microbiology, New York, pp 209–229
Gomez I, Snchez J, Miranda R, Bravo A, Soberon M (2002) Cadherin like receptor binding facilitates proteolytic cleavage of helix α-1 in domain I and oligomer pre-pore formation of *Bacillus thuringiensis* Cry1Ab toxin. FEBS Lett 513:242–246
Gonzales JM, Carlton BC (1980) Patterns of plasmid DNA in crystalliferous strains of *B. thuringiensis*. Plasmid 3:92–98
Gordon RE, Haynes WC, Pang CHP (1973) The genus *Bacillus*. Department of Agriculture Handbook no. 427. U.S. Department of Agriculture, Washington, DC
Gould F (1998) Sustainability of transgenic insecticidal cultivars: integrating pest genetics and ecology. Annu Rev Entomol 43:701–726
Griffitts JS, Whitacre JL, Stevens DE, Aroian RV (2001) *Bt* toxin resistance from loss of a putative carbohydrate-modifying enzyme. Sci 293:860–864
Griffitts JS, Huffman DL, Whitacre JL, Barrows BD, Marroquin LD, Muller R, Brown JR, Hennet T, Esko JD, Aroian RV (2003) Resistance to a bacterial toxin is mediated by removal of a conserved glycosylation pathway required for toxin–host interactions. J Biol Chem 278:45594–45602
Grochulski P, Masson L, Borisova S, Pusztai-Carey M, Schwartz JL, Brousseau R, Cygler M (1995) *B. thuringiensis* CryIA(a) insecticidal toxin: crystal structure and channel formation. J Mol Biol 254:447–464
Groulx N, McGuire H, Laprade R, Schwartz JL, Blunck R (2011) Single molecule fluorescence study of the *Bacillus thuringiensis* toxin Cry1Aa reveals tetramerization. J Biol Chem 286:42274–42282
Guerchicoff A, Delécluse A, Rubinstein CP (2001) The *Bacillus thuringiensis cyt* genes for the hemolytic endotoxins constitute a gene family. Appl Environ Microbiol 67:1090–1096
Gunning RV, Dang HT, Kemp FC, Nicholson IC, Moores GD (2005) New resistance mechanism in *Helicoverpa armigera* threatens transgenic crops expressing *Bacillus thuringiensis* Cry1Ac toxin. Appl Environ Microbiol 71:2558–2563

Guo S, Ye S, Liu Y, Wei L, Xue J, Wu H, Song F, Zhang J, Wu X, Huang D, Rao Z (2009) Crystal structure of *Bacillus thuringiensis* Cry8Ea1: an insecticidal toxin toxic to underground pests, the larvae of *Holotrichia parallela*. J Struct Biol 168:259–266

Hastowo S, Lay BW, Ohba M (1992) Naturally occurring *B. thuringiensis* in Indonesia. J Appl Bacteriol 73:108–113

Hautea DM, Taylo LD, Masanga APL, Sison MLJ, Narcisco JO, Quilloy RB, Hautea RA, Shotkoski FA, Shelton AM (2016) Field performance of *Bt* eggplants (*Solanum melongena* L.) in the Philippines: Cry1Ac expression and control of the eggplant fruit and shoot borer (*Leucinodes orbonalis* Guenée). PLoS One 11(6):e0157498

Head GP, Carroll MW, Evans SP, Rule DW, Willse AR, Clark TL, Storer NP, Flannagan RD, Samuel LW, Meinke LJ (2017) Evaluation of SmartStax and SmartStax PRO maize against western corn rootworm and northern corn rootworm: efficacy and resistance management. Pest Manag Sci 73(9):1883–1899

Helgason E, Okstad OA, Caugant DA, Johansen HA, Fouet A, Mock M, Hegna I, Kolsto AB (2000) *Bacillus anthracis, Bacillus cereus*, and *Bacillus thuringiensis* one species on the basis of genetic events. Appl Environ Microbiol 66:2627–2630

Hickey LT, Hafeez AN, Robinson H, Jackson SA, Leal-Bertioli SCM, Tester M, Gao C, Godwin ID, Hayes BJ, Wulff BBH (2019) Breeding crops to feed 10 billion. Nat Biotechnol 37(7):744-754

Hofman C, Vanderbruggen H, Höfte H, Van-Rie J, Jansen S, Van Melleart H (1988) Specificity of *Bacillus thuringiensis* delta-endotoxins is correlated with the presence of high affinity binding sites in the brush border membrane of target insect midguts. Proc Natl Acad Sci U S A 85:7844–7848

Hofte H, Whiteley HR (1989) Insecticidal crystal proteins of *B. thuringiensis*. Microbiol Rev 53:242–255

Huang F, Ghimire MN, Leonard BR, Daves C, Levy R, Baldwin J (2012) Extended monitoring of resistance to *Bacillus thuringiensis* Cry1Ab maize in *Diatraea saccharalis* (Lepidoptera: Crambidae). GM Crops 3:245–254

Hui F, Scheib U, Hu Y, Sommer RJ, Aroian RV, Ghosh P (2012) Structure and glycolipid binding properties of the nematicidal protein Cry5B. Biochemistry 51:9911–9921

Iatsenko I, Nikolov A, Sommer RJ (2014) Identification of distinct *Bacillus thuringiensis* 4A4 nematicidal factors using the model nematodes *Pristionchus pacificus* and *Caenorhabditis elegans*. Toxins 6:2050–2063

ISAAA (2017) Global status of commercialized biotech/GM crops in 2017: biotech crop adoption surges as economic benefits accumulate in 22 years. ISAAA brief no. 53. ISAAA, Ithaca, NY

Jenkins JL, Lee MK, Valaitis AP, Curtiss A, Dean DH (2000) Bivalent sequential binding model of a *Bacillus thuringiensis* toxin to gypsy moth aminopeptidase N receptor. J Biol Chem 275:14423–14431

Jouzani GS, Valijanian E, Sharafi R (2017) *Bacillus thuringiensis*: a successful insecticide with new environmental features and tidings. Appl Microbiol Biotechnol 101:2691–2711

Juarez-Hernández EO, Casados-Vazquez LE, del Rincon-Castro MC, Salcedo-Hernández R, Bideshi DK, Barboza-Corona JE (2015) *Bacillus thuringiensis* subsp. *israelensis* producing endochitinase ChiA74Dsp inclusions and its improved activity against *Aedes aegypti*. J Appl Microbiol 119:1692–1699

Jurat-Fuentes JL, Adang MJ (2001) Importance of Cry1 δ-endotoxin domain II loops for binding specificity in *H. virescens* (L). Appl Environ Microbiol 67:323–329

Jurat-Fuentes JL, Adang MJ (2006) The *Heliothis virescens* cadherin protein expressed in Drosophila S2 cells functions as a receptor for *Bacillus thuringiensis* Cry1A but not Cry1Fa toxins. Biochemist 45:9688–9695

Karaminejadranjbar M, Eckermann KN, Ahmed HMM, CHM S, Dippel S, Marshall JM, Wimmer EA (2018) Consequences of resistance evolution in a Cas9-based sex conversion suppression gene drive for insect pest management. Proc Natl Acad Sci U S A 115:6189–6194

Karlova R, Weeman-Hendriks M, Naimov S, Ceron J, Dukiandjiev S, de Maagd RA (2005) *Bacillus thuringiensis* δ-endotoxin Cry1Ac domain–III enhances activity against *Heliothis virescens* in some, but not all Cry 1-Cry1Ac hybrids. J Invertebr Pathol 88:169–172

Khanna HK, Raina SK (2002) Elite indica transgenic plants expressing modified cry 1Ac endotoxin of *Bacillus thuringiensis* show enhanced resistance to yellow stem borer (*Scirpophaga incertulas*). Transgenic Res 11:411–423

Kido S, Doi Y, Kim F, Morishita E, Narita H, Kanaya S, Ohkubo T, Nishikawa K, Yao T, Ooi T (1995) Characterization of vitelline membrane outer layer protein I, VMO-I: amino acid sequence and structural stability. J Biochem 117:1183–1191

Knowles BH (1994) Mechanism of action of *Bacillus thuringiensis* insecticidal δ-endotoxins. Adv Insect Physiol 24:275–308

Knowles BH, Knight PJ, Ellar DJ (1991) N-Acetylgalactosamine is a part of the receptor in the insect gut epithelia that recognizes an insecticidal protein from *Bacillus thuringiensis*. Proc R Soc Lond B 245:31–35

Koller CN, Bauer LS, Hollingworth RM (1992) Characterization of the pH-mediated solubility of *B. thuringiensis* var. *sandiego* native δ-endotoxin crystals. Biochem Biophys Res Commun 184:692–699

Kour D, Rana KL, Yadav AN, Yadav N, Kumar M, Kumar V et al (2020) Microbial biofertilizers: bioresources and eco-friendly technologies for agricultural and environmental sustainability. Biocatal Agric Biotechnol 23:101487. https://doi.org/10.1016/j.bcab.2019.101487

Koziel MG, Beland GL, Bowman C, Carozzi NB, Crenshaw R, Crossland L, Dawson J, Desai N, Hill M, Kadwell S, Launis K, Maddox D, McPherson K, Meghji MR, Merlin R, Rhodes R, Warren GW, Wright M, Evola SV (1993) Field performance of elite transgenic maize plants expressing an insecticidal protein derived from *B. thuringiensis*. Bio Technol 11:194–200

Krieg VA, Huger AM, Longenbruch GA, Schnetter W (1983) *B. thuringiensis* var. *tenebrionis*: a new pathotype effective against larvae of coleoptera. Z Angew Entomol 96:500

Kumar H, Kumar V (2004) Tomato expressing Cry1Ab insecticidal proteins from *Bacillus thuringiensis* protected against tomato fruit borer *Helicoverpa armigera* (Hubner) (Lepidoptera: Noctuidae) damage in laboratory, greenhouse and field. Crop Prot 23:135–139

Kumar GVS, Swamy SVSG (2014) A duo-decennium of Bt cotton adoption in India: an overview. Curr Biotica 8:322–340

Landen R, Bryne M, Abdel-Hameed A (1994) Distribution of *B. thuringiensis* strains in southern Sweden. World J Microbiol Biotechnol 10:45–50

Lecadet MM, Frachon E, Dumanoir VC, Ripouteau H, Hamon S, Laurent P, Thiey I (1999) Updating the H-antigen classification of *B. thuringiensis*. J Appl Microbiol 86:660–672

Lee MK, Milne RE, Ge AZ, Dean DH (1995) Location of a *Bombyx mori* receptor binding region on a *Bacillus thuringiensis* δ-endotoxin. J Biol Chem 267:3115–3121

Lee MK, Walters FS, Hart H, Palekar N, Chen JS (2003) The mode of action of the *Bacillus thuringiensis* vegetative insecticidal protein Vip3A differs from that of Cry1Ab δ-endotoxin. Appl Environ Microbiol 69(8):4648–4657

Lereclus D, Mahillon J, Menou G, Lecadet MM (1986) Identification of Tn 4430, a transposon of *Bacillus thuringiensis* functional in *Escherichia coli*. Mol Gen Genet 204:52–57

Levinson BL (1990) High performance liquid charamotography analysis of two β-exotoxins produced by some *Bacillus thuringiensis* strains. In: Hickle LA, Fitch WL (eds) Analytical chemistry of *Bacillus thuringiensis*. American Chemical Society, Washington, DC, pp 114–136

Li J, Caroll J, Ellar DJ (1991) Crystal structure of insecticidal δ-endotoxin from *Bacillus thuringiensis* at 2.5 Å resolution. Nature 353:815–821

Li J, Koni PA, Ellar DJ (1996) Structure of the mosquitocidal δ-endotoxin CytB from *Bacillus thuringiensis* ssp. *kyushuensis* and implications for membrane pore formation. J Mol Biol 257:129–152

Li H, Gonzalez-Cabrera J, Opert B, Ferre J, Higgins RA, Suschman LL, Radke GA, Zhu KY, Huang F (2004) Binding analysis of *Cry1Ab* and *Cry1Ac* with membrane vesicles from *Bt* resistant and susceptible *Ostrinia nubilalis*. Biochem Biophys Res Commun 323:52–57

Lin Y, Fang G, Cai F (2008) The insecticidal crystal protein Cry2Ab10 from *Bacillus thuringiensis:* cloning, expression, and structure simulation. Biotechnol Lett 30:513–519

Lin X, Parthasarathy K, Surya W, Zhang T, Mu Y, Torres J (2014) A conserved tetrameric interaction of cry toxin helix 3 suggests a functional role for toxin oligomerization. Biochim Biophys Acta 1838:1777–1784

Liu X, Ruan L, Peng D, Li L, Sun M, Yu Z (2014) Thuringiensin: a thermostable secondary metabolite from *Bacillus thuringiensis* with insecticidal activity against a wide range of insects. Toxins 6:2229–2238

Liu Q, Hallerman E, Peng Y, Li Y (2016) Development of Bt rice and Bt maize in China and their efficacy in target pest control. Int J Mol Sci 17:E1561

Liu YL, Wang YL, Shu CL, Lin KJ, Song FP, Bravo A, Sobero'n M, Zhang J (2018) Cry64Ba and Cry64Ca, two ETX/MTX2-type *Bacillus thuringiensis* insecticidal proteins active against hemipteran pests. Appl Environ Microbiol 84:e01996-e01917. (doi:https://doi.org/10.1128/AEM.01996-17)

Lopez-pazos SA, Ceron-salamanca JA (2007) Three-dimensional structure of *Bacillus thuringiensis* toxins: a review. Acta Biol Colomb 12(2):19–32

Lovgren AM, Zang Y, Engstöm A, Dalhammar G, Landen R (1990) Molecular characterization of immune inhibitor a, a secreted virulence protease from *Bacillus thuringiensis*. Mol Microbiol 4:2137–2146

MacIntosh SC, Kishore GM, Perlak FJ, Marrone PG, Stone TB, Sims SR, Fuchs RL (1990) Potentiation of *Bacillus thuringiensis* insecticidal activity by serine protease inhibitors. J Agricult Food Chem 38:1145–1152

Mahillon J, Rezsöhazy R, Ballet B, Delcour J (1994) *IS231* and other *Bacillus thuringiensis* transposable elements: a review. Genetica 93:13–26

Marroquin LD, Elyssnia D, Griffitts JS, Feitelson JS, Aroian RV (2000) *Bacillus thuringiensis* (*Bt*) toxin susceptibility and isolation of resistance mutants in the nematode *Ceanorhabditis elegans*. Genetics 155:1693–1699

Martinez C, Caballero P (2002) Contents of *cry* genes and insecticidal toxicity of *B. thuringiensis* strains from terrestrial and aquatic habitats. J Appl Microbiol 92:745–752

Melo AL, Soccol VT, Soccol CR (2016) *Bacillus thuringiensis*: mechanism of action, resistance, and new applications: a review. Crit Rev Biotechnol 36:317–326

Mendez-Lopez I, Basurto-Ríos R, Ibarra JE (2003) *B. thuringiensis* serovar *israelensis* is highly toxic to the coffee berry borer, *Hypothenemus hamper* Ferr. (Coleoptera: Scolytidae). FEMS Microbiol Lett 11131:1-5

Moellenbeck DJ, Peters ML, Bing JW, Rouse JR, Higgins LS, Sims L, et al. (2001) Insecticidal proteins from *Bacillus thuringiensis* protect corn from corn root worms. Nature Biotechnol 19:668–672

Morse RJ, Yamamoto T, Stroud RM (2001) Structure of Cry2Aa suggests an unexpected receptor binding epitope. Structure 9:409–417

Nava PE, García GC, Camacho BJR, Vázquez MEL (2012) Bioplaguicidas: Una opción para el control biológico de plagas. Ra Ximhai 8:17–29

Oerke E, Dehne H (1997) Global crop production and the efficacy of crop protection-current situation and future trends. Eur J Plant Pathol 103:203–215

Oppert B, Kramer KJ, Beeman RW, Johnson D, McGaughey WH (1997) Proteinase mediated insect resistance to *B. thuringiensis* toxins. J Biochem 272:23473–23476

Ounjai P, Unger VM, Sigworth FJ, Angsuthanasombat C (2007) Two conformational states of the membrane-associated *Bacillus thuringiensis* Cry4Ba -endotoxin complex revealed by electron crystallography: implications for toxin-pore formation. Biochem Biophys Res Commun 361:890–895

Palma L, Muñoz D, Berry C, Murillo J, Caballero P (2014) *Bacillus thuringiensis* toxins: an overview of their Biocidal activity. Toxins 6:3296–3325

Pardo-Lopez L, Gomez I, Rausell C, Snchez J, Soberon M, Bravo A (2006) Structural changes of the Cry1Ac oligomeric pre-pore from *Bacillus thuringiensis* induced by N-acetylgalactosamine facilitates toxin membrane insertion. Biochemistry 45:10329–10336

Pardo-Lopez L, Soberon M, Bravo A (2013) *Bacillus thuringiensis* insecticidal three-domain cry toxins: mode of action, insect resistance and consequences for crop protection. FEMS Microbiol Rev 37:3–22

Perlak FJ, Fuchs RL, Dean DA, McPherson SL, Fischhnff DA (1991) Modification of the coding sequence enhances plant expression of insect control protein genes. Proc Natl Acad Sci U S A 88:3324–3328

Perlak FJ, Stone TB, Muskopf YM, Petersen LJ, Parker GB, McPherson SA, Yman J, Love S, Reed G, Biever D (1993) Genetically improved potatoes: protection from damage by Colorado potato beetles. Plant Mol Biol 22:313–321

Peyronnet O, Vachon V, Schwartz JL, Laprade R (2001) Ion channels induced in planar lipid bilayers by the *Bacillus thuringiensis* toxin Cry1Aa in the presence of gypsy moth (*Lymantria dispar*) brush border membrane. J Membr Biol 184:45–54

Pimentel D, Burgess M (2014) Environmental and economic costs of the application of pesticides primarily in the United States. In: Pimentel D, Peshin R (eds) Integrated Pest management. Springer, Dordrecht, pp 47–71

Pohare MB (2010) Cloning of *cry2a* gene from a new isolate *of Bacillus thuringiensis* and its expression in *E. coli* (Master's thesis). Tamil Nadu agricultural University, Coimbatore

Pohare MB, Akita M (2017) A rapid and simple, recombination-based cloning method in *Escherichia coli*. Biosci Biotechnol Res Asia 14(1):43–52

Pohare MB, Bhor SA, Patil PK, Gakkula M (2017) Integrated and sustainable initiative towards better farm returns led by rural youth. Popular Kheti 5(3):113–115

Pohare MB, Sharma M, Wagh SG (2019) CRISPR/Cas9 genome editing and its medical potential. In: Kumar S (ed) Advances in biotechnology and biosciences. NavNik Publications, Pusa, pp 69–90

Portugal L, Gringorten JL, Caputo GF, Mario Soberón M, Muñoz-Garay C, Bravo A (2014) Toxicity and mode of action of insecticidal Cry1A proteins from *Bacillus thuringiensis* in an insect cell line, CF-1. Peptides 53:292–299

Rajamohan N (1998) Pesticides resistance -A global scenario. Pesticide Wld 3:34–40

Ramachandran S, Buntin GD, Tabashnik JN, Reymer BE, Adang PL, Pulliam MJ, Steward DA (1998) Survival, development and oviposition of resistant diamondback moth (Lepidoptera: Plutellidae) on transgenic canola toxin. J Econ Entomol 91:1239–1244

Ramalakshmi A, Udayasuriyan V (2010) Diversity of *Bacillus thuringiensis* isolated from Western Ghats of Tamil Nadu state, India. Curr Microbiol 61:13–18

Ranjithkumar L, Patil BV, Ghante VN, Bheemanna M, Arunkumar H (2013) Baseline sensitivity of brinjal shoot and fruit borer, Leucinodes orbonalis (guene'e) in South India to Cry1Ac insecticidal protein of *Bacillus thuringiensis*. Curr Sci 105:366–370

Raybould A, Vlachos D (2011) Non-target organism effects tests on Vip3A and their application to the ecological risk assessment for cultivation of MIR162 maize. Transgenic Res 20:599–611

Reinoso PY, Del Rincón CMC, Ibarra EJ (2016) Characterization of a highly toxic strain of *Bacillus thuringiensis* serovar *kurstaki* very similar to the HD-73 strain. FEMS Microbiol Rev 363:1–6

Reyaz AL, Balakrishnan N, Udayasuriyan V (2019) Genome sequencing of *B. thuringiensis* isolate T414 toxic to pink bollworm and its insecticidal genes. Microb Pathog 134:103553

Sahay H, Yadav AN, Singh AK, Singh S, Kaushik R, Saxena AK (2017) Hot springs of Indian Himalayas: potential sources of microbial diversity and thermostable hydrolytic enzymes. 3. Biotech 7:1–11

Sajid M, Geng C, Li M, Wang Y, Liu H, Zheng J, Peng D, Sun M (2018) Whole genomic analysis of *Bacillus thuringiensis* revealing partial genes as a source of novel cry toxins. Appl Microbiol Biotechnol 84:e00277–e00218

Sasaki J, Asano S, Hashimoto N, Lay BW, Hastowo S, Bando H, Jizuka T (1997) Characterization of a *cry2A* gene cloned from an isolate of *B. thuringiensis* serovar *sotto*. Curr Microbiol 35:1–8

Saxena AK, Yadav AN, Rajawat M, Kaushik R, Kumar R, Kumar M et al (2016) Microbial diversity of extreme regions: an unseen heritage and wealth. Indian J Plant Genet Resour 29:246–248

Saxena AK, Padaria JC, Gurjar GT, Yadav AN, Lone SA, Tripathi M et al. (2020) Insecticidal formulation of novel strain of *Bacillus thuringiensis* AK 47. Indian Patent 340541

Schnepf HE, Whiteley HR (1981) Cloning and expression of the *B. thuringiensis* protein gene in *E. coli*. Proc Natl Acad Sci U S A 78:2989–2897

Schnepf HE, Crickmore N, Van Rie J, Lerecurs D, Baum J, Feitelson J, Zeigler JDR, Dean DH (1998) *B. thuringiensis* and its pesticidal crystal proteins. Microbiol Mol Biol Rev 62:775–806

Shao E, Lin L, Chen C, Chen H, Zhuang H, Wu S, Sha L, Guan X, Huang Z (2016) Loop replacements with gut-binding peptides in Cry1Ab domain II enhanced toxicity against the brown planthopper, *Nilaparvata lugens* (Stal). Sci Rep 6:20106

Shelake RM, Pramanik D, Kim JY (2019a) Exploration of plant-microbe interactions for sustainable agriculture in CRISPR era. Microorganisms 7(8):269

Shelake RM, Pramanik D, Kim JY (2019b) Evolution of plant mutagenesis tools: a shifting paradigm from random to targeted genome editing. Plant Biotechnol Rep 13:423–445

Shelton AM, Zhao JZ, Roush RT (2002) Economic, ecological, food safety and social consequences of the deployment of *Bt* transgenic plants. Annu Rev Entomol 47:845–881

Shelton AM, Hossain MJ, Paranjape V, Azad AK, Rahman ML, Khan ASMMR, et al. (2018) *Bt* eggplant project in Bangladesh: history, present status, and future direction. Front Bioeng Biotechnol 6(106):1–6

Soberon M, Perez RV, Nunez-Valdez ME, Lorence A, Gomez I, Sanchez J, Bravo A (2000) Evidence for intermolecular interaction as a necessary step for pore-formation activity and toxicity of *Bacillus thuringiensis* Cry1Ab toxin. FEMS Microbiol Lett 191:221–225

Soberon M, Pardo-Lopez L, Lopez I, Gomez I, Tabashnik B, Bravo A (2007) Engineering modified *Bt* toxins to counter insect resistance. Sci 318:1640–1642

Soufiane B, Cote JC (2009) Discrimination among *Bacillus thuringiensis* H serotypes, serovars and strains based on 16S *rRNA*, *gyrB* and *aroE* gene sequence analyses. Ant Van Leeuwen 95(1):33–45

Stabb EV, Jacobson LM, Handelsman J (1994) Zwittermycin A-producing strains of *B. cereus* from diverse soils. Appl Environ Microbiol 60:4404–4412

Stahly DP, Andrews RE, Yousten AA (1991) The genus *Bacillus*- insect pathogens. In: Balows A, Truper HG, Dworkin M, Harder W, Schleifer KH (eds) The prokaryotes, 2nd edn. Springer Science and Business Media, pp 1697–1745

Tabashnik BE, Carrière Y (2017) Surge in insect resistance to transgenic crops and prospects for sustainability. Nat Biotechnol 35(10):926

Tabashnik B, Finson EN, Chilcutt CF, Cushing NL, Johnson M (1994) Increasing efficiency of bioassays: evaluation of resistance to *Bacillus thuringiensis* in diamondback moth (Lepidoptera: Plutellidae). J Econ Entomol 86:635–644

Tabashnik BE, Biggs RW, Higginson DM, Henderson S, Unnithan DC, Unnithan GC, Elers-Kirk C, Sisterson MS, Dennehy TJ, Carriere Y, Shai M (2005) Association between resistance to *Bt* cotton and cadherin genotype in pink bollworm. J Econ Entomol 98:635–644

Tabashnik BE, Huang F, Ghimire MN, Leonard BR, Siegfried BD, Rangasamy M, Yang Y, Wu Y, Gahan LJ, Heckel DG, Bravo A, Soberon M (2011) Efficacy of genetically modified *Bt* toxins against insects with different genetic mechanisms of resistance. Nat Biotechnol 29:1128–1131

Tabashnik BE, Brevault T, Carriere Y (2013) Insect resistance to Bt crops: lessons from the first billion acres. Nat Biotechnol 31:510–521

Tan F, Zhu J, Tang J, Tang X, Wang S, Zheng S, Li P (2009) Cloning and characterization of two novel crystal protein genes, *cry54Aa1* and *cry30Fa1*, from *Bacillus thuringiensis* strain *Bt*MC28. Curr Microbiol 58:654–659

Tang H, Chen G, Chen F, Han L, Peng Y (2018) Development and relative fitness of Cry1C resistance in *Chilo suppressalis*. Pest Manag Sci 74:590–597

Tanwar R, Dureja P, Rathore H (2012) Bio-pesticides. In: Rathore H, Nollet L (eds) Pesticides, evaluation of environmental pollution. CRC Press, Boca Raton, FL, pp 587–603

Thakore Y (2006) The bio-pesticide market for global agricultural use. Ind Biotechnol 2:194–208

Thomas WE, Ellar DJ (1983) Mechanism of action of *Bacillus thuringiensis* var *israelensis* insecticidal delta-endotoxins. FEBS Letter 154:362–368

Thomas DJ, Morgan AW, Whipps JM, Saunders JR (2001) Plasmid transfer between *Bacillus thuringiensis* subsp. *israelensis* strains in laboratory culture, river water and dipteran larvae. Appl Environ Microbiol 67:330–338

Thompson MA, Schnepf HE, Feitelson JS (1995) Structure, function and engineering of *Bacillus thuringiensis* toxins. In: Setlow JK (ed) Genetic engineering: principles and methods. Plenum Press, New York, pp 99–117

Tigue NJ, Jacoby J, Ellar DJ (2001) The alpha-helix 4 residue, Asn135, is involved in the oligomerization of Cry1Ac1 and Cry1Ab5 *Bacillus thuringiensis* toxins. Appl Environ Microbiol 67:5715–5720

Tikar S, Prakash S (2017) Fly ash-based *Bacillus thuringiensis israelensis* formulation: an eco-friendly approach. Indian J Med Res 146(6):680

Torres-Quintero M, Gómez I, Pacheco S, Sánchez J, Flores H, Osuna J, Mendoza G, Soberón M, Bravo A (2018) Engineering *Bacillus thuringiensis* Cyt1Aa toxin specificity from dipteran to lepidopteran toxicity. Sci Rep 8:4989

Udayasuriyan V, Nakamura A, Mori A, Masaki H, Uozumi T (1994) Cloning of a new *crylA*(a), gene from *B. thuringiensis* strain FU-2-7 and analysis of chimeric *cry1A*(a) proteins of toxicity. Biosci Biotechnol Biochem 58:830–835

Udayasuriyan V, Indra Arulselvi P, Balasubramani V, Sudha DR, Balasubramanian P, Sangeetha P (2006) Construction of new chimeric *cry2AX1* gene of *B. thuringiensis* encoding protein with enhanced insecticidal activity. Indian Patent 244427

Vadlamudi RK, Weber E, Ji I, Ji TH, Bulla LA (1995) Cloning and expression of a receptor for an insecticidal toxin of *B. thuringiensis*. J Biol Chem 270:5490–5494

Verma P, Yadav AN, Kazy SK, Saxena AK, Suman A (2014) Evaluating the diversity and phylogeny of plant growth promoting bacteria associated with wheat (*Triticum aestivum*) growing in central zone of India. Int J Curr Microbiol Appl Sci 3:432–447

Verma P, Yadav AN, Khannam KS, Panjiar N, Kumar S, Saxena AK et al (2015) Assessment of genetic diversity and plant growth promoting attributes of psychrotolerant bacteria allied with wheat (*Triticum aestivum*) from the northern hills zone of India. Ann Microbiol 65:1885–1899

Verma P, Yadav AN, Khannam KS, Kumar S, Saxena AK, Suman A (2016) Molecular diversity and multifarious plant growth promoting attributes of bacilli associated with wheat (*Triticum aestivum* L.) rhizosphere from six diverse agro-ecological zones of India. J Basic Microbiol 56:44–58

Verma P, Yadav AN, Kumar V, Singh DP, Saxena AK (2017) Beneficial plant-microbes interactions: biodiversity of microbes from diverse extreme environments and its impact for crop improvement. In: Singh DP, Singh HB, Prabha R (eds) Plant-microbe interactions in agro-ecological perspectives, Microbial interactions and agro-ecological impacts, vol 2. Springer, Singapore, pp 543–580. https://doi.org/10.1007/978-981-10-6593-4_22

Verma P, Yadav AN, Khannam KS, Mishra S, Kumar S, Saxena AK et al (2019) Appraisal of diversity and functional attributes of thermotolerant wheat associated bacteria from the peninsular zone of India. Saudi J Biol Sci 26:1882–1895. https://doi.org/10.1016/j.sjbs.2016.01.042

Vie V, Van-Mau N, Pomarede P, Dance C, Schwartz JL, Laprade R, Frutos R, Rang C, Masson L, Heitz CF, Grimellec L (2001) Lipid-induced pore formation of the *Bacillus thuringiensis* Cry1Aa insecticidal toxin. J Membr Biol 180:195–203

Wagh SG, Pohare MB (2019) Current and future prospects of plant breeding with CRISPR/Cas. Current Journal of Applied Science and Technology 38(3):1–17

Wang J, Boets A, Van Rie J, Ren G (2003) Characterization of *cry1, cry2* and *cry9* genes in *B. thuringiensis* isolates from China. J Invertebr Pathol 82:63–71

Wang Y, Liu Y, Zhang J, Crickmore N, Song F, Gao J, Shu C (2018) Cry78Aa, a novel *Bacillus thuringiensis* insecticidal protein with activity against *Laodelphax striatellus* and *Nilaparvata lugens*. J Invertebr Pathol 158:1–5

Watrud LS, Perlak FJ, Tran M, Kusano K, Mayer EJ, Miller-Wideman MA, Obukowicz MG, Nelson DR, Kreitinger JP, Kaufman RJ (1985) Cloning of the *B. thuringiensis* subsp. *kurstaki* delta-endotoxin gene into *P. fluorescens*: molecular biology and ecology of an engineered microbial pesticide. In: Halverson Pramer HO, Rogul M (eds) Engineered organisms in the environment scientific issues. American Society for Microbiology, Washington, DC, pp 40–44

Wei J, Yuyuan Guo Y, Gemei Liang G, Kongming Wu K, Jie Zhang J, Bruce E, Tabashnik BE, Xianchun Li X (2015) Cross-resistance and interactions between *Bt* toxins Cry1Ac and Cry2Ab against the cotton bollworm. Sci Rep 5

Whitehouse MEA, Wilson LJ, Davies AP, Cross D, Goldsmith P, Thompson A, Harden S, Baker G (2014) Target and non-target effects of novel "triple-stacked" *Bt* transgenic cotton 1: canopy arthropod communities. Environ Entomol 43:218–241

Wilcks A, Jayaswal N, Lereclus D, Andrup L (1998) Characterization of plasmid pAW63, a second self-transmissible plasmid in *Bacillus thuringiensis* subsp. *Kurstaki HD 73*. Microbiol 144:1263–1270

Wu D, Aronson AI (1992) Localized mutagenesis defines regions of the *B. thuringiensis* δ-endotoxin involved in toxicity and specificity. J Biol Chem 267:2311–2317

Wu X, Huang F, Leonard BR, Moore SH (2007) Evaluation of transgenic *Bacillus thuringiensis* corn hybrids against Cry1Ab-susceptible and resistant sugarcane borer (Lepidoptera: Crambidae). J Econ Entomol 100:1880–1886

Xiao Y, Wu K (2019) Recent progress on the interaction between insects and *Bacillus thuringiensis* crops. Phil Trans R Soc B 374:20180316

Xu X, Yu L, Wu Y (2005) Disruption of a cadherin gene associated with resistance to Cry1Ac δ-endotoxin of *Bacillus thuringiensis* in *Helicoverpa armigera*. Appl Environ Microbiol 71:948–954

Yadav AN, Saxena AK (2018) Biodiversity and biotechnological applications of halophilic microbes for sustainable agriculture. J Appl Biol Biotechnol 6:48–55

Yadav AN, Sachan SG, Verma P, Saxena AK (2015a) Prospecting cold deserts of north western Himalayas for microbial diversity and plant growth promoting attributes. J Biosci Bioeng 119:683–693. https://doi.org/10.1016/j.jbiosc.2014.11.006

Yadav AN, Verma P, Kumar M, Pal KK, Dey R, Gupta A et al (2015b) Diversity and phylogenetic profiling of niche-specific bacilli from extreme environments of India. Ann Microbiol 65:611–629

Yadav AN, Sachan SG, Verma P, Kaushik R, Saxena AK (2016a) Cold active hydrolytic enzymes production by psychrotrophic bacilli isolated from three sub-glacial lakes of NW Indian Himalayas. J Basic Microbiol 56:294–307

Yadav AN, Sachan SG, Verma P, Saxena AK (2016b) Bioprospecting of plant growth promoting psychrotrophic bacilli from cold desert of north western Indian Himalayas. Indian J Exp Biol 54:142–150

Yamamoto T, Mclanghlin RE (1981) Isolation of a protein from the parasporal crystal of *Bacillus thuringiensis* var. k*urstaki* toxic to the mosquito larva, *Aedes taeniorhynchus*. Biochem. Biophys. Res Commun 103:414–421

Yamamoto T, Powell GK (1993) *Bacillus thuringiensis* crystal proteins: recent advances in understanding its insecticidal activity. In: Kim L (ed) Advanced engineered pesticides. Marcel Dekker Inc, New York, pp 3–42

Yu HL, Li YH, Li XJ, Romeis J, Wu KM (2013) Expression of Cry1Ac in transgenic Bt soybean lines and their efficiency in controlling lepidopteran pests. Pest Manag Sci 69:1326–1333

Zhang MY, Lövgren A, Low MG, Landen R (1993) Characterization of an avirulent pleiotropic mutant of the insect pathogen *Bacillus thuringiensis*: reduced expression of flagellin and phospholipases. Infect Immun 61:4947–4954

Zhang X, Candas M, Griko NB, Taissing R Jr, Bulla LA (2006) A mechanism of cell death involving an adenylylcyclase/PKA signaling pathway is induced by the Cry1Ab toxin of *Bacillus thuringiensis*. Proc Natl Acad Sci U S A 103:9897–9902

Zhang H, Yin W, Zhao J, Jin L, Yang Y, Wu S, Tabashnik BE, Wu Y (2011) Early warning of cotton bollworm resistance associated with intensive planting of *Bt* cotton in China. PLoS One 6(8):e22874

Zhang TT, Coates BS, Wang YQ, Wang YD, Bai SX, Wang ZY, He KL (2017) Down-regulation of aminopeptidase N and ABC transporter subfamily G transcripts in Cry1Ab and Cry1Ac resistant Asian corn borer, Ostrinia furnacalis (Lepidoptera: Crambidae). Int J Biol Sci 13:835–851

Zhao M, Yuan X, Wei J, Zhang W, Wang B, Khaing MM, Liang G (2017) Functional roles of cadherin, aminopeptidase-N and alkaline phosphatase from *Helicoverpa armigera* (HUbner) in the action mechanism of *Bacillus thuringiensis* Cry2Aa. Sci Rep 7:46555

Zheng A, Zhu J, Tan F, Guan P, Yu X, Wang S, Deng Q, Li S, Liu H, Li P (2010) Characterisation and expression of a novel haplotype *cry2A*-type gene from *Bacillus thuringiensis* strain JF19-2. Ann Microbiol 60:129–134

Zhong C, Ellar DJ, Bishop A, Johnson C, Lin S, Hart ER (2000) Characterization of *B. thuringiensis* δ-endotoxin which is toxic to insects in three orders. J Invertebr Pathol 76:131–139

Zhu J, Zheng A, Wang S, Liu H, Li P (2010) Characterization and expression of *cry4Cb1* and *cry30Ga1* from *Bacillus thuringiensis* strain HS18-1. J Invertebr Pathol 103:200–202

Zhuang M, Oltean DI, Gomez I, Pullikuth AK, Soberion M, Bravo A, Gill SS (2002) *Heliothis virescens* and *Manduca sexta* lipid rafts are involved in Cry1A toxin binding to the midgut epithelium and subsequent pore formation. J Biol Chem 277:13863–13872

Zotti M, Santos EAD, Cagliari D, Christiaens O, Taning CNT, Smagghe G (2018) RNA interference technology in crop protection against arthropod pests, pathogens and nematodes. Pest Manag Sci 74:1239–1250

Entomopathogenic Microbes for Sustainable Crop Protection: Future Perspectives

19

P. S. Soumia, Ram Krishna, Durgesh Kumar Jaiswal, Jay Prakash Verma, Janardan Yadav, and Major Singh

Abstract

Food security for a growing world population is a challenging task under limited agriculture land and increasing incidences of pest infestation. In the pursuit of this challenging task, modern agriculture has seen a major escalation in the use of agrochemicals like chemical pesticides and fertilizer by the farmers. Repeated application of pesticides to agricultural soil not only leads to the development of pesticide-resistant pests but also imposes their toxicity in environmental systems and human health. Annually, ~40% of global crop loss occurs due to the inability of farmers in managing the pest attack and due to the resistance shown by these pests against different pesticides. Also, maintaining the sustainability of agro-ecosystems during agricultural production has become an important issue worldwide. Therefore, it is imperative to identify and develop biopesticides, which can provide protection to crops against a broad range of pests and insects' infestations as well as be cost-effective and environmentally friendly at the same time. In the development of biocontrol agents (BAC), entomopathogenic microorgan-

P. S. Soumia · M. Singh
ICAR-Directorate of Onion and Garlic Research, Pune, Maharashtra, India

R. Krishna
ICAR-Directorate of Onion and Garlic Research, Pune, Maharashtra, India

Institute of Environment and Sustainable Development, Banaras Hindu University, Varanasi, Uttar Pradesh, India

D. K. Jaiswal (✉) · J. P. Verma
Institute of Environment and Sustainable Development, Banaras Hindu University, Varanasi, Uttar Pradesh, India
e-mail: durgesh.jaiswal1@bhu.ac.in

J. Yadav
Department of Soil Science, Institute of Agricultural Sciences, Banaras Hindu University, Varanasi, Uttar Pradesh, India

A. N. Yadav et al. (eds.), *Current Trends in Microbial Biotechnology for Sustainable Agriculture*, Environmental and Microbial Biotechnology,
https://doi.org/10.1007/978-981-15-6949-4_19

ism (EPM) plays an important role as ecofriendly biopesticides and serves as an alternative to synthetic pesticides.

EPMs consist of bacteria, fungi, nematodes, and Viruses, which can facilitate in developing a potential biological technology to suppress the pests and promote sustainable agriculture. In this chapter, we focus on the important roles and application of entomopathogenic microorganism and their types for pest and insect control.

Keywords

Biopesticides · Entomopathogenic microbes · Entomopathogenic bacteria · Entomopathogenic fungi · Entomopathogenic nematodes · Entomopathogenic virus · Sustainable crop protection

19.1 Introduction

Plants are key converters of solar energy into chemical energy, which gives about 90% and 80% of calories and protein respectively to the global human population. Rather than human, all the animals get their foods directly or indirectly from the plants. For food, humans utilize nearly 3000 plant species; however, presently the global population depends on their major calorie requirements on 20 species of crops only; of which 50% are from eight species of cereal crops (Krishna et al. 2019). Thirty species of vegetables and fruits provide minerals and vitamins. As per estimation earth can sustain 15 billion of vegetarian people or 5 billion of mixed diet population, but the world population by 2050 will reach up to 10 billion (Krishna et al. 2019). The plants in natural agroecosystem are challenged by various abiotic (drought, heat, salinity, waterlogging, etc.) and biotic (virus, bacteria, fungi, and insects) stresses (Karkute et al. 2019; Prasanna et al. 2015). The biotic stresses affect almost all the growth and developmental stages of the crop including postharvest and storage. The biotic factors including viruses, bacteria, fungi, and insects drastically reduce the crop yield and quality. Insects belong to the class Insecta of phylum Arthropoda, which comprises the largest group of invertebrates. Generally, insect pest reduces 5–25% of yield in major staple grains like wheat, maize, and rice (Deutsch et al. 2018).

In changing climatic scenario, plants are readily exposed to different abiotic and biotic stresses, which reduce the global crop yield by 20–40% annually (FAO; 2012). Moreover, yield loss will increase globally by 10–25% per degree of global mean temperature; this is because of (1) increased temperature will increase the insect metabolic rate consequently require more food and (2) change in pest insect population growth rate (Dillon et al. 2010; Irlich et al. 2009). In the near future, insect pests will become more destructive and challenging for agriculture. The prevention of losses due to these biotic factors is of paramount importance. Although chemical pesticides stand as "Miracle Weapon" in the front line of defense, agroecosystems are not free from problems. Sole reliance on an array of inorganic chemicals has led to several problems associated with pest resistance to insecticides; effects on nontarget organisms like predators, parasitoids and pollinators, environmental residues, and pollution, etc. (Soumia et al. 2015). Therefore, a more sustainable and eco-friendly pest management strategy is required to overcome the above problem. In this regard, the use of successful biological control strategies has gained special attention, as they try to strike a balance in ecological safety and enhance the natural pest control either uniquely or can be integrated with various pest management programs.

Biopesticides are biochemical pesticides which are natural in origin and generally originated from bacteria, fungi, animals, plants, and certain minerals used for the management of pests (Saxena et al. 2020; Thakur et al. 2020). At present, only 2% of biopesticides cover for the plant protection globally; which is derived either as active ingredient or products (microbial products, phytochemicals, or semiochemicals). As part of bio-intensive integrated pest management, biopesticides have an essential role in crop protection together with conventional insecticides. This approach has led to the use of various entomopathogenic microbial species such as viruses, bacteria, fungi, and nematodes as biopesticides (Vega and Kaya 2012). These naturally occurring entomopathic microbes act as bioinsecticides, fungicides, and bio-herbicides (Gupta and Dikshit 2010).

19.2 Entomopathogenic Bacteria

Bacteria are found ubiquitously in nature. Most bacterial species inhibit insect bodies and developed mutualistic relationships with insects as they have coevolved with insects (Feldhaar 2011), while some act as pathogens of insect. The latter has evolved a multiplicity of strategies to infect the host, to conquer its immune responses, to infect, and to kill it. In general, Entomopathogenic bacteria (EPB) is broadly divided into two categories (Fig. 19.1). Under unfavorable conditions, these spore-forming bacteria produce endospores along with proteinaceous parasporal bodies. Whereas, non-spore-forming bacteria do not produce any spores, lack stability in storage.

19.2.1 Classification of Entomopathogenic Bacteria

19.2.1.1 Spore-Forming Bacteria

With the discovery of entomopathogenic bacteria (EPB), belonging to genus *Bacillus* has resulted in the paradigm shift in the biological control program (Glare and O'callaghan 2000). At the outset, Japanese beetle, *Popillia japonica* Newman (Steinhaus 1975) was managed by *Paenibacillus* sp. (former *Bacillus popilliae* Dutky). The bacterium caused milky disease in larvae of *P. japonica*. However, the pathogenicity of EPB was proved with the discovery of Gram-positive, a rod-shaped and sporulating bacterium, *Bacillus thuringiensis* (*Bt*) showing high degree species specificity. Ishiwata was first to isolate *B. thuringiensis* from sick silkworms during

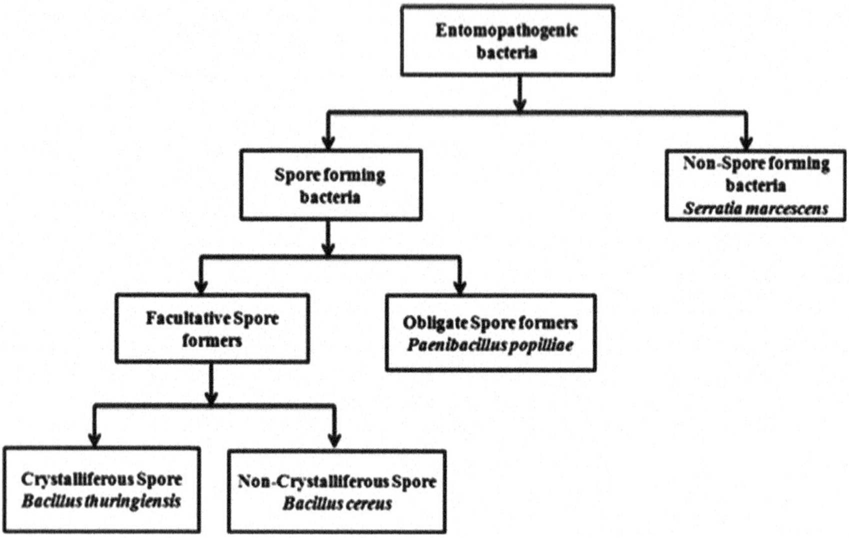

Fig. 19.1 Spore forming bacteria belonging to genera *Bacillus*, *Paenibacillus*, and *Lysinibacillus* and non-spore-forming bacteria belonging to genera Serratia. Further spore-forming bacteria is classified as facultative spore formers (Crystalliferous: *B. thuringiensis* and non-crystalliferous: *B. cereus*) and obligate spore formers (Paenibacilluspopilliae)

1901 in Japan, however, Ernst Berliner properly explained and named *Bt* after isolating from sick larvae of Mediterranean flour moth (*Ephestia kuehniella*) (Berliner 1915). Consequently, isolation and monitoring efforts have resulted in the recognition of new *Bt* strains capable to infest different insect species belonging to six different taxonomic orders (van Frankenhuyzen 2009). *Bt* subspecies *kurstaki* (*Btk*) and *aizawai* (*Bta*) against lepidopteran pest (de Barjec and Lemille 1970), *Bt* subspecies *israelensis* (*Bti*) and *Lysinibacillus* (*Bacillus*) *sphaericus* against dipteran pest (Goldberg and Margalit 1977), and *Bt biovar. Tenebrionis* against coleopteran pests (Krieg et al. 1983).

Undoubtedly, *B. thuringiensis* is the species that are exploited the most as commercial biopesticides. In terms of efficacy and costs of production, *Bt* formulations provided high toxicity as compared to conventional insecticides. A potential decrease of leaf miner, *Tuta absoluta* in tomato, was recorded by formulations of *Bt* in Spain (González-Cabrera et al. 2011). The formulations of *Bt* are also recommended to control various lepidopteran pests infecting blueberry, grape, and strawberry (Haviland 2014; Zalom et al. 2014; Varela et al. 2015).

19.2.1.2 Non-spore-Forming Entomopathogenic Bacteria

Besides *Bacillus*, other Entomopathogenic bacteria like *Clostridium bifermentans* serovar *Malaysia* against blackflies and mosquitoes (Nicolas et al. 1990), *Pseudomonas entomophila* trigger a response of systemic immune in *Drosophila melanogaster* (Vodovar et al. 2005) have shown the potential for insect pest management. Another group of Gamma proteobacteria members includes two genera

Xenorhabdus and *Photorhabdus,* characterized by endosymbionts of entomopathogenic nematodes. The first is naturally related with nematodes in the genus *Steinernema*, while the later colonizes the genus *Heterorhabditis*. Once a susceptible host is invaded by a nematode, symbiotic bacteria are released into the hemocoel. Subsequently, proliferate within the hemocoel and produce various antimicrobial compounds to restrain the growth of other microorganisms. Further, they impair the insect immune system and kill the host. Moreover, *Serratia entomophila* (Enterobacteriaceae) against grass grub, *Costelytra zealandica* (White) (Jackson et al. 1992) produces insecticidal toxins viz., *Sep* proteins (*SepA*, *SepB*, *SepC*) showing similarities to the insecticidal toxins of *Photorhabdus luminescens.* Similarly, *Yersinia entomophaga,* a non-spore-forming bacteria, produces toxic complex *Yen-Tc*, containing chitinases and toxins (Landsberg et al. 2011).

19.2.2 Mode of Action of Entomopathogenic Bacteria

The virulence and species specificity of entomopathogenic bacteria like *B. thuringiensis (Bt)* generally depend on the produced toxins within the parasporal crystal body at the time of stationary phase of growth. Toxins are of two types: *Cry* (from Crystal) and *Cyt* (from Cytolytic) toxins, present in the parasporal bodies. Whereas, vegetative *Bt* cells secrete *Vip* toxins (from vegetative insecticidal protein) and production of Bin toxins by *L. sphaericus* (Djukic et al. 2011; Adang et al. 2014). The classification and naming of *Cry*, *Cyt,* and *Vip* toxins are based on the similarity in the sequences of proteins (Crickmore et al. 1998); >300 *Cry*, 11 *Cyt*, and 32 *Vip* toxin holotypes have been described (Crickmore et al. 2015) (Fig. 19.2).

Fig. 19.2 Entomopathogenic microbes: (1) NPV, (2) Bt, and (3) EPF

19.2.2.1 Crystal Proteins or *Cry* Toxins

During sporulation, bacterial cells produce crystal proteins within parasporal crystalline inclusions; broadly known as the δ-endotoxins which consist of both *Cry* and *Cyt* toxins. The presence of crystal toxins is not distinctive to the *Bt* parasporal body; however, identified as secretory proteins (Varani et al. 2013) or the product of cryptic genes (Crickmore et al. 1994) from other bacteria (Barloy et al. 1996). On the basis of amino acid sequence similarity, *Cry* toxins are named and classified; with about 70 *Cry* groups having <40% identity are described at the primary rank represented with definite numerals (e.g., *Cry1*, *Cry2*, *Cry3*, etc.) (Crickmore et al. 2015). Based on <70% identity, these *Cry* groups are further subdivided and represented with different alphabets (i.e., *Cry1A*, *Cry1B*, *Cry2A*, etc.). Usually, the size of the *Cry* proteins ranges, from 40 to 148 kDa, nevertheless, a division of five conserved domain sequence is found within most *Cry* proteins (Höfte and Whiteley 1989); essential for their structural stability and functionality of the toxin (Adang et al. 2014). It consists of three main domains. Domain I or perforating domain is made up of α-helices (seven or eight) among them five number α-helix is hydrophobic and being located in the center; responsible for toxin insertion to the membrane and formation of pore. Whereas, domain II (central or middle domain) made up of three antiparallel β-sheets, which result in the interactions of toxin and receptor, while the domain III (galactose-binding domain) is a β-sandwich composed of two antiparallel β-sheets compressed into a "jelly roll" topology, showing structural similarity to carbohydrate-binding lectins and aid in determining specificity. It is also participated in binding to the receptor and formation of pore.

When parasporal *Bt* crystal ingested by a susceptible host gets easily solubilized due to alkaline condition in the gut (pH 8–11). Later on, the proteases in gut fluid activate the dormant crystal proteins (i.e., Protoxin) of ~120 kDa into a toxin core (delta endotoxin) of 65- to 55 kDa by digesting about 500 amino acids of the C terminus and 43 of the N-terminus from the protoxin. *Cry* protoxins of small size lack C-terminus; hence generally processing takes place at the N-terminus. The N-terminus processing is crucial for the generation of oligomers having insecticide activity (Bravo et al. 2007). The specificity is influence by both solubilization and activation (Walters et al. 2008), and their changes may lead to resistance development (Oppert et al. 1997). Certain proteins and glycoconjugates receptor functionality are provided for cadherin-like proteins (Tsuda et al. 2003), aminopeptidase-N (Sivakumar et al. 2007), alkaline phosphatase (Fernandez et al. 2006), and proteins in the ATP-binding cassette (ABC) transporter family (Tanaka et al. 2013).

The activated insecticidal δ-endotoxins recognize specific binding receptors on the brush-border midgut epithelial cells and create pores. As a result, the integrity of midgut epithelial cells is disturbed, osmoregulation lost, paralysis of midgut, and consequently cell lysis occurs which facilitates bacterial invasion into a body cavity (hemocoel). Once *Bt* cells invade the hemocoel, utilizes the host's resources to grow and multiply, causes septicemia, and then enters the sporulation phase (Raymond et al. 2010), finally the host insect death. Insects displaying various types of responses to *Bt* toxins. The response types given below are based on the caterpillars to susceptibility *Bt* toxins.

Type I response: Once delta endotoxin is ingested, midgut paralysis and septicemia occur within 24–48 hours. General indications consist of feeding cessation, an increase in hemolymph pH, diarrhea, sluggishness, and vomiting, e.g., Type I response showed by silkworm and tobacco hornworm.

Type II response: Upon ingestion of delta endotoxin by a susceptible host, midgut paralysis and septicemia occur within 24–72 hours. General paralysis of the host does not occur. e.g.: Alfalfa caterpillar and cabbage butterfly.

Type III response: The paralysis of midgut takes place after delta endotoxin is ingestion followed by feeding cessation; no usual paralysis. Mortality takes place in 48–96 hours, e.g., Corn earworm, gypsy moth, Mediterranean flour moth, and spruce budworm.

Type IV response: Insects, in general, are resistant to infection by nature; however, younger ones are more susceptible than old instars. The paralysis of midgut takes place after delta endotoxin is ingestion followed by feeding cessation, but no usual paralysis. The mortality takes place in 72–96 hours or more, e.g., Armyworms and cutworms. Dissimilar to the caterpillars, response in mosquitoes is different whereupon ingestion of *Bt* subsp. *israelensis* delta endotoxin, the larva of mosquito is killed within 20–30 min.

19.2.2.2 Cytotoxic Proteins or *Cyt* Toxins

In general, some *Bt* crystals also contain *Cyt* proteins which exhibit cytolytic (hemolytic) activity, mostly found toxic against mosquitoes and black flies (Soberon et al. 2013). *Cyt1Aa* (Cohen et al. 2011) and *Cyt2Ba* (Cohen et al. 2008) have a single domain of three-layer alpha–beta proteins, while *Cyt1Ca* the domain shows homology to the carbohydrate-binding domain of ricin but has no larvicidal or hemolytic activity. Two different action modes have been suggested for the *Cyt* group proteins: first, pore-formation model where it binds to the midgut membrane lipids and forms pores; whereas, the second suggests an imprecise detergent-like action mechanism resulting in cytolysis (Butko 2003; de Maagd et al. 2003; Soberon et al. 2013).

19.2.2.3 Vegetative Insecticidal Protein or *Vip* Proteins

When a susceptible host ingests *Vip* proteins or binary *Cry* toxins which are produced by vegetative cells target hemocoel mostly during the later stages of infection. *Vip* toxins comprise a group of three toxins namely *Vip1*, *Vip2,* and *Vip3*, wherein *Vip1* and *Vip2* act as A/B binary toxins (Barth et al. 2004), in which *Vip1* protein functions as binding and translocation (or "B") component, while the latter corresponds to the toxic (or "A") component showing similarity with actin–ADP–ribosylating toxins (Jucovic et al. 2008). While *Vip3* protein results in cytotoxicity through pore formation (Lee et al. 2003). *Vip1* protein is reported to bind the insect midgut epithelial cells (Sattar and Maiti 2011) and result in pore creation (Leuber et al. 2006). Subsequently, the pore thus formed allows *Vip2* to pierce the cell and occupy ribosylation activity to prevent polymerization of actin (Han et al. 1999). While *Vip3* exhibits a similar mode of action as that of *Cry* toxins, which includes activation through midgut proteases, binding to receptors, and pore formation (Lee et al. 2003; Liu et al. 2011). In comparison to the Cry proteins, *Vip3* toxins seem to lack a protease-resistant toxin core.

19.2.2.4 Binary Toxins or Bin Toxins

L. sphaericus produce spherical endospore toxic against mosquitoes. The parasporal crystals are produced within the exosporium and are closely associated with the endospore. A single parasporal crystal contains two equimolar proteins of 42 kDa (*BinA* or *P42*) and 51 kDa (*BinB* or *P51*) represented as binary toxins/*Bin* toxins (Broadwell et al. 1990). Both the toxins exhibit low similarity within sequences but share several identical regions which are necessary for toxicity (Clark and Baumann 1991). On the basis of variability in sequence *Bin* toxins are differentiated into four groups represented with different numbers as *Bin1*, *Bin2*, *Bin3,* and *Bin4* (Humphrey and Berry 1998). *BinB* protein structure has a lectin-like N-terminal domain and a C-terminal domain with the similarity to aerolysin-type toxins (Srisucharitpanit et al. 2014); the former is crucial for midgut receptor binding (Singkhamanan et al. 2010) while the latter is responsible for interacting with *BinA* (Oei et al. 1992). BinA/BinB protein binding to receptors leads in pore creation with the help of a β-barrel structure in the larval gut (Boonserm et al. 2006) and allows internalization of *BinA* to exert cytotoxicity. Furthermore, the vegetative cells of some *L. sphaericus* strains synthesize toxins lethal for mosquito (*Mtx* proteins).

19.3 Entomopathogenic Fungi

Among all the biological control agents, entomopathogenic fungi (EPF) are most important due to its broad host range, route of pathogenicity, and its ability to control sap-sucking pests (Qazi and Khachatourians 2005; Thomas and Read 2007; Fan et al. 2007) as well as chewing pests (Hajek and St. Leger 1994; de Faria and Wraight 2007). More than 700 species of entomopathogenic fungi from about 90 genera are reported (Khachatourians and Qazi 2008); of which a fair share belongs to *Ascomycota* and *Zygomycota*. EPF may be applied either as conidia or mycelium which sporulates after application. Among EPF, *Beauveria bassiana, Metarhizium anisopliae,* and *Verticillium lecanii* are intensively studied (de Faria and Wraight 2007). Most EPFs infect almost all insect orders (Ramanujam et al. 2014) e.g., *B. bassiana* and *M. anisopliae* taint in excess of 700 species in a few insect orders. While some have limited their host go, for example, *Aschersonia aleyrodis* taints whiteflies and *Nomuraea rileyi* infects lepidopteran hatchlings as it were (Table 19.1).

During the late nineteenth century, *B. bassiana* is the first fungal bioinsecticide to be used that caused white muscadine disease in insects. Apart from *B. bassiana*, *B. brongniartii* strains are also found pathogenic against wide range of insects, exhibits varying degree of virulence. These fungi colonize numerous plants endophytically and induce systemic resistance against various plant pathogens like *Pythium, Rhizoctonia,* and *Fusarium* (Ownley et al. 2010). *B. bassiana* formulation has resulted

Table 19.1 Entomopathogenic microbes to insect pests of agricultural crops

Entomopathogenic microbes		Target pest	Experiment assays	Study	References
Entomopathogenic bacteria (EPB)	*Brevibacillus laterosporus*	*Chrysoperla agilis*	Laboratory bioassays	Safety evaluation of the entomopathogenic bacterium *Brevibacillus laterosporus* for the green lacewing *Chrysoperla agilis* (Neuroptera: Chrysopidae)	Ruiu et al. 2020
	B.aryabhattai	*Cassida vittata*	Field trials	Entomopathogenic Bacteria, *Bacillus Aryabhattai* strain B8w 22, isolated from *Cassida vittata* Vill. And its pathogenicity against, this insect in Egyptian sugar beet fields	Bazazo et al. 2019
	Metarhizium anisoplia	*Callosobruchus chinensis*	Chickpea	Synergistic effect of entomopathogenic fungi and Bacteria against pulse beetle, *C. chinensis*	Iqbal et al. 2019
	Bacillus sp. ABP14	*Anticarsia gemmatalis*	Laboratory bioassays	Genome analysis of Entomopathogenic *Bacillus* sp. ABP14 isolated from a lignocellulosic compost	Andreazza et al. 2019
	Serratia marcescens PS-1	*Bactrocera dorsalis*	Laboratory bioassays	Intestinal responses of the oriental fruit fly *Bactrocera dorsalis* (Hendel) after ingestion of an entomopathogenic bacterium strain	Zeng et al. 2019
	Enterobacter cloacae, Serratia marcescens, Serratia liquefaciens, P. aeruginosa, Kocuria kristinae	*D. micans, T. pityocampa, L. dispar*	Laboratory bioassays	Potential of five non-spore-forming bacteria, originated from the European cockchafer, *M. melolontha* on three economic insect pests	Ertürk and Yaman 2019
	B. thuringiensis, B. brevis B. sphaericus	*Ricania simulans*	Laboratory bioassays	Prospects of Entomopathogenic Bacteria and Fungi for biological control of *Ricania simulans* (Walker 1851) (*Hemiptera: Ricaniidae*)	Gokturk et al. 2018
	Serratia marcescens	*Polyphylla olivieri*	Laboratory bioassays	Isolation and characterization of a *Serratia marcescens* with insecticidal activity from *Polyphylla olivieri* (Col.: Scarabaeidae)	Sajnaga et al. 2018
	Staphylococcus vitulinus	*Spodoptera litura* fab	Cotton	Larvicidal effect of *Staphylococcus vitulinus* bacteria against *Spodoptera litura* fab	Kalane and Pardeshi 2018

(continued)

Table 19.1 (continued)

Entomopathogenic microbes		Target pest	Experiment assays	Study	References
	Bacillus cereus, P. pulmonis	*Brahmina coriacea*	Laboratory bioassays	Pathogenicity of entomopathogenic bacteria isolated from white grub *Brahmina coriacea* on its life stages	Sharma et al. 2018
	B. thuringiensis	*Spodoptera litura*	Laboratory bioassays	Isolation and toxicity test of *Bacillus thuringiensis* from Sekayu region soil, South Sumatra on *Spodoptera litura*	Afriani et al. 2018
	B. thuringiensis	*Hypothenemus hampei*	Laboratory bioassays	Isolation, morphological, and molecular characterization of *B. thuringiensis* strains against *Hypothene mushampei* Ferrari (Coleoptera: Curculionidae: Scolytinae)	Zorzetti et al. 2018
	B. thuringiensis	*Rhynchophorus ferrugineus*	Laboratory bioassays	An entomopathogenic bacterium strain, *B. thuringiensis*, as a biological control agent against the red palm weevil, *R. ferrugineus* (Coleoptera: Curculionidae)	Pu et al. 2017
	Alcaligenes aquatilis, Alcaligenes faecalis, E.mundtii, P. protegens, S. nematodiphila, S. marcescens, S. maltophilia	*L. dispar, M. neustria*	Laboratory bioassays	Oral insecticidal activity of new bacterial isolates against insects in two orders	Ruiu et al. 2017
	Serratia marcescens	*Nilaparvata lugens*	Laboratory bioassays	Identification of a bacterium isolated from the diseased brown planthopper and determination of its insecticidal activity	Niu et al. 2016
	S. aureus, S. sciuri, B. circulans, B. thuringiensis	*Crepidodera aurata*	Laboratory bioassays	Insecticidal effects of entomopathogenic bacteria from the willow flea beetle, *C. aurata.*	Yaman and Ertürk 2016

	Pseudomonas sp., *Pseudomonas* sp. *Enterobacter* sp., *Bacillus* sp.	*Locusta migratoria*	Laboratory bioassays	Isolation and identification of entomopathogenic bacteria from Algerian desert soil and their effects against the migratory locust, *Locusta migratoria* (Linnaeus, 1758)	Oulebsir-Mohandkaci et al. 2015
Entomopathogenic fungi (EPF)	*P. taiwanensis*, *P. elaphomyceticola*	*Meloidogyne graminicola*	Laboratory bioassays	In vitro inferred interactions of selected entomopathogenic fungi from Taiwan and eggs of *Meloidogyne graminicola*	Yang et al. 2020
	B. bassiana, *M. anisopliae*	*C. vittata*	Laboratory bioassays	Isolation, mass production, and application of entomopathogenic fungi for insect pests control	Abdel-Raheem 2020
	B. brongniartii, *N. rileyi*	*P. operculella*	Potato		
	M. robertsii, *B. bassiana*	*T. urticae*	Strawberry	Root inoculation of strawberry with the entomopathogenic fungi *M. robertsii* and *B. bassiana* reduces incidence of the two spotted spider mite and selected insect pests and plant diseases in the field	Canassa et al. 2020
	Isaria javanica BE01	*Hyphantria cunea*	Laboratory bioassays	Isolation and identification of entomopathogenic fungi and an evaluation of their actions against the larvae of the fall webworm, *H. cunea*	Wang et al. 2019
	Beauveria bassiana, *Metarhizium robertsii*	*Lobesia botrana*	Laboratory bioassays	Characterization of entomopathogenic fungi from vineyards in Argentina with potential as biological control agents against the European grapevine moth *L. botrana*	Plantey et al. 2019
	B. bassiana deb4, *P. verrucosum* Nlg1, *F. equiseti* khr4	*Aphis gossypii*	Laboratory bioassays	Biological control of papaya aphid (*Aphis gossypii* glover) using entomopathogenic fungi	Mukherjee et al. 2019
	B. bassiana *M. anisoplidae*	*Sitophilus granarius*	Laboratory bioassays	Efficacy of two species of entomopathogenic fungi against the stored-grain pest, *S. granarius* L. via oral ingestion	Batta 2018
	M. anisopliae ICIPE 30	*Spoladea recurvalis*	Laboratory bioassays	Effects of entomopathogenic fungi and *Bacillus thuringiensis*-based biopesticides on *S. recurvalis*	Opisa et al. 2018

(continued)

Table 19.1 (continued)

Entomopathogenic microbes		Target pest	Experiment assays	Study	References
	B. bassiana, *M. anisopliae*	*R. ferrugineus*	Laboratory bioassays	Isolation, molecular identification, and under lab evaluation of the entomopathogenic fungi *M. anisopliae* and *B. bassiana* against the red palm Weevil *R. ferrugineus* in Gaza strip	El Kichaoui et al. 2017
	I. farinose, *B. bassiana*	*Thaumetopoea pityocampa*	Laboratory bioassays	Entomopathogenic fungi isolated from *T. pityocampa* and their efficacies against its larvae	Akinci et al. 2017
	B. bassiana SD15, *M. anisopliae* SD3	*Myzuspersicae Botrytis cinerea*	Laboratory bioassays	Entomopathogenic fungi as dual control agents against both the pest *Myzuspersicae* and Phytopathogen *Botrytis cinereal*	Yun et al. 2017
	B. Bassiana *P. lilacinus*	*Galleria mellonella*	Laboratory bioassays	Selection of entomopathogenic fungi as biocontrol agent against the greater wax moth, *Galleria mellonella* L.	Ibrahim et al. 2016
	P. xylostella	*Plutella xylostella (L.)*	Laboratory bioassays	Potential of entomopathogenic fungi as biological control agents of diamondback moth (*Lepidoptera*: *Plutellidae*) and compatibility with chemical insecticides	Duarte et al. 2016
	B. medogensis sp. nov.	*Helicoverpa armigera Tenebrio molitor*	Laboratory bioassays	*B. medogensis* sp. nov., a new fungus of the entomopathogenic genus from China	Imoulan et al. 2016
	M. anisopliae Gg-12	*Gryllotalpa gryllotalpa*	Laboratory bioassays	Characterization and virulence of entomopathogenic fungi from *Gryllotalpa gryllotalpa*	Sönmez et al. 2016
	M. anisopliae FCM AR *B. bassiana* GAR 17 B3	*Planococcus citri*	Laboratory bioassays	Screening of entomopathogenic fungi against citrus mealybug, *Plannococcus citri*	Jaihan et al. 2016
	P. lilacinum SD17	*M. persicae*	Laboratory bioassays	Evaluation of entomopathogenic fungi against the green peach aphid, *Myzuspersicae*, using multiple tools	Lee et al. 2015

	M. robertsii MTL151 and GW461	*Cydiapomonella*	Laboratory bioassays	Pathogenicity of some south African entomopathogenic fungi (Ascomycota) against eggs and larvae of *Cydiapomonella*	AbaajehAR 2015
	Purpureocillium lilacinum	*L. decemlineata*, *P. operculella*	Laboratory bioassays	Identification of Turkish isolate of the entomopathogenic fungi, *P. lilacinum* and its effect on potato pests, Lepidoptera: Gelechiidae and Leptinotarsa decemlineata	Kepenekci et al. 2015
Entomopathogenic nematodes (EPN)	*S. carpocapsae*, *S. riobrave*	*Limonius californicus*	Laboratory bioassays	Efficacy of entomopathogenic nematodes against the sugarbeet wireworm, *Limonius californicus* (Mannerheim) (Coleoptera: Elateridae)	Sandhi et al. 2020
	S. pakistanense , *S. bifurcatum*, *S. affinae*, *S. cholashanense*	*Tribolium confusum*, *Rhyzopertha dominica*	Laboratory bioassays	Biocontrol potential of entomopathogenic nematode species against *Tribolium confusum* (Jac.) and *Rhyzopertha dominica* (fab.) under laboratory conditions	Javed et al. 2020
	S.feltiae	*B. odoriphaga*	Laboratory bioassays	Integrated management of chive gnats (*Bradysia odoriphaga* Yang & Zhang) in chives using entomopathogenic nematodes and low-toxicity insecticides	Yan et al. 2019
	Steinernema feltiae	*Bradysia odoriphaga*	Laboratory bioassays	Bioefficacy and searching effect of entomopathogenic nematode *Steinernema feltiae* SF-SN treated with thiamethoxam on *Bradysia odoriphaga* (Diptera: Sciaridae) larvae	Wu et al. 2019
	S. carpocapsae	*Leucopholis burmesterie*	Laboratory bioassays	Management of white grub (*Leucopholis burmesterie*) infesting Arecanut through entomopathogenic nematodes under field conditions	Patil et al. 2019
	S.riobrave	*Naupactus godmani*	Field trials	A biological control approach to reducing *Naupactus godmani* (Curculionidae) populations in citrus using entomopathogenic nematodes	Gulcu et al. 2019

(continued)

Table 19.1 (continued)

Entomopathogenic microbes	Target pest	Experiment assays	Study	References
S. huense, *H. bacteriophora* *H. amazonensis.*	*Rhizoglyphus robini*	Laboratory bioassays	Entomopathogenic nematodes as natural enemies for control of *Rhizoglyphus robini* (Acari: Acaridae)	Nermuť et al. 2019
H. bacteriophora *S. carpocapsae*	*Tuta absoluta*	Greenhouse	New insight into the management of the tomato leaf miner, *Tuta absoluta* (Lepidoptera: Gelechiidae) with entomopathogenic nematodes	Kamali et al. 2018
H. noenieputensis, *S. yirgalemense*	*Ceratitis capitata*	Laboratory bioassays	Surveying and screening south African entomopathogenic nematodes for the control of the Mediterranean fruit fly, *Ceratitis capitata*	James et al. 2018
S. pakistanense 94–1 *H.indica* 212–2	*Phyllotreta striolata*	Laboratory bioassays	Characterization of biological and biocontrol traits of entomopathogenic nematodes promising for control of striped flea beetle	Yan et al. 2018
H. indica.	*Galleria mellonella*	Laboratory bioassays	Isolation and identification of indigenous entomopathogenic nematode (EPN) isolate from Egyptian fauna	Lulamba et al. 2018
S. feltiae, *X. bovienii,* *P. lilacinum*	*Meloidogyne javanica,* *Meloidogyne incognita*	Field trials	Application methods of *Steinernema feltiae*, *Xenorhabdus bovienii*, and *Purpureocillium lilacinum* to control root-knot nematodes in greenhouse tomato systems	Abdel-Razek et al. 2018
H.amazonensis, *M.rainai,* *O.tipulae* *S.rarum*	*Galleria mellonella*	Laboratory bioassays	Entomopathogenic nematodes in agricultural areas in Brazil	De Brida et al. 2017
S. jeffreyense, *S.feltiae,* *H. bacteriophora*	*Cydia pomonella*	Field trials	Entomopathogenic nematodes for the control of the codling moth (*Cydia pomonella* L.) in field and laboratory trials	Odendaal et al. 2016
Steinernemafeltiae, *S. carpocapsae,* *H.bacteriophora*	*Tuta absoluta*	Laboratory bioassays	Efficacy of entomopathogenic nematodes against larvae of *Tuta absoluta* in the laboratory	Van et al. 2016

	H. zealandica	*Thaumatotibia leucotreta*	Laboratory bioassays	Evaluation of local entomopathogenic nematodes for the control of false codling moth, *T. leucotreta* (Meyrick, 1913), in a citrus orchard in South Africa	Malan and Moore 2016
	Steinernema spp., *Heterorhabditis* spp.	*Phyllotreta cruciferae*	Field spray trials	Efficacy of entomopathogenic nematodes and sprayable polymer gel against crucifer flea beetle (Coleoptera: Chrysomelidae) on canola	Antwi and Reddy 2016
	Steinernema feltiae	*Agriotes lineatus*	Laboratory bioassays	Immune defenses of *Agriotes lineatus* larvae against entomopathogenic nematodes	Rahatkhah et al. 2015
	S. feltiae, *H. bacteriophora*	*Trialeurodes vaporariorum*	Greenhouse trials	Pathogenicity of two species of entomopathogenic nematodes against the greenhouse whitefly, *Trialeurodes vaporariorum*, in laboratory and greenhouse experiments	Rezaei et al. 2015
Entomopathogenic virus (EPV)	*Baculoviruses* SfCH32 and SfCH15	*Spodoptera frugiperda*	Laboratory bioassays	Molecular and morphological characterization of multiple nucleopolyhedrovirus from Mexico and their insecticidal activity against *Spodoptera frugiperda* (Lepidoptera: Noctuidae)	Wang et al. 2020
	Red pine sawfly nuclear polyhedrosis virus, Lymantria nuclear polyhedrosis virus	*Cydalima prespectalis*	Laboratory bioassays	Virus application experience to protect boxwood against *Cydalima prespectalis* caterpillars	Gninenko et al. 2018
	Nucleopolyhedrovirus (NPV), *Granulovirus* (GV)	*Darnatrima*, *S.asigna*, *Setoranitens*	Laboratory bioassays	The efficacy of entomopathogenic virus for the control of oil palm nettle caterpillar	Loong et al. 2017

in 93% control of *Tetranychus urticae* in greenhouse tomato (Chandler et al. 2005). Besides exoenzymes, *B. bassiana* produces low-molecular weight cyclic peptides and *Cyclosporins* A and C (toxic compounds), insecticidal cyclic peptides like beauvericin, enniatins, bassianolide (Roberts 1981; Vey et al. 2001), antibacterial activity Oosporein and cyclic peptides with immunosuppressive activities.

M. anisopliae represents another well-exploited fungal species for biocontrol of insects. It causes Green muscadine disease in insects. Some of the *Metarhizium* spp are generalists, e.g., *M. anisopliae* against mosquitoes, termites, and cattle ticks, while some are specialists, e.g., *M. acridum* infects only Acrididae family (locusts and grasshoppers) (Gao et al. 2011). A variety of insecticidal toxins and virulence factors produced by *M. anisopliae* of which Destruxins A, B, C, D, E, F are produced during pathogenesis. Among which destruxins A and E are more insecticidal that suppresses the cellular and humoral immune response of the host.

Apart from these other EPFs are *Lecanicelium lecanii* and *Hirsutella thompsonii* controls sucking pests such as whiteflies, aphids, thrips, brown planthopper, scale insects, mealy bugs, red mites, etc., *N. rileyi* against pod borers, cutworms, cabbage borers, etc., while *Paecelomyces fumosoroseus* controls yellow and red mite, whiteflies, etc., *Lecanicellium muscarium* formulation has resulted in greenhouse whitefly (*Trialeurodes vaporariorum*) population reduction by 76–96% in tomato (Fargues et al. 2005).

Aside from crop protection, EPFs additionally showed positive impact on drought resilience and plant growth development in cabbage, strawberry, and estranges plant pathogens (Dara 2013; Dara et al. 2016; Dara et al. 2017). However, the main hitch of using these EPFs as biopesticides is their mode of action, i.e., conidia or spores have to come in direct contact with the host and moreover moderate temperatures and high relative humidity are required for spore germination.

19.3.1 Mode of Action of Entomopathogenic Fungi

Entomopathogenic fungi (EPF) infection in general comprises of six stages: adhesion, germination, appressorium formation, penetration, colonization of hemolymph, and extrusion and sporulation. After attaching to the host body, infection starts with the germination of conidia or spores. Subsequently, due to the enzymatic and mechanical action, EPF penetrate into the host body through specialized structures called penetration pegs or appressoria and proliferates internally by draining the insect nutrients thereby eventually kills the host. During vegetative growth phase, the fungus produces different metabolites that favor its growth and also act as virulence factors or toxins. Whereas at later stage, different types of conidia or spores are produced outside the infected host which ensures its spread. Therefore, unlike bacterial and viral pathogens of insects, fungal pathogens infect insects by breaching the host cuticle and do not need to be consumed by the host to cause infection. Use of these EPF as an alternative to chemical pesticide or their combined applications with insecticide could be useful in insecticide-resistant management (Table 19.1; Fig. 19.3).

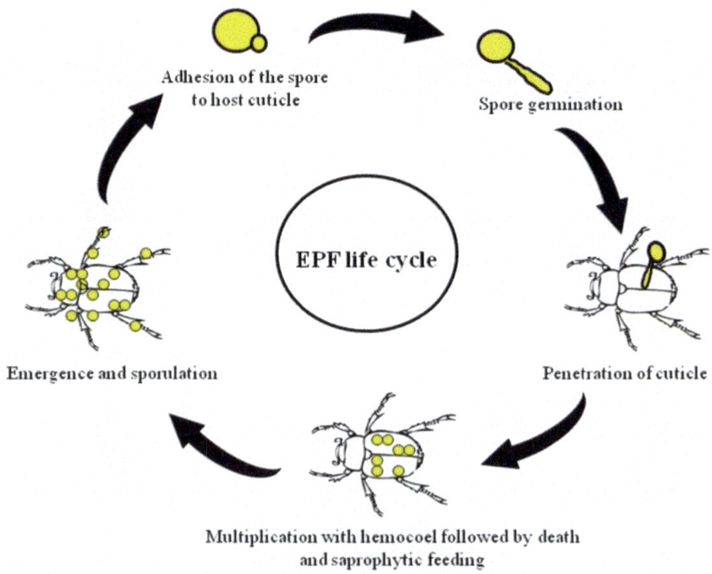

Fig. 19.3 Generalized life cycle of an entomopathogenic fungi

19.4 Entomopathogenic Nematodes

Entomopathogenic nematodes (EPNs) is tiny, soil-abiding worms that are parasitic to insects. The most broadly examined EPNs are *Steinernema* and *Heterorhabditis*. These nematodes harbor symbiotic bacteria (*Xenorhabdus* spp. for *Steinernematidae* and *Photorhabdus*spp. for *Heterorhabditidae*) in their digestive tract. The infectious phase of these nematodes is the third juvenile stag frequently alluded to as the J3 stage or "dauer" larvae. Nematodes in this stage get by without taking care of in clammy soil and similar habitats, here and there for expanded periods. After finding a powerless host, infective juveniles infiltrate into the larval body pit by means of characteristic openings like mouth, butt, or some other hypersensitive body part. Juveniles of *Heterorhabditis* nematodes can enter through the intersegmental layers of insect cuticle. Once inside the host body, the symbiotic microorganisms from the nematode are discharged into the insect hemocoel. Further, the nematode–bacterial complex would increase and quickly slaughter the host by causing septicemia. Nematodes feed on duplicating microorganisms, develop into grown-ups, replicate, and afterward rise as infective juveniles from the host body to look for new larvae in the soil. The EPN *Steinernema feltiae* decreased raspberry crown borer (*Penni setia marginata*) populaces by 33–67% (Capinera et al. 1986). In like manner, *S. carpocapsae* is generally suggested for branch and twig borer (*Melagus confertus*) the board in grapes (Varela et al. 2015).

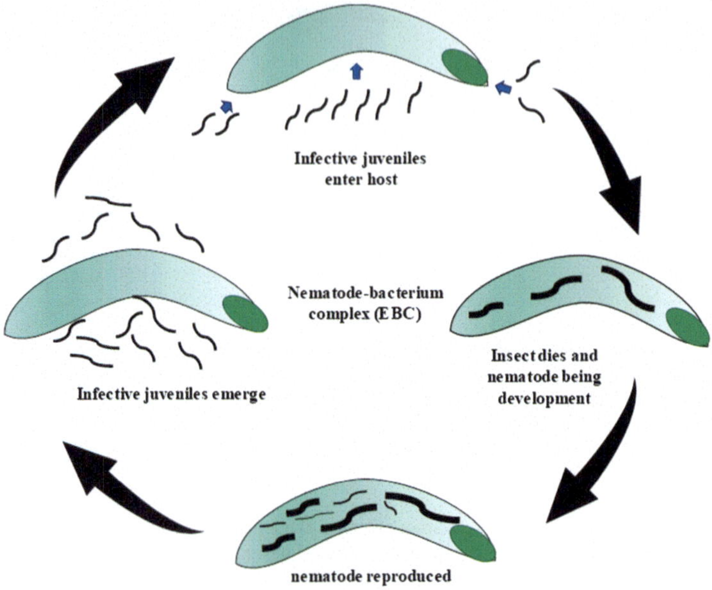

Fig. 19.4 Generalized life cycle of entomopathogenic nematode (EFN)

19.4.1 Mode of Action of Entomopathogenic Nematodes

In field application, immediately after release infective juveniles of entomopathogenic nematodes start searching for their host insects. Subsequently, infective juveniles enter the larval body through natural openings such as mouth, anus, and spiracles. In the body cavity (hemocoel), EPNs releases the symbiotic bacteria (*Xenorhabdus* spp. of *Steinernematidae* and *Photorhabdus* spp. of *Heterorhabditidae*) from their gut. The nematode–bacterium complex (NBC) starts augmentation, causes septicemia, and later executes the insect host have inside 48 hours after disease. In life pattern of nematode–bacterium complex, nematodes benefit from duplicating bacteria, develop into grown-ups, imitate, and afterward rise as infective juveniles from the insect cadaver looking for new hatchlings either in the dirt or yields (Table 19.1; Fig. 19.4).

19.5 Entomopathogenic Viruses

Entomopathogenic viruses (EPV) are obligate intracellular parasites having either DNA or RNA encapsulated into a protein coat called capsid to form virions or nucleocapsids. Similar to the mode of action of EPB, entomopathogenic viruses also needed to be ingested by the host and exhibit a high level of specificity, thereby making them ideal for controlling chewing pests. More than 1100 entomopathogenic viruses are reported to infect over 20 different insects' families, among which over 600 belongs to family *Baculoviridae* (Miller 1997; Eberle et al. 2012a, 2012b). Since the sixteenth-century, disease brought about by entomopathogenic viruses is reported; for example, *jaundice grasserie* of *Bombyx mori* currently identified as a nucleopolyhedrovirus (NPV); shaped inside nuclei. Maestri and Cornalia in 1856 recognized the occlusion bodies of the silkworm nucleopolyhedrovirus, and Paillot described the granuloviruses (GVs) just because by 1926. Similarly, another kind of polyhedrosis from silkworm whose OBs are framed inside the cytoplasm was accounted for Ishimori in 1934 as cypoviruses. With the appearance of electron microscopy, rod-shaped capsids in the OBs of baculoviruses by Bergold.

19.5.1 Classification of Entomopathogenic Viruses

According to the norms of International Committee on Taxonomy of Viruses (ICTV), EPV is a categorization dependent on the sort of hereditary material (e.g., single- or double-stranded DNA, singe- or double-stranded RNA, positive or negative strand), virion morphology and size (e.g., icosahedral, rod-shaped, etc.), nearness of an envelope encompassing the virion, nearness of an impediment body inundating the virions and host range, and so forth. Abbreviations are given dependent on the host it taints and viral gathering it has a place, for instance, the *Helicoverpa armigera* different nucleopolyhedrovirus is named as HaMNPV. In this way, all nucleopolyhedroviruses are named NPV, granuloviruses as GV, entomopoxviruses as EPV, iridoviruses as IV, and cytoplasmic polyhedrosis viruses (cypoviruses) as CPV.

19.5.1.1 Baculovirus (Nucleopolyhedrovirus [NPV]/Granulovirus [GV])

Belongs to the family Baculoviridae, wherein the virion is enveloped, occluded either within polyhedral or granular OBs and has rod-shaped viral particles. It contains circular dsDNA of 80 and 130 Kb extensively exploited as biological control agents. The family Baculoviridae contains four genera, Alpha-, Beta-, Gamma-, and Deltabaculovirus. Previously Alpha-, Gamma-, and Deltabaculovirus are collectively known as nucleopolyhedrovirus (NPV) while Betabaculovirus as Granulovirus (GV). International Committee on Taxonomy of Viruses (ICTV) just perceives 30 species inside two genera: NPVs and GVs. Further dependent on various virions per envelope; NPVs are partitioned into two: single nucleopolyhedroviruses or SNPVs having one virion for each envelope, e.g., *BmSNPV* infecting *B. mori*; and multiple nucleopolyhedroviruses or MNPVs with several virions per envelope e.g., *AcMNPV* infecting *Autographa californica*. With respect to NPV, virions replicate only within the nuclei and the size of their polyhedral OBs ranges from 1 to 15 µm; hence clearly visible under light microscopy; whereas for GV, virions replicate within the cytoplasm; always enveloped singly and their granular

OBs are very small (0.2–0.5 μm). Moreover, another virus species similar to baculoviruses are nudivirus or non-occluded baculovirus (NOB) specific to coconut rhinoceros beetle, *Oryctes rhinoceros*; rod-shaped virions but not occluded within an OB. To date, GVs have been isolated only from Lepidoptera, whereas NPVs from Lepidoptera, Hymenoptera, and Diptera.

During baculovirus infections, two phenotypes of virions are formed: (1) occlusion-derived virion (ODV) and (2) budded virion (BV). Within the nuclei of the susceptible host, the ODV is occluded in a crystalline protein matrix during the last phase of virus replication, whereas BV is produced as nucleocapsids bud through the plasma membrane of infected cells. BVs are responsible for systemic infection as they spread throughout the host tissues, while ODVs are responsible for horizontal transmission of baculovirus upon the ingestion of the OB by a susceptible host.

19.5.1.2 Entomopoxvirus

Entomopoxvirus (EPV) has a place with the family *Poxviridae*, show wide host extend. Virions are allantoid to brick formed, impeded inside ovoid OBs called spheroids, and contain dsDNA going from 270 to 320 kb. Virions are probably the biggest in the virus world, measuring 400 nm long and 250 nm in width. Like CPV, EPV additionally recreates inside the cytoplasm susceptible cells. EPVs have been disconnected from 27 species from various insect orders like orthoptera, lepidoptera, diptera, and coleoptera. The subfamily *Entomopoxvirinae* incorporates three genera called as *Entomopoxvirus* A (taints the coleoptera), *Entomopoxvirus* B (taints both lepidotpera and coleoptera), and *Entomopoxvirus* C (taints diptera).

19.5.1.3 Cypovirus (Cytoplasmic Polyhedrosis Virus)

Cytoplasmic polyhedrosis virus (CPV) has a place with the family Reoviridae, where Icosahedral virions with 12 parallel projections are impeded inside enormous isometrical polyhedra (up to 10 μm in size) and contain 10–12 segments of dsRNA of 12–32 kb. Taints a profoundly various host go and imitate inside the cytoplasm of susceptible cells. All insect explicit reoviruses are amassed inside the genus Cypovirus, and ICTV perceives 70 species facilitated by lepidoptera, e.g., *B. mori* cytoplasmic polyhedrosis infection or BmCPV. Be that as it may, 20 additional species are accounted for from diptera.

ICTV just perceives 30 species inside two genera: NPVs and GVs. Further dependent on various virions per envelope, NPVs are partitioned into two: single nucleopolyhedroviruses or SNPVs having one virion for each envelope.

19.5.1.4 Polydnavirus

Belongs to the family *Polydnaviridae*, where the virion is non-occluded, ovoid in shape, and possess multipartite dsDNA of 75–200 kb. Infects exclusively endoparasitic Hymenoptera. Polydnavirus two genera: *Ichneovirus* infecting Icheumonids and *Bracovirus* infecting Braconids.

19.5.1.5 Ascovirus

Belongs to the family *Ascoviridae*, where the virion is enveloped and has vesicle-like OBs specifically infecting the order Lepidoptera. Virions are bacilliform, ovoid, or allantoid in shape and their genomes contain circular dsDNA of 100–180 kb.

19.6 Safety and Ecotoxicology

Screening of entomopathogenic microbes (EPM) against spineless creatures gives data on dangers as well as uncovers much about their versatility and potential utility against a more extensive scope of pests or more vulnerable connections in the lifecycle of a susceptible host. Subsequently, there is a need to segregate among apparent and genuine risk. For instance, EPF is typically used to control adult and larval phases of insect yet have likewise indicated ovicidal action or decreased fertility in red spider mite and mosquitoes (Rocha et al. 2015). EPF can control the development of egg mass of nonhosts, for example, *Biomphalaria glabrata* (Duarte et al. 2015). Shockingly it has been seen that the predatory mites demonstrate resilience to EPF, along these lines permitting them to be utilized in the mix to control thrips and red spider mites successfully (Jacobson et al. 2001; Seiedy et al. 2012).

Many researchers have performed several biosafety reviews (Lacey et al. 2015; Ruiu et al. 2013; Rohlfs and Churchill 2011; Zimmermann 2007) which showed that EPM is highly safe with minimal risk of human beings and the environment. Because of the negative impact of chemical pesticide on human health and ecosystem, it is necessary to accelerate the testing of biopesticides prior to its release (Toriello 2003). Globally many international organizations have encouraged and developed the registration of EPM as biopesticides. Registration of these EPMs follows stringent rules like analysis of the product, residue, toxicology, impact on nontarget organisms, the fate of the environment, efficiency, and functionality (Siegel 1997). Insight into the host–pathogen interaction may provide information regarding host specificity. Whereas identification of certain biochemical/molecular markers will resolve the risk posed to nontarget organisms.

19.7 Future Prospects

The study of a microbial biopesticide is as yet viewed as a novel and exploring; comparatively new crop protection option in both organic and conventional production systems that attracts much of the global attention. Some of these microbial biopesticides are currently underdeveloped; hitherto prove as excellent alternatives to the chemical pesticides. Hence, in-depth research is needed in many areas such as its production, formulation, delivery, and commercialization. A public–private sector approach to the development, manufacturing, and sale of these microbial biopesticides is the need of the day. Moreover, continuous search for new biomolecules and improving the efficiency of existing biopesticides, recombinant DNA technology is also being deployed. Novel fusion proteins are being designed to develop next-generation biopesticides. Several other innovative approaches are also being applied to develop biopesticides as effective, efficient, and acceptable pest control measures. The future for biopesticides is unquestionably looking splendid, with the expanding withdrawal of agrochemical, low quantities of new synthetic substances being created, and open weight for safe pesticides. In any case, for the possibility to be acknowledged, operators including entomopathogenic microorganisms should meet the prerequisites of modern agriculture.

Acknowledgments The authors are thankful to Director, ICAR-DOGR, Pune for providing necessary facilities and timely guidance in improving the manuscript. Also express their heartfelt gratitude to Head (Dept. of ESD) and Director of Institute of Environment & Sustainable Development, Banaras Hindu University, for providing the necessary facilities during research and constant encouragement. Also, special thanks to the reviewers for their valuable suggestions to improve the manuscript.

References

AbaajehAR NF (2015) Isolation and pathogenicity of some South African entomopathogenic fungi (Ascomycota) against eggs and larvae of *Cydia pomonella* (Lepidoptera: Tortricidae). Biocontrol Sci Tech 25:828–842

Abdel-Raheem M (2020) Isolation, mass production and application of Entomopathogenic Fungi for insect pests control. In: Cottage industry of biocontrol agents and their applications. Springer, Cham, pp 231–251

Abdel-Razek AS, Hussein M, Shehata I (2018) Isolation and identification of indigenous entomopathogenic nematode (EPN) isolate from Egyptian fauna. Arch Phytopathol Plant Prot 51:197–206

Adang MJ, Crickmore N, Jurat-Fuentes JL (2014) Diversity of *Bacillus thuringiensis* crystal toxins and mechanism of action. In Adv Insect Physiol Academic Press 47:39–87

Afriani SR, Pujiastuti Y, Irsan C, Damiri N, Nugraha S, Sembiring ER (2018) Isolation and toxicity test of *Bacillus thuringiensis* from Sekayu region soil, South Sumatra on Spodopteralitura. In: In IOP conference series: earth and environmental science, vol 102:1. IOP Publishing, p 012066

Akinci HA, Ozman-Sullivan SK, Diler H, Celik N, Sullivan GT, Karaca G (2017) Entomopathogenic fungi isolated from *Thaumetopoea pityocampa* and their efficacies against its larvae. Fresenius Environ Bull 26(8):5251–5257

Andreazza AP, Cardoso RL, Cocco J, Guizelini D, Faoro H, Tadra-Sfeir MZ, Souza EM (2019) Genome analysis of entomopathogenic *Bacillus* sp. ABP14 isolated from a lignocellulosic compost. Genome Biol Evol1 1(6):1658–1662

Antwi FB, Reddy GV (2016) Efficacy of entomopathogenic nematodes and sprayable polymer gel against crucifer flea beetle (Coleoptera: Chrysomelidae) on canola. J EconEntomo 109(4):1706–1712

Barloy F, Delécluse A, Nicolas L, Lecadet MM (1996) Cloning and expression of the first anaerobic toxin gene from *Clostridium bifermentans* subsp. *Malaysia*, encoding a new mosquitocidal protein with homologies to *Bacillus thuringiensis* delta-endotoxins. *J* Bacteriol 178(11):3099–3105

Barth H, Aktories K, Popoff MR, Stiles BG (2004) Binary bacterial toxins: biochemistry, biology, and applications of common *Clostridium* and *Bacillus* proteins. Microbiol Mol Biol Rev 68(3):373–402

Batta YA (2018) Efficacy of two species of entomopathogenic fungi against the stored-grain pest, *Sitophilus granaries* L.(Curculionidae: Coleoptera), *via* oral ingestion. Egypt J Biol Pest Co 28(1):44

Bazazo KG, Ekram AA, El-Sheikh MF (2019) New record of entomopathogenic bacteria, *Bacillus aryabhattai* strain b8w 22, isolated from *Cassidavittata*vill. and its pathogenicity against this insect in Egyptian sugar beet fields. Zagazig J Agri Res 46(6): 2247–2254

Berliner E (1915) Über die Schlaffsuchtder Mehlmottenraupe (*Ephestiakühniella* Zell.) und ihren Erreger *Bacillus thuringiensis* n. sp. Z Angew Entomol 2(1):29–56

Boonserm P, Moonsom S, Boonchoy C, Promdonkoy B, Parthasarathy K, Torres J (2006) Association of the components of the binary toxin from *Bacillus sphaericus* in solution and with model lipid bilayers. Biochem Bioph Res Co 342(4):1273–1278

Broadwell AH, Baumann L, Baumann P (1990) Larvicidal properties of the 42 and 51 kilodalton *Bacillus sphaericus* proteins expressed in different bacterial hosts: evidence for a binary toxin. Curr Microbiol 21(6):361–366

Bravo A, Gill SS, Soberón M (2007) Mode of action of *Bacillus thuringiensis* Cry and Cyt toxins and their potential for insect control. Toxicon 49(4):423–435

Butko P (2003) Cytolytic toxin Cyt1A and its mechanism of membrane damage: data and hypotheses. Appl Environ Microbiol 69(5):2415–2422

Canassa F, Esteca FC, Moral RA, Meyling NV, Klingen I, Delalibera I (2020) Root inoculation of strawberry with the entomopathogenic fungi *Metarhizium robertsii* and *Beauveria bassiana* reduces incidence of the twospotted spider mite and selected insect pests and plant diseases in the field. J Pest Sci 93(1):261–274

Capinera JL, Cranshaw WS, Hughes HG (1986) Suppression of raspberry crown borer, *Pennisetia marginata* (Harris)(Lepidoptera: Sesiidae) with soil applications of *Steinernema feltiae* (Rhabditida: Steinernematidae). J Invertebr Pathol 48(2):257–258

Chandler D, Davidson G, Jacobson RJ (2005) Laboratory and glasshouse evaluation of entomopathogenic fungi against the two-spotted spider mite, *Tetranychus urticae* (Acari: Tetranychidae), on tomato, *Lycopersicon esculentum*. Biocon Sci Tech 15(1):37–54

Clark MA, Baumann PAUL (1991) Modification of the *Bacillus sphaericus* 51-and 42-kilodalton mosquitocidal proteins: effects of internal deletions, duplications, and formation of hybrid proteins. Appl Environ Microbiol 57(1):267–271

Cohen S, Dym O, Albeck S, Ben-Dov E, Cahan R, Firer M, Zaritsky A (2008) High-resolution crystal structure of activated Cyt2Ba monomer from *Bacillus thuringiensis* subsp. *Israelensis*.J. Mol Biol 380(5):820–827

Cohen S, Albeck S, Ben-Dov E, Cahan R, Firer M, ZaritskyA DO (2011) Cyt1Aa toxin: crystal structure reveals implications for its membrane-perforating function. J Mol Biol 413(4):804–814

Crickmore N, Baum J, Bravo A, Lereclus D, Narva K, Sampson K, Zeigler DR (2015) *Bacillus thuringiensis* toxin nomenclature. Available at: http://www.lifesci.sussex.ac.uk/Home/Neil_Crickmore/Bt/. Accessed on 14

Crickmore N, Wheeler VC, Ellar DJ (1994) Use of an operon fusion to induce expression and crystallisation of a *Bacillus thuringiensis* δ-endotoxin encoded by a cryptic gene. Mol Gen Genet 242(3):365–368

Crickmore N, Zeigler DR, Feitelson J, Schnepf E, Van Rie J, Lereclus D, Dean DH (1998) Revision of the nomenclature for the *Bacillus thuringiensis* pesticidal crystal proteins. Microbiol Mol Biol Rev 62(3):807–813

Dara SK (2013) Entomopathogenic fungus *Beauveria bassiana* promotes strawberry plant growth and health. UCANR eJournal Strawberries and Vegetables:30

Dara SK, Dara SS, Dara SS (2016) First report of entomopathogenic fungi, *Beauveria bassiana*, *Isaria fumosorosea* and *Metarhizium brunneum* promoting the growth and health of cabbage plants growing under water stress. UCANR eJournal Strawberries and Vegetables:19

Dara SSR, Dara SS, Dara SK, Anderson T (2017) Fighting plant pathogenic fungi with entomopathogenic fungi and other biologicals. CAPCA Adviser 20(1):40–44

de Barjec H, Lemille F (1970) Presence of flagellar antigenic subfactors in serotype 3 of *Bacillus thuringiensis*. J Invertebr Pathol 15(1):139–140

De Brida AL, Rosa JMO, De Oliveira CMG, BMDC C, Serrão JE, Zanuncio JC, SRS W (2017) Entomopathogenic nematodes in agricultural areas in Brazil. Sci Rep-Uk 7:45254

de Maagd RA, Bravo A, Berry C, Crickmore N, Schnepf HE (2003) Structure, diversity, and evolution of protein toxins from spore-forming entomopathogenic bacteria. Annu Rev Genet 37(1):409–433

Deutsch CA, Tewksbury JJ, Tigchelaar M, Battisti DS, Merrill SC, Huey RB, Naylor RL (2018) Increase in crop losses to insect pests in a warming climate. Science 361(6405):916–919

Dillon ME, Wang G, Huey RB (2010) Global metabolic impacts of recent climate warming. Nat 467(7316):704

Djukic M, Poehlein A, Thürmer A, Daniel R (2011) Genome sequence of *Brevibacillus laterosporus* LMG 15441, a pathogen of invertebrates. J Bacteriol 193:5535–5536

Duarte GF, Rodrigues J, Fernandes ÉK, Humber RA, Luz C (2015) New insights into the amphibious life of *Biomphalaria glabrata* and susceptibility of its egg masses to fungal infection. J Invertebr Patho 125:31–36

Duarte RT, Gonçalves KC, Espinosa DJL, Moreira LF, De Bortoli SA, Humber RA, Polanczyk RA (2016) Potential of entomopathogenic fungi as biological control agents of diamondback moth (Lepidoptera: Plutellidae) and compatibility with chemical insecticides. J Econ Entomol 109(2):594–601

Eberle KE, Jehle JA, Huber J (2012a) 10 microbial control of crop pests using insect viruses. Integrated pest management: principles and practice 281

Eberle KE, Wennman JT, Klespies RG, Jehle JA (2012b) Basic techniques in insect virology. Manual of techniques in invertebrate pathology, 2nd edn. Academic Press, London, UK, pp 15–74

El Kichaoui AY, Bara'a A, El Hindi M (2017) Isolation, molecular identification and under lab evaluation of the entomopathogenic fungi *M. anisopliae* and *B. bassiana* against the Red Palm Weevil *R. ferrugineus* in Gaza Strip. 7(01)

Ertürk Ö, Yaman M (2019) Potential of five non-spore-forming bacteria, originated from the European cockchafer, *Melolontha melolontha* (Linnaeus, 1758)(Coleoptera: Scarabaeidae), on three economic insect pests. Egypt J Biol Pest Co 29(1):1–5

Fargues J, Smits N, Rougier M, Boulard T, Ridray G, Lagier J et al (2005) Effect of microclimate heterogeneity and ventilation system on entomopathogenic hyphomycete infection of *Trialeurodes vaporariorum* (Homoptera: Aleyrodidae) in Mediterranean greenhouse tomato. Biol Control 32(3):461–472

de Faria MR, Wraight SP (2007) Mycoinsecticides and mycoacaricides: a comprehensive list with worldwide coverage and international classification of formulation types. Biol Control 43(3):237–256

Fan Y, Fang W, Guo S, Pei X, Zhang Y, Xiao Y, Pei Y (2007) Increased insect virulence in *Beauveria bassiana* strains overexpressing an engineered chitinase. Appl Environ Microbiol 73(1):295–302

Feldhaar H (2011) Bacterial symbionts as mediators of ecologically important traits of insect hosts. Ecol Entomol 36(5):533–543. https://doi.org/10.1111/j.1365-2311.2011.01318.x

Fernandez LE, Aimanova KG, Gill SS, Bravo A, Soberón M (2006) A GPI-anchored alkaline phosphatase is a functional midgut receptor of Cry11Aa toxin in Aedes aegypti larvae. Biochem J 394(1):77–84

Gao Q, Jin K, Ying SH, Zhang Y, Xiao G, Shang Y et al (2011) Genome sequencing and comparative transcriptomics of the model entomopathogenic fungi *Metarhizium anisopliae* and *M. acridum*. PLoS Genet 7:1

Glare TR, O'callaghan M (2000) *Bacillus thuringiensis* Biology, Ecology and safety (No. 632.951 G5)

Gninenko YI, Ponomarev VL, Nesterenkova A (2018) Virus application experience to protect boxwood against *Cydalima prespectalis* caterpillars. Recent advances in the researches and application of viruses and entomophages in forest health protection 40

Gokturk T, Tozlu E, Kotan R (2018) Prospects of entomopathogenic bacteria and fungi for biological control of *Ricania simulans* (Walker 1851)(Hemiptera: Ricaniidae). Pak J Zool 50(1)

Goldberg LJ, Margalit J (1977) A bacterial spore demonstrating rapid larvicidal activity against *Anopheles sergentii, Uranotaenia unguiculata, Culex univitattus, Aedes aegypti* and *Culexpipiens*. Mosq News 37(3):355–358

González-Cabrera J, Mollá O, Montón H, Urbaneja A (2011) Efficacy of *Bacillus thuringiensis* (Berliner) in controlling the tomato borer, *Tuta absoluta* (Meyrick)(Lepidoptera: Gelechiidae). BioControl 56(1):71–80

Gupta S, Dikshit AK (2010) Biopesticides: an ecofriendly approach for pest control. J Biopest 3(1):186–188

Hajek AE, St Leger RJ (1994) Interactions between fungal pathogens and insect hosts. Annu Rev Entomol 39(1):293–322

Han S, Craig JA, Putnam CD, CarozziNB TJA (1999) Evolution and mechanism from structures of an ADP-ribosylating toxin and NAD complex. Nat Struct sBiol 6(10):932–936

Haviland DR (2014) UC IPM Pest Management Guidelines: Blueberry. UCANR Pub. 3542

Höfte H, Whiteley HR (1989) Insecticidal crystal proteins of *Bacillus thuringiensis*. Microbiol Mol Biol R53(2):242–255

Humphrey SMJ, Berry C (1998) Variants of the *Bacillus sphaericus* binary toxins: implications for differential toxicity of strains. J Invertebr Pathol 71(2):184–185

Ibrahim AA, Mohamed HF, El-Naggar SEM, Swelim MA, Elkhawaga OE (2016) Isolation and selection of entomopathogenic fungi as biocontrol agent against the greater wax moth, *Galleria mellonella* L.(Lepidoptera: Pyralidae). Egypt J Biol Pest Co 26(2):249

Imoulan A, Wu HJ, Lu WL, Li Y, Li BB, Yang RH et al (2016) *Beauveria medogensis* sp. *nov*., a new fungus of the entomopathogenic genus from China. J Invertebr Pathol 139:74–81

Iqbal M, Shaheen FA, Mahmood R, Rafique MK, Bodlah I, Naz F, Raja MU (2019) Synergistic effect of Entomopathogenic Fungi and Bacteria against pulse beetle, *Callosobruchus chinensis*. Pak J Zool 51(5)

Irlich UM, Terblanche JS, Blackburn TM, Chown SL (2009) Insect rate-temperature relationships: environmental variation and the metabolic theory of ecology. Nat 174(6):819–835

Jackson TA, Pearson JF, O'callaghan M, Mahanty HK, Willocks MJ (1992) Pathogen to product-development of *Serratia entomophila* Enterobacteriaceae) as a commercial biological control agent for the New Zealand grass grub (*Costelytra zealandica*). Use of pathogens in scarab pest management Ed. by TrevorAJ, Travis RG

Jaihan P, Sangdee K, Sangdee A (2016) Selection of entomopathogenic fungus for biological control of chili anthracnose disease caused by *Colletotrichum* spp. Eur J Plant Pathol 146(3):551–564

James M, Malan AP, Addison P (2018) Surveying and screening south African entomopathogenic nematodes for the control of the Mediterranean fruit fly, *Ceratitis capitata* (Wiedemann). Crop Prot 105:41–48

Jucovic M, Walters FS, Warren GW, Palekar NV, Chen JS (2008) From enzyme to zymogen: engineering Vip2, an ADP-ribosyltransferase from *Bacillus cereus*, for conditional toxicity. Protein Eng Des Sel 21(10):631–638

Kalane VG, Pardeshi AB (2018) Larvicidal effect of *Staphylococcus vitulinus* bacteria against *Spodoptera litura* fab. J Pharmacogn Phytochem 7(3):3054–3057

Kamali S, Karimi J, Koppenhöfer AM (2018) New insight into the management of the tomato leaf miner, *Tuta absoluta* (Lepidoptera: Gelechiidae) with entomopathogenic nematodes. J Econ Entomol 111(1):112–119

Karkute SG, Krishna R, Ansari WA, Singh B, Singh PM, Singh M, Singh AK (2019) Heterologous expression of the AtDREB1A gene in tomato confers tolerance to chilling stress. Biol Plant 63(1):268–277

Kepenekci I, Atay T, Oksal E, Saglam HD, Tulek A, Evlice E (2015) Identification of Turkish isolate of the Entomopathogenic Fungi, *Purpureocillium lilacinum* (syn: *Paecilomyces lilacinus*) and its effect on potato pests, *Phthorimaea operculella* (Zeller)(Lepidoptera: Gelechiidae) and *Leptinotarsa decemlineata* (say)(Coleoptera: Chrysomelidae). Egypt J Biol Pest Co 25(1)

Khachatourians GG, Qazi SS (2008) Entomopathogenic fungi: biochemistry and molecular biology. In: Human and animal relationships. Springer, Heidelberg, pp 33–61

Krieg AV, Huger AM, Langenbruch GA, Schnetter W (1983) *Bacillus thuringiensis* var. *tenebrionis*: einneuer, gegenüberLarven von ColeopterenwirksamerPathotyp. Z Angew Entomo 96(1–5):500–508

Krishna R, Karkute SG, Ansari WA, Jaiswal DK, Verma JP, Singh M (2019) Transgenic tomatoes for abiotic stress tolerance: status and way ahead. 3 Biotech 9(4):143

Lacey LA, Grzywacz D, Shapiro-Ilan DI, Frutos R, Brownbridge M, Goettel MS (2015) Insect pathogens as biological control agents: back to the future. J Invertebr Pathol 132:1–41

Landsberg MJ, Jones SA, Rothnagel R, Busby JN, Marshall SDG, Simpson RM, Lott JS, Hankamer B, Hurst MRH (2011) 3D structure of the *Yersinia entomophaga* toxin complex and implications for insecticidal activity. Proc Natl Acad Sci USA 108:20544–20549

Lee MK, Walters FS, Hart H, Palekar N, Chen JS (2003) The mode of action of the *Bacillus thuringiensis* vegetative insecticidal protein Vip3A differs from that of Cry1Ab δ-endotoxin. Appl Environ Microbiol 69(8):4648–4657

Lee WW, Shin TY, Bae SM, Woo SD (2015) Screening and evaluation of entomopathogenic fungi against the green peach aphid, *Myzuspersicae*, using multiple tools. J Asia Pac Entomol 18(3):607–615

Leuber M, Orlik F, Schiffler B, Sickmann A, Benz R (2006) Vegetative insecticidal protein (Vip1Ac) of *Bacillus thuringiensis* HD201: evidence for oligomer and channel formation. Biochemist 45(1):283–288

Liu JG, Yang AZ, Shen XH, Hua BG, Shi GL (2011) Specific binding of activated Vip3Aa10 to *Helicoverpa armigera* brush border membrane vesicles results in pore formation. J Invertebr Pathol 108(2):92–97

Loong CY, Shamsudin SH, Chong TC (2017) The efficacy of entomopathogenic virus for the control of oil palm nettle caterpillar. Proceeding of Agriculture, Biotechnology and Sustainability Conference, Kuala Lumpur 121–124

Lulamba TE, Green E, Serepa-Dlamini MH (2018) Entomopathogenic nematodes, potential industrial pest control agents: a south African perspective. J Entomol Nematol 11(1):1–12

Malan AP, Moore SD (2016) Evaluation of local entomopathogenic nematodes for the control of false codling moth, *Thaumatotibia leucotreta* (Meyrick, 1913), in a citrus orchard in South Africa. Afr Entomol 24(2):489–501

Miller LK (1997) The Baculoviruses. Plenum Press, New York, p 477

Mukherjee A, Debnath P, Ghosh SK, Medda PK (2019) Biological control of papaya aphid (*Aphis gossypii* glover) using entomopathogenic fungi. Vegetosp 1-10

Nermuť J, Zemek R, Mráček Z, Palevsky E, Půža V (2019) Entomopathogenic nematodes as natural enemies for control of *Rhizoglyphus robini* (Acari: acaridae)? Biol Control 128:102–110

Nicolas L, Lecroisey A, Charles JF (1990) Role of the gut proteinases from mosquito larvae in the mechanism of action and the specificity of the *Bacillus sphaericus* toxin. Can J Microbiol 36(11):804–807

Niu H, Liu B, Li Y, Guo H (2016) Identification of a bacterium isolated from the diseased brown planthopper and determination of its insecticidal activity. Biocontrol Sci Tech 26(2):217–226

Odendaal D, Addison MF, Malan AP (2016) Entomopathogenic nematodes for the control of the codling moth (*Cydia pomonella* L.) in field and laboratory trials. J Helminthol 90:615–623

Oei C, Hindley J, Berry C (1992) Binding of purified *Bacillus sphaericus* binary toxin and its deletion derivatives to *Culex quinquefasciatus* gut: elucidation of functional binding domains. Microbiol 38(7):1515–1526

Opisa S, Du Plessis H, Akutse KS, Fiaboe KKM, EkesiS (2018) Effects of Entomopathogenic fungi and *Bacillus thuringiensis* based biopesticides on *Spoladea recurvalis* (Lepidoptera: Crambidae). J Appl Entomol 142(6): 617–626

Oppert B, Kramer KJ, Beeman RW, Johnson D, McGaughey WH (1997) Proteinase-mediated insect resistance to Bacillus thuringiensis toxins. J Biol Chem 272(38):23473–23476

Oulebsir-Mohandkaci H, Khemili-Talbi S, Benzina F, Halouane F (2015) Isolation and identification of Entomopathogenic Bacteria from Algerian Desert soil and their effects against the migratory locust, *Locusta migratoria* (Linnaeus, 1758)(Orthoptera: Acrididae). Egypt J Biol Pest Co 25(3):739

Ownley BH, Gwinn KD, Vega FE (2010) Endophytic fungal entomopathogens with activity against plant pathogens: ecology and evolution. Biol Control 55(1):113–128

Plantey RL, Papura D, Couture C, Thiéry D, Pizzuolo PH, Bertoldi MV, Lucero GS (2019) Characterization of entomopathogenic fungi from vineyards in Argentina with potential as biological control agents against the European grapevine moth *Lobesia botrana*. BioControl 64(5):501–511

Prasanna HC, Sinha DP, Rai GK, Krishna R, Kashyap SP, Singh NK et al (2015) Pyramiding T y2 and T y3 genes for resistance to monopartite and bipartite tomato leaf curl viruses of India. Plant Pathol J 64(2):256–264

Pu YC, Ma TL, Hou YM, Sun M (2017) An entomopathogenic bacterium strain, *Bacillus thuringiensis*, as a biological control agent against the red palm weevil, *Rhynchophorus ferrugineus* (Coleoptera: Curculionidae). Pest Manag Sci 73(7):1494–1502

Qazi SS, Khachatourians GG (2005) Insect pests of Pakistan and their management practices: prospects for the use of entomopathogenic fungi. Biopestic Int 1:13–24

Rahatkhah Z, Karimi J, Ghadamyari M, Brivio MF (2015) Immune defenses of *Agriotes lineatus* larvae against entomopathogenic nematodes. BioControl 60(5):641–653

Ramanujam B, Rangeshwaran R, Sivakmar G, Mohan M, Yandigeri MS (2014) Management of Insect Pests by microorganisms. Proc Natl Acad Sci U S A 80(2):455–471

Raymond B, Johnston PR, Nielsen-LeRouxC LD, Crickmore N (2010) *Bacillus thuringiensis*: an impotent pathogen? Trends Microbiol 18(5):189–194

Rezaei N, Karimi J, Hosseini M, Goldani M, Campos-Herrera R (2015) Pathogenicity of two species of entomopathogenic nematodes against the greenhouse whitefly, *Trialeurodes vaporariorum* (Hemiptera: Aleyrodidae), in laboratory and greenhouse experiments. J Nematol 47(1):60

Roberts DW (1981) Toxins of entomopathogenic fungi. In: Microbial control of pests and plant diseases 1970–1980. In: Burges HD (ed)

Rocha LFN, Sousa NA, Rodrigues J, AML C, Marques CS, Fernandes ÉK, Luz C (2015) Efficacy of *Tolypocladium cylindrosporum* against *Aedes aegypti* eggs, larvae and adults. J Appl Microbiol 119(5):1412–1419

Rohlfs M, Churchill AC (2011) Fungal secondary metabolites as modulators of interactions with insects and other arthropods. Fungal Genet Biol 48(1):23–34

Ruiu L, Satta A, Floris I (2013) Emerging entomopathogenic bacteria for insect pest management. Bull Insectol 66(2):181–186

Ruiu L, Virdis B, Mura ME, Floris I, Satta A, Tarasco E (2017) Oral insecticidal activity of new bacterial isolates against insects in two orders. Biocontrol Sci Tech 27(7):886–902

Ruiu L, Lazzeri AM, Nuvoli MT, Floris I, Satta A (2020) Safety evaluation of the entomopathogenic bacterium *Brevibacillus laterosporus* for the green lacewing *Chrysoperla agilis* (Neuroptera: Chrysopidae). J Invertebr Pathol 169:107281

Sajnaga E, Kazimierczak W, Skowronek M, Lis M, Skrzypek T, Waśko A (2018) *Steinernema poinari* (Nematoda: Steinernematidae): a new symbiotic host of entomopathogenic bacteria *Xenorhabdus bovienii*. Arch Microbiol 200(9):1307–1316

Sattar S, Maiti MK (2011) Molecular characterization of a novel vegetative insecticidal protein from *Bacillus thuringiensis* effective against sap-sucking insect pest. J Microbiol Biotechnol 21(9):937–946

Saxena AK, Padaria JC, Gurjar GT, Yadav AN, Lone SA, Tripathi M et al. (2020) Insecticidal formulation of novel strain of *Bacillus thuringiensis* AK 47. Indian Patent 340541

Sharma A, Thakur DR, Chandla VK (2018) Pathogenicity of entomopathogenic bacteria isolated from white grub *Brahmina coriacea* on its lifestages. Ind J Entomol 80(3):668–676

Siegel JP (1997) Testing the pathogenicity and infectivity of entomopathogens to mammals. In: Manual of techniques in insect pathology. Academic Press, pp 325–336

Singkhamanan K, Promdonkoy B, Chaisri U, Boonserm P (2010) Identification of amino acids required for receptor binding and toxicity of the *Bacillus sphaericus* binary toxin. FEMS Microbiol Lett 303(1):84–91

Sivakumar S, Rajagopal R, Venkatesh GR, Srivastava A, Bhatnagar RK (2007) Knockdown of aminopeptidase-N from *Helicoverpa armigera* larvae and in transfected Sf21 cells by RNA

interference reveals its functional interaction with *Bacillus thuringiensis* insecticidal protein Cry1Ac. J Biol Chem 282(10):7312–7319
Soberon M, Lopez-Diaz JA, Bravo A (2013) Cyt toxins produced by *Bacillus thuringiensis*: a protein fold conserved in several pathogenic microorganisms. Peptides 41:87–93
Sönmez E, Sevim A, Demirbağ Z, Demir İ (2016) Isolation, characterization and virulence of entomopathogenic fungi from *Gryllotalpa gryllotalpa* (Orthoptera: Gryllotalpidae). Appl Entomol Zool 51(2):213–223
Soumia PS, Srivastava C, Dikshit HK, Guru Pirasanna Pandi G (2015) Screening for resistance against pulse beetle, *Callosobruchus analis* (F.) in Greengram (*Vigna radiata* (L.) Wilczek) accessions. Proc Natl Acad Sci India Sect B Biol Sci 87(2):551–558
Srisucharitpanit K, Yao M, Promdonkoy B, Chimnaronk S, Tanaka I, Boonserm P (2014) Crystal structure of B inB: a receptor binding component of the binary toxin from *Lysinibacillus sphaericus*. Proteins: Struct Funct Bioinf 82(10):2703–2712
Steinhaus EA (1975) Disease in a minor chord: being a semihistorical and semibiographical account of a period in science when one could be happily yet seriously concerned with the diseases of lowly animals without backbones, especially the insects. The OhioState University Press
Tanaka S, Miyamoto K, Noda H, Jurat-Fuentes JL, Yoshizawa Y, Endo H, Sato R (2013) The ATP-binding cassette transporter subfamily C member 2 in *Bombyx mori* larvae is a functional receptor for cry toxins from *Bacillus thuringiensis*. FEBS J 280(8):1782–1794
Thakur N, Kaur S, Tomar P, Thakur S, Yadav AN (2020) Microbial pesticides: current status and advancement for sustainable environments. In: Rastegari AA, Yadav AN, Yadav N (eds) Trends of microbial biotechnology for sustainable agriculture and biomedicine systems: diversity and functional perspectives. Elsevier, Amsterdam, pp 243–282. https://doi.org/10.1016/B978-0-12-820526-6.00016-6
Thomas MB, Read AF (2007) Can fungal biopesticides control malaria? Nat Rev Microbiol 5(5):377–383
Tsuda Y, Nakatani F, Hashimoto K, Ikawa S, Matsuura C, Fukada T, Himeno M (2003) Cytotoxic activity of *Bacillus thuringiensis* Cry proteins on mammalian cells transfected with cadherin-like Cry receptor gene of *Bombyx mori* (silkworm). Biochem J 369(3):697–703
Van Frankenhuyzen K (2009) Insecticidal activity of *Bacillus thuringiensis* crystal proteins. J Invertebr Pathol 101(1):1–16
Van Damme VM, Beck B, Berckmoes E, Moerkens R, Wittemans L, De Vis R et al. (2016) Efficacy of entomopathogenic nematodes against larvae of *Tuta absoluta* in the laboratory. Pest Manag Sci 72(9): 1702–1709
Varani AM, Lemos MV, Fernandes CC, Lemos EG, Alves EC, Desidério JA (2013) Draft genome sequence of *Bacillus thuringiensis var. thuringiensis* strain T01-328, a Brazilian isolate that produces a soluble pesticide protein, Cry1Ia. Genome Announc 1(5):e00817–e00813
Varela LG, Haviland DR, Bentley WJ, Zalom FG, Bettiga LJ, Smith RJ, Daane KM (2015) UC IPM Pest Management Guidelines: Grape. UCANR Pub. 3448
Vega FE, Kaya HK (2012) Insect pathology. Academic press, London
Vey A, Hoagland R, Butt TM (2001) Toxic metabolites of fungal biocontrol agents. In: Butt TM, Jackson CW, Magan N (eds) Fungi as biocontrol agents: progress, problems and potential. CAB International, Wallingford, pp 311–346
Vodovar N, Vinals M, Liehl P, Basset A, Degrouard J, Spellman P et al (2005) *Drosophila* host defense after oral infection by an entomopathogenic *Pseudomonas* species. PNAS 102(32):11414–11419
Walters FS, Stacy CM, Lee MK, Palekar N, Chen JS (2008) An engineered chymotrypsin/cathepsin G site in domain I renders *Bacillus thuringiensis* Cry3A active against western corn rootworm larvae. ApplEnviron Microbiol 74(2):367–374
Wang W, Zhou L, Dong G, Chen F (2019) Isolation and identification of entomopathogenic fungi and an evaluation of their actions against the larvae of the fall webworm, *Hyphantria cunea* (Drury)(Lepidoptera: Arctiidae). BioControl:1–11
Wang P, Yang G, Shi N, Huang B (2020) Molecular characterization of a new partitivirus, MbPV1, isolated from the entomopathogenic fungus *Metarhizium brunneum* in China. Arch of Virol:1–5

Yaman M, Ertürk Ö (2016) Isolation, identification and insecticidal effects of entomopathogenic bacteria from the willow flea beetle, *Crepidodera aurata* (Coleoptera; Chrysomelidae). Prog Plant Prot 56(2):225–229

Yan X, Lin Y, Huang Z, Han R (2018) Characterisation of biological and biocontrol traits of entomopathogenic nematodes promising for control of striped flea beetle (*Phyllotreta striolata*). Nematology 20(6):503–518

Yang JI, Stadler M, Chuang WY, Wu S, Ariyawansa HA (2020) In vitro inferred interactions of selected entomopathogenic fungi from Taiwan and eggs of *Meloidogyne graminicola*. Mycol Prog 19(1):97–109

Yun HG, Kim DJ, Gwak WS, Shin TY, Woo SD (2017) Entomopathogenic fungi as dual control agents against both the pest *Myzuspersicae* and phytopathogen *Botrytis cinerea*. Mycobiol 45(3):192–198

Zalom FG, Bolda MP, Dara SK, Joseph S (2014) UC IPM Pest Management Guidelines: Strawberry. UCANR Pub 3468

Zorzetti J, Ricietto APS, Fazion FAP, Meneghin AM, Neves PMOJ, Vilas-Boas LA, Vilas-Bôas GT (2018) Isolation, morphological and molecular characterization of *Bacillus thuringiensis* strains against *Hypothenemus hampei* Ferrari (Coleoptera: Curculionidae: Scolytinae). Rev Bras Entomol 62(3):198–204

Soil Microbes as Biopesticides: Agricultural Applications and Future Prospects

20

Arun Karnwal and Dhriti Kapoor

Abstract

Conventional biological pesticides have been dominated over time by synthetic chemicals probably due to their efficacy, reliability and quick knock-down effect. In the past 50 years, crop safety has relied extensively on synthetic pesticides, but now, due to recent legal regulations and resistance developed in pests, is being restricted as a result. Thus acceptable alternative strategies of pest control are needed. Biopesticides are plant, microorganism and insect derivatives. They have been documented to control pests and are used worldwide. Biopesticides sources are readily accessible, biodegradable, exhibit a variety of different modes of operation, are affordable and less or no toxic to human and non-target organisms. Microbial biopesticides comprise a diverse group of microorganisms including fungi, bacteria, Baculoviruses or nematode-associated microbes working against the pests in agroecosystems. The biopesticide market is growing significantly, and various innovations have been made in the form of new biopesticide products, which boost this market worldwide. Bacterial entomopathogens include several *Bacillaceae*, *Serratia*, *Pseudomonas*, *Yersinia*, *Burkholderia*, *Chromobacterium*, *Streptomyces*, and *Saccharopolyspora* species, while fungi comprise different strains of *Beauveria bassiana*, *Beauveria brongniartii*, *Metarhizium anisopliae*, *Verticillium*, *Lecanicillium*, *Hirsutella*, *Paecilomyces*, and *Isaria* species. Entomopathologic nematodes (EPNs) predominantly contain organisms of the *Heterorhabditis* and *Steinernema* genera related in the *Photorhabdus* and *Xenorhabdus* symbiotic mutualistic bacteria. However,

A. Karnwal (✉)
Department of Microbiology, School of Bioengineering and Biosciences, Lovely Professional University, Phagwara, Punjab, India

D. Kapoor
Department of Botany, School of Bioengineering and Biosciences, Lovely Professional University, Phagwara, Punjab, India

A. N. Yadav et al. (eds.), *Current Trends in Microbial Biotechnology for Sustainable Agriculture*, Environmental and Microbial Biotechnology,
https://doi.org/10.1007/978-981-15-6949-4_20

biopesticides are faced with challenges of formulation, registration, commercialization, acceptance, and adoption. There are still significant technological obstacles to making the biopesticides more efficient. This chapter explores the insecticide properties and possible effectiveness of microbes in pesticide control.

Keywords

Baculoviruses · Biodegradable · Entomopathogenic nematodes · Integrated pest management · Legal regulation · Sustainability

20.1 Introduction

Farming has faced the detrimental actions of various pests such as fungus, weed and insects on food production as overall agricultural productivity are decreased by 20–40% yearly due to pests and diseases (Berg 2009). The problem has been largely overcome with the introduction of synthetic pesticides (Savoie and Mata 2016). However, the excessive use on synthetic pesticides and their subsequent carefree use have demanded replacements primarily for environmental safety concerns (Leng et al. 2011; Nawaz et al. 2016). Although the percentage of biopesticides worldwide cover around 1% for all crop safety products (Berg 2009). Over the past two decades, growth rate in the production of these types of products was increased and about 175 new biopesticides and 700 products reported worldwide. *Beauveria bassiana*, nuclear polyhedrosis viruses (NPV), *Bacillus thuringiensis* (Bt), *Metarhizium* spp., and *Trichoderma viride* are commonly reported and applied worldwide plant protectants for protection of plants (Blanford et al. 2005; Cawoy et al. 2011; Cantrell et al. 2012; Rastegari et al. 2020). The first and most common insect pathogen reported in pest management is the *B. thuringiensis* (Bt) that currently accounts for around 2% total insecticidal industry (Biswas 2009).

The most commonly identified bacterial spp. as insect pathogens are the sub species or types of *B. thuringiensis*. Each species or strain of Bt synthesize a different combination of crystal protein for specifically killing one or several related insect species, like, *B. thuringiensis* subspecies *kurstaki* and *aizawai* kill larvae of lepidopteric (Geden 2012; Ruiu 2018), and *B. thuringiensis* subspecies *tenebrionis* kill larvae of coleopteric larvae (Damalas and Koutroubas 2018). Some strains of *B. thuringiensis* subspecies *israelensis* kill mosquito-specific (Geden 2012). The most important insect control viruses for Lepidoptera and Diptera particularly belong to baculovirus family including, NPV and GV. NPVs were widely applied on cotton, vegetable, and fruit crops across various countries to control destructive pests including Heliothis and Spodoptera species (Gasic and Tanovic 2013; Hubbard et al. 2014; Olson 2015).

Fungal entomopathogens, such as *Metarhizium, Isaria, Beauveria,* and *Lecanicillium* species, were established as killing machine for different insect pests (Ruiu 2018). There are many agricultural products, globally available for the biocontrol of insect–pests in farming, including fungi, bacteria, and viruses (Kaaya and Hassan 2000; Yadav et al. 2020d). Over the next 10 years, the development of the bio-pesticides market is estimated at 10–15% per year opposed to 2–3% of synthetic pesticides (Chandler et al. 2011; Bautista et al. 2018). The key strengths of such biocontrol product are (1) its specific nature towards specific pests, (2) its protection for the non-target species, (3) its non-harmful nature for environmental and human health, (4) its use as an optimal tool for integrated pesticide (IPM) management and (5) its sensitivity for those pests that developed resistance to chemical pesticides. In present chapter, we have discussed about soil microbes as biopesticides and their applications for agricultural and environmental sustainability.

20.2 Need of Biopesticides

Biologic materials are used by a number of methods to manage pests, infections, pathogens and plant diseases (Sudakin 2003). Control of pathogens or parasites that attack the targeted plant may be included in microbial biocontrol methods. They could also serve as opponents or inducers of resistance to the plant host. Biological bio-controls may also be implemented via a number of methods. Many of them deal with the growth, nutrition, propagation or development of a disease or pathogen. Still, plenty of other bio-controls can be considered to create a surface defence so that they can serve as a regulator of nutrition or infection (Heinig et al. 2015; Olson 2015; Karimi et al. 2019; Singh and Yadav 2020).

Plant derivatives are probably the oldest agriculture biocontrol since nicotine is documented in literature during late seventeenth century for suppressing plum beetles (Walia et al. 2017). Biochemical pest management studies in the agricultural sector date back to 1835, when Agostino Bassi revealed the application of *B. bassiana* (a white fungus) for the control of a deadly disease in silkworm (Biswas 2009). In the nineteenth-century studies have also been documented the use of mineral oils as plant protection agent. An increasing number of studies and ideas regarding biocontrol were established during the gradual systematic growth of agriculture science at the beginning of the twentieth century. In 1901, the first microbial biocontrol agent, *B. thuringiensis* was obtained by Japanese biologist Shigetane Ishiwata from the infected silk worm (Buchholz et al. 2006) and was the most extensively utilized bio-controls to produce *B. thuringiensis* spores (Kramer and Muthukrishnan 1997) and after 10 years in a sick caterpillar of flour-moth, Ernst Berliner found it once again in Germany, (Thuringen) (Meshram 2010). The *B. thuringiensis* was first time classified and characterized as a *B. thuringiensis* in 1911 and is the biocontrol method most commonly utilized up to the time. *B. thuringiensis* became the synthetic pesticide in the early 1920s, in France. In France, in 1938, the first commercially produced product of *B. thuringiensis*, Sporeine, was announced (Rosas García 2014; Hernández-Fernández 2016). The French began using Bt as a natural biological control agent for insects in the early 1920s. In the 1950s, a large-scale overuse of biocontrol started to take over in the United States, when a group of scientists reported on BT effectiveness.

During twentieth century, the application of least expensive but poisonous chemically synthesized pesticides appeared on a widespread level that harm environment at higher level (Dubey and Juwarkar 2001, Medina et al. 2014, Martinez-Medina et al. 2014). Throughout this period, newer safe products have been formulated and introduced; particularly in regional markets where chemical-based products have not been reported, not active or are not affordable. In 1956, the Pacific-Yeast Company introduced a commercial method called as submerged fermentation that enabled Bt to be produced at a large-scale level (Whalon and Wingerd 2003).

In 1973, Heliothis moth-based nuclear polyhedrosis virus (NPV) was excluded from resistance and Elcar became the first virus-based insecticide to be branded in 1975 (Senthil-Nathan 2015). *B. thuringiensis* toxic to flies, var. *israelensis*, was identified in 1977, and the var. *tenebrion* which is toxic for beetles, was identified in 1983. Environmental Protection Agency (EPA), in 1979, has licensed the very first insect pheromone in the United States, to be used in the widespread capturing for *Popillia japonica* (Japanese beetles) (Potter and Held 2002). Biochemical advances in the management of crop pathogens have experienced a complete transformation. At the beginning of twentieth century, soil microbiology and biodiversity research contributed to the discovery of several distinct microbes that function as rivals or hyper-parasites for parasites and pests. A variety of such have been demonstrated being effective in ground-scale inoculations, but very few have been widely produced due to the massive introduction of synthetic pesticides over that timeframe (Blanford et al. 2005; Olson 2015; Vea et al. 2018). Major breakthrough in the 1980s and 1990s are those of use of *Agrobacterium radiobacter* in formulations for the control of crown-gall in crops and *Pseudomonas fluorescens* in orchards (Biswas 2009; Suprapta 2012), where overused use of streptomycin and widespread immune pathogenic populations of fire blight was reported. Products carrying a number of microbes that reduced soil-borne infections were launched commercially for the nursery and potting-mix industries.

20.3 Biopesticides

Biopesticides are such kinds of artificial/biological pesticides, as described by the United States Environmental Protection Agency (EPA), obtained from natural resources such as zoological, botanical, microbes and certain minerals (Zazouli et al. 2015). Commercially, biopesticides usually involve microbes that regulate various types of infection causing pests (microbial pesticides), natural substances that regulate pests (biochemical pesticides) and plants materials-based pesticides with additional genetic material (PIPs: plant-incorporated protectants) (Rodgers 1993; Khater 2012; Beas-Catena et al. 2014; Yadav et al. 2020b, c). Biopesticides are used in cultivation for everything from the objectives of controlling weeds, biological infection, pests, nematodes and provide better crop propagation and development (Fig. 20.1). The Environmental Protection Agency has divided biopesticides into three main groups, depending on the composition used like, biochemical metabolites, PIPs and microorganism-based products (Seiber et al. 2014). There are

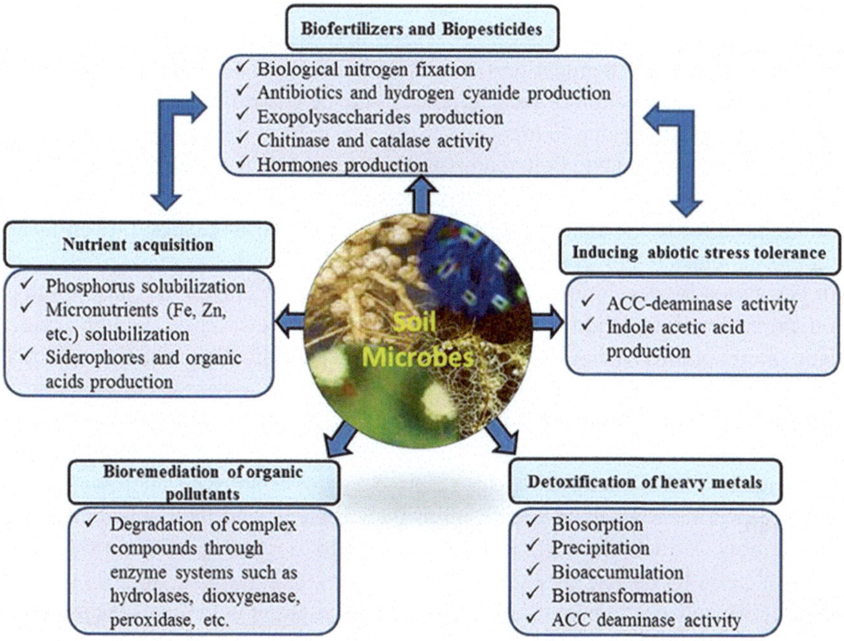

Fig. 20.1 Soil microbes and their potential applications for agricultural and environmental sustainability. Adapted from Ahmad et al. (2018)

different kinds of products within each group and each product has its own mechanism of action. The following properties underlie the benefits about the use of biopesticides:

- Economic and environmental advantage; least damaging and least stress for the environment.
- Specificity of the target; engineered to directly influence one possibly two specified species.
- Environmental friendly; mostly very active and decomposes easily, leading to cause least microbial exposures to environment and minimizing the issue of toxicity.
- Suitability; biopesticides can make a major contribution as an integrated pest management (IPM) component.

20.3.1 Microbial Pesticides

Microorganisms-based pesticides are often considered as biological control agents (Robles-Gonzalez et al. 2008; Abraham and Silambarasan 2015; Karmakar et al. 2015). Microbial pesticides are composted of variety of microorganisms like fungi, virus, bacteria, algae or protozoans those are spontaneously or genetically reconstructed (Céspedes et al. 2014). There are different methods for development of microbial biopesticides (Fig. 20.2). These may be practical and effective as a solution to synthetic/chemical pesticides (Table 20.1). In case of microbial pesticides, the pesticidal action of these products may be either due to metabolites produced by organism or due to organism itself. These have the benefits of enhanced selectivity and less or no toxicity compared to traditional chemical-based pesticides (Olson 2015). The most widely utilized microbial biopesticides are live organisms that are harmful to the concerned pest. These include biofungicides, bioherbicides and biopesticides (Deravel et al. 2014; Thakur et al. 2020; Yadav et al. 2020a). The kill pets either by developing pest-specific toxic metabolic products, causing infection, preventing interaction of certain beneficial microbes with pest through antagonistic or competition behaviours or through another undefined mechanisms (Rosell et al. 2008; Lovett and St. Leger 2018).

Earlier work has shown that pathogens induce essential pathogenesis through microbial toxins in pests (Gupta and Dikshit 2010; Mazid 2011; Kachhawa 2017). The majority of the microbial toxins released by the established microbial pathogens are peptides, but their composition, their toxicity and their specificity vary considerably. A microbial toxin is usually described as a biological toxin substance either released by bacteria or fungi (Mondal et al. 2020; Verma et al. 2017). These are effective and safe to humans as well as most non-target individuals (Mazid et al. 2011; Satyanarayana et al. 2012). These pesticides left very little or zero traces in food or crops. Therefore, microbial agents remain pretty selective to target pest species, while allowing beneficial pests to live in treated crops without causing any

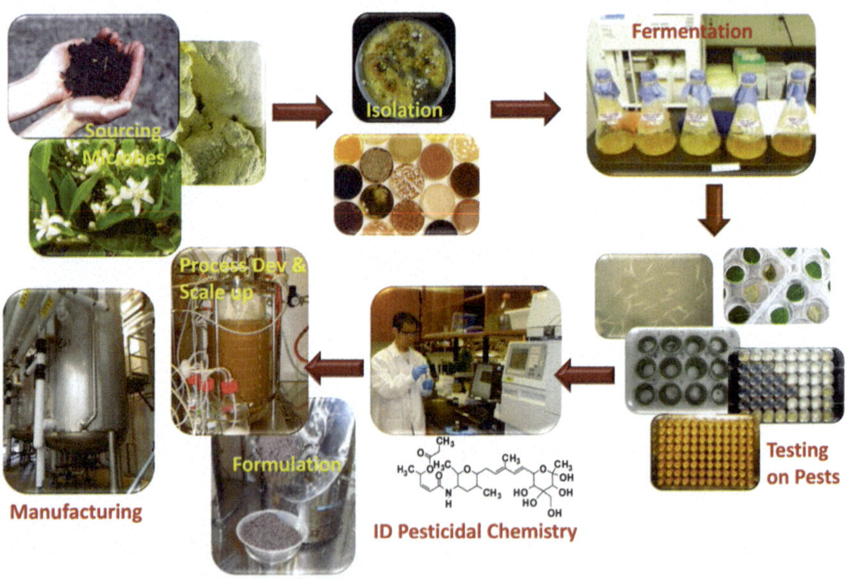

Fig. 20.2 The process of discovery and development of microbial biopesticides. Adapted with permission from Marrone (2019)

Table 20.1 Microbial Insecticides: A summary of products and their uses

Biopesticides microbes	Commercially marketed product	Infectable host	Application and other information
Bacteria			
Bacillus thuringiensis var. *kurstaki* (*Bt*)	Bactur®, Bactospeine®, bioworm®, Caterpillar killer®, Dipel®, Futura®, javelin®, SOK-Bt®, Thuricide®, topside®, Tribactur®, worthy attack®	Caterpillars (larvae of moths and butterflies)	Effective for foliage-feeding caterpillars (and Indian meal moth in stored grain). Deactivated rapidly in sunlight; apply in the evening or on overcast days and direct some spray to lower surfaces or leaves. Does not cycle extensively in the environment.
Bacillus thuringiensis var. israelensis (*Bt*)	Aquabee®, Bactimos®, Gnatrol®, LarvX®, mosquito attack®, Skeetal®, Teknar®, Vectobac®	Larvae of *Aedes* and *Psorophora* mosquitoes, black flies and fungus gnats	Effective against larvae only. Active only if ingested. *Culex* and *Anopheles* mosquitoes are not controlled at normal application rates. Does not cycle extensively in the environment.
Bacillus thuringiensis var. *tenebrinos*	Foil®M-one®M-track®,Novardo®trident®	Larvae of Colorado potato beetle, elm leaf beetle adults	Effective against Colorado potato beetle larvae and the elm leaf beetle. Like other *Bt*s, it must be ingested. It is subject to breakdown in ultraviolet light and does not cycle extensively in the environment.
Bacillus thuringiensis var. *aizawai*	Certan®	Wax moth caterpillars	Used only for control of was moth infestations in honeybee hives.
Bacillus popilliae and *Bacillus lentimorbus*	Doom¨, Japidemic¨,® milky spore disease, grub attack®	Larvae (grubs) of Japanese beetle	The main Illinois lawn grub (the annual white grub, *Cyclocephala* sp.) is NOT susceptible to milky spore disease.
Bacillus sphaericus	Vectolex CG®,Vectolex WDG®	Larvae of *Culex, Psorophora* and *Culiseta* mosquitos, larvae of some *Aedes* spp.	Active only if ingested, for use against *Culex, Psorophora,* and *Culiseta* species; also effective against *Aedes vexans*. Remains effective in stagnant or turbid water

(continued)

Table 20.1 (continued)

Biopesticides microbes	Commercially marketed product	Infectable host	Application and other information
Fungi			
Beauveria bassiana	Botanigard®,Mycotrol®,Naturalis®	Aphids, fungus gnats, mealy bugs, mites, thrips, whiteflies	Effective against several pests. High moisture requirements, lack of storage longevity and competition with other soil microorganisms are problems that remain to be solved.
Lagenidium giganteum	Laginex®	Larvae of most pest mosquito species	Effective against larvae of most pest mosquito species; remains infective in the environment through dry periods. A main drawback is its inability to survive high summertime temperatures.
Protozoa			
Nosema locustae	NOLO bait®, grasshopper attack®	European cornborer caterpillars, grasshoppers and mormon crickets	Useful for rangeland grasshopper control. Active only if ingested. Not recommended for use on a small scale, such as backyard gardens, because the disease is slow acting and grasshoppers are very mobile. Also effective against caterpillars.
Viruses			
Gypsy moth nuclear polyhedrosis (NPV)	Gypchek® virus	Gypsy moth caterpillars	All of the viral pesticides used for control of forest pests are produced and used exclusively by the U.S. Forest Service.
Tussock moth NPV	TM Biocontrol-1®	Tussock moth caterpillars	
Pine sawfly NPV	Neochek-S®	Pine sawfly larvae	

(continued)

Table 20.1 (continued)

Biopesticides microbes	Commercially marketed product	Infectable host	Application and other information
Codling moth granulosis virus (GV)	(see comments)	Codling moth caterpillars	Commercially produced and marketed briefly, but no longer registered or available. Future re-registration is possible. Subject to rapid breakdown in ultraviolet light.
Entomogenous nematodes			
Steinernema feltiae (*Neoaplectana carpocapsae*) *S. riobravis*, *S. carpocapsae* and other *Steinernema* species	Biosafe®,Ecomask®, Scanmask®, also sold generically (wholesale and retail),vector®	Larvae of a wide variety of soil-dwelling and boring insects	*Steinernema riobravis* is the main nematode species marketed retail in the United States. Because of moisture requirements, it is effective primarily against insects in moist soils or inside plant tissues. Prolonged storage or extreme temperatures before use may kill or debilitate the nematodes.
Heterorhabditis heliothidis	Currently available on a wholesale basis for large-scale operations	Larvae of a wide variety of soil-dwelling and boring insects	Not commonly available by retail in the United States; this species is used more extensively in Europe. Available by wholesale or special order for research or large-scale commercial uses.
Pathogen			
steinernema scapterisci	Nematac®S	Late nymph and adult stages of mole crickets	*S. scapterisci* is the main nematode species marketed to target the tawny and southern mole cricket. Best applied where irrigation is available. Irrigate after application.

Source: Usta (2013)

harm to them. This is one of the primary reasons for developing microbial pesticides during the last three decades as biological controlling agent (Debode et al. 2007; Bailey et al. 2010; Asari et al. 2016; Melo et al. 2016).

20.3.1.1 Bacteria

Bacteria are tiny, single-cell species that live within and outside of millions of individuals throughout every environment. Few bacteria are dangerous, but many of them utilized for a useful purpose. They sustain multiple life forms, both in crops and animals, and are employed successfully in commercial, agricultural and pharmaceutical applications (Singh et al. 2020). It has been believed, around 4 billion years ago, that bacteria were the first creatures to exist on earth. Bacteria can utilize various types of compounds as a food and can survive even in extreme conditions.

The Bacillaceae, Pseudomonadaceae, Enterobacteriaceae, Streptococcaceae and Micrococcaceae bacterial families are the most common pathogenic bacterial species for various insects (Khetan 2000; Berg 2009). Family Bacillaceae members, especially *Bacillus* spp. are counted as microbial controllers and have achieved the highest attention during last three decades. The most severe form of biological pesticides which work in many different ways is bacterial biopesticides. They are generally employed as pesticides, but can also be applied to regulate phyto-pathogenic bacterial and fungal pathogens (Wani et al. 2016). These are usually specific to individual inset species like moths, butterflies, flies, mosquito. To become as an effective pesticide, it is necessary for bacterial pathogen to be ingested by target pest through infection or feed supply (Han et al. 2011). Bacteria damage the digestive tract of insects through releasing of endotoxins which are also specific to that of the individual insect disease. When applied to manage phyto-pathogenic bacterial or fungal spp., the bacterial biopesticide crowds the plant and compete with pathogenic species (Hall et al. 2003; Ruiu 2018).

The first and most commonly identified bacterial pesticides are subspecies and variants of *B. thuringiensis* (Bt), responsible for nearly 90% of the biopesticide industry throughout the United States. Bt has been commonly utilized for managing insect pests of significance in farming, forestry and therapy. The main attribute is the production of crystal protein during sporulation phase of growth, containing inclusions known as delta-endotoxins or Cry-proteins with insect-killing characteristics. *B. thuringiensis* and Cry-Protein are effective, secure and sustainable substitutes to synthetic pesticides for managing insect pests, because to its high environmental safety and target specificity (Hubbard et al. 2014). The main families comprising pesticide bacteria are belonging to the Enterobacteriaceae, Micrococcaceae, Pseudomonadaceae and Bacillaceae. In several bacteria-based pesticides, the active ingredient is cylindrical type bacteria of *Bacillus* genus. Around 100 spp. have been known for attacking insects and 4 of them (*B. thuringiensis*, *B. popilliae*, *B. lentimorbus* and *B. sphaericus*) as insect-control agents are being thoroughly examined. The main benefit of bacterial spp. is their targeted selection of hosts and their lack of toxicity on non-target organisms, like humans and natural insect rivals.

In 1948, the outbreak of *Bacillus* has been recorded as 'doom' or 'milky spore infection' in Japanese beetles (Beard 1945). A milky sporeal disease, consisting of two bacterial spp., *lentimorbus* and *popilliae*, is available in market and second sp. was *thuringiensis* var. that was marketed individually in market. The most effective bioinsecticide against caterpillar is *kurstaki* (BTK) (Dixit et al. 2018; Nautiyal et al. 2008). Subsequently, several *Bacillus* strains have been observed for the regulation of various insects and formulated industrially.

The above-mentioned four *Bacillus* spp. are usually located throughout soil and form spores. The spore is the phase of bacteria tidying over a favourable period and is capable of confronting degradations, for at least a short period of time, by ultraviolet light, drought and other adverse ambient conditions. Protein crystals are also formed by *B. thuringiensis* and *Bacillus popilliae* during their sporulative cells. Such crystals are toxic substances for insects. Bacteria not producing spores will not remain long enough to regulate pests until added to a crop; thus these bacteria spp. counted as poor individuals as pesticides (Whalon and Wingerd 2003; Loth et al. 2015).

The term Thuringiensis (Bt) derives from Thuringia, a town located in Germany, where it was reported in the diseased moths of Mediterranean flour, in 1911. Bt is a strong lepidopteran larval insecticide. The Cry toxin produced by the bacterium can be isolated and used as direct insecticide in field. Bt often exists generally both in the intestines of moth and butterfly caterpillars and on the dark surface of plants. There were at least 35 Bt types, each targeting specific insect hosts via the toxic protein crystals (Marrone 2008). The amount of *Bacillus* strains researched for potential production of pesticides for specific use has raised farmer's direct interest of using biopesticides. Some companies have hundreds or more cultures those can be used for formulation of biopesticides. The first drug of industrial Bt was introduced in 1958 as bioinsecticide. It is derived from the Kurstaki (BTK) type and only acts with the insect order Lepidoptera against a larval stage of moths and butterflies (Whalon and Wingerd 2003; Gupta and Dikshit 2010; Olson 2015).

Across the years, a number of drugs were produced with enhanced BTK strains. The BT var. *Israelensis* (BTI) in 1980, which destroys insects in the insecticide category of Dipter, Bt Var. *san diego* and *tenebrionis*, which are lethal to some insect beetles in the Coleptera insect order, are additional Bt stains were discovered. Bt tends to be small white, brown or reddish orange granules composed of spores or poisonous and harmless crystals (Buchholz et al. 2006; Krieg et al. 1987; Milner 1994).

20.3.1.2 Fungi

Entomopathogenic fungi are known to serve a crucial game changer in insect populations control as a biocontrol tool (Meikle et al. 2012; Senthil-Nathan 2015). A wide range of fungal species present in nature belong to various categories that can kill or reduce pathogenicity of insects. Such pathogenic insects' killers are present in a broad array of habitats, including infection skills such as facultative and strict pathogens (Berg 2009; Kour et al. 2019b). In many insect species, the transmission of fungal diseases is a normal process during their life span, although other insects might not have been affected due to fungal infection. The first pathogenic insect management studies were conducted in the 1980s, with the goal of finding ways to control the silkworm disease (Milner 1994; Kunimi 2007).

Bassi proposed the germ theory for the first time in 1835, employing white–muscardine fungi on the silkworm against muscardine disease caused by arthropods in silkworm, which was then called *B. bassiana* in honour of Bassi (James and Li 2012; Olson 2015). Gilbert and Gill (2010) outlined the concept about the use of insect pest control by using infection fungi as a consequence of the silkworm disease. Currently, about 90 genera and roughly more than 700 species of insect infecting fungi have been known to comprise all major classes (Chandler 2017). An insect-killing community of fungi is named entomopathogen fungi, as they invade and destroy their insect host. The primary method of entomopathogen entry is via insect integument that can also harm the insect by absorption or damaging of insect trachea.

Entomopathogens provide an enormous capacity as pest biocontrol agent, as these represent a community of more than 750 fungal spp. and induce infections in insect populations when released in the atmosphere (Rodgers 1993; Kramer and Muthukrishnan 1997; Arthurs and Dara 2019). The infective cycle of these fungi starts as spores on the integument surface persist and germinate, after germination the fungi continue to secrete digestive enzymes like proteases, chitinases, quitobiases, Upases and lipoxygensases. Such excreted fungal enzymes disintegrate the cuticle of the insect and help to penetrate it through a mechanical pressure induced by the particular structure developed throughout a germination tube, appressorium (Ambethgar 2009). Once inside, the fungus grows into hyphal bodies and propagates throughout the haemocoel and attack different insect muscle tissues, fatty bodies, malpighian channels, mitochondria and hemocytes that induce bug death within 3–14 days since penetration (Kaaya and Hassan 2000; Bianciotto et al. 2001). After the insect dies and several nutrients have been depleted, fungal species start to propagate and penetrate the host's organs. Lastly, the hyphae penetrate the cuticle from insect interior and appear at the surface, initiating the emergence of spores in suitable ambient conditions.

Biotrophic fungus uses their hosts' living cells. Some of them grow under commensals association which receives food from the insect's digestive system (Mendgen and Hahn 2002). This fungal class is generally used in several areas but not widely used in pesticide monitoring, because it is not possible to observe the fungi asymptomatically in insects or the changes caused by pathogens (Koeck et al. 2011).

Necrotrophic fungus survives on alive host cells and must destroy their hosts before absorbing them (Mendgen and Hahn 2002; Srivastava et al. 2016). Fungi in this community are potential biological insect control agents and are highly efficient in their action. A variety of host comprises the Coleoptera, Lepidoptera, Hymenoptera, Hemiptera, Orthoptera, Homoptera and Diptera orders (Sanchez-Vallet et al. 2010; Salas-Marina et al. 2011). Attack can happen in a number of life cycle stages. Various fungal molecules produced by fungal spp. like *Beauveria,*

Metarhizium, Nomuraea, Aspergillus, Verticillium, Paecilomyces, Isaria, Fusarium, Cordyceps and *Entomophthora* have been extracted having insecticidal properties.

Phenotypically, fungus may appear either as individual (yeast) or as networked (hyphae) filaments forming sheets and they may replicate through sexually or asexually methods. Sexual reproduction requires a merger of some kind of material between two gametes or hyphae. The conidial spore is the key, frequently employed fungal microbial pesticide infectious phase (Jaber and Ownley 2018). Certain growth structures like mycelial fragments and blastospores, have been studied, but not having any commercial application as biopesticide. Commercial biopesticides based on fungi (conidia) are predominantly belonging to Deuteromycota. Most fungi do not invade hosts through the intestine, even if they are consumed with conidia.

Entomopathogenic fungal spores typically enter the host via the integument. The fungal host ranges differ from small to large, but a number of species with a broad range of pathotypes may be more common (Biswas 2009). Infections of the fungus are very complex, but for Zygomycota and Deuteromycota the following overview is common. After conidia or other infectious processes fungal infections begin spontaneously to reach a susceptible host by way of a direct application to the target after delivery by water, rain or animals or, in the case of biopesticides (Razinger et al. 2014).

The conidia on the host cuticulum must be adhered to and germinated after interaction. The fungi may emerge as hyphae, yeasts and protoplasts without having walls. Protoplast helps to bypass the host barriers because the immune system is not noticed the presence of parasite. These operations are influenced by physicochemical characteristics of host cuticle and directly affect the insect ranges of the fungal parasite. Mucilaginous products are often used to facilitate conidia adhesion. The conidium, after it was attached on the insect cuticle and under sufficient moisture, creates a germ. The hypha penetration (germ tube) exerts pressure onto a location which has been partially degraded by the preceding release of hydrolyzing enzymes. Fully hard cuticle presents a larger barrier than newly developed cuticle to fungal penetration, which makes insects more susceptible to moulting.

20.3.1.3 Nematodes

Entomopathogenic nematodes are non-segmented circular worms which are obligatory or sometimes strict insect parasites (Ruiu 2018; Brownbridge and Buitenhuis 2019). Entomopathologic nematodes naturally occur in the soil ecosystem and search the recipient or host for carbon dioxide, movement as well as other biochemical markers. Species have been successfully used for biological pesticides in pest management schemes belong to two nematodes groups (Heterorhabditidae and Steinernematidae) (Geden 2012; Senthil-Nathan 2015; Ruiu 2018). Entomopathogens are fitting well into integrated pest management systems or IPM, since these are known as non-toxic to humanity, are extremely relevant to the targeted pests and could be utilized with conventional pesticides equipments. Entomopathogens nematodes have been removed from the regulation of pesticides by the United States Environmental Protection Agency (EPA). Personal protection equipment and re-entry controls are not mandatory in case of entomopathogen nematodes (Sarwar 2015; Senthil-Nathan 2015). Problems with insect resistance are improbable. The only free stage of entomopathogenic nematodes is the infective juvenile stage (IJ). The young juvenile reaches inside the host insect through spiracles, mouth and anus or through the cuticle's cross-segment membranes in some species and then reaches the haemocoel. All Heterorhabditis and *Steinernema* are related to the genera of bacteria *Photorhabdus* and *Xenorhabdus* (Ruiu 2018). The juvenile stage releases cells bearing their symbiotic bacteria from their small intestine into the haemocoel of insect. In the insect haemolymph, the bacteria spread and grow in higher amount and typically the contaminated host dies within 24–48 hours. Upon the host's death, nematodes continue to feed, evolve and reproduce on the host tissue. The offspring nematodes mature into the adult from four juvenile stages. Based on the existing resources, several generations of infectious juveniles could be emitted throughout the environment to harm multiple hosts and maintain its life cycle.

20.3.1.4 Viruses

A virus is a not reproducible, smallest parasite. Moreover, when a sensitive cell is infected, a virus could capture the host cell system to create additional viruses. Most viruses have single- or double-stranded genetic materials either RNA or DNA. The whole viral particle, referred as a virion, has been made of single type of nucleic acid but also an external protein shell (Kalawate 2014; Maitani et al. 2006; Pierre et al. 2012).

Around 1000 insect species have been reported infected by viruses and comprising of at least 13 distinct insects' orders. The entomogenous viruses were classified into two groups, including non-inclusion viruses those note create inclusion bodies in host and inclusion bodies those produce inclusion bodies in host (Kaaya and Hassan 2000; Biswas 2009). The inclusion viruses are sub-classified into polyhedron (polyhedroses) viruses, which contain polyhedral bodies and granulosis viruses those contain granular bodies (Johne et al. 2011).

Polyhedrosis (plural polyhedroses) is one of the infectious found in insect larvae triggered due to infection of polyhedral virus, i.e. Baculoviridae family. Polyherdrosis bodies, that are featured for the breakdown of host tissues and deposition of viral comprising granules in the body fluid of insect, that are double-band DNA viruses of the families Baculoviridae or Reoviridae (Cypovirus)(Rizvi et al. 2009; Senthil-Nathan 2015; Shafiq Ansari et al. 2012). Viruses/polyhedroses those inhabit in nucleus are called NPV (nuclear polyhedrosis viruses), as well as in similar manner, viruses/polyhedroses those inhabit in cytoplasm are called CPV (cytoplasmic polyhedrosis virus) (Ince et al. 2007; Senthil-Nathan 2015).

Over 1600 distinct viruses invade 1100 insects and mites. More than 10% of all insect infections contain a specific group of viruses, named baculovirus (family Baculoviridae), to which some 100 insect species are susceptible (Kramer and Muthukrishnan 1997). Family Baculoviridae are used for the production of various agricultural virus-incorporated biopesticides. Viruses of this family are considered safe for chordates and no incidents of baculovirus pathogenicity to the chordates have been recorded. Baculoviruses are supercoiled dsDNA-based rod-shaped, insect-specific, enveloped and circular virus. Genome size ranges from 80 to 180 kbp. In research, it was documented that more than 600 types of Baculoviruses caused specific infection in butterflies, moths, sawflies and mosquitoes (Hall et al. 2003). Baculoviruses are contagious and have effective transmission through horizontal channels in environment. If insect-consume occlusion bodies orally, the high pH (alkaline) midgut environment causes polyhedral occlusion bodies to disintegrate and virions release into the lumen of mid-gut of insect (Beas-Catena et al. 2014). In the midgut cell nucleus, viruses multiply within a nucleus of sensitive tissue cells, and their susceptibility varies greatly between viruses and type of cell, although certain NPVs destroy all forms of cells in a tissue and many of the GVs are tissue-specific (only fat body cells) (Gupta and Dikshit 2010). The bloomed new virions cause infection in the haemolymph tissues, like nerve system cells, fat bodies and haemocytes. The infected tissues or cells that are exposed to virus during the latter round in the insect larva produce polyhedral occluded viral particles in nucleus and also produce bloomed virion particles (Kalawate 2014; Senthil-Nathan 2015). Polyhedral virus particle aggregation in the arthropod takes place until the host is completely filled with virus. The insect liquefies and develops polyhedral particles during the final stage of invasion; these newly formed polyhedral particles can be used when it infects other insects.

One caterpillar will have more than 109 occlusion bodies in an initial dose of 1000 when it is dead. Until resorting to the virus infection, the contaminated larvae lead to negative geotropism, encouraging extensive diffusion of killer virus in insect community (Kramer and Muthukrishnan 1997; Rodgers 1993; Szewczyk et al. 2006). The risk of mortality of insect in a community is partially influenced by the environmental factors. The targeted pests could be destroyed in 3–7 days in optimal conditions, but mortality could be induced within 3–4 weeks if conditions are not suitable.

20.3.1.5 Protozoa

Entomopathogens belong to protozoans are a number of organisms comprised of about 1000 species that target invertebrates, especially insect varieties, and are typically considered as microsporidians. These appear to be host unique and sluggish, contributing to chronic infections with weakening of host. The protozoan-formed spore has been the contagious phase and must be consumed for pathogenicity by the insect host (Nawaz et al. 2016). The spores germinate in the midgut of the insect body and released sporoplasm invades the host cells that cause the host to become infected. Only few species have been moderately successful. The effectiveness of *N. locustae* as a pest-management biological agent in the grasshopper appears uncertain, though, because the efficacy of a moving organism is exceedingly difficult to evaluate (Lewis et al. 2009). *Nosema pyrausta* is also an effective microsporidian that decreases productivity and length of adult insects and also destroys European corn-borer larvae.

20.4 Potential Applications of Soil Microbes as Biopesticides

The main purpose of using biopesticides in agriculture is to regulate the growth and functioning of plants along with controlling weeds, nematodes, insects and consequently the diseases spread by them. Biopesticides are categorized by the EPA into three foremost classes which are founded on the kind of vigorous ingredient utilized, such as biochemicals, defensive substances in plants and pesticides produced inside plant body (Clemson 2007). Different products have been synthesized by their specific mechanisms inside each soil microbes. Apart from acting as pesticides, soil microbes may act as biocontrol agents.

The functional component of these microbes exists either naturally or produced by the process of genetic engineering. The mode of actions of soil microbes as a pesticide is generated inside the organism itself or by the action of certain precursors. These substances are specifically used and as compared to other chemical pesticides; these have been proved less toxic (MacGregor 2006). Generally, living organisms like soil microbes are majorly used as biopesticides, which kill the actinomycetes, protozoa, bacteria, nematodes etc. Examples of these organisms are biofungicides (*Trichoderma, Pseudomonas and Bacillus*), bioherbicides (*Phytophthora*) and bioinsecticides (*Bt*) (Gupta and Dikshit 2010). According to the reports, certain soil microbes also include mycoplasma, virus and rickettsia which act as biopesticides and having functional substance in their body regarding that. These contribute in suppressing different types of pests by giving rise to toxic metabolites production which is particular to the pest, leading to certain diseases, trigger the prevention of microbe establishment, via competing, or by different other strategies (Clemson 2007).

20.4.1 Bacteria as Biopesticides

Bacteria are prokaryotes and single-celled organisms and their size may range from 1 μm to certain longer lengths. Mostly the bacteria, which causes disease belongs to certain families including Rhizobiaceae, Enterobacteriaceae, Bacillaceae and Micrococcaceae. *Bacillus spp.*, belonging Bacillaceae family is considered as the most significantly important pest control species (Tanada and Kaya 2012). Soil bacterial species are majorly used as biopesticides which contribute in different ways, though these can be utilized in killing the insects as well as retarding the growth of bacterial and fungal species which cause diseases (Kour et al. 2019a; Yadav et al. 2018). These bacterial microbes affect specifically to certain species of organisms like beetle, mosquitoes, moths and butterflies etc., hence play role as insecticides. For the effective functioning, these biocontrol agents come in contact with that particular pest and finally ingested by them. Digestive system of the insects is interrupted by certain endotoxins produced by the bacteria which are mainly explicated to the specific insect pest. For controlling the pathogens like fungi and bacteria, application of soil bacterial species inhabits the plants and contributes to crowding out those pathogens (O'Brien et al. 2009).

Certain species of *Bacillus* produce spores and live proficiently on earth. The well-known species utilized as a biopesticide is *B. thuringiensis*. These bacterial species are certified also for organic gardening and synthesizing a toxic substance, called Bt, which kills the insects. *B. thuringiensis* is most common bacterial species, which is being used as pesticides and contributing to about 90% of the biopesticide market (Chattopadhyay et al. 2004). Insects' pests are mainly regulated by Bt and therefore intensively used in agricultural practices, medical field and forestry (Keswani et al. 2016).

Bacillus subtilis is also applied as biofungicides. It synthesizes antibiotics, restricts the plant pathogen spore germination and also triggers an immune response in plants. Another species acting as a soil inhabitant and used as a soil inoculant is *Bacillus amyloliquefaciens*. It can grow on roots and gives rise to a physical barrier preventing various other in colonizing the roots. Thus, it can be helpful in providing resistance in plants (O'Brien et al. 2009). *Streptomyces* is a bacterium which is capable of producing mycelium, a thread-like structure. Certain species like *Streptomyces lividicus*, are sold in market as biofungicides. This species is functional in rhizospheric zone in overcoming the pathogenic fungal species by causing deficiency of nutritional content for the fungi (MacGregor 2006).

20.4.2 Fungi as Biopesticide

Insect pathogens exist in broad adjustments range and their disease-causing capabilities include obligate and facultative pathogens (Pucheta et al. 2016). The first pathogenic studies related to insects were done in 1980s and this study was mainly focused on the control of the disease caused by silkworms. With the application of white muscardine fungi on the silkworm in 1835, Bassi first time framed the germ theory and it was considered as *B. bassiana*. According to Gilbert and Gill (2010) explained this particular disease of silkworm gave an idea about utilizing soil fungi as a biocontrol agent for the insect pest management. There are various such products world widely accessible that can be synthesized by the use of approximately ten fungal species (Copping and Menn 2000). Integuments are the major path for the entry of these entomopathogen and further the insects are digested by this ingestion method or via another openings like trachea, wounds etc. (Holder and Keyhani 2005).

Certain soil fungi have a great capability of acting as biocontrol agent and these entomopathogenic fungi create a group with almost 750 species and these stimulate the infections inside the insects when they spread in the environment. Fungus species start producing certain enzymes like chitinases, proteases and lipoxygenases. Insect cuticle is degraded by the production of these enzymes and penetration process is started by the appressorium by applying mechanical pressure and then a structure is formed in the germinative tube. In the body of insect, hyphae are developed by fungal species which spread via haemocoel and attack different muscle tissues, fatty bodies, Malpighian tubes, mitochondria and haemocytes etc. consequently triggers the death of the insect between 3 and 14 days after contamination.

After the death of insects and deprivation of different nutrients, fungal species lead to the formation of growth of micelles and conquer various organs of the host. Conclusively, hyphae bodies enter into the cuticle inside the insect and appear at the surface, which leads to the formation of spores in the suitable conditions of environment (Pucheta et al. 2016).

These fungi live inside the soil and also plant roots. *Trichoderma* has many benefits in comparison to other soil inoculants as it can grow on plant roots. Besides, these fungal species play significant role in acting as a biopesticide as well as enhancing the plant growth (Gilbert and Gill 2010).

20.4.3 Nematodes as Biopesticide

Entomopathogenic nematodes are having soft bodies, unsegmented roundworms which are obligate or also facultative parasites of insects. These nematodes exist in nature, mainly in edaphic conditions and search the host with change according to the different abiotic conditions like chemicals, carbon dioxide, vibration etc. (Grewal et al. 2005). Among soil microbes, specifically in nematodes mainly belong to families like Heterorhabditidae and Steinernematidae are efficiently utilized as bio-insecticides in pest management programs. These soil nematodes can be used in integrated pest management programs as these are not toxic to human beings, comparatively precise to the target pests, and may be treated with standard pesticide equipment (Shapiro-Ilan et al. 2006). The juvenile stage of nematodes enters the host insect through the openings like mouth, spiracles, anus, or may be through intersegmental membranes of the cuticle and finally penetrates into the haemocoel.

Heterorhabditis and *Steinernema*are symbiotically related to the bacteria of the genera *Photorhabdus* and *Xenorhabdus*, respectively (Ferreira and Malan 2014). Cells of mutually associated bacteria are released at the juvenile stage from their intestines which further enters the haemocoel. These microbes increase their number in the insect haemolymph, and then the infected host leads to death within 24–48 hours. Nematodes take their food from the host even after their death, then these organisms mature and undergo the reproduction process. Nematode progeny develops after passing through the four juvenile stages to the adult. According to the accessible resources, few generations might occur inside the host dead body and various numbers of infective juveniles are ultimately free into environment for infecting different other hosts and endure their life cycle (Tofangsazi et al. 2015).

20.4.4 Virus as Biopesticide

A large number of viruses used as biopesticides have now been commercialized. Some DNA-containing viruses like baculoviruses (BVs), nucleopolyhedrosis viruses (NPVs), granuloviruses (GVs), acoviruses, iridoviruses, parvoviruses, polydnaviruses and poxviruses and the RNA-containing reoviruses, cytoplasmic polyhedrosis viruses, nodaviruses, picrona-like viruses and tetraviruses are used for the control of insects. However, NPVs and GVs are the main categories which have been used in the management of pests. These viruses are very effectively used against plant-chewing insects and are widely used to control vegetable and field crop pests. The use of these viruses has a great impact on forest habitats against gypsy moths, pine sawflies, Douglas fir tussock moths and pine caterpillars (Shukla et al. 2019).

Cydia pomonella GVs control codling moth on fruit trees (Lacey et al. 2008) and in stored potatoes, potato tubeworm is controlled by *Phthorimaea operculella* GVs. (Arthurs et al. 2008). For corn earworms, cotton leafworms, cabbage moths, and bollworms, beet armyworms, tobacco budworms and celery loopers a number of virus-based products are also available. Baculoviruses generally are target specific and have the ability to infect and destroy a large number of important plant pests. They are effective against various pests particularly against lepidopterous pests of crops like cotton, rice and some vegetables. The use of these pests is restricted to small areas as their large-scale production causes some difficulties. They are being synthesized on a small scale by various agencies like IPM centres and state agricultural departments as they are commercially unavailable in India (Keswani et al. 2015).

Baculoviruses are well-known arthropod viruses and these are having capability to act as biopesticides in the field of agriculture, forestry and horticulture. Baculoviruses mainly cause infection in the larvae of insects. After eating the foliage of plants, plant debris inadvertently uptake polyhedral larvae. Then transport of polyhedral occurs towards the midgut after ingestion by an insect, polyhedrin protein coat is dissolved due to the alkaline conditions and nucleocapsid is released, which is infectious in nature. The nucleocapsids are transported to the cytoplasm and then towards the nucleus, where the nucleocapsid has been detached. Replication of viruses occurs in nucleus and the assembly of nucleocapsid. Subsequently, the infection blow outs from one cell to another through the insect. One of the best examples of insect regulation by baculovirus is the control of the *Gilipinia hercyniae* population in eastern Canada (Au et al. 2013).

20.4.5 Protozoa as Biopesticides

An extremely diverse group of organisms, the entomopathogenic protozoans consist of around 1000 species of microsporidians (species which attack invertebrates along with insect species) (Solter and Becnel 2007). These are slow-acting and host-specific and produce chronic infections and general host debilitation of the host. For pathogenicity the spore which is the infectious stage of the protozoan formed by it has to be ingested by the insect host. On germination of spore in the mid gut, the sporoplasm is released which invades the target cells and causes the infection of the host. The result of infection is reduced feeding, vigour, fertility and durability of the insect host as wrongly applied microbial control agents. But there are hardly a few species which have been moderately successful (Rosell et al. 2008). However, because of the difficulty in the assessment of a highly mobile insect the use of locusate as a biocontrol agent against grasshopper remains questionable (Ranga Rao et al. 2007). One of the beneficial microorganisms is *N. pyrausta* which lowers the fertility and longevity of the adults and the larvae of European corn borer also face mortality because of this (Tofangsazi et al. 2015).

20.5 Conclusion and Future Prospects

Biocontrol has renewed attributable to both biological benefits as part of international environmental biodiversity and human health. Throughout every area of the world market for biopesticides is steadily rising. When it's used in integrated systems for pest control, the effectiveness of biopesticides, in particular for crops such as berries, veggies and flowers, can be equal or better than traditional pesticides. Biopesticides are successful by balancing efficiency with protection, thus providing versatility to minimal application constraints, superior residue control and resistance reduction ability as well as advantages for human and environmental health. In a study which found that Bs binary crystal toxins were sensitive to extracellular proteases of aquatic microbes, it was also shown that environmental microorganisms are also affected by their extracellular proteases, which naturally are released in the area of toxin applications. In order to influence the Bs toxin efficacy in the control of mosquitoes, in particular *Culex* spp. The susceptibility of target species to the microbial entamopathogens, in particular, is commonly considered genetic potential of insects. In further research, the forms and characteristics of successful proteases released into the application areas of Bs toxins can be established. So that the Bs toxin protein's preventive manipulations or other genetic derivative protein toxins may be well established so that a toxin cannot affect certain proteases, while a toxin can still kill mosquito spp. In the future, their involvement throughout agriculture and forestry is very likely to be more significant. Biopesticides have a distinctly potential role for designing future sustainable pesticide management strategies. We expect that further practical solution to biopesticides will be implemented slowly in the short term and that the fate of biopesticide will not be determined by the real-term profits of synthetic pesticides.

References

Abraham J, Silambarasan S (2015) Plant growth promoting bacteria *Enterobacter asburiae*JAS5 and *Enterobacter cloacae*JAS7 in mineralization of endosulfan. Appl Biochem Biotechnol 175:3336–3348

Ahmad M, Pataczek L, Hilger TH, Zahir ZA, Hussain A, Rasche F et al (2018) Perspectives of microbial inoculation for sustainable development and environmental management. Front Microbiol. https://doi.org/10.3389/fmicb.2018.02992

Ambethgar V (2009) Potential of entomopathogenic fungi in insecticide resistance management (IRM): a review. J Biopest 2:177–193

Arthurs S, Dara SK (2019) Microbial biopesticides for invertebrate pests and their markets in the United States. J Invertebr Pathol 165:13–21

Arthurs SP, Lacey LA, De La Rosa F (2008) Evaluation of a granulovirus (PoGV) and *Bacillus thuringiensis* subsp kurstaki for control of the potato tuberworm (Lepidoptera: Gelechiidae) in stored tubers. J Econ Entomol 101:1540–1546

Asari S, Tarkowska D, Rolcik J, Novak O, Palmero DV, Bejai S, Meijer J (2016) Analysis of plant growth-promoting properties of *Bacillus amyloliquefaciens* UCMB5113 using *Arabidopsis thaliana* as host plant. Planta 245:15–30

Au S, Wu W, Panté N (2013) Baculovirus nuclear import: open, nuclear pore complex (NPC) sesame. Viruses 5(7):1885–1900

Bailey KL, Boyetchko SM, Längle T (2010) Social and economic drivers shaping the future of biological control: a Canadian perspective on the factors affecting the development and use of microbial biopesticides. Biol Control 52(3):221–229

Bautista E, Mesa L, Gómez M (2018) Alternatives for the production of microbial biopesticides based on fungi: the case of Latin America and the Caribbean. Sci Agropecu 9(4):585–604

Beard RL (1945). Studies on the milky disease of Japanese beetle larvae (No. Folleto 1355)

Beas-Catena A, Sánchez-Mirón A, García-Camacho F, Contreras-Gómez A, Molina-Grima E (2014) *Baculovirus* biopesticides: an overview. J Anim Plant Sci 24(2):362–373

Berg G (2009) Plant-microbe interactions promoting plant growth and health: perspectives for controlled use of microorganisms in agriculture. Appl Microbiol Biotechnol 84:11–18

Bianciotto V, Andreotti S, Balestrini R, Bonfante P, Perotto S (2001) Extracellular polysaccharides are involved in the attachment of Azospirillum brasilense and rhizobium leguminosarum to arbuscular mycorrhizal structures. Eur J Histochem 2001:39–50

Biswas A (2009) Role of biopesticides in crop protection: present status and future thrust. Green Farming 13(1 (Special 1)):933–937

Blanford S, Chan BHK, Jenkins N, Sim D, Turner RJ, Read AF, Thomas MB (2005) Fungal pathogen reduces potential for malaria transmission. Science 308(5728):1638–1641

Brownbridge M, Buitenhuis R (2019) Integration of microbial biopesticides in greenhouse floriculture: the Canadian experience. J Invertebr Pathol 165:4–12

Buchholz S, Neumann P, Merkel K, Hepburn HR (2006) Evaluation of *Bacillus thuringiensis* Berliner as an alternative control of small hive beetles, *Aethina tumida* Murray (Coleoptera: Nitidulidae). J Pest Sci 79(4):251

Cantrell CL, Dayan FE, Duke SO (2012) Natural products as sources for new pesticides. J Nat Prod 75(6):1231–1242

Cawoy H, Bettiol W, Fickers P, Onge M (2011) *Bacillus*-Based Biological Control of Plant Diseases. Pesticides in the Modern World - Pesticides Use and Management: 273–302

Céspedes CL, Salazar JR, Ariza-Castolo A, Yamaguchi L, Ávila JG, Aqueveque P, Kubo I, Alarcón J (2014) Biopesticides from plants: *Calceolaria integrifolia* s.l. Environ Res 132:391–406

Chandler D (2017) Basic and applied research on Entomopathogenic Fungi. In: Microbial control of insect and mite pests: from theory to practice. Academic Press, pp 69–89

Chandler D, Bailey AS, Mark Tatchell G, Davidson G, Greaves J, Grant WP (2011) The development, regulation and use of biopesticides for integrated pest management. Philos Trans R Soc B Biol Sci 366(1573):1987–1998

Chattopadhyay A, Bhatnagar NB, Bhatnagar R (2004) Bacterial insecticidal toxins. Crit Rev Microbiol 30(1):33–54

Clemson HGIC (2007) Organic pesticides and biopesticides, Clemson extension, home and garden information center. Clemson University, Clemson

Copping LG, Menn JJ (2000) Biopesticides: a review of their action, applications and efficacy. Pest Manag Sci 56(8):651–676

Damalas CA, Koutroubas SD (2018) Current status and recent developments in biopesticide use. Agriculture 8(13):1–6

Debode J, De Maeyer K, Perneel M, Pannecoucque J, De Backer G, Hofte M (2007) Biosurfactants are involved in the biological control of *Verticillium microsclerotia* by *Pseudomonas* spp. J Appl Microbiol 103:1184–1196

Deravel J, Krier F, Jacques P (2014) Les biopesticides, compléments et alternatives aux produits phytosanitaires chimiques (synthèse bibliographique). Biotechnol Agron Soc Environ 18(2):220–232

Dixit R, Agrawal L, Singh SP, Prateeksha SPC, Prasad V, Chauhan PS (2018) *Paenibacillus lentimorbus* induces autophagy for protecting tomato from *Sclerotiumrolfsii* infection. Microbiol Res 215:164–174

Dubey K, Juwarkar A (2001) Distillery and curd whey wastes as viable alternative sources for biosurfactant production. World J Microbiol Biotechnol 17(1):61–69

Ferreira T, Malan AP (2014) *Xenorhabdus* and *Photorhabdus*, bacterial symbionts of the entomopathogenic nematodes *Steinernema* and *Heterorhabditis* and their in vitro liquid mass culture: a review. African Entomology 22(1):1–15

Gasic S, Tanovic B (2013) Biopesticide formulations, possibility of application and future trends. Pestic i fitomedicina 28(2):97–102

Geden CJ (2012) Status of biopesticides for control of house flies. J Biopest 5:1–6

Gilbert LI, Gill SS (2010) Insect control: biological and synthetic agents. Academic Press

Grewal PS, Ehlers RU, Shapiro-Ilan DI (2005) Nematodes as biocontrol agents. CABI

Gupta S, Dikshit AK (2010) Biopesticides: an ecofriendly approach for pest control. J Biopest 3(1 Special Issue):186–188

Hall FR, Menn JJ, Boyetchko S, Pedersen E, Punja Z, Reddy M (2003) Formulations of biopesticides. In: Biopesticides, pp 487–508

Han JI, Choi HK, Lee SW, Orwin PM, Kim J, Laroe SL, Kim TG, O'Neil J, Leadbetter JR, Lee SY, Hur CG, Spain JC, Ovchinnikova G, Goodwin L, Han C (2011) Complete genome sequence of the metabolically versatile plant growth-promoting endophyte *Variovorax paradoxus* S110. J Bacteriol 193:1183–1190

Heinig RL, Paaijmans KP, Hancock PA, Thomas MB (2015) The potential for fungal biopesticides to reduce malaria transmission under diverse environmental conditions. J Appl Ecol 52(6):1558–1566

Hernández-Fernández J (2016) *Bacillus thuringiensis*: a natural tool in insect pest control. In: Gupta et al (eds) The handbook of microbial bioresources. CABI publishers, UK

Holder DJ, Keyhani NO (2005) Adhesion of the entomopathogenic fungus *Beauveria* (Cordyceps) *bassiana* to substrata. Appl Environ Microbiol 71(9):5260–5266

Hubbard M, Hynes RK, Erlandson M, Bailey KL (2014) The biochemistry behind biopesticide efficacy. Sustain Chem Process 2(1):18

Ince IA, Demir I, Demirbag Z, Nalcacioglu R (2007) A cytoplasmic polyhedrosis virus isolated from the pine processionary caterpillar, *Thaumetopoea pityocampa*. J Microbiol Biotechnol 17(4):632

Jaber LR, Ownley BH (2018) Can we use entomopathogenic fungi as endophytes for dual biological control of insect pests and plant pathogens? Biol Control 116:36–45

James RR, Li Z (2012) From silkworms to bees BT - insect pathology. Insect Pathology 1:425–459

Johne R, Buck CB, Allander T, Atwood WJ, Garcea RL, Imperiale MJ, Major EO, Ramqvist T, Norkin LC (2011) Taxonomical developments in the family Polyomaviridae. Arch Virol 156(9):1627–1634

Kaaya GP, Hassan S (2000) Entomogenous fungi as promising biopesticides for tick control. Exp Appl Acarol 24(12):913–926

Kachhawa D (2017) Microorganisms as a biopesticides. J Entomol Zool Stud 5(3):468–473

Kalawate AS (2014) Microbial viral insecticides. In: Basic and applied aspects of biopesticides. Springer, New Delhi, pp 47–68

Karimi J, Dara SK, Arthurs S (2019) Microbial insecticides in Iran: history, current status, challenges and perspective. J Invertebr Pathol 165:67–73

Karmakar K, Rana A, Rajwar A, Sahgal M, Johri BN (2015) Legume-rhizobia symbiosis under stress. In: Plant microbes Symbiosis: applied facets. Springer, New Delhi, pp 241–258

Keswani C, Bisen K, Singh V, Sarma BK, Singh HB (2016) Formulation technology of biocontrol agents: present status and future prospects. In: Bioformulations: for sustainable agriculture. Springer, New Delhi, pp 35–52

Khater (2012) Prospects of botanical biopesticides in insect pest management. J Appl Pharm Sci 3(12):641–656

Khetan S (2000) Microbial Pest Control. CRC Press, New Delhi

Koeck M, Hardham AR, Dodds PN (2011) The role of effectors of biotrophic and hemibiotrophic fungi in infection. Cell Microbiol 13(12):1849–1857

Kour D, Rana KL, Yadav N, Yadav AN, Kumar A, Meena VS et al (2019a) Rhizospheric microbiomes: biodiversity, mechanisms of plant growth promotion, and biotechnological applications for sustainable agriculture. In: Kumar A, Meena VS (eds) Plant growth promoting Rhizobacteria for agricultural sustainability : from theory to practices. Springer Singapore, Singapore, pp 19–65. https://doi.org/10.1007/978-981-13-7553-8_2

Kour D, Rana KL, Yadav N, Yadav AN, Singh J, Rastegari AA et al (2019b) Agriculturally and industrially important fungi: current developments and potential biotechnological applications. In: Yadav AN, Singh S, Mishra S, Gupta A (eds) Recent advancement in white biotechnology through Fungi, Perspective for value-added products and environments, vol 2. Springer International Publishing, Cham, pp 1–64. https://doi.org/10.1007/978-3-030-14846-1_1

Kramer KJ, Muthukrishnan S (1997) Insect chitinases: molecular biology and potential use as biopesticides. Insect Biochem Mol Biol 27(11):887–900

Krieg A, Huger AM, Schnetter W (1987) Bacillus thuringiensis var. san diego Stamm M-7 ist identisch mit dem zuvor in Deutschland isolierten käferwirksamen B. thuringiensis subsp. tenebrionis Stamm BI 256-82. J Appl Entomol 104(1–5):417–424

Kunimi Y (2007) Current status and prospects on microbial control in Japan. J Invertebr Pathol 95(3):181–186

Lacey LA, Headrick HL, Arthurs SP (2008) Effect of temperature on long-term storage of codling moth granulovirus formulations. J Econ Entomol 101(2):288–294

Leng P, Zhang Z, Pan G, Zhao M (2011) Applications and development trends in biopesticides. African J Biotechnol 10(86):19864–19873

Lewis LC, Bruck DJ, Prasifka JR, Raun ES (2009) Nosema pyrausta: its biology, history, and potential role in a landscape of transgenic insecticidal crops. Biol Control 48(3):223–231

Loth K, Costechareyre D, Effantin G, Rahbé Y, Condemine G, Landon C, Da Silva P (2015) New Cyt-like δ -endotoxins from Dickeya dadantii: structure and aphicidal activity. Sci Rep 5(1):1–0

Lovett B, St. Leger RJ (2018) Genetically engineering better fungal biopesticides. Pest Manag Sci 74(4):781–789

MacGregor JT (2006) Genetic toxicity assessment of microbial pesticides: needs and recommended approaches. Intern Assoc Environ Mutagen Soc: 1–17

Maitani Y, Yano S, Hattori Y, Furuhata M, Hayashi K (2006) Liposome vector containing biosurfactant-complexed DNA as herpes simplex virus thymidine kinase gene delivery system. J Liposome Res 16:359–372

Marrone PG (2008) Barriers to adoption of biological control agents and biological pesticides. In: Integrated Pest management: concepts, tactics, strategies and case studies. Cambridge University Press, Cambridge, UK, pp 163–178

Marrone PG (2019) Pesticidal natural products – status and future potential. Pest Manag Sci 75:2325–2340. https://doi.org/10.1002/ps.5433

Martinez-Medina A, Del Mar AM, Pascual JA, Van Wees SC (2014) Phytohormone profiles induced by trichoderma isolates correspond with their biocontrol and plant growth-promoting activity on melon plants. J Chem Ecol 40:804–815

Mazid S (2011) A review on the use of biopesticides in insect pest management. Int J Sci Adv Technol 1(7):169–178

Mazid S, Kalita J, Rajkhowa R (2011) A review on the use of biopesticides in insect pest management paper subtitle: biopesticides -a safe alternative to chemical control of pests. Int J Sci Adv Technol 1(7):169–178

Meikle WG, Sammataro D, Neumann P, Pflugfelder J (2012) Challenges for developing pathogen-based biopesticides against *Varroa destructor* (Mesostigmata: Varroidae). Apidologie 43(5):501–514

Melo J, Carolino M, Carvalho L, Correia P, Tenreiro R, Chaves S, Meleiro AI, de Souza SB, Dias T, Cruz C, Ramos AC (2016) Crop management as a driving force of plant growth promoting rhizobacteria physiology. Springerplus 5:1574

Mendgen K, Hahn M (2002) Plant infection and the establishment of fungal biotrophy. Trends Plant Sci 7(8):352–356

Meshram PB (2010) Role of some biopesticides in management of some forest insect pests. J Biopest 3(Special Issue):250

Milner RJ (1994) History of *Bacillus thuringiensis*. Agric Ecosyst Environ 49(1):9–13

Mondal S, Halder SK, Yadav AN, Mondal KC (2020) Microbial consortium with multifunctional plant growth promoting attributes: future perspective in agriculture. In: Yadav AN, Rastegari AA, Yadav N, Kour D (eds) Advances in plant microbiome and sustainable agriculture, Functional annotation and future challenges, vol 2. Springer, Singapore, pp 219–254. https://doi.org/10.1007/978-981-15-3204-7_10

Nautiyal CS, Govindarajan R, Lavania M, Pushpangadan P (2008) Novel mechanism of modulating natural antioxidants in functional foods: involvement of plant growth promoting Rhizobacteria NRRL B-30488. J Agric Food Chem 56:4474–4481

Nawaz M, Mabubu JI, Hua H (2016) Current status and advancement of biopesticides: microbial and botanical pesticides. J Entomol Zool Stud 4(2):241–246

O'Brien KP, Franjevic S, Jones J (2009) Green chemistry and sustainable agriculture: the role of biopesticides advancing green chemistry. Ecology 90:2223–2232

Olson S (2015) An analysis of the biopesticide market now and where it is going. Outlooks Pest Manag 26(5):203–206

Pierre L, Cécile L and Patrick DM (2012) The Complex World of Polysaccharides. In: A Complex Biophysical World, The Complex World of Polysaccharides, Desiree Nedra Karunaratne, IntechOpen, https://doi.org/10.5772/51213

Potter DA, Held DW (2002) Biology and Management of the Japanese Beetle. Annu Rev Entomol 47(1):175–205

Pucheta DM, Macias AF, Navarro SR (2016) Mechanism of action of Entomopathogenic Fungi. Interciencia 156(12):2164–2171

Ranga Rao GV, Rupela OP, Rao VR, Reddy YVR (2007) Role of biopesticides in crop protection: present status and future prospects. Indian J Plant Prot 35(1):1–9

Rastegari AA, Yadav AN, Yadav N (2020) New and future developments in microbial biotechnology and bioengineering: Trends of microbial biotechnology for sustainable agriculture and biomedicine systems: diversity and functional perspectives. Elsevier, Amsterdam

Razinger J, Lutz M, Schroers HJ, Urek G, Grunder J (2014) Evaluation of insect associated and plant growth promoting fungi in the control of cabbage root flies. J Econ Entomol 107:1348–1354

Rizvi PQ, Choudhury RA, Ali A (2009) Recent advances in biopesticides. In: Microbial strategies for crop improvement springer. Heidelberg, Berlin, pp 185–203

Robles-Gonzalez IV, Fava F, Poggi-Varaldo HM (2008) A review on slurry bioreactors for bioremediation of soils and sediments. Microb Cell Factories 7:5

Rodgers PB (1993) Potential of biopesticides in agriculture. Pestic Sci 39(2):117–129

Rosas García NM (2014) *Bacillus thuringiensis*: una aplicación de la ciencia. Rev Colomb Biotecnol 16(2):5–6

Rosell G, Quero C, Coll J, Guerrero A (2008) Biorational insecticides in pest management. J Pestic Sci 33:103–121

Ruiu L (2018) Microbial biopesticides in agroecosystems. Agronomy 8(11):235

Salas-Marina MA, Silva-Flores MA, Cervantes-Badillo MG, Rosales-Saavedra MT, Islas-Osuna MA, Casas-Flores S (2011) The plant growth-promoting fungus *Aspergillus ustus* promotes growth and induces resistance against different lifestyle pathogens in *Arabidopsis thaliana*. J Microbiol Biotechnol 21:686–696

Sanchez-Vallet A, Ramos B, Bednarek P, López G, Piślewska-Bednarek M, Schulze-Lefert P, Molina A (2010) Tryptophan-derived secondary metabolites in *Arabidopsis thaliana* confer non-host resistance to necrotrophic *Plectosphaerella cucumerina* fungi. Plant J 63(1):115–127

Sarwar M (2015) Biopesticides: an effective and environmental friendly insect-pests inhibitor line of action. Int J Eng Adv Res Technol 1(2):10–15

Satyanarayana T, Johri BN, Prakash A (2012) Microorganisms in sustainable agriculture and biotechnology. Springer, Dordrecht; New York

Savoie JM, Mata G (2016) Growing Agaricus bisporus as a contribution to sustainable agricultural development. In: Mushroom biotechnology: developments and applications. Academic Press

Seiber JN, Coats J, Duke SO, Gross AD (2014) Biopesticides: state of the art and future opportunities. J Agric Food Chem 62(48):11613–11619

Senthil-Nathan S (2015) A review of biopesticides and their mode of action against insect pests. In: Environmental sustainability: role of green technologies. Springer, New Delhi

Shafiq Ansari M, Ahmad N, Hasan F (2012) Potential of biopesticides in sustainable agriculture. In: Environmental protection strategies for sustainable development. Springer, Dordrecht

Shapiro-Ilan DI, Gouge DH, Piggott SJ, Fife JP (2006) Application technology and environmental considerations for use of entomopathogenic nematodes in biological control. Biol Control 38(1):124–133

Shukla N, Singh EANA, Kabadwa BC, Sharma R, Kumar J (2019) Present status and future prospects of bio-agents in agriculture. Int J Curr Microbiol App Sci 8(4):2138–2153

Singh J, Yadav AN (2020) Natural bioactive products in sustainable agriculture. Springer, Singapore

Singh A, Kumar R, Yadav AN, Mishra S, Sachan S, Sachan SG (2020) Tiny microbes, big yields: microorganisms for enhancing food crop production sustainable development. In: Rastegari AA, Yadav AN, Yadav N (eds) Trends of microbial biotechnology for sustainable agriculture and biomedicine systems: diversity and functional perspectives. Elsevier, Amsterdam, pp 1–15. https://doi.org/10.1016/B978-0-12-820526-6.00001-4

Solter LF, Becnel JJ (2007) Entomopathogenic microsporidia in field manual of techniques in invertebrate pathology. Springer, Dordrecht, pp 199–221

Srivastava S, Trivedi PK, Nautiyal CS, Chauhan PS, Singh PC, Asif MH, Bist V, Srivastava S (2016) Unraveling aspects of *Bacillus amyloliquefaciens* mediated enhanced production of Rice under biotic stress of Rhizoctonia solani. Front Plant Sci 7:587

Sudakin DL (2003) Biopesticides. Toxicol Rev 22(2):83–90

Suprapta DN (2012) Potential of microbial antagonists as biocontrol agents against plant fungal pathogens. J Int Soc Southeast Asian Agric Sci 18(2):1–8

Szewczyk B, Hoyos-Carvajal L, Paluszek M, Skrzecz I, Lobo De Souza M (2006) *Baculoviruses* - Re-emerging biopesticides. Biotechnol Adv 24(2):143–160

Tanada Y, Kaya HK (2012) Insect pathology. Academic Press

Thakur N, Kaur S, Tomar P, Thakur S, Yadav AN (2020) Microbial biopesticides: current status and advancement for sustainable agriculture and environment. In: Rastegari AA, Yadav AN, Yadav N (eds) Trends of microbial biotechnology for sustainable agriculture and biomedicine systems: diversity and functional perspectives. Elsevier, Amsterdam, pp 243–282. https://doi.org/10.1016/B978-0-12-820526-6.00016-6

Tofangsazi N, Arthurs SP, Davis RMG (2015) Entomopathogenic Nematodes (Nematoda: Rhabditida: families Steinernematidae and Heterorhabditidae) one of a series of the Entomology and Nematology Department, UF/IFAS Extension, 2015, 1–5

Usta C (2013) Microorganisms in Biological Pest Control — A Review (Bacterial Toxin Application and Effect of Environmental Factors). In: Current Progress in Biological Research: 287–317

Vea EB, Romeo D, Thomsen M (2018) Biowaste valorisation in a future circular bioeconomy. Procedia CIRP 69:591–596

Verma P, Yadav AN, Kumar V, Singh DP, Saxena AK (2017) Beneficial plant-microbes interactions: biodiversity of microbes from diverse extreme environments and its impact for crop improvement. In: Singh DP, Singh HB, Prabha R (eds) Plant-microbe interactions in agro-ecological perspectives, Microbial interactions and agro-ecological impacts, vol 2. Springer, Singapore, pp 543–580. https://doi.org/10.1007/978-981-10-6593-4_22

Walia S, Saha S, Tripathi V, Sharma KK (2017) Phytochemical biopesticides: some recent developments. Phytochem Rev 16(5):989–1007

Wani ZA, Mirza DN, Arora P, Riyaz-Ul-Hassan S (2016) Molecular phylogeny, diversity, community structure, and plant growth promoting properties of fungal endophytes associated with the corms of saffron plant: an insight into the microbiome of Crocus sativus Linn. Fungal Biol 120:1509–1524

Whalon ME, Wingerd BA (2003) Bt: mode of action and use. Arch Insect Biochem Physiol 54(4):200–211

Yadav AN, Verma P, Kumar S, Kumar V, Kumar M, Singh BP et al (2018) Actinobacteria from Rhizosphere: molecular diversity, distributions and potential biotechnological applications. In: Singh B, Gupta V, Passari A (eds) New and future developments in microbial biotechnology and bioengineering, Amsterdam, pp 13–41. https://doi.org/10.1016/B978-0-444-63994-3.00002-3

Yadav AN, Kour D, Kaur T, Devi R, Gukeria G, Rana KL et al (2020a) Microbial biotechnology for sustainable agriculture: current research and future challenges. In: Rastegari AA, Yadav AN, Yadav N (eds) Trends of microbial biotechnology for sustainable agriculture and biomedicine systems: diversity and functional perspectives. Elsevier, Amsterdam, pp 331–343. https://doi.org/10.1016/B978-0-12-820526-6.00020-8

Yadav AN, Rastegari AA, Yadav N, Kour D (2020b) Advances in plant microbiome and sustainable agriculture: diversity and biotechnological applications. Springer, Singapore

Yadav AN, Rastegari AA, Yadav N, Kour D (2020c) Advances in plant microbiome and sustainable agriculture: functional annotation and future challenges. Springer, Singapore

Yadav AN, Singh J, Rastegari AA, Yadav N (2020d) Plant microbiomes for sustainable agriculture. Springer, Cham

Zazouli MA, Asgharnia H, Cherati JY, Hezarjeribi HZ, Ahmadnezhad A (2015) Evaluation of cow manure effect as bulking agent on concentration of heavy metals in municipal sewage sludge vermicomposting. J Maz Univ Med Sci 25(124):152–169

Biofertilizers for Agricultural Sustainability: Current Status and Future Challenges

21

Arpita Roy

Abstract

Due to the continuous rise in worldwide population and reduction in natural resources, various environmental and agricultural problems have been growing. To overcome this situation, there is a requirement of sustainable agriculture and environment. The innovative method is essential to fulfilling the current food demands and sustainable production of crops. Optimum growth of plants, enhanced productivity, balanced, and adequate nutrient supply are important for sustainable agriculture. In developing countries, soil infertility is one the major problem which hinders the crop productivity. Use of chemical fertilizers enhanced the productivity but extensive uses not only deteriorate soil quality but also influence groundwater quality. This problem can be overcome by the use of green method, i.e., use of biofertilizers. Biofertilizers are a mixture of growth-specific nutrients which improve soil fertility and simultaneously enhanced crop productivity without causing any environmental problems. Industrial-scale production of biofertilizers not only influences the agricultural sectors but also change the farmer's life. This chapter provides information about different types of biofertilizers, its market demand, methods of applications, methods for its use, advantages, commercial production, and limitations.

Keywords

Advantages · Biofertilizers · Commercial methods · Production · Types · Production

A. Roy (✉)
Department of Biotechnology, School of Engineering & Technology, Sharda University, Greater Noida, India

A. N. Yadav et al. (eds.), *Current Trends in Microbial Biotechnology for Sustainable Agriculture*, Environmental and Microbial Biotechnology,
https://doi.org/10.1007/978-981-15-6949-4_21

21.1 Introduction

Global population is growing day by day which leads to creating pressure on the agricultural system. Majority of the populace depends on the agriculture sector for their food and other essential resources such as wood, fiber, gums, and medicinal products. Exponential growth of human population demands simultaneous production and sufficient food supply. There is a requirement of a sustainable agriculture system to fulfill the present demand (Herve et al. 2016). Scientists are developing a suitable method for sustainable agriculture which can effectively fulfill the hunger necessities (Singh et al. 2014). Traditional farming techniques involve production of food and other resources for domestic fulfillment and these methods are generally restricted to the farmer's families and their local societies (Pandey 2018). Advanced scientific methods can be used to enhance the production of crops. The idea of sustainable agriculture is not just to produce the maximum number of crops, but at the same time preservation of ecology is also taken under consideration (Barragan-Ocana and Rivera 2016). Various hormones, synthetic minerals, and chemical fertilizers are used to increase crop production and among these, fertilizers are majorly used. Fertilizer is defined as any substance that is utilized for enhancing soil productivity and crop production by addition of nutrients in the soil which further improves the growth of the plant.

Fertilizers help in the proper supply of vital macronutrients and micronutrients to plants such as nitrogen, phosphorus, calcium, potassium, magnesium, zinc, sulfur, iron, copper, and molybdenum (Alley and Vanlauwe 2009). There is an increasing demand for standard fertilizers such as NPK fertilizers that supply nitrogen (N), phosphorus (P), and potassium (K). Micro-enriched fertilization involves micronutrient addition to the standard fertilizers, which enhanced bio-fortification and improve micronutrient deficiencies such as iron, iodine, zinc, selenium, copper, and fluorine. Although the production of crop enhanced with application of fertilizers but excess use can deplete vital minerals and nutritional factors. Generation of secondary pollutant may enter into the food chain and finally goes into the human body. Secondary pollutants may continue to be present in the atmosphere for a longer period (Uosif et al. 2014). Production and transport of fertilizers also need the burning of fossil fuels which result in the release of carbon dioxide and nitrogen and leads to cause air pollution. Additionally, excess use of chemical fertilizers to soil often causes deterioration of plant roots and abiotic stress such as leaching, acidification, volatilization, and denitrification (Bargaz et al. 2018). Increased concentration of nitrate and phosphorus in water bodies leads to eutrophication (Yang et al. 2008) and emissions of greenhouse gases (Aloe et al. 2014). These fertilizers possess positive effect on crop production but they do have negative impacts on quality of soil, surface water, groundwater, and plant health, which further impact biodiversity and ecosystem functioning (Campos et al. 2018). Therefore, it is necessary to attain sustainable agriculture but without hampering the environmental condition. It is important to lessen the utilization of these substances and find an alternative solution to this problem. Use of biologically derived fertilizers in place of chemical one provides an alternative opportunity due to its harmless and renewable nature.

Bacteria, algae, fungi, and diverse metabolites that are extracted from them can enhance soil productivity and crop yield (Bargaz et al. 2018; Chittora et al. 2020) and they act as a biofertilizer.

Biofertilizers are efficient in supplying essential nutrients to the crops without hampering the natural environment. They possess a vital role in enhanced crop production and sustainability of soil and agriculture (Giri et al. 2019; Kaur et al. 2020; Mondal et al. 2020). They are cost-efficient, environmentally friendly, and renewable sources and possess efficiency to replace chemical fertilizers for sustainable production of crops. In case of chemical fertilizers, plants are only able to use 10–40% as nutrients and the rest gets lost in various forms such as NH_4, CH_4, and CO_2. Biofertilizers can support slow, stable, and continuous nutrients release by their metabolism (Bargaz et al. 2018; Kour et al. 2020a). The requirement of biofertilizers has increased mainly due to the two reasons, i.e., chemical fertilizers although increase fertility of soil and crop productivity, however, extensive use leads to increase various problems and the other reason is that biofertilizers are economical as well as environment friendly. Therefore, in this chapter, use of biofertilizers, their current status, market demand, and further challenges for sustainable agriculture has been discussed.

21.2 Biofertilizers and its Types

Biofertilizers are organic substances that enhance crop productivity and can be produced from bacterial source, fungal source, and algal source. Various microbes that are used as a biofertilizer include Arbuscular mycorrhizal, *Azolla, Azospirillum, Azotobacter,* blue–green algae, and *Rhizobium* (Yadav et al. 2020a, b). These organisms produce metabolites in response to certain conditions that are used by the plants for their various biochemical reactions and further leads to the effective production of crops (Kaur and Purewal 2019; Yadav et al. 2020c). Metabolites that are produced by the microbes help in the breakdown of complex soil minerals to a simple form which acts as a growth enhancer for a particular crop (Fig. 21.1).

Plant growth promotion occurs by alteration of microbes present in the root area through the production of various substances such as growth hormone, siderophores, hydrogen cyanide, ammonia, and exopolysaccharides (Backer et al. 2018). Mostly, bacteria promote plant growth either directly facilitating plant growth hormone levels or indirectly by declining the inhibitory effects of various pathogens.

21.2.1 Bacterial Biofertilizers

Bacterial biofertilizers contain a helpful bacterium that promotes the growth of plants. Bacteria that are used as biofertilizers include *Anabaena, Azolla, Azotobacter, Azospirillum, Bacillus subtills, Bacillus mucilaginosus, Nostoc, Pseudomonas striata, and Rhizobium* (Thomas and Singh 2019; Kour et al. 2020e; Mondal et al. 2020). When bacterial cultures are applied to the surfaces of seed, plant, or soil, they

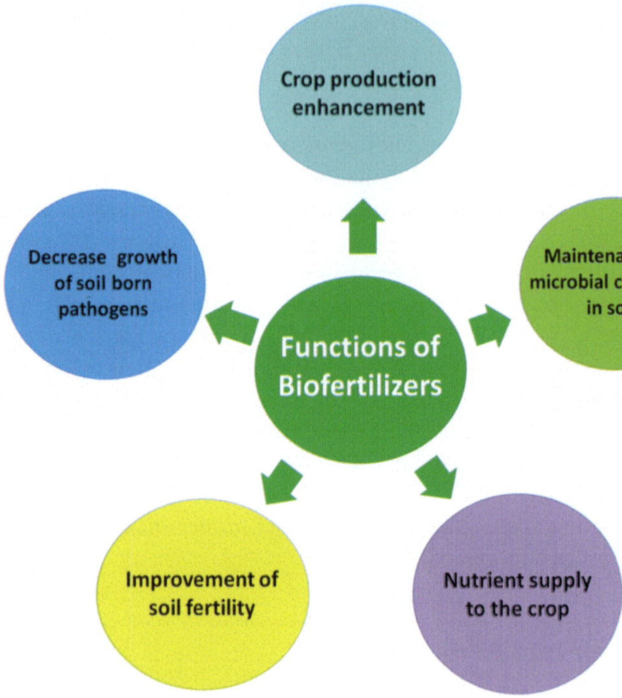

Fig. 21.1 Functions of Biofertilizers

are colonizing either in the interior of the host plant or at rhizosphere. They enhance the accessibility, supply, and nutrient uptake in the host and promote host growth. Majority of bacteria possess symbiotic relationships with the host plants. Bacterial biofertilizers are grouped based on their nature of work such as nitrogen-fixing, phytohormone production, phosphorus solubilizing, and micronutrients supplier (Fig. 21.2). A list of different bacterial biofertilizers has been mentioned in Table 21.1.

21.2.1.1 Nitrogen-Fixing Bacteria

Rhizobium

They are gram-negative, nitrogen-fixing bacteria belong to Rhizobiaceae family, which involves in symbiotic relationship with the host. These are nontoxic and cost-effective biofertilizers which do not possess any adverse health effects on human (Naseer et al. 2019). This bacterium infects plant roots and forms-specific root nodules (de Bruijn 2020). Within the root nodules, these bacteria decrease the molecular nitrogen to ammonia which is subsequently used by the plant for vitamins, proteins, and other nitrogenous compound syntheses (Nyoki and Ndakidemi 2018; Belhadi et al. 2018). Different *Rhizobium* species that act as effective biofertilizers include *Rhizobium japonicum, R. lupine, R. melliloti, R. Phaseoli,* and *R. trifoli*

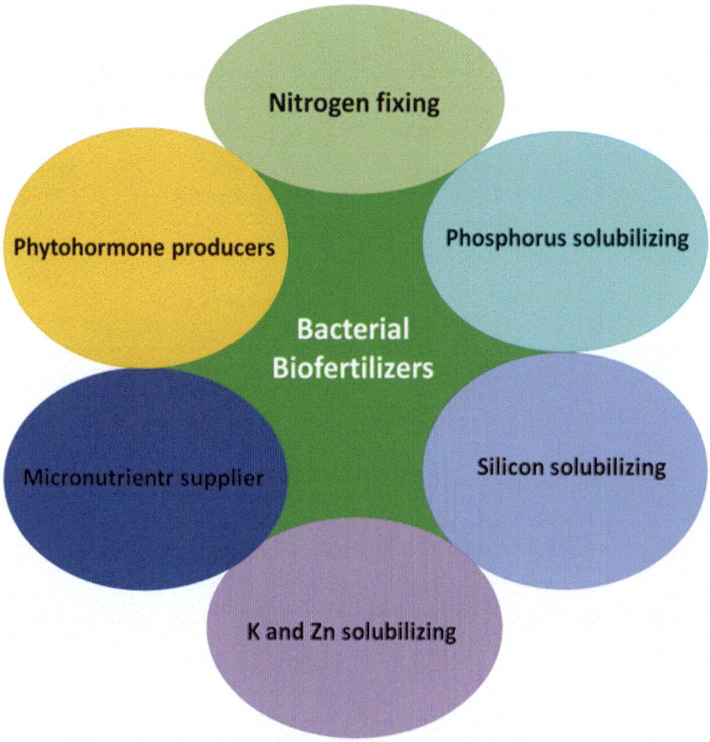

Fig. 21.2 Types of bacterial biofertilizers

(Kaur and Purewal 2019). Utilization of *Rhizobium* in the specific host helps in the maintenance of plant growth and crop productivity (Sahu et al. 2018). In a study, two strains of *Rhizobium* were isolated and these strains enhanced the plant growth by increasing the root and shoot length, height of the plants, and biomass (Kumari et al. 2018).

In another study, *Rhizobium* was inoculated in *Medicago sativa* plant which regulated the synthesis of phytochelatin and expression of MT-related gene and protects plants from excess copper stress (Chen et al. 2018). Bacterial inoculation improved inhibition of copper-induced growth and in seedlings, increased nitrogen concentration was observed. A substantial enhancement in copper uptake was detected in case of *Rhizobium*-inoculated plants as compared to control one (Chen et al. 2018). Verma and Yadav (2019) used *Rhizobium* biofertilizers for *Cicer arietinum* L and found that the application of biofertilizer yields higher biomass as compared to the untreated one. Barakzai et al. (2020) also used *Rhizobium* biofertilizer in case of *Vigna radiata* L. and a substantial enhancement in the plant growth such as plant height, trifoliate leaves, branch number, leaf area, and biomass was observed.

Table 21.1 Bacterial biofertilizers and its roles in plant growth promotion

Bacterial species	Plant	Role	References
Mesorhizobium loti	*Brassica campestris*	Production of siderophores and IAA, solubilization of phosphorus	Maheshwari et al. (2007)
Rhizobium sp.	*Arabidopsis thaliana*, *Brassica napus*	Production of exopolysaccharides	Santaella et al. (2008)
Azorhizobium sp.	*Oryza sativa*	Solubilization of phosphorus, zinc, production of IAA, ammonia, and siderophores	Islam et al. (2009)
Sinorhizobium sp.	*Cajanus cajan*	Production of siderophores and IAA, chitinase and glucanase activity, solubilization of phosphorus	Kumar et al. (2010)
Bradyrhizobium sp.	*Arachis hypogaea*	Production of IAA, siderophores, and HCN solubilization of phosphorus	Badawi et al. (2011)
Rhizobium leguminosarum	*Lactuca sativa*, *Daucus carota*	Production of IAA, siderophores, nitrogen fixation, plant growth enhancement	Flores-Felix et al. (2012)
Rhizobium leguminosarum	*Psoralea corylifolia*	Production of IAA, siderophores, psoralen, solubilization of phosphorus, plant growth enhancement	Prabha et al. (2013)
Rhizobium sp.	–	Solubilization of phosphorus, nitrogen fixation	Karpagam and Nagalakshmi (2014)
Rhizobium nepotum	*Vicia faba L*	Solubilization of phosphorus	Rfaki et al. (2015)
Rhizobium leguminosarum	–	Solubilization of phosphorus	Singha et al. (2016)
Rhizobium sp.	–	Production of IAA	Marczak et al. (2017)
Rhizobium sp.	*Psoralea corylifolia*	Solubilization of phosphorus, production of IAA, nitrogen fixation	Shengepallu et al. (2018)
Bacillus amyloliquefaciens, *Burkholderia cepacia*	*Arachis hypogaea*	Solubilization of phosphorus	Pradhan et al. (2019)
Acidovoraxvalerianellae, *Sinorhizobiumfredii*	*Cajanus cajan*	Plant growth enhancement	Arya et al. (2020)
Bacillus sp.	*Zea mays* L.	Enhanced growth, yield, and quality	Hussain et al. (2020)

Azospirillum

Azospirillum is another type of biofertilizer, a Gram-negative bacterium belongs to Rhodospirillaceae family. They are involved in an associated relationship with the host and able to perform nitrogen fixation, production of phytohormones, or fungicides (Fukami et al. 2018). Different *Azospirillum* species that act as effective biofertilizers are *Azospirillum brasilense, A. halopraeferens, A. irakense,* and *A. lipoferum.* They can produce various phytohormones such as auxins, salicylic acid, indole-3-butaric acid, and indole-3-acetic acid (Egamberdieva et al. 2017). They protect crops from different stresses and enhance nutrient and moisture uptake, which leads to the overall improvement in productivity (Egamberdieva et al. 2017; Fukami et al. 2018). Utilization of *Azospirillum* biofertilizers in plants showed notable physiological and morphological changes such as enhancement of shoot growth and grains with high nitrogen content (Cassan and Diaz-Zorita 2016). Reddy et al. (2018) reported the use of *Azospirillum* treatment in *Lycopersicon esculentum* and found that soil fertility and plant growth were significantly enhanced. In a study *Azospirillum* biofertilizer was used for *Zea mays* L., where it was found that plant growth was promoted with high chlorophyll content as compared to the untreated one (Hekmat et al. 2019). In another study, *Azospirillum* biofertilizer was used for *Arachis hypogeae* L and *Sesamum indicum* L and it was found that treatment with *Azospirillum* for 30 days increases the chlorophyll and nitrogen content where in case of seeds vitamin E content was enhanced (Gayathri and Aiswariya 2020).

Azotobacter

Azotobacter is an aerobic, Gram-negative bacterium that belongs to *Azotobacteraceae* family. They are non-symbiont diazotrophs which play various beneficial roles in crop productivity. They help plants to maintain a healthy life along with the highest production. They can improve the fertility of soil, nitrogen fixation, yield improvement, and plant growth enhancement (Bargaz et al. 2018; Shirinbayan et al. 2019; Subrahmanyam et al. 2020). Various reports suggested that use of *Azotobacter* increased the dry matter of plant and bioactive compound production (Bhardwaj et al. 2014; Mahato and Kafle 2018). During nitrogen fixation, *Azotobacter* produces specific pigments which vary from yellowish-green to dark brown or purple colors. These pigments protect the nitrogenase from harmful effects of oxygen. Different *Azotobacter* species that act as effective biofertilizers include *Azotobacter chroococum, A. salinestris, A. vinelandii,* etc. In a study *Azotobacter* was used for *Allium cepa L* and results showed that there was a significant enhancement in germination, bulb weight, bulb diameter, dry weight, height, leaf number, chlorophyll content, and carotenoid production (Kurrey et al. 2018). Another study reported use of *Azotobacter* for enhancement of growth, quality, and yield of strawberry plant. They found that there was an enhancement of height, leaf number, fruit number, and fruit weight (Soni et al. 2018). Effect of *Azotobacter* on height, straw, and crop yield was evaluated for *Triticum aestivum* L and it was found that apart from an increase in height, yield of straw, and grain, there was also an increase in nitrogen, phosphorus, and potassium content (Khandare et al. 2020).

21.2.1.2 Phosphorus-Solubilizing Microbes

Phosphorus is an important macronutrient that controls signal transduction, respiration, synthesis of proteins, and nitrogen fixation in plants (Naseer et al. 2019). It is available in the soil as an insoluble form; therefore, plants are not able to utilize it, therefore conversion of it from complex to free form is required for proper consumption (Thomas and Singh 2019; Kour et al. 2019a, b; Kumar et al. 2019a; Singh et al. 2020b). Bacteria that can solubilize phosphate are abundant in nature and depending on the region and soil types from where they are isolated influence their number (Awais et al. 2017). *Aereobacter, Agrobacterium, Bacillus megaterium, B. circulans, B. subtilis, Flavobacterium, Microccocus, and P. striata* can effectively solubilize phosphorus. *B. megaterium,* a Gram-positive bacterium with rod shape, can increase inorganic phosphorus and able to solubilize zinc, iron, potassium, and manganese (Kang et al. 2014; Rastegari et al. 2020a, b; Singh and Yadav 2020). In another study, *B. megaterium* var. *phosphaticum* was used for sugarcane and it enhanced the yield of sugar and quality of juice (Sundara et al. 2002). Sharma et al. (2007) isolated two bacteria, i.e., *Pseudomonas fluorescens* and *B. megaterium* which have phosphate-solubilizing capacity and used as biofertilizer for *C. arietinum* seeds. They found that bacteria treatment enhanced the plumule length and radicle. Dastager et al. (2010) isolated *Micrococcus sp.* which showed potential phosphate-solubilizing capability and production of siderophores. Zhao et al. (2014) reported that in maize plant a bacterium *Burkholderia cepacian* was able to solubilize phosphate and enhanced growth as well as salt concentration. Studies reported that *Bacillus and Pseudomonas* showed the capability to increase phosphorus uptake by the plants (Otieno et al. 2015; Radhakrishnan et al. 2017). Ribaudo et al. (2020) reported that in barley plant *Enterobacter ludwigii* was able to solubilize phosphate and promote plant growth. There are many reports on P-solubilization Indian Himalayas region from different habitats including cold deserts (Yadav et al. 2015a, 2016; Yadav et al. 2015b) and cereal crops (Kour et al. 2020b; Kour et al. 2020d; Rana et al. 2020). Phosphorus-solubilizing attributes and mechanisms of P-solubilization by halophilic archaea have been reported by Yadav et al. (2015c).

21.2.1.3 Phytohormones-Producing Microbes

Apart from nitrogen fixation and solubilization of phosphorus, numerous bacteria are able to produce chemicals that promote growth of the plants and used as biofertilizers. In a study, it was reported that *Bacillus polymyxa* showed various beneficial roles such as antibiotics production, cytokinins, chitinase, and other hydrolytic enzymes production, phosphorus solubilization, nitrogen fixation, and also improves the porosity of soil (Timmusk et al. 1999; Tiwari et al. 2020). In another study, two bacteria, i.e., *Bacillus licheniformis* and *Bacillus pumilus* were able to produce gibberellin which promotes plant growth (Gutierez-Mañero et al. 2001). Audenaert et al. (2002) reported that *Pseudomonas aeruginosa* was able to produce siderophores pyochelin, salicylic acid, and pyoverdine, which protect tomato plants against *Botrytis cinereal*. Rhizobacteria are also able to produce some antimicrobial metabolites such as siderophores and antibiotics, which inhibit the growth of fungal cells and protect the plants (Beneduzi et al. 2012). Ali et al. (2010) reported that

Escherichia, Micrococcus, Pseudomonas, and *Staphylococcus* able to produce IAA hormone which enhanced the shoot length, pod number, and weight in case of *V. radiata*. Wani et al. (2013) reported that *Azotobacter* was able to produce some antibiotics which inhibit the growth of soilborne pathogens and prevent its spread in plants. A study reported that *Pseudomonas pseudoalcaligenes* and *Pseudomonas putida* were able to enhance *Coriandrum sativum* growth and increased salt tolerance capacity (Al-Garni et al. 2019). Xia et al. (2020) reported that bacterial consortium was significantly able to increase plant growth and biomass of salt grass. Yadav et al. (2019a) reported the phytohormones producing haloarchaea isolated from halophytic plants growing in hypersaline region of Kutch, India.

21.2.1.4 Mineral-Solubilizing Microbes

Various soil bacteria provide nutrients such as iron, zinc, potassium, and copper to the plants and these are also essential for plant growth. Some rhizobacteria able to solubilize insoluble forms of potassium which can further be taken up by the plants (Jakobsen et al. 2005). Bacteria that can solubilize potassium include *B. mucilaginosus, B. circulans, B. edaphicus, Paenibacillus* sp., and *Acidothiobacillus ferrooxidans*. These bacteria able to enhance the uptake of potassium in numerous plants such as wheat, black pepper, eggplant, pepper, cucumber, etc., and promote biomass production (Etesami et al. 2017). Similarly, some bacteria can also hydrolyze silicates that are required for plant growth. A study reported that *Bacillus sp.* was able to solubilize silicate due to which enhanced growth of rice production was observed (Cakmakci et al. 2007). Rajawat et al. (2020) reported bacteria in stressed environments capable of weather potassium aluminosilicate mineral from different extreme habitats of India. Zinc is another important mineral that is present in a limited amount in the Earth's crust and therefore external supply is required. Some bacterial species, i.e., *Thiobacillus thiooxidan, Bacillus subtilis, Bacillus edaphicus, Paenibacillus glucanolyticus* can solubilized insoluble zinc compounds such as zinc carbonate, zinc oxide, and zinc sulfide (Ansori and Gholami 2015). The mineral solubilizing and mobilizing microbes associated with wheat crops from different agroecological zone of India have been characterized for different abiotic stresses including low temperature (Verma et al. 2015), high temperature (Verma et al. 2014), salinity (Kumar et al. 2019b; Verma et al. 2016), drought (Kour et al. 2020c; Verma et al. 2019), acidic and alkaline habitats (Verma et al. 2013; Verma et al. 2016).

21.2.2 Fungal Biofertilizers

Fungi are abundant in nature and some of them do have useful effects on plants. Fungal inoculum enhanced crop production by increasing uptake of essential nutrients, stimulating plant growth hormone, and decomposing the organic residues (Rashid et al. 2016). Various fungi that act as biofertilizers include *Acaulospora* sp., *Gigaspora* sp., *Glomus* sp., *Laccaria* sp., *Pisolithus* sp., *Pezizellaericae, Rhizoctonia solani,* and *Scutellospora* sp.

21.2.2.1 Arbuscular Mycorrhizal Fungi

They are one of the important biofertilizers, which works in the symbiotic association with plant roots and help in the uptake of nutrient and supports various enzymatic reactions in plants (Begum et al. 2019). They are glomus, intercellular, and nonspecific obligates and plays various growth-promoting roles such as nutrition improvement, increased resistance to drought and insects, and controlled soil structure (Berruti et al. 2015). Arbuscular mycorrhizal fungi are phosphorus mobilizers or phosphate absorbers and nutrients transfer from fungi to host generally happens at the arbuscular interface. They also enhance the uptake of insoluble nutrients such as phosphorus, copper, calcium, and zinc in plants by increasing the root surface area (Singh and Giri 2017). These fungi form obligate or facultative relations with more than 80% of plants and they are dependent on the host for energy and photosynthates and provides various benefits to the host plant (Thakur and Singh 2018).

Additionally, they also improve soil aeration, soil quality, water dynamics, and tolerance to heavy metal and drought and makes the plant less vulnerable to pathogens (Thakur and Singh 2018). Arbuscular mycorrhizal fungi are used in the form of spores, fragments of colonized roots, or combination of both (Klironomos and Hart 2002). They also improve sulfur and potassium nutrition in the host plant. Pellegrino and Bedini (2014) reported that inoculation of arbuscular mycorrhizal fungi improved the chickpea growth as well as grain nutrition in terms of iron, protein, and zinc content. Battini et al. (2017) reported that arbuscular mycorrhizal fungi not only able to mobilize soil phosphates but also enhanced growth of maize plants as compared to control one. Nafady et al. (2018) reported that use of arbuscular mycorrhizal fungi in *Viciafaba* and found that there was increased growth of the plant. Tahjib-Ul-Arif et al. (2018) reported that arbuscular mycorrhizal fungi enhanced the nutrient contents in *Basella alba* and *Amaranthus tricolor*. A study reported that inoculation of arbuscular mycorrhizal fungi with yeast in sunflower plants enhanced the nutrient availability and growth of the plant (Nafady et al. 2019). Another study reported the role of arbuscular mycorrhizal fungi in nitrogen fixation, nutrient uptake in faba bean and wheat plant (Ingraffia et al. 2019). In case of *Panax quinquefolius,* arbuscular mycorrhizal fungi improved the plant growth and at the same time inhibit the soilborne pathogens (Liu et al. 2020).

21.2.2.2 Other Fungi

Ectomycorrhizal fungi form a mutual relationship with several trees and do not penetrate root cells but only surround them. They are naturally present in association with numerous forest trees such as spruce, larch, pine, willow, hemlock, poplar, and eucalyptus (Kaewchai et al. 2009). Various ectomycorrhizal fungi that act as biofertilizers include *Amanita* sp., *Cenococcum* sp., *Elaphomyces* sp., *Lactarius* sp., *Pisolithus* sp., *Rhizopogon* sp., and *Tuber* sp. These fungi support the growth and development of trees by root colonization which enables absorption and accumulation of nitrogen, potassium, phosphorus, calcium, etc. They are able to break down the organic substances and complex minerals that are present in the soil and transfer these to the plants. Apart from these they also provide drought resistance to the plant and prevent plants from soil toxins. These fungi are used for large-scale inoculum such as nursery or forestry plantations (Anderson and Cairney 2007). *Pisolithustinctorius* are the most commonly used ectomycorrhizal fungi that act as a potential biofertilizers and another potential ectomycorrhizal fungus is *Piriformospora indica* (Yadav et al. 2019b, c).

Aspergillus sp. and *Pencillium* sp. are important phosphate-solubilizing fungi which are utilized as biofertilizers. Symbiotic nitrogenous rhizobia and *Arthrobotrys oligospora* have also shown phosphate-solubilizing activity (Khan et al. 2009). Kapri and Tewari (2010) reported the use of *Trichoderma* sp. in case of *C. arietinum* and they found that there was an increase in length of shoot and root as well as biomass. Yadav et al. (2011) treated *C. arietinum* with *Aspergillus niger* and *Trichoderma harzianum* and found that these two fungi were able to solubilize phosphates and enhanced the plant growth. In a study, *Avena sativa* was inoculated with *Paecilomyces carneus* and *Geomyces pannorum* and it was found that these fungi were capable of phosphate solubilization as well as inhibits the nematode growth (Lima-Rivera et al. 2016). Another study reported that *Claroideoglomus etunicatum* and *Rhizophagus clarus* improved the growth of coffee plants as compared to control plants (Moreira et al. 2019).

21.2.3 Algal Biofertilizers

Algae are unicellular or multicellular organisms which belong to a large group of photosynthetic eukaryotic or prokaryotic organisms. They are a rich source of bioactive compounds and used as alginate, carrageenan, agar, food supplements, and functional foods (Ścieszka and Klewicka 2019). Algal bio-fertilizer has shown various advantages and various reports suggest its valuable effects on plant growth (Khan et al. 2009; Ścieszka and Klewicka 2019). Algal fertilizers enhance the micronutrient, macronutrients, vitamins, growth regulators, and amino acids content which further helps in growth and development of plants (Ronga et al. 2019; Malyan et al. 2020). Algal biofertilizers are easier to handle and suitable for use, they are cost-effective, possess extended shelf life, retains soil moisture, and enhanced uptake of nutrients. They also improve the plant resistance against various diseases, i.e., such as insects, pests, nematodes, and prevent against various stresses such as frost, salinity, and drought. A list of different algal biofertilizers has been mention in Table 21.2.

21.2.3.1 Blue-Green Algae

Blue-green algae are filamentous in nature and have the capability to fix nitrogen. They establish symbiotic relationships with ferns, fungal strains, and flowering plants for nitrogen fixation (Kaur and Purewal 2019). They are also able to solubilize zinc, phosphorous, sulfur, potassium, and other micronutrients (Renneberg et al. 2017; Adeniyi et al. 2018). These symbiotic relationships occur either extracellularly or intracellularly. Amongst various symbiotic association, *Azolla-Anabaena azollae* association with blue-green algae makes a unique mutual relationship which possesses both ecological and economic benefits and used as

Table 21.2 Algal biofertilizers and its roles in plant growth promotion

Algal species	Plant	Role	References
Nostoc	–	Improved stability and mineral content of saline soil	Malam Issa et al. (2007)
Blue-green algae	*Oryza sativa*	Improved growth, yield, and mineral composition	Tripathi et al. (2008)
Anabaena iyengarii var. tenuis, Nostoc commune, N. linckia, and *Nostoc* sp.	*Oryza sativa*	Enhanced yield and quality	Pereira et al. (2009)
Nostocentophytum, Oscillatoria angustissima	*Pisum sativum*	Enhanced metabolic activities, growth, and yield	Osman et al. (2010)
Chroococcidiopsis and *Anabaena*	*Triticum aestivum*	Enhanced shoot length, spike length, lateral root, grain weight in wheat plant	Hussain and Hasnain (2011)
Botryococcusbraunii	–	Improved soil stability	Weiss et al. (2012)
Dunaliellasalina, Phaeodactylum Tricornutum	*Capsicum annuum*	Enhanced germination rate in seeds	Guzman-Murillo et al. (2013)
Ascophyllum nodosum	*Vitis vinifera*	Growth enhancement	Popescu and Popescu (2014)
Stoechospermum marginatum	*Solanum melongena*	Growth and yield enhancement	Ramya et al. (2015)
Dunaliella salina	*Triticum aestivum*	Enhanced seed germination	El Arroussi et al. (2016)
Chlorella sp.	*Triticum aestivum, Hordeum vulgare*	Enhanced germination rate	Odgerel and Tserendulam (2017)
Chlorella, Scenedesmus sp.	–	Improved growth in leafy vegetables, wheat, and tomato	Das et al. (2018)
Ulva lactuca, Cystoseira sp., *Gelidium crinale*	*Brassica napus* L.	Productivity and salt stress tolerance	Hashem et al. (2019)
Spirulina platensis, Chlorella vulgaris	*Allium cepa*	High growth and yield	Dineshkumar et al. (2020)

fertilizer. *Azolla* is commonly known as duckweed and belongs to *Salviniaceae* family, it generally forms symbiotic association with *Anabaena*. Yao et al. (2018) reported that *Azolla* was able to provide up to ten tons of protein and other vital minerals for rice growth. Bandonill et al. (2017) reported that blue-green algae were able to improve the rice plant growth in terms of the height of plant, number of panicles/m^2, and number of spikelets/m^2. A study evaluated the role of blue-green algae on growth of rice plants and found that there was an increase in yield of grain about 2–3 quintal/hectare as compared to the control one (Verma and Srivastava

2018). Another study reported that blue-green algae along with vermi compostable to increase the nitrate reductase activity in *Capsicum annuum* and enhanced the growth of the plant (Sundaram and Sundaram 2019). Tomar et al. (2020) evaluated the role of blue-green algae on the mustard plant and found that blue-green algae were able to improve the fertility of soil as well as plant growth.

21.2.3.2 Cyanobacteria

Cyanobacteria are an autotrophic organism, generally found in marine and freshwater. They are small, usually unicellular, and frequently grow in large colonies. They have nitrogen fixation capability and therefore can be utilized as a biofertilizer for economically important crop cultivation like rice, beans, wheat, etc. *Anabaena variabilis Aulosirafertilisima*, *Calothrix* sp., *Scytonema* sp., and *Tolypothrix* sp. are some of the most efficient nitrogen-fixing cyanobacteria which are found in the rice cultivation area (Prasad and Prasad 2001; Singh and Yadav 2020; Yadav et al. 2020c). Cyanobacteria do not require any host for the growth, development, and production of valuable organic products. They play an important role in soil fertility maintenance (Song et al. 2005), which can degrade various pollutants and possess diverse roles in the soil ecosystem (Chittora et al. 2020). A study reported that cyanobacteria able to enhanced the production of *Salix viminalis* L and enhanced the quantity of nitrogen, potassium, and phosphorus in plants (Grzesik et al. 2017). Chittapun et al. (2018) reported that cyanobacteria were able to increase the total number and weight of grain in case of *Oryza sativa*. Singh et al. (2019) reported that cyanobacteria can fix the atmospheric nitrogen and effectively act as biofertilizers. Ma et al. (2019) reported cyanobacterial use on the paddy field for nitrogen fixation and its supply to the plant. Now biomass of different cyanobacteria is also explored for soil quality improvement and plant growth.

21.3 Production of Biofertilizers

For the production of biofertilizers, various things should be undertaken such as microbe type, growth pattern of microbe, optimum conditions for growth, and inoculum preparation. Application method and product storage are also crucial for market consistency. Generally, six major steps are involved in production of biofertilizer (Mohammadi and Sohrabi 2012) and they are

1. Selection of microbe.
2. Isolation of a particular microbe.
3. Appropriate method.
4. Best proliferation method.
5. Evaluation at a smaller scale (trials).
6. Evaluation at a larger scale.

Identifying a specific strain for nitrogen fixation or solubilization of potassium, phosphate, and zinc which are able to survive in various environmental conditions, soils, and crops are required. Use of modern as biotechnological methods for strain development can also be used for development of better quality strains.

21.4 Methods Used for the Application of Biofertilizers

Seed treatment is the most widely used method. In this method initially, slurry is prepared by mixing 200 g of biofertilizers with a double amount of water, i.e., 1:2 ratio. The amount is enough to treat 10–12 kg seeds. Prepared slurry is sprayed on the seeds and uniformly mixed for a thin coating. This method is generally applied to the crop like rice, wheat, sorghum, maize, sunflower, groundnut, safflower, mustard, gram, soybean, etc.

Another method is set treatment which is generally used for sets or pieces of sugarcane, banana, and potato. In this method initially, slurry is prepared by mixing 500 g of biofertilizers with 25 L of water, i.e., 1:50 ratio. Then seeds are immersed in the slurry for at least 30 min and then dried for 1 h under shade. After planting the seeds, the field should be irrigated within 24 h.

Seedling treatment is also used for crops like chilly, onion, rice, cabbage, tomato, cauliflower, etc. In this method, slurry is prepared by mixing 1 kg of biofertilizers with 10–15 L of water, i.e., 1:10 or 1:15 ratio. Seedling roots are then dipped in the mixture for 20–30 min and instantly planted in the field without drying. This method can also be used for flowers and ornamental plants (Shabbir et al. 2019).

Soil treatment is another method used for biofertilizer application. This method varies from one crop to another depending upon its duration. In case of short duration crop (6 months), for one-acre land, 2–3 kg of biofertilizer mixed with 40–60 kg of manure. Then the mixture is directly used into the soil. In case of long duration crops, twofold amount of biofertilizers is required (Shabbir et al. 2019).

21.5 Precautions for Biofertilizers Applications

For proper use of biofertilizers in order to get the best results, there is some precautionary measure which should be followed

- Biofertilizer packets should be kept at a shady and cool place.
- Storage should not be above 35 °C and below 0 °C.
- Utilized the material before expiry date.
- Biofertilizer should only be used for a particular crop that is mentioned on the packet.
- Cultures should not be kept in warm or hot water which may destroy the microbes.
- Mixing of nitrogenous fertilizers with other biofertilizers should not be done.
- Fungicides should not be used along with biofertilizers.

21.6 Advantages of Biofertilizers

Biofertilizers are playing an important role in improving soil fertility and crop production. They form a mutual relationship with the host plants as they grow in the soil. They possess several advantages such as

- Reduction in chemical fertilizer use.
- Reduction in environmental pollution, eco-friendly.
- Enhance nutrients which are easily absorbed.
- Enhance overall properties of soil.
- Cost-effective therefore poor farmers can also use.
- Increase the decomposition rate in compost pit.
- Improve root propagation.
- Permit plant growth and better survival under various stress conditions.

21.7 Commercial Production and Release of Biofertilizers

One of the major restrictions for improved production of crops is the availability of quality biofertilizers in the market. Biofertilizer quality differs from one production unit to another and also on the method of preparation. Before commercial-scale production, there are some prerequisites which need to be undertaken such as estimation of inoculum, field suitability, feasible of production, adequate staff training, quality control measures, availability of proper microbiological facilities, continuous supply of microbial culture, and requirement of equipment which can maintain healthy life cycle of microbes and provides maximum production.

Currently, in the market numerous biofertilizers are available and their quality and quantity depend on a particular unit. However, before market release, a biofertilizer should have some qualities such as

- It should be easily available.
- There should be less transportation cost.
- It should be stable under different conditions.
- Quality should be remaining same for longer period.
- It should be required in minimal amount and effectively provide the nutrients required for crops.
- It should be water-soluble so that can be easily applied in huge field area.
- It should supply nutrients immediately without posing any negative impact on plants.
- User-friendly and do not cause any harmful health effect on the farmers.
- Cost-effective and independent of seasonal variations.

Steps that are involved in commercial production to market release of biofertilizers are mention in Fig. 21.3. Biofertilizers that are presently available in the market are listed in Table 21.3.

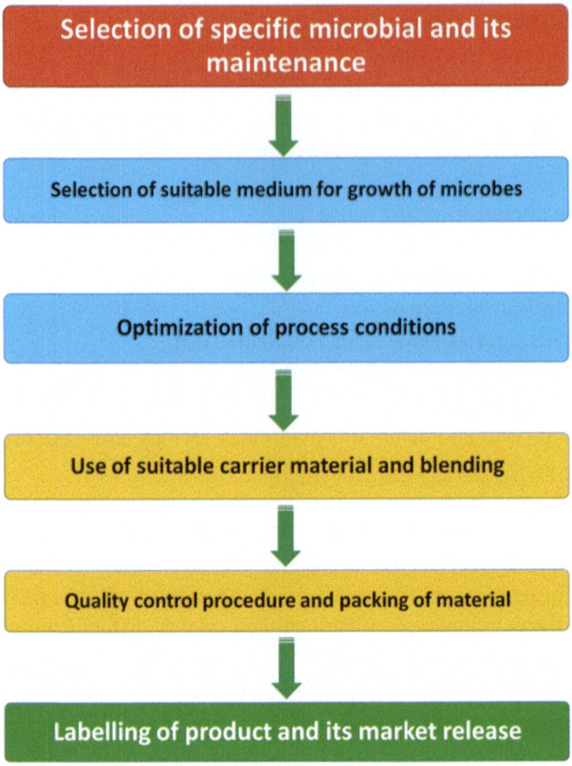

Fig. 21.3 Steps involved in commercial production of biofertilizers

21.8 Biotechnological Role of Biofertilizers

Biofertilizer possess various biotechnological applications such as amino acid synthesis, photosynthesis, bioremediation of pollutants, prevention from abiotic stresses, and biofortification. Growth of plants depends on the photosynthesis rate, a study found that when certain *Rhizobia* strains were inoculated with rice substantial improvement in overall photosynthetic rate was observed (Mia and Shamsuddin 2010). Another study reported that water stress conditions, *Bacillus lentus*, *Pseudomonas* sp., and *A. brasilense* enhanced the antioxidant activity and photosynthetic capacity of basil plants (Heidari and Golpayegani 2012). Cohen et al. (2015) studied the morphological, physiological, and biochemical changes in *Arabidopsis thaliana* when it was inoculated with *A. brasilense* sp. 245 strain under drought and controlled condition. It was observed that the bacterial strain triggers the photosynthetic and photoprotective pigment synthesis. Heavy metals are one of the major pollutants which accumulate soil biosphere and they are nondegradable in nature (Akhtar et al. 2013). There are various soil microbes and plant growth-promoting rhizobacteria such as *Azotobacter, Brevibacillus, Kluyvera, Mesorhizobium, Ochrobactrum, Pseudomonas, Ralstonia, Rhizobium, Sinorhizobium, Variovox,* and

Table 21.3 Various biofertilizers available in market and their application for agricultural sustainability

Trade name	Organism name	Applications
Bioazoto	Various *Azotobacter* strains such as *Azotobacter chroococcum*	For all crops like wheat, maize, paddy, sorghum, barley, mustard, sunflower, sugarcane, banana, grapes, papaya, watermelon, onion, potato, tomato, cauliflower, sesamum, cotton, chilly, ladyfinger, rapeseed, linseed, tobacco, mulberry, coconut, spices, fruits, flowers, plantation crops, and forest plants
Bhoomi Rakshak,		
Kisaan Azotobacter culture		
Azonik		
Biospirillum	Various *Azospirillum* strains such as *Azospirillum lipoferum*, *Azospirillum brasilense*,	For paddy and other crops, normal, acidic and dry soils
Green plus		
Bio-N		
Azo-S		
ROM		
Spironik		
Grow agro	Various *Rhizobium* sp.	For pulse crops like gram, lentil, pea, pigeon pea, soybean, green gram, black gram, etc. and other leguminous crops groundnut, berseem, and lucerne
Samridhi bio Rhizo		
Iffco *Rhizobium*		
VBL- Rhizo		
Jaiveek *Rhizobium*		
Biobium		
Kisaan rhizobium culture		
Rhizo-enrich,		
Rhizoteeka,		
Green earth reap N4		
Rhizonik		
Biophos, get-Phos	*Bacillus megaterium*, mycorrhizal fungi	Phosphorus-solubilizing and mobilizing for all types of crop plant
MYCORISE		
Kisaan P.S.B. culture		
MycoRhiz,		
Reap P		
Phosphonive		
BIO-NPK	*Bacillus mucilagenosus*	Potassium-mobilizing for all types of crop plant
BioPotash		
Potash-cure		
Green earth reap K		

(continued)

Table 21.3 (continued)

Trade name	Organism name	Applications
Biozinc	Zinc-solubilizing bacteria such as *Pantoea dispersa, Pantoea agglomerans, Pseudomonas fragi, Rhizobium* sp.	Zinc-solubilizing for crops like wheat, pulses, paddy, ginger, citrus, pomegranate, etc.
Zinc-cure		
Zinc activator		
Zinc extra		
MicroZ-109		
Biosulf	Sulfur-solubilizing microbes such as *Thiobacillus thioxidans*	For cereals, pulses, millets, oilseeds, fiber crops, sugar crops, forage crops, plantation crops, spices, flowers, medicinal crops, etc.
Sulphonik		
S sol B®		
MicroS-109		
Silica-cure	Silica-solubilizing microbes such as *Burkholderia eburnean, Rhizobium* sp., etc.	For crops like cereals, sugar, cane, onions, leafy greens, legumes, cucumber, pumpkin, and gourd
BioSilica		
Silica-109		

Xanthomonas which have shown potential in remediation of heavy metals contaminants (Shinwari et al. 2015; Malyan et al. 2019; Sharaff et al. 2020). Pesticides are other pollutants which are regularly used in agriculture for management of pests. They possess negative impact on environment as well as on plant health. Bacterial species such as *Azospirillum, Bacillus, Gordonia, Klebsiella, Pseudomonas,* and *Serratia* showed potential ability in reducing toxic effects caused by pesticides (Shaheen and Sundari 2013; Rai et al. 2020; Singh et al. 2020a; Thakur et al. 2020). Microbes are able to produce various enzymes such as esterases and hydrolases which are in majorly involved in pesticide degradation (Kour et al. 2019a, b; Devi et al. 2020). Biofertilizers also help plants to survive in abiotic stress. Salt-tolerant microbes such as *P. extremorientalis, P. fluorescens, S. plymuthica,* and *S. rhizophila* are able to produce plant growth hormones. In a study, it was found that biofertilizer-treated wheat plants showed 78% higher biomass under drought conditions as compared to control one (Timmusk et al. 2014). Thus, biofertilizers possess potential biotechnological role for sustainable environment and agriculture.

21.9 Challenges of Biofertilizers

Although biofertilizers are eco-friendly, cost-effective, and holds various advantages, it does have some limitations in its application at a larger scale. Several restrictions are there which affect the impact of this technique in production, marketing, or practicality.

- They act slowly as compared to chemical fertilizer.
- Storage problem due to its high sensitivity toward temperature and moisture.
- Often it is difficult to find a vendor selling biofertilizers in rural or remote areas.
- They can't completely replace the other fertilizers but can complement them.

- Lacks of particular microbial strain or efficient growth medium lessen the availability of some biofertilizers.
- There are no appropriate facilities for production of biofertilizers which results in low-quality production.
- Limited availability of trained labor for production unit.
- Biofertilizers demand is seasonal based due to the microorganism's activity.

These limitations cause doubt among investors about its potential applications. Additionally, there can be restrictions about application of biofertilizers which can influence production, marketing, and usage.

21.10 Conclusion and Future Prospects

Due to the increase in worldwide populace, there is a requirement of sustainable agricultural system which can efficiently fulfill the food, feed, and other needs. In modern-day rigorous cropping systems requires more energy inputs to obtain higher production. Fertilizers are used for enhanced crop productivity, however, prolonged use of them can possess various negative impacts on human health, soil quality, and environment such as land degradation and poor soil quality. Therefore, eco-friendly sustainable agricultural production method is required to meet food demand and simultaneous conserve environment. Biofertilizers are one of the alternatives to overcome this problem which can solve the agro-industrial challenges in a better way. They provide nutrition to the crop plant and at the same time maintains the ecological balance. Biotech businesses in developing countries have attained notable success in biofertilizers development and supply. However, some factors are associated with the biofertilizers usage, i.e., development, detailed knowledge of production, application method, and promotion of biofertilizers is required for proper agriculture. It is important to understand production method, utilization, and storage conditions for sustainable use. Additionally, farmers should be properly trained for the proper usage of biofertilizers. Biofertilizers can solve the current challenges present in the agriculture industries and creates new openings for farmers, business, academia, and other sectors. As completely dependent on chemical fertilizers not only contaminates ecosystem but also increases farming cost which eventually leads to crises among the farmers. Therefore, efficient utilization of biofertilizers for crop production is a vital parameter for sustainable agriculture. Biofertilizers have various advantages, though they also possess certain limitations such as slow release of mineral elements in the fields which limits their flexibility in present agriculture practices. This can be rectified by the use of modern biotechnology and therefore further research work is required to make this technique more efficient.

References

Adeniyi OM, Azimov U, Burluka A (2018) Algae biofuel: current status and future applications. Renew Sust Energ Rev 90:316–335

Akhtar MS, Chali B, Azam T (2013) Bioremediation of arsenic and lead by plants and microbes from contaminated soil. Res Plant Sci 1:68–73

Al-Garni SM, Khan MMA, Bahieldin A (2019) Plant growth-promoting bacteria and silicon fertilizer enhance plant growth and salinity tolerance in *Coriandrum sativum*. J Plant Interact 14(1):386–396

Ali B, Sabri AN, Hasnain S (2010) Rhizobacterial potential to alter auxin content and growth of *Vigna radiata* (L.). World J Microbiol Biotechnol 26(8):1379–1384

Alley MM, Vanlauwe B (2009) The role of fertilizers in integrated plant nutrient management. International Fertilizer Industry Association, Paris, p 59

Aloe AK, Bouraoui F, Grizzetti B, Bidoglio G, Pistocchi A (2014) Managing nitrogen and phosphorus loads to water bodies: characterisation and Solutions.Towards macro-regional integrated nutrient management. Joint Research Centre, JRC-Ispra

Anderson IC, Cairney JW (2007) Ectomycorrhizal fungi: exploring the mycelial frontier. FEMS Microbiol Rev 31(4):388–406

Ansori A, Gholami A (2015) Improved nutrient uptake and growth of maize in response to inoculation with *Thiobacillus* and mycorrhiza on an alkaline soil. Commun Soil Sci Plant Anal 46:2111–2126

Arya R, Pandey C, Dhiman S, Aeron A, Dubey RC, Maheshwari DK et al (2020) Fertilizer adaptive bacteria *Acidovorax valerianellae* and *Sinorhizobium fredii* in integrated nutrient management of pigeon pea (*Cajanus cajan* L.).S Afr J bot. https://doi.org/10.1016/j.sajb.2020.03.018

Audenaert K, Pattery T, Cornelis P, Höfte M (2002) Induction of systemic resistance to *Botrytis cinerea* in tomato by *Pseudomonas aeruginosa* 7NSK2: role of salicylic acid, pyochelin, and pyocyanin. MPMI 15:1147–1156

Awais M, Tariq M, Ali A, Ali Q, Khan A, Tabassum B et al (2017) Isolation, characterization and inter-relationship of phosphate solubilizing bacteria from the rhizosphere of sugarcane and rice. Biocatal Agric Biotechnol 11:312–321

Backer R, Rokem JS, Ilangumaran G, Lamont J, Praslickova D, Ricci E et al (2018) Plant growth-promoting rhizobacteria: context, mechanisms of action, and roadmap to commercialization of biostimulants for sustainable agriculture. Front Plant Sci 23(9):1473

Badawi FSF, Biomy AMM, Desoky AH (2011) Peanut plant growth and yield as influenced by co-inoculation with *Bradyrhizobium* and some rhizo-microorganisms under sandy loam soil conditions. AOAS 56(1):17–20

Bandonill EH, Cacerez JCA, Malabayabas MD, Martinez-Goss MR (2017) Blue-green algae: a promising biofertilizer for rice cultivation. Phil J Crop Sci 42(S1):62–63

Barakzai KR, Dhar S, Khalili A, Rasooli M, Tanha J (2020) Effect of sources of nutrient and biofertilizers on growth and yield of mungbean (*Vigna radiata* L.). IJCS 8(1):555–559

Bargaz A, Lyamlouli K, Chtouki M, Zeroual Y, Dhiba D (2018) Soil microbial resources for improving fertilizers efficiency in an integrated plant nutrient management system. Front Microbiol 9:1606

Barragan-Ocana A, Rivera MC (2016) Rural development and environmental protection through the use of biofertilizer's in agriculture: an alternative for underdeveloped countries. Technol Soc 46:90–99

Battini F, Grønlund M, Agnolucci M, Giovannetti M, Jakobsen I (2017) Facilitation of phosphorus uptake in maize plants by mycorrhizosphere bacteria.Sci. Rep 7(1):1–11

Begum N, Qin C, Ahanger MA, Raza S, Khan MI, Ahmed N et al (2019) Role of Arbuscular Mycorrhizal Fungi in plant growth regulation: implications in abiotic stress tolerance. Front Plant Sci 10:1068

Belhadi D, Lajudie P, Ramdani N, Roux C, Boulila F, Tisseyre P et al (2018) *Vicia faba* L. in the Bejaia region of Algeria is nodulated by *Rhizobium leguminosarum* sv. Viciae, *Rhizobium laguerreae* and two new genospecies. Syst Appl Microbiol 41:122–130

Beneduzi A, Ambrosini A, Passaglia LM (2012) Plant growth-promoting rhizobacteria (PGPR): their potential as antagonists and biocontrol agents. Genet Mol Biol 35(4):1044–1051

Berruti A, Lumini E, Balestrini R, Bianciotto V (2015) Arbuscular mycorrhizal Fungi as natural biofertilizers: let's benefit from past successes. Front Microbiol 6:1559

Bhardwaj D, Ansari MW, Sahoo RK, Tuteja N (2014) Biofertilizers function as key player in sustainable agriculture by improving soil fertility, plant tolerance and crop productivity. Microb Cell Factories 13(1):66

de Bruijn FJ (2020) Signaling and early infection events in the rhizobium–legume symbiosis: introduction. In: de Bruijn F (ed) The Model Legume *Medicago truncatula*. Wiley, pp 432–433. https://doi.org/10.1002/9781119409144.ch53

Cakmakci R, Dönmez MF, Erdoğan U (2007) The effect of plant growth promoting rhizobacteria on barley seedling growth, nutrient uptake, some soil properties, and bacterial counts. Turk J Agric For 31:189–199

Campos EVR, Proenca PLF, Oliveira JL, Bakshi M, Abhilash PC, Fraceto LF (2018) Use of botanical insecticides for sustainable agriculture: future perspectives. Ecol Indic

Cassan F, Diaz-Zorita M (2016) *Azospirillum* sp. in current agriculture: from the laboratory to the field. Soil Biol Biochem 103:117–130

Chen J, Liu YQ, Yan XW, Wei GH, Zhang JH, Fang LC (2018) *Rhizobium* inoculation enhances copper tolerance by affecting copper uptake and regulating the ascorbate-glutathione cycle and phytochelatin biosynthesis-related gene expression in *Medicago sativa* seedlings. Ecotoxicol Environ Saf 162:312–323

Chittapun S, Limbipichai S, Amnuaysin N, Boonkerd R, Charoensook M (2018) Effects of using cyanobacteria and fertilizer on growth and yield of rice, Pathum Thani I: a pot experiment. J Appl Phycol 30(1):79–85

Chittora D, Meena M, Barupal T, Swapnil P (2020) Cyanobacteria as a source of biofertilizers for sustainable agriculture. Biochem Biophys Rep 22:100737

Cohen AC, Bottini R, Pontin M, Berli FJ, Moreno D, Boccanlandro H et al (2015) *Azospirillum brasilense* ameliorates the response of *Arabidopsis thaliana* to drought mainly *via* enhancement of ABA levels. Physiol Plant 153:79–90

Das P, Quadir MA, Thaher MI, Alghasal GSHS, Aljabri HMSJ (2018) Microalgal nutrients recycling from the primary effluent of municipal wastewater and use of the produced biomass as bio-fertilizer. Int J Environ Sci Technol 16(7):3355–3364

Dastager SG, Deepa CK, Pandey A (2010) Isolation and characterization of novel plant growth promoting *Micrococcus* sp NII-0909 and its interaction with cowpea. Plant Physiol Biochem 48(12):987–992

Devi R, Kaur T, Guleria G, Rana K, Kour D, Yadav N et al (2020) Fungal secondary metabolites and their biotechnological application for human health. In: Rastegari AA, Yadav AN, Yadav N (eds) Trends of microbial biotechnology for sustainable agriculture and biomedicine systems: perspectives for human health. Elsevier, Amsterdam, pp 147–161. https://doi.org/10.1016/B978-0-12-820528-0.00010-7

Dineshkumar R, Subramanian J, Arumugam A, Rasheeq AA, Sampathkumar P (2020) Exploring the microalgae biofertilizer effect on onion cultivation by field experiment.Waste and. Biomass Valorization 11(1):77–87

Egamberdieva D, Wirth SJ, Alqarawi AA, AbduAllah EF, Hashem A (2017) Phytohormones and beneficial microbes: essential components for plants to balance stress and fitness. Front Microbiol 8:2104

El Arroussi H, Elbaouchi A, Benhima R, Bendaou N, Smouni A, Wahby I (2016) Halophilic microalgae *Dunaliella salina* extracts improve seed germination and seedling growth of *Triticum aestivum* L. under salt stress. In: Acta horticulturae. International Society for Horticultural Science (ISHS), Leuven, pp 13–26

Etesami H, Emami S, Alikhani HA (2017) Potassium solubilizing bacteria (KSB): mechanisms, promotion of plant growth, and future prospects, a review. J Soil Sci Plant Nutr 17:897–911

Flores-Felix JD, Menendez E, Rivera LP (2012) Use of *Rhizobium leguminosarum* as a potential biofertilizer for *Lactuca sativa* and *Daucus carota* crops. J Plant Nutr Soil Sci 176:876–882

Fukami J, Cerezini P, Hungria M (2018) *Azospirillum*: benefits that go far beyond biological nitrogen fixation. AMB Express 8(1):73

Gayathri V, Aiswariya K (2020) Effect of different bio-fertilizers on the chlorophyll, nitrogen and vitamin E content in *Arachis hypogeae* L. and *Sesamum indicum* L.Agricultural science digest-a. Res J 40(1):49–52

Giri B, Prasad R, Wu QS, Varma A (2019) Biofertilizers for sustainable agriculture and environment. Springer, Cham

Grzesik M, Romanowska-Duda Z, Kalaji HM (2017) Effectiveness of cyanobacteria and green algae in enhancing the photosynthetic performance and growth of willow (*Salix viminalis* L.) plants under limited synthetic fertilizers application. Photosynthetica 55(3):510–521

Gutierez-Mañero FJ, Ramos-Solano B, Probanza A, Mehouachi J, Tadeo FR, Talon M (2001) The plant-growth-promoting rhizobacteria *Bacillus pumilus* and *Bacillus licheniformis* produce high amounts of physiologically active gibberellins. Physiol Plant 111:206–211

Guzman-Murillo MA, Ascencio F, Larrinaga-Mayoral JA (2013) Germination and ROS detoxification in bell pepper (*Capsicum annuum* L.) under NaCl stress and treatment with microalgae extracts. Protoplasma 250:33–42

Hashem HA, Mansour HA, El-Khawas SA, Hassanein RA (2019) The potentiality of marine macro-algae as bio-fertilizers to improve the productivity and salt stress tolerance of canola (*Brassica napus* L.) plants. Agronomy 9(3):146

Heidari M, Golpayegani A (2012) Effects of water stress and inoculation with plant growth promoting rhizobacteria (PGPR) on antioxidant status and photosynthetic pigments in basil (*Ocimum basilicum* L.). J Saudi Soc Agric Sci 11:57–61

Hekmat AW, Mohammadi NK, Ghosh G (2019) Effect of NPK, biofertilizer and zinc foliar nutrition on growth and growth attributes of babycorn (*Zea mays* L.). IJCS 7(4):2432–2436

Herve M, Albert CH, Bondeau A (2016) On the importance of taking into account agricultural practices when defining conservation priorities for regional planning. J Nat Conserv 33:76–84

Hussain A, Hasnain S (2011) Phytostimulation and biofertilization in wheat by cyanobacteria. J Ind Microbiol Biotechnol 38:85–92

Hussain A, Zahir ZA, Ditta A, Tahir MU, Ahmad M, Mumtaz MZ, Hussain S (2020) Production and implication of bio-activated organic fertilizer enriched with zinc-solubilizing bacteria to boost up maize (*Zea mays* L.) production and biofortification under two cropping seasons. Agronomy 10(1):39

Ingraffia R, Amato G, Frenda AS, Giambalvo D (2019) Impacts of arbuscular mycorrhizal fungi on nutrient uptake, N_2 fixation, N transfer, and growth in a wheat/faba bean intercropping system. PLoS One 14(3):e0213672

Islam MR, Madhaiyan M, Deka HPB, Yim W, Lee G, Saravanan VS et al (2009) Characterization of plant growth-promoting traits of free-living diazotrophic bacteria and their inoculation effects on growth and nitrogen uptake of crop plants. J Microbiol Biotechnol 19(10):1213–1222

Jakobsen I, Leggett ME, Richardson AE (2005) Rhizosphere microorganisms and plant phosphorus uptake. In: Sims JT, Sharpley AN (eds) Phosphorus, agriculture and the environment. Am Soc Agronomy, Madison, pp 437–494

Kaewchai S, Soytong K, Hyde KD (2009) Mycofungicides and fungal biofertilizers. Fungal Divers 38:25–50

Kang SM, Radhakrishnan R, You YH, Joo GJ, Lee IJ, Lee KE et al (2014) Phosphate solubilizing *Bacillus megaterium* mj1212 regulates endogenous plant carbohydrates and amino acids contents to promote mustard plant growth. Indian J Microbiol 54(4):427–433

Kapri A, Tewari L (2010) Phosphate solubilization potential and phosphatase activity of rhizospheric *Trichoderma* spp. Braz J Microbiol 41(3):787–795

Karpagam T, Nagalakshmi PK (2014) Isolation and characterization of phosphate solubilizing microbes from agricultural soil. J Curr Microbiol App Sci 3(3):601–614

Kaur P, Purewal SS (2019) Biofertilizers and their role in sustainable agriculture. In: Giri B, Prasad R, Wu QS (eds) Varma ABiofertilizers for sustainable agriculture and environment. Springer, Cham, pp 285–300. https://doi.org/10.1007/978-3-030-18933-4

Kaur T, Rana KL, Kour D, Sheikh I, Yadav N, Kumar V et al (2020) Microbe-mediated biofortification for micronutrients: present status and future challenges. In: Rastegari AA, Yadav AN, Yadav N (eds) Trends of microbial biotechnology for sustainable agriculture and biomedicine systems: perspectives for human health. Elsevier, Amsterdam, pp 1–17. https://doi.org/10.1016/B978-0-12-820528-0.00002-8

Khan V, Jilani MS, Akhtar SMS, Naqvi RM (2009) Phosphorus solubilizing bacteria: occurrence, mechanisms and their role in crop production. J Agric Biol Sci 1:48–58

Khandare RN, Chandra R, Pareek N, Raverkar KP (2020) Carrier-based and liquid bioinoculants of *Azotobacter* and PSB saved chemical fertilizers in wheat (*Triticum aestivum* L.) and enhanced soil biological properties in Mollisols. J PlantNutr 43(1):36–50

Klironomos JN, Hart MM (2002) Colonization of roots by arbuscular mycorrhizal fungi using different sources of inoculum. Mycorrhiza 12:181–184

Kour D, Rana KL, Yadav AN, Yadav N, Kumar M, Kumar V et al (2019a) Microbial biofertilizers: bioresources and eco-friendly technologies for agricultural and environmental sustainability. Biocatal Agric Biotechnol 28:101487

Kour D, Rana KL, Yadav AN, Yadav N, Kumar V, Kumar A et al (2019b) Drought-tolerant phosphorus-solubilizing microbes: biodiversity and biotechnological applications for alleviation of drought stress in plants. In: Sayyed RZ, Arora NK, Reddy MS (eds) Plant growth promoting Rhizobacteria for sustainable stress management, Rhizobacteria in abiotic stress management, vol 1. Springer, Singapore, pp 255–308. https://doi.org/10.1007/978-981-13-6536-2_13

Kour D, Kaur T, Devi R, Rana KL, Yadav N, Rastegari AA et al (2020a) Biotechnological applications of beneficial microbiomes for evergreen agriculture and human health. In: Rastegari AA, Yadav AN, Yadav N (eds) Trends of microbial biotechnology for sustainable agriculture and biomedicine systems: perspectives for human health. Elsevier, Amsterdam, pp 255–279. https://doi.org/10.1016/B978-0-12-820528-0.00019-3

Kour D, Rana KL, Kaur T, Sheikh I, Yadav AN, Kumar V et al (2020b) Microbe-mediated alleviation of drought stress and acquisition of phosphorus in great millet (*Sorghum bicolour* L.) by drought-adaptive and phosphorus-solubilizing microbes. Biocatal Agric Biotechnol 23:101501. https://doi.org/10.1016/j.bcab.2020.101501

Kour D, Rana KL, Sheikh I, Kumar V, Yadav AN, Dhaliwal HS et al (2020c) Alleviation of drought stress and plant growth promotion by *Pseudomonas libanensis* EU-LWNA-33, a drought-adaptive phosphorus-solubilizing bacterium. Proc Natl Acad Sci India B. https://doi.org/10.1007/s40011-019-01151-4

Kour D, Rana KL, Yadav AN, Sheikh I, Kumar V, Dhaliwal HS et al (2020d) Amelioration of drought stress in foxtail millet (*Setaria italica* L.) by P-solubilizing drought-tolerant microbes with multifarious plant growth promoting attributes. Environ Sustain 3:23–34. https://doi.org/10.1007/s42398-020-00094-1

Kour D, Rana KL, Yadav AN, Yadav N, Kumar M, Kumar V et al (2020e) Microbial biofertilizers: bioresources and eco-friendly technologies for agricultural and environmental sustainability. Biocatal Agric Biotechnol 23:101487. https://doi.org/10.1016/j.bcab.2019.101487

Kumar H, Bajpai VK, Dubey RC (2010) Wilt disease management and enhancement of growth and yield of *Cajanus cajan* (L) var. Manak by bacterial combinations amended with chemical fertilizer. Crop Protect 29:591–598

Kumar M, Saxena R, Rai PK, Tomar RS, Yadav N, Rana KL et al (2019a) Genetic diversity of Methylotrophic yeast and their impact on environments. In: Yadav AN, Singh S, Mishra S, Gupta A (eds) Recent advancement in white biotechnology through Fungi, Perspective for sustainable environments, vol 3. Springer International Publishing, Cham, pp 53–71. https://doi.org/10.1007/978-3-030-25506-0_3

Kumar V, Joshi S, Pant NC, Sangwan P, Yadav AN, Saxena A et al (2019b) Molecular approaches for combating multiple abiotic stresses in crops of arid and semi-arid region. In: Singh SP,

Upadhyay SK, Pandey A, Kumar S (eds) Molecular approaches in plant biology and environmental challenges. Springer, Singapore, pp 149–170. https://doi.org/10.1007/978-981-15-0690-1_8
Kumari P, Meena M, Gupta P, Dubey MK, Nath G, Upadhyay RS (2018) Plant growth promoting rhizobacteria and their biopriming for growth promotion in mung bean (*Vigna radiata* (L.) R. Wilczek). Biocatal Agric Biotechnol 16:163–171
Kurrey DK, Lahre MK, Pagire GS (2018) Effect of *Azotobacter* on growth and yield of onion (*Allium cepa* L). J Pharmacogn Phytochem 7:1171–1175
Lima-Rivera DL, Lopez-Lima D, Desgarennes D, Velazquez-Rodriguez AS, Carrion G (2016) Phosphate solubilization by fungi with nematicidal potential. J Soil Sci Plant Nutr 16(2):507–524
Liu N, Shao C, Sun H, Liu Z, Guan Y, Wu L, Zhang B (2020) Arbuscular mycorrhizal fungi biofertilizer improves American ginseng (*Panax quinquefolius* L.) growth under the continuous cropping regime. Geoderma 363:114155
Ma J, Bei Q, Wang X, Lan P, Liu G, Lin X et al (2019) Impacts of Mo application on biological nitrogen fixation and diazotrophic communities in a flooded rice-soil system. Sci Total Environ 649:686–694
Mahato S, Kafle A (2018) Comparative study of *Azotobacter* with or without other fertilizers on growth and yield of wheat in Western hills of Nepal. Ann Agric Sci 16(3):250–256
Maheshwari DK, Chandra S, Choure K, Dubey RC (2007) Rhizosphere competent *Mesorhizobium loti* mp6 induces root hair curling, inhibits *Sclerotinia sclerotiorum* and enhances growth of Indian mustard (*Brassica campestris*). BJM 38:124–130
Malam Issa O, Défarge C, Le Bissonnais Y, Marin B, Duval O, Bruand A, D'Acqui LP, Nordenberg S, Annerman M (2007) Effects of the inoculation of cyanobacteria on the microstructure and the structural stability of a tropical soil. Plant Soil 290:209–219
Malyan SK, Kumar A, Baram S, Kumar J, Singh S, Kumar SS et al (2019) Role of fungi in climate change abatement through carbon sequestration. In: Yadav AN, Singh S, Mishra S, Gupta A (eds) Recent advancement in white biotechnology through Fungi, Perspective for sustainable environments, vol 3. Springer, Cham, pp 283–295. https://doi.org/10.1007/978-3-030-25506-0_11
Malyan SK, Singh S, Bachheti A, Chahar M, Sah MK, Narender et al. (2020) Cyanobacteria: a perspective paradigm for agriculture and environment. In: Rastegari AA, Yadav AN, Yadav N (eds) Trends of microbial biotechnology for sustainable agriculture and biomedicine systems: diversity and functional perspectives. Elsevier, Amsterdam pp 215–224. https://doi.org/10.1016/B978-0-12-820526-6.00014-2
Marczak M, Mazur A, Koper P, Żebracki K, Genes AS (2017) Synthesis of rhizobial exopolysaccharides and their importance for symbiosis with legume plants. Genes 8(12):360
Mia MB, Shamsuddin Z (2010) Nitrogen fixation and transportation by rhizobacteria: a scenario of rice and banana. Int J Bot 6:235–242
Mohammadi K, Sohrabi Y (2012) Bacterial biofertilizers for sustainable crop production: a review. ARPN J Agric Biol Sci 7(5):307–316
Mondal S, Halder SK, Yadav AN, Mondal KC (2020) Microbial consortium with multifunctional plant growth promoting attributes: future perspective in agriculture. In: Yadav AN, Rastegari AA, Yadav N, Kour D (eds) Advances in plant microbiome and sustainable agriculture, Functional annotation and future challenges, vol 2. Springer, Singapore, pp 219–254. https://doi.org/10.1007/978-981-15-3204-7_10
Moreira SD, França AC, Grazziotti PH, Leal FDS, Silva EDB (2019) Arbusc ular mycorrhizal fungi and phosphorus doses on coffee growth under a non-sterile soil. Revista Caatinga 32(1):72–80
Nafady NA, Hassan EA, Abd-Alla MH, Bagy MMK (2018) Effectiveness of eco-friendly arbuscular mycorrhizal fungi biofertilizer and bacterial feather hydrolysate in promoting growth of *Vicia faba* in sandy soil. Biocatal Agric Biotechnol 16:140–147
Nafady NA, Hashem M, Hassan EA, Ahmed HA, Alamri SA (2019) The combined effect of arbuscular mycorrhizae and plant-growth-promoting yeast improves sunflower defense against *Macrophomina phaseolina* diseases. Biol Control 138:104049
Naseer I, Ahmad M, Nadeem SM, Ahmad I, Zahir ZA (2019) Rhizobial inoculants for sustainable agriculture: prospects and applications. In: Giri B, Prasad R, Wu QS, Varma A (eds)

Biofertilizers for sustainable agriculture and environment. Springer, Cham, pp 245–283. https://doi.org/10.1007/978-3-030-18933-4

Nyoki D, Ndakidemi PA (2018) *Rhizobium* inoculation reduces P and K fertilization requirement in corn-soybean intercropping. Rhizosphere 5:51–56

Odgerel B, Tserendulam D (2017) Effect of *Chlorella* as a biofertilizer on germination of wheat and barley grains. Proc Mong Acad Sci 56:26

Osman MEH, El-Sheekh MM, El-Naggar AH, Gheda SF (2010) Effect of two species of cyanobacteria as biofertilizers on some metabolic activities, growth, and yield of pea plant. Biol Fertil Soils 46(8):861–875

Otieno N, Lally RD, Kiwanuka S, Lloyd A, Ryan D, Germaine KJ et al (2015) Plant growth promotion induced by phosphate solubilizing endophytic *Pseudomonas* isolates. Front Microbiol 6:745

Pandey G (2018) Challenges and future prospects of Agri-nanotechnology for sustainable agriculture in India. Environ Technol Innov 11:299–307

Pellegrino E, Bedini S (2014) Enhancing ecosystem services in sustainable agriculture: biofertilization and biofortification of chickpea (*Cicer arietinum*L.) by arbuscular mycorrhizal fungi. Soil Biol Biochem 68:429–439

Pereira I, Ortega R, Barrientos L, Moya M, Reyes G, Kramm V (2009) Development of a biofertilizer based on filamentous nitrogen-fixing cyanobacteria for rice crops in Chile. J Appl Phycol 21(1):135–144

Popescu GC, Popescu M (2014) Effect of the brown alga *Ascophyllum nodosum* as biofertilizer on vegetative growth in grapevine (*Vitis vinifera* L.). Curr Trends Nat Sci 3(6):61–67

Prabha C, Maheshwari DK, Bajpai VK (2013) Diverse role of fast-growing rhizobia in growth promotion and enhancement of psoralen content in *Psoralea corylifolia*. Phcog Mag 9:57–65

Pradhan M, Dhali S, Sahoo RK, Pradhan C, Mohanty S (2019) Effect of P solubilizing bacteria and P fertilizer on inorganic P fractions of acid soil and its influence on P uptake in groundnut (*Arachis hypogaea* L). Legum Res-Int J 42(5):694–698

Prasad RC, Prasad BN (2001) Cyanobacteria as a source biofertilizer for sustainable agriculture in Nepal. J Plant Sci Bot Orient 1:127–133

Radhakrishnan R, Hashem A, AbdAllah EF (2017) *Bacillus*: a biological tool for crop improvement through bio-molecular changes in adverse environments. Front Physiol 8:667

Rai PK, Singh M, Anand K, Saurabhj S, Kaur T, Kour D et al (2020) Role and potential applications of plant growth promotion rhizobacteria for sustainable agriculture. In: Rastegari AA, Yadav AN, Yadav N (eds) Trends of microbial biotechnology for sustainable agriculture and biomedicine systems: diversity and functional perspectives. Elsevier, Amsterdam, pp 49–60. https://doi.org/10.1016/B978-0-12-820526-6.00004-X

Rajawat MVS, Singh R, Singh D, Yadav AN, Singh S, Kumar M et al (2020) Spatial distribution and identification of bacteria in stressed environments capable to weather potassium aluminosilicate mineral. Braz J Microbiol 51:751–764. https://doi.org/10.1007/s42770-019-00210-2

Ramya SS, Vijayanand N, Rathinavel S (2015) Foliar application of liquid biofertilizer of brown alga *Stoechospermum marginatum* on growth, biochemical and yield of *Solanum melongena*. Int J Recycl Org Waste Agric 4(3):167–173

Rana KL, Kour D, Kaur T, Sheikh I, Yadav AN, Kumar V et al (2020) Endophytic microbes from diverse wheat genotypes and their potential biotechnological applications in plant growth promotion and nutrient uptake. Proc Natl Acad Sci India B. https://doi.org/10.1007/s40011-020-01168-0

Rashid MI, Mujawar LH, Shahzad T, Almeelbi T, Ismail IM, Oves M (2016) Bacteria and fungi can contribute to nutrients bioavailability and aggregate formation in degraded soils. Microbiol Res 183:26–41

Rastegari AA, Yadav AN, Yadav N (2020a) New and future developments in microbial biotechnology and bioengineering: trends of microbial biotechnology for sustainable agriculture and biomedicine systems: diversity and functional perspectives. Elsevier, Amsterdam

Rastegari AA, Yadav AN, Yadav N (2020b) New and future developments in microbial biotechnology and bioengineering: trends of microbial biotechnology for sustainable agriculture and biomedicine systems: perspectives for human health. Elsevier, Amsterdam

Reddy S, Singh AK, Masih H, Benjamin JC, Ojha SK, Ramteke PW et al (2018) Effect of *Azotobacter* sp and *Azospirillum* sp on vegetative growth of tomato (*Lycopersicon esculentum*).J. Pharmacogn Phytochem 7(4):2130–2137

Renneberg R, Berkling V, Loroch V (2017) Green biotechnology. In: Biotechnology for beginners, 2nd edn. Academic Press, Cambridge, MA, pp 235–279. https://doi.org/10.1016/B978-0-12-801224-6.00007-2

Rfaki A, Nassiri L, Ibijbijen J (2015) Isolation and characterization of phosphate solubilizing bacteria from the rhizosphere of faba bean (*Vicia faba* L.) in Meknes region, Morocco. BMRJ 6(5):247–254

Ribaudo C, Zaballa JI, Golluscio R (2020) Effect of the phosphorus-solubilizing bacterium *Enterobacter Ludwigii* on barley growth promotion. ASRJETS 63(1):144–157

Ronga D, Biazzi E, Parati K, Carminati D, Carminati E, Tava A (2019) Microalgal biostimulants and biofertilisers in crop productions. Agronomy 9(4):192

Sahu PK, Singh DP, Prabha R, Meena KK, Abhilash PP (2018) Connecting microbial capabilities with the soil and plant health: options for agricultural sustainability. Ecol Indic 105:601–612

Santaella C, Schue M, Berge O, Heulin T, Achouak W (2008) The exopolysaccharide of *Rhizobium* sp. YAS34 is not necessary for biofilm formation on *Arabidopsis thaliana* and Brassica napus roots but contributes to root colonization. Environ Microbiol 10:2150–2163

Ścieszka S, Klewicka E (2019) Algae in food: a general review. Crit Rev Food Sci Nutr 59(21):3538–3547

Shabbir RN, Ali H, Nawaz F, Hussain S, Areeb A, Sarwar N et al (2019) Use of biofertilizers for sustainable crop production. In: Agronomic crops. Springer, Singapore, pp 149–162

Shaheen S, Sundari K (2013) Exploring the applicability of PGPR to remediate residual organophosphate and carbamate pesticides used in agriculture fields. Int J Agric Food Sci Technol 4:947–954

Sharaff MS, Subrahmanyam G, Kumar A, Yadav AN (2020) Mechanistic understanding of root-microbiome interaction for sustainable agriculture in polluted soils. In: Rastegari AA, Yadav AN, Yadav N (eds) Trends of microbial biotechnology for sustainable agriculture and biomedicine systems: diversity and functional perspectives. Elsevier, Amsterdam, pp 61–84. https://doi.org/10.1016/B978-0-12-820526-6.00005-1

Sharma K, Dak G, Agrawal A, Bhatnagar M, Sharma R (2007) Effect of phosphate solubilizing bacteria on the germination of *Cicer arietinum* seeds and seedling growth. J Herb Med Toxicol 1(1):61–63

Shengepallu MD, Gaikwad RT, Chavan VA, Anand YR (2018) Isolation and characterization of nitrogen fixing bacteria from babchi (*Psoralea corylifolia* L.) and testing them for plant growth promotion traits in vitro. Int J Curr Microbiol App Sci 7:441–447

Shinwari KI, Shah AU, Afridi MI, Zeeshan M, Hussain H, Hussain J et al (2015) Application of plant growth promoting rhizobacteria in bioremediation of heavy metal polluted soil. Asian J Multidiscip Stud 3:179–185

Shirinbayan S, Khosravi H, Malakouti MJ (2019) Alleviation of drought stress in maize (*Zea mays*) by inoculation with *Azotobacter* strains isolated from semi-arid regions. Appl Soil Ecol 133:138–145

Singh I, Giri B (2017) Arbuscular mycorrhiza mediated control of plant pathogens. In: Mycorrhiza-Nutrient uptake, biocontrol, ecorestoration. Springer, Cham, pp 131–160

Singh J, Yadav AN (2020) Natural bioactive products in sustainable agriculture. Springer, Singapore

Singh S, Singh BK, Yadav SM, Gupta AK (2014) Potential of biofertilizer's in crop production in Indian agriculture. Am J Plant Nutr Fertil Technol 4:33–40

Singh JS, Kumar A, Singh M (2019) Cyanobacteria: A sustainable and commercial bioresource in production of bio-fertilizer and bio-fuel from waste waters. Environmental and Sustainability Indicators, 100008

Singh A, Kumar R, Yadav AN, Mishra S, Sachan S, Sachan SG (2020a) Tiny microbes, big yields: microorganisms for enhancing food crop production sustainable development. In: Rastegari AA, Yadav AN, Yadav N (eds) Trends of microbial biotechnology for sustainable agriculture and biomedicine systems: diversity and functional perspectives. Elsevier, Amsterdam, pp 1–15. https://doi.org/10.1016/B978-0-12-820526-6.00001-4

Singh B, Boukhris I, Pragya KV, Yadav AN, Farhat-Khemakhem A et al (2020b) Contribution of microbial phytases to the improvement of plant growth and nutrition: a review. Pedosphere 30:295–313. https://doi.org/10.1016/S1002-0160(20)60010-8

Singha B, Mazumder PB, Pandey P (2016) Characterization of plant growth promoting rhizobia from root nodule of *Crotalaria pallida* grown in Assam. IJBT 15:210–216

Song T, Mårtensson L, Eriksson T, Zheng W, Rasmussen U (2005) Biodiversity and seasonal variation of the cyanobacterial assemblage in a rice paddy field in Fujian, China.FEMS. Microbiol Ecol 54(1):131–140

Soni S, Kanawjia A, Chaurasiya R, Singh P (2018) Effect of organic manure and biofertilizers on growth, yield and quality of strawberry (*Fragaria X ananassa* Duch) cv. Sweet Charlie J J Pharmacogn Phytochem 2:128–132

Subrahmanyam G, Kumar A, Sandilya SP, Chutia M, Yadav AN (2020) Diversity, plant growth promoting attributes, and agricultural applications of rhizospheric microbes. In: Yadav AN, Singh J, Rastegari AA, Yadav N (eds) Plant microbiomes for sustainable agriculture. Springer, Cham, pp 1–52. https://doi.org/10.1007/978-3-030-38453-1_1

Sundara B, Natarajan V, Hari K (2002) Influence of phosphorus solubilizing bacteria on the changes in soil available phosphorus and sugarcane and sugar yields. Field Crops Res 77(1):43–49

Sundaram SS, Sundaram SP (2019) Nitrate reductase activities on *Capsicum annuum* L. by treating vermi compost and blue green algae. J Med Plant 7(6):144–146

Tahjib-Ul-Arif M, Ghosh A, Chamely SG, Haque MR, Rahman MM (2018) Arbuscular mycorrhizal fungi inoculation with organic matter and phosphorus supplementation enhance nutrient contents of *Amaranthus tricolor* L. and *Basella alba* L. by improving nutrients uptake. Trop Plant Res 5(3):375–384

Thakur P, Singh I (2018) Biocontrol of soilborne root pathogens: an overview. In: Giri B, Prasad R, Wu QS, Varma A (eds) Root biology, soil biology. Springer, pp 181–220

Thakur N, Kaur S, Tomar P, Thakur S, Yadav AN (2020) Microbial biopesticides: current status and advancement for sustainable agriculture and environment. In: Rastegari AA, Yadav AN, Yadav N (eds) Trends of microbial biotechnology for sustainable agriculture and biomedicine systems: diversity and functional perspectives. Elsevier, Amsterdam, pp 243–282. https://doi.org/10.1016/B978-0-12-820526-6.00016-6

Thomas L, Singh I (2019) Microbial biofertilizers: types and applications. In: Giri B, Prasad R, Wu QS, Varma A (eds) Biofertilizers for sustainable agriculture and environment. Springer, Cham, pp 1–19. https://doi.org/10.1007/978-3-030-18933-4

Timmusk S, Nicander B, Granhall U, Tillberg E (1999) Cytokinin production by *Paenobacillus polymyza*. Soil Biol Biochem 31:1847–1852

Timmusk S, El-Daim IA, Copolovici L, Tanilas T, Kännaste A, Behers L et al (2014) Drought-tolerance of wheat improved by rhizosphere bacteria from harsh environments: enhanced biomass production and reduced emissions of stress volatiles. PLoS One 9(5):e96086

Tiwari P, Bajpai M, Singh LK, Mishra S, Yadav AN (2020) Phytohormones producing fungal communities: metabolic engineering for abiotic stress tolerance in crops. In: Yadav AN, Mishra S, Kour D, Yadav N, Kumar A (eds) Agriculturally Important Fungi for Sustainable Agriculture, Perspective for diversity and crop productivity, vol 1. Springer, Cham, pp 1–25. https://doi.org/10.1007/978-3-030-45971-0_8

Tomar P, Thakur H, Sudarsan JS, Nithiyanantham S, Prathap MG (2020) Effect of blue green algae on plant growth and improving soil quality. Test Eng Manag 82:3069–3073

Tripathi RD, Dwivedi S, Shukla MK, Mishra S, Srivastava S, Singh R, Gupta DK (2008) Role of blue green algae biofertilizer in ameliorating the nitrogen demand and fly-ash stress to the growth and yield of rice (*Oryza sativa* L.) plants. Chemosphere 70(10):1919–1929

Uosif MAM, Mostafa AMA, Elsaman R, Moustafa ES (2014) Natural radioactivity levels and radiological hazards indices of chemical fertilizers commonly used in upper Egypt. J Radiat Res Appl Sci 7:430–437

Verma DC, Srivastava DK (2018) Response of mixed blue green algal bio fertilizer on yield of Rice (*Oryza sativa*) of Uttar Pradesh. Biotech Today 8(1):73–75

Verma G, Yadav DD (2019) Effect of fertility levels and biofertilizers on the productivity and profitability of chickpea (*Cicer arietinum*). Ind J Agronomy 64(1):138–141

Verma P, Yadav AN, Kazy SK, Saxena AK, Suman A (2013) Elucidating the diversity and plant growth promoting attributes of wheat (*Triticum aestivum*) associated acidotolerant bacteria from southern hills zone of India. Natl J Life Sci 10:219–227

Verma P, Yadav AN, Kazy SK, Saxena AK, Suman A (2014) Evaluating the diversity and phylogeny of plant growth promoting bacteria associated with wheat (*Triticum aestivum*) growing in central zone of India. Int J Curr Microbiol Appl Sci 3:432–447

Verma P, Yadav AN, Khannam KS, Panjiar N, Kumar S, Saxena AK et al (2015) Assessment of genetic diversity and plant growth promoting attributes of psychrotolerant bacteria allied with wheat (*Triticum aestivum*) from the northern hills zone of India. Ann Microbiol 65:1885–1899

Verma P, Yadav AN, Khannam KS, Kumar S, Saxena AK, Suman A (2016) Molecular diversity and multifarious plant growth promoting attributes of bacilli associated with wheat (*Triticum aestivum* L.) rhizosphere from six diverse agro-ecological zones of India. J Basic Microbiol 56:44–58

Verma P, Yadav AN, Khannam KS, Mishra S, Kumar S, Saxena AK et al (2019) Appraisal of diversity and functional attributes of thermotolerant wheat associated bacteria from the peninsular zone of India. Saudi J Biol Sci 26:1882–1895. https://doi.org/10.1016/j.sjbs.2016.01.042

Wani SA, Chand S, Ali T (2013) Potential use of *Azotobacter chroococcum* in crop production: an overview. Curr Agric Res 1:35–38

Weiss TL, Roth R, Goodson C, Vitha S, Black I, Azadi P et al (2012) Colony organization in the green alga *Botryococcus braunii* (race B) is specified by a complex extracellular matrix. Eukaryot Cell 11:1424–1440

Xia M, Chakraborty R, Terry N, Singh RP, Fu D (2020) Promotion of saltgrass growth in a saline petroleum hydrocarbons contaminated soil using a plant growth promoting bacterial consortium. Int Biodeterior Biodegradation 146:104808

Yadav J, Verma JP, Tiwari KN (2011) Plant growth promoting activities of fungi and their effect on chickpea plant growth.Asian. J Biol Sci 4(3):291–299

Yadav AN, Sachan SG, Verma P, Saxena AK (2015a) Prospecting cold deserts of north western Himalayas for microbial diversity and plant growth promoting attributes. J Biosci Bioeng 119:683–693. https://doi.org/10.1016/j.jbiosc.2014.11.006

Yadav AN, Sachan SG, Verma P, Tyagi SP, Kaushik R, Saxena AK (2015b) Culturable diversity and functional annotation of psychrotrophic bacteria from cold desert of Leh Ladakh (India). World J Microbiol Biotechnol 31:95–108. https://doi.org/10.1007/s11274-014-1768-z

Yadav AN, Sharma D, Gulati S, Singh S, Dey R, Pal KK et al (2015c) Haloarchaea endowed with phosphorus Solubilization attribute implicated in phosphorus cycle. Sci Rep 5:12293. https://doi.org/10.1038/srep12293

Yadav AN, Sachan SG, Verma P, Saxena AK (2016) Bioprospecting of plant growth promoting psychrotrophic bacilli from cold desert of north western Indian Himalayas. Indian J Exp Biol 54:142–150

Yadav AN, Gulati S, Sharma D, Singh RN, Rajawat MVS, Kumar R et al (2019a) Seasonal variations in culturable archaea and their plant growth promoting attributes to predict their role in establishment of vegetation in Rann of Kutch. Biologia 74:1031–1043. https://doi.org/10.2478/s11756-019-00259-2

Yadav AN, Singh S, Mishra S, Gupta A (2019b) Recent advancement in white biotechnology through Fungi. Volume 2: perspective for value-added products and environments. Springer International Publishing, Cham

Yadav AN, Singh S, Mishra S, Gupta A (2019c) Recent advancement in white biotechnology through Fungi. Volume 3: perspective for sustainable environments. Springer International Publishing, Cham

Yadav AN, Rastegari AA, Yadav N, Kour D (2020a) Advances in plant microbiome and sustainable agriculture: diversity and biotechnological applications. Springer, Singapore

Yadav AN, Rastegari AA, Yadav N, Kour D (2020b) Advances in plant microbiome and sustainable agriculture: functional annotation and future challenges. Springer, Singapore

Yadav AN, Singh J, Rastegari AA, Yadav N (2020c) Plant microbiomes for sustainable agriculture. Springer, Cham

Yang XE, Wu X, Hao HL, He ZL (2008) Mechanisms and assessment of water eutrophication.J. Zhejiang Univ Sci B 9(3):197–209

Yao Y, Zhang M, Tian Y, Zhao M, Zeng K, Zhang B et al (2018) *Azolla* biofertilizer for improving low nitrogen use efficiency in an intensive rice cropping system. Field Crop Res 216:158–164

Zhao K, Penttinen P, Zhang X, Ao X, Liu M, Yu X et al (2014) Maize rhizosphere in Sichuan, China, hosts plant growth promoting *Burkholderia cepacia* with phosphate solubilizing and antifungal abilities. Microbiol Res 169:76–82

Current Trends in Microbial Biotechnology for Agricultural Sustainability: Conclusion and Future Challenges

22

Abd El-Latif Hesham, Tanvir Kaur, Rubee Devi, Divjot Kour, Shiv Prasad, Neelam Yadav, Chhatarpal Singh, Joginder Singh, and Ajar Nath Yadav

Abstract

Microbial biotechnology is an emerging field with greater applications in diverse sectors involving food security, human nutrition, plant protection, and overall basic research in the agricultural sciences. The environment has been sustaining the burden of mankind since decades and indiscriminate use of the resources has led to the degradation of the environment, loss of soil fertility, and has created a need for sustainable strategies. The major focus in the coming decades would be on a green and clean environment by utilizing the plant-associated beneficial

A. E.-L. Hesham (✉)
Genetics Department, Faculty of Agriculture, Beni-Suef University, Beni-Suef, Egypt
e-mail: hesham_egypt5@agr.bsu.edu.eg; hesham_egypt5@aun.edu.eg

T. Kaur · R. Devi · D. Kour · A. N. Yadav
Department of Biotechnology, Dr. Khem Singh Gill Akal College of Agriculture, Eternal University, Baru Sahib, Sirmour, Himachal Pradesh, India

S. Prasad
Centre for Environment Science & Climate Resilient Agriculture, ICAR-Indian Agricultural Research Institute, New Delhi, India

N. Yadav
Veer Bahadur Singh Purvanchal University, Jaunpur, Uttar Pradesh, India

C. Singh
School for Environmental Sciences, Babasaheb Bhimrao Ambedkar University, Lucknow, Uttar Pradesh, India

J. Singh
Department of Microbiology, Climate Mitigation and Sustainable Agriculture Research lab (CMaSAR), Division of Research and Development, Lovely Professional University, Phagwara, Punjab, India

A. N. Yadav et al. (eds.), *Current Trends in Microbial Biotechnology for Sustainable Agriculture*, Environmental and Microbial Biotechnology,
https://doi.org/10.1007/978-981-15-6949-4_22

microbial communities. These beneficial microbial communities represent a novel and promising solution for a sustainable environment. Microbial communities possess a huge sink of mechanisms by which they act as biofertilizers, bioprotectants, and biostimulants as well as the alleviators of abiotic stress conditions. Thus, utilizing plant-associated microbiomes will surely support sustainable agriculture thereby reducing the production costs and environmental pollution. The present chapter exclusively concluded the horizon covered book content of microbial biotechnology for sustainable agriculture.

Keywords

Beneficial microbiomes · Microbial biotechnology · Plant growth promotion · Soil fertility · Sustainable agriculture

This book contains current knowledge about beneficial microbes including soil and plant microbiomes, recent microbial technologies, and potential applications of microbes for agro-environmental sustainability. The agriculturally and environmentally important microbiomes are the key components of soil–plant systems, where they are engaged in an intense network of diverse interactions. Soil and plant microbiomes (rhizospheric, endophytic, and epiphytic) with plant growth-promoting (PGP) attributes have emerged as an important and promising tool for sustainable agriculture and environment. These beneficial microbes plant the plant growth and enhanced soil fertility by diverse mechanisms that include releasing plant growth regulators; solubilization of phosphorus, potassium, and zinc; biological nitrogen fixation, or by producing siderophores, ammonia, HCN, and other secondary metabolites. The aim of the present book is to collect and compile the current developments in the understanding of microbes, plant and soil interaction, and their biotechnological applications for agro-environmental sustainability. The book will be highly useful to the faculty, researchers, and students associated with microbiology, biotechnology, agriculture, molecular biology, environmental biology, and related subjects.

The increasing food demand and sufficient food production with an increasing population are one of the major concerns for the twenty first century. At the end of 2033, the increased human population will create demands for food and shelter. This poses a great challenge to the agricultural system to solve the problem of food demand (Mosttafiz et al. 2012). It has been estimated by the United Nations Population Fund that the global human population may reach 10 billion by 2050. If the food production globally is to keep pace with a growing population, greater efforts will be required by modern societies to boost up agricultural productivity in an environmentally sustainable manner (Morrissey et al. 2004). Agriculture in developed countries already leads to serious environmental issues through the excessive use of pesticides, salinization, and the depletion of water resources. Thus, to achieve a clean and green environment, finding efficient ways for controlling pests, increasing the tolerance of the plants to abiotic stress conditions, and recycling of the nutrients is of major importance (Umesha et al. 2018).

Soil health is defined as functional ability within agro-ecosystem boundaries to support biological productivity, promote plant and animal fitness, and sustain environmental quality (Singh and Yadav 2020). Healthy soils function to support organic matter, water and nutrients cycling decaying, inactivate toxic compounds, suppress pathogens, and enhance the sustainability of production system (Rastegari et al. 2020a, b; Yadav et al. 2020d, e). The soil health directly or indirectly impacts plant health, environmental health, and food safety and quality. The soil serves as a biological filter for removing unwanted solids and gaseous constituents from air and water. Healthy soils produce nutritious crops that, in turn, nourish humans and animals. Soil microbes have an immense impact on relations between soil and plant and microbe and play a vital role in sustaining soil fertility (Yadav et al. 2020f). Nutrient cycling is the most significant of these relationships (Fig. 22.1).

Soil and plant microbiome are key players to achieve sustainability and ultimately leading to the conservation of the biosphere. Diverse PGPMs have been reported to belong to all three domains namely Archaea, Bacteria, and Eukarya and dominant of genus viz, *Acinetobacter, Arthrobacter*, *Aspergillus*, *Azospirillum*, *Bacillus*, *Brevibacillus*, *Burkholderia*, *Enterobacter*, *Flavobacterium*, *Halococcus, Klebsiella*, *Methylobacterium*, *Micrococcus*, *Paenibacillus*, *Penicillium*, *Pseudomonas*, *Rhizobium*, *Serratia*, *Staphylococcus,* and *Stenotrophomonas* (Kour et al. 2019c; Rana et al. 2019; Singh et al. 2020b; Yadav et al. 2018a; Yadav et al. 2018b). These microbiomes play a major role in maintaining plant health, stimulating growth, increasing the productivity, and stress tolerance of plants by different direct and indirect mechanisms (Yadav et al. 2020a, b).

Interestingly, PGPMs play fascinating roles which influence growth and development of plants ranging from the production of the indole acetic acid, cytokinins, and gibberellins (Tiwari et al. 2020), nitrogen fixation (Rana et al. 2020),

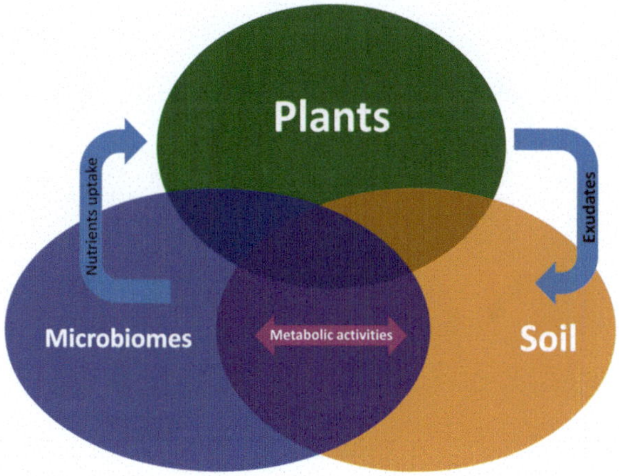

Fig. 22.1 Interactions between plants, microbiota, and soil

production of low molecular weight iron-chelating compounds (Subrahmanyam et al. 2020), solubilization of phosphorus (Singh et al. 2020a; Yadav et al. 2016), potassium (Kour et al. 2020b), zinc and manganese (Kaur et al. 2020), thus acting as biofertilizers (Kaur et al. 2019; Mondal et al. 2020). PGPMs also protect the plants from phytopathogens thereby acting as biocontrol agents (Yadav et al. 2017). Furthermore, recent researches focusing on the use of PGPMs in the agricultural sector has highlighted their role in the amelioration of drought, salinity, low temperature, high temperature, and heavy metal stress in plants by diverse mechanisms (Kour et al. 2020a). These mechanisms include the accumulation of osmolytes like proline and glycine betaine preventing plants from osmotic stress, production of reactive oxygen species scavenging enzymes, decreasing lipid peroxidation thereby maintaining membrane integrity and reducing the increased levels of ethylene by producing ACC deaminase enzyme (Kour et al. 2019b). PGPMs can be used as bioinoculants for increasing the fertility of the soil, yield, as well as for healthy food production both in normal and stressed environments. Furthermore, they possess the capability to reduce the use of agrochemicals thus protecting the environment and ultimately the ecosystem. Thus, these microbiomes pave the way for the next agricultural green revolution and can be potent bioresources for a sustainable environment. Beneficial plant and soil microbiomes play a major role in maintaining crop and soil health through nutrient cycling and uptake (Fig. 22.2).

The research on microbe's mode of action for the enhancement of plant growth is increasing at a rapid rate, so that the microbes can be used according to their abilities of promoting plant growth. The mechanism of plant growth promotion has been investigated for many years, as in the late 1990s a report by Holguin and Patten (1999) suggested that biological nitrogen fixation, nutrients solubilization, production of siderophores and phytohormones, lowering of ethylene levels, and induction

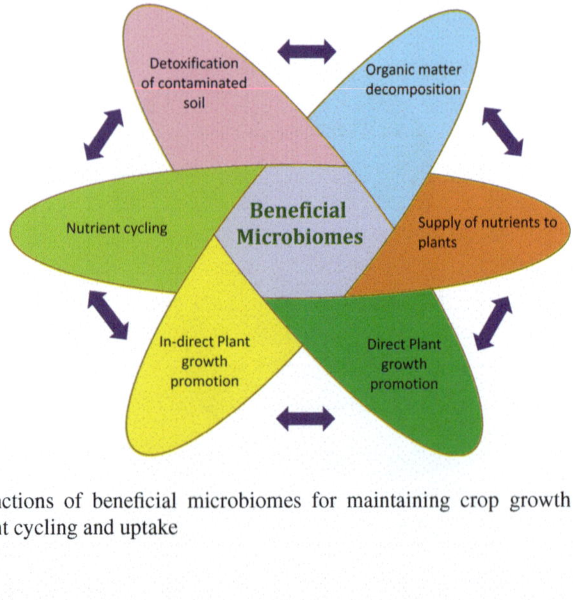

Fig. 22.2 Functions of beneficial microbiomes for maintaining crop growth and soil health through nutrient cycling and uptake

of pathogen resistance are the general mechanisms of microbes. All these mechanisms were later categorized under two modes of actions namely direct plant growth promotion and indirect plant growth promotion mechanisms (Rai et al. 2020; Sharaff et al. 2020).

The mode of action, direct plant growth promotion was the action that stimulates the growth of plants directly by providing nutrients and growth stimulators. This type of mode of action stimulates growth by mechanisms like fixation of atmospheric nitrogen; solubilization of macro and micronutrients; production of various growth hormones like auxin, cytokinin, and gibberellins, enzymes, siderophores (Van Loon 2007). Direct mechanism of plant growth promotion by microbes mostly takes place when the microbes interact with the roots part of the plant, especially the provision of nutrients through the mechanism of solubilization of nutrients (Gamalero and Glick 2011). Soil contains most of the nutrients like macronutrients like phosphorous, potassium, and micronutrients like zinc, iron, magnesium, calcium, and selenium. But all these nutrients in the soil are available in an insoluble form that cannot be utilized by the plants. So, microbes having the ability of nutrient solubilization change the form of the nutrients from insoluble to soluble, which is readily available for the utilization of plants. This mechanism of microbes solves the basic requirement of the plant's nutrients along with the enhancement of soil fertilization as the use of chemical fertilizers has ruined the soil nutrients completely (Kour et al. 2020d).

Another essential mechanism of microbes that directly facilitates plant growth is biological nitrogen fixation. Eventually, this mechanism of microbes is represented as an ecological and economically beneficial alternative to the chemical nitrogen fertilizer as agriculture is heavily dependent upon the nitrogen fertilizer prepared through an expensive chemical reaction that has a hazardous effect on the environment and humans (Fig. 22.3). In the atmosphere, nitrogen is abundantly present in the form of complex diatom N_2 that is not used by the plants as such. It is necessary to reduce the nitrogen into ammonia before the plant metabolizes it and form integral components like protein, nucleic acid, and some other essential biomolecules. Nitrogen-fixing microbes, mainly bacteria that have enzyme nitrogenase fix the atmospheric nitrogen. These bacteria have a special structural gene known as *nif* gene, which is a nitrogenase biosynthesis gene that helps in the activation of required proteins and regulates the enzymes for the fixation process. Naturally, large number of *Rhizobia*, the associative bacteria, fixes the huge amount of nitrogen but only for leguminous plants and whereas *Azospirillum* bacterial species are the free-living soil nitrogen fixer (James and Olivares 1998; Prasad et al. 2019).

Iron, another essential mineral or micronutrient, is a cofactor for proteins and is involved in most important process like photosynthesis and respiration in plants. On the earth's crust, it is the fourth abundant element but unfortunately, this much quantity of iron is not available to plants as they are present in the form of ferric ions which is not assimilated by the plants. To overcome this difficulty, microbes produce a low molecular weight organic molecule known as siderophores under the iron-limiting conditions. This organic molecule helps in the sequestering of iron from the soil and converting it into ferrous ions, which are assimilated by the plants.

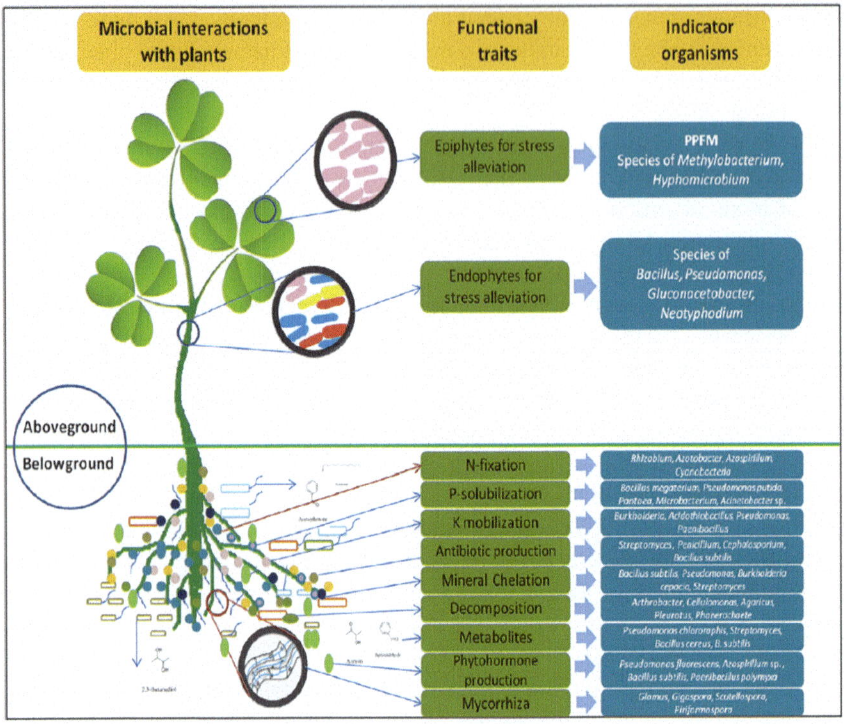

Fig. 22.3 Microbial associations with plants and soils and their inherent functional attributes that make them potential indicators for plant and soil health management (Sahu et al. 2019)

Using this mechanism, deficiency of iron in plants can be overcome (Saha et al. 2016).

The production of plant growth hormones is the other mechanism of this mode of action. Normally, the plant itself produces phytohormones like auxin, cytokinin, gibberellins, and abscisic acid for their proper growth and development of plant parts like roots and shoots but is the harsh conditions where plants had to undergo the climate stress, and hence were not able to produce the sufficient amount of phytohormones (Kour et al. 2020c; Verma et al. 2019). These conditions of the plants weaken the plant growth. On the same, microbes that have the capability of producing plant growth hormones can be used as phytostimulators to enhance the growth and functionality of the plant (Rademacher 2015).

The indirect method, the other mode of action is basically related to the biocontrol of pathogens and pest through the production of antibiotics, lytic enzymes, hydrogen cyanide, induced systemic resistance, and siderophore production (Thakur et al. 2020; Van Loon 2007). All these mechanisms suppress the growth of the phytopathogens by competing with pathogens for nutrient or by releasing of some toxic chemical. Mechanism of antibiotic production is one of the most common mechanisms of PGPMs (Devi et al. 2020). They release various types of antibiotics like

2,4-diacetylphloroglucinol, agrocin 84 and 434, herbicolin, cyclic lipopeptides, oomycin, pyoluteorin, pyrrolnitrin, and phenazines that control the growth of pathogens by targeting the electron transport chain (Raaijmakers et al. 2009; Rademacher 2015). Hydrogen cyanide is another toxic compound that is produced by the microbes. This compound is so toxic that a very small amount can kill the most phytopathogens.

Lytic enzymes are the cell wall–degrading enzymes that degrade the cell wall of the pathogens and suppress their growth. A number of lytic enzymes have been known that are produced by the microbes like b-1,3 glucanase, cellulose, chitinase, protease, and lipase (Kour et al. 2019a; Singh et al. 2016). Some of the microbes use induced systemic resistance (ISR) mechanism to biocontrol the pathogens. In this mechanism, the defense mechanism of the plant is activated by ISR signaling. Jasmonate and ethylene are the two signaling compounds released by the microbes and the host plant stimulates the response to kill the pathogens (Glick 2012).

Limiting the nutrients for the pathogens is also another method of suppressing the growth of phytopathogens. Siderophore production can also help in the biocontrol of pathogens as this iron chelator, chelates the iron from the soil that is not available for the pathogens. This mechanism competes with the pathogens on nutrients and suppresses their growth (Saha et al. 2016).

Another mechanism of the indirect method is the production of 1-amino cyclopropane-1-carboxylic acid (ACC) deaminase. This mechanism helps in the biocontrol by lowering the level of ethylene in plants. Ethylene is the known phytohormone which is produced by the plants as it is required for normal growth, but if its concentration increases it becomes a poison for plants as it suppresses the growth of the plant. ACC deaminase production mechanism is useful in lowering the level of increased ethylene in plants (mainly increase in stress condition) by using ethylene as a precursor to cleave ACC into ammonia along with α-ketobutyrate. This mechanism even helps plants to grow normally even in stressful conditions (Glick et al. 2007; Suman et al. 2016).

Abiotic stress is the stress exerted by the environment on the plant like drought, salinity, heavy metals, temperature extremes, and flood stress. These stresses disturb the plant's optimal functionality and also affect the productivity of the crop plants. Half of the global production of the crop plants is being destroyed by the stress exerted by the environment. Globally, to alleviate the abiotic stresses for plants, microbial biotechnology is widely studied and being practiced (Kumar et al. 2019d; Verma et al. 2017). Microbial utilization for the abiotic stress mitigation strengthens the plant and increase the productivity of the plants even under the harsh and unfavorable conditions (Yadav et al. 2020c).

Around the world, the foremost abiotic stress that affects the agricultural crop production is drought or water deficit condition as 64% of the land around the world is water deficit and it is expected that drought condition will increase by twofold by the year 2050 (Falkenmark 2013; Mittler and Blumwald 2010). In the drought conditions, there is a decrease in the water potential of the leaf, size of the leaf reduces, there is a reduction in the viability, number, and size of the seed, root growth, flowering and fruiting of the plant are delayed and it lowers the productivity of the plant

(Xu et al. 2016). To mitigate the drought stress microbes from various conditions have been reported like *Klebsiella* sp., *Flavobacterium* sp., *Enterobacter ludwigii, Pseudomonas putida, Bacillus* sp., *Paenibacillus beijingensis, Bacillus subtilis, Acinetobacter calcoaceticus,* and *Penicillium* sp. (Gontia-Mishra et al. 2016; Li et al. 2019; Vurukonda et al. 2016; Woo et al. 2020).

Salinity of soil is also one of the major stresses for the crop plants and it is reported that worldwide, 6% of the land have a high concentration of salt. Plant growth under the condition of high salt concentration has low osmotic potential due to the imbalance of salt. Salinity in plants can cause damage to proteins and even DNA. To alleviate the stress of salinity microbes like *Pseudomonas fluorescens, Pseudomonas putida* (Jalili et al. 2009), *Mortierella* sp., *Glomus aggregatum, Glomus mosseae* (Zhang et al. 2011), *Azotobacter chroococcum* (Rojas-Tapias et al. 2012), *Methylobacterium oryzae, Glomus etunicatum* (Lee et al. 2015), and *Curtobacterium albidum* (Vimal et al. 2019) have been reported.

Another abiotic stress that affects plant growth is temperature extremes. Temperature of the earth is changing drastically due to global warming. The growth of the plant is affected by both the extremes temperature that is high temperature and low temperature. In this condition, almost all the processes of the plants are affected including growth, germination, and reproduction (McClung and Davis 2010). Particularly, in high-temperature conditions, plants have several damages like leaf burning, abscission and senescence, inhibition in the growth of root and shoot, discoloration of fruit along with hormonal changes, and imbalance (Vollenweider and Günthardt-Goerg 2005). Microbes like *Pseudomonas putida* (Srivastava et al. 2008) have been reported. Whereas, in the case of low temperature, plants have cold injuries due to extremely low temperatures. To alleviate the low temperature, numerous microbes have been reported like *Pseudomonas lurida, Pseudomonas jessani, Arthrobacter, Flavobacterium, Massilia,* and *Pedobacter* (Subramanian et al. 2016; Verma et al. 2015). Stress due to heavy metal in the soil is also another major stress (Kumar et al. 2019a; Kumar et al. 2019c). This stress is generally experienced by the plant growing in the mines area. Heavy metal high (Cd, Sb, Cr, Hg, Pb) concentration in soil affects photosynthesis, respiration, blocks the protein structure and growth of the plants. To mitigate the heavy metal stress microbes like *Pseudomonas* sp. (Gupta et al. 2018) and *Arthrocnemum macrostachyum* (Navarro-Torre et al. 2016) have been reported.

Modern intensive farming techniques improve crop production but are associated with more problems over time, causing adverse environmental contamination to human health and eventually, low crop production and the increasing world population's threat to food security. Sustainable agriculture growth is therefore the biggest challenge facing the growing population's enormous demand for food grain in a sustainable environmental and cost-effective way (Reddy et al. 2020). In modern agriculture, the availability and affordability of chemical fertilizers based on fossil fuels at the farm level have only been ensured through imports and subsidy. There are different groups of microbes associated with plants as plant microbiome (Epiphytic, endophytic, rhizospheric). These microbes play a significant role in plant growth promotion, nutrient cycling, and soil health. The endophytic microbes

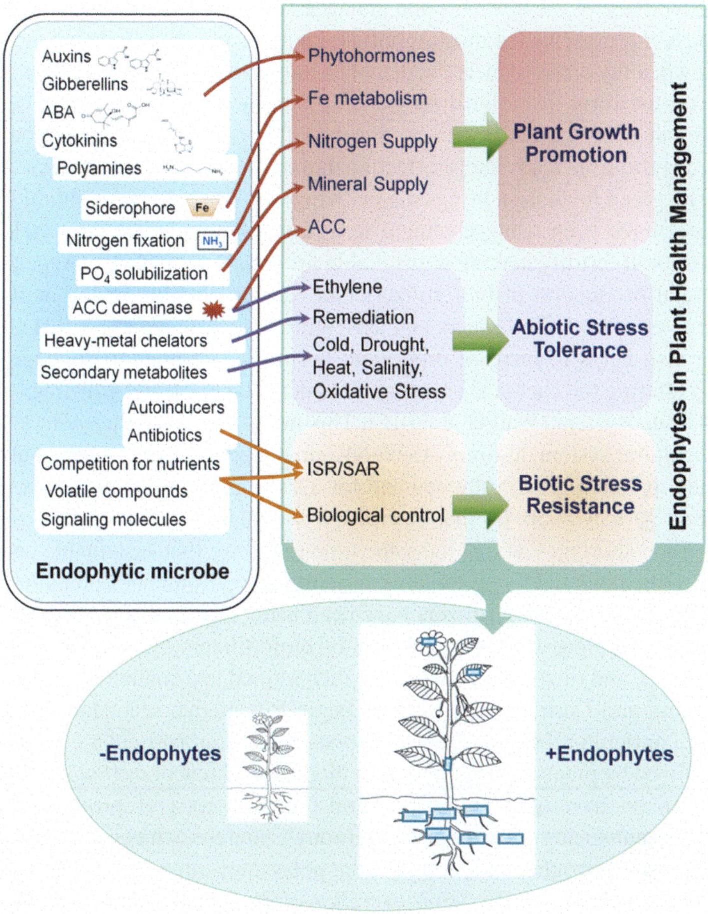

Fig. 22.4 Role and potential applications of endophytic microbiomes for sustainable agriculture and environments, Adapted with permission from Waghunde et al. (2017)

present inside the plant tissues promote the plant growth by different plant growth-promoting mechanisms (Fig. 22.4). The concept of biofertilizers ("bio means life and fertilizer means the substances used to provide available plant nutrients") was first introduced in 1834 when a French farmer named J.B. Boussingault documented the buildup of nitrogen from soil by cultivating legumes (Bhattacharyya and Jha 2012).

Today, sustainability has become the core research agenda in worldwide agricultural land resources as it focuses primarily on working out the urgent needs of humans such as surface water, groundwater, air pollution, suppressed ecosystem function, as well as deterioration of soil quality and reduced biodiversity that has adverse effects on human health and the environment as well as improving agricultural production to feed the growing population. In the same way, we can improve crop productivity of the nutrients and enhance soil fertility by using biofertilizers. Plant nutrient represents the most important components of sustainable farming by producing healthy crops to meet the demands of the world population. This depends entirely on the form of biofertilizers used to supply the plants with all necessary nutrients, but excess reliability on chemical fertilizers is damaging the biodiversity of the ecosystem and adversely affecting human health (Bhattacharyya and Jha 2012). The green revolution brings major changes to food and agricultural life.

Farmers have been using chemical fertilizers for increasing crop yield but it reduces the soil fertility making it unsuitable for raising crop plants (Aggani 2013). Commercial production of biofertilizers and easy availability in the market may alter the lives of both the farmers and agricultural sectors. We know that chemical fertilizer also helps to increase crop yield but they contribute to spoilage of soil health by altering the chemical composition, microbial flora, and affecting biodiversity and its ecosystem (Wall et al. 2015). This poses a huge challenge to a sustainable agricultural system in solving the food demand problem (Mosttafiz et al. 2012). According to Barea (2015), the demand for agricultural production is expected to rise at least 70% by 2050. By that time people would be more aware of agricultural demand, food shortages, and at the same time to realize that sustainable practice is essential to the potential for sustainable agriculture, environment, and human health (Umesha et al. 2018). Biofertilizers have been using living organisms such as bacteria, fungi, and algae for the production of biofertilizers by isolating them from soil, water, air, and rhizosphere, which are then refined to a condensed form for field uses (Mishra and Dash 2014). Microbes begin to generate secondary metabolites reaction of agricultural significance in response to certain particular conditions and can be utilized by plants to sustain various biochemical reactions (Salar et al. 2017).

Biofertilizers have been used for several functions for crop production system significantly enhancing crop production through numerous mechanisms including phosphorus solubilization and mobilization, potassium solubilization, biofertilizers for micronutrients, nitrogen fixation, excretion of growth hormones, and also reducing the hazardous influences of inorganic fertilizers on crops and soil productivity (Hamid and Ahmad 2012; Rajawat et al. 2020). Several types of biofertilizers are present that we can use to enhance crop productivity like *Azospirillum*, *Azotobacter*, *Arbuscular Mycorrhizal Azolla* and Blue-Green Algae, *Cyanobacteria*, *Diazotrophs*, *Pseudomonas*, Phosphate-Solubilizing Microbe, *Rhizobium* and silicon-solubilizing microbe (Kaur and Purewal 2019). We can use advantageous microbes as a biofertilizer has become minor in the agricultural sector because of its probable role in food safety and sustainable crop production.

Globally, biopesticides have been rising annually while the number of conventional pesticides is decreasing (Bailey et al. 2010). Biopesticides are naturally occurring substance including microbes, insects, and environmentally safe pesticides, and amidst its numerous types like the bacterial biopesticides that are used extensively and are of prodigious importanance. These substances found in plants and animals were used to control disease in crops, animals, and humans. Such factors have contributed to its use in the world for the plant control system, rather than chemical pesticides. Biopesticides are extracted from animals, plants, and other natural materials including algae, bacteria, fungi, viruses, nematodes, and protozoa and their bioactive compounds (Sharma et al. 2019; Verma et al. 2018). However, environment and human health issues due to excessive dependence on chemical pesticides, improvements in residue requirements, and increased demand for organically grown products have led in recent years to substantial growth in their uses (Arthurs and Dara 2019). Despite the many problems that have been faced in the adoption of biopesticides, the use of synthetic chemical pesticides has raised many concerns because of its negative impact on the climate, human health, natural enemies, and the balance maintain of the ecosystem. Some of the active ingredients available in synthetic pesticides were found to be carcinogenic and may pose a danger to human life. The biopesticidal based materials are readily available in the market at an inexpensive rate (Fig. 22.5).

In the field of agriculture, advanced research and development of biopesticides greatly reduce environmental emissions caused by residues of chemical synthetic and encourage sustainable agricultural production (Nawaz et al. 2016). However, due to their toxicity and contamination synthetic pesticides have become a health threat to humans and the environment (Lengai and Muthomi 2018). The global population must grow to 10.12 billion by 2100 to meet the increasing population's food demand; higher and more advanced competitive agricultural materials are required. The highest yield is based on improved variety, effective management of pests and diseases, and prescribed fertilization. Proper management of pests is an essential factor for safe and high yielding crops, which can provide food for the growing population. Adequate pest management is a key necessity today to produce maximum food from less. Acute or chronic pesticides-related poisoning is a problem in

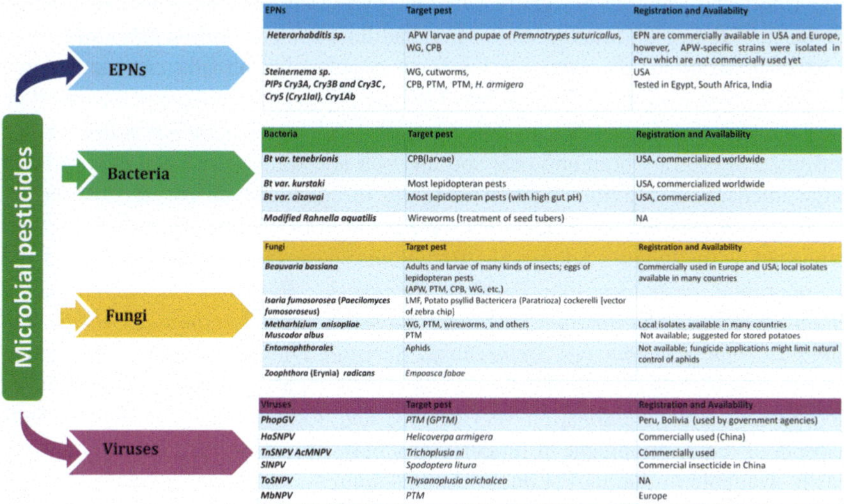

Fig. 22.5 Different biopesticides used for controlling different pest along with their commercial availability, Adapted with permission from Thakur et al. (2020)

many countries around the world, especially in developing countries (Casida and Durkin 2013; Green et al. 2013).

The demand for microbial biopesticides accounts for around 90% of total biopesticides and there is sufficient room for further growth in agricultural and public health, although challenges still exist. Biopesticides currently account for about 5% of the Indian pesticide market with at least 15 microbial species and 970 microbial formulations registered by the Central Insecticides Board and Registration Committee (CIBRC) (Kumar et al. 2019b). The same as of 2017 there are 200 products available based on nematicidal fungi (*Pochonia chlamydosporia* and *Purpureocillium lilacinum*) and entomopathogenic fungi (*Hirsutella thompsonii*, *Lecanicillium lecanii*, *Metarhizium anisopliae*, *B. brongniartii*, and *Beauveria bassiana*) have been documented for use against several arthropods and plant-parasitic nematodes, as well as 30 products have been commercialized of the bacteria species based on *Bacillus thuringiensis* (Bt) (kurstaki) that are registered against bollworms, lepidopterans, and other loopers and other 12 species based on Bt subsp. *Israelensis* and four entomopathogenic species are sold in the Indian market.

Bacillus thuringinesis strain of Lipdoptera is the most famous product, but new bacterial strains and its metabolites are produced for use in the fruits, vegetables, and ornamental crops against a wide range of arthropods. Ten fungal species are published for crop production in nursery, greenhouse and field crop against thrips, whiteflies, aphids, or other sucking pests. Berić et al. (2012) documented *Bacillus* sp. showed antagonistic activity against rice pathogens *Xanthomonas Oryzae p.v. Oryzae* and the activity was accredited to the making of a bacteriocin by the bacterial species wherein we can treat rice and wheat plants among the concentration of *Chaetomium globosum* to decrease the harshness of rice blast (*Magnaporthe grisea*) and wheat rust (*Puccinia recondite*) by up to 80%.

Arthurs and Dara (2019) reported that the pyrethrum (*Chrysanthemum cinerariifolium*), Neem (*Azadirachta indica*), sabadilla (*Schoenocaulon officinale*), and tobacco (*Nicotiana tabacum*) are the most popular and already commercialized botanical pesticides and other sources of botanical pesticides are pepper, euphorbia, citrus, and garlic. In another investigation, *Trichoderma harzianum* is used in the formulation of gram bran and peat soil and soil and water have been exposed to a large intensity of activities against dumping of eggplant and its seed caused by *Sclerotium rolfsii* and these activities were credited to the huge quantity of spores produced by the fungus (Adan et al. 2015. Some species have been commercialized as microbial pesticides like *Bacillus*, *Beauveria*, *Burkholderia*, *Enterobacter*, *Heterarhabditis*, *Metarhizium*, *Pythium*, *Pseudomonas*, *Paecilomyces*, *Penicillim* and *Fusarium*, *Streptomyces*, *Steinernama*, *Serratia*, *Trichoderma*, *Verticillium*, *and Xanthomonas*, however, biopesticides are faced with challenges of adaptation, acceptance, commercialization registration, and formulation. Although various aspects of the development of biopesticides have been described but they are not easily available in the market due to limited sources, commercialization, efficacy, formulation, production, and role of sustainable agriculture. Therefore researchers should work with government and industrial engineers, as well as the farmers to provide safe and lasting formulation of the biopesticides.

Plant-associated microbiomes greatly influence the productivity of the plants and also provides them a physiological and environmental advantage. Recently, a sustainable environment has become the need of the hour for the world to bring out the ultimate solution to reduce the use of chemical fertilizers and pesticides and to mitigate the undesirable effects of climate change. Applying knowledge about beneficial microbiomes may allow us to increase food production and simultaneously reduce the stress on the environment and global biodiversity. All future biotechnological developments in achieving a clean environment whether based on genetic-engineering technologies must take into consideration the role of plant-associated microbiomes. Further studies should be focused on beneficial microbial communities along with taking into account multidisciplinary research combining applications in agro-biotechnology, biotechnology, material science, and nanotechnology and bringing together functional biological and ecological approaches which will provide opportunities and open a new door for researchers, industries, and ultimately farmers with vast potential.

References

Adan MJ, Baque MA, Rahman MM, Islam MR, Jahan A (2015) Formulating of *Trichhoderma* based biopesticede for controlling damping off pathogen of eggplant seedling. Univer J Agr Res 3:106–113

Aggani SL (2013) Development of bio-fertilizers and its future perspective. Schol Acad J Pharma 2:327–332

Arthurs S, Dara SK (2019) Microbial biopesticides for invertebrate pests and their markets in the United States. J Invert Pathol 165:13–21

Bailey K, Boyetchko S, Längle T (2010) Social and economic drivers shaping the future of biological control: a Canadian perspective on the factors affecting the development and use of microbial biopesticides. Biol Con 52:221–229

Barea J (2015) Future challenges and perspectives for applying microbial biotechnology in sustainable agriculture based on a better understanding of plant-microbiome interactions. J Soil Sci Plant Nutr 15:261–282

Berić T, Kojić M, Stanković S, Topisirović L, Degrassi G, Myers M et al (2012) Antimicrobial activity of *Bacillus* sp. natural isolates and their potential use in the biocontrol of phytopathogenic bacteria. Food Technol Biotechnol 50:25–31

Bhattacharyya PN, Jha DK (2012) Plant growth-promoting rhizobacteria (PGPR): emergence in agriculture. World J Microbiol Biotechnol 28:1327–1350

Casida JE, Durkin KA (2013) Neuroactive insecticides: targets, selectivity, resistance, and secondary effects. Annu Rev Entomol 58:99–117

Devi R, Kaur T, Guleria G, Rana K, Kour D, Yadav N et al (2020) Fungal secondary metabolites and their biotechnological application for human health. In: Rastegari AA, Yadav AN, Yadav N (eds) Trends of microbial biotechnology for sustainable agriculture and biomedicine systems: perspectives for human health. Elsevier, Amsterdam, pp 147–161. https://doi.org/10.1016/B978-0-12-820528-0.00010-7

Falkenmark M (2013) Growing water scarcity in agriculture: future challenge to global water security. Philos Trans R Soc A Math Phys Eng Sci 371:20120410

Gamalero E, Glick BR (2011) Mechanisms used by plant growth-promoting bacteria. In: Maheshwari DK (ed) Bacteria in agrobiology: plant nutrient management. Springer, Berlin Heidelberg, Berlin, Heidelberg, pp 17–46. https://doi.org/10.1007/978-3-642-21061-7_2

Glick BR (2012) Plant growth-promoting bacteria: mechanisms and applications. Scientifica 2012:963401. https://doi.org/10.6064/2012/963401

Glick BR, Todorovic B, Czarny J, Cheng Z, Duan J, McConkey B (2007) Promotion of plant growth by bacterial ACC deaminase. Crit Rev Plant Sci 26:227–242

Gontia-Mishra I, Sapre S, Sharma A, Tiwari S (2016) Amelioration of drought tolerance in wheat by the interaction of plant growth-promoting rhizobacteria. Plant Biol 18:992–1000

Green BT, Welch KD, Panter KE, Lee ST (2013) Plant toxins that affect nicotinic acetylcholine receptors: a review. Chem Res Toxicol 26:1129–1138

Gupta P, Rani R, Chandra A, Kumar V (2018) Potential applications of *Pseudomonas* sp. (strain CPSB21) to ameliorate Cr6+ stress and phytoremediation of tannery effluent contaminated agricultural soils. Sci Rep 8:1–10

Hamid A, Ahmad L (2012) Soil phosphorus fixation chemistry and role of phosphate solubilizing bacteria in enhancing its efficiency for sustainable cropping-a review. J Pure Appl Microbiol 66:1905–1911

Holguin G, Patten CL (1999) Biochemical and genetic mechanisms used by plant growth promoting bacteria. World Scientific. https://doi.org/10.1142/p130

Jalili F, Khavazi K, Pazira E, Nejati A, Rahmani HA, Sadaghiani HR et al (2009) Isolation and characterization of ACC deaminase-producing fluorescent pseudomonads, to alleviate salinity stress on canola (*Brassica napus* L.) growth. J Plant Physiol 166:667–674

James EK, Olivares FL (1998) Infection and colonization of sugar cane and other graminaceous plants by endophytic diazotrophs. Crit Rev Plant Sci 17:77–119

Kaur P, Purewal SS (2019) Biofertilizers and their role in sustainable agriculture. In: Giri B, Prasad R, Wu Q-S, Varma A (eds) Biofertilizers for sustainable agriculture and environment. Springer International Publishing, Cham, pp 285–300. https://doi.org/10.1007/978-3-030-18933-4_12

Kaur T, Devi R, Rana KL, Kour D, Yadav AN (2019) Microbes with multifarious plant growth promoting attributes for sustainable agriculture. EU Voice 5:11–13

Kaur T, Rana KL, Kour D, Sheikh I, Yadav N, Kumar V et al (2020) Microbe-mediated biofortification for micronutrients: present status and future challenges. In: Rastegari AA, Yadav AN, Yadav N (eds) Trends of microbial biotechnology for sustainable agriculture and biomedicine systems: perspectives for human health. Elsevier, Amsterdam, pp 1–17. https://doi.org/10.1016/B978-0-12-820528-0.00002-8

Kour D, Rana KL, Kaur T, Singh B, Chauhan VS, Kumar A et al (2019a) Extremophiles for hydrolytic enzymes productions: biodiversity and potential biotechnological applications. In: Molina G, Gupta VK, Singh B, Gathergood N (eds) Bioprocessing for biomolecules production, pp 321–372. https://doi.org/10.1002/9781119434436.ch16

Kour D, Rana KL, Yadav AN, Yadav N, Kumar V, Kumar A et al (2019b) Drought-tolerant phosphorus-solubilizing microbes: biodiversity and biotechnological applications for alleviation of drought stress in plants. In: Sayyed RZ, Arora NK, Reddy MS (eds) Plant growth promoting Rhizobacteria for sustainable stress management, Rhizobacteria in abiotic stress management, vol 1. Springer, Singapore, pp 255–308. https://doi.org/10.1007/978-981-13-6536-2_13

Kour D, Rana KL, Yadav N, Yadav AN, Kumar A, Meena VS et al (2019c) Rhizospheric microbiomes: biodiversity, mechanisms of plant growth promotion, and biotechnological applications for sustainable agriculture. In: Kumar A, Meena VS (eds) Plant growth promoting Rhizobacteria for agricultural sustainability: from theory to practices. Springer, Singapore, pp 19–65. https://doi.org/10.1007/978-981-13-7553-8_2

Kour D, Kaur T, Devi R, Rana KL, Yadav N, Rastegari AA et al (2020a) Biotechnological applications of beneficial microbiomes for evergreen agriculture and human health. In: Rastegari AA, Yadav AN, Yadav N (eds) Trends of microbial biotechnology for sustainable agriculture and biomedicine systems: perspectives for human health. Elsevier, Amsterdam, pp 255–279. https://doi.org/10.1016/B978-0-12-820528-0.00019-3

Kour D, Rana KL, Kaur T, Devi R, Yadav N, Halder SK et al (2020b) Potassium solubilizing and mobilizing microbes: biodiversity, mechanisms of solubilization and biotechnological implication for alleviations of abiotic stress. In: Rastegari AA, Yadav AN, Yadav N (eds) Trends of microbial biotechnology for sustainable agriculture and biomedicine systems: diversity and functional perspective. Elsevier, Amsterdam, pp 177–202. https://doi.org/10.1016/B978-0-12-820526-6.00012-9

Kour D, Rana KL, Sheikh I, Kumar V, Yadav AN, Dhaliwal HS et al (2020c) Alleviation of drought stress and plant growth promotion by *Pseudomonas libanensis* EU-LWNA-33, a drought-adaptive phosphorus-solubilizing bacterium. Proc Natl Acad Sci India B. https://doi.org/10.1007/s40011-019-01151-4

Kour D, Rana KL, Yadav AN, Yadav N, Kumar M, Kumar V et al (2020d) Microbial biofertilizers: bioresources and eco-friendly technologies for agricultural and environmental sustainability. Biocatal Agric Biotechnol 23:101487. https://doi.org/10.1016/j.bcab.2019.101487

Kumar A, Chaturvedi AK, Yadav K, Arunkumar KP, Malyan SK, Raja P et al (2019a) Fungal phytoremediation of heavy metal-contaminated resources: current scenario and future prospects. In: Yadav AN, Singh S, Mishra S, Gupta A (eds) Recent advancement in white biotechnology through Fungi, Perspective for sustainable environments, vol 3. Springer, Cham, pp 437–461. https://doi.org/10.1007/978-3-030-25506-0_18

Kumar KK, Sridhar J, Murali-Baskaran RK, Senthil-Nathan S, Kaushal P, Dara SK et al (2019b) Microbial biopesticides for insect pest management in India: current status and future prospects. J Invert Pathol 165:74–81

Kumar M, Saxena R, Rai PK, Tomar RS, Yadav N, Rana KL et al (2019c) Genetic diversity of methylotrophic yeast and their impact on environments. In: Yadav AN, Singh S, Mishra S, Gupta A (eds) Recent advancement in white biotechnology through Fungi, Perspective for sustainable environments, vol 3. Springer International Publishing, Cham, pp 53–71. https://doi.org/10.1007/978-3-030-25506-0_3

Kumar V, Joshi S, Pant NC, Sangwan P, Yadav AN, Saxena A et al (2019d) Molecular approaches for combating multiple abiotic stresses in crops of arid and semi-arid region. In: Singh SP, Upadhyay SK, Pandey A, Kumar S (eds) Molecular approaches in plant biology and environmental challenges. Springer, Singapore, pp 149–170. https://doi.org/10.1007/978-981-15-0690-1_8

Lee Y, Krishnamoorthy R, Selvakumar G, Kim K, Sa T (2015) Alleviation of salt stress in maize plant by co-inoculation of arbuscular mycorrhizal fungi and *Methylobacterium oryzae* CBMB20. J Kor Soc Appl Biol Chem 58:533–540

Lengai GM, Muthomi JW (2018) Biopesticides and their role in sustainable agricultural production. J Biosci Med 6:7

Li Y, Shi H, Zhang H, Chen S (2019) Amelioration of drought effects in wheat and cucumber by the combined application of super absorbent polymer and potential biofertilizer. PeerJ 7:e6073

McClung CR, Davis SJ (2010) Ambient thermometers in plants: from physiological outputs towards mechanisms of thermal sensing. Curr Biol 20:R1086–R1092

Mishra P, Dash D (2014) Rejuvenation of biofertilizer for sustainable agriculture and economic development. Consilience pp:41–61

Mittler R, Blumwald E (2010) Genetic engineering for modern agriculture: challenges and perspectives. Ann Rev Plant Biol 61:443–462

Mondal S, Halder SK, Yadav AN, Mondal KC (2020) Microbial consortium with multifunctional plant growth promoting attributes: future perspective in agriculture. In: Yadav AN, Rastegari AA, Yadav N, Kour D (eds) Advances in plant microbiome and sustainable agriculture, Functional annotation and future challenges, vol 2. Springer, Singapore, pp 219–254. https://doi.org/10.1007/978-981-15-3204-7_10

Morrissey JP, Dow JM, Mark GL, O'Gara F (2004) Are microbes at the root of a solution to world food production? EMBO Rep 5:922–926

Mosttafiz SB, Rahman M, Rahman MM (2012) Biotechnology: role of microbes in sustainable agriculture and environmental health. Int J Microbiol 10:1–6

Navarro-Torre S, Mateos-Naranjo E, Caviedes M, Pajuelo E, Rodríguez-Llorente I (2016) Isolation of plant-growth-promoting and metal-resistant cultivable bacteria from *Arthrocnemum macrostachyum* in the Odiel marshes with potential use in phytoremediation. Mar Poll Bull 110:133–142

Nawaz M, Mabubu JI, Hua H (2016) Current status and advancement of biopesticides: microbial and botanical pesticides. J Entomol Zool Stud 4:241–246

Prasad M, Srinivasan R, Chaudhary M, Choudhary M, Jat LK (2019) Plant growth promoting rhizobacteria (PGPR) for sustainable agriculture: perspectives and challenges. In: Kumar A, Singh PK (eds) Singh AK. Woodhead Publishing, PGPR Amelioration in Sustainable Agriculture, pp 129–157. https://doi.org/10.1016/B978-0-12-815879-1.00007-0

Raaijmakers JM, Paulitz TC, Steinberg C, Alabouvette C, Moënne-Loccoz Y (2009) The rhizosphere: a playground and battlefield for soilborne pathogens and beneficial microorganisms. Plant Soil 321:341–361

Rademacher W (2015) Plant growth regulators: backgrounds and uses in plant production. J Plant Growth Reg 34:845–872

Rai PK, Singh M, Anand K, Saurabhj S, Kaur T, Kour D et al (2020) Role and potential applications of plant growth promotion rhizobacteria for sustainable agriculture. In: Rastegari AA, Yadav AN, Yadav N (eds) Trends of microbial biotechnology for sustainable agriculture and biomedicine systems: diversity and functional perspectives. Elsevier, Amsterdam, pp 49–60. https://doi.org/10.1016/B978-0-12-820526-6.00004-X

Rajawat MVS, Singh R, Singh D, Yadav AN, Singh S, Kumar M et al (2020) Spatial distribution and identification of bacteria in stressed environments capable to weather potassium aluminosilicate mineral. Braz J Microbiol 51:751–764. https://doi.org/10.1007/s42770-019-00210-2

Rana KL, Kour D, Sheikh I, Dhiman A, Yadav N, Yadav AN et al (2019) Endophytic fungi: biodiversity, ecological significance and potential industrial applications. In: Yadav AN, Mishra S, Singh S, Gupta A (eds) Recent advancement in white biotechnology through Fungi, Diversity and enzymes perspectives, vol 1. Springer, Switzerland, pp 1–62

Rana KL, Kour D, Kaur T, Devi R, Yadav AN, Yadav N et al (2020) Endophytic microbes: biodiversity, plant growth-promoting mechanisms and potential applications for agricultural sustainability. Antonie Van Leeuwenhoek. https://doi.org/10.1007/s10482-020-01429-y

Rastegari AA, Yadav AN, Yadav N (2020a) New and Future Developments in Microbial Biotechnology and Bioengineering: Trends of Microbial Biotechnology for Sustainable Agriculture and Biomedicine Systems: Diversity and Functional Perspectives. Elsevier, Amsterdam

Rastegari AA, Yadav AN, Yadav N (2020b) New and Future Developments in Microbial Biotechnology and Bioengineering: Trends of Microbial Biotechnology for Sustainable Agriculture and Biomedicine Systems: Perspectives for Human Health. Elsevier, Amsterdam

Reddy GC, Goyal R, Puranik S, Waghmar V, Vikram K, Sruthy K (2020) Biofertilizers toward sustainable agricultural development. In: Plant microbe Symbiosis. Springer, Cham, pp 115–128

Rojas-Tapias D, Moreno-Galván A, Pardo-Díaz S, Obando M, Rivera D, Bonilla R (2012) Effect of inoculation with plant growth-promoting bacteria (PGPB) on amelioration of saline stress in maize (*Zea mays*). Appl Soil Ecol 61:264–272

Saha M, Sarkar S, Sarkar B, Sharma BK, Bhattacharjee S, Tribedi P (2016) Microbial siderophores and their potential applications: a review. Environ Sci Pollut Res 23:3984–3999

Sahu PK, Singh DP, Prabha R, Meena KK, Abhilash P (2019) Connecting microbial capabilities with the soil and plant health: options for agricultural sustainability. Ecol Indic 105:601–612

Salar RK, Purewal SS, Sandhu KS (2017) Bioactive profile, free-radical scavenging potential, DNA damage protection activity, and mycochemicals in *Aspergillus awamori* (MTCC 548) extracts: a novel report on filamentous fungi. 3. Biotech 7:164

Sharaff MS, Subrahmanyam G, Kumar A, Yadav AN (2020) Mechanistic understanding of root-microbiome interaction for sustainable agriculture in polluted soils. In: Rastegari AA, Yadav AN, Yadav N (eds) Trends of microbial biotechnology for sustainable agriculture and biomedicine systems: diversity and functional perspectives. Elsevier, Amsterdam, pp 61–84. https://doi.org/10.1016/B978-0-12-820526-6.00005-1

Sharma S, Kour D, Rana KL, Dhiman A, Thakur S, Thakur P et al (2019) *Trichoderma*: biodiversity, ecological significances, and industrial applications. In: Yadav AN, Mishra S, Singh S, Gupta A (eds) Recent advancement in white biotechnology through Fungi, Diversity and enzymes perspectives, vol 1. Springer, Cham, pp 85–120. https://doi.org/10.1007/978-3-030-10480-1_3

Singh J, Yadav AN (2020) Natural bioactive products in sustainable agriculture. Springer, Singapore

Singh RN, Gaba S, Yadav AN, Gaur P, Gulati S, Kaushik R et al (2016) First, high quality draft genome sequence of a plant growth promoting and cold active enzymes producing psychrotrophic *Arthrobacter agilis* strain L77. Stand Genomic Sci 11:54. https://doi.org/10.1186/s40793-016-0176-4

Singh A, Kumar R, Yadav AN, Mishra S, Sachan S, Sachan SG (2020a) Tiny microbes, big yields: microorganisms for enhancing food crop production sustainable development. In: Rastegari AA, Yadav AN, Yadav N (eds) Trends of microbial biotechnology for sustainable agriculture and biomedicine systems: diversity and functional perspectives. Elsevier, Amsterdam, pp 1–15. https://doi.org/10.1016/B978-0-12-820526-6.00001-4

Singh B, Boukhris I, Pragya KV, Yadav AN, Farhat-Khemakhem A et al (2020b) Contribution of microbial phytases to the improvement of plant growth and nutrition: a review. Pedosphere 30:295–313. https://doi.org/10.1016/S1002-0160(20)60010-8

Srivastava S, Yadav A, Seem K, Mishra S, Chaudhary V, Nautiyal C (2008) Effect of high temperature on *Pseudomonas putida* NBRI0987 biofilm formation and expression of stress sigma factor RpoS. Curr Microbiol 56:453–457

Subrahmanyam G, Kumar A, Sandilya SP, Chutia M, Yadav AN (2020) Diversity, plant growth promoting attributes, and agricultural applications of rhizospheric microbes. In: Yadav AN, Singh J, Rastegari AA, Yadav N (eds) Plant microbiomes for sustainable agriculture. Springer, Cham, pp 1–52. https://doi.org/10.1007/978-3-030-38453-1_1

Subramanian P, Kim K, Krishnamoorthy R, Mageswari A, Selvakumar G, Sa T (2016) Cold stress tolerance in psychrotolerant soil bacteria and their conferred chilling resistance in tomato (*Solanum lycopersicum* mill.) under low temperatures. PLoS One 11(8):e0161592. https://doi.org/10.1371/journal.pone.0161592

Suman A, Yadav AN, Verma P (2016) Endophytic microbes in crops: diversity and beneficial impact for sustainable agriculture. In: Singh DP, Singh HB, Prabha R (eds) Microbial inoculants in sustainable agricultural productivity, Research perspectives, vol 1. Springer, New Delhi, pp 117–143. https://doi.org/10.1007/978-81-322-2647-5_7

Thakur N, Kaur S, Tomar P, Thakur S, Yadav AN (2020) Microbial biopesticides: current status and advancement for sustainable agriculture and environment. In: Rastegari AA, Yadav AN, Yadav N (eds) Trends of microbial biotechnology for sustainable agriculture and biomedicine systems: diversity and functional perspectives. Elsevier, Amsterdam, pp 243–282. https://doi.org/10.1016/B978-0-12-820526-6.00016-6

Tiwari P, Bajpai M, Singh LK, Mishra S, Yadav AN (2020) Phytohormones producing fungal communities: metabolic engineering for abiotic stress tolerance in crops. In: Yadav AN, Mishra S, Kour D, Yadav N, Kumar A (eds) Agriculturally important Fungi for sustainable agriculture, Perspective for diversity and crop productivity, vol 1. Springer, Cham, pp 1–25. https://doi.org/10.1007/978-3-030-45971-0_8

Umesha S, Singh PK, Singh RP (2018) Microbial biotechnology and sustainable agriculture. In: Biotechnology for sustainable agriculture. Elsevier, pp 185–205

Van Loon L (2007) Plant responses to plant growth-promoting rhizobacteria. In: New perspectives and approaches in plant growth-promoting rhizobacteria research. Springer, Cham, pp 243–254

Verma P, Yadav AN, Shukla L, Saxena AK, Suman A (2015) Alleviation of cold stress in wheat seedlings by *Bacillus amyloliquefaciens* IARI-HHS2-30, an endophytic psychrotolerant K-solubilizing bacterium from NW Indian Himalayas. Natl J Life Sci 12:105–110

Verma P, Yadav AN, Kumar V, Singh DP, Saxena AK (2017) Beneficial plant-microbes interactions: biodiversity of microbes from diverse extreme environments and its impact for crop improvement. In: Singh DP, Singh HB, Prabha R (eds) Plant-microbe interactions in agro-ecological perspectives, Microbial interactions and agro-ecological impacts, vol 2. Springer, Singapore, pp 543–580. https://doi.org/10.1007/978-981-10-6593-4_22

Verma P, Yadav AN, Kumar V, Khan MA, Saxena AK (2018) Microbes in termite management: potential role and strategies. In: Khan MA, Ahmad W (eds) Termites and sustainable management: volume 2 - economic losses and management. Springer International Publishing, Cham, pp 197–217. https://doi.org/10.1007/978-3-319-68726-1_9

Verma P, Yadav AN, Khannam KS, Mishra S, Kumar S, Saxena AK et al (2019) Appraisal of diversity and functional attributes of thermotolerant wheat associated bacteria from the peninsular zone of India. Saudi J Biol Sci 26:1882–1895. https://doi.org/10.1016/j.sjbs.2016.01.042

Vimal SR, Patel VK, Singh JS (2019) Plant growth promoting Curtobacterium albidum strain SRV4: an agriculturally important microbe to alleviate salinity stress in paddy plants. Ecol Indic 105:553–562

Vollenweider P, Günthardt-Goerg MS (2005) Diagnosis of abiotic and biotic stress factors using the visible symptoms in foliage. Environ Poll 137:455–465

Vurukonda SSKP, Vardharajula S, Shrivastava M, SkZ A (2016) Multifunctional *Pseudomonas putida* strain FBKV2 from arid rhizosphere soil and its growth promotional effects on maize under drought stress. Rhizosphere 1:4–13

Waghunde RR, Shelake RM, Shinde MS, Hayashi H (2017) Endophyte microbes: a weapon for plant health management. In: Panpatte DG, Jhala YK, Vyas RV, Shelat HN (eds) Microorganisms for Green revolution, Microbes for sustainable crop production, vol 1. Springer Singapore, Singapore, pp 303–325. https://doi.org/10.1007/978-981-10-6241-4_16

Wall DH, Nielsen UN, Six J (2015) Soil biodiversity and human health. Nature 528:69–76

Woo O-G, Kim H, Kim J-S, Keum HL, Lee K-C, Sul WJ et al (2020) *Bacillus subtilis* strain GOT9 confers enhanced tolerance to drought and salt stresses in Arabidopsis thaliana and Brassica campestris. Plant Physiol Biochem 148:359–367

Xu Z, Jiang Y, Jia B, Zhou G (2016) Elevated-CO2 response of stomata and its dependence on environmental factors. Front Plant Sci 7:657

Yadav AN, Rana KL, Kumar V, Dhaliwal HS (2016) Phosphorus solubilizing endophytic microbes: potential application for sustainable agriculture. EU Voice 2:21–22

Yadav AN, Kumar R, Kumar S, Kumar V, Sugitha T, Singh B et al (2017) Beneficial microbiomes: biodiversity and potential biotechnological applications for sustainable agriculture and human health. J Appl Biol Biotechnol 5:45–57. https://doi.org/10.7324/JABB.2017.50607

Yadav AN, Kumar V, Dhaliwal HS, Prasad R, Saxena AK (2018a) Microbiome in crops: diversity, distribution, and potential role in crop improvement. In: Crop improvement through microbial biotechnology. Elsevier, Amsterdam, pp 305–332. https://doi.org/10.1016/B978-0-444-63987-5.00015-3

Yadav AN, Verma P, Kumar S, Kumar V, Kumar M, Kumari Sugitha TC et al (2018b) Actinobacteria from rhizosphere: molecular diversity, distributions, and potential biotechnological applications. In: Singh BP, Gupta VK, Passari AK (eds) New and future developments in microbial biotechnology and bioengineering. Elsevier, Amsterdam, pp 13–41. https://doi.org/10.1016/B978-0-444-63994-3.00002-3

Yadav AN, Mishra S, Kour D, Yadav N, Kumar A (2020a) Agriculturally important Fungi for sustainable agriculture, volume 1: perspective for diversity and crop productivity. Springer International Publishing, Cham

Yadav AN, Mishra S, Kour D, Yadav N, Kumar A (2020b) Agriculturally important Fungi for sustainable agriculture, volume 2: functional annotation for crop protection. Springer International Publishing, Cham

Yadav AN, Rastegari AA, Yadav N (2020c) Microbiomes of extreme environments: biodiversity and biotechnological applications. CRC Press, Taylor & Francis, Boca Raton

Yadav AN, Rastegari AA, Yadav N, Kour D (2020d) Advances in plant microbiome and sustainable agriculture: diversity and biotechnological applications. Springer, Singapore

Yadav AN, Rastegari AA, Yadav N, Kour D (2020e) Advances in plant microbiome and sustainable agriculture: functional annotation and future challenges. Springer, Singapore

Yadav AN, Singh J, Rastegari AA, Yadav N (2020f) Plant microbiomes for sustainable agriculture. Springer, Cham

Zhang H, Wu X, Li G, Qin P (2011) Interactions between arbuscular mycorrhizal fungi and phosphate-solubilizing fungus (*Mortierella* sp.) and their effects on Kostelelzkya virginica growth and enzyme activities of rhizosphere and bulk soils at different salinities. Biol Fert Soils 47:543

The manufacturer's authorised representative in the EU is Springer Nature Customer Service Centre GmbH, Europaplatz 3, 69115 Heidelberg, Germany. If you have any concerns regarding our products, please contact ProductSafety@springernature.com

Printed and bound by CPI Group (UK) Ltd, Croydon, CR0 4YY
30/09/2026
02225328-0012